In Sight of the Suture

GEOLOGICAL SOCIETY SPECIAL PUBLICATION NO. 160

In Sight of the Suture: the Palaeozoic geology of the Isle of Man in its Iapetus Ocean context

EDITED BY

N. H. WOODCOCK
Department of Earth Sciences, University of Cambridge, UK

D. G. QUIRK
School of Construction and Earth Sciences, Oxford Brookes University, UK
(Present address: Burlington Resources (Irish Sea) Limited, London, UK)

W. R. FITCHES
Robertson Research International, Llandudno, UK

R. P. BARNES
British Geological Survey, Edinburgh, UK

1999
Published by
The Geological Society
London

THE GEOLOGICAL SOCIETY

The Geological Society of London was founded in 1807 and is the oldest geological society in the world. It received its Royal Charter in 1825 for the purpose of 'investigating the mineral structure of the Earth' and is now Britain's national society for geology.

Both a learned society and a professional body, the Geological Society is recognized by the Department of Trade and Industry (DTI) as the chartering authority for geoscience, able to award Chartered Geologist status upon appropriately qualified Fellows. The Society has a membership of 8600, of whom about 1500 live outside the UK.

Fellowship of the Society is open to persons holding a recognized honours degree in geology or a cognate subject and who have at least two years' relevant postgraduate experience, or not less than six years' relevant experience in geology or a cognate subject. A Fellow with a minimum of five years' relevant postgraduate experience in the practice of geology may apply for chartered status. Successful applicants are entitled to use the designatory postnominal CGeol (Chartered Geologist). Fellows of the Society may use the letters FGS. Other grades of membership are available to members not yet qualifying for Fellowship.

The Society has its own Publishing House based in Bath, UK. It produces the Society's international journals, books and maps, and is the European distributor for publications of the American Association of Petroleum Geologists (AAPG), the Society for Sedimentary Geology (SEPM) and the Geological Society of America (GSA). Members of the Society can buy books at considerable discounts. The Publishing House has an online bookshop (*http://bookshop.geolsoc.org.uk*).

Further information on Society membership may be obtained from the Membership Services Manager, The Geological Society, Burlington House, Piccadilly, London W1V 0JU (Email: *enquiries@geolsoc.org.uk*; tel: +44 (0)171 434 9944).

The Society's Web Site can be found at *http://www.geolsoc.org.uk/*. The Society is a Registered Charity, number 210161.

Published by The Geological Society from:
The Geological Society Publishing House
Unit 7, Brassmill Enterprise Centre
Brassmill Lane
Bath BA1 3JN, UK

(*Orders*: Tel. +44 (0)1225 445046
Fax +44 (0)1225 442836)
Online bookshop: *http://bookshop.geolsoc.org.uk*

First published 1999

The publishers make no representation, express or implied, with regard to the accuracy of the information contained in this book and cannot accept any legal responsibility for any errors or omissions that may be made.

British Library Cataloguing in Publication Data
A catalogue record for this book is available from the British Library.

ISBN 1-86239-046-0
ISSN 0305-8719

Typeset by E & M Graphics, Midsomer Norton, Bath, UK.

Printed by The Alden Press, Osney Mead, Oxford, UK.

Distributors

USA
AAPG Bookstore
PO Box 979
Tulsa
OK 74101-0979
USA
Orders: Tel. + 1 918 584-2555
Fax +1 918 560-2652
Email *bookstore@aapg.org*

Australia
Australian Mineral Foundation Bookshop
63 Conyngham Street
Glenside
South Australia 5065
Australia
Orders: Tel. +61 88 379-0444
Fax +61 88 379-4634
Email *bookshop@amf.com.au*

India
Affiliated East-West Press PVT Ltd
G-1/16 Ansari Road, Daryaganj,
New Delhi 110 002
India
Orders: Tel. +91 11 327-9113
Fax +91 11 326-0538

Japan
Kanda Book Trading Co.
Cityhouse Tama 204
Tsurumaki 1-3-10
Tama-shi
Tokyo 206-0034
Japan
Orders: Tel. +81 (0)423 57-7650
Fax +81 (0)423 57-7651

Contents

Tectonics and metamorphism

Regional comparisons

Bibliography

References to this volume

It is recommended that reference to all or part of this book should be made in one of the following ways:

WOODCOCK, N. H., QUIRK, D. G., FITCHES, W. R. & BARNES, R. P. (eds) 1999. *In Sight of the Suture: the Palaeozoic geology of the Isle of Man in its Iapetus Ocean context*. Geological Society, London, Special Publication, **160**.

PIPER, J. D. A., BIGGIN, A. J. & CROWLEY, S. F. 1999. Magnetic survey of the Poortown Dolerite, Isle of Man. *In*: WOODCOCK, N. H., QUIRK, D. G., FITCHES, W. R. & BARNES, R. P. (eds) *In Sight of the Suture: the Palaeozoic geology of the Isle of Man in its Iapetus Ocean context*. Geological Society, London, Special Publication, **160**, 155–164.

In sight of the suture: the early Palaeozoic geological history of the Isle of Man

N. H. WOODCOCK,[1] D. G. QUIRK,[2] W. R. FITCHES,[3] & R. P. BARNES[4]

[1]*Department of Earth Sciences, University of Cambridge, Cambridge CB2 3EQ, UK*

[2]*Department of Geology, Oxford Brookes University, Gipsy Lane, Oxford OX3 0BP, UK*

Present address: Burlington Resources (Irish Sea) Ltd, 1 Canada Square, Canary Wharf, London E14 5AA, UK

[3]*Robertson Research International, Llanrhos, Llandudno, North Wales, LL30 1SA, UK*

[4]*British Geological Survey, Murchison House, West Mains Road, Edinburgh EH9 3LA, UK*

Abstract: The pre- and syn-Caledonian rocks of the Isle of Man are now known to comprise three distinct units: the early Ordovician Manx Group, the mid-Silurian Dalby Group and the ?late Silurian–early Devonian Peel Sandstones.

The Manx Group is dominated by Arenig deep-marine turbidites and debrites deposited in oxygenated basins on the northwest-facing margin of Avalonia. Its organization into a sand-rich lower part and a mud-rich upper part invites comparison with the Skiddaw Group (Lake District) and Ribband Group (Leinster) and points to control by margin-wide events, in part eustatic sea-level changes. Episodes of mass-wasting and Fe–Mn fluid exhalation also correlate along the margin. A mid–late Ordovician volcanic arc is missing above the Manx Group, although parts of its intrusive substructure may be preserved.

The Dalby Group comprises northwest-derived turbidites, sedimented into an anoxic basin during Wenlock (mid-Silurian) time. These turbidites were deposited in a successor basin above the Iapetus suture zone. The Dalby Group sits with a tectonic contact on the Manx Group.

No evidence has been found of a pre-Silurian cleavage. The main Caledonian D1 and D2 shortening phases are post-Wenlock, comparable in age with those further along the margin in the Lake District and Leinster. The Peel Sandstones preserve a Lower 'Old Red Sandstone' sequence, mostly removed by post-Caledonian erosion elsewhere along this outboard part of the Avalonian margin. The unit does not host a definite Caledonian cleavage, and it must have been deposited late in the deformation history. The granitic intrusions into the Manx Group range from early in D1 to late in D2. The intrusions generate only local aureoles, and the high metamorphic grade in parts of the Manx Group may be enhanced by favourable protolith compositions.

The Isle of Man enjoys a unique geographical position, lying as it does in the Irish Sea within sight of Wales, England, Scotland and Ireland (Fig. 1). However, its geological setting is no less special. Although now part of a horst block surrounded by Mesozoic basins, it lies tantalizingly close to the surface trace of that most important of regional Palaeozoic structures, the Iapetus Suture. Geophysical evidence (Soper *et al.* 1992) suggests that this boundary, between the former Avalonian microcontinent to the south and the Laurentian continent to the north, skirts the northwestern edge of the island (Fig. 1). Over most of the British Isles, the surface trace of the suture is hidden by Upper Palaeozoic rocks. Only in eastern Ireland and the Isle of Man do Lower Palaeozoic rocks crop out at, or close to, the suture. The difficulty in deciphering the eastern Irish evidence across the suture (Harper & Murphy 1989; Todd *et al.* 1991; Owen *et al.* 1992, Vaughan & Johnston 1992) highlights the need for more information from the Isle of Man. The results from Lower Palaeozoic rocks reported in this volume promise to augment substantially our knowledge of the geology of the Iapetus Suture Zone and of the outboard edge of the Avalonian margin. The Upper Palaeozoic and Mesozoic geology of the surrounding Irish Sea has been summarized recently in the volumes edited by Meadows *et al.* (1997) and a thematic issue of the *Journal of Petroleum Geology* (1999) edited by D. G. Quirk.

Research past and present

The Lower Palaeozoic rocks of the Isle of Man, until recently all assigned to the Manx Group, have

From: WOODCOCK, N. H., QUIRK, D. G., FITCHES, W. R. & BARNES, R. P. (eds) 1999.
In Sight of the Suture: the Palaeozoic geology of the Isle of Man in its Iapetus Ocean context.
Geological Society, London, Special Publications, **160**, 1–10. 1-86239-046-0/99/$15.00

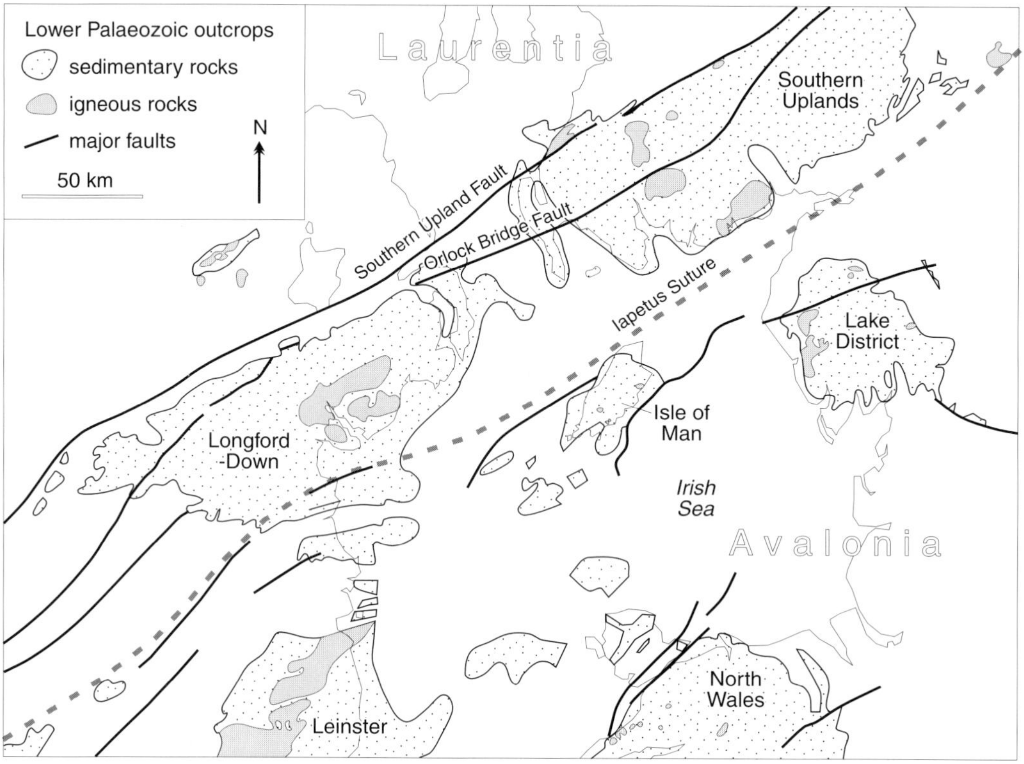

Fig. 1. Location of the Isle of Man in relation to other Lower Palaeozoic regions around the Irish Sea.

a long history of investigation. This history is detailed in the present volume by **Ford *et al.*** and listed in a comprehensive bibliography of the island by **Wilson**. Despite this diversity of past research, our present view of the geology of the Manx Group has been predominantly formed by the work of two people: G. W. Lamplugh, who mapped the island for the British Geological Survey at the end of the last century, and A. Simpson, who studied the Manx slates in the 1960s. The work of both geologists pre-dates, of course, ideas about the crucial plate tectonic setting of the island. The regional context of the Manx Group has been built up instead from work in related areas, particularly the Lake District of England, the Welsh Basin and the Leinster Basin of Ireland. On this evidence, the Manx Group is seen as part of the early Ordovician sediment prism on the outboard edge of the Avalonian segment of the Gondwana continent, continuous with the Skiddaw Group of the Lake District and the Ribband Group of Leinster (Cooper *et al.* 1995). The polyphase deformation history established by Simpson (1963) for the Manx Group is assumed to be predominantly late Caledonian in age, and generally due to the Silurian–Early Devonian impingement of Avalonia with the Laurentian continent (Soper *et al.* 1987, 1992).

Recent interest in the Manx Group was rekindled through biostratigraphic work by Molyneux (1979), metamorphic studies by Roberts *et al.* (1990) and a field guidebook by Ford (1993). New research (e.g. Rushton 1993; Quirk & Kimbell 1997; Stone & Evans 1997) was eventually focused into a multidisciplinary field-based project carried out on the island between 1995 and 1998. This volume reports many of the results of this new wave of research. The papers are organized into sections covering the main themes in the deposition of the Lower Palaeozoic sedimentary rocks of the Isle of Man and their subsequent deformation, metamorphism, intrusion and mineralization. The stratigraphical focus of each of the papers is shown on Fig. 2. This introductory review sets these papers (denoted by bold type) within an interpretative summary of Palaeozoic geological history of the island, highlighting current debates and the scope for future work.

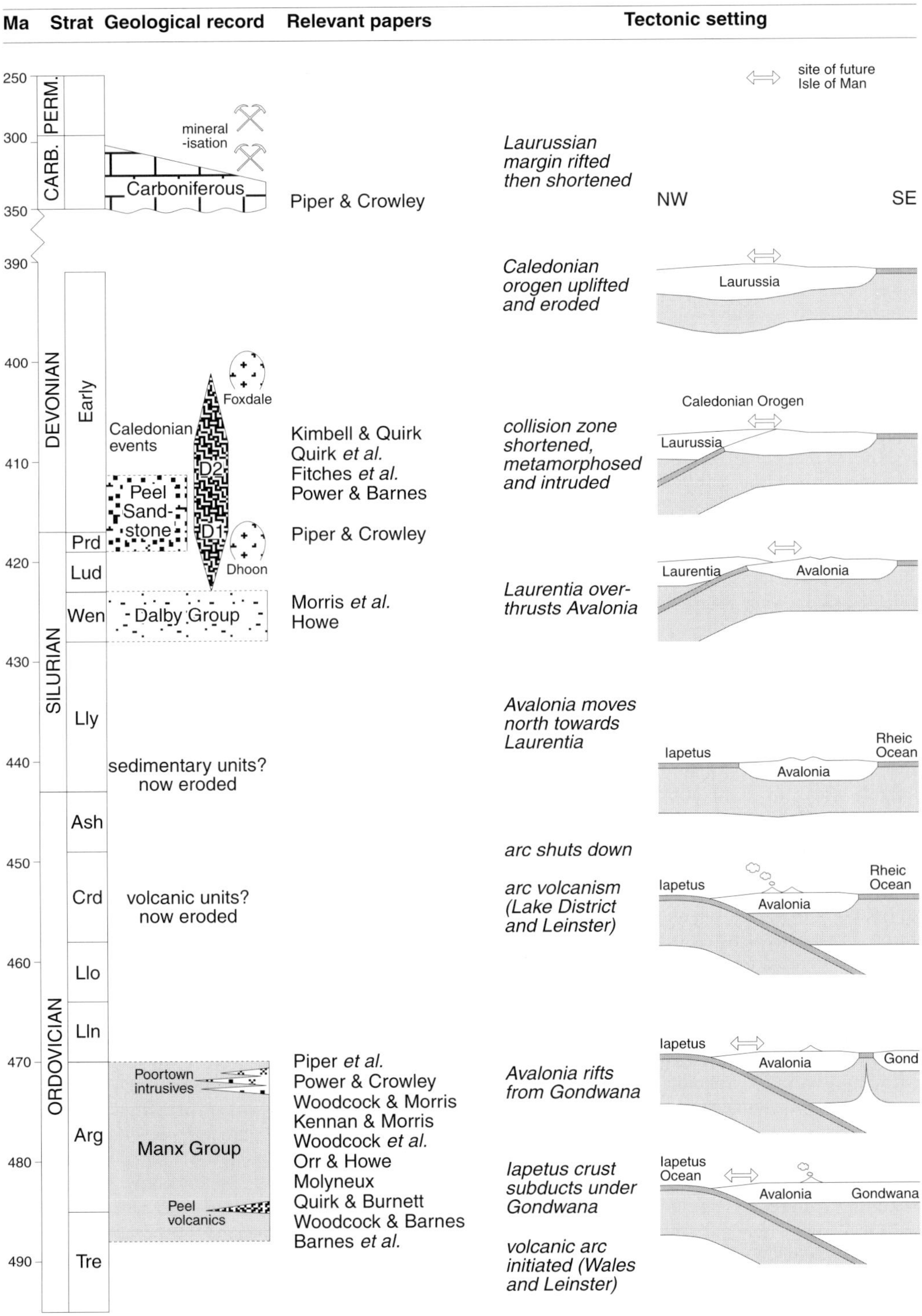

Fig. 2. Stratigraphical location of the studies reported in this volume, plotted on a geochronological diagram of the Palaeozoic geological history of the Isle of Man.

Stratigraphic and structural framework

The prevailing structural view of the Manx Group throughout this century has been of a major syncline, involving repetition of sandstone-rich units on the northwest and southeast coasts and cored by mudstone-rich units along the northeast–southwest spine of the island. However, the stratigraphic scheme required by this synclinal model, outlined by Lamplugh (1903) and embellished by Simpson (1963), has proved to be inconsistent with more recent biostratigraphic data (Molyneux 1979; Cooper *et al.* 1995). In this volume, the correlation of the two sandstone units, the Lonan and Niarbyl Flags, across the syncline is shown to be untenable on lithological grounds by **Morris *et al.***, **Quirk & Burnett** and **Woodcock & Barnes**. More conclusively still, **Howe** has identified mid-Silurian (Wenlock) graptolites from the Niarbyl Flags, and **Molyneux** and **Orr & Howe** have confirmed the early Ordovician (Arenig) faunas from the Lonan flags. These conflicting data have provoked **Morris *et al.*** to remove the formally redefined Niarbyl Formation from the Manx Group and to place it in a distinct Dalby Group, separated from the underlying Manx Group by the Niarbyl Thrust. The considerable regional implications of this relationship are discussed later.

The simplicity of a major D1 syncline in the Manx Group has also been questioned by structural observations summarized by **Fitches *et al.*** The lithological repetition across the island is due, in part, to significant, probably reverse, faults that separate tracts of continuous stratigraphy. Other tracts have suspect boundaries, which may prove to be stratigraphic or faulted. Only a few of the faults are well exposed on the coast. However, the existence of important faults in the less well-exposed part of the central and northwestern Manx Group is suggested by northeast–southwest and east–west oriented aeromagnetic lineaments noted by **Quirk *et al.***

Uncertainty over the large-scale structure has encouraged **Woodcock *et al.*** to reorganize Manx Group stratigraphy in terms of a series of seven northeast–southwest trending tectonostratigraphic tracts. The approach is analogous to the successful tract strategy employed in the Southern Uplands of Scotland (e.g. Barnes *et al.* 1989). Faults or high-strain zones form some tract boundaries, but the possibility cannot be excluded that some tract boundaries are essentially stratigraphic. Some of the 12 component formations may be repeated in another tract, but have been separately named until such correlations can be better founded. Because type sections are designated for the first time, the new stratigraphic scheme should provide a firm basis for future work. That such work will require further refinement of the scheme is suggested by the lithofacies mapping of **Quirk & Burnett**. Using a detailed facies characterisation of the Manx Group lithologies, they have revealed the internal complexity of some of the formations. They have also begun to delimit possible formation boundaries in the northern segment of the Manx Group left undefined by **Woodcock *et al.***

Poor biostratigraphic zonation of the Manx Group continues to hamper its interpretation. **Orr & Howe** report new graptolite finds, but the material is indifferently preserved and merely serves to confirm the Arenig age of the upper part of the Lonan Flags, now formally defined as the Santon Formation. More encouragingly, the trace fossil assemblage described by Orr & Howe closely matches that from the Skiddaw Group, strengthening the correlation both with the deep-water setting of the Skiddaw Group and its Tremadoc–Arenig age. However, it is the acritarch assemblages that still provide the best zonation of the Manx Group itself. **Molyneux** reassigns the available data to the new lithostratigraphic scheme, confirming that the Manx Group includes definite early, mid and late Arenig assemblages. Further biostratigraphic work on both micro- and macrofossils is essential for future understanding of the depositional history of the Manx Group.

Early Arenig sand–turbidite fan deposition

The base of the Manx Group, like its correlative Skiddaw and Ribband Groups, is not seen at outcrop. Comparison with the structural model of Hughes *et al.* (1993) for the Lake District, or that of Max *et al.* (1990) for Leinster, suggests that these units might, in any case, have been tectonically detached from their original basement. The earliest evidence of Manx Group sedimentation comes from poorly preserved Tremadoc or early Arenig graptolites from the Cronk Sumark Slates, within the area mapped by **Quirk & Burnett**. A volcanic arc had already been initiated in Wales and Leinster by this time (Kokelaar 1988; McConnell & Morris 1997), and subduction of Iapetus oceanic crust must already have started beneath this segment of the Gondwana margin. Unless dramatic reorganization of marginal crustal slivers has since occurred, the Manx Group was presumably deposited in the forearc of this active margin (Fig. 2). Another glimpse of the early Manx Group may be provided by the lower Arenig 'Peel Volcanics' (Simpson 1963; briefly discussed by **Morris *et al.***) and by widespread early felsitic intrusions. How these rocks relate to the arc volcanism awaits further study.

The lower Arenig components of the Manx Group seem to be dominated by sandstone-rich turbidite successions. **Quirk & Burnett** chart the facies variation in these successions, ranging from very thin-bedded silt–mud couplets through thin- and medium-bedded sand–mud couplets, to thick-bedded sandstones. **Orr & Howe** confirm from trace fossils that these successions were deposited in a deep-marine environment. **Woodcock & Barnes** show the sedimentological complexity of the lower Arenig sandstones, with flows travelling downslope to the northwest but with their concentrated sandy basal parts tending to be deflected along-slope to the southwest, possibly by a fault-structured topography on the margin.

A feature of these lower Arenig sandstone turbidites is the compositional variation from lithic wackes to arenites, in places highly quartzose. **Woodcock & Barnes** find this variation too marked to be produced solely by intrabasinal sorting or flow-stripping, and follow Cooper *et al.* (1995) in invoking a distinct supply of clean sand to the Gondwana shelf at this time. This sand was plausibly sourced from the continental sheet sands preserved in Europe as the Armorican quartzite, implying a continued connection between Avalonia and Gondwana during early Arenig time. Geochemical studies by **Barnes *et al.*** confirm the compositional distinction between a silica-rich and silica-poor group of lower Arenig sandstones. The silica-poor sandstones resemble the Loweswater Formation of the Skiddaw Group. The silica-rich sandstones match only the enigmatic Redmain Sandstone of the Lake District (Cooper *et al.* 1995) rather than any unit in the main Skiddaw Group. The ordering of the lower Arenig Manx Group sandstone units is only partially constrained by the available field data. Several possibilities are outlined by **Barnes *et al.***

Later Arenig mud-prone slope deposition

The relatively good biostratigraphic control in the Skiddaw Group, summarized by Cooper *et al.* (1995) and by **Stone *et al.***, reveals a prominent mid-Arenig transition from the sandstones of the Loweswater Formation to the mudstones of the Kirkstile Formation. A similar transition occurs within both the Manx Group (**Woodcock *et al.***) and the Ribband Group (**McConnell *et al.***), a transition that the available biostratigraphy also allows to be of mid-Arenig age. The rapid decrease in the rate of sand supply to the Avalonian segment of the Gondwana margin suggests a relative sea-level rise. However, the continued availability of limited volumes of clean sand implies that the Avalonian connection with Gondwana was not yet completely severed (Fig. 2; **Woodcock *et al.***).

Quirk & Burnett and **Woodcock *et al.*** describe the typical mix of facies within the probable middle and upper Arenig successions of the Manx Group: mudstones, quartzose sandstones and pebbly mudstones. **Woodcock & Morris** interpret one example of such a succession, the Lady Port Formation. They suggest that slumping and debris flow on the muddy margin slope created an irregular topography, which in turn trapped the sandy basal parts of downslope turbidity flows. This phase of down-margin mass wasting might simply reflect the mud-rich nature of the contemporary slope. Alternatively, it might record seismic triggering during the decisive rifting of Avalonia from Gondwana (Cooper *et al.* 1995; **Woodcock & Morris**; **Stone *et al.***; Fig. 2), a timing consistent with regional evidence (Prigmore *et al.* 1997).

These syn-rift Manx Group successions also record evidence for a high rate of exhalation of iron and manganese-rich fluids into the late Arenig ocean, as deduced by **Kennan & Morris** from the manganese carbonate rocks in the Creggan Mooar and Lady Port Formations. They also suggest that such rocks are the protolith to the Mn–garnet-bearing 'coticule' that characterizes many higher grade Ordovician rocks bordering the former Iapetus Ocean.

The probability that Iapetus oceanic crust had started to subduct under the Avalonian margin as early as the Tremadoc (Fig. 2) implies that the Manx Group lay on an active margin during its deposition. Deformation of the Manx Group soon after its deposition is a possibility in such a setting and Max *et al.* (1990) have suggested that the analogous Ribband Group formed a thrust-dominated accretionary complex. The evidence for deformation timing in the Manx Group (**Fitches *et al.***) suggests that the penetrative D1 event is post-Wenlock (mid-Silurian) in age. If Arenig folds and fabrics are preserved, they have yet to be discriminated from the overprinting D1 and D2 events. The evidence therefore matches that from the Skiddaw Group, where earlier proposals of a major mid-Ordovician cleavage-forming deformation are now discounted (Cooper *et al.* 1993).

Mid-Ordovician to early Silurian history

In both the Lake District and Leinster, the predominantly sedimentary early Ordovician successions are overlain by later Ordovician volcanic arc successions of the Borrowdale and Duncannon Groups, respectively (Max *et al.* 1990; Cooper *et al.* 1993). Any such volcanic superstructure is missing in the Isle of Man. However, a suggestion that it may have accumulated comes from the Poortown intrusive complex in the north-western Manx Group. The magnetic survey of **Piper *et al.*** shows that this complex comprises basaltic sheets, probably sills. **Power & Crowley** describe a compositional range from tholeiitic basalt to basaltic andesite, and a chemistry consistent with a volcanic-arc setting. The sills were intruded at a high level into probable upper Arenig rocks. Although the intrusive complex could therefore be of any subsequent age, the possibility that it represents the substructure to a late Ordovician volcanic succession is supported by estimates of its palaeolatitude, derived from the palaeomagnetic work of **Piper *et al.*** Common minor intrusions elsewhere in the Manx Group (Lamplugh 1903) may belong to the same suite as the Poortown Complex.

If an Ordovician arc was indeed constructed above the Manx Group, comparison with the Borrowdale Group suggests that it would have been mostly of Caradoc age and substantially subaerial. The volcanics, and any post-Caradoc sedimentary cover analogous to the Windermere Supergroup, must have been subsequently eroded. This erosion must have occurred at latest before the overthrusting of the Dalby Group on to the Manx Group, probably in early Devonian time (Fig. 2).

Silurian deposition of the Dalby Group and Peel Sandstone

A new and intriguing factor in Manx geological history arises from the demonstration by **Howe** and **Morris *et al.*** that the Niarbyl Formation is of mid-Silurian rather than early Ordovician age. These arc-derived turbidites, now assigned to the Dalby Group, were supplied southeastward into an anoxic marine basin during Wenlock, probably late Wenlock, time. They were subsequently faulted into their present position above the western Manx Group.

The tectonic significance ascribed to the Dalby Group depends crucially on the postulated site of their original deposition. **Barnes *et al.*** favour correlation with the mid-Silurian turbidites of the Windermere Supergroup (Lake District). This hypothesis implies that the Dalby Group was originally deposited on the Manx Group, probably above an intervening succession of upper Ordovician to lower Silurian rocks, which are now cut out by a normal-faulted contact. In contrast, **Morris *et al.*** favour correlation of the Dalby Group with the Riccarton Group of the Scottish Southern Uplands, a match also allowed by the geochemical data of **Barnes *et al.*** The second hypothesis implies that the Niarbyl Formation represents the toe end of the Southern Uplands turbidite prism, either onlapping or overthrusted on to the outboard edge of the Avalonian continental margin. The contrast between these two hypotheses is, in fact, rather small if the Wenlock and Ludlow rocks of the Windermere Supergroup are themselves regarded as the diachronous southward advance of Southern Uplands deposits in a successor basin across the, now essentially closed, Iapetus Ocean (Barnes *et al.* 1989; Kneller 1991; Fig. 2).

Faulted against both the Manx and Dalby Groups are the red alluvial Peel Sandstones (Crowley 1985). These rocks contain derived Ashgill and Wenlock shelly fossils, but otherwise their age has been unconstrained. Now **Piper & Crowley** have measured a remanent magnetism in these beds that includes a pre-folding component interpreted as late Silurian in age. This evidence favours deposition of the Peel Sandstones in one of the 'Lower Old Red Sandstone' basins that accompanied the Late Caledonian deformation event (Allen & Crowley 1983) in the culminating collision zone between Avalonia and Laurentia. However, a late Silurian age begs several important questions, yet to be satisfactorily answered. Were the continental Peel Sandstones deposited in conformity with underlying marine Silurian rocks, including the Dalby Group, or in a separate unconformable basin fill? Why are the Peel Sandstones only locally folded and cleaved? Did their deposition post-date at least the D1 event, or was the Peel Sandstones basin simply at too high a crustal level to be affected?

In any case, uplift of the Iapetus collision zone had already begun by the time the Peel Sandstones were deposited, prior to presumed exhumation and erosion during an early Devonian peak of orogenic activity (Fig. 2).

Late Caledonian orogenic events

Simpson's (1963) three-stage deformation history of the Manx Group is essentially verified by **Fitches *et al.***, although the third phase (D3) is regarded as locally developed and of possible Variscan or later age. A steeply dipping penetrative first cleavage and a more gently dipping second crenulation cleavage, both with geometrically congruent folds, appear in both the Manx and Dalby Groups. This relationship suggests that D1 and D2 are essentially post-Wenlock in age, although **Morris *et al.*** urge caution in this geometric correlation, particularly of D1 structures. The localized cleavage in the Peel Sandstones cannot be matched with one in the Manx Group, and is regarded as Variscan by **Fitches *et al.*** and by **Quirk & Kimbell**.

Power & Barnes show that the main metamorphic mineral growth in the Manx Group coincided with, or just post-dated, D1, whereas mineral growth during D2 was more limited. The intrusion of the Foxdale Granite, tentatively dated by Rb–Sr methods at *c.* 400 Ma (Crowley & Power, pers. comm.), overlapped or postdated D2. In summary then, it is most likely that both the D1 and D2 phases on the Isle of Man occurred in latest Silurian or early Devonian time, between *c.* 420 and 400 Ma. This is essentially the same age as that deduced for the main deformation in the Lake District (Soper *et al.* 1987; Hughes *et al.* 1993) and slightly younger than the southernmost deformation in the Southern Uplands (Barnes *et al.* 1989). The deformation in the Isle of Man is presumed to be driven by the culminating collision of Avalonia with Laurentia (Fig. 2), although Soper *et al.* (1992) have suggested that the impingement of Armorica on the southern edge of Avalonia may have been an important factor.

Whilst the geometry of the minor structures produced by the D1 and D2 events is now well established, the nature and kinematic significance of the related major structures is more debatable. **Fitches *et al.*** note that the largest D1 folds often verge southeastward, consistent with a southeast directed overthrusting, but that some large D1 structures in the southeast of the Manx Group verge northwestward. Some major high-strain zones, such as the Niarbyl Shear Zone, preserve evidence of D1 sinistral shear (**Fitches *et al.***) and possible southeast-directed overthrusting (**Morris *et al.***). The sinistral strike-slip component is also suggested by the patterns of shallow aeromagnetic lineaments identified by Quirk & Kimbell (1997) and **Quirk & Kimbell**. The kinematic pattern of the D2 deformation locally has a southeast directed component (**Morris *et al.***), but in most places resolves into a vertical flattening (**Fitches *et al.***).

The southeast directed theme during the Late Caledonian deformation of the Isle of Man is consistent with a regional control in which Laurentian outboard terranes attempted to overthrust the Avalonian margin during the final closure of the Iapetus Ocean (Fig. 2). The preservation, as the Dalby Group, of a possible slice of the Laurentian terranes raises the possibility that a strand of the Iapetus Suture itself passes through the island. The Niarbyl Thrust and the subjacent shear zone, below the Dalby Group, comprise a candidate for such a strand (**Morris *et al.***). However, a search for the Iapetus Suture is likely to prove inconclusive amongst the plexus of related shear zones at this high level in the crust. For instance, **Kimbell & Quirk** show that a major contrast in basement character now occurs on a northwest dipping zone that comes to surface further southeast, entirely within the Manx Group at the present erosion level.

Deformation zones within the Isle of Man may indeed have localized displacement in the continental convergence zone for some of its history. It is now important to add to the kinematic knowledge of these zones for comparison with analogous settings bordering the suture zone in the Lake District (Hughes *et al.* 1993; **Stone *et al.***), eastern Ireland (Vaughan & Johnston 1992; **McConnell *et al.***) and southern Scotland (**Barnes & Stone**).

The metamorphic and igneous history associated with the Late Caledonian deformation is also proving a particular challenge to unravel. Roberts *et al.* (1990) suggested that the high metamorphic grade along the central northeast–southwest spine of the island records the enhanced thermal gradient due to granite emplacement. However, **Power & Barnes** show that the two main granitic bodies have different time relations to the main deformation, with the Foxdale intrusion postdating D2, and the Dhoon intrusion pre-dating D2 and possibly overlapping D1. These authors demonstrate a strong lithological control on metamorphic mineralogy and therefore apparent grade. For instance, the high content of iron and aluminium in the Barrule Formation favours the formation of chloritoid. Nevertheless, the metamorphic conditions indicate that a substantial missing overburden has been removed from above the Manx Group since early Devonian time.

Post-Caledonian events

Mid- and Late Devonian events are not represented in the preserved geological record of the Isle of Man. Uplift and erosion of the thickened collision zone presumably dominated this time interval (Fig. 2). **Piper& Crowley** attempt to constrain the end of this interval by palaeomagnetic dating of the Langness Conglomerate Formation, which lies inconformably on the Manx Group in the southeast of the island, beneath the sequence of Lower Carboniferous limestones. However, they have recovered only an Early Permian magnetization, similar to that which pervades many red bed successions in Britain and Ireland. The Peel Sandstones also record this component. Late Carboniferous–Permian hydrothermal activity was also responsible for the mineralization that is concentrated in fractures cutting the Manx Group.

The Carboniferous and Permo-Triassic strata preserved offshore are the remnants of two, once much more extensive, rift-thermal sag basins, each partially removed by 1–3 km of uplift in the early Permian and early Tertiary (e.g. Green *et al.* 1997). The Peel and Solway Basins, lying west and north of the Isle of Man, respectively, have been a recent exploration target, after the successful hydrocarbon finds in the southeast Irish Sea Basin. The same Sherwood Sandstone reservoir is capped by saliferous mudstones of the Mercia Mudstone Group. but the lower Namurian Holywell Shale source rock is apparently absent in the Peel and Solway Basins (e.g. Armstrong *et al.* 1997). Wells have penetrated Dinantian limestone directly below the base Permian unconformity (Newman 1999). Recent seismic analysis has indicated that the Holywell Shale may be preserved on the undrilled downthrown side of major normal faults (Quirk 1999) and geochemical work also suggests that Dinantian source rocks may remain (Clayton *et al.* 1999; Racey *et al.* 1999), so prospects may still exist.

Many of the faults bounding the offshore basins, including those defining the present shape of the Isle of Man, display kilometres of normal offset due to early Permian, ?late Jurassic and early Tertiary extensional episodes (Quirk & Kimbell 1997). Some of these, such as the southeast dipping Lagman Fault to the northeast of the Isle of Man, probably represent reactivated Caledonian structures. Others, particularly north–south faults defining, for example, the western side of the Peel Basin, were formed in the Mesozoic. Whether northeast–southwest trending faults, such as the Central Valley Lineament crossing the Isle of Man and the Keys Fault east of the Isle of Man, are Caledonian or younger is as yet uncertain.

Scope for further work

The new wave of work on Manx geology reported in this volume has greatly enhanced the knowledge of local relationships and provides a firmer basis for regional comparisons. However, a number of old problems persist and new ones have been identified for future attention.

The most pressing need is for better biostratigraphic control of the sedimentary succession. The identification of Silurian rocks on the island has been an important breakthrough, but the calibration and intertract correlation of the Manx Group remains poorly constrained. Equally crucial are better radiometric ages of igneous and metamorphic events. Palaeomagnetic dating is proving helpful in the absence of other methods.

The stratigraphic organization of the Manx Group needs further refinement. The tectonostratigraphic tract scheme provides a testable model for future studies, but removes the predictability of previous synclinal models. Detailed interpretation of each individual tract is now required. Further structural studies are needed, particularly to establish the magnitude and kinematic pattern of displacement on the tract-bounding structures. The correlation of deformation phases between Manx Group tracts, and between them and the Dalby and Peel tracts, becomes a debatable issue, and the possibility of a pre-Silurian cleavage-forming deformation in parts of the Manx Group has yet to be conclusively disproved.

Sedimentological studies of the turbidites and debrites in the Manx and Dalby Groups have revealed a number of generally important features. The apparent dual supply of clean and dirty sand to the Gondwana margin will repay further sedimentological and geochemical investigation, particularly to test the link with remote sources in the Gondwana interior. Trace fossil assemblages clearly hold further information on the deep-water environmental conditions on the margin.

Refinement of the chronology of the Caledonian metamorphism and magmatism depends partly on the acquisition of radiometric dates, but partly on further studies of textural relationships with respect to the main deformation phases. Whether the high-grade 'spine' to the Manx Group is entirely explained by compositional control will repay further work.

The need for a comprehensive remapping of the Isle of Man has frequently been voiced (e.g. Roberts *et al.* 1990; Rushton 1993; Ford 1993). Invaluable though this would be, the difficulty and cost of such an exercise should not be underestimated. The inaccessibility of many of the coastal exposures and the thick drift cover inland would demand a considerable effort to improve on the century-old results of Lamplugh.

Regional conclusions

The results reported in this volume fill a major gap in understanding the northern margin of Avalonia, and its subsequent deformation, and metamorphism. The main conclusions of regional importance are:

- the Manx Group mainly comprises Arenig (Lower Ordovician) deep-marine turbidites and debrites, supplied from the southeast into oxygenated basins on the northwest facing margin of Avalonia;
- the gross organization of the Manx Group into a sand-rich lower part and mud-rich upper part echoes that in the Skiddaw Group (Lake District) and Ribband Group (Leinster); this organization points to control by margin-wide events, in part eustatic sea-level changes;
- episodes of down-margin mass wasting and Fe–Mn fluid exhalation also correlate along the margin, and may relate to rifting of Avalonia from Gondwana;
- a mid–late Ordovician volcanic arc is missing above the Manx Group, although remnants of its intrusive substructure may be preserved;
- a succession of northwest derived turbidites,

sedimented into an anoxic basin during Wenlock (mid-Silurian) time, sits with a tectonic contact on the Manx Group; these Silurian sediments were deposited in a successor basin above the Iapetus Suture Zone;

- no evidence of Ordovician cleavage-forming deformation has been found; the main D1 and D2 shortening phases are post-Wenlock, comparable in age with those further along the margin in the Lake District and Leinster;
- the Peel Sandstones preserve a Lower 'Old Red Sandstone' sequence, mostly removed by post-Caledonian erosion elsewhere along this outboard part of the Avalonian margin; they do not host a definite Caledonian cleavage, and must have been deposited late in the deformation history;
- the granitic intrusions into the Manx Group range from early in D1 to late in D2; generating only local aureoles, and the high metamorphic grade in parts of the Manx Group may be due rather to lithological control.

The authors particularly thank Padhraig Kennan, John Morris, Mike Howe, Greg Power and Dave Burnett for numerous discussions on Manx geology. RPB publishes with permission of the Director, British Geological Survey. This work was funded by NERC research grant GR9/01834.

References

ALLEN, J. R. L. & CROWLEY, S. F. 1983. Lower Old Red Sandstone fluvial dispersal systems in the British Isles. *Transactions of the Royal Society of Edinburgh*, **74**, 61–68.

ARMSTRONG, J. P., SMITH, J., D'ELIA, V. A. A. & TRUEBLOOD, S. P. 1997. The occurrence and correlation of oils and Namurian source rocks in the Liverpool Bay–North Wales area. *In*: MEADOWS, N., TRUEBLOOD, S., COWAN, G. & HARDMAN, M. (eds) *Petroleum Geology of the Irish Sea and Adjacent Areas*. Geological Society, London, Special Publications, **124**, 195–211.

BARNES, R. P., LINTERN, B. C. & STONE, P. 1989. Timing and regional implications of deformation in the Southern Uplands of Scotland. *Journal of the Geological Society, London*, **146**, 905–908.

CLAYTON, G., FERNANDEZ, P. & RACEY, A. 1999. Hydrocarbon source rocks in the Carboniferous succession of eastern Ireland: implications for Irish Sea exploration. *Journal of Petroleum Geology*, in press.

COOPER, A. H., MILLWARD, D., JOHNSON, E. W. & SOPER, N. J. 1993. The early Palaeozoic evolution of northwest England. *Geological Magazine*, **130**, 711–724.

——, RUSHTON, A. W. A., MOLYNEUX, S. G., HUGHES, R. A., MOORE, R. M. & WEBB, B. C. 1995. The stratigraphy, correlation, provenance and palaeogeography of the Skiddaw Group (Ordovician) in the English Lake District. *Geological Magazine*, **132**, 185–211.

CROWLEY, S. F. 1985. Lithostratigraphy of the Peel Sandstones, Isle of Man. *Mercian Geologist*, **10**, 73–76.

FORD, T. D. 1993. *The Isle of Man*. Geologists' Association, London, 1–94.

GREEN, P. F., DUDDY, I. R. & BRAY, R. J. 1997. Variation in thermal history styles around the Irish Sea and adjacent areas: implications for hydrocarbon occurrence and tectonic evolution. *In*: MEADOWS, N., TRUEBLOOD, S., COWAN, G. & HARDMAN, M. (eds) *Petroleum Geology of the Irish Sea and Adjacent Areas*. Geological Society, London, Special Publications, **124**, 73–93.

HARPER, D. A. T. & MURPHY, F. C. 1989. The Iapetus Suture in the British Isles – comment on its position in eastern Ireland. *Geological Magazine*, **126**, 723–724.

HUGHES, R. A., COOPER, A. H. & STONE, P. 1993. Structural evolution of the Skiddaw Group (English Lake District) on the northern margin of eastern Avalonia. *Geological Magazine*, **130**, 621–629.

KNELLER, B. C. 1991. A foreland basin on the southern margin of Iapetus. *Journal of the Geological Society, London*, **148**, 207–210.

KOKELAAR, B. P. 1988. Tectonic controls of Ordovician arc and marginal basin volcanism in Wales. *Journal of the Geological Society, London*, **145**, 759–776.

LAMPLUGH, G. W. 1903. *The Geology of the Isle of Man*. Memoir of the Geological Survey, UK. HMSO.

MAX, M. D., BARBER, A. J. & MARTINEZ, J. 1990. Terrane assemblage of the Leinster Massif, SE Ireland, during the Lower Palaeozoic. *Journal of the Geological Society, London*, **147**, 1035–1050.

MCCONNELL, B. & MORRIS, J. 1997. Initiation of Iapetus subduction under Irish Avalonia. *Geological Magazine*, **134**, 213–218.

MEADOWS, N., TRUEBLOOD, S., COWAN, G. & HARDMAN, M. (EDS) 1997. *Petroleum Geology of the Irish Sea and Adjacent Areas*. Geological Society, London, Special Publications, **124**.

MOLYNEUX, S. G. 1979. New evidence for the age of the Manx Group, Isle of Man. *In*: HARRIS, A. L., HOLLAND, C. H. & LEAKE, B. E. (eds) *Caledonides of the British Isles: reviewed*. Geological Society, London, Special Publications, **8**, 415–421.

NEWMAN, P. 1999. The geology and hydrocarbon potential of the Peel and Solway Basins. *Journal of Petroleum Geology*, in press.

OWEN, A. W., HARPER, D. A. T. & ROMANO, M. 1992. The Ordovician biogeography of the Grangegeeth terrane and the Iapetus suture zone in eastern

Ireland. *Journal of the Geological Society, London*, **149**, 3–6.

PRIGMORE, J. K., BUTLER, A. J. & WOODCOCK, N. H. 1997. Rifting during separation of Eastern Avalonia from Gondwana: Evidence from subsidence analysis. *Geology*, **25**, 203–207.

QUIRK, D. G. 1999. Hydrocarbon potential of the Irish Sea. *Journal of Petroleum Geology*, in press.

—— & KIMBELL, G. S. 1997. Structural evolution of the Isle of Man and central part of the Irish Sea. *In*: MEADOWS, N., TRUEBLOOD, S., COWAN, G. & HARDMAN, M. (eds) *Petroleum Geology of the Irish Sea and Adjacent Areas*, Geological Society, London, Special Publications, **124**, 135–160.

RACEY, A., GOODALL, J. G. S. & QUIRK, D. G. 1999. Palynological and geochemical analysis of borehole and outcrop samples from the Carboniferous of the Isle of Man. *Journal of Petroleum Geology*, in press.

ROBERTS, B., MORRISON, C. & HIRONS, S. 1990. Low grade metamorphism of the Manx Group, Isle of Man: a comparative study of white mica 'crystallinity' techniques. *Journal of the Geological Society, London*, **147**, 271–277.

RUSHTON, A. W. A. 1993. Graptolites from the Manx Group. *Proceedings of the Yorkshire Geological Society*, **49**, 259–262.

SIMPSON, A. 1963. The stratigraphy and tectonics of the Manx Slates Series. *Quarterly Journal of the Geological Society, London*, **119**, 367–400.

SOPER, N. J., WEBB, B. C. & WOODCOCK, N. H. 1987. Late Caledonian (Acadian) transpression in North West England: timing, geometry and geotectonic significance. *Proceedings of the Yorkshire Geological Society*, **46**, 175–192.

——, ENGLAND, R. W., SNYDER, D. B. & RYAN, P. D. 1992. Iapetus suture zone in England Scotland and eastern Ireland: a reconciliation of geological and deep seismic data. *Journal of the Geological Society, London*, **149**, 697–700.

——, STRACHAN, R. A., HOLDSWORTH, R. E., GAYER, R. A. & GREILING, R. O. 1992. Sinistral transpression and the Silurian closure of Iapetus. *Journal of the Geological Society, London*, **149**, 871–880.

STONE, P. & EVANS, J. A. 1997. A comparison of the Skiddaw and Manx groups (English Lake District and Isle of Man) using neodymium isotopes. *Proceedings of the Yorkshire Geological Society*, **51**, 343–347.

TODD, S. P., MURPHY, F. C. & KENNAN, P. S. 1991. On the trace of the Iapetus suture in Ireland and Britain. *Journal of the Geological Society, London*, **148**, 869–880.

VAUGHAN, A. P. M. & JOHNSTON, D. J. 1992. Structural constraints on closure geometry across the Iapetus suture in eastern Ireland. *Journal of the Geological Society, London*, **149**, 65–74.

Previous ideas and models of the stratigraphy, structure and mineral deposits of the Manx Group, Isle of Man

T. D. FORD,[1] EVA WILSON[2] & D. J. BURNETT[3]
[1]*Geology Department, University of Leicester, Leicester LE1 7RH, UK*
[2]*The Lifeboat House, Castletown, Isle of Man IM9 1LD, UK*
[3]*Geology, Oxford Brookes University, Oxford 0X3 0EP, UK*

Abstract: Since Bishop Wilson's [Wilson, Bishop T. (1772) *The Works of Bishop T. Wilson*, Cruttwell, Bath] account of the geology of the Isle of Man there has been a succession of increasingly sophisticated accounts of the Manx slates. Outlines were provided by Woods [Woods, G. (1811) *An account of the Past and Present State of the Isle of Man, Including ... a Sketch of its Geology*, Baldwin, London], Berger [Berger, J. F. (1814) *Transactions of the Geological Society of London*, **2**; (1824) *Annals of Philosophy, Series 2*, **viii**] and MacCulloch [MacCulloch, J. (1819) *A Description of the Western Isles of Scotland, Including the Isle of Man*, Constable, London] in the early nineteenth century but the first attempt at a stratigraphic subdivision was by Cumming in 1846 [Cumming, J. G. (1846) *Quarterly Journal of the Geological Society*, **31**]. Lamplugh's map and memoir at the turn of the century provided the first full account [Lamplugh, G. W. (1903) *The Geology of the Isle of Man*, HMSO]. Later, Gillott [e.g. Gillott, J. E. (1955) *Geological Magazine*, **92**] examined the Manx slate in more detail but it was not until the 1960s that Simpson [Simpson, A. (1963) *Quarterly Journal of the Geological Society*, **119**] defined an 11 unit stratigraphic succession and delineated a refolded synclinal structure. Subsequently, discoveries of graptolites and acritarchs have necessitated revisions to both Simpson's succession and structure. Correlation with some parts of the Skiddaw Group is now well established.

Although the Manx 'slates' outcrop over more than three-quarters of the island, their supposed monotonous character and apparent lack of fossils meant that they have attracted little attention from geologists until the last few decades. Of a bibliography of 572 entries related to Manx geology (Wilson, 1999) only 29 were concerned with the Manx slates. Their apparent correlation with the Skiddaw slates similarly meant that they were merely considered as an extension of the Lower Palaeozoic of the Lake District.

The purpose of this contribution is to review the growth of knowledge concerning the Manx slates as a historical background for the following papers in this volume.

Before Lamplugh

The Isle of Man's geology first came to notice with brief accounts by Bishop Wilson (1772) and Woods (1811). Their comments were soon followed by those of Jean Berger (1814), a Swiss political refugee who had been a student of Gottlieb Werner at Freiburg and who became a protege of MacCulloch. Berger's descriptions include tables of barometric heights. He noted the uplands as 'a rising en masse of the land, a foundation on which rest several mountains ... disposed in some regular order'. Granite and slate on the highest ground were regarded as 'primitive' rocks in Werner's system, whilst the coarser grained rocks of the lower ground were 'transition' rocks. The Carboniferous limestone and later rocks were classified as 'secondary'. Berger (1814) observed that there were three transverse valleys, the central valley, the Douglas–Port Erin Tract and Calf Sound. He deduced that, at some time, there had been a subsidence to the south which caused the three dislocations and that the same subsidence may have produced the general dip of the stratified rock to the south.

William Smith's famous map of 1815 included what must have been secondhand knowledge of the geology of the Isle of Man with an area of Carboniferous limestone stretching from the Calf of Man to Onchan, much larger than is actually present, and the rest of the island shown, surprisingly, as Old Red Sandstone!

Only five years after Berger, MacCulloch himself included an account of the Isle of Man in his Description of the Western Isles of Scotland (MacCulloch 1819). In it he moved away from the

From: WOODCOCK, N. H., QUIRK, D. G., FITCHES, W. R. & BARNES, R. P. (eds) 1999.
In Sight of the Suture: the Palaeozoic geology of the Isle of Man in its Iapetus Ocean context.
Geological Society, London, Special Publications, **160**, 11–21. 1-86239-046-0/99/$15.00

Wernerian theory and followed Hutton in regarding the granites as intrusive, but otherwise he supported Berger's ideas. He was followed by John Henslow, who brought from Cambridge what must have been one of the first ever field classes to study a specific area. Henslow's account (1821) included severe criticism of Berger but he added an improved map. The argument continued in 1824 with a reply from Berger and further comments from Henslow (1824).

From 1846 onwards the Reverend Cumming published a series of papers in the *Quarterly Journal of the Geological Society*. He was Vice Principal of King William's College at Castletown from 1841 to 1858, when he moved to Birmingham to become Professor of both Classical Literature and of Geology. Most of these papers (e.g. Cumming 1848) were concerned with philosophical matters or with the Quaternary, including land bridges, variations in sea level and glacial erratics, not so long after the recognition of glaciation in Britain. Cumming (1846, 1847, 1861) also studied the Manx slates; the cited papers represent a first attempt at applying stratigraphic terms to subdivisions of the Manx slates. He also suggested that it was 'convenient at present to include (them) under the general term Cambro-Silurian, intending thereby all the rocks under the Lower Silurian of Sir Roderick Murchison' (Cumming 1861). However, Grindley (1862) disagreed and assigned the Manx slates to the Cambrian. Harkness & Nicholson (1866) accepted Cumming's Cambro-Silurian age as they misinterpreted the green colouration of the Lonan Flags on Clay Head as indicating that they were part of the Borrowdale Volcanic Series. Later, Ward (1880) disagreed with this view, although he accepted a correlation of the Manx and Skiddaw slates as proposed by Harkness & Nicholson (1866). Horne (1874) argued that the Manx slates were older than the 'Silurian' rocks of the Southern Uplands.

Possible worm trails and dubio-fossils were found in the Manx slates at various localities and caused Taylor (1864) to refer the slates to the Lower Cambrian. Poorly preserved dendroid graptolites were found in a loose block on Cronk Sumark (also spelled Shamerk or Shumark) in 1893 (Bolton 1899) and were taken to indicate a Tremadoc age for all the underlying beds, effectively the whole of the Manx slates. Doubts remained, as these dendroid graptolites could not be found in the solid rock, and the possibility arose that they might have been part of a a glacial erratic.

Lamplugh

George W. Lamplugh was an officer of the Geological Survey of England and Wales, whose previous experience of mapping had been in Lincolnshire. Although the Isle of Man was strictly not part of the United Kingdom, he was, in effect, seconded to the Isle of Man Government to make a geological map of the island and he spent five field seasons there, 1892–1897. Fair copies of his 6 inch:1 mile field maps are held in the Manx Museum. Lamplugh's 1 inch:1 mile map was published in 1898 and his memoir followed in 1903 (Lamplugh 1903). At 620 pages it was one of the largest memoirs ever published by the Geological Survey. The Economic Geology section of the memoir was also issued separately as an Economic Memoir in the same year. In hindsight Lamplugh's effort was magnificent in recording details, but he was working long before modern techniques of sedimentological and structural analysis were known. His sketches suggest that he was not fully aware of the significance of some sedimentary structures and he omitted much of what would be regarded as diagnostic today. His memoir provided a comprehensive summary of all previous literature and was packed with his observations so that it became the obvious starting point for all later investigations.

Lamplugh (1903) saw the Manx 'slates' as a sequence of marine sedimentary rocks deposited in a subsiding area of moderate depth at no great distance from land. A source area provided all grades from coarse sand to mud. He recognized lateral facies changes and stated that his boundaries were only approximate owing to the difficulty of tracing boundaries inland. He assigned the Manx slates to the Cambrian, though with some reservations (the Tremadoc was then regarded as the topmost subdivision of the Cambrian). He divided the succession into three units but left large areas as unassigned (Fig. 1). The so-called 'crush conglomerates' were locally developed between the two highest units:

3. Barrule Slate (crush conglomerates);
2. Agneash Grits, plus other grits;
1. Lonan and Niarbyl Flags.

Lamplugh's failure to find suitable marker horizons precluded more detailed subdivision. It is, however, notable that he correlated the Lonan Flags with the Niarbyl Flags, thereby indicating a structure with repetition on the northwest and southeast limbs. Although he made this correlation he knew that there were differences, with the Niarbyl Flags being more calcareous (Lamplugh 1903). The absence of the Agneash Grits on the west side was explained by a thrust fault at Niarbyl. The Agneash Grits around Agneash were impure flaggy quartzites with thin argillaceous partings. This lithology appeared at various other localities, such as Santon Head and

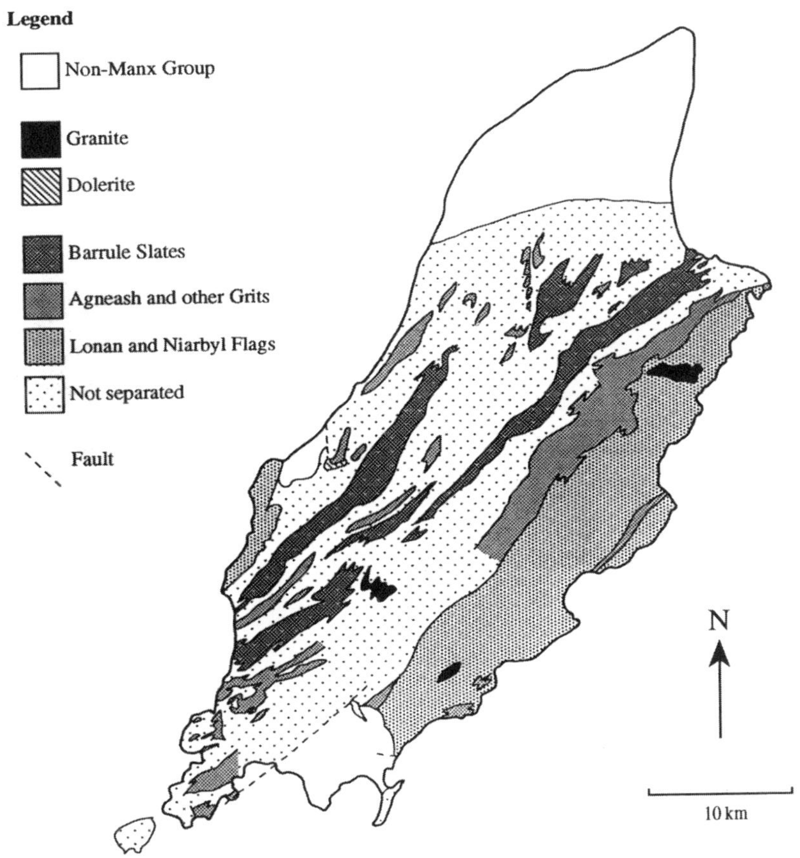

Fig. 1. Sketch map of the solid geology of the Isle of Man after Lamplugh's 1 inch:1 mile Geological Survey map, showing his subdivisions of the Manx Slates.

Mull Hill, but Lamplugh found difficulty in including these in his stratigraphic interpretation. The highest unit was the Barrule Slate, a fine-grained slate belt which is a prominent feature along the axis of the island, apparently repeated in places and interpreted as indicating the synclinorial structure. The slate seemed to pinch-out downwards and this was taken to confirm the basic synclinal structure as shown in Figs 2 and 3. Beds consisting of alternating laminae of argillaceous and arenaceous material occurred between the main units and were regarded as transitional between the three main divisions and left unassigned. The crush conglomerates were thought by Lamplugh (1903) to be produced by brecciation during compression of juxtaposed different strata, mainly the Barrule Slates and the Agneash Grits.

According to current ideas, Lamplugh (1903) seems to have grossly underestimated the thicknesses of his units. He suggested that the Barrule Slate was 200 ft thick before deformation, 2000 ft after, the ten-fold increase being due to repetition by isoclinal folding. The Agneash Grits were thought to be only 100 ft thick before deformation, whilst the Lonan Flags were 1000 ft thick before deformation. The total thickness of the Manx slates was therefore estimated as a mere 1500–2000 ft, but with concertina-like repetition of strata (Fig. 2). Lamplugh noted that the cleavage seemed to dip outwards from the northeast–southwest trending central spine of the island, although his sections indicate a more consistent westerly dip of the cleavage and associated structures (Fig. 3).

Lamplugh deduced a succession of repeated episodes of folding and cleavage formation and was therefore well ahead of his time. He proposed the following sequence of events:

1. folding – the main period of deformation – with strain-slip and brecciation;

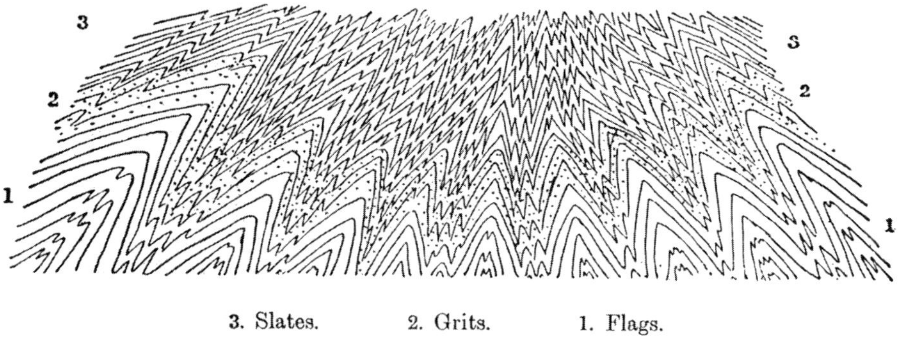

Fig. 2. Idealized diagram of the Manx synclinorium: a diagrammatic northwest–southeast cross-section across the Manx Slate Massif. From Lamplugh's Geological Survey memoir (Lamplugh 1903).

2. intrusion of basic dykes and quartz veins;
3. folding with formation of cleavage and mineralization;
4. intrusion of granite;
5. renewed folding and cleavage formation.

Palaeontologically, Lamplugh (1903) had little new to report, and repeated much of Bolton's (1899) paper, which provides a useful summary of all previous records. Lamplugh was sceptical over most of the alleged fucoids, burrows and a trilobite, but he accepted the worm trails *Palaeochorda major*, *P.minor* and *Chondrites*. He regarded the record of *Dictyonema* and other dendroid graptolites as of doubtful value as they were on a loose block.

On regional correlations, Lamplugh was guarded. He could find little in the way of comparison with the Cambrian of North Wales or Ireland and he was uncertain of correlation with the Skiddaw slates, although if there was one it was with the lower part thereof, i.e. Upper Cambrian.

After Lamplugh

Only two years after the publication of Lamplugh's (1903) seminal memoir, Blake (1905) challenged his interpretation of the stratigraphic succession. Blake presented arguments, supported by one cross-section of the north-central part of the island (Fig. 4) and a scatter of local sections, from which he deduced the the sequence was:

5. Lonan and Sulby Flags unconformable on;
4. Schistose Breccia;
3. Barrule Slates;
2. Snaefell Laminated Slates;
1. Agneash Grits.

Blake apparently regarded the Lonan and Sulby Flags as equivalents resting unconformably on the older strata – a marked contrast to both Lamplugh's (1903) and Simpson's (1961, 1963*a*) works. The Niarbyl Flags were shown on Blake's section as Lonan Flags, indicating that he accepted the correlation between flaggy beds on east and west coasts.

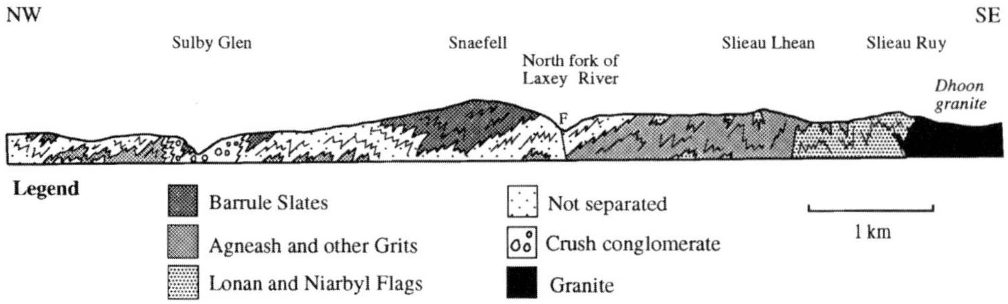

Fig. 3. Part of one of Lamplugh's (1903) sections across the Manx Slate Massif to show the synclinorial structure and westerly dopping cleavage. from Geological Survey memoir.

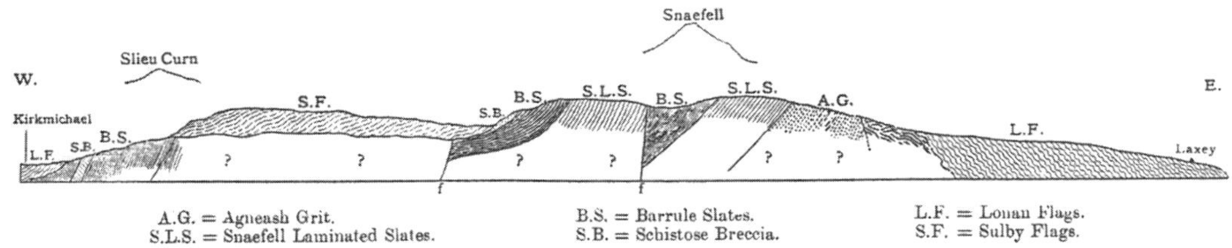

Fig. 4. Cross-section of the Manx Slate Massif by Blake (1905) showing repetition of the Barrule Slates by faulting, and the alleged unconformable relationship of the Lonan and Sulby Flags (reproduced from the *Quarterly Journal of the Geological Society*).

Blake (1905) argued that the two parallel Barrule Slate belts were simply repetitions by faulting rather than the synclinorium concept of Lamplugh (1903). Blake showed his units 1–4 dipping regularly to the northwest, with the Lonan Flags regarded as a younger unit resting on the eroded ends of the Agneash Grits.

For the next 50 years little research was attempted on the Manx Group. The stratigraphic and sedimentological part of Gillott's PhD thesis (1954) mostly remained unpublished, and we have to be satisfied with very brief summaries of his stratigraphic succession in three short papers (Gillott 1955, 1956*a*, *b*). Without full description, the logic of the stratigraphic sequence is difficult to interpret, although he presented useful structural data. Gillott listed the sequence within the Manx Group as:

7. Cronk Sumark Slate;
6. Sulby Flags;
5. Breccia;
4. Barrule Slates;
3. Banded Strata;
2. Agneash Grits;
1. Lonan Flags.

Gillott's (1956*a*) simple map of the northeastern part of the outcrop implied that he regarded the basic structure as anticlinal, with subsidiary folds superimposed upon it. Indeed, he referred to northwesterly dips in the northwest and southeasterly dips in the southeast, having at times apparently confused cleavage directions with bedding dips. The oldest unit, the Lonan Flags, apparently lay in the core of the anticline despite its main outcrop being along the east coast. Gillott (1956*a*) also recognized that there had been at least two periods of folding and cleavage. The first had acute folds and axial-plane flow cleavage whilst the second had more open folds and axial-plane 'fracture' cleavage. Gillott (1955) regarded the higher grade metamorphic axial zone of the island as being due to its depth within the core of the anticline. He recognized the distribution of garnet and cordierite as indicative of amphibolite facies metamorphism but did not distinguish the thermal (contact) effects of the Foxdale Granite from the regional metamorphism. The axial climax of metamorphic effects was, however, ascribed to a buried granite mass.

Simpson

Based on his PhD thesis, Simpson (1961) provided a comprehensive synthesis of the Manx slate stratigraphy, structure, metamorphism and intrusions (Simpson 1963 *a*, *b*, 1964 *a*, *b*, 1965 *a*, *b*, 1968). He provided the first comprehensive map of the Manx slate (Fig. 5) subdivided into 11 units (with a twelfth volcanic unit of uncertain relationship):

	Thicknesses (ft)
11. Glen Dhoo Flags	3200
10. Cronkshamerk Slates	1580–1850
9. Sulby Flags	2100
8. Sulby Slump–Breccias	900–1500
7. Slieugh Managh Slates	1100
6. Injebreck Banded Group	2000–2900
5. Barrule Slates	530–3400
4. Maughold Banded Group	2600–6000
3. Lonan and Niarbyl Flags	2000–10 000
2. Bailanayre Slump–Breccia	20–500
1. Lady Port Banded Group	400

This succession totalled *c.* 25 000 ft (*c.* 8000 m) and was thought to have been deposited in a geosyncline during the Cambrian, based on the graptolite dating of the Cronk Sumark Slate as Tremadoc (then regarded as the topmost series of the Cambrian period but now regarded as lowermost Ordovician). It should be noted that Simpson's work just pre-dated acceptance of plate tectonic concepts, and thus the Iapetus Ocean and its suture were unknown to him.

Simpson (1961, 1963*a*) recognized four lithofacies:

- flaggy formations – siltsone, greywacke and subordinate pelite;

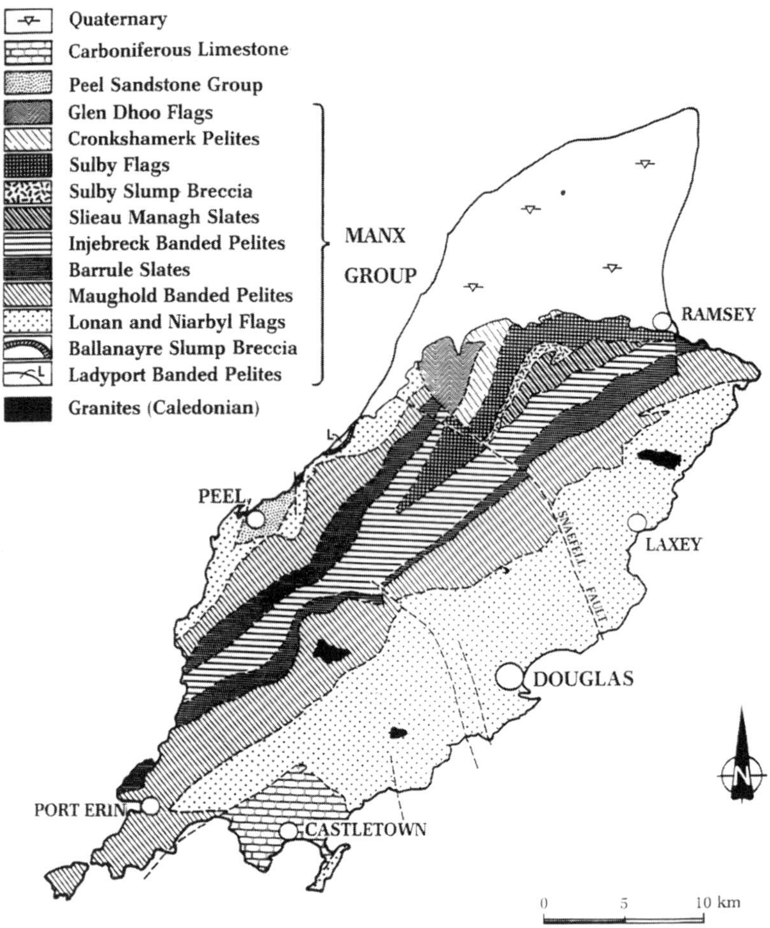

Fig. 5. Simplified version of Simpson's 1963 map of the subdivisions of the Manx Slate Series (from Ford 1993).

- banded formations – rapid alternations of siltstone and pelite, with occasional thicker beds;
- pelitic formations – uniform pelites with occasional silty bands;
- slump breccias – pelitic groundmass with clasts of slate, siltstone and greywacke, with occasional interbeds of undisturbed banded sediment.

Two small areas of andesitic extrusive rocks were also noted.

Although there was considerable variation in the character and thicknesses of these formations, his interpretation was, in part, similar to Lamplugh's (1903) with a basic synclinal structure. Simpson (1961, 1963*a*) also argued for a correlation between the Lonan Flags on the east and the Niarbyl Flags on the west.

Within the main synclinal arrangement, Simpson recognized a series of structural episodes:

- F1 – folds resulting from northwest–southeast compression with northeast–southwest axes and plunging at 35° to the southwest (the Isle of Man syncline fell within this category);
- S1 – cleavage dipping between 35° to the northwest and vertical; synchronous with and parallel to F1 fold axes;
- F2 – folds due to nearly vertical compression with axes trending northeast–southwest and with axial planes dipping gently northwest causing F1 folds to be 'crumpled': the folds included the Peel–Ballaugh antiform, structurally the highest; the Manx synform, which refolded the Isle of Man syncline; and the Mull Hull antiform, at the lowest level;
- S2 – cleavage, due to slippage parallel to F2 fold axes;
- F3 – folding resulting from east-northeast–west-

southwest compression, almost at right angles to previous structures, with axes trending north-northwest and with axial planes dipping east-northeast;
- S3 – strain-slip cleavage causing small displacements of all previous structures.

A detailed map of the fold axes and faults was provided by Simpson (1963*a*). Simpson, together with Helm *et al.* (1963), also identified the three folding phases, F1–F3, in the Lake District, and elsewhere in the Caledonides.

A map of the axial metamorphic zone was also provided by Simpson (1964*a*). He noted the presence of cordierite, garnet and chloritoid porphyroblasts confined to the axial zone. He deduced that the most intense metamorphic effects were in a zone dipping at a low angle to the northwest, so that rocks on higher ground were, in places, at a higher metamorphic grade than those on adjacent lower ground. He also noted that the Foxdale Granite was surrounded by a considerable contact metamorphic aureole containing garnet and cordierite, as well as metasomatic and vein tourmaline. Abundant quartz veining in the axial zone was linked to the F2 phase of folding (Simpson 1963*b*) .

Simpson's (1961) research on the Manx Slate Series was a seminal piece of work which has stood for 30 years. It was summarized by Robinson & McCarroll (1990) and provided the basis for Ford's guidebook (Ford 1993), where the names of the units were defined as formations in accordance with the Rules of Stratigraphic Nomenclature. Recently, however, problems with Simpson's (1963*a*) model have become apparent, and a revised stratigraphy and structure are presented in the papers which follow.

Post-Simpson

The first step towards revision of Simpson's interpretation of the stratigraphic sequence was that of Molyneux (1980). Following Downie & Ford (1966), who reported late Tremadoc–early Arenig acritarchs in the Lonan Flags, Molyneux (1980) found acritarchs at some 20 other localities, from which he deduced that Simpson's units 1 and 4 were Arenig age whilst units 3 and 11 were Tremadoc age, an apparent reversal of parts of Simpson's (1963*a*) stratigraphic succession.

The discovery of Arenig graptolites in the Lonan Flags (Rushton 1993) confirmed a correlation with the northern belt of the Skiddaw slates in the Lake District (Cooper & Molyneux 1990). As a result, a completely revised stratigraphic sequence was proposed, which preserved as much of Simpson's (1963*a*) order of succession as possible while putting the biostratigraphically dated formations in their correct superpositional order (Cooper *et al.* 1995), as shown below:

9. Sulby Flags;
8. Sulby Slump–Breccia/Ballanayre Slump–Breccia;
7. Managh Slieau Slates;
6. Lady Port Banded Group;
5. Injebreck Banded Group;
4. Barrule Slates;
3. Maughold Banded Group;
2. Lonan Flags/Glen Dhoo Flags (Niarbyl Flags);
1. Cronk Sumark Slates.

This revised sequence contradicted Simpson's (1963*a*) interpretation of the synclinal structure and necessitated the presence of a series of strike faults, or even thrusts.

One of the few other attempts to follow up Simpson's work was that of Roberts *et al.* (1990) – who applied the new technique of mica crystallinity analysis to the metamorphic rocks – which came to much the same conclusion regarding the distribution of metamorphic grades and argued that these confirmed the presence of a large buried granite intrusion, as proposed by Cornwell (1972). Roberts *et al.* (1990) also indicated the need for previously unrecognized faults to explain the relationships on the west coast.

By analogy with the structures seen in the Skiddaw slates, a rigid block comparable to the Borrowdale Volcanic Group was thought necessary to the southeast of the Isle of Man to explain the structures in the Manx Group, but no geophysical evidence has yet been put forward (Hughes *et al.* 1993).

Slump Breccias

Lamplugh's (1895, 1903) interpretation of these pebbly mudstones as 'crush conglomerates', caused by tectonic crushing together of different rock types, was challenged by Blake (1905), who thought these 'schistose breccias' had been deposited as sedimentary rocks with subsequent schistosity imposed upon them. Gillott (1956*b*) agreed that they were of sedimentary origin but went further in explaining the brecciation as being due to slumping or sliding. As noted above, Simpson (1963*a*) agreed with Gillott's (1956*b*) interpretation of an origin by sedimentary slumping, but he also used them as stratigraphic units. Similar rocks are present within the Skiddaw slates and both those and the Manx slump breccias are now regarded as examples of debris flows or olistostromes, i.e. massive sedimentary slumps (Cooper *et al.* 1995).

Granites

The positions and characters of the three small granitic intrusions at Foxdale, Dhoon and Oatlands were well established by Lamplugh (1903). For a long time, the Foxdale Granite failed to attract any subsequent investigation but the Dhoon Granite was shown by Nockolds (1931) to be a complex with varying degrees of contamination by absorption of country rock. Similarly, the poorly exposed Oatlands Intrusion was found to be a complex intrusion with a gabbroic core containing abundant, partly digested, xenoliths (White 1909; Taylor & Gamba 1933). The Foxdale Granite was shown by both Lamplugh (1903) and Simpson (1965*a*) to have a much larger subsurface extent than its 1 km^2 area of outcrop, as its aureole was up to 3 km wide. Mining records showed that granite was at shallow depths in the Glen Rushen and Cornelly Mines, indicating that the granite intrusion had gently sloping sides.

The gravity survey by Cornwell (1972) suggested that the small isolated granite masses on the Isle of Man were linked at depth and he deduced that a major granite intrusion lay beneath the whole axial region of the island.

The Foxdale Granite was dated by K–Ar techniques at 374 ± 7 Ma (Brown *et al.* 1969; Ineson & Mitchell 1979) and the Dhoon Granite at 370 Ma (Harper 1966). The Manx slate itself was dated by K–Ar methods at 411 Ma (Harper 1966). Unmetamorphosed mudstone from the east coast was dated at 415 Ma. However, argon loss is a well-known problem with K–Ar dating and the above dates may be unreliable. None the less, it seems clear that the granites and metamorphism were late Caledonian events.

Simpson (1965*a*) deduced that the emplacement of the Foxdale Granite was syn-tectonic with the F2 phase of folding, whilst Harper (1966) thought that the Dhoon Granite was early F2 in age. Pegmatite veins with local concentrations of beryl cut the Foxdale Granite (Dawson 1966), whilst quartz veins with muscovite, tourmaline and ilmenite cut both granite and pegmatite veins.

An intrusive 'porphyritic feldspar-rich rock' was noted by Lamplugh (1903) at Lhergydhoo, east-northeast of Peel. Subsequent quarrying has revealed an augite-rich basic intrusion, known locally as the Poortown Gabbro although it is actually a coarse dolerite. Contact relationships with surrounding rocks are largely obscured by glacial deposits and until recently it seems to have escaped further scientific investigation, despite being the island's main source of roadstone.

Minor intrusions

Lamplugh (1903) mapped some 80 dykes or 'greenstones', mostly on coastal sections. These are altered-basic or intermediate rocks with biotite-, hornblende- or augite-rich varieties, with some classified as lamprophyres. Several small elvans or porphyritic microgranite intrusions were noted by Lamplugh and were more thoroughly investigated by Simpson (1964*b*). They were regarded as coeval with the Foxdale and Dhoon Granites.

Mineral deposits

The mines at Foxdale and Laxey were worked from at least the 18th century and were active until 1919. However, there is little in Lamplugh's memoir (1903), or any other literature, giving detailed descriptions of the deposits or of hypotheses concerning their origin. The Geological Survey's Reports on Mineral Resources (e.g. Carruthers & Strahan 1923; Dewey & Eastwood 1925) do little more than repeat Lamplugh's memoir, whilst Skelton's (1956) review was more concerned with mining history. However, Mackay & Schnellmann's (1963) report to the Isle of Man Government was a more thorough description and included much material from mining archives in the Manx Museum, although, again, little comment was made on metallogenesis. Uraniferous hydrocarbons (thucolite) found in the Laxey Mine were investigated by Parnell (1988), who deduced that the veins had once extended up into a Carboniferous cover and hence that the mineralization was no older than Carboniferous in age, not 'Caledonian' as previously thought. Isotopic investigations by Crowley & Bottrell (1997) have indicated mineralizing events at 300–280 and 210–190 Ma, corresponding to Hercynian inversion and Mesozoic basin formation, respectively. Their arguments have been applied to the Laxey lode (Ford 1999).

Palaeontology

As long ago as 1864, Taylor (1864) recorded *Orthoceras* on 'Mount Craig', thought to be 'The Creg' in the axial belt of highest metamorphic grade, which seems a most unlikely place for any fossil to survive. A variety of worm trails were also noted by Taylor (1864), Grindley (1863) and Binney (1877), the details of which were summarized in Bolton (1899) and Lamplugh, (1903). Dendroid graptolites, including *Dictyonema socialis*(?), were found in a loose block on Cronk Sumark (Bolton 1893, 1899) and were thought to prove a Tremadoc (then regarded as Upper Cambrian) confirming correlation with the

Skiddaw slates. The identity of the graptolites was uncertain owing to their poor preservation, and also as dendroids range up into the Carboniferous their use in dating has its limits.

Acritarchs have been found in at least four of Simpson's (1963*a*, *b*) units and confirm a Tremadoc–Arenig age, as noted above.

An assemblage of rather poorly preserved graptolites was found in the Lonan Flags at Baltic Rock on the Santon coast and again indicated an Arenig age (Rushton 1993).

As reported elsewhere in this volume, both graptolites and orthocones have recently been found in the Niarbyl Flags, indicating a middle Silurian age for this unit, now reassigned to the Dalby Group and no longer correlated with the Lonan Flags.

The wider context

Soon after Simpson (1963) published his research the concept of plate tectonics, with such processes as sea-floor spreading and subduction, threw new light on the early Palaeozoic evolution of Britain. In particular, a hypothesis of closure of an ancient ocean, Iapetus, was proposed with eventual collision between northern Britain, as part of the ancient continent Laurentia, with southern Britain, as part of Avalonia (a fragment of the northern margin of a southern continent, Gondwana). Final collision occurred during Early Devonian times, along a suture lying close to the English–Scottish border and extending across the Irish Sea to Leinster (McKerrow & Soper 1989). The suture is generally thought to lie in the northern Irish Sea not far north of the Isle of Man (Soper *et al.* 1992).

Within this plate tectonic context, the Manx and the Skiddaw Groups, and probably the Ribband Group of Leinster, are regarded by most investigators as parts of the same depositional regime and the most likely palaeogeographical position of the Manx Group was on the northwestern margin of eastern Avalonia (Cooper *et al.* 1995; Quirk & Kimbell 1997).

Petrographic and geochemical work by Moore (1992) suggested that the Manx and Skiddaw Groups may have shared a common depositional basin with a heterolithic, recycled sediment, source. Neodymium isotope results, as reported by Stone & Evans (1997), also support the likelihood of a common sediment source for most of the Skiddaw Group and at least part of the Manx Group. However, they also provide evidence for an influx of juvenile volcanic debris into the Manx basin earlier than seen in the Skiddaw Group, indicating that sedimentation along the northern margin of Avalonia was not uniform. Stone & Evans' (1997) results are, however, compatible with the back-arc basin depositional setting for the Manx Group proposed by Quirk & Kimbell (1997).

Conclusion

Over 200 years of geological research on the Isle of Man have seen the development of a variety of ideas and theories. A consensus has yet to be reached but the papers presented in this volume go a long way towards arriving at a coherent story. Although the Manx Group succession proposed by Simpson (1963*a*) has proved to be flawed, his work has provided a basis for all subsequent study on the stratigraphy, sedimentology and structure of the island. In the post-Simpson period, research has focused on revising the stratigraphic succession, with the recognition of Silurian rocks within the Manx massif and presence of large faults. Other research has investigated the provenance and transport of the sediments, the timing of the granite and other intrusions, both thermal and regional metamorphism, and the mineralization process. As these lines of research progress, so further questions arise which are addressed in some of the following papers.

References

BERGER, J. F. 1814. Mineralogical account of the Isle of Man. *Transactions of the Geological Society of London.*, **2**, 29–65.

—— 1824. Reply to Mr Henslow's obeservations on Dr Berger's account of the Isle of Man. *Annals of Philosophy, Series 2*, **viii**, 367.

BINNEY, E. W. 1877. A notice of some organic remains from the schists of the Isle of Man. *Proceedings of the Manchester Literary & Philosophical Society*, **16**, 1–8.

BLAKE, J. F. 1905. On the order of succession in the Manx Slates. *Quarterly Journal of the Geological Society*, **61**, 358-373.

BOLTON, H. 1893. Observations on the Skiddaw Slates of the North of the Isle of Man. *Report of the British Association for 1893*, 770–771.

—— 1899. The palaeontology of the Manx Slate of the Isle of Man. Memoirs & *Proceedings of the Manchester Literary & Philosophical Society*, **43**, 1–15.

BROWN, P. E., MILLER, J. A. & GRASTY, R. L. 1969. Isotopic ages of the late Caledonian granitic intrusions in the British Isles. *Proceedings of the Yorkshire Geological Society*, **36**, 261–276.

CARRUTHERS, R. G. & STRAHAN, A. 1923. Lead and zinc ores of Durham, Yorkshire, Derbyshire and the Isle of Man. *Geological Survey Special Report on Mineral Resources on the Isle of Man*. **26**, 89–91.

COOPER, A. H. & MOLYNEUX, S. G. 1990. The age and correlation of the Skiddaw Group (early Ordovician) sediments in the Cross Fell inlier, northern England. *Geological Magazine*, **127**, 147–157, (IOM 156).

——, RUSHTON, A. W. A., MOLYNEUX, S. G., HUGHES, R. A., MOORE, R. M. & WEBB, B. C. 1995. The stratigraphy, correlation, provenance and palaeogeography of the Skiddaw Group (Ordovician) in the English Lake District. *Geological Magazine*, **132**, 185–211. (IOM 206–207).

CORNWELL, J. D. 1972. A gravity survey of the Isle of Man. *Proceedings of the Yorkshire Geological Society*, **39**, 93–106.

CROWLEY, S. F. & BOTTRELL, S. H. 1997. An isotopic investigation of base metal mineralization on the Isle of Man and its relationship to fluid flow events and geotectonic evolution of the Irish Sea: initial results. *Geo-fluids Conference, Belfast*, **Abs**.

CUMMING, J. G. 1846. On the geology of the Isle of Man. *Quarterly Journal of the Geological Society*, **2**, 317–348.

—— 1847. On the geology of the Calf of Man. *Quarterly Journal of the Geological Society*, **3**, 179–185.

—— 1848. *The Isle of Man: history, physical, ecclesiastical, civil and legendary*. Van Voorst, London.

—— 1861. *A guide to the Isle of Man*. Stanford, London.

DAWSON, J. 1966. Beryllium in the Foxdale granite. *Bulletin of the Geological Survey of Great Britain* **25**, 55–58.

DEWEY, H. & EASTWOOD, T. 1925. Copper ores. *Geological Survey Special Reports on the Isle of Man*, **30**, 78–82

DOWNIE, C. & FORD, T. D. 1966. Microfossils from the Manx Slate Series. *Proceedings of the Yorkshire Geological Society*, **35**, 307–322.

FORD, T. D. 1993. *The Isle of Man*. Geologists Association Guide, **46**.

—— 1999. The geology of Laxey Mine. *Mercian Geologist*, **14**, 125–134.

GILLOTT, J. E. 1954. *Sedimentary petrology, metamorphism, and structure of the Manx Slates*. PhD Thesis, University of Liverpool.

—— 1955. Metamorphism in the Manx Slates. *Geological Magazine*, **92**, 141–154.

—— 1956*a*. Structural geology of the Manx Slate. *Geological Magazine*, **93**, 301–313.

—— 1956*b*. Breccias in the Manx Slates: their origins and stratigraphic relations. *Liverpool & Manchester Geological Journal*, **1**, 370–380.

GRINDLEY, T. 1862. Geology of the Isle of Man. *Geologist*, **5**, 171–183.

—— 1863. Footprints in the Cambrian? slates. *Geologist*, **6**, 315.

HARKNESS, R. & NICHOLSON, H. A. 1866. On The lower Silurian rocks of the Isle of Man. *Quarterly Journal of the Geological Society*, **22**, 488–491.

HARPER, C. T. 1966. Potassium–argon ages of slates from the southern Caledonides of the British Isles. *Nature*, **212**, 1339–1341.

HELM, D. G., ROBERTS, B. & SIMPSON, A. 1963. Polyphase folding in Caledonides south of the Scottish Highlands. *Nature*, **200**, 1060–1062.

HENSLOW, J. S. 1821. Supplementary observations to Dr Berger's account of the Isle of Man. *Transactions of the Geological Society of London*, **5**, 482–505.

—— 1824. Remarks on Dr Berger''s reply to Mr Henslow's observations. *Annals of Philosophy, Series 2*, **viii**, 407.

HORNE, J. 1874. A sketch of the geology of the Isle of Man. *Transactions of the Edinburgh Geological Society*, **2**, 323–347.

HUGHES. R. A., COOPER, A. H. & STONE, P. 1993. Structural evolution of the Skiddaw Group (English Lake District) on the northern margin of eastern Avalonia. *Geological Magazine*, **130**, 621–629.

INESON, P. R. & MITCHELL, J. G. 1979. K–Ar ages from the ore deposits and related rocks of the Isle of Man. *Geological Magazine*, **116**, 117–128.

LAMPLUGH, G. W. 1895. The crush conglomerates of the Isle of Man. *Quarterly Journal of the Geological Society*, **51**, 564–597.

—— 1903. *The geology of the Isle of Man*. Memoir of the Geological Survey, United Kingdom. HMSO.

MACCULLOCH, J. 1819. *A Description of the Western Isles of Scotland, Including The Isle of Man*. Constable, London.

MACKAY, L. & SCHNELLMANN, G. A. 1963. *The mines and minerals of the Isle of Man*. Report to the Isle of Man Government.

MCKERROW, W. S. & SOPER, N. J. 1989. The Iapetus Suture in the British Isles. *Geological Magazine*, **126**, 1–8.

MOLYNEUX, S. G. 1980. New evidence for the age of the Manx Group, Isle of Man. *In*: HARRIS, A. L., HOLLAND, C. H. & LEAKE, B. E. (eds) *The Caledonides of the British Isles – Reviewed*. Geological Society, London, Special Publications, **8**, 415–421.

MOORE, R. M. 1992. *The Skiddaw Group of Cumbria: early Ordovician Turbidite sedimentation and provenance on an evolving micro-continental margin*. PhD Thesis, University of Leeds.

NOCKOLDS, S. R. 1931. The Dhoon (Isle of Man) granite: a study in contamination. *Mineralogical Magazine*, **22**, 494–509.

PARNELL, J. 1988. Mineralogy of uraniferous hydrocarbons in Carboniferous-hosted mineral deposits, Great Britain. *Uranium*, **4**, 197–218.

QUIRK, D. G. & KIMBELL, G. S. 1997. Structural evolution of the Isle of Man and central part of the Irish Sea. *In*: MEADOWS, N. S., TRUEBLOOD, S. P., HARDMAN, M. & COWAN, G. (eds) *Petroleum Geology of the Irish Sea and Adjacent Areas*. Geological Society, London, Special Publications, **124**, 135–159.

ROBERTS, B., MORRISON, C. & HIRONS, S. 1990. Low grade metamorphism of the Manx Group, Isle of Man: a comparative study of white mica crystallinity techniques. *Journal of the Geological Society*, **147**, 271–277.

ROBINSON, V. & MCCARROLL, D. 1990. *The Isle of Man: celebrating a sense of place*. Liverpool University Press.

RUSHTON, A. W. A. 1993. Graptolites from the Manx

Group. *Proceedings of the Yorkshire Geological Society*, **49**, 259–262.

SIMPSON, A. 1961. *The stratigraphy and tectonics of the Manx Slate*. PhD Thesis, University of London.

—— 1963*a*. The stratigraphy and tectonics of the Manx - Slate Series, Isle of Man. *Quarterly Journal of the Geological Society*, **119**, 367–400.

—— 1963*b*. F2 quartz veining in the Manx Slate Series. *Nature*, **199**, 900–901.

—— 1964*a*. The metamorphism of the Manx Slate Series. *Geological Magazine*, **101**, 20–36.

—— 1964*b*. Deformed acid intrusions in the Manx Slate Series, Isle of Man. *Geological Journal*, **4**, 189–206.

—— 1965*a*. The syn-tectonic Foxdale–Archallagan granite and its metamorphic aureole. *Geological Journal*, **4**, 415–434.

—— 1965*b*. F1 cross-folding in the Manx Slate Series, Isle of Man. *Geological Magazine*, **102**, 440–444.

—— 1968. The Caledonian history of the northeastern Irish Sea region and its relation to surrounding areas. *Scottish Journal of Geology*, **4**, 135–163, (discussion 375–385).

SKELTON, R. H. 1956. The Manx mines. *Mining Magazine*, 9–18.

SOPER, N. J., ENGLAND, R. W. & SNYDER, D. B. 1992. The Iapetus suture in England, Scotland and eastern Ireland: a reconciliation of geological and deep seismic data. *Journal of the Geological Society*, **149**, 697–700.

STONE, P. & EVANS, J. A. 1997. A comparison of the Skiddaw and Manx Groups (English Lake District and Isle of Man) using Neodymium isotopes. *Proceedings of the Yorkshire Geological Society*, **51**, 343–347.

TAYLOR, J. E. 1864. The Cambrian strata of the Isle of Man. *Transactions of the Manchester Geological Society*, **4**, 70–78.

TAYLOR, J. H. & GAMBA, E. A. 1933. The Oatlands Igneous complex. *Proceedings of the Geologists' Association*, **44**, 355–377.

WARD, J. C. 1880. Notes on the geology of the Isle of Man. *Geological Magazine, Decade 2*, **7**, 1–9.

WHITE, F. W. 1909. The complex of igneous rocks at Oatlands, Santon, Isle of Man. *Proceedings of the Yorkshire Geological Society*, **17**, 110–131.

WILSON, BISHOP T. 1772. A new survey and description. An account of the Isle of Man. *In*: CRUTTWELL, C. & CRUTTWELL, R. (eds) *The Works of Bishop T. Wilson*. Cruttwell, Bath.

WILSON, E. 1999. A bibliography of the geology of the Isle of Man. *This volume.*

WOODS, G. 1811. *An Account of the Past and Present State of the Isle of Man, Including ... a Sketch of its Geology*. Baldwin, London.

A reassessment of Manx Group acritarchs, Isle of Man

S. G. MOLYNEUX

British Geological Survey, Keyworth, Nottingham NG12 5GG, UK

Abstract: The biostratigraphy of acritarch assemblages from the Manx Group, Isle of Man, has been reassessed in the light of evidence from the Skiddaw Group, rocks in the English Lake District that are the Manx Group's equivalents in terms of both age and lithofacies. Assemblages from the Santon Formation (formerly the Lonan Flags) and the Glen Dhoo Unit (formerly the Glen Dhoo Flags) are considered to be of early Arenig rather than the previously suggested latest Tremadoc age, and indicate that these two formations probably correlate with each other, and with the Hope Beck Formation of the Skiddaw Group. An acritarch assemblage associated with volcanic rocks at Peel is reassessed as being of early Arenig age, rather than late Arenig as previously thought. The age of an assemblage from the Glen Rushen Formation (formerly the Maughold Banded Group) is now considered to be of mid-Arenig age and to indicate correlation with the Loweswater Formation of the Skiddaw Group, while that from the Lady Port Formation is regarded as being an upper Arenig assemblage indicating correlation with the Kirk Stile Formation. An assemblage from Glenfaba Brickworks was previously assigned to the Niarbyl (Flags) Formation, but is now considered to be from undifferentiated Manx Group rocks. It is a poor assemblage that is not younger than early Arenig, but could be as old as Late Cambrian, although its proximity to the Peel locality makes an early Ordovician age more likely.

Acritarchs recorded from the Isle of Man by Downie & Ford (1966) and Molyneux (1979) showed Simpson's (1963) interpretation of Manx Group stratigraphy to be incorrect. Molyneux (1979) assigned ages ranging from Tremadoc to late Arenig to five of Simpson's formations, plus an additional formation that he erected for volcanic rocks at Peel. The acritarch ages showed that some of Simpson's formations, placed at different levels in the succession, were probably equivalents, whereas other formations were in the wrong superpositional order. Molyneux (1979) concluded, for example, that the Lonan and Glen Dhoo Flags, placed in the lower part and at the top of Simpson's succession (Table 1), respectively, probably correlated, while the Lady Port Banded Group, placed at the base of the succession in Simpson's scheme, yielded the youngest of the acritarch microfloras from the Manx Group.

No new Manx acritarch assemblages of stratigraphical significance have come to light since 1979, although some stratigraphically important taxa have been added to the list of those present, but a considerable amount of data has been obtained from the Skiddaw Group, the lithostratigraphical and lithofacies equivalent of the Manx Group in the English Lake District (Cooper *et al.* 1995). Understanding of lower Ordovician acritarch biostratigraphy throughout the region has improved as a result. The purpose of this paper is to reassess and, if necessary, revise the Manx Group ages based on acritarch evidence, amplifying the information given by Cooper *et al.* (1995) and applying the results to the revised lithostratigraphical scheme of Woodcock *et al.* (1999).

Acritarch biostratigraphy in the Skiddaw Group

Molyneux (in Cooper *et al.* 1995) documented a succession of acritarch assemblages in the Skiddaw Group of the Northern Fells Belt in the Skiddaw Inlier. At the base of the succession, the *Cymatiogalea messaoudensis–Stelliferidium trifidum* assemblage comprises a distinctive microflora associated with graptolites of the *Araneograptus murrayi* and *Tetragraptus phyllograptoides* Biozones (Fig. 1). The boundary between these Lake District graptolite zones has been correlated with the boundary between the *Hunnegraptus copiosus* and *T. phyllograptoides* Biozones of the Scandinavian succession (Cooper *et al.* 1995), which has in turn been equated with the Tremadoc–Arenig series boundary (Williams *et al.* 1994). Successive microfloras comprise the *Stelliferidium trifidum–Coryphidium bohemicum* assemblage, the *Coryphidium bohemicum* assemblage, the *Stelliferidium* aff. *pseudoornatum*

From: Woodcock, N. H., Quirk, D. G., Fitches, W. R. & Barnes, R. P. (eds) 1999.
In Sight of the Suture: the Palaeozoic geology of the Isle of Man in its Iapetus Ocean context.
Geological Society, London, Special Publications, **160**, 23–32. 1-86239-046-0/99/$15.00

Table 1. *Lithostratigraphy of the Manx Group*

Manx Group stratigraphy according to Simpson (1963)	Equivalent lithostratigraphical units in the revised scheme (Woodcock *et al.* 1999)*	Acritarch-based ages (this account)
Glen Dhoo Flags	**Glen Dhoo Unit** (informal unit)	Early Arenig
Cronkshamerk Slates Sulby Flags Sulby Slump Breccia Slieau Managh Slates	Not yet redefined as formal lithostratigraphical units	
Injebreck Banded Group	Injebreck Formation (tracts 4 and 5)	
Barrule Slates	**Glen Rushen Formation** (tract 5) Barrule Formation (tract 4)	Mid-Arenig
Maughold Banded Group	Creggan Mooar Formation (tract 6) Maughold Formation (tract 3) Creg Agneash Formation (tract 3) Mull Hill Formation (tract 2) Port Erin Formation (tract 2)	
Lonan Flags	Ny Garvain Formation (tract 3) **Santon Formation** (tract 1) Lonan Formation (tract 1)	 Early Arenig
Niarbyl Flags	Niarbyl Formation (Dalby Group, Silurian) Glion Cam Unit (tract 7) (possibly including rocks at **Peel** (volcanic locality) and **Glenfaba Brickworks**)	 Early Arenig (Peel) Early Ordovician? (Glenfaba)
Ballanayre Slump Breccia Lady Port Banded Group	**Lady Port Formation** (tract 7)	Late Arenig

* Acritarch evidence applies to those units in **bold** type.

assemblage and the *Frankea hamata–Striatotheca rarirrugulata* assemblage. The *bohemicum* assemblage is associated with graptolites of the *varicosus* Biozone, while the *S.* aff. *pseudoornatum* assemblage spans the *simulans* Biozone, from the top of the *varicosus* Biozone into the lower part of the *gibberulus* Biozone. The *hamata–rarirrugulata* assemblage appears in the *gibberulus* Biozone and also correlates with the highest Arenig *hirundo* Biozone. It is not clear whether it extends into the Llanvirn, but there are significant differences between Arenig and Llanvirn acritarch assemblages which suggest that its top lies close to the series boundary.

Acritarch assemblages from the Manx Group

Acritarchs have been obtained from the Santon, Glen Rushen and Lady Port Formations of the revised lithostratigraphical scheme (Woodcock *et al.* 1999), from Manx Group rocks of uncertain stratigraphical position at Peel and Glenfaba, and from the Glen Dhoo Unit defined by Quirk & Burnett (1999). Correspondence between these assemblages and those listed by Molyneux (1979) is indicated below (see also Table 1), and localities are shown in Fig. 2. The determinations of acritarchs recorded from the Manx Group have also been revised. Taxa are listed in Table 2 and are cross-referenced to the taxa listed by Molyneux (1979, table 1).

Santon Formation

The Santon Formation is equivalent to part of the Lonan Flags in Simpson's (1963) lithostratigraphical scheme (Table 1) and the acritarch assemblage was discussed under the heading 'Lonan Flags' by Molyneux (1979). The assemblage is based on the material described by Downie & Ford (1966) from Wallberry Hill [SC 370 735] and on specimens from samples of buff-coloured siltstones collected from the Marine Drive [SC 3715 7353], *c.* 200 m east of Wallberry Hill (Fig. 2). It should be noted, however, that Quirk & Burnett (1999) indicate the lithofacies in this area to be unrepresentative of the Santon Formation further south and consider the section to have been faulted in from slightly lower in the succession, possibly representing the Lonan Formation of the revised lithostratigraphy or even older rocks.

Molyneux (1979) drew attention to the fact that

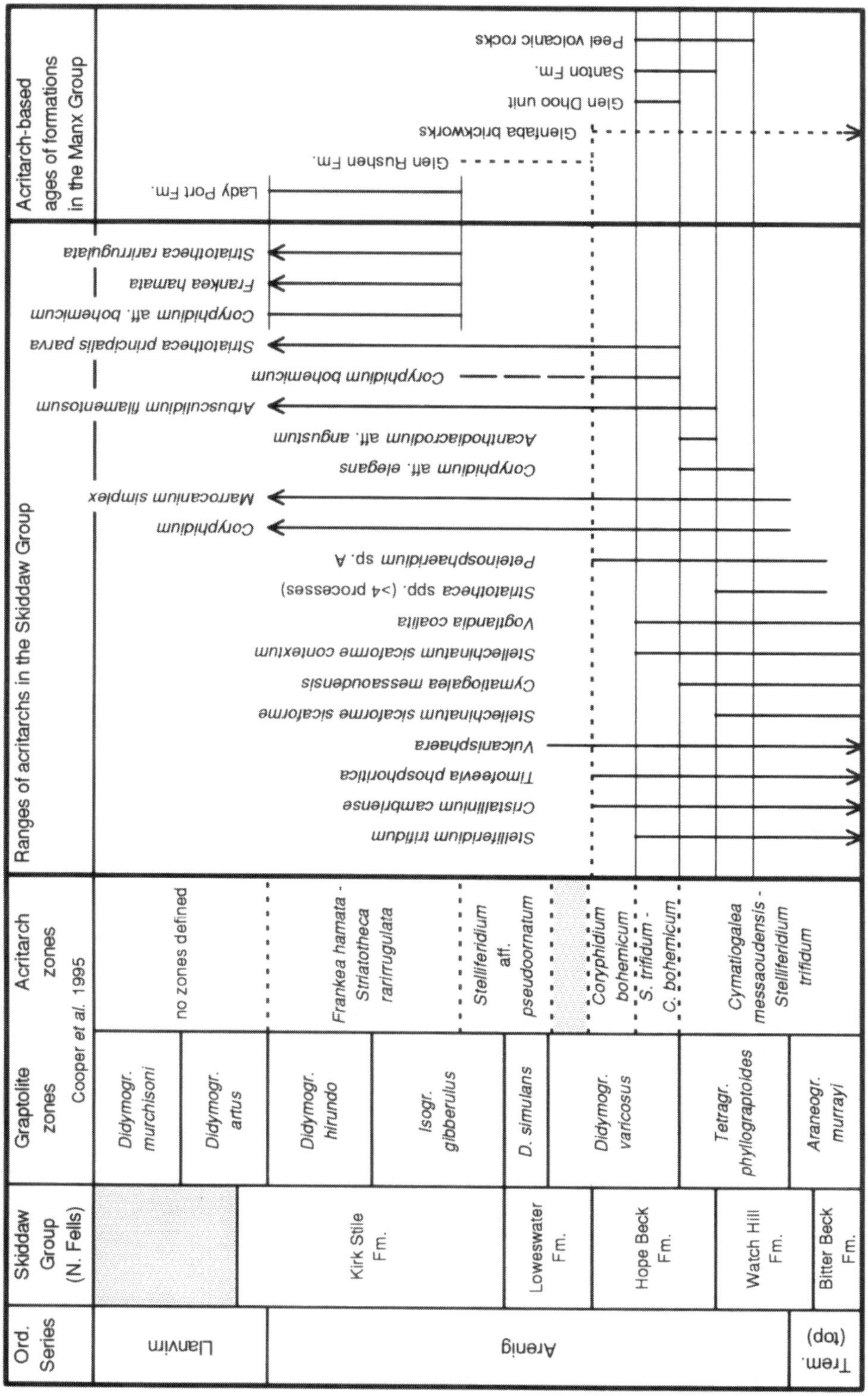

Fig. 1. Ranges in the Skiddaw Group, English Lake District, of selected acritarch taxa recorded from the Manx Group and acritarch-based ages of formations in the Manx Group.

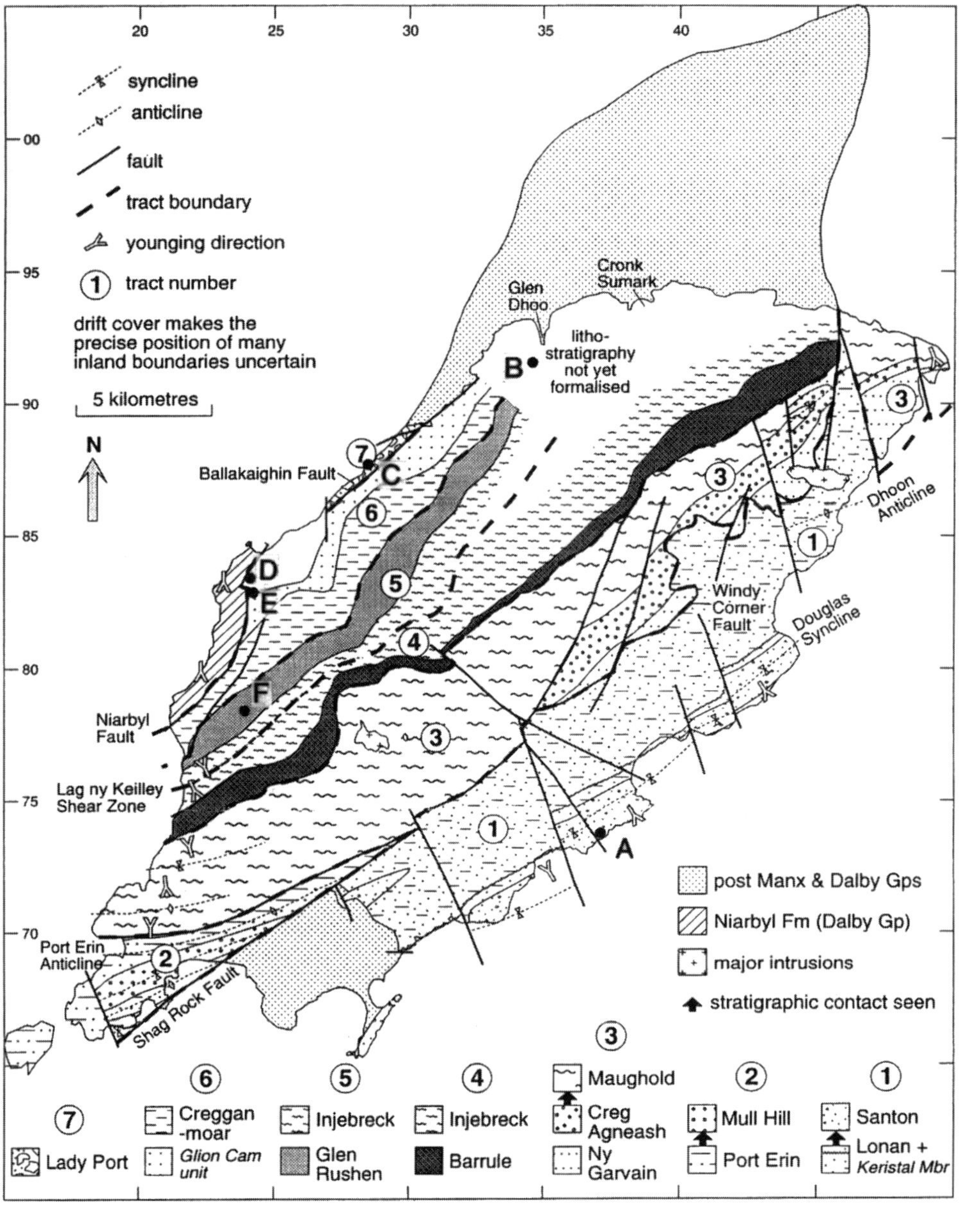

Fig. 2. Geological map of the Isle of Man [after Woodcock *et al.* (1999, fig. 1)]. Acritarch localities (●) are labelled A–F: A, Santon Formation, Wallberry Hill [SC 370 735] and Marine Drive [SC 3715 7353]; B, Glen Dhoo Unit, Slieau Curn [SC 3463 9156]; C, Lady Port Formation, Lady Port [SC 2883 8785] to Ooig Mooar [SC 2919 8825]; D, Peel, volcanic locality [SC 2386 8334]; E, Glenfaba Brickworks *c.* [SC 242 828]; F, Glen Rushen Formation, Glen Mooar [SC 2413 7929].

the assemblage contained forms known to occur in Tremadoc rocks, together with taxa that were not known to occur below the Arenig. In the absence of comparative empirical data, Molyneux (1979) concluded that the assemblage might occur close to the Tremadoc–Arenig series boundary, where some overlap in ranges might be expected, and arbitrarily assigned a late Tremadoc age to the assemblage. This conclusion has been borne out, to some extent, by work in the Lake District, where ranges of characteristic Tremadoc and younger acritarchs overlap across the Tremadoc–Arenig series boundary, in the Bitter Beck, Watch Hill and Hope Beck Formations. The Lake District work has added more detailed information, however, which makes it possible to refine the age of the assemblage.

Figure 1 shows the ranges of selected acritarch taxa in the Skiddaw Group. Of particular importance are *Stelliferidium trifidum* (Fig. 3a) and *Coryphidium bohemicum* (Fig. 3b), the ranges of which overlap in the *trifidum–bohemicum* assemblage (Cooper *et al.* 1995, fig. 5). The acritarch-based age of the formation, indicated by the solid vertical bar on Fig. 1, extends below the *trifidum–bohemicum* level to accommodate *Acanthodiacrodium* aff. *angustum*, a species that is characteristic of the highest part of the *messaoudensis–trifidum* assemblage. Nevertheless, the acritarchs are now considered to indicate an early Arenig rather than latest Tremadoc age for the Santon Formation and correlation with the Hope Beck Formation of the Lake District.

Rushton (1993) reported graptolites of Arenig age from the Santon Formation at Baltic Rock [SC 3270 7023], *c.* 5 km southwest of the Marine Drive locality. He suggested that these might indicate correlation with the middle or upper part of the Arenig succession in the Lake District (Loweswater and Kirk Stile Formations), although with a considerable amount of uncertainty because of their poor preservation. The revised acritarch age brings the two pieces of biostratigraphical evidence more in line with each other, suggesting an early–middle Arenig age for the Santon Formation (cf. Woodcock *et al.* 1999, fig. 9).

Glen Dhoo Unit

Green–grey mudstones and interbedded grey siltstones of the Glen Dhoo Unit, equivalent to Simpson's (1963) Glen Dhoo Flags (Table 1), yielded acritarchs *c.* 1 km northeast of the summit of Slieau Curn [SC 3463 9156] (Fig. 2). The lithofacies of the Glen Dhoo Unit is similar to that of the beds exposed at Wallberry Hill and along the Marine Drive (see 'Santon Formation' above), and the assemblage resembles that from the Santon Formation in the sense that it contains both 'Tremadoc' and younger taxa. As with the Santon Formation, Molyneux (1979) inferred an overlap in species ranges near the Tremadoc–Arenig series boundary to suggest a late Tremadoc age.

Stelliferidium trifidum (Fig. 3a) and *Striatotheca principalis parva* (Fig. 3c), both present in the Glen Dhoo assemblage, occur together in the *trifidum–bohemicum* assemblage of the Skiddaw Group. They indicate a probable early Arenig rather than late Tremadoc age for the Glen Dhoo Unit, and correlation with the Hope Beck Formation of the Lake District (Fig. 1). The other acritarch taxa recorded from the Glen Dhoo Unit range through this interval. The revised age of the Glen Dhoo Unit is in line with that of the Santon Formation, confirming the view of Molyneux (1979) that these formations [i.e. the Glen Dhoo Flags and Lonan Flags of Simpson (1963)] are probably equivalent, but reassigning both to the lower Arenig.

Glen Rushen Formation

A single sample of dark grey mudrock, interbedded with sparse paler grey siltstones, from a roadside exposure in Glen Mooar [SC 2413 7929], 780 m southeast of the road junction in Glenmaye (Fig. 2), yielded an acritarch assemblage discussed under the heading 'Maughold Banded Group' by Molyneux (1979). Rocks at the locality have been reassigned to the Glen Rushen Formation (Woodcock *et al.* 1999), equivalent to the western outcrop of Simpson's (1963) Barrule Slates (Table 1).

The acritarch assemblage is of low diversity, dominated by simple acanthomorph acritarchs of *Polygonium* type (Fig. 3i) and species of *Stelliferidium*, accompanied by species of *Veryhachium* (Fig. 3j). Molyneux (1979) could only assign an Arenig age to this microflora, but similar assemblages are known from the Lake District where they characterize the middle part of the Arenig succession. Cooper *et al.* (1995) described an interzonal assemblage from the lower part of the Loweswater Formation as being of low diversity with a dominant component of simple acanthomorph acritarchs, including *Polygonium*, while the *Stelliferidium* aff. *pseudoornatum* assemblage in the upper part of the Loweswater Formation and the base of the Kirk Stile Formation is similar, but with the addition of the eponymous *Stelliferidium* species. In the Buttermere area, a low diversity assemblage, with *Polygonium* as a characteristic component, occurs at Buttermere Quarry, where the graptolites indicate a level close to the *varicosus–simulans* zonal boundary (Molyneux & Rushton 1988; Cooper *et al.* 1995, p. 201). Perhaps significantly, the representatives of the *Veryhachium lairdii* group recorded from Buttermere include large, long-processed forms, similar to specimens that occur in the Glen Rushen assemblage (Fig. 3j). The similarities between the Loweswater and Glen Rushen assemblages suggest a possible middle Arenig age for the latter, indicated by the broken vertical bar on Fig. 1. The difference in facies between the Loweswater Formation, comprising turbidite sandstones (Cooper *et al.* 1995), and the Glen Rushen

Table 2. *Acritarch and equivalent (Molyneux 1979, table 1) taxa from the Manx Group*

Revised list of acritarch taxa from the Manx Group	Santon Fm	Glen Dhoo Unit	Glen Rushen Fm	Lady Port Fm	Peel volcanic location	Glenfaba Brickworks	Equivalent taxa
Acanthodiacrodium aff. *angustum* (Downie) Combaz	×						*A. angustum*
Acanthodiacrodium simplex Combaz				×			
Acanthodiacrodium spp.	×	×	×	×	×	?	*A. costatum*; *A. ignoratum*; *A. rectinerve*; *A.* cf. *ubui*; *A. uniforme*; *A.* spp.
acanthomorph acritarchs	×	×	×	×	×	×	'*A. zalesskyi*'; '*Archaeohystricho.*' spp.
Adorfia prolongata Burmann					×		*Multiplici.* cf. *maroquense*
Arbusculidium filamentosum (Vavrdová) Vavrdová					×		*A. filamentosum*
Aureotesta clathrata Vavrdová				?			
Coryphidium bohemicum Vavrdová	×				×		*C. bohemicum*
Coryphidium aff. *bohemicum*				×			*C. bohemicum*
Coryphidium aff. *elegans* Cramer *et al.*					×		*C. bohemicum*
Coryphidium spp.	×	?	×	×			*C. elegans*, *C. bohemicum*
Cristallinium cambriense (Slaviková) Vanguestaine		×				×	*D. cambriense*
Cymatiogalea messaoudensis Jardiné *et al.*					?		*Multiplici.* sp. A
	×						
Cymatiogalea spp.	×	?			×		*C. cristata*; *C. cuvillieri*; *C.* cf. *granulata*; *C.* spp.
Frankea hamata Burmann				×			*F. hamata*
Frankea sartbernardensis (Martin) Colbath				×	×		*F. sartbernardensis*
Frankea? sp.				×			
Leiofusa sp. A		×					*Leiofusa* sp. A
Marrocanium simplex Cramer *et al.*					×		
Micrhystridium spp.	×	×		×	×	×	*M. shinetonense*; *M.* spp.
Multiplicisphaeridium spp.	?			×			*M.* spp.
Peteinosphaeridium spp.			×	?	×		*P.* cf. *paucifurcatum*; *P.* spp.
Polygonium spp.	×	×	×	×	×		'*A.* cf. *pungens*'; *P. gracile*; *P.* spp.
Priscotheca complanata Deunff	×						*P. complanata*
Schizodiacrodium cf. *ramiferum* Burmann					×		*S.* cf. *ramiferum*
sphaeromorph acritarchs	×	×		×	×	×	*Protoleiosphaeridium* spp.
Stellechinatum sicaforme contextum Servais & Molyneux					×		
Stellechinatum sicaforme sicaforme Molyneux					×		
Stellechinatum uncinatum (Downie) Molyneux	×	×		×	?		*B. uncinatum*
Stelliferidium trifidum (Rasul) Fensome *et al.*	×	×				?	*S.* sp. A
Stelliferidium spp.	×	×	×	×	×		*S.* cf. *cortinulum*; ?*S. redonensis*; *S.* cf. *striatulum*; *S.* spp.
Stephanodiacrodium stephanum (Vavrdová) Vavrdová					×		*Arbusculidium stephanum*
Striatotheca principalis parva Burmann	?	×	×	×	×		*S. principalis parva*
Striatotheca principalis principalis Burmann				×			
Striatotheca rarirrugulata (Cramer *et al.*) Eisenack *et al.*				×			*S. rarirrugulata*

Table 2. *continued*

Revised list of acritarch taxa from the Manx Group	Santon Fm	Glen Dhoo Unit	Glen Rushen Fm	Lady Port Fm	Peel volcanic location	Glenfaba Brickworks	Equivalent taxa
Striatotheca spp.			×		×		*S.* spp.
Timofeevia phosphoritica Vanguestaine	×	×				×	*T. phosphoritica*
Veryhachium lairdii Deflandre *ex* Loeblich *s.l.*	?		×	×	×		*V. lairdii*
Veryhachium minutum Downie	×	×	?	×	?		*V. minutum*
Veryhachium trispinosum (Eisenack) Stockmans & Willière *s.l.*			×	×	?		*V. trispinosum* group
Vogtlandia coalita Martin					×		*Multiplici.* cf. *maroquense*
Vulcanisphaera africana Deunff - *V. cirrita* Rasul	×				×		*V. cirrita*
Vulcanisphaera sp. A		×				×	*V.* sp. A

Formation, predominantly an anoxic hemipelagic deposit (Woodcock *et al.* 1999), suggests that the apparent reduction in acritarch diversity in the middle Arenig represents a stratigraphic event rather than a local response to conditions of sedimentation.

Lady Port Formation

Dark grey and blue–grey mudstones with pale grey siltstone bands of the Lady Port Formation, exposed in the cliff and on the shore between Lady Port [SC 2883 8785] and a point north of Ooig Mooar [SC 2919 8825] (Fig. 2), yielded poorly preserved acritarchs that Molyneux (1979) used to suggest a late Arenig–early Llanvirn age.

The presence of *Coryphidium* aff. *bohemicum* (Fig. 3g), *Frankea hamata* (Fig. 3d) and *Striatotheca rarirrugulata* (Fig. 3e) provides a clear indication of the *hamata–rarirrugulata* assemblage. Although it is not known whether the *hamata–rarirrugulata* assemblage ranges into the base of the Llanvirn Series, there are sufficient differences between upper Arenig and Llanvirn acritarch assemblages in the Skiddaw Group to suggest that the highest occurrence of the *hamata–rarirrugulata* assemblage lies close to the Arenig–Llanvirn series boundary. There is nothing in the Lady Port assemblage to indicate a Llanvirn age, so the Lady Port Formation is now considered to be of late Arenig age and to correlate with the Kirk Stile Formation of the Skiddaw Group (Fig. 1).

The other acritarchs recorded from the Lady Port Formation are consistent with this conclusion and some, notably *Acanthodiacrodium simplex*, *Aureotesta clathrata* and *Striatotheca principalis principalis*, have their first appearances in the upper part of the Skiddaw Group, either in the *Stelliferidium* aff. *pseudoornatum* assemblage or in the *hamata–rarirrugulata* assemblage (Cooper *et al.*, fig. 3).

Peel

The volcanic rocks on the southwest outskirts of Peel [mentioned by Simpson (1963) and termed the Peel Volcanic Formation by Molyneux (1979)] crop out in a flooded quarry [SC 2386 8334] on the west bank of the River Neb (Fig. 2). The position of these rocks in the revised lithostratigraphical scheme remains uncertain. Woodcock *et al.* (1999, fig. 9) show the Peel volcanic rocks beneath the Glion Cam Unit, which in turn underlies the Creggan Mooar Formation, but the stratigraphical contacts between these units have not been seen. The same authors note a predominantly northward sense of younging in the Creggan Mooar Formation, however, implying that it youngs towards the Glion Cam Unit, all of which adds to the uncertainties surrounding the lithostratigraphy in this part of the Manx Group [tract 6 of Woodcock *et al.* (1999)], and therefore to the position of the Peel volcanic rocks.

Sedimentary intercalations of unindurated grey mudstone yielded abundant and reasonably well-preserved acritarchs which Molyneux (1979) considered to indicate a late Arenig–early Llanvirn age, based in part on the first appearance datum of *Acanthodiacrodium costatum* and other species of *Acanthodiacrodium*. Other forms that are diagnostic of the upper Arenig or Llanvirn are absent,

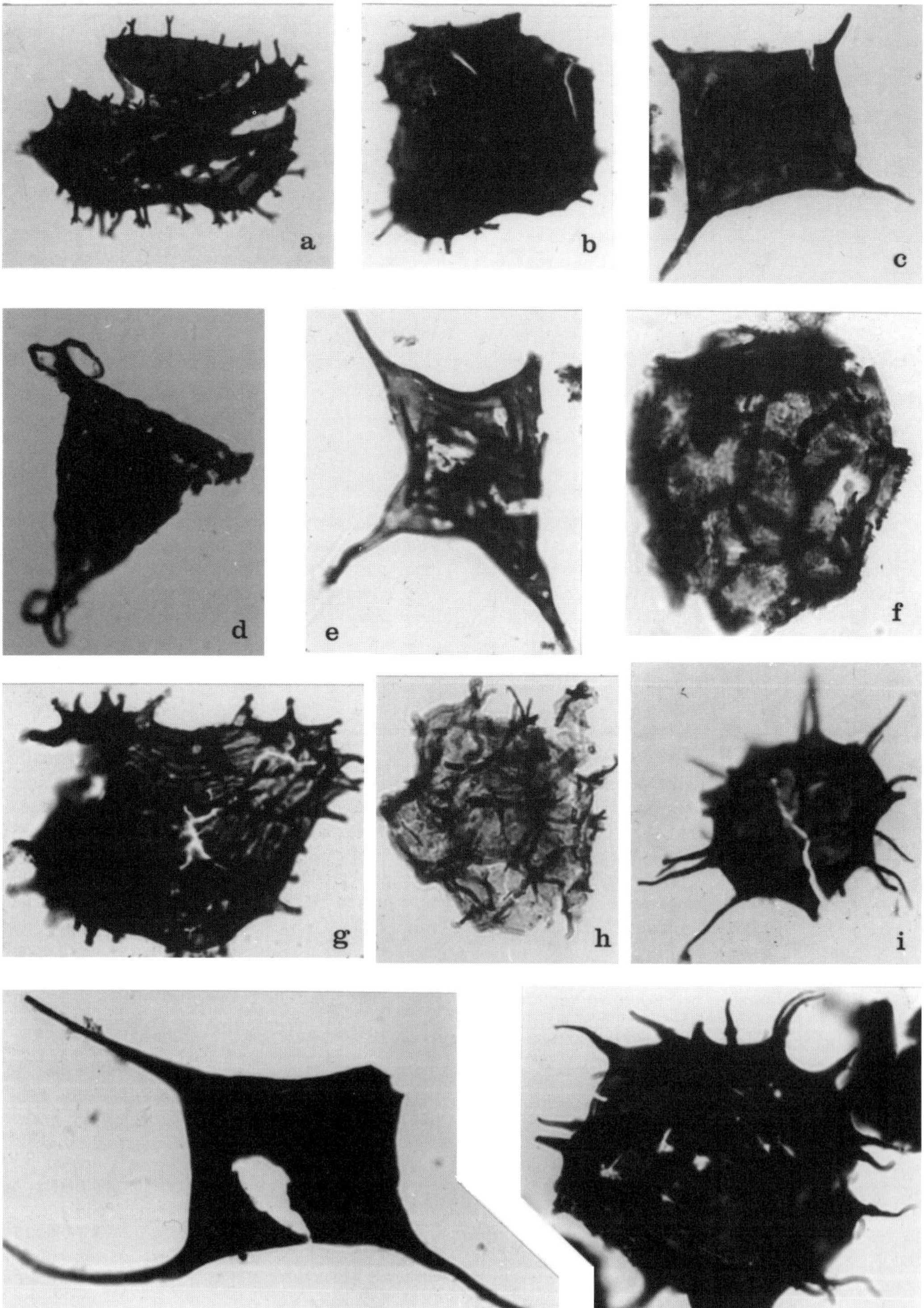
a
b
c
d
e
f
g
h
i
j
k

however, and the assemblage does not bear comparison with upper Arenig or Llanvirn assemblages from the Lake District.

In contrast, there are a number of species that have restricted upper Tremadoc–lower Arenig ranges in the Skiddaw Group, notably *Coryphidium* aff. *elegans* sensu Molyneux & Leader (1997), *Stellechinatum sicaforme contextum*, *Stellechinatum sicaforme sicaforme* (Fig. 3k), species of *Striatotheca* with more than four processes and *Vogtlandia coalita* (Fig. 1). *Cymatiogalea messaoudensis* is also questionably present. The stratigraphical ranges in the Skiddaw Group of these and other taxa recorded from Peel suggest an early Arenig age, equivalent to the upper part of the *messaoudensis–trifidum* assemblage or the *trifidum–bohemicum* assemblages (Fig. 1).

It is difficult to determine the age of the assemblage with more precision because certain species in the Peel assemblage, e.g. *Arbusculidium filamentosum*, *Coryphidium bohemicum* and *Striatotheca principalis parva*, appear above the last occurrences of others in the Skiddaw Group (e.g. *Stellechinatum sicaforme sicaforme*; Fig. 1). This might indicate unfilled ranges in the Skiddaw Group, or possible reworking in the Manx Group. Nevertheless, the evidence for an early Arenig age is now considered to outweigh that used in the previous interpretation.

Most of the other taxa in the revised list from Peel (Table 2) are consistent with the reinterpretation, and range through the upper *messaoudensis–trifidum* and *trifidum–bohemicum* intervals in the Lake District. Only the presence of *Frankea sartbernardensis* is inconsistent, since it appears much higher in the Lake District succession in the upper Arenig *hamata–rarirrugulata* assemblage (Cooper *et al.* 1995, fig. 5). Even so, it is only represented by a single specimen in the Peel assemblage, in contrast to its much more common occurrence in the *hamata–rarirrugulata* assemblage.

Glenfaba Brickworks

Molyneux (1979) discussed the acritarch assemblage from Glenfaba Brickworks [SC 242 828] under the heading 'Niarbyl Flags' and the locality lies on the outcrop of that formation on Simpson's (1963) map. The lithologies exposed in the brickworks, comprising soft, unindurated, blue–grey mudstones with pale grey and buff silty laminae, are unlike those of the Niarbyl Formation, however, and the locality lies to the east of the faulted contact between the Dalby and Manx Groups (Morris *et al.* 1999). The position of the beds exposed at Glenfaba in the revised lithostratigraphical scheme (Woodcock *et al.* 1999) is uncertain. The locality lies on the outcrop of the Glion Cam Unit as currently mapped, but assignment of the beds at Glenfaba to that unit has to be confirmed, as does the position of the Glion Cam Unit in the revised succession (see 'Peel' above).

The presence of ?*Acanthodiacrodium* spp. and '*Archaeohystrichosphaeridium* cf. *zalesskyi*' was reported by Molyneux (1979) to indicate a possible Tremadoc age, but the evidence is weak. The presence of *Cristallinium* (Fig. 3f), *Timofeevia* and *Vulcanisphaera* (Fig. 3h) suggests that the Glenfaba assemblage is no younger than early Arenig (Fig. 1), based on their last appearances in the Skiddaw Group. Combined with a single, questionable specimen of *Stelliferidium trifidum*, they would indicate a late Tremadoc–early Arenig age. If the latter is discounted, however, because of the uncertainty surrounding its determination, only a broad age within the interval from the Late Cambrian–early Arenig can be suggested. The

Fig. 3. Selected acritarchs from the Manx Group. All specimens are ×1200 and are held in the micropalaeontological collections of the British Geological Survey, Keyworth, Nottingham (MPK series). (**a**) *Stelliferidium trifidum* (Rasul) Fensome *et al.* Glen Dhoo Unit, slide MS43/6-3, England Finder coordinate F40/1. Specimen number MPK 10848. The species also occurs in the assemblage from the Santon Formation and a single questionable specimen was found in a sample from Glenfaba Brickworks. (**b**) *Coryphidium bohemicum* Vavrdová. Santon Formation, slide MS14/2-1, coordinate N29/0. MPK 10849. (**c**) *Striatotheca principalis parva* Burmann. Glen Rushen Formation, slide MS50/1-3, coordinate M49/2. MPK 10850. The species occurs in other Manx acritarch assemblages (Table 2), including that from the Glen Dhoo Unit where its occurrence with *Stelliferidium trifidum* indicates an early Arenig age. (**d**) *Frankea hamata* Burmann. Lady Port Formation, slide MS5/13-4, co-ordinate C50/2. MPK 10851. (**e**) *Striatotheca rarirrugulata* (Cramer *et al.*) Eisenack *et al.* Lady Port Formation, slide MS5/19-3, coordinate V41/2. MPK 10852. (**f**) *Cristallinium cambriense* (Slaviková) Vanguestaine. Glenfaba Brickworks, slide MS71/4-2, coordinate W41/2. MPK 10853. (**g**) *Coryphidium* aff. *bohemicum* Vavrdová *sensu* Molyneux & Leader (1997). Lady Port Formation, slide MS5/19-4, coordinate K51/1. MPK 10854. (**h**) *Vulcanisphaera*. Glenfaba Brickworks, slide MS71/4-4, coordinate L50/0. MPK 10855. (**i**) *Polygonium*. Glen Rushen Formation, slide MS50/1-7, coordinate M46/4. MPK 10856. (**j**) *Veryhachium lairdii* Deflandre *ex* Loeblich *sensu lato*. Glen Rushen Formation, slide MS50/1-1, coordinate T53/0. MPK 10857. This large, long-processed form resembles specimens obtained from the middle Arenig of the English Lake District (see text). (**k.**) *Stellechinatum sicaforme sicaforme* Molyneux. Peel, slide MS6/13-2, coordinate L37/2. MPK 10858.

proximity of the Glenfaba locality to the Peel volcanic locality, dated herein as early Arenig, suggests that an early Ordovician age is perhaps more likely.

Conclusions

All of the acritarch-based ages suggested for the Manx Group by Molyneux (1979) have been revised in the light of data obtained from the Skiddaw Group. The assemblages from the Santon Formation (formerly the Lonan Flags) and the Glen Dhoo Unit (formerly the Glen Dhoo Flags) are still considered to indicate correlation of these units, but they are judged to be of early Arenig age rather than the latest Tremadoc age suggested previously. Both are considered to correlate with the Hope Beck Formation of the Skiddaw Group.

The assemblage from the Glen Rushen Formation (formerly assigned to the Maughold Banded Group) is suggested to indicate a mid-Arenig age, based on its similarity to middle Arenig assemblages from the Skiddaw Group, rather than the undifferentiated Arenig age suggested previously. The age of the Lady Port assemblage is similarly refined and is now considered to be late Arenig rather than late Arenig–early Llanvirn.

The most radical revision concerns the Peel volcanic rocks, previously thought to be late Arenig but now revised to early Arenig. There remain some difficulties over the age of this assemblage, notably the presence of acritarchs of the *Acanthodiacrodium costatum* group and *Frankea sartbernardensis*, taxa that are more typical of the upper Arenig (or higher), but they are associated with diagnostic lower Arenig species. Reworking of the lower Arenig forms would account for this association but is considered unlikely. In the first place, the assemblage is unlike upper Arenig assemblages from the Lake District, even those in which reworking can be recognized, lacking the common and diagnostic upper Arenig species. Furthermore, *Frankea sartbernardensis* is represented by a single specimen, in contrast to its more common occurrence at higher levels, and *Acanthodiacrodium costatum* is not considered to be diagnostic of upper Arenig assemblages throughout the region because it is rare in Lake District assemblages and so its range is uncertain.

The poorest Manx acritarch assemblage, from Glenfaba, is no longer assigned to the Niarbyl Flags but to Manx Group rocks in faulted contact with them. The Glenfaba assemblage is no younger than early Arenig but could be much older, embracing the possible Tremadoc age suggested by Molyneux (1979). In contrast to the other assemblages, the possible age of the Glenfaba assemblage has been broadened rather than refined. Given its proximity to the Peel volcanic assemblage, however, an early Ordovician age seems most likely.

This paper is published by permission of the Director, British Geological Survey, NERC.

References

COOPER, A. H., RUSHTON, A. W. A., MOLYNEUX, S. G., HUGHES, R. A., MOORE, R. M. & WEBB, B. C. 1995. The stratigraphy, correlation, provenance and palaeogeography of the Skiddaw Group (Ordovician) in the English Lake District. *Geological Magazine*, **132**, 185–211.

DOWNIE, C. & FORD T. D. 1966. Microfossils from the Manx Slate Series. *Proceedings of the Yorkshire Geological Society*, **35**, 307–322.

MOLYNEUX, S. G. 1979. New evidence for the age of the Manx Group, Isle of Man. *In*: HARRIS, A. L., HOLLAND, C. H. & LEAKE, B. E. (eds) *The Caledonides of the British Isles – Reviewed*. Geological Society, London, Special Publications, **8**, 415–421.

—— & LEADER, R. U. 1997. Morphological variation in *Coryphidium* from the Arenigian Series (Lower Ordovician) of northwestern England. *Review of Palaeobotany and Palynology*, **98**, 81–94.

—— & RUSHTON, A. W. A. 1988. The age of the Watch Hill Grits (Ordovician), English Lake District: structural and palaeogeographical implications. *Transactions of the Royal Society of Edinburgh: Earth Sciences*, **79**, 43–69.

MORRIS, J. H., WOODCOCK, N. H. & HOWE, M. P. A. 1999. The Silurian succession of the Isle of Man: the late Wenlock Niarbyl Formation, Dalby Group. *This volume*.

QUIRK, D. G. & BURNETT, D. J. 1999. Lithofacies of Lower Palaeozoic deep-marine sediments in the Isle of Man: a new map and stratigraphic model of the Manx Group. *This volume*.

RUSHTON, A. W. A. 1993. Graptolites from the Manx Group. *Proceedings of the Yorkshire Geological Society*, **49**, 259–262.

SIMPSON, A. 1963. The stratigraphy and tectonics of the Manx Slate Series, Isle of Man. *Quarterly Journal of the Geological Society, London*, **119**, 367–400.

WILLIAMS, S. H., BARNES, C. R., O'BRIEN, F. H. C. & BOYCE, W. D. 1994. A proposed global stratotype for the second series of the Ordovician System: Cow Head Peninsula, western Newfoundland. *Bulletin of Canadian Petroleum Geology*, **42**, 219–231.

WOODCOCK, N. H., MORRIS, J. H., QUIRK, D. G. *ET AL.* 1999. Revised lithostratigraphy of the Manx Group, Isle of Man. *This volume*.

Macrofauna and ichnofauna of the Manx Group (early Ordovician), Isle of Man

P. J. ORR[1] & M. P. A. HOWE[2]

[1]*Department of Earth Sciences, University of Oxford, Parks Road, Oxford OX1 3PR and Department of Geology, National University of Ireland, Galway, Ireland*

[2]*Department of Geology, University of Leicester, University Road, Leicester LE1 7RH, UK*

Abstract: Within the Manx Group, the key elements of the limited macrofauna include a *Rhabdinopora*-like graptoloid from Cronk Sumark and a didymograptid–dichograptid fauna from Baltic Rock. The ichnofaunal assemblage is dominated by fodinichnia (mainly examples of *Phycodes*). This is atypical of Phanerozoic deep-marine environments as a whole, but is a feature of other early Ordovician deep-marine ichnofaunal assemblages, notably that within the Skiddaw Group of the English Lake District. *Dictyodora zimmermani* and *Glockerichnus radiatus* occur within the Manx Group; these ichnotaxa are present in other ichnofaunal assemblages of early Ordovician age which were also emplaced at high southerly palaeolatitudes.

This conference volume comes just a century after the publication of the first 'modern' palaeontological and ichnological study of the Manx Group, that of Bolton (1899). Here, both body and trace fossils were accurately figured and described, and, most importantly, deposited in museum collections (where they can still be recognized from their figures) and, in most instances, from their published accession numbers.

Since 1899, the graptoloids have been reviewed on several occasions (see Rushton 1993) but the ichnofauna has been ignored. This contribution therefore concentrates on the latter, but the occurrence and stratigraphical implications of the graptoloid (and possible trilobite) macrofauna are first summarized briefly. The important recent discovery of a middle Wenlock (*lundgreni* Biozone) graptoloid fauna from the Niarbyl Formation near Peel is described elsewhere (Howe 1999) because the Niarbyl Formation is now assigned to the Dalby Group and not to the Manx Group.

The material used in this study is housed in the following institutions: British Geological Survey, Keyworth, Nottinghamshire (BGS); Manx Museum, Douglas, Isle of Man (IOMMM); Manchester Museum (MM); Sedgwick Museum, University of Cambridge (SM A).

All map references refer to British National Grid square SC. Eight figure references are generally given for recently collected material, six figure references for older material where the precise locality is less certain.

Macrofaunas – graptoloids

Cronk Sumark

Bolton (1893*b*) recorded the discovery of *Dictyonema* and *Dendrograptus* from the slates in the quarry [SC 391 941] on Cronk Sumark – the Cronk Sumark Formation of Woodcock *et al.* (1999), although not formally defined. He subsequently figured them (Bolton 1899, plate 1 figs 1 and 2; MM L4491, L.4494) as *Dictyonema sociale* (Salter) and *Dendrograptus flexuosus* (Hall). Lamplugh (1903, p. 94) dismissed Bolton's interpretation, largely because of his belief that the 'crush breccias' on Cronk Sumark indicated exceptionally high strain levels. These 'crush breccias' are now known to be of sedimentary origin [e.g. see Woodcock & Morris (1999)]. Lamplugh also cited as evidence the absence of *Dictyonema* from the Skiddaw Slates, which he considered the precise correlative of the Manx Group.

Rushton (1993) recorded the discovery of some additional graptoloid stipe fragments (BGS RX4098-4100), reviewed the subsequent literature and elegantly demonstrated, using a variable *XY*-zoom photocopier, that only one species was present. The variation in appearance was purely due to the deformation. Because of the absence of a well preserved proximal end, he was unable to distinguish between the nema-bearing form (*Rhabdinopora*, typical of the Tremadoc) and the rooted form [e.g. *Dictyonema* (Cambrian–Carboniferous)]. However, he concluded that a

From: WOODCOCK, N. H., QUIRK, D. G., FITCHES, W. R. & BARNES, R. P. (eds) 1999. *In Sight of the Suture: the Palaeozoic geology of the Isle of Man in its Iapetus Ocean context*. Geological Society, London, Special Publications, **160**, 33–44. 1-86239-046-0/99/$15.00

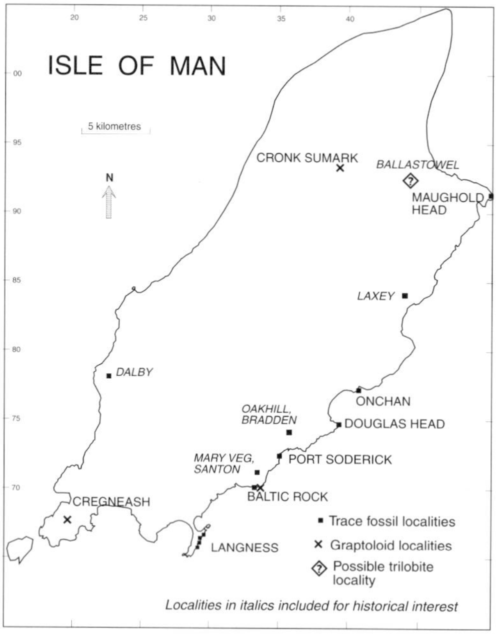

Fig. 1. Locality map for the Manx Group macrofauna and ichnofauna. The four marked localities on Langness are: (from northeast to southwest) Gullet Buigh; Horse Gullet; Martha Gullet; Creg Custane.

Tremadoc, or possibly Arenig age, was most likely.

Baltic Rock

Graptolite fragments (BGS RX4104-4160) were discovered at Baltic Rock, west of Santon Head and 1 km east of Port Grenaugh [SC 3270 7023], during a field excursion in 1992 (Rushton 1993). The fragments were collected from a 15 cm thick, horizontally laminated sandstone bed, interpreted as a Bouma B unit, overlying massive units, 1–1.5 m thick, of coarse grained sandstone. This falls within the Santon Formation of Woodcock *et al.* (1999).

During the current study, the locality was re-sampled and a further graptolitic horizon was identified 2 cm above the main band. The sandstone bed also contained ripple-laminated horizons (Bouma C), representing either several amalgamated events or one event with considerable variation in current strength, allowing the graptolite debris to accumulate at several discrete horizons.

Rushton (1993) figured eight specimens, now in the collections of the BGS GSM, and determined a fauna of:

- dichograptid fragments (possibly *Clonograptus* or *Orthodichograptus*);
- *Didymograptus* and/or *Xiphograptus* spp. (including a few proximal ends of slightly declined forms);
- *Tetragraptus*(?) spp.

He considered that they represented an Arenig age; the declined *Didymograptus* possibly suggesting a correlation with the Loweswater or Kirkstile Formations of the Skiddaw Group.

A considerable quantity of additional material has been collected during the present study but, because of the absence of well preserved proximal ends, the conclusions of Rushton (1993) cannot be improved upon.

Cregneash

During the 1992 field excursion, undeterminable graptolite fragments (BGS RX4101, RX4102) were collected from the quarry south of Mull Hill, Cregneash [SC 1896 6737], which exposes the Mull Hill Formation of Woodcock *et al.* (1999). During the present study, a further fragment was collected but this too was indeterminate. However, it is clear that this locality merits further study in the future.

Macrofaunas – trilobites

A single specimen, collected by H. Bolton in 1892 (Bolton 1893*a*) from a small quarry at Ballastowell, near Ramsey, was identified at the time as either *Asaphus* or *Aeglina*. Bolton (1893*a*, p. 30) describes the locality as 'Ballastowel, a hill crest overlooking Ramsey on the south-west, and lying midway between the familiar Albert Tower and Elfin Glen.' A more detailed description is given by Lamplugh (1903, p. 66) and the locality would appear to be at [SC 4522 9335], within the Injebreck Formation of Woodcock *et al.* (1999).

The specimen was subsequently figured by Bolton (1899, plate 1, fig. 12) (MM L1169). However, given that Bolton (1893*a*, p. 30) records that 'the specimen is too distorted and fragmentary' to allow a precise determination, its re-examination has not been a priority of this study.

Macrofaunas – stratigraphical conclusions

The biostratigraphy of the Manx Group is based largely upon the acritarch faunas, as detailed by Molyneux (1979, 1999) and Cooper *et al.* (1995). The Arenig graptolite faunas from Baltic Rock support the more precise acritarch biozonation. If the Tremadoc age, perhaps suggested by the dendroid fauna from the Cronk Sumark Formation, is correct, it fits well with the low Arenig age determined with acritachs from the presumed overlying Glen Dhoo Formation (Cooper *et al.* 1995).

Ichnofauna – general

The trace fossils (ichnofauna) preserved within the Manx Group have received little study since Bolton (1899) and are in need of revision. The early Ordovician age of the Manx Group is particularly important in this respect as early Palaeozoic, flysch-hosted, trace fossil assemblages have been relatively understudied in comparison to their Mesozoic and Cenozoic counterparts (McCann & Pickerill 1988). This has implications for models of behavioural evolution and the rate at which the deep-marine environment was colonized during the early Phanerozoic. The early Ordovician is recognized as a interval in which there is a marked increase in the diversity of deep-marine ichnofaunal assemblages (Pickerill 1980; Frey & Seilacher 1980; Crimes *et al.* 1992; Orr 1996). Pascichnia (systematically meandering or spiralling grazing traces, e.g. *Nereites* MacLeay in Murchison 1839) and agrichnia (polygonal or petaloid networks of open burrows, e.g. *Paleodictyon* Meneghini in Murchison 1850) are important, often the dominant, behaviour patterns in many deep-marine ichnofaunal assemblages; e.g. both are common in strata of Llandovery age in the Welsh Basin (Crimes & Crossley 1980, 1991; McCann 1990, 1993; Orr 1995). Orr (1996, p. 212–213) noted that, in contrast, the ichnofaunal assemblage of the Skiddaw Group (early Ordovician) of the Lake District of England is dominated by fodinichnia (trace fossils combining deposit feeding and dwelling); agrichnia are absent and only a single unequivocal pasichnion has been recorded. The significance of this observation for models of behavioural evolution within early Phanerozoic deep-marine environments cannot be fully assessed until more analyses of Ordovician deep-marine ichnofaunal assemblages are made.

This study is a reassessment of the trace fossils from the Manx Group using the collections of BGS GSM, SM A and new material collected by D. J. Quirk & N. H. Woodcock, now deposited at IOMMM.

Ichnofauna – previous research

Although earlier authors (e.g. Cumming 1848) had alluded to the occurrence of 'fucoids' in the Manx Group, the first description of possible trace fossils appears to have been by Taylor (1862). Taylor interpreted ovate structures found in quarries at Dalby as imprints (each 200 × 100 mm, based on his illustration) and compared them with *Protichnites* Owen 1852. However, doubts were expressed by several authors (see Bolton 1899, p. 4), including the editor of the journal in which the original study was published [see footnote to Taylor (1862)]. Although specimens were collected, their present whereabouts are unknown. The locality is uncertain and it is even possible that the specimens came from the Dalby Group, and are thus Silurian (Morris *et al.* 1999). Further interpretation is not attempted.

Harkness & Nicholson (1866, p. 488) recorded '*Palaeochorda major*' 'along the coast from Douglas to about a mile north-east thereof.' No descriptions or illustrations were provided, nor do the specimens remain. As there is uncertainty as to the validity of '*Palaeochorda*' as an ichnotaxon, and what it represents [see discussion in Orr (1996, p. 207–209; 1999, fig. 17)], no reinterpretation can be attempted. In contrast, it has been possible to reinterpret material from the Manx Group that was assigned to '*Palaeochorda*' by Bolton (1899) (see below).

Binney (1877, p. 107–109, plates 1 and 2, fig. 1) erected *Nemerites* (*sic. Nemertites* MacLeay in Murchison 1839) *monensis* and *Nereites monensis*. The ichnogenus '*Nemertites*' has been reinterpreted as part of a *Dictyodora* burrow (Häntzschel 1975), almost certainly a horizontal cross-section through the vertical 'wall' structure; see reconstructions of *Dictyodora* in Benton & Trewin (1980, fig. 1A) and Benton (1982, fig. 10). '*Nemerites*' *monensis*, 'from Oakhill, in Bradden' (Binney 1877, p. 110), [SC 352 740], is probably a morphologically simple burrow of *Planolites* Nicholson 1873*a*,*b*–*Palaeophycus* Hall 1847 affinity. The holotype of *Nereites monensis* (SM A10892) is 'from the slate quarry of Mary Veg, in Santon' (Binney 1877, p. 110), [SC 330 711]. It lacks any of the diagnostic features of the ichnogenus *Nereites* (Orr & Pickerill 1995, p. 396) and is reinterpreted herein as a morphologically simple burrow of *Planolites–Palaeophycus* affinity.

Bolton (1899) reviewed the work of previous authors and provided descriptions and figures of additional material; almost all of the latter is now in the BGS collections. M'Coy (1848) erected '*Chondrites informis*' for what is actually a network of closely spaced and overlapping burrows of *Planolites–Palaeophycus* affinity; the 'holotype' from the Skiddaw Group is specimen SM A40367. '*Chondrites informis*' has been rejected as a nomen dubium (Orr 1996, p. 199). Three of Bolton's additional specimens (BGS GSM92719, BGS GSM92723 and BGS GSM92720) were assigned to '*Chondrites informis*' (Bolton 1899, p. 11, plate 1, figs 8–10); these have been reinterpreted herein (see below).

Ichnofauna – description

None of the specimens in the BGS collection has information as to how they were oriented relative to bedding prior to collection. In some, it can be deduced by assuming laminae within the adjacent or adhering matrix were oriented parallel, or subparallel, to bedding. The way-up of the specimens, however, remains uncertain. This necessitates the use of 'observed ... in semi-relief' in the following descriptions, rather than epirelief or hyporelief, unless either of the latter can be deduced.

Dictyodora zimmermani *Hundt 1913* *(Fig. 2a–c)*

Material. *IOMMM 98-86.*

Locality and horizon. Port Soderick [SC 3508 7292] [Lonan Formation of Woodcock *et al.* (1999)] *c.* 1 m above Keristal Member (Woodcock, pers. comm.).

Description. The narrow (*c.* 1 mm wide) structure in specimen IOMMM 98-86 (Fig. 2a and b) has a meandering course; individual meanders have high amplitude and short wavelength. Superimposed upon these meanders is a secondary sinuosity in which the amplitude of the meanders is less than their wavelength. Vertical sections (Fig. 2c) indicate that this narrow structure continues vertically below the subhorizontal plane of splitting upon which it is exposed; it can be identified where its fill contrasts in both colour and grain size with that of the laminae it cross-cuts (at arrow in Fig. 2c).

Remarks. The ichnogenus *Dictyodora* Weiss 1884 comprises a basal burrow from the mid-line of the dorsal side of which a narrow 'wall' structure extends vertically upwards; see reconstructions in Benton & Trewin (1980, fig. 1A) and Benton (1982, fig. 10). Development of the burrow system is dominantly bedding parallel; the 'wall' structure therefore forms narrow, meandering lines on horizontal planes of splitting that it intersects. Specimen IOMMM 98-86 is interpreted as a horizontal (and, where sectioned, a vertical) cross-section through the 'wall' structure of an example

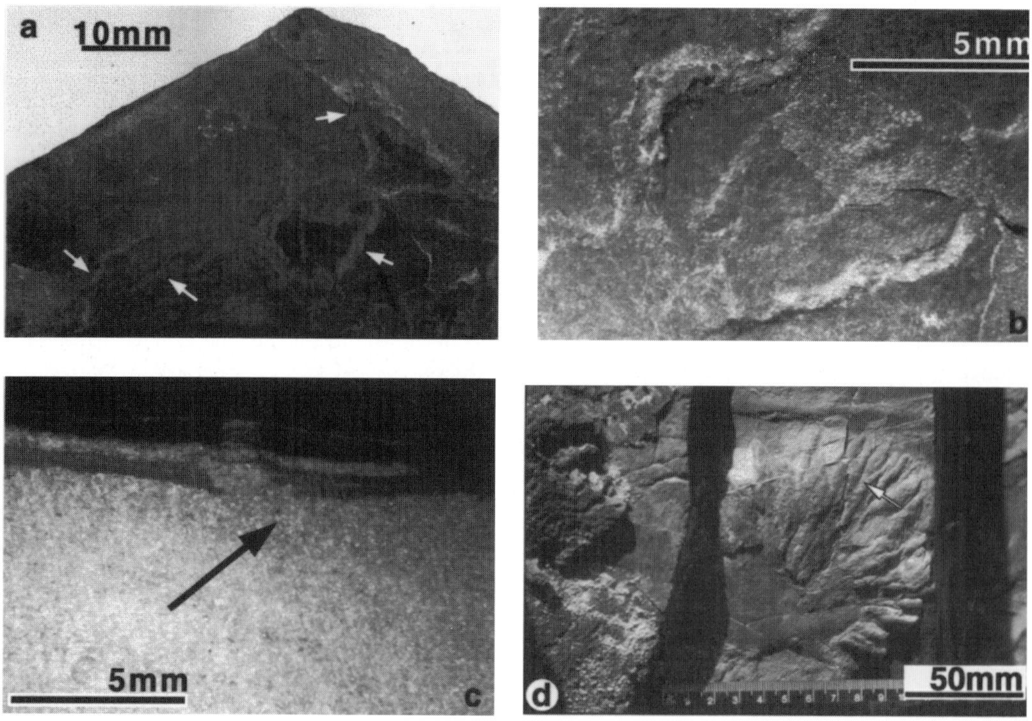

Fig. 2. (**a**)–(**c**) *Dictyodora zimmermani* Hundt 1913. IOMMM 98-86; (a) And (b) are horizontal sections, (c) vertical section. (**d**) *Glockerichnus radiatus* (Etheridge 1876).

of *Dictyodora*. Two ichnospecies of *Dictyodora* have secondary sinuosity superimposed upon the primary meander pattern: *Dictyodora tenuis* (M'Coy 1851) and *Dictyodora zimmermani*. The secondary sinuosity is more irregular in *Dictyodora tenuis* than it is in both *Dictyodora zimmermani* and specimen IOMMM 98-86; c.f. Fig. 2a and b with examples of *Dictyodora tenuis* in Benton & Trewin (1980, fig. 5). Specimen IOMMM 98-86 is thus identified as an example of *Dictyodora zimmermani*.

Glockerichnus radiatus *(Etheridge 1876) (Fig. 2d)*

Material. Specimen not collected; interpretation on basis of field photograph by N. H. Woodcock.

Locality and horizon. [SC 3900 7469] near Douglas Head Lighthouse; base of turbidite sand (Woodcock, pers. comm.). Santon Formation of Woodcock *et al.* (1999).

Description. Preservation is in positive hyporelief. The diameter of the specimen is *c.* 140 mm. A series of closely spaced shafts radiate outwards from a poorly defined centre. Although incomplete, it is evident that the outline of the specimen was originally circular with the shafts arranged in a stellate pattern. Individual shafts are subequal in length and curved in the vertical plane such that their proximal and distal ends are above their middle portion. Each shaft appears to be unbranched; the appearance of branching (at the arrow in Fig. 2d) is more likely due to partial overlap of two individual shafts.

Macaronichnus segregatis *Clifton & Thompson 1978 (Fig. 3a and b)*

Material. *BGS GSM92721*, partim.

Locality and horizon details. 'Lonan Flags; north side of Martha Gullet, Langness' (Bolton 1899, p. 14), [SC 289 662]. Lonan Formation of Woodcock *et al.* (1999).

Fig. 3. (**a**) And (**b**) *Macaronichnus segregatis* Clifton & Thompson 1978. BGS GSM92721, *partim*. (**c**)–(?**e**) *Phycodes palmatus* (Hall 1852); (c) BGS GSM92722; (d) BGS GSM105366; (?e) BGS GSM105365.

1899 *Palaeochorda minor* (Bolton, p. 9, plate 1, fig. 5).

Description. Specimen BGS GSM92721 is a complex network of overlapping and interpenetrating burrows of variable size. The burrows to which the following description applies are identified by the arrows in Fig. 3a and are shown enlarged in Fig. 3b. These 3–4 mm wide burrows are straight to curved on the plane of splitting. The burrow fill is coarser grained than the matrix, from which it is separated by a thin, fine-grained wall lining; the wall lining is indicated by the arrows in Fig. 3b. The fill is, at least locally, meniscate (see below). Individual menisci are shallow and arcuate. Adjacent menisci define meniscate segments (rather than meniscate packets); the meniscate fill as a whole is homogeneous [terminology of Keighley & Pickerill (1994)].

Remarks. The menisci are consistently oriented transverse to the margins of the burrow, although the course of the burrow changes along its length. They are therefore not an artefact of the rocks having a well-developed cleavage fabric. Following Clifton & Thompson (1978), Curran (1985), Fillion (1989), Fillion & Pickerill (1990) and Keighley & Pickerill (1995), lined burrows which are actively filled (i.e. filled by the producer) are referred to the ichnogenus *Macaronichnus*. Assignment of specimen BGS GSM92721 *partim* to *Macaronichnus segregatis* is based on the similar overall diameters, curving to sinuous paths and the absence of branching in the examples of the latter illustrated by Clifton & Thompson (1978, figs 1–3).

Phycodes palmatus *Hall 1852 (Fig. 3c–e)*

Material. BGS GSM92722, BGS GSM105366, ?BGS GSM105365.

Locality and horizon details. BGS GSM92722, '65 yards north of Creg Custane , Langness' (Bolton 1899, p. 15) [SC 289 661]; GSM 105366, Langness shore, south side of Horse Gullet [SC 291 665]; GSM 105365, Langness shore, 65 yards north of Creg Custane [SC 289 661] (sources: BGS records). All Lonan Formation of Woodcock *et al.* (1999).

1899 *Palaeochorda major* (Bolton, p. 10, plate 1, fig. 7).

Description. Overall development of the burrow system is subparallel to bedding (identified from laminated matrix in BGS GSM92722; Fig. 3c). A series of shafts (?11 in BGS GSM92722; at least 11 in BGS GSM105366) that radiate from a common origin are preserved in positive semi-relief. Individual shafts are asymmetrically curved in the vertical plane; the medial parts are at a slightly different level than the more gently inclined proximal parts and the more steeply sloping, recurving, distal parts. In the horizontal plane, medially sited shafts are straight in specimen BGS GSM105366; the more marginal shafts are slightly laterally curved (Fig. 3d). In BGS GSM92722, each shaft is straight or slightly sinuous in the horizontal plane (Fig. 3c). Proximally, the large number of shafts are accommodated by the vertical superimposition of adjacent shafts. These are not points of branching; individual shafts can be traced back towards the origin in both the horizontal and vertical planes (e.g. at arrows in Fig. 3c). Changes in the diameter of shafts along their length are more obvious in specimen BGS GSM105366 (Fig. 3d); they have a clavate outline, broadening distally (up to 10 mm diameter) and terminating in a rounded outline. This correlates with their being inclined to the single sub-horizontal plane on which they are exposed. Specimen BGS GSM105365 (Fig. 3e) comprises a series of straight, unbranched shafts (maximum diameter *c.* 5 mm) arranged in a digitate pattern; the proximal part is missing.

Remarks. Specimens BGS GSM105366 and BGS GSM92722 compare closely to the specimen of *Phycodes palmatus* illustrated by Orr (1996, fig. 6c). As specimen BGS GSM105365 is incomplete it is only tentatively identified as *Phycodes palmatus*. Use of the ichnospecific epithet *palmatus* follows Fillion & Pickerill (1990).

Planolites montanus *Richter 1937 (Fig. 4a and b)*

Material. BGS GSM7655; BGS GSM7656.

Locality and horizon details. 'Lonan Flags; cliff south-west of Onchan Harbour, near Douglas' (Bolton 1899, p. 14), [SC 404 774]. Santon Formation of Woodcock *et al.* (1999).

1899 *Palaeochorda minor* (Bolton, p. 9, plate 1, figs 3 and 4).

Description. Specimen BGS GSM7655 comprises two burrows and specimen BGS GSM7656 a number of burrows that are preserved in positive semi-relief. Each burrow is narrow [diameter *c.* 1 mm (BGS GSM7656) or 2–3 mm (BGS GSM7655)], straight or slightly curved, subhorizontal to bedding (see below) and unbranched (see below). A wall lining is not present. Their infill is structureless and coarser than the pelitic matrix;

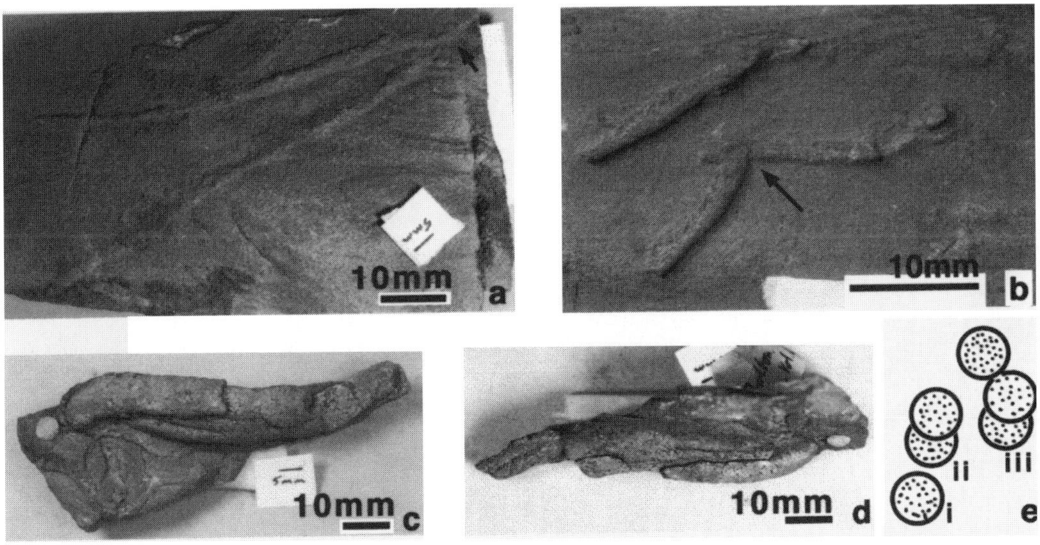

Fig. 4. (**a**) and (**b**) *Planolites montanus* Richter 1937. (a) BGS GSM7655; (b) BGS GSM7656. (**c**) And (**d**) ?*Phycodes* isp 1. BGS GSM92723. (**e**) Reconstruction of sequential emplacement of shafts in ?*Phycodes* isp. 1, including at stage ii, partial reworking of previous shaft.

this is more pronounced in specimen BGS GSM7656.

Remarks. In specimen BGS GSM7656 the matrix comprises thin alternations of pelitic and coarser grained laminae. These are parallel to the plane of splitting on which the burrows are exposed. The latter is thus assumed to originally have had a horizontal orientation. The way up of either specimen cannot be identified. The appearance of branching (at arrows in Fig. 4a and b) arises from the overlap of two individual burrows; in specimen BGS GSM7655 (Fig. 4a) each burrow can be traced either side of the point indicated by the arrow.

Following Pemberton & Frey (1982), the absence of a wall lining, the smooth exterior and the small overall diameter of the burrows identifies them as *Planolites montanus*.

Ichnogenus Phycodes *Richter 1850*

?Phycodes *isp 1 (Fig. 4c and d)*

Material. *BGS GSM92723.*

Locality and horizon details: 'north side of Martha Gullet, Langness' (Bolton 1899, p. 15), [SC 289 662]. Lonan Formation of Woodcock *et al.* (1999). 1899 *Chondrites informis* (Bolton, p. 11, plate 1, fig. 9).

Description. The specimen comprises three discrete shafts, each with a diameter of 6–7 mm. In what is assumed to be the horizontal plane (Fig. 4c;see below) each shaft is sinuous. In the vertical plane (Fig. 4d) each shaft is asymmetrically curved; the more steeply sloping end is interpreted as the proximal. The shortest shaft has a crescentic vertical cross-section suggesting that it pre-dates, and was partially reworked during the emplacement of, that adjacent to it (see reconstruction in Fig. 4e). In vertical cross-section, the other two shafts abut but retain subcircular outlines. The structureless infill of the shafts is coarser grained than the matrix.

Remarks. Specimen BGS GSM92723 is preserved in full relief but its orientation prior to collection is unknown. Variation in the length of the shafts is almost certainly due to weathering prior to collection. The bundled arrangement of the shafts is strikingly similar to that of *Phycodes* and the overall morphology is reminiscent of the example of *Phycodes circinatus* Richter 1853, figured by Fillion & Pickerill (1990, plate 11, fig. 10). In the latter, the shafts are sinuous in the horizontal plane. As specimen BGS GSM92723 is incomplete and its orientation can only be suggested by analogy, it is only tentatively identified as an example of *Phycodes.*

Ichnogenus Phycodes *Ritcher 1850*

?Phycodes *isp 2 (Fig. 5a and b)*

Material. *IOMMM 98-87.*

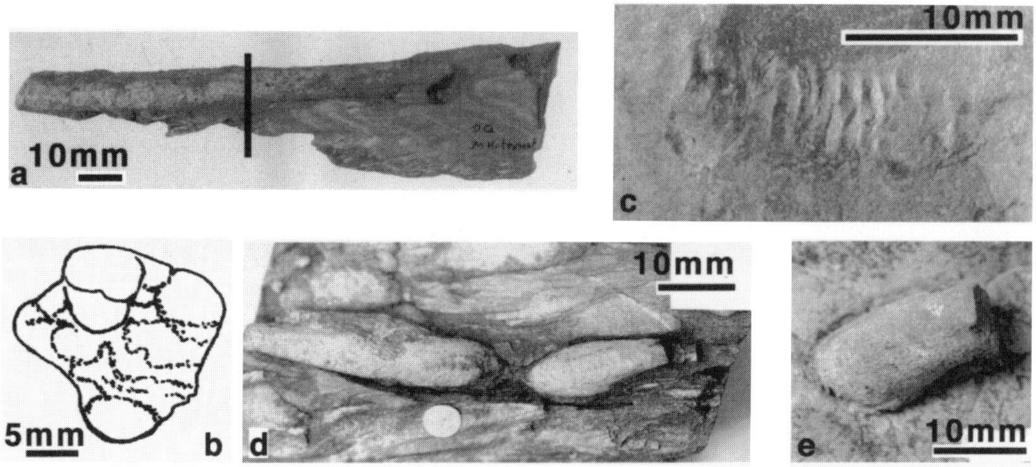

Fig. 5. (**a**) And (**b**) ?*Phycodes* isp 2. IOMMM 98-87. (**c**) Arthropod?-produced repichnion. IOMMM 98-88. (**d**) Burrow 1. BGS GSM92720. (**e**) Burrow 2. BGS GSM92719.

Locality and horizon details. Base of bed of thin- to medium-bedded quartz arenite, hanging wall of Maughold Head thrust [SC 4960 9231] (Quirk, pers. comm.). Creg Agneash Formation of Woodcock *et al.* 1999.

Description. Specimen IOMMM 98-87 is preserved in full relief (Fig. 5a). It has a coarser grain size than, and has weathered positively with respect to, the pelitic matrix. In cross-section (Fig. 5b; along the line indicated in Fig. 5a) the structure comprises a pair of laterally offset burrows with complete, subcircular outlines that each surmount a crude spreite (Fig. 5b).

Remarks. Some forms of *Phycodes* consist of a few main branches showing a spreite-like structure that distally give rise to numerous free branches (Fillion & Pickerill 1990). The combination of the two adjacent burrows and spreiten in specimen IOMMM 98-87 is reminiscent of this and forms the basis for the tentative assignment of the specimen to *Phycodes*.

Arthropod?-produced repichnion (Fig. 5c)

Material. *IOMMM 98-88.*

Locality and horizon details. Baltic Rock [SC 3270 7027] (Quirk, pers. comm.). Santon Formation of Woodcock *et al.* (1999).

Description. Specimen IOMMM 98-88 is preserved in positive semi relief; it comprises two parallel series of narrow (1 mm wide) ridges arranged in a ladder-like pattern. Each ridge is elongate normal to the longest axis of the overall structure. Successive ridges are spaced 1 mm apart on each side; the ridges are out of phase between the left- and right-hand sides. The overall breadth of the structure is 7 mm; as each ridge is between 4 and 5 mm in length, successive ridges overlap medially.

Remarks. As the original attitude of specimen IOMMM 98-88 is unknown, and only a short length (20 mm) is preserved, more formal nomenclature is unwarranted. If an epirelief view, the ridges could represent sediment pushed backwards during locomotion; alternatively, if a hyporelief view, the ridges could represent casts of imprints into the sediment surface during locomotion. In both scenarios the trace fossil would be considered as a repichnion (a locomotion trace); the ridges would indicate discrete steps, rather than continuous locomotion. This would favour it having been produced by an arthropod.

Unassigned material

For completeness, the morphology of material that cannot be assigned, largely because of its fragmentary nature, is briefly summarized and accession details provided. The specimens are simple burrows, probably of *Planolites–Palaeophycus* affinity. The surfaces of the specimens have clearly been extensively weathered prior to their collection. The primary ichnotaxobase used to distinguish *Planolites* from *Palaeophycus* (the absence or presence of a wall lining, respectively) cannot therefore be employed (Pemberton & Frey 1982; Keighley & Pickerill 1995).

Burrow 1 (Fig. 5d)

Material. *BGS GSM92720.*

Locality and horizon details. 'South side of Gullet Buigh, Langness' (Bolton 1899, p. 15), [SC 292 668]. Lonan Formation of Woodcock *et al.* (1999).

1899 *Chondrites informis* (Bolton, p. 11, plate 1, fig. 10).

Description. Specimen BGS GSM92720, a straight, unbranched, burrow with a structureless infill, is preserved in full relief. The diameter varies between 4 and 6 mm along the course of the burrow. The pelitic matrix in which specimen BGS GSM92720 occurs provides no indication as to the original orientation. The weathered appearance accounts for the burrow having a variable diameter along its length.

Burrow 2 (Fig. 5e)

Material. *BGS GSM92719.*

Locality and horizon details. 'Foreshore north of Creg Custane, Langness' (Bolton 1899, p. 15), [SC 289 661]. Lonan Formation of Woodcock *et al.* (1999)

1899 *Chondrites informis* (Bolton, p. 11, plate 1, fig. 8).

Description. Specimen BGS GSM92719 is preserved in full relief. It is a short fragment of a 7 mm diameter burrow; the structureless infill is coarser grained than the silt-grade matrix. The burrow is inclined to the plane of splitting; there is no indication as to whether this plane is bedding parallel or a cleavage plane.

Palaeogeographical and stratigraphical distribution of *Dictyodora zimmermani* and *Glockerichnus radiatus*

Dictyodora zimmermani also occurs in the Barrancos Shale of eastern Portugal and the Skiddaw Group of the Lake District of England. The Ordovician stratigraphy of the Barrancos area is still poorly understood (Romano 1982). Perdigão (1967) reported *Didymograptus hirundo* from the Barrancos Shale at the 'Pedreira de Mestre André' (Delgado 1908), indicating that, at least here, it has an Arenig age. This quarry, near the Portuguese–Spanish border, is also known as 'la carrière de Mestre Andre' (Delgado 1910). One of us (PJO) has observed *Dictyodora zimmermani* in this quarry. Although no information on their toponomy is available, other probable examples of *Dictyodora zimmermani* from this locality are illustrated by Delgado (1910, plate XXXI, figs 2, 3, 5 and 6; plate XXXII, fig. 4; plate XXXIII, fig. 3). These specimens, as is that from the Manx Group, are interpreted as horizontal cross-sections through the 'wall' structure that surmounts the basal burrow. Examples of *Dictyodora zimmermani* from the Northern Fells Belt of the Skiddaw Group in the Lake District of England are figured by Orr (1996, figs 4a–c, and possibly figs 4d and e).

There are strong morphological similarities between the specimen of *Glockerichnus radiatus* from the Manx Group and the material described and figured as *Bifasciculus radiatus* by Crimes & Crossley (1968, plate X, fig. D; plate XI, fig. A) and Crimes (1970, plate 1a), and, subsequently, as *Glockerichnus glockeri* by Crimes *et al.* (1992, fig. 2B–D). The latter occurs in the Ribband Group (early Ordovician) in southeastern Ireland. *Glockerichnus radiatus* also occurs at a number of localities in the Northern Fells Belt of the Skiddaw Group (Orr 1996, fig. 5); see Orr (1996, figs 1 and 2) for specific locality details.

A number of other well-defined ichnotaxa recur in early Ordovician shallow- and deep-marine strata from different geographic locations (Fig. 6a); see Orr (1996) for further details. The occurrence of *Dictyodora zimmermani* and *Glockerichnus radiatus* in the Manx Group supports the suggestion by Orr (1996, p. 211–212) that the distribution of certain ichnotaxa may be geographically restricted to high southerly palaeolatitudes in the early Ordovician (Fig. 6b).

Ichnofauna – conclusions

- As in the Skiddaw Group, the ichnofauna of the Manx Group is dominated by examples of fodinichnia (e.g. *Phycodes*, *Glockerichnus*). The absence of unequivocal pascichnia (with the exception of *Dictyodora zimmermani*) and agrichnia in both assemblages is notable. The available evidence suggests that the extensive development of pascichnia and agrichnia within deep-marine environments is restricted to strata of post-early Ordovician age.
- The ichnotaxa of the Manx Group known to date are mainly morphologically simple forms (e.g. *Planolites*) and more complex forms (e.g. *Phycodes*) that recur throughout the Phanerozoic in a variety of depositional environments. Such

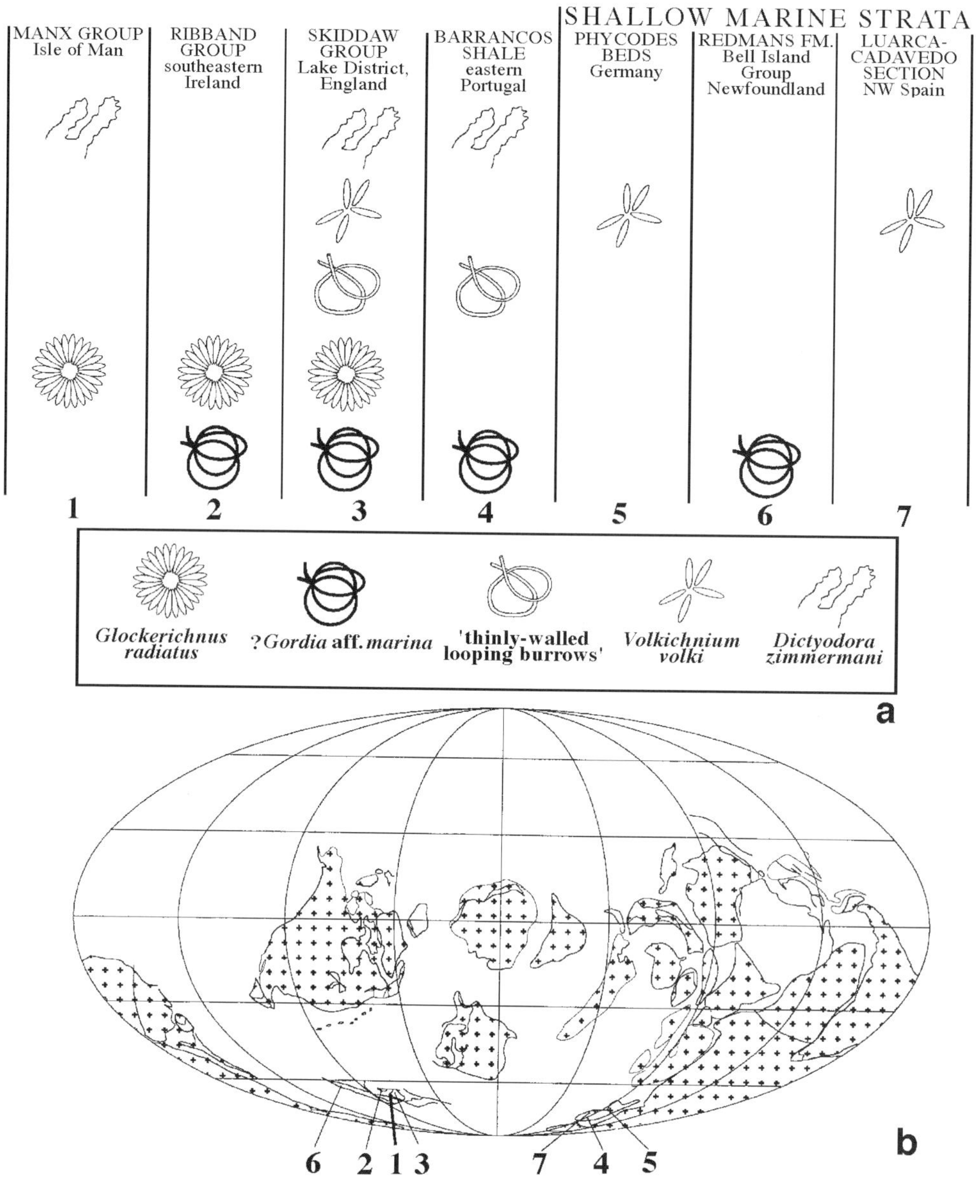

Fig. 6. Elements of the ichnofaunal assemblage in the Manx Group are also present in other ichnofaunal assemblages of early Ordovician age (**a**), all of which were emplaced at high southerly palaeolatitudes (**b**). Modified from Orr (1996, fig. 10).

ichnotaxa neither constrain the age nor the depositional environment of the Manx Group.

- The occurrence of *Dictyodora zimmermani* and *Glockerichnus radiatus* in the Manx Group may be more significant. These morphologically complex and taxonomically well-defined ichnotaxa repeatedly occur together in deep-marine strata of early Ordovician age deposited at high southerly palaeolatitudes.

Dave Quirk and Nigel Woodcock provided material and valuable discussion; Steve Tunnicliff assisted with BGS Collections and Trevor Ford supplied some locality information. Sören Jensen and Jan Zalasiewicz provided thorough and constructive reviews. PJO acknowledges funding from the Department of Education for Northern Ireland, the Royal Society and the Royal Irish Academy. MPAH acknowledges assistance from NERC Research Grant GR9/01834 towards partial defrayment of field expenses.

References

BENTON, M. J. 1982. Trace fossils from Lower Palaeozoic ocean-floor sediments of Southern Uplands; Scotland. *Transactions of the Royal Society of Edinburgh, Earth Sciences*, **73**, 67–87.

—— & TREWIN, N. H. 1980. *Dictyodora* from the Silurian of Peebleshire, Scotland. *Palaeontology*, **23**, 501–513.

BINNEY, E. W. 1877. A notice of some organic remains from the schists of the Isle of Man. *Proceedings of the Literary and Philosophical Society of Manchester*, **16**, 102-110.

BOLTON, H. 1893*a*. On the occurrence of a trilobite in the Skiddaw Slate of the Isle of Man. *Geological Magazine*, **10**, 29–31.

—— 1893*b*. *Observations on the Skiddaw Slates of the Isle of Man*. British Association Reports, Nottinghamshire.

—— 1899. The palaeontology of the Manx Slates of the Isle of Man. *Memoirs and Proceedings of the Manchester Literary and Philosophical Society*, **43**, 1–15.

BROMLEY, R. G. 1990. *Trace Fossils: Biology and Taphonomy. Special topics in paleontology*, **3**. Unwin & Hyman.

CLIFTON, H. E. & THOMPSON, J. K. 1978. *Macaronichnus segregatis*: a feeding structure of shallow marine polychaetes. *Journal of Sedimentary Petrology*, **48,** 1293–1301.

COOPER, A. H., RUSHTON, A. W. A., MOLYNEUX, S. G., HUGHES, R. A. & WEBB, B. C. 1995. The stratigraphy, correlation, provenance and palaeogeography of the Skiddaw Group (Ordovician) in the English Lake District. *Geological Magazine*, **132**, 185–211.

CRIMES, T. P. 1970. The significance of trace fossils in sedimentology, stratigraphy and palaeoecology with examples from Lower Palaeozoic strata. *In*: CRIMES, T. P. & HARPER, J. C. (eds) *Trace Fossils*, Geological Journal, Special Issue, **3**, 101–126.

—— & CROSSLEY, J. D. 1968. The stratigraphy, sedimentology, ichnology and structure of the Lower Palaeozoic rocks of part of north-eastern Co. Wexford. *Proceedings of the Royal Irish Academy*, **67B**, 185–215.

—— & —— 1980. Inter-turbidite bottom current orientation from trace fossils, with an example from the Silurian flysch of Wales. *Journal of Sedimentary Petrology*, **50,** 821–830.

—— & —— 1991. A diverse ichnofauna from Silurian flysch of the Aberystwyth Grits Formation, Wales. *Geological Journal*, **26,** 27–64.

——, GARCIA HIDALGO, J. F. & POIRE, D. G. 1992. Trace fossils from Arenig flysch sediments of Eire and their bearing on the early colonisation of the deep seas. *Ichnos*, **2**, 61–77.

CUMMING, J. G. 1848. *The Isle of Man: Its History, Physical, Ecclesiastical, Civil and Legendary.* Van Voorst.

CURRAN, H. A. 1985. The trace fossil assemblage of a Cretaceous nearshore environment: Englishtown formation of Delaware, U.S.A. *In*: CURRAN, H. A. (ed.) *Biogenic Structures: Their Use in Interpreting Depositional Environments*. Society of Economic Paleontologists and Mineralogists, Special Publications, **35**, 261–276.

DELGADO, J. F. N. 1908. *Systéme Silurique du Portugal. Étude de Stratigraphie Paléontologique*. Commission du Service Géologique du Portugal. Imprimerie Nationale, Lisbonne.

—— 1910. *Terrains Paléozoïques du Portugal. Étude sur les Fossiles des Schistes à Néréites de San Domingos et des Schistes à Néréites et à Graptolites de Barrancos*. Commission du Service Géologique du Portugal. Imprimerie Nationale, Lisbonne.

ETHERIDGE, R. 1876. *Appendix A. Description of new fossils occurring in the Arenig or Skiddaw slates. In:* WARD, J. (ed.)*The Geology of the Northern Part of the English Lake District*. Memoir of the British Geological Survey (England and Wales), 108–112.

FILLION, D. 1989. Les critères discriminants à l'intérieur du triptyque *Palaeophycus–Planolites–Macaronichnus*. Essai de synthèse d'un usage critique. *Comptes rendus de l'Academie des Sciences de Paris, Série 2*, **309**, 169–172.

—— & PICKERILL, R. K. 1990. Ichnology of the Upper Cambrian(?) to Lower Ordovician Bell Island and Wabana groups of eastern Newfoundland, Canada. *Palaeontographica Canadiana*, **7**, 1–119.

FREY, R. W. & SEILACHER, A. 1980. Uniformity in marine invertebrate ichnology. *Lethaia*, **13**, 183–207.

HALL, J. 1847. *Palaeontology of New York. Volume I. Containing Descriptions of the Organic Remains of the Lower Division of the New York System (Equivalent of the Lower Silurian Rocks of Europe)*. C. van Benthuysen.

—— 1852. *Palæontology of New York. Volume II. Containing Descriptions of the Organic Remains of the Lower Middle Division of the New York System (Equivalent of the Middle Silurian Rocks of Europe)*. C. van Benthuysen.

HÄNTZSCHEL, W. 1975. *Trace fossils and Problematica. In:* TEICHERT, C. (ed.) *Treatise on Invertebrate Palaeontology, part W, Miscellanea; Supplement 1*. Geological Society of America, Boulder, Colorado and The University of Kansas.

HARKNESS, R. & NICHOLSON, H. 1866. On the Lower Silurian Rocks of the Isle of Man. *Quarterly Journal of the Geological Society, London*, **22**, 488–491.

HOWE, M. P. A. 1999. The Silurian fauna (graptolite and nautiloid) of the Niarbyl Formation, Isle of Man. *This volume*.

HUNDT, R. 1913. Eine Ergänzung zu 'Organische Reste aus dem Untersilur des Hüttchenberges bei Wünschendorf an der Elster'. *Zentralblatt für*

Mineralogie, Geologie, und Paläontologie, **1913**, 180–181.

KEIGHLEY, D. G. & PICKERILL, R. K. 1994. The ichnogenus *Beaconites* and its distinction from *Ancorichnus* and *Taenidium*. *Palaeontology*, **37**, 305–337.

—— & —— 1995. The ichnotaxa *Palaeophycus* and *Planolites*: historical perspectives and recommendations. *Ichnos*, **3**, 301–309.

LAMPLUGH, G. W. 1903. *The Geology of the Isle of Man*. Memoir of the Geological Survey, United Kingdom. HMSO.

M'COY, F. 1848. Note on some Skiddaw slate fossils. *In:* SEDGWICK, A. (ed.) *On the Organic Remains Found in the Skiddaw Slate, with some Remarks on the Classification of the Older Rocks of Cumberland and Westmoreland*. Quarterly Journal of the Geological Society of London, **4**, 223–225.

—— 1851–1855. A systematic description of the British Palaeozoic fossils in the Geological Museum of the University of Cambridge. *In*: SEDGWICK A. (ed.) *A Synopsis of the Classification of the British Palaeozoic Rocks*. J. W. Parker, 1–184 (1851), 185–406 (1852), 407–661 (1855).

MACLEAY, W. S. 1839. Note on the annelida. *In:* MURCHISON, R. I. (ed.) *The Silurian System, Part II: Organic Remains*, J. Murray, 699–701.

MCCANN, T. 1990. Distribution of Ordovician–Silurian ichnofossil assemblages in Wales–implications for Phanerozoic ichnofaunas. *Lethaia*, **23**, 243–255.

—— 1993. A *Nereites* ichnofacies from the Ordovician–Silurian Welsh Basin. *Ichnos*, **2**, 1–18.

—— & PICKERILL, R. K. 1988. Flysch trace fossils from the Cretaceous Kodiak Formation of Alaska. *Journal of Paleontology*, **62**, 345–366.

MOLYNEUX, S.G. 1979. New evidence for the age of the Manx Group, Isle of Man. *In*: HARRIS, A. L., HOLLAND, C. H. & LEAKE, B. E. (eds) *The Caledonides of the British Isles – Reviewed*. Special Publication of the Geological Society, London, **8**, 415–421.

—— 1999. A reassessment of Manx Group acritarchs, Isle of Man. *This volume*.

MORRIS, J. H., WOODCOCK, N. H. & HOWE, M. P. A., 1999. The Silurian succession of the Isle of Man: the late Wenlock Niarbyl Formation, Dalby Group. *This volume*.

MURCHISON, R. I. 1839. *The Silurian System. Part I. Founded on Geological Researches in the Counties of Solop, Hereford, Radnor, Montgomery, Caermarthen, Brecon, Pembroke, Monmouth, Gloucester, Worcester, and Stafford; With Descriptions of the Coal-fields and Overlying Formations. Part II. Organic Remains*. J. Murray.

—— 1850. *Memoria sulla struttura geologica delle Alpi, delle Apennini e dei Carpazi*. Stamperia granucale, Firenze.

NICHOLSON, H. A. 1873*a*. Contributions to the study of the errant annelides of the older Palaeozoic rocks. *Proceedings of the Royal Society of London*, **21**, 288–290.

—— 1873*b*. Contributions to the study of the errant annelides of the older Palaeozoic rocks. *Geological Magazine*, **10**, 309–310.

ORR, P. J. 1995. A deep-marine ichnofaunal assemblage from Llandovery strata of the Welsh Basin, west Wales, UK. *Geological Magazine*, **132**, 267–285.

—— 1996. The ichnofauna of the Skiddaw Group (early Ordovician) of the Lake District, England. *Geological Magazine,* **133**, 193–216.

—— 1999. Quantitative approaches to the resolution of taxonomic problems in invertebrate ichnology. *In*: HARPER, D. A. T. (ed.) *Numerical Palaeobiology: Analysing and Modelling Fossils and their Distributions*. John Wiley, in press.

—— & PICKERILL, R. K. 1995. Trace fossils from early Silurian flysch of the Waterville Formation, Maine, U.S.A. *Northeastern Geology and Environmental Sciences*, **17**, 394–414.

OWEN, R. 1852. Description of the impressions and footprints of the Protichnites from the Potsdam sandstone of Canada. *Quarterly Journal of the Geological Society, London*, **8**, 214–225.

PEMBERTON, S. G. & FREY, R. W. 1982. Trace fossil nomenclature and the *Planolites–Palaeophycus* dilemma. *Journal of Paleontology*, **56**, 843–881.

PERDIGÃO, J. C. 1967. Estudos geológicos na pedreira do Mestre André (Barrancos). *Serviços Geológicos de Portugal, Comunicações,* **52**, 55–64.

PICKERILL, R. K. 1980. Phanerozoic flysch trace fossil diversity – observations based on an Ordovician flysch ichnofauna from the Aroostook–Matapedia Carbonate Belt of northern New Brunswick. *Canadian Journal of Earth Sciences*, **17**, 1259–1270.

RICHTER, REINHARD. 1850. Aus der thüringischen Grauwacke. *Deutsche Geologische Gesellschaft, Zeitschrift*, **2**, 198–206.

———— 1853. *Gaea von Salfeld*. Programm d. Realsch. Saalfeld, 3–32

RICHTER, RUDOLF. 1937. Marken und spuren aus allen Zeiten I–II. *Senckenbergiana,* **19**, 150–169.

ROMANO, M. 1982. The Ordovician biostratigraphy of Portugal – A review with new data and re-appraisal. *Geological Journal,* **17**, 89–110.

RUSHTON, A. W. A. 1993. Graptolites from the Manx Group. *Proceedings of the Yorkshire Geological Society*, **49**, 259–262.

TAYLOR, J. 1862. Supposed imprints in the Lower Cambrian beds of the Isle of Man. *The Geologist*, **5,** 321–323.

VYALOV, O. S. & GOLEV, B. T. 1960. K sistematike *Paleodictyon. Doklady, Akademiya Nauk USSR,* **134**, 175–178.

WEISS, E. 1884. *Vorlegung des Dictyophytum liebeanum Gein. aus der Gegend von Gera. Gesellschaft Naturforschender Freunde Berlin, Sitzungsberichte,* **1884**, 17.

WOODCOCK, N. H. & MORRIS, J. H. 1999. Debris flows on the Ordovician margin of Avalonia: Lady Port Formation, Manx Group, Isle of Man. *This volume*.

——, ——, QUIRK, D. G., BARNES, R. P., BURNETT, D., FITCHES, W. R., KENNAN, P. S. & POWER, G. M. 1999. Revised lithostratigraphy of the Manx Group, Isle of Man. *This volume*.

Revised lithostratigraphy of the Manx Group, Isle of Man

N. H. WOODCOCK[1], J. H. MORRIS[2], D. G. QUIRK[3], R. P. BARNES[4], D. J. BURNETT[3], W. R. FITCHES[5], P. S. KENNAN[6] & G. M. POWER[7]

[1]*Department of Earth Sciences, University of Cambridge, Cambridge CB2 3EQ, UK*

[2]*Geological Survey of Ireland, Beggars Bush, Haddington Road, Dublin 4, Ireland*

[3]*Department of Geology, Oxford Brookes University, Gipsy Lane, Oxford OX3 0BP, UK*

[4]*British Geological Survey, Murchison House, West Mains Road, Edinburgh EH9 3LA, UK*

[5]*Robertson Research International, Llanrhos, Llandudno, North Wales, LL30 1SA, UK*

[6]*Department of Geology, University College, Belfield, Dublin 4, Ireland*

[7]*Department of Geology, University of Portsmouth, Portsmouth PO1 3QL, UK*

Abstract: The existing lithostratigraphy of the Manx Group (lower Ordovician) of the Isle of Man is formalized as far as is possible given a paucity of biostratigraphical and structural control. Coherent packages of strata can be recognized within northeast striking structural tracts. Some tract boundaries are faulted but poor exposure leaves doubt about the nature of others. Some structural repetition of successions is likely. However, the long-held equivalence of the sandstone units on either side of the island can no longer be maintained, because the Niarbyl Formation in the west is now known to be of Silurian age. Repetition of Manx Group units across a syncline along the axis of the island is therefore in doubt. Different outcrops of superficially similar units are accordingly named separately on each flank of the island. The stratigraphy of part of the northern Manx Group is left unrevised awaiting further work.

The southeastern, sandstone-dominated units of the Manx Group comprise three tracts divided into the Lonan and Santon (lower Arenig) Formations, the Port Erin and Mull Hill Formations, and the Ny Garvain and Creg Agneash Formations. The Creg Agneash Formation is stratigraphically overlain by the mudstone-dominated Maughold Formation. The next two tracts to the northwest both contain a black mudstone unit, the possibly correlative Barrule and Glen Rushen (middle Arenig) Formations, each structurally overlain by laminated silty mudstone of the Injebreck Formation. A further unit of laminated siltstone and mudstone, the Creggan Mooar Formation, is defined in tract 6, west of which an enigmatic unit contains lowest Arenig fragmental volcanic rocks. In the most northwesterly tract, the Lady Port Formation is a varied assemblage of black mudstone, turbidites, pebbly mudstone and quartz arenite. Lithological correlation of these different Manx Group successions is attempted, but is constrained by only four biostratigraphical control points.

The Manx Group stratigraphy has long been correlated with that of the Skiddaw Group of the Lake District. A tentative correlation of events on this part of the early Ordovician margin of Gondwana can be made assuming some synchroneity of Early Arenig lowstand turbidite fans, of a mid-Arenig transgression, and of a late Arenig mass-wasting event.

The Manx Group comprises a Lower Palaeozoic, probably entirely lower Ordovician, succession of deep-marine sedimentary rocks, cropping out over about three-quarters of the Isle of Man (Fig. 1). Successive studies, reviewed in the next section, have struggled to make stratigraphic sense of this crucial succession, facing the same frustrations disconsolately recorded by Blake (1905, pp. 358–359): 'The determination of the order of succession of the various members of the Manx-Slate Series is a matter, on the whole, of considerable difficulty: owing, in the first place, to the relative paucity of exposure over wide areas, the solid rocks being much concealed by the abundance of their own debris, which may be more or less mixed and removed from their original site; and, in the second place, to the similarity in many of their aspects of portions of the different members themselves, when seen otherwise than in mass. And when, to these difficulties, we add the easy confusion so often introduced by cleavage of a complicated character, and the indubitable contortion to which parts are

From: Woodcock, N. H., Quirk, D. G., Fitches, W. R. & Barnes, R. P. (eds) 1999.
In Sight of the Suture: the Palaeozoic geology of the Isle of Man in its Iapetus Ocean context.
Geological Society, London, Special Publications, **160**, 45–68. 1-86239-046-0/99/$15.00

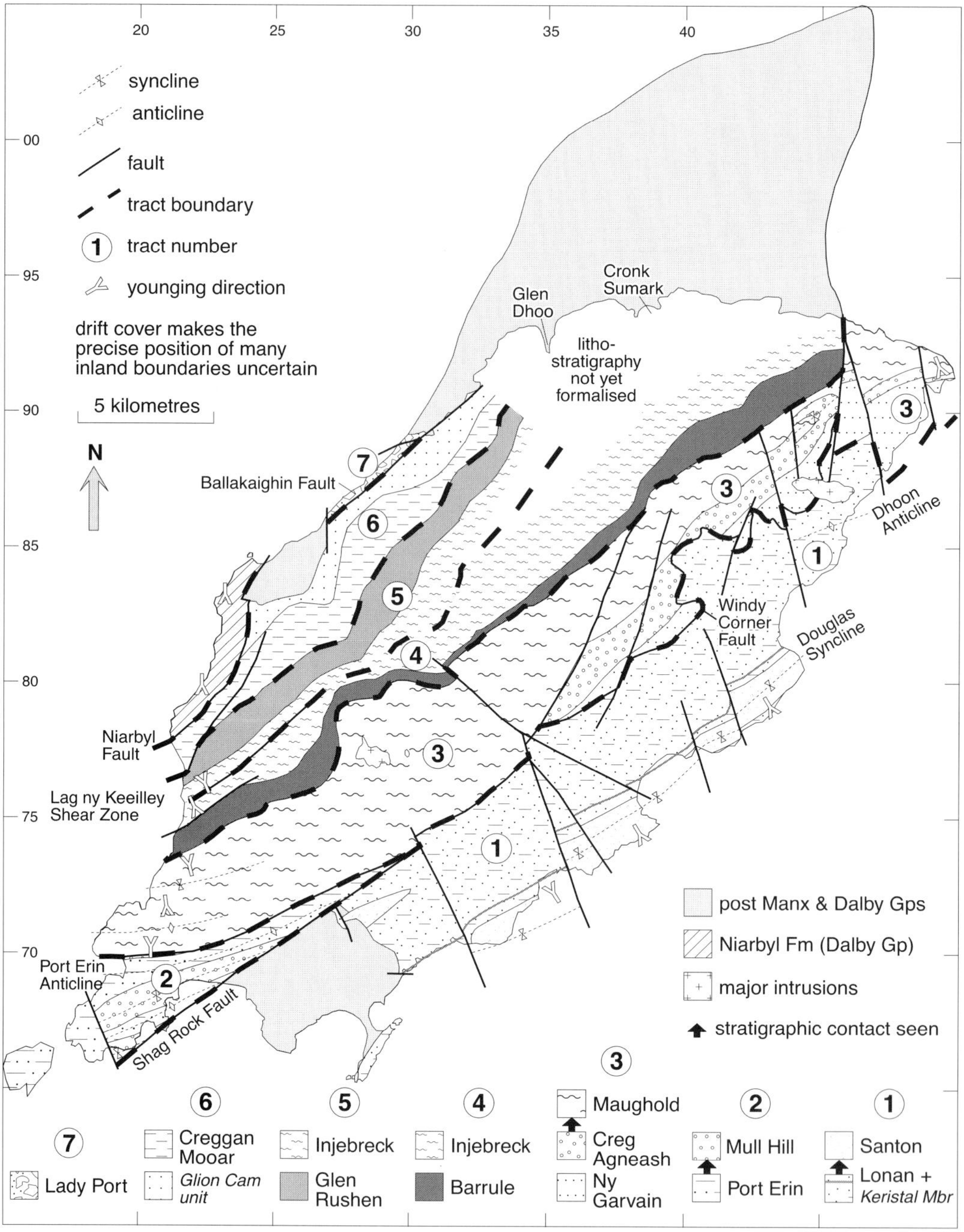

Fig. 1. Geological map of the Manx Group showing the extent of the lithostratigraphic units formally defined in this paper and the boundaries of the structural tracts that contain them.

subject, we have all the elements united which might lead us to despair of ever solving the problem.'

In our recent re-examination of the Manx Group, three specific barriers to further unravelling its stratigraphy have been tackled.

First, all the accessible coastal outcrop of the Manx Group has been remapped at a scale of

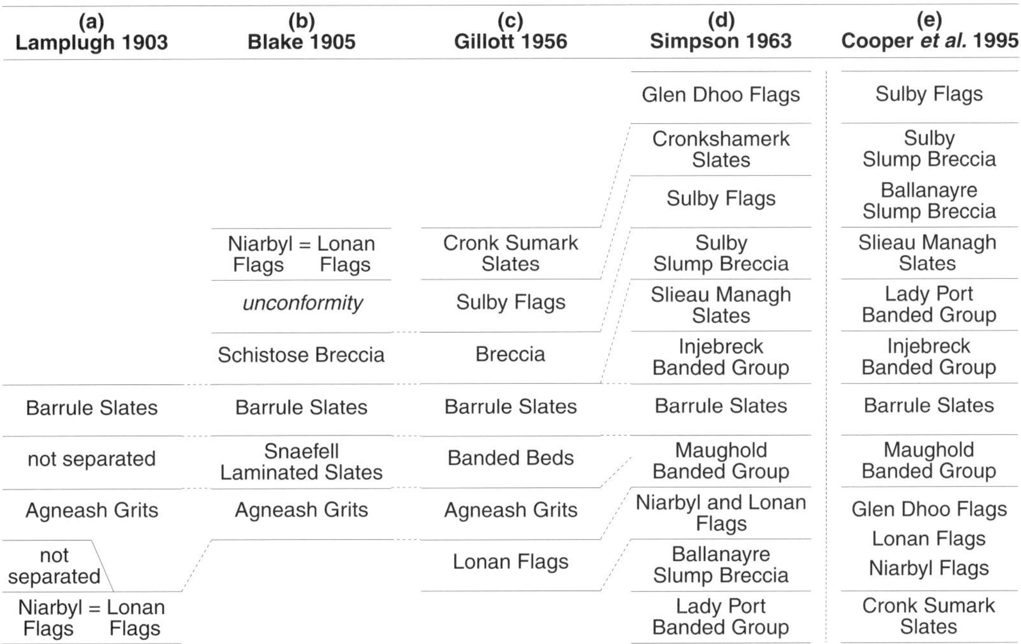

Fig. 2. Previous stratigraphic schemes for the Manx Group. 'Pecked' lines indicate approximate correlations where possible.

1:10 000 or better. The need for re-mapping has been widely recognized (e.g. Roberts *et al.* 1990, p. 277; Rushton 1993, p. 262) and the present project has answered some crucial questions of lithostratigraphic sequence and correlation. The available resources – for about 40 person-weeks of fieldwork – permitted only selective inland mapping, mainly to extrapolate coastal stratigraphy and structure. However, this experience has confirmed that only a much more costly effort would substantially improve on the excellent mapping of Lamplugh, carried out between 1892 and 1897. As a substitute, Lamplugh's meticulous six-inch scale field slips have been consulted in the present study, as well as key inland traverses recently carried out by Quirk & Burnett (1999).

Secondly, an attempt to address the poor biostratigraphic control within the Manx Group has been made. Better knowledge of the depositional environment of each sedimentary unit has allowed a well-constrained search for graptolites and other macrofossils. The number of specimens from the Manx Group has been increased but the preservation quality of the material is such that no significantly enhanced biostratigraphic control has been achieved (Orr & Howe 1999).

Thirdly, and the main purpose of the present paper, formal lithostratigraphic definition of component formations of the Manx Group has been started. A major barrier to progress has been the loose lithological definition of existing units, without reference to type sections or discussion of contact relationships (Simpson 1963). The confusion is compounded by the use of the same names for units supposed to correlate across the central axis of the island. Several of these correlations can now be shown to be untenable, and the others to be premature. Nevertheless, a conservative approach has been taken to the formalized lithostratigraphy, retaining old names wherever possible and leaving some tracts, mainly poorly exposed inland areas, with the existing informal nomenclature pending further work.

The Manx Group: previous classifications

During the later part of the nineteenth century, the older rocks of the Isle of Man were habitually referred to as Skiddaw Slates, following Harkness & Nicholson's (1866) recognition of their general similarity to these oldest rocks of the Lake District. However, G. W. Lamplugh's comprehensive re-mapping of the island failed to confirm a close correlation with the Skiddaw Slates. With the publication of his geological map (Geological Survey of United Kingdom 1898) and accompanying memoir (Lamplugh 1903) the Manx Slate Series was separately designated, although the understanding of a general equivalence to the Skiddaw Group was retained.

Lamplugh used the three lithological descriptors: 'grits' for sandstone-rich units, 'flags' for mixed sandstone–mudstone units and 'slates' for mudstone units. On this basis, he distinguished four named lithostratigraphic units in the Manx Series, intercalated with undifferentiated units designated as 'not separated' (Fig. 2a). The Lonan Flags of the southeast coast were tentatively correlated with the Niarbyl Flags of the northwest coast, and regarded as the base of the Manx succession. The Agneash Grits overlay the Lonan Flags, either directly or with an intervening 'not separated' unit. Other grits, regarded by Lamplugh either as infolds of the Agneash or, more likely, as an older horizon, were shown intercalated with the Lonan Flags. Overlying the Agneash Grits, Lamplugh mapped a second 'not separated' unit, itself overlain by the Barrule Slates. Additionally, Lamplugh mapped horizons of 'crush-conglomerate', regarded by him (Lamplugh 1903, pp. 55–71; Lamplugh & Watts 1895) as of tectonic origin and therefore without primary stratigraphic significance. He ascribed their preferential distribution above the Agneash Grits to a lithological control on deformation.

Explicit in Lamplugh's (1903) stratigraphic scheme was its repetition across a 'synclinorium' axis following the central northeast–southwest spine of the island. This repetition explained the two main outcrop tracts of Barrule Slates and allowed the correlation of the Lonan and Niarbyl Flags but, as Lamplugh (1903, pp. 47–48) observed: '... it remains altogether doubtful whether there be any equivalent of the Agneash Grits to the westward of the central axis...'

Blake (1905) proposed that the 'crush-conglomerate' of Lamplugh (1903) was a primary fragmental deposit. Blake recognized that these 'schistose breccias' are typically associated with a host lithology of dark mudstone, and had a limited stratigraphic range, believed to be above the Barrule Slates (Fig. 2b). He proposed that the upward succession of Agneash Grits, Snaefell Laminated Slates (the lower of Lamplugh's unseparated units), the Barrule Slates and the Schistose Breccias was repeated twice on the northwest of the island's spine by large reverse faults rather than by a syncline. However, Blake placed the Lonan and Niarbyl Flags as an unconformable cover to this faulted assemblage, a view with which Lamplugh [in discussion of Blake (1905)] strongly disagreed, and which has not been confirmed by any later work.

Gillott (1956) used way-up evidence in the Lonan Flags to deduce that this unit passed up into the Agneash Grits and was therefore at the base of the succession, as Lamplugh had maintained (Fig. 2c). Gillott substantiated the pre-tectonic, submarine slide origin of the breccias and arranged them above the Barrule Slates as had Blake (1905). Gillott recognized two higher units in the north of the island, the Sulby Flags and the Cronk Sumark Slates. However, he did not address how they might relate to the Niarbyl Flags, nor to the conflicting synclinal and reverse-faulted hypotheses for Manx Series structure.

Simpson (1963) produced structural evidence in favour of the synclinal hypothesis, invoking a large but relatively simple Isle of Man Syncline rather than Lamplugh's (1903) more complicated synclinorium. However, Simpson's strict application of this structural model spawned a number of new stratigraphic units (Fig. 2d). A Lady Port Banded Group and Ballanayre Slump Breccia supposedly lay below the Niarbyl Flags on the northwest coast, the Injebreck Banded Group and Slieau Managh Slates were required below the Cronk Sumark Slates (Cronkshamerk Slates of Simpson) in the north of the island. Ford (1993) informally recognized Simpson's units as having formation status. He used the lithological descriptor 'pelites' for the Lady Port, Maughold, Injebreck and Cronkshamerk Units, but retained Simpson's stratigraphic succession.

The early finds of fossils in the Manx Slates, detailed by Lamplugh (1903, pp. 89–95) proved to be mostly either of dubious organic origin or trace fossils of little value for dating. The main find that has stood up to later re-examination (Rushton 1993; Orr & Howe 1999) is an early Ordovician (Tremadoc or possibly Arenig) graptolite, allegedly from the Cronk Sumark Slates (Bolton 1899). However, only with the analysis of microfossil assemblages (Downie & Ford 1966; Molyneux 1979) has any biostratigraphical guide emerged to the ordering of units within the Manx Slates (Manx Group; Molyneux 1979). These palaeontological constraints, including a new Arenig graptolite find from the Lonan Flags reported by Rushton (1993), have been summarized by Cooper *et al.* (1995). They found it necessary to re-order Simpson's lithological succession (Fig. 2e), correlating the three units of flags, placing the Cronk Sumark Slates at the base of the group, and moving the Lady Port and Ballanayre Units higher up. The low metamorphic grade of the Lady Port and Ballanayre Units is also consistent with a high stratigraphic position (Roberts *et al.* 1990). However, this low-grade zone also include the Niarbyl Flags, prompting Roberts *et al.* (1990) to suggest that all three units may have been emplaced along a major fault, the Niarbyl Thrust of Lamplugh (1903), against the higher grade mass of the Manx Group.

Recently, Quirk & Kimbell (1997) have presented geophysical evidence suggesting that the Manx Group is compartmentalized into a number of fault-bound slices.

A new lithostratigraphic framework

The new lithostratigraphic scheme for the Manx Group recognizes the following realities:

- The island contains several tracts, apparently separated from each other by major faults, in which coherent stratigraphy may be preserved but between which no continuity can, at present, be demonstrated (Table 1). Different lithostratigraphic names are considered advisable within tracts that cannot be correlated, although existing names have been retained where possible (Fig. 3). The new formation names can later be suppressed if correlations are proved. Quirk & Burnett (1999) do not recognize the Windy Corner Fault, regarded here as as tract boundary. Conversely, they have mapped a number of faults, additional to the tract boundaries, which affect the stratigraphy. These include an east–west mineralized fault traversing Maughold Head (e.g. [SC 497 915]), an east-northeast–west-southwest shear zone which cuts off the northern end of the Douglas Syncline (e.g. [SC 442 808]) and a number of north–south and east-northeast–west-southwest lineaments in the northern inland area.
- Stratigraphic contacts between formations within each tract cannot always be demonstrated, often due to faulting in the coastal sections. Figure 1 and the following text are explicit about which contacts are demonstrably stratigraphic and which are of uncertain status. Formations with uncertain contacts can later be severed from each other if a tectonic contact is demonstrated.
- The depositional way-up across contacts cannot always be observed, particularly in fine-grained facies. Tight early folding also makes the gross younging direction within such units difficult to determine. Doubt must attach, therefore, to the relative age of formations across any contact lacking local way-up evidence (Fig. 1).
- The northern part of the Manx Group, only exposed inland between Kirk Michael and Ramsey, and containing the upper five of Simpson's (1963) units, cannot yet be reliably assigned to any of the stratigraphic tracts. Rather than forcing these units into the new scheme, they are left as informal units awaiting further work (Fig. 1). Quirk & Burnett (1999) offer new evidence in this area.

Seven tectonostratigraphic tracts are distinguished in the new scheme for the Manx Group (Fig. 1). A further zone lies northwest of a north-east–southwest fault and shear zone, the Niarbyl Thrust of Lamplugh (1903) or Niarbyl Slide of Simpson (1963), cropping out between Niarbyl and Peel (Fig. 1). This tract is entirely composed of the Niarbyl Formation, the Niarbyl Flags of previous workers (Fig. 3). There is strong evidence against correlating this unit with any other unit in the Manx Group, and for removing the Niarbyl Formation to a Dalby Group of Silurian age (Morris *et al.* 1999).

Quirk & Burnett (1999) have adopted an alternative approach to interpreting the Manx Group, which has involved defining a simplified set of lithofacies associations and mapping their occurrence. Their results are mostly incorporated in this paper, with specific differences of interpretation noted in the text. In general, Quirk & Burnett (1999) recognize additional units on the basis of groupings of lithofacies associations and are more conservative in their interpretation of the lateral extent of formations. Few of their units are shown to continue uninterrupted across the island; e.g. their Injebreck Formation is confined to the north of the island and their Keristal Member is regarded, at least in part, as representing similar events of different ages.

Table 1. *Approximate equivalence of formally defined units*

This paper	Simpson (1963)
Informal units	Sulby Flags Sulby Slump Breccia Slieau Managh Slates Glen Dhoo Flags Cronkshamerk Slates
Niarbyl Fm (now in Dalby Group)	Niarbyl Flags (SW)
Lady Port Fm	Ballanayre Slump Breccia Lady Port Banded Group
Glion Cam Unit Creggan Mooar Fm	Niarbyl Flags (NE) Maughold Banded Gp (SE)
Injebreck Fm Glen Rushen Fm	Injebreck Banded Gp (NW) Barrule Slates (NW)
Injebreck Fm Barrule Fm	Injebreck Banded Gp (SE) Barrule Slates (SE)
Maughold Fm Creg Agneash Fm Ny Garvain Fm	Part Maughold Banded Group (SE) Lonan Flags (NE)
Mull Hill Fm Port Erin Fm	Part Maughold Banded Group (SE)
Santon Fm Keristal Mbr Lonan Fm	Lonan Flags (SW)

In the following descriptions, the lower bounds of bed thickness classes are 1 (very thin), 3 (thin), 10 (medium), 30 (thick) and 100 cm (very thick). The term mudstone designates rocks containing both silt- and clay-grade material, unless these are specifically distinguished. The boundary between wacke and arenite is taken at 15% matrix.

Manx Group

The term 'Manx Group' was introduced by Simpson (1968, p. 135) as the precise equivalent of the 'Manx Slate Series' of Lamplugh (1903), Simpson (1963) and other earlier workers. It is here specified as containing the following formations: Lonan, Santon, Port Erin, Mull Hill, Ny Garvain, Creg Agneash, Maughold, Barrule, Injebreck, Glen Rushen, Creggan Mooar and Lady Port. It also contains the other, as yet informally designated, strata included within the Manx Slate Series by Lamplugh (1903) and Simpson (1963), and distinguished on Fig. 1, except that it specifically excludes the redefined Niarbyl Formation.

The Manx Group contains minor volcanic rocks, the Peel Volcanics (Lamplugh 1903; fig. 8). The group also hosts intrusions, some of which show evidence of early introduction into semi-consolidated sediment and most of which pre-date the main tectonic deformation of the Manx Group. Some of these intrusions may belong within the group but are excluded until a temporal link is demonstrated. The earlier intrusions include the Dhoon Granodiorite, which was intruded before the deformation: spots in the contact aureole were flattened in the first cleavage, which is also present in associated granitic dykes (Power & Barnes 1999). However, the Foxdale Granite was probably intruded late in, or after, the deformation (Simpson 1965; Power & Barnes 1999).

The Manx Group specifically excludes the rocks now recognized to be of Silurian age, the Dalby Group of Morris *et al.* (1999) and the Carboniferous rocks around Castletown, which are clearly separated from the Manx Group by an angular unconformity. The Peel Sandstones are also excluded because, although their observable contacts with the Manx Group are faulted, they are of contrasting facies and are thought to be of upper Silurian or lower Devonian age (Crowley 1985; Piper & Crowley 1999).

Lonan Formation

Name and equivalence

The Lonan Formation is named after the parish district of Lonan in the east of the island. The formation is the lower part only of the Lonan Flags of Lamplugh (1903) and Simpson (1963), the upper part being renamed the Santon Formation. The Lonan Formation also excludes the northeastern part of the Lonan Flags, lying within tract 3 (Figs 1 and 3), which is assigned to the Ny Garvain Formation. Quirk & Burnett (1999) address facies variation in the Lonan Formation, Woodcock & Barnes (1999) describe its sedimentology and Barnes *et al.* (1999) report on its geochemical provenance.

Type area and subsidiary localities

The type area is designated in Laxey Bay, in Lonan district (between [SC 4400 8344] and [SC 4476 8313]), which displays the lithology that Lamplugh regarded as typical of the Lonan Flags. The variability of the formation is exposed in coastal sections between Dhoon Bay [SC 4617 8645] and Garwick Bay [SC 4378 8125], in Douglas Bay [SC 4010 7726] to [SC 3876 7651] and from Keristal Bay [SC 3578 7300] to Port Soderick [SC 3473 7205]. Similar facies exposed between Purt Veg [SC 3242 7030] and Cass ny Hawin [SC 2980 6926], and on the east side of Langness [SC 2930 6735] to [SC 2788 6534] are included in the Lonan Formation by Woodcock & Barnes (1999) but equated with the Port Erin Formation by Quirk & Burnett (1999).

Lithological characteristics

The Lonan Formation mainly comprises thin-bedded or very thin-bedded fine sandstone and mudstone (Fig. 3a). Each bed typically has a sharp base and grades from light grey cross-laminated fine wacke through parallel-laminated silt–mudstone up to dark grey mudstone, sometimes bioturbated; the sandstone component is 30–50%. Convolute lamination is common in the cross-laminated divisions. Bedding surfaces often show straight crested to undulatory, asymmetric current ripples. The graded beds of the type Lonan Formation are interpreted as the product of deposition from low-concentration turbidity currents in an oxygenated deep-marine environment.

The thin-bedded facies varies in sandstone proportion from as low as 25% (e.g. Cass ny Hawin [SC 2980 6934]) to as high as 80% (e.g. Keristal [SC 3490 7280]). Intervals of very thin-bedded, or even thickly laminated, turbidites occur, particularly in the core of the Dhoon Anticline, e.g. in Dhoon Bay [SC 4616 8648; Fig. 3b] and above Bulgham Bay [SC 4554 8600]. These beds comprise very fine sandstone or siltstone grading up to mudstone. The formation also contains

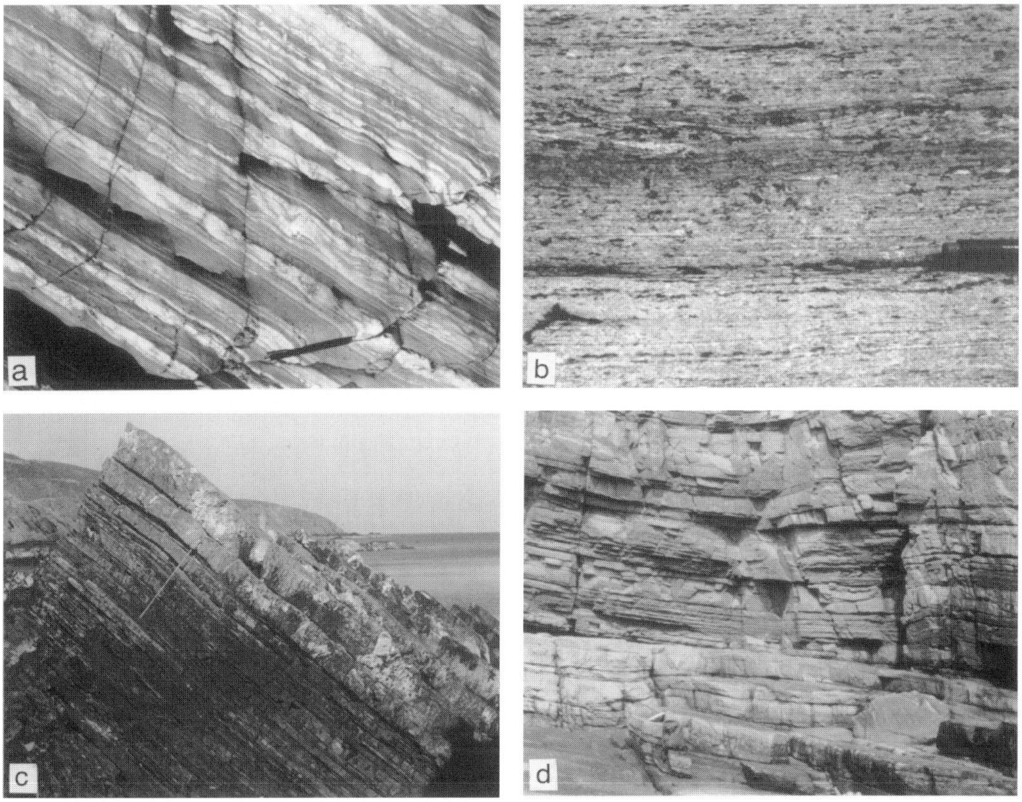

Fig. 3. Typical lithologies in tract 1. (a) Thin- and very thin-bedded turbidites in the Lonan-type area (pen length, 13 cm); Laxey Bay [SC 4400 8344]. (b) Laminated to very thin-bedded Lonan facies (pen diameter,1 cm), Dhoon Bay [SC 4616 8648]. (c) Base of thickly to medium-bedded quartz arenite packet (staff length, 1 m), Keristal Member, Keristal Bay [SC 3505 7296]. (d) Medium-bedded turbidites at the type section of the Santon Formation (height of section, 7 m), east of Santon Head [SC 3278 7028].

sporadic medium-bedded units grading from fine-grained wacke to mudstone. A distinctive packet of medium- to thick-bedded quartz arenites, the Keristal Member, occurs in the upper part of the formation.

Thickness and relationship with adjoining units

The base of the Lonan Formation does not occur at outcrop. The very thin-bedded facies in the core of the Dhoon Anticline is probably the lowest exposed level in the formation. Thickness estimates are difficult because of frequent folding. About 1800 m of strata occur on a relatively uncomplicated limb between the Dhoon Anticline at Bulgham Bay [SC 4575 8580] and the base of the Santon Formation east of Garwick Bay [SC 4378 8125]. The formation may be even thicker in the vicinity of Douglas Harbour [SC 3832 7520] but the inland control on fold position and geometry are poor here. The top of the Lonan Formation is defined by the base of the overlying Santon Formation.

Keristal Member

Name and equivalence

The Keristal Member is named after Keristal Bay [SC 3525 7300], northeast of Port Soderick. This unit has not been previously named although Lamplugh (1903, pp. 46–47) explicitly recorded and mapped some of its outcrops as ‘other Eastern grits’, in particular at Coolebegad [SC 3539 7295], the eastern headland to Keristal Bay. However, he did not separate the ‘grits’ assigned here to the Keristal Member from those now assigned to the stratigraphically higher Santon Formation.

Type area and subsidiary localities

The type section is defined at the west side of Keristal Bay [SC 3505 7296] where a south-southeast dipping succession includes the complete thickness, here 7 m, of the Keristal Member (Fig. 3c). Immediately to the north, the member is repeated over an anticline–syncline pair. The same fold pair affects the member at the eastern side of the bay [SC 3536 7300], where it is faulted southward on to the headland of Coolebegad. Seven localities, all recognized by Lamplugh (1903), provide additional exposures attributable to the Keristal Member. There are coastal exposures of the member at Cass ny Hawin Head [SC 3000 6930], Port Soldrick [SC 3057 6970], Port Greenaugh [SC 3183 7036], Port Jack [SC 3997 7730] and Garwick Bay [SC 4362 8130]. Inland exposures occur at the Nunnery [SC 3719 7543] and in Bibaloe Glen [SC 4162 7894] and [SC 4173 7950].

Lithological characteristics

The Keristal Member comprises a discrete packet of between six and 14 light grey to white fine sandstone beds, typically medium- to thick-bedded, but including some very thick beds (Fig. 3c). The sandstones comprise quartz arenite or quartz wacke with up to 25% matrix, compared with the more matrix-rich wacke that typifies the bulk of the Lonan Formation. The sandstone occurs in non-graded or weakly graded beds, often massive or with parallel lamination, but sometimes with ripple cross-lamination in thinner beds. The beds within the Keristal Member are sometimes organized in either a thinning-up or a thickening-up sequence. In thinning-up sequences, the basal bed is often strongly erosional into underlying thin-bedded turbidites and has basal flute marks.

The Keristal Member is interpreted by Woodcock & Barnes (1999) as the product of a short-lived system of high- to medium-concentration turbidity currents, that tapped a source of clean quartz sand, distinct from the more muddy sands of the typical Lonan turbidity flows (cf. Barnes *et al.* 1999). Quirk & Burnett (1999) regard the Keristal Member as, at least in part, representing similar events of different ages.

Thickness and relationship with adjoining units

The thickness of the Keristal Member varies from *c.* 2 m at Port Jack to 9 m at Cass ny Hawin. The member is underlain and overlain by thin-bedded or, in the south, very thin-bedded turbidites of the Lonan Formation. The boundaries of the member are sometimes locally faulted due to the strong mechanical contrast with the surrounding strata. The member occurs *c.* 150–300 m below the base of the Santon Formation.

Santon Formation

Name and equivalence

The formation is named after Santon parish and specifically after Santon Head [SC 3328 7022], around which coastal sections typical of the Santon Formation occur. The Santon Formation was previously included within the Lonan Flags, although Lamplugh (1903) recognized the distinctive medium- to thick-bedded nature of some of its outcrop, e.g. at Clay Head, Douglas Head and, particularly, St Ann's (Santon) Head.

Type area and subsidiary localities

The type section of the Santon Formation is taken between Baltic Rock [SC 3290 7031] and Santon Head. However, no one section can display the full facies variation in the formation. A continuous but atypical section of the basal 150 m of the Santon Formation is exposed between Purt Veg [SC 3236 7030] and Baltic Rock. More typical sections occur northeast of Santon Head to Gob Lhiack [SC 3475 7200], on Onchan Head [SC 4026 7729] and on the northeast side of Onchan Harbour [SC 4079 7753]. The cuttings along the old Marine Drive tramway between The Whing [SC 3603 7330] and Douglas Head [SC 3887 7468] display strata assigned to the Santon Formation by Woodcock & Barnes (1999), although certain differences between the lithofacies exposed in this area and the rest of the Santon Formation have been highlighted by Quirk & Burnett (1999).

Lithological characteristics

The Santon Formation is characterized by medium- or thick-bedded sandstones intercalated with the thin-bedded facies typical of the Lonan Formation. In the type section (Fig. 4d), typical beds grade from fine or very fine sandstone to mudstone. Each bed comprises 60–90% of wacke, light grey or light greenish grey in colour. Parallel and ripple cross lamination are common, and some bed bases preserve flute marks and horizontal burrows. The thicker graded beds are interpreted as the product of deposition from high- to medium-concentration turbidity currents within a background of low-concentration events.

The wackes that dominate the formation become particularly quartzose in places along the Marine Drive, e.g. at Port Skillion Lighthouse [SC 3900 7470] and on Little Ness [SC 3661 7293]. Yet another variant is seen below Wallberry Hill [SC 3700 7349] and at The Whing [SC 3603 7330], where isolated or weakly grouped medium to thick beds of quartzose arenite punctuate a background of medium-bedded wacke.

Very thick-bedded sandstones dominate the Santon Formation in one section only, on the coast either side of Purt Veg [SC 3255 7037]. Here, thin-bedded turbidites of the Lonan Formation are abruptly overlain by *c.* 30 m of very thick-bedded, parallel-laminated medium sandstones, with lenses of granules or coarse sand at their bases, probably deposited from high-concentration turbidity currents. These pass up into 20 m of very thick- or thick-bedded, medium- to fine-grained sandstones, with intercalated thin-bedded sand–mud turbidites. Thick- or very thick-bedded sandstones are intercalated with medium-bedded turbidites through at least another 50 m of the succession (Fig. 4e), northwards to past Baltic Rock [SC 3320 7027] before typical Santon Formation lithofacies predominate. The Purt Veg succession is interpreted (Woodcock & Barnes 1999) as the fill of a trunk distributary channel in the Santon turbidite system.

Thickness and relationship with adjoining units

About 200 m of Santon Formation can be directly measured above its sharp base at Purt Veg. A continuous fold limb south of Pistol Castle [SC 3395 7140] exposes a thickness of *c.* 300 m and a limb at Clay Head *c.* 400 m. The possible Santon Formation between Little Ness and The Whing totals as much as 600 m. Except at Purt Veg, the base of the Santon Formation is gradational above thin-bedded Lonan Formation, and is arbitrarily defined where medium-bedded sandstone forms > 25% of the unit. The base is visible dipping southeastwards between Port Soderick and Gob Lhiack [SC 3475 7220] and on Onchan Head, and dipping northwestward at Brither Clip Gut [SC 4110 7738] and south of Port Groudle [SC 4208 7806]. The Santon Formation crops out in the core of the Douglas Syncline and its top is above the present level of erosion into the syncline. According to Quirk & Burnett (1999), it is equivalent to the Ny Garvain Formation, but in this paper a correlation with the Creg Agneash Formation (Fig. 9) is suggested, on the basis of arguments in Barnes *et al.* (1999) and Woodcock & Barnes (1999).

Port Erin Formation

Name and equivalence

The formation is named after the town of Port Erin [SC 1970 6900]. The unit was included as part of the Maughold Banded Group by Simpson (1963) and as undifferentiated Manx Group by Lamplugh (1903).

Type area and subsidiary localities

The type section is taken at Traie Vane [SC 1950 6943] in Port Erin Bay. The formation is also exposed for much of the way around the coast of the Cregneash Peninsula and Calf of Man. It is easily accessible at Little Sound [SC 1730 6662], Perwick Bay [SC 2050 6725] and Port St Mary [SC 2105 6778].

Lithological characteristics

The Port Erin Formation is dominated by thin- or very thin-bedded sandstone–mudstone couplets (Fig. 4a), similar to those of the Lonan Formation in tract 1. However, the Port Erin Formation includes a higher proportion, *c.* 50% rather than 10%, of thickly laminated siltstone–mudstone couplets. Both facies of the formation are interpreted as the deposits of low-concentration turbidity flows.

Thickness and relationship with adjoining units

The formation may be at least 600 m thick in Port Erin Bay, although folding makes this estimate unreliable. The base of the formation is not exposed; the top is a rapid upward transition into the Mull Hill Formation. The Port Erin Formation is similar in facies to the thinner bedded parts of the Lonan Formation. Strata on Langness [SC 2930 6735] to [SC 2788 6534], and between Purt Veg [SC 3242 7030] and Cass ny Hawin [SC 2980 6926], are equated with the Port Erin Formation by Quirk & Burnett (1999) and with the Lonan Formation by Woodcock & Barnes (1999). Lateral equivalence of the Port Erin Formation with the Lonan Unit is speculatively suggested by a possible occurrence of the Keristal Member in the Port Erin Formation in Perwick Bay [SC 060 6715] (Woodcock & Barnes 1999).

Mull Hill Formation

Name and equivalence

The formation is named after Mull Hill [SC 1897 6757] on which numerous natural exposures of the

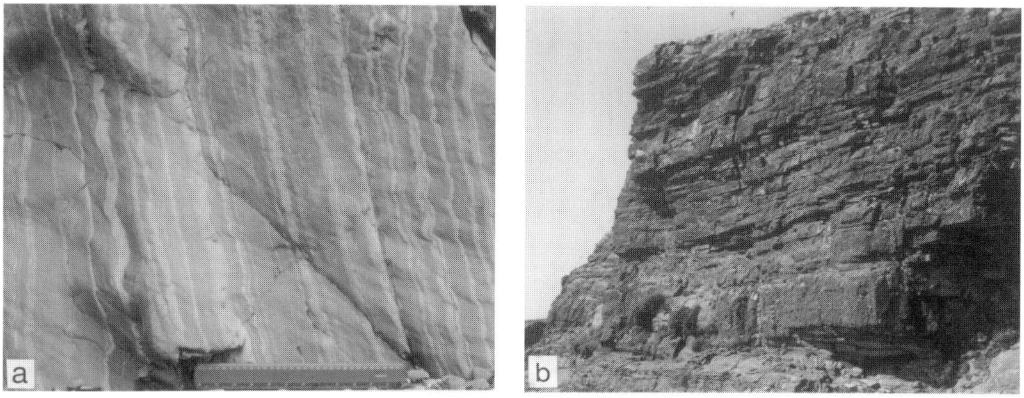

Fig. 4. Typical lithologies in tract 2. (a) Thin- to very thin-bedded sandstone–mudstone couplets (ruler, 31 cm), Port Erin Formation, Traie Vain [SC 1950 6943]. (b) Medium- to very thick-bedded quartzose sandstones (height of section, 11 m), Mull Hill Formation, near the top of the Chasms [SC 1934 6638].

unit occur. Lamplugh (1903, pp. 42–43) used the informal term Mull Hill Grits for these rocks and grouped them under the more formal term Agneash Grits, regarding the Mull Hill unit as a lateral equivalent of the principal outcrop of the Agneash. The two units do occur at a comparable structural level but cannot be mapped into each other, so that their original depositional continuity is unproven. For this reason, the Mull Hill Formation is defined separately here.

Simpson (1963) recognized the Mull Hill Quartzite and included it within that part of the Maughold Banded Group reassigned here to the Port Erin Formation (Fig. 1). He also recognized the, supposedly lower, Chasms Quartzite cropping out on the coast south of Mull Hill [SC 1915 6642], ignoring Lamplugh's conclusion that this mass is a faulted or folded repetition of the Mull Hill Unit. Recent mapping has confirmed this repetition, showing that the Mull Hill and Chasms outcrops occur on two gently dipping limbs of an anticline–syncline pair (Fig. 1), the intervening steep to overturned limb of which is exposed in Chapel Bay (e.g. [SC 2105 6795]). The formalized Mull Hill Formation therefore includes Simpson's Chasms Quartzite. It tentatively includes quartzose arenites at Spaldrick [SC 1940 6963], although their structural continuity with the main outcrop cannot be demonstrated.

Type area and subsidiary localities

The type section is taken on the west side of Chapel Bay, Port St Mary [SC 2100 6800], where a steeply dipping, north younging succession exposes *c.* 130 m of typical Mull Hill Formation above a gradational base. This succession becomes progressively inverted along-strike to the northeast, and is exposed both on the eastern side of Chapel Bay [SC 2130 6806] and on the west side of Bay ny Carrickey [SC 2155 6830]. The southern limb of the formation is continuously exposed in coastal cliffs from Bay Stacka [SC 1982 6644] to the west side of Perwick Bay [SC 2011 6689]. The northern limb of the formation is sporadically exposed on Mull Hill, with a good section near its top at Cregneash Quarry [SC 1912 6742]. Further north, over the Port Erin Anticline, a possible section of the Mull Hill Formation is exposed at Spaldrick [SC 1940 6957].

Lithological characteristics

The formation is characterized by light grey to white, medium- to very thick-bedded quartz arenite beds, contrasting strongly with the thin- or very thin-bedded facies of the underlying Port Erin Formation. Each sandstone bed grades up from medium sand to very fine sand or silt, with any mud top being thin or absent (Fig. 4b). Sporadic beds have coarse to very coarse sand bases with mudstone rip-up clasts. Beds are typically massive to parallel laminated at the base with an overlying cross-laminated division. The beds are interpreted as the product of deposition from high- and medium-concentration turbidity currents in a deep-marine environment. The section at the Chasms [SC 1936 6637] suggests that the average bed thickness in the Mull Hill Formation increases upwards, with a predominance of thick and very thick beds at the top of this section (Fig. 4b).

Near the base, top and lateral margins of the formation, packets of medium to thick sandstone beds are intercalated with thin-bedded turbidites, typical of the Port Erin Formation.

Thickness and relationship with adjoining units

Continuous sections through the Mull Hill Formation measure 130 m at Chapel Bay and *c.* 100 m at the Chasms. However, the total thickness of the formation is estimated at between 350 and 400 m across Mull Hill. The quartzose sandstone packet at Spaldrick, interpreted as a lateral margin of the Mull Hill turbidite system, totals no more than 40 m.

The Mull Hill Formation overlies the Port Erin Formation with a gradational depositional contact. In the type section at Chapel Bay, and in its continuation to the east, the basal transition occurs over *c.* 25 m. The base of the Mull Hill Formation is taken at the horizon [SC 2104 6790] where two thick beds mark the start of this transition zone. The formation base is also exposed in Bay Stacka and on the western side of Perwick Bay. On the eastern side of Perwick Bay [SC 2060 6715], two discrete quartzose sandstone packets, *c.* 2.6 and 1.8 m thick, occur *c.* 100 m below the faulted base of the Mull Hill Formation. The possibility that these packets might correlate with the Keristal Member of the Lonan Formation has already been noted, implying a correlation of the Mull Hill Formation with the Santon Formation.

The upper contact of the Mull Hill Formation is nowhere unequivocally exposed. Cregneash Quarry is near the top of the formation, but is still included within it on the basis of common intercalations of medium- to thick-bedded sandstones. Along-strike to the northeast, the Mull Hill Formation is faulted out against Carboniferous rocks.

Ny Garvain Formation

Name and equivalence

The Ny Garvain Formation is named after the headland of Gob ny Garvain [SC 4885 8986]. The formation was included within the Lonan Flags by both Lamplugh (1903) and Simpson (1963).

Type area and subsidiary localities

The type area is the coastal section between Port Cornaa [SC 4738 8787] and Traie Farkan [SC 4958 9128], most easily accessible at Port Mooar [SC 4880 9100].

Lithological characteristics

The type section of the Ny Garvain Formation comprises two parts, informally termed lower and upper by Quirk & Burnett (1999). Between Port Cornaa and Gob ny Garvain the succession is sand dominated. Medium to thick beds of fine- to medium-grained sandstone are separated by very thin mudstone partings. The sandstones are massive, parallel laminated or ripple cross-laminated. Sandstone-rich packets are interspersed with intervals of thin-bedded, cross-laminated, sandstone–mudstone couplets. Further north, between Gob ny Garvain and Traie Farkan, the succession is dominated by thin-bedded sandstone–mudstone couplets, but with packets of medium-bedded sandstone–mudstone couplets locally comprising up to 30% of the unit. Typical beds are sharp based, with a lower division of green–grey ripple cross-laminated fine sandstone passing abruptly up into interlaminated sand and dark grey mudstone (Fig. 5a). Convolute lamination is common in the cross-laminated divisions, which show current ripples on bedding surfaces. The main part of the Ny Garvain Formation was mostly deposited from low- to medium-concentration turbidity currents.

At two localities, Port Cornaa [SC 4728 8778] and north of Gob ny Garvain [SC 4880 9015], the wacke sequence is punctuated by intervals of quartz arenites. At Port Cornaa these take the form of thick- to very thick, irregular beds in a discrete package, not dissimilar to the style of the Keristal Member in the Lonan Formation. The quartz arenite at Gob ny Garvain occurs in medium to thin beds interspersed over a section of several tens of metres.

Thickness and relationship with adjoining units

Much of the coastal exposure in the formation occurs in a folded succession with a shallow sheet dip, in which thickness determinations are consequently unreliable. About 250 m of continuous succession is exposed in a steeply dipping limb near Port Cornaa, but the overall thickness of the unit may be several times this estimate. The base of the formation is not exposed, its contact with the Lonan Formation to the south being taken at a north-northwest striking steep fault, just south of Port Cornaa. Inland, a gently dipping contact of the Ny Garvain Formation over the Lonan Formation can be mapped on the basis of bed thickness, but its geometry suggests a fault. With no biostratigraphical control, the original relationship of the Ny Garvain and Lonan Formations therefore remains uncertain.

The upper contact of the Ny Garvain Formation on the coast south of Maughold Head is also complicated by a minor northwest dipping thrust,

Fig. 5. Typical lithologies in tract 3. (a) Sandy thin- and very thin-bedded turbidites (hammer length, 30 cm), Ny Garvain Formation, Port Mooar [SC 4883 9100]. (b) Thin- to medium-bedded quartzose sandstone turbidites (staff length, 1 m), Creg Agneash Formation, Traie Carn [SC 4977 9154]. (c) Contact between thin-bedded sandstone–mudstone couplets (Creg Agneash Formation) and laminated siltstones and mudstones (Maughold Formation) (staff length, 1 m); Traie Carn [SC 4970 9158], (d) Bioturbated laminated siltstone–mudstone couplets (pen length, 13 cm); Maughold Formation, Port Lewaigue [SC 4663 9310].

and by northwest striking, steeply dipping faults. However, a continuous upward-thinning Ny Garvain succession is evident, with the appearance, over several tens of metres, of intercalated thin to medium quartz arenite beds marking a stratigraphic transition to the Creg Agneash Formation. Inland evidence, particularly in the River Laxey and around Agneash, is also compatible with a stratigraphic order of the Creg Agneash Formation over the Ny Garvain Formation.

Creg Agneash Formation

Name and equivalence

The formation is named after Creg Agneash [SC 4295 8705], one of the exposures chosen by Lamplugh (1903, p. 39) as typical of his Agneash Grits. The formation approximates to Lamplugh's principal outcrop of Agneash Grits stretching from Maughold Head [SC 4986 9142] to Mount Rule [SC 3546 7934] (Fig. 1). The formation excludes other 'grits' correlated by Lamplugh with the Agneash Grits, specifically those assigned here to the Keristal Member, Santon Formation and Mull Hill Formation, or included as part of the Maughold Formation. Simpson (1963) included the Agneash Grits within his Maughold Banded Group.

Type area and subsidiary localities

The type area for the Creg Agneash Formation is chosen at Maughold Head. Its surrounding bays of Traie Carn [SC 4978 9156], Traie Foillan [SC 4979 9131] and Traie Farkan [SC 4959 9131] provide cleaner and more accessible exposures than those inland at Creg Agneash, both of the typical facies and of the base and top of the formation. There are many natural inland crag exposures of the Creg Agneash Formation, well detailed by Lamplugh (1903, pp. 39–42). The more informative of these are, in addition to Creg Agneash, above Corrany [SC 4496 8943], on Cronk y Vaare [SC 4114 8631], Carn Gerjoil [SC 3936 8405] and Gob y Creggagh [SC 3820 8428]. Quarry exposures are available at Dreemskerry [SC 4760 9112] and Windy Corner [SC 3900 8400], and an accessible stream section occurs along the River Laxey (between [SC 4183 8670] and [SC 4088 8716]).

Lithological characteristics

The formation is characterized by light grey or white, thin- or medium-bedded quartzose arenites, with occasional thick beds (Fig. 5b). Near the base of the formation, the quartz arenite is interbedded within a very thin-bedded facies that passes upwards into very thin but persistent beds of dark grey silty mudstone. The sandstone beds grade upwards from medium or fine sand to very fine sand or silt, sometimes with a very thin mud top. Beds have weakly defined, often upward-thinning, parallel lamination, sometimes with a thin overlying ripple cross-laminated division. The very thin-bedded facies comprises graded silt–mud couplets, often with a basal lamina of very fine sand. The Creg Agneash Formation is interpreted as the depositional product of medium-concentration turbidity currents, punctuating a persistent background of low-concentration events in a deep-marine environment.

The frequency of quartzose sandstones decreases near the top and bottom of the Creg Agneash Formation. In Traie Carn [SC 4970 9158], the formation top is a zone in which sandstone beds progressively decrease from 60 to 30% of the succession before rapidly giving way to very thin-bedded or laminated facies (Fig. 5c). This transition zone is included within the Creg Agneash Formation and the base of the Maughold Formation taken where only very sporadic sandstones remain interbedded with the silty mudstone. A similar transitional facies occurs below the typical Creg Agneash Formation in Traie Foillan [SC 4983 9130], but the stratigraphic continuity of this section is disrupted by a thrust fault below the sandstone-rich part of the Creg Agneash Formation.

Thickness and relationship with adjoining units

About 120 m of sandstone-rich Creg Agneash Formation can be directly measured above a thrusted base on the south side of Traie Carn. This succession is faulted against, but was probably originally succeeded by, 60 m of marginal facies on the north side of Traie Carn [SC 4970 9158]. Allowing for a similar marginal zone at the base of the Creg Agneash Formation, the minimum total thickness of the formation in its type area is 240 m. The outcrop width of the Creg Agneash Formation increases southwestwards to the Laxey Valley. In this region, the formation could be as much as 750 m thick, although there may be some fold and fault repetition here.

The base of the Creg Agneash Formation is almost certainly a depositional contact with the underlying Ny Garvain Formation. At Maughold Head, and apparently inland, the Creg Agneash Formation passes gradationally up into the Maughold Formation (Fig. 5c). The Creg Agneash and Mull Hill Formations have some lithological and geochemical similarities (Barnes *et al.* 1999; Quirk & Burnett 1999), and occur at an analogous structural level in the Manx Group. However, the outcrop of both formations fails towards the centre of the island and their equivalence must therefore remain speculative.

Maughold Formation

Name and equivalence

The unit is named after the village of Maughold [SC 4920 9170], north and east of which the coast displays good sections through the formation. The unit was mostly mapped by Lamplugh (1903) as undivided Manx Group southeast of the Barrule Formation, although parts containing significant quartz arenite were incorporated within his Agneash Grit. In the southwest of the island, Lamplugh also distinguished horizons of 'crush conglomerate', here described as pebbly mudstones. Simpson (1963) included most of the unit in his Maughold Banded Group.

Type area and subsidiary localities

The type section is the coast west of Port Lewaigue [SC 4630 9328] to [SC 4682 9304]. The formation is exposed from here as far as Maughold Head, and is particularly accessible at Port e Vullen [SC 4738 9281] and Traie Carn [SC 4973 9158]. Comparative inland sections occur on Mullagh Ouyr [SC 3982 8616] and in the adjacent Laxey Valley [SC 4058 8710]. The southwestern facies are exposed around Bradda Head [SC 1847 6990] and on the prominent cliff-top crags from Fleshwick Bay [SC 2020 7140] to the north flank of Gob ny Beinn [SC 2135 7320], where the distinctive pebbly mudstones and quartzites are best displayed. The coast section here is inaccessible, other than by boat, apart from a few hundred metres at Fleshwick Bay.

An alternative interpretation of the section between Port Lewaigue and Maughold Head has been made by Quirk & Burnett (1999), who assign it to the Barrule Formation. The Maughold Formation is strongly variable along-strike to the southwest, and the name Fleshwick Unit has instead been informally adopted by Quirk & Burnett (1999) where it occurs at outcrop between Bradda Head [SC 1930 6970] and The Slock [SC 2150 7330], as it also appears to overlie the Barrule Formation.

Lithological characteristics

Three intergradational mudstone-rich facies dominate the type section. Laminated mudstones are predominantly dark grey, with subordinate pale grey faint silt laminae. Silt-filled bioturbation spots occur sporadically. The second facies has inter-laminated siltstone and mudstone in about equal proportions (Fig. 5d). The silt laminae are up to 3 mm thick, occasionally graded but more usually diffuse. In the third facies, the siltstone is dominant, occurring in distinct graded beds up to 2 cm thick. Bioturbation is common as spots, or as discordant silt- or mud-filled burrows.

The mudstones at the western end of the type section contain a locally fault-bounded unit of pebbly mudstone, estimated to be *c.* 12 m thick. Clasts are in the 0.2–3 cm range, occasionally up to 35 cm, supported in a medium to dark grey mud matrix. The clasts are mostly of shale, siltstone or fine sandstone, all of which could have been supplied from within the Maughold Formation.

In the western coastal exposures of the formation, between Port Erin and Gob ny Beinn, pebbly mudstone units increase markedly in abundance and comprise *c.* 65% of the succession. As in the type section, these pebbly mudstones are matrix supported, with intraformational clasts occupying between 1 and 25% of the rock volume. The primary mudstone lithofacies is correspondingly less common on the west coast. Moreover, both the mudstone and pebbly mudstone successions are punctuated by packets of quartzose sandstone, and quartzite is only sparsely developed at the eastern end of the type section. The sandstone is off-white to pale or very pale greyish-white, fine to coarse grained, and occurs in beds 10–100 cm thick, averaging 20–40 cm. The sandstone beds are commonly graded with parallel-laminated and cross-laminated divisions, with or without pelitic interbeds. The quartzite beds are not discernibly graded. They show sharp top and bottom contacts, commonly contain tabular cross-sets through the thickness of each bed, and may contain dark grey pelitic interbeds with or without very thin, flaser-bedded quartzite lenses. Beds occur in packets, 7–25 m thick, in which they are amalgamated or separated by very thin dark shaly or very fine sandy interbeds.

The distribution of sandstones within the inland outcrop of the Maughold Formation is imperfectly known. In the Laxey Valley, quartz arenite beds are distributed through most of the succession, essentially continuing the depositional character of the Creg Agneash Formation. However, mudstones predominate here, some of them manganiferous (e.g. on the north slope of Mullagh Ouyr [SC 3982 8616]). A prominent sandstone packet occurs towards the top of the succession in the Laxey Valley, above which arenite beds are rare or absent. North of Windy Corner, a packet of sandstone several tens of metres in thickness can be mapped along-strike but, from Lamplugh's mapping of the relatively poor exposure further southwest, quartz arenite appears sparse or absent until the southwestern end of the outcrop.

The mudstones in the Maughold Formation are interpreted as the product of hemipelagic fall-out and low-concentration turbidity flows into a deep-marine, periodically oxygenated basin. Basin instability resedimented the bedded facies as pebbly mudstones, particularly in the southwest of the formation. Here too, packets of quartzose sandstones accumulated by deposition from the medium- to high-concentration parts of turbidity flows.

Thickness and relationship with adjoining units

Uncertainty about the structure of the type section precludes accurate thickness estimates. The abnormal outcrop width of 4 km in this region is undoubtedly due to tectonic repetition. Inland, the mapped outcrop widths suggest a thickness of some hundreds of metres. The moderately dipping succession on the southwest coast, north of Fleshwick Bay, represents a maximum thickness of 920 m. However, recognition of at least one tight fold implies that the succession may be considerably thinner, perhaps no more than 500–600 m.

The basal gradational contact of the Maughold Formation with the underlying Creg Agneash Formation is exposed at Traie Carn [SC 4973 9158] (Fig. 5c); the base to the Maughold Formation is not exposed in the southwest of the island. The formation possibly overlies the Mull Hill Formation there, but the present contact, e.g. at the north side of Port Erin Bay, is across a zone of high strain suggesting a faulted boundary between tracts 2 and 3.

The northwestern contact of the Maughold Formation with the Barrule Formation is nowhere exposed, although mapping in the northeast indicates that it truncates major fold structures in the adjacent Maughold Formation and is thus likely to be tectonic. In the southwest, the Maughold Formation youngs southeastward away from the Barrule but the nature of the contact is uncertain.

Barrule Formation

Name and equivalence

The Barrule Formation is equivalent to the southeastern outcrop of the Barrule Slates of Lamplugh (1903) and Simpson (1963). The northwestern outcrop of lithologically similar material, previously assigned to the Barrule Slate, is renamed the Glen Rushen Formation until such correlation can be confirmed.

Type area and subsidiary localities

The Barrule Formation is defined on the summit of South Barrule [SC 2580 7590]. Its uniform lithology can be observed at intervals along the outcrop, notably in the extensive cliff-top crags northeast of Burroo Mooar [SC 2160 7350], west and north of Cronk ny Arrey Laa [SC 2240 7472], at South Barrule Quarry [SC 2700 7693], near Black Hut [SC 4040 8850], on Clagh Ouyr [SC 4128 8883] and on North Barrule [SC 4430 9090].

Lithological characteristics

The formation is remarkably homogeneous, consisting of medium to dark grey or bluish-grey, occasionally black, dense slaty pelite (Fig. 6a). The pelite is massive or, more commonly, contains a faint, but persistent, plane-parallel lamination unaffected by bioturbation. The lamination is defined by lighter grey laminae, probably originally silt, spaced on a submillimetre scale. The proportion of silt is typically < 2%. Pale grey sandstone beds < 5 cm thick occur in a number of localities close to the southern margin of the outcrop. Rocks of the Barrule Formation are interpreted as hemipelagic and fine-grained turbiditic deposits into anoxic bottom water.

Thickness and relationship with adjoining units

The thickness of the formation is conjectural in the absence of evidence of way-up and internal structure. Simpson's (1963) estimate was in the 160–1030 m range, assuming that the succession is homoclinal. However, small-scale, tight folds are visible locally.

The boundaries of the Barrule Formation are not exposed inland and the unit is not thought to reach the northeast coast. The boundaries may occur in inaccessible southwestern cliff sections. Thin sandstone beds present near the southern edge of the outcrop of the formation in the southwest may hint at a gradational stratigraphic contact with the Maughold Formation. However, a prominent north-northeast trending topographic lineament along the edges of the crags at Gob ny Beinn [SC 2125 7305] suggests a fault at this contact. Northeast of the Central Valley, exposures near to this southern contact are highly sheared, again suggesting a fault. In the northeast of the island, the southern margin

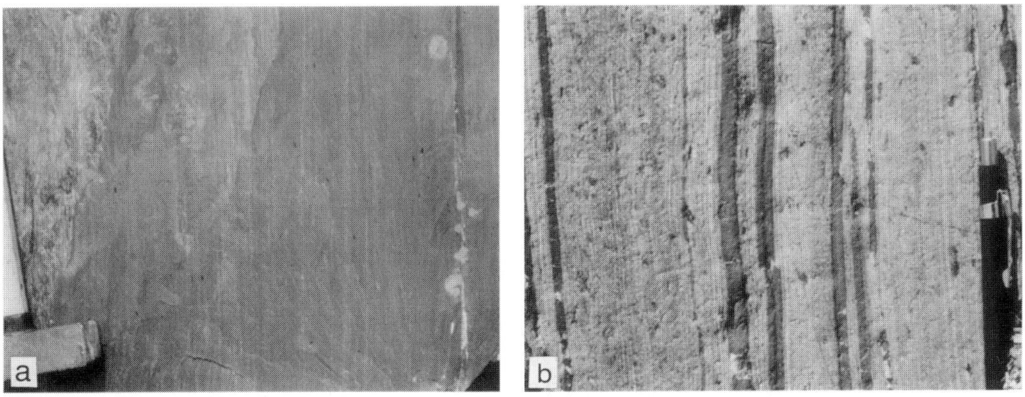

Fig. 6. Typical lithologies in tracts 4 and 6. (a) Laminated mudstones (hammer head width, 3 cm), Barrule Formation, South Barrule Quarry [SC 2690 7690]. (b) Very thin-bedded and thickly laminated manganiferous sandstone–mudstone couplets (pen length, 13 cm), Creggan Mooar Formation, southwest of Creggan Mooar [SC 2163 7684].

of the Barrule Formation cuts across and is unaffected by structures in the Maughold Formation, strongly suggesting a faulted contact. It maps out as a northwest dipping structure, parallel to bedding in the Barrule Formation.

A further topographic lineament parallels the northwestern margin of the Barrule Formation in the southwest and a fault with the Injebreck Formation is suggested by relationships in the cliffs around Lag ny Keeilley Chapel [SC 2158 7453]. Nothing is known about this contact in the northeast of the island.

Injebreck Formation

Name and equivalence

The Injebreck Formation is named after the Injebreck Valley [SC 3560 8480]. It comprises much of the Injebreck Banded Group of Simpson (1963), created from what Lamplugh (1903) had previously categorized as 'Manx Slate Series – not separated'.

Type area and subsidiary localities

The type area is defined in foreshore sections on the west coast, from below Lag ny Keeilley [SC 2158 7453] north to Con Shellagh [SC 2165 7495]; and from Da Leura [SC 2175 7530] north to Gob ny Ushtey [SC 2152 7565]. Quartzites and other lithologies are also well exposed in the south facing crags above Da Leura [SC 2195 7545]. A further coastal section south from Dul Ushtey [SC 2156 7600] to The Stack [SC 2160 7633] is also assigned here to the Injebreck Formation, following Lamplugh (1903), although it was mapped within the Barrule Slates by Simpson (1963). Inland exposures occur around Injebreck [SC 3520 8580].

Two successions occur in the type area, separated by a zone of highly strained rocks termed the Lag ny Keeilley Shear Zone exposed on the foreshore at Ooig ny Seyu [SC 2162 7470]. To the south, the succession is right way-up and north younging, whereas the northern succession is generally overturned and south younging, at least for the 1 km north to Gob ny Ushtey. The Lag ny Keeilley Shear Zone is interpreted as a significant tectonic boundary, separating tracts 4 and 5 (Fig. 1) (see Fitches *et al.* 1999).

Lithological characteristics

In its type area, the southern succession of the Injebreck Formation consists of a thickening- and coarsening-up succession of quartzose sandstone and mudstone. The succession starts in very thin- to thin-bedded pale grey sandstone–mudstone couplets, in which the mudstone predominates. The sandstone beds show parallel and ripple cross-lamination, and increase in proportion upwards until the succession becomes predominantly medium bedded. In the wave-cut rock platform just south of its truncation by the Lag ny Keeilley Shear Zone, the succession is dominated by thick- or very thick-bedded, amalgamated, medium- to coarse-grained, stratified quartz sandstones. Many of these beds exceed 1 m in thickness and a number contain prominent conglomerates up to 20 cm thick.

The northern succession is notably more pelitic than the southern succession, albeit punctuated by quartzose sandstones and siltstones, and occasional thin pebbly mudstone horizons. The mudstones resemble those of the Barrule and Glen Rushen Formations. They are medium to dark grey, almost black, in colour and they may contain dispersed, very thin, pale grey silty laminae. Thin-bedded, pale to medium grey sandstone–mudstone couplets are particularly prevalent in exposures north of Gob ny Ushtey, where they occur intercalated with slaty pelite. The typical grain size at the base of each couplet is silt or fine sand, with sand proportions ranging between 20 and 70%, and the beds show parallel or cross-lamination. Pebbly mudstone horizons are < 1 m thick and consist of dispersed, intraformational, pale grey laminated silt or sandstone clasts, up to 10 cm in diameter, in a medium blue–grey to almost black pelitic matrix. Pale grey to cream–buff, thin- to thick-bedded quartzose sandstone occurs in a single, horizon *c.* 35 m thick, in the Glion ny Goayr crags [SC 2190 7555] above Da Leura. Thin-bedded packets up to 1.3 m thick, with *c.* 40% sandstone, are intercalated with packets of the thick-bedded facies, comprising up to 70% sandstone. Thin beds show parallel and ripple cross-lamination, whilst thick beds are massive or parallel laminated.

The Injebreck Formation is interpreted as the product of deep-marine turbidity flows, some of which were highly quartzose, together with occasional debris flows. The paucity of inland exposures makes it difficult to be sure how typical the coastal exposures of the Injebreck Formation are of the unit as a whole. Simpson (1963) suggests that the proportion of siltstone–mudstone couplets increases northeastward from the coast, at the expense of the mudstones and pebbly mudstones. This is confirmed by exposures in the northern tributary of Glen Helen [SC 3248 8450] to [SC 3337 8418].

Thickness and relationship with adjoining units

The exposed thickness of the southern succession is *c.* 120 m, although inaccessible sections further south undoubtedly add to this thickness. In contrast, the northern succession is, with due allowance for repetition by folding, estimated to be *c.* 500 m thick.

The southern, probably faulted, boundary of the Injebreck Formation with the Barrule Formation has already been discussed. Near the northern boundary, the intercalation of thin-bedded turbidites with slaty pelite, just south of The Stack [SC 2160 7633], combined with the absence of any obvious tectonic break, suggests that the Glen Rushen Formation passes stratigraphically up into the Injebreck Formation.

Quirk & Burnett (1999) question the validity of the correlation between the southern and northern parts of the Injebreck Formation on lithofacies and structural grounds. Instead, they assign the section between Lag ny Keeilley and The Stack on the west coast to the informally named Eary Cushlin Unit.

Glen Rushen Formation

Name and equivalence

The Glen Rushen Formation, named after the extensive, disused slate quarries in Glen Rushen [SC 2445 7845], equates with the northwestern outcrop strip of the Barrule Slates of both Lamplugh (1903) and Simpson (1963). Molyneux (1979) described an Arenig acritarch fauna from a road cutting in Glen Mooar [SC 241 793], originally thought to be from the Maughold Banded Group but now referred to the Glen Rushen Formation.

Type area and subsidiary localities

The type area for the formation is the Glen Rushen Quarries, but a valuable reference section is the coastal exposure at Fheustal [SC 2173 7654] where the lithological variation in the formation is better exposed. The formation is also exposed on Slieau Whallian [SC 2645 8040] and in Glen Helen [SC 2952 8432].

Lithological characteristics

Two distinct, though intergradational, lithologies are evident at Fheustal: laminated mudstone and laminated silty mudstone. Laminated mudstone predominates, consisting of medium to dark grey or bluish-grey mudstone with indistinct very pale greyish–white silt laminae < 1 mm thick. This lithology is similar to that in the Barrule Formation (Fig. 6a). Bioturbation is absent and the silt laminae are laterally persistent across exposure widths, although the percentage of siltstone is < 20%. The subordinate laminated silty mudstone typically contains 40–60% silt, in laminae 0.1–0.3 mm thick. A single, 1 m thick horizon of pebbly mudstone occurs within this facies, composed of isolated, intraformational lenticular siltstone clasts in grey mudstone.

Elsewhere, there is more lithological variation in the Glen Rushen Formation than in the Barrule Formation. Interbedded thin sandstones are not uncommon in the southern outcrop and the laminated silty mudstone is locally abundant. The Glen Rushen Formation is interpreted as an anoxic hemipelagic facies with occasional input of low-concentration turbidity flows.

Thickness and relationship with adjoining units

A maximum thickness of 250–500 m is suggested for the Glen Rushen Formation, based on the outcrop width, corrected for the regional dip and the estimated effects of steeply inclined D1 isoclinal folds and recumbent D2 folds.

The southern contact of the Glen Rushen Formation, gradational up into the Injebreck Formation, has been discussed above. The northern contact of the formation is a late north-northwest trending fault at Fheustal. A strike-parallel fault is inferred in Glen Maye [SC 2370 7967] between phyllites assigned to the Creggan Mooar and Glen Rushen Formations to the east, at the boundary between tracts 5 and 6 (Fig. 1). However, an increased proportion of silt laminae in the Glen Rushen Formation towards the late fault at Fheustal, followed by laminated siltstones in the Creggan Mooar Formation, suggests the possibility of an originally gradational stratigraphic contact. The predominantly northward sense of younging in the Creggan Mooar Formation implies that it would overlie the Glen Rushen Formation. The outcrop of the Glen Rushen Formation would then lie in the core of a tight anticline, younging out into more sandy formations on either flank. In Glen Helen the southeastern (?lower) contact is sharp but may be faulted (Quirk & Burnett 1999).

Creggan Mooar Formation

Name and equivalence

The formation is named after the hamlet of Creggan Mooar [SC 2195 7711], just inland from the main coastal exposure. It is a new formation, broadly equivalent to the northwestern outcrop of the Maughold Banded Group of Simpson (1963) and to part of the 'unseparated Manx Series' of Lamplugh (1903). The formation is fully described by Kennan & Morris (1999).

Type area and subsidiary localities

The type area is defined in low cliffs and the adjoining wave-cut platform at Gob ny Gamera [SC 2170 7650]. The formation is, however, well displayed in almost continuous coastal exposures from a point near the bottom of the footpath down the cliff at Fheustal [SC 2173 7654] northward to Niarbyl [SC 2118 7758]. The formation becomes increasingly more deformed north of Traie Vrish [SC 2131 7743]. The Creggan Mooar Formation is also well exposed in the Glen Maye Stream section, from the Mona Erin water wheel pit [SC 2302 7983] to the road bridge in Glen Maye [SC 2361 7969], and particularly at the waterfall and gorge at [SC 2354 7976].

Lithological characteristics

The formation is dominated by very thin- or thin-bedded couplets, each grading upwards from pale to medium grey or greenish-grey, exceptionally reddish-brown siltstone or very fine sandstone to dark grey mudstone. The beds are parallel laminated or, in rare medium beds, cross-laminated. They commonly have bioturbated tops comprising spots and lenses either parallel or discordant to bedding. This sand-rich facies, comprising 60–70% sandstone, is intercalated with pale to medium grey, greenish-grey or brownish-grey, plane-parallel laminated siltstone. This siltstone characteristically occurs in thick, monotonous, apparently non-organized sequences, although intercalated, very thin-bedded siltstone–mudstone couplets are occasionally discernible. Other less common lithofacies include laminated silt–mudstone, identical to that described in the Glen Rushen Formation, and sporadic pale grey, faintly laminated, brown, buff weathering, 3–15 cm thick quartzite beds. The Creggan Mooar succession is interpreted as a mix of deep-marine redeposited facies, mostly the product of low-concentration turbidity flows into oxygenated bottom waters.

A minor, though distinctive, lithofacies in the Creggan Mooar Formation comprises bright reddish brown, black to dark brown weathering siliceous beds, ranging in thickness from 2 mm to *c.* 1 cm, occasionally to *c.* 2 cm (Fig. 6b). The weathering colours may occur either irregularly dispersed or very notably banded into 'tram track' black margins and reddish-brown centres. This lithofacies is particularly rich in manganese and is regarded by Kennan & Morris (1999) as the possible protolith for coticule horizons, which are manganiferous garnet-rich rocks occurring in higher grade Arenig successions worldwide. The manganiferous beds occur, intermittently or clustered, in all principal background lithofacies throughout the formation, though they are most prevalent in the laminated siltstone facies.

Thickness and relationship with adjoining units

Assuming a constant sheet dip of 55° and a doubling of the Creggan Mooar Formation thickness across tight folds, and disregarding the effects of thrusts, a thickness in the order of 350 m is estimated for the formation from its faulted southern contact to the north end of the Traie Vrish Beach. North of that point, the formation is so complexly deformed that no reliable thickness estimate can be made.

The southeastern contact with the Glen Rushen Formation has been discussed above. The northwestern contact of the Creggan Mooar Formation with the Glion Cam Unit is an unexposed boundary of uncertain nature. On the west coast, this boundary has been truncated by the west-northwest dipping brittle fault at the tectonic base of the Silurian Niarbyl Formation. This fault is exposed at Niarbyl and again in Glen Maye, where it is interpreted to host a felsite intrusion. At both locations, the underlying Creggan Mooar Formation is characterized by high-strain fabrics of the 140 m wide Niarbyl Shear Zone (Roberts *et al.* 1990), although the stratigraphic and structural implications of this zone are unknown (see Fitches *et al.* 1999).

Lady Port Formation

Name and equivalence

The formation is named after the bay of Lady Port [SC 2882 8786], which contains fine-grained lithofacies typical of the formation. The formation includes both the Lady Port Banded Group and the Ballanayre Slump Breccia of Simpson (1963). It also includes a small part of the turbidite sandstone and siltstone units grouped by Simpson with the Niarbyl Flags.

Type area and subsidiary localities

The type area is the coastal section between Will's Strand [SC 2696 8609] and *c.* 200 m southwest of Glen Mooar [SC 3008 8917]. Additional exposures of the formation occur in the old railway cutting north of Ballacarnane [SC 3036 8908], but poor exposure prevents the unit being mapped further inland to the northeast.

Lithological characteristics

The formation is heterogeneous, but the most

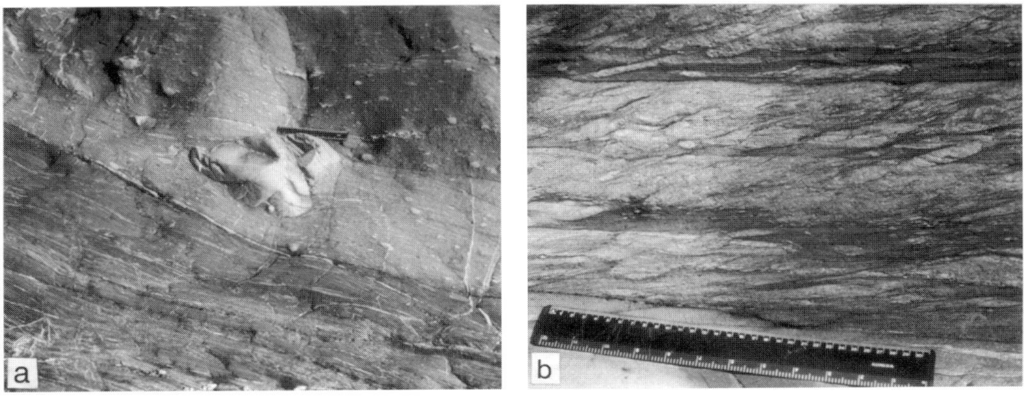

Fig. 7. Typical lithologies in the Lady Port Formation, tract 7. (a) Laminated mudstones overlain by a pebbly mudstone unit (pen length, 13 cm), base of Glion Thooar [SC 2955 8862]. (b) Bioturbated fine sand–mud turbidites (ruler, 30 cm), Ballanayre Strand [SC 2768 8683].

common lithology is medium grey to black mudstone, thinly to thickly laminated due to alternation of siltstone and clay-rich mudstone (Fig. 7a). The silt–clay couplets may be sharp based and graded, but are more commonly diffuse. Laminated mudstone dominates the coast section from Lynague Beach [SC 2806 8706] to beyond Ooig Beg [SC 2944 8847]. At Lynague Beach the mudstones contain reddish-brown manganiferous siltstone beds similar to those that characterize the Creggan Mooar Foramtion. The various mudstones of the Lady Port Formation are intruded by pre-cleavage doleritic and granodioritic sills and dykes. Some of the mudstones show strongly disrupted bedding, partly due to tectonic deformation and partly due to soft-sediment slumping.

Light grey quartzose sandstone punctuates the mudstone successions in places, typically in thin or very thin beds; examples occur north of Ooig Beg. South of Gob ny Creggan Glassey [SC 2962 8871] an isolated 5 m thick packet of quartzose sandstone comprises medium and thick beds. Each bed is sharp based, graded from fine sand to mud, typically massive but with occasional ripple cross-lamination near the bed top.

A third lithofacies comprises light grey to greenish-grey wacke sandstones, in thin to medium beds. Each bed grades upwards from fine or very fine sand up to mud. Parallel and ripple cross-lamination are preserved, but both these divisions are often conspicuously bioturbated (Fig. 7b). This turbidite facies is best seen at Ballanayre Strand [SC 2763 8678] and below Buggane Mooar [SC 2747 8663]. However, a finer grained variant of the same facies is exposed north of Gob ny Creggan Glassey [SC 2973 8890] to [SC 3008 8917]. Here, dark and light greenish-grey, thin-bedded, bioturbated silt–mud turbidites dominate the succession. Some very thin-bedded intervals resemble the typical laminated mudstone facies of the Lady Port Formation, suggesting a gradational relationship between the two lithofacies.

The fourth major lithofacies of the Lady Port Formation is pebbly mudstone and disrupted laminated mudstone (Fig. 7a). Clast diameters are typically in the 1–5 cm range, but outsize clasts up to 20 cm are not uncommon. Sporadic rafts of bedded sediment occur, reaching 50 m long in one extreme example ([SC 2778 8693]). Associated laminated mudstones and turbidite sandstones show all stages of disruption whilst partially lithified (Woodcock & Morris 1999).

The Lady Port Formation is interpreted as the product of deposition in a mudstone-prone deep marine sub-basin that was subject to repeated slumping and debris flow. The sub-basin received at least one episode of medium-concentration turbidity flows carrying mainly quartzose sand (Woodcock & Morris 1999) and a separate episode of muddy sand turbidite deposition.

Thickness and relationship with adjoining units

The total thickness of the formation is impossible to calculate accurately because of complicated thrust repetitions within the succession. About 100 m of black mudstones and turbidites can be measured below Buggane Mooar, and these are overlain by a comparable, or possible greater, thickness of pebbly mudstones. About 150 m of black mudstones, quartzose sandstone and pebbly mudstones occur in sequence south of Gob ny Creggan Glassey.

The Lady Port Formation seems to be separated from the rest of the Manx Group by a northeast striking fault several hundred metres inland from the coast section. Turbidites just southeast of this putative Ballakaighin Fault, at Glion Cam [SC 2908 8772], cannot be matched with the Lady Port Formation. Moreover, neither the Glion Cam rocks or the turbidites in the Lady Port Formation resemble the Niarbyl Formation, with which they were both correlated by Simpson (1963). The closest lithological matches to the Lady Port lithologies are some of the muddy facies with the Glen Rushen Formation, the quartzose-punctuated mudstone facies with the Creggan Mooar Formation, and the pebbly mudstones with the Sulby Slump Breccia (Simpson 1963). Any firm correlations await further work. Quirk & Burnett (1999) indicate that the Lady Port Formation is the youngest in the Manx Group succession and does not correlate with other areas.

Other units

The Manx Group contains a number of other units not formally defined in this paper and awaiting further inland mapping. The **Glion Cam Unit**, in tract 6 (Fig. 1), comprises mudstones and muddy sandstones assigned by Simpson (1963) to the Niarbyl Flags. However, they lie east of the Niarbyl Fault and are lithologically distinguished from the Niarbyl Formation by Morris *et al.* (1999). They are also distinct from the bioturbated turbidites in the Lady Port Formation in tract 7 (Woodcock & Morris 1999). Also poorly exposed within tract 6 are the pyroclastic **Peel Volcanics** mentioned by Lamplugh (1903) and Simpson (1963), and which yielded lower Arenig acritarchs (Molyneux 1979, 1999).

Several of Simpson's (1963) units are exposed inland in the north of the Manx Group (Fig. 1) and, despite partial remapping by Quirk & Burnett (1999), cannot yet be unequivocally matched with the units formally defined in this paper. They include the **Sulby Flags** and **Sulby Slump Breccia**, tentatively assigned to tract 4, and the **Cronk Sumark Slates**, **Glen Dhoo Flags**, and **Slieu Managh Slates**, tentatively placed within tract 5. Quirk & Burnett (1999) assign these to the informally named Glen Dhoo unit.

The **Niarbyl Flags** of Simpson (1963) and earlier workers have been shown to contain a Silurian (?late Wenlock) graptolite fauna (Howe 1999) and thus are explicitly excluded from the Manx Group. This unit has been redefined as the Niarbyl Formation within a newly erected Dalby Group (Morris *et al.* 1999).

Biostratigraphic constraints

The biostratigraphic control on Manx Group lithostratigraphy has been reviewed most recently by Cooper *et al.* (1995), Molyneux (1999) and Orr & Howe (1999). Existing zonal assignments have been retained here but have been transposed on to the revised lithostratigraphy arranged by structural tract (Fig. 8). Acritarch or graptolite control is available for the Lady Port Formation, the Glen Rushen Formation [the Maughold Banded Group site of Molyneux (1979)], and the Santon Formation [the Lonan Flags sites of Molyneux (1979) and Rushton (1993)]. The informal Peel Volcanics, Glen Dhoo Flags and Cronkshamerk (properly Cronk Sumark) Slates can also be assigned a chronostratigraphic position (Fig. 8). New mapping (Morris *et al.* 1999) has established that the assemblage assigned by Molyneux (1979) to the Niarbyl Formation lies instead in tract 6.

The units without biostratigraphic control have been positioned on the chart in harmony with their observed stratigraphic sequence or, failing that, structural sequence within each tract. The scheme assumes that the black mudstone-dominated successions of the Barrule and Glen Rushen Formations correlated with each other and, more tentatively, with the mudstone-rich lower parts of both the Lady Port Formation and the Maughold Formation. Above, the pebbly mudstones of the Lady Port Formation could correlate with the Sulby Slump Breccia, as suggested by Cooper *et al.* (1995), and with similar pebbly mudstones in the Maughold Formation, although this correlation is unconstrained. The manganiferous siltstones found in the Lady Port, Creggan Mooar and Maughold Formations may provide a further lithological tie below the level of the main pebbly mudstones.

The sand-rich formations of tracts 1–3 occur below the mudstone-rich intervals on this scheme, although they include two compositionally different sandstone types (Barnes *et al.* 1999; Woodcock & Barnes 1999). They may then plausibly correlate with the sandstones of the Glen Dhoo Flags in tract 4, as suggested by Cooper *et al.* (1995).

The new stratigraphic chart (Fig. 8) therefore confirms the views of Molyneux (1979), Roberts *et al.* (1990) and Rushton (1993) that Simpson's (1963) stratigraphic succession is flawed, and builds on the attempt of Cooper *et al.* (1995) to begin a revision. However, the formally redefined Manx Group units presently contain only four biostratigraphical control points. Correlation, both within the island and regionally is, therefore, very uncertain and a wide range of alternatives is possible (cf. Barnes *et al.* 1999; Fitches *et al.* 1999; Quirk & Burnett 1999).

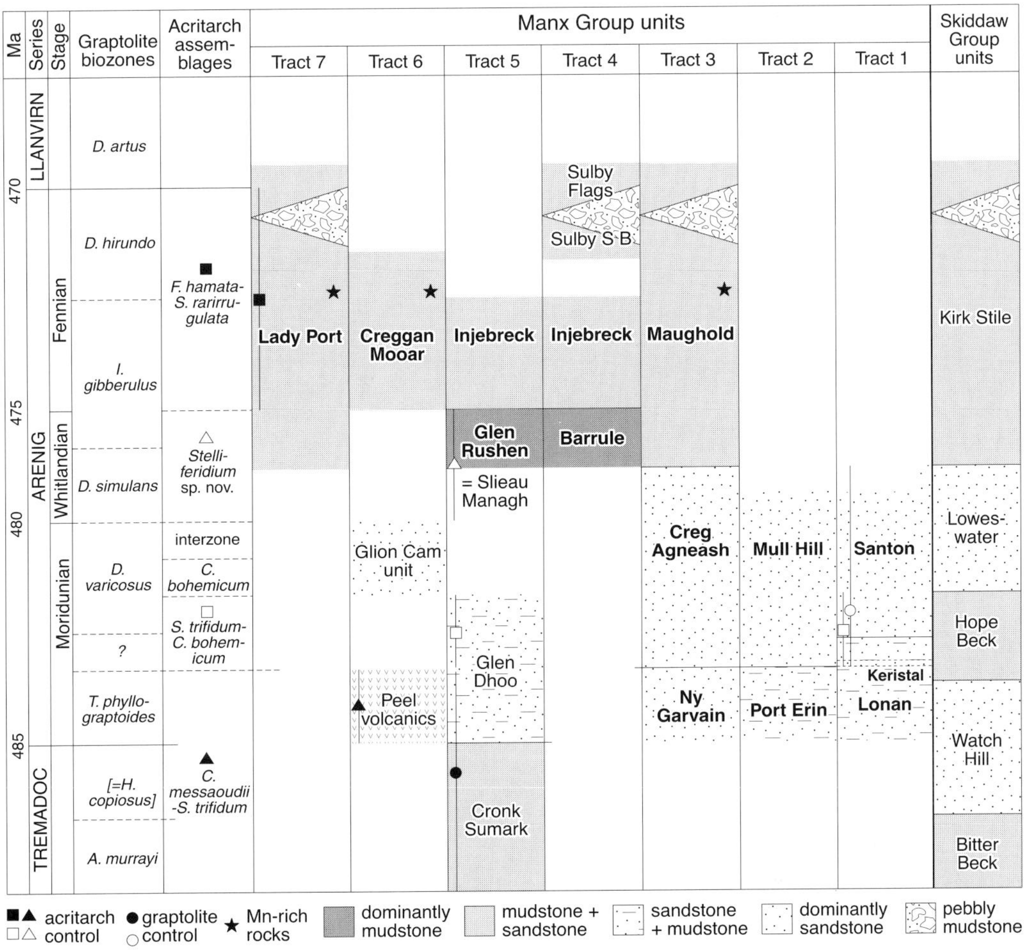

Fig. 8. Biostratigraphic control on, and possible regional correlation of, the Manx Group lithostratigraphy. Biostratigraphic zonation and Skiddaw Group stratigraphy from Cooper *et al.* (1995), except that the Whitlandian is taken to include the *D. simulans* Biozone, following Cooper *et al.* (1993) and Fortey *et al.* (1995).

Regional correlations

Comparison of the revised Manx Group stratigraphy with that of the Skiddaw Group shows some tentative correlation, but also some mismatch, of sandstone-rich and sandstone-poor units (Fig. 9). At the base of the exposed succession, the mudstones of the Cronk Sumark Slates could be coeval, within the Tremadoc, with those of the Bitter Beck Formation. The Lower Arenig succession is then dominated by sandstone-rich units in both the Skiddaw and Manx Groups. However, the mudstone intercalation of the Hope Beck Formation in the Skiddaw Group cannot yet be matched in the Manx Group. Indeed, the acritarch evidence points to the sand-rich Glen Dhoo Unit and, particularly, the Santon Formation being deposited at this time. Barnes *et al.* (1999) provide geochemical evidence for possible correlations of the Lower Arenig sandstones

The conspicuously abrupt upward loss of sandstones from the Creg Agneash into the Maughold Formation may correlate with that from the Loweswater into the Kirk Stile Formation in Mid-Arenig time. The predominance of mudstones in the Upper Arenig successions is common to both the Manx and Skiddaw Groups. Manganiferous horizons have yet to be firmly identified in the Skiddaw Group. However, both the Buttermere Formation olistostrome and pebbly mudstones and slumps in the upper part of Kirk Stile Formation correlate well with the pebbly mudstones in the

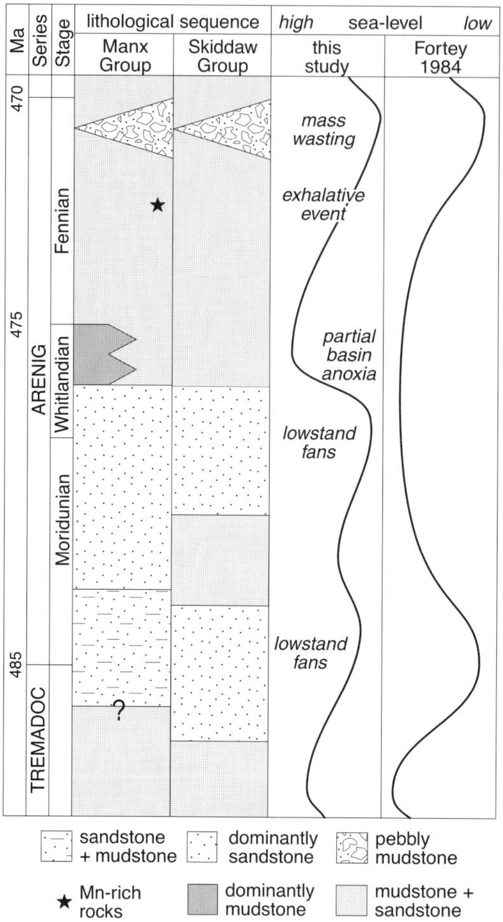

Fig. 9. Generalized Manx and Skiddaw Group lithological successions with their implied relative sea-level curve, compared with the proposed global curve of Fortey (1984).

Lady Port Formation and possibly others in the Manx Group.

If the tentative correlations between the Manx Group and the Skiddaw Group could be confirmed bistratigraphically, they would suggest the possibility of a consistent event stratigraphy along this part of the Avalonian margin (Fig. 9). A local curve of relative sea-level change is derived on the assumptions that the major influxes of turbidite sandstone record lowstand fans and that the mudstone intervals, particularly those with evidence of anoxia, record subsequent transgressions. Comparison of this curve with proposed global sea-level curves for late Tremadoc and Arenig time (e.g. Fortey 1984; Ross & Ross 1992; Nicoll *et al.* 1992) is hampered by problems of detailed chronostratigraphic correlation. Only the generalized curve of Fortey (1984) is shown.

A highstand in the late Tremadoc may have coincided with the lowest mudstones of the Skiddaw succession. The sandstone turbidite successions of Early Arenig (Moridunian) time suggest a time of relatively low sea level following a latest Tremadoc regression. However, evidence from platform successions in Britain (Fortey 1984), North America (Ross & Ross 1992), Scandinavia and Australia (Nicoll *et al.* 1992) suggests that this lowstand lasted only through the early part of the Moridunian. The persistence and indeed, in the Manx Group, increase in sand supply to the Avalonian margin throughout the Early Arenig is incompatible with the world wide transgression expected at this time, without additional regional influences on relative sea levels or sediment supply. The acme of sandstone deposition in late Moridunian–earliest Whitlandian (Mid-Arenig) time could, however, correspond to a marked regression well displayed on the North American Platform (Ross & Ross 1992) and the North China Platform (Meng *et al.* 1997). The abrupt switch to mudstone-dominated successions would mark the succeeding transgression. The persistence of mudstone-rich successions through the Late Arenig (Fennian) is consistent with globally raised sea levels at this time. The widely recognized regressive event at the end of the Arenig might plausibly be a factor in the instability of the Avalonian slope. However, the size and frequency of slump sheets and debrites at this time argues for additional tectonic factors. The most important regional tectonic event during the Arenig was the rifting of Avalonia from its parent Gondwana continent. The Late Arenig mass wasting on the Iapetus-facing margin of Avalonia may therefore date this continental break-up (Cooper *et al.* 1995; Prigmore *et al.* 1997).

Phil Stone and Stewart Molyneux are thanked for their careful reviews of a long manuscript, Mike Howe and Trevor Ford for helpful additional discussions on Manx Group stratigraphy, and Bob Holdsworth for editorial assistance. This work was funded by NERC Research Grant GR9/01834 to NHW, DGQ and WRF.

References

Barnes, R. P., Power, G. M. & Cooper, D. C. 1999. The definition of sandstone-bearing formations in the Isle of Man and correlation with adjacent areas – evidence from sandstone chemistry. *This volume*.

Blake, J. F. 1905. On the order of succession of the Manx

Slates in their northern half, and its bearing on the origin of the schistose breccia associated therewith. *Quarterly Journal of the Geological Society, London*, **61**, 358–373.

BOLTON, H. 1899. The palaeontology of the Manx Slate of the Isle of Man. *Memoirs and Proceedings of the Manchester Literary and Philosophical Society*, **43**, 1–15.

COOPER, A. H., MILLWARD, D., JOHNSON, E. W. & SOPER, N. J. 1993. The early Palaeozoic evolution of northwest England. *Geological Magazine*, **130**, 711–724.

——, RUSHTON, A. W. A., MOLYNEUX, S. G., HUGHES, R. A., MOORE, R. M. & WEBB, B. C. 1995. The stratigraphy, correlation, provenance and palaeogeography of the Skiddaw Group (Ordovician) in the English Lake District. *Geological Magazine*, **132**, 185–211.

CROWLEY, S. F. 1985. Lithostratigraphy of the Peel Sandstones, Isle of Man. *Mercian Geologist*, **10**, 73–76.

DOWNIE, C. & FORD, T. D. 1966. Microfossils from the Manx Slate Series. *Proceedings of the Yorkshire Geological Society*, **35**, 307–322.

FITCHES, W. R., BARNES, R. P. & MORRIS, J. H. 1999. Geological structure and tectonic evolution of the Lower Palaeozoic rocks of the Isle of Man. *This volume.*.

FORD, T. D. 1993. *The Isle of Man*. Geologists' Association, London, 1–94.

FORTEY, R. A. 1984. Global earlier Ordovician transgressions and regressions and their biological implications. *In:* BRUTON, D. L. (ed.) *Aspects of the Ordovician System*. Universitetsforlaget, 37–50.

——, HARPER, D. A. T., INGHAM, J. K., OWEN, A. W. & RUSHTON, A. W. A. 1995. A revision of Ordovician series and stages from the historical type area. *Geological Magazine*, **132**, 15–30.

GEOLOGICAL SURVEY OF UNITED KINGDOM, 1898. Isle of Man. *1:63 360 Geological Map*. **Sheets 36, 45, 46, 56 and 57**.

GILLOTT, J. E. 1956. Breccias in the Manx Slates: their origin and stratigraphic relationships. *Liverpool and Manchester Geological Journal*, **1**, 370–380.

HARKNESS, R. & NICHOLSON, H. A. 1866. On the Lower Silurian rocks of the Isle of Man. *Quarterly Journal of the Geological Society, London*, **22**, 488–491.

HOWE, M. P. A. 1999. The Silurian fauna (graptolite and nautiloid) of the Niarbyl Formation, Isle of Man. *This volume.*

KENNAN, P. S. & MORRIS, J. H. 1999. Manganiferous ironstones in the early Ordovician Manx Group, Isle of Man: a protolith of coticule. *This volume.*

LAMPLUGH, G. W. 1903. *The Geology of the Isle of Man.* Memoir of the Geological Survey, United Kingdom. HMSO..

—— & WATTS, W. W. 1895. The crush-conglomerates of the Isle Of Man. *Quarterly Journal of the Geological Society, London*, **51**, 563–599.

MENG, X., GE, M. & TUCKER, M. E. 1997. Sequence stratigraphy,sea-level changes and depositional systems in the Cambro–Ordovician of the North China carbonate platform. *Sedimentary Geology*, **114**, 189–222.

MOLYNEUX, S. G. 1979. New evidence for the age of the Manx Group, Isle of Man. *In:* HARRIS, A. L., HOLLAND, C. H. & LEAKE, B. E. (eds) *Caledonides of the British Isles – reviewed.* of the Geological Society, London, Special Publications, **8**, 415–421.

—— 1999. A reassessment of Manx Group acritarchs, Isle of Man. *This volume.*

MORRIS, J. H., WOODCOCK, N. H. & HOWE, M. P. A. 1999. The Silurian succession of the Isle of Man: the late Wenlock Niarbyl Formation, Dalby Group. *This volume.*

NICOLL, R. S., LAURIE, J. R., SHERGOLD, J. H. & NIELSEN, A. T. 1992. Preliminary correlation of latest Cambrian to Early Ordovician sea level events in Australia and Scandinavia. *In:* WEBBY, B. & LAURIE J. R. (eds) *Global Perspectives on Ordovician Geology*. Balkema, 381–394.

ORR, P. J. & HOWE, M. P. A. 1999. Macrofauna and ichnofauna of the Manx Group (early Ordovician), Isle of Man. *This volume.*

PIPER, J. D. A. & CROWLEY, S. F. 1999. Palaeomagnetism of (Palaeozoic) Peel Sandstones and Langness Conglomerate Formation, Isle of Man: implications for the age and regional diagenesis of Manx red beds. *This volume.*

POWER, G. M. & BARNES, R. P. 1999. Relationships between metamorphism and structure on the northern edge of eastern Avalonia in the Manx Group, Isle of Man. *This volume.*

PRIGMORE, J. K., BUTLER, A. J. & WOODCOCK, N. H. 1997. Rifting during separation of Eastern Avalonia from Gondwana: Evidence from subsidence analysis. *Geology*, **25**, 203–207.

QUIRK, D. G. & BURNETT, D. 1999. Lithofacies of Lower Palaeozoic deep marine sediments in the Isle of Man: a new map and stratigraphic model for the Manx Group. *This volume.*

—— & KIMBELL, G. S. 1997. Structural evolution of the Isle of Man and central part of the Irish Sea. *In:* MEADOWS, N., TRUEBLOOD, S., COWAN, G. & HARDMAN, M. (eds) *Petroleum Geology of the Irish Sea and Adjacent Areas*. Geological Society, London, Special Publications, **124**, 135–160.

ROBERTS, B., MORRISON, C. & HIRONS, S. 1990. Low grade metamorphism of the Manx Group, Isle of Man: a comparative study of white mica 'crystallinity' techniques. *Journal of the Geological Society, London*, **147**, 271–277.

ROSS, J. R. & ROSS, C. A. 1992. Ordovician sea-level fluctuations. *In:* WEBBY, B. & LAURIE, J. R. (eds) *Global Perspectives on Ordovician Geology*, Balkema, 327–335.

RUSHTON, A. W. A. 1993. Graptolites from the Manx Group. *Proceedings of the Yorkshire Geological Society*, **49**, 259–262.

SIMPSON, A. 1963. The stratigraphy and tectonics of the Manx Slates Series. *Quarterly Journal of the Geological Society, London*, **119**, 367–400.

—— 1965. The syn-tectonic Foxdale–Archallagan granite and its metamorphic aureole. *Geological Journal*, **4**, 189–206.

—— 1968. The Caledonian history of the north-eastern Irish Sea region and its relation to surrounding areas. *Scottish Journal of Geology*, **4**, 135–163.

WOODCOCK, N. H. & BARNES, R. P. 1999. An early Ordovician turbidite system on the Gondwana margin: the southeastern Manx Group, Isle of Man. *This volume*.

—— & MORRIS, J. H. 1999. Debris flows on the Ordovician margin of Avalonia: Lady Port Formation, Manx Group, Isle of Man. *This volume*.

Lithofacies of Lower Palaeozoic deep-marine sediments in the Isle of Man: a new map and stratigraphic model of the Manx Group

D. G. QUIRK[1] & D. J. BURNETT

Geology, Oxford Brookes University, Gipsy Lane, Oxford OX3 0BP, UK

[1]Present address: Burlington Resources (Irish Sea) Limited, 1 Canada Square, Canary Wharf, London E14 5AA, UK

Abstract: A classification scheme for lithofacies in the Lower Palaeozoic Manx Group has been designed based on a simple assessment of the sandstone:mudstone ratio, bed thickness and sedimentary structures. The scheme was used to acquire high-resolution, standardized field data over areas of good exposure on the Isle of Man. These data were then correlated in order to produce a lithostratigraphic framework and a simplified geological map. Overall, the succession appears to young northwestwards from early Arenig, sand-rich sediments on the southeast coast to mud-rich sediments of mid–late Arenig age inland and on the northwest coast. A number of alternative structural reconstructions have been made which fit to a greater or lesser degree with limited biostratigraphic data. The favoured model involves a compromise between inferred duplication of stratigraphic units by reverse faults and repetition of similar depositional cycles or sequences within the succession. This model implies that the Manx Group is between 5400 and 9250 m thick and contains within it both a lateral facies change (from proximal to distal character east–west) and an overall fining-upwards stratigraphic signature. The sediments are interpreted to have been rapidly deposited on the west dipping margin of a basin situated between the Isle of Man and southeast Ireland, which forms an embayment on the northwest edge of Eastern Avalonia. The main depositional processes were turbidity currents and debris flows, with evidence of possible current deflection or reworking of the upper parts of turbidites by northwards flowing contour currents. During the Caledonian orogeny, the basin was inverted with the development of northeast–southwest trending thrusts and east–west dextral faults. Tentative evidence also exists for the presence of an older north-northeast–south-southwest normal fault trend, although whether it was active during the Arenig is uncertain. It is recommended that future stratigraphic work should concentrate on testing possible correlations between and within the mixed sand–mud Lonan, Port Erin, Injebreck and Creggan Mooar Formations.

During recent years it has become clear, mainly from micropalaeontological evidence (e.g. Molyneux 1979), that the stratigraphy of the Isle of Man proposed by Simpson (1963) is flawed. Little explanation is given by Simpson (1963) of the criteria used to identify and classify his units, making it difficult to use his stratigraphic framework in the field and casting doubt on his structural model (cf. Fitches *et al.* 1999). For example, Quirk & Kimbell (1997) and Quirk *et al.* (1999*b*) have shown that the Isle of Man is traversed by major northeast–southwest and east–west faults, unrecognized by Simpson (1963), making the great lateral persistence of some of his stratigraphic units seem untenable (see Ford *et al.* 1999). A solution to these problems was to resurvey the distribution of different rock types in the Manx Group, rather than the more subjective approach of assigning the rocks to formations and mapping their boundaries. However, although it appears that all the sediments are marine (and probably deep marine) in origin (Orr & Howe 1999), it was found that existing classification schemes for these sort of sediments provide insufficient resolution to subdivide the Manx Group into more than a few distinctive rock types. For example, the lithofacies scheme of Pickering *et al.* (1989) is useful for sedimentological analysis but not ideal for mapping (cf. Woodcock & Barnes 1999; Woodcock & Morris 1999). Instead, a new lithofacies scheme was developed specifically for the Manx Group based on simple sedimentological characteristics such as sandstone type, mudstone content, sedimentary structures and bed thicknesses. The

From: Woodcock, N. H., Quirk, D. G., Fitches, W. R. & Barnes, R. P. (eds) 1999.
In Sight of the Suture: the Palaeozoic geology of the Isle of Man in its Iapetus Ocean context.
Geological Society, London, Special Publications, **160**, 69–88. 1-86239-046-0/99/$15.00

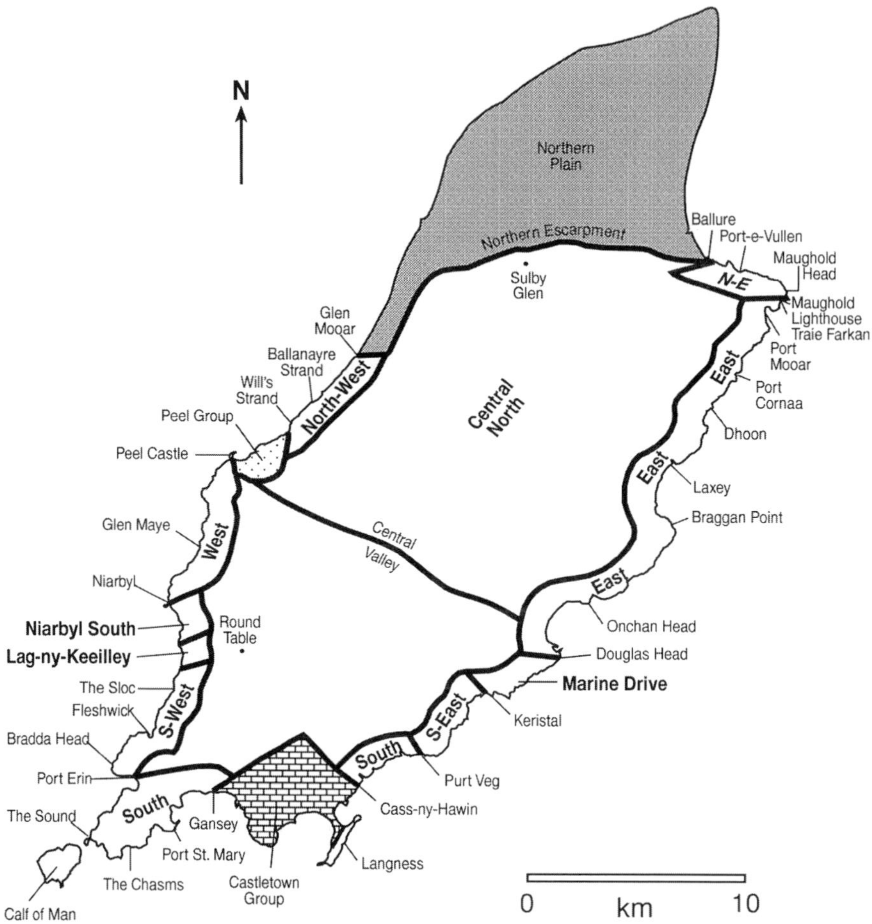

Fig. 1. Map of the Isle of Man showing coastal regions and localities referred to in the text.

scheme is summarized in Table 1, illustrated in Fig. 2 and described in detail in a Supplementary Publication.[1]

The lithofacies nomenclature that has been adopted reflects the sandstone:mudstone ratio of the sediments: e.g. a lower limit of 55% mudstone defines lithofacies class M (mudstone) suffixed by the letters V, H, I and L to indicate very high (95–100%), high (90–95%), intermediate (70–90%) and low (55–70%), respectively. In addition to these standard mudstone lithofacies, lithofacies M_P and M_{IC} are two distinctive types that comprise pebbly mudstone and mudstones with iron–manganese carbonate bands, respectively (Table 1). Sandstone-bearing lithofacies are divided into three field-based classes, and further subdivided in a similar way to the mudstones. Sandstones (S_L, S_I and S_H) comprise quartz arenites containing moderate to low amounts of interbedded mudstone. Quartzites (Q_H and Q_V) are coarser grained, recrystallized quartz arenites with no visible matrix. Wackes (W_H and W_U) are lithofacies containing varying amounts of lithic fragments and feldspar grains.

After a pilot study, the scheme was applied in mapping outcrops in detail along coastal exposures and inland transects. The results were then used to correlate the succession and to determine the gross lithostratigraphic and structural relationships, based on large-scale trends in the dominant lithofacies

[1]Supplementary Publication No SUP 18134 (12 pages) contains detailed descriptions of each lithofacies. It is available from the Society Library or the British Library Document Supply Centre, Boston Spa, Wetherby, West Yorkshire LS23 7BQ, UK.

Table 1. *Main field characteristics of lithofacies in the Manx Group, described in detail in the Supplementary Publication No. SUP 18134*

Lithofacies	Type section	Sand-stone (%)	Sandstone bed thickness (cm)	Distinguishing characteristics	Bouma units	Pickering class*	Lithostratigraphic unit in which commonly found	Field term used by Lamplugh (1903)	Associated lithofacies
$\mathbf{M_V}$	Round Table	< 5	< 0.1	Dark grey, massive	Te	E1.1, E1.2	Barrule Fm, Glen Rushen Fm	Barrule slate	M_H, M_P (, M_I)
$\mathbf{M_H}$	Maughold Head	5–10	< 0.5 (0.1 av)	Laminated	Te	E2.2, D2.3	Barrule Fm	Barrule slate	M_V (, M_P, M_I)
$\mathbf{M_I}$	Sulby Glen	10–30	< 1 (0.2 av)	Laminated, slumping common	Tde	C2.3, D2.3	Injebreck Fm (lower unit)	Unassigned	M_L (, M_P, M_H, Q_H)
$\mathbf{M_{IC}}$	Niarbyl south	5–20	< 2 (0.15 av)	Contains bands of Fe–Mn carbonate	Tde	C2.3, D2.3	Creggan Mooar Fm	Unassigned	M_P, M_V (, M_I)
$\mathbf{M_L}$	Port Erin	30–45	< 3 (1 av)	Planar (-ripple) laminated	Tde	C2.3	Port Erin Fm	Mostly unassigned	M_I, S_L, Q_H, Q_V
$\mathbf{M_P}$	Ballanayre Strand	< 30		Conglomeratic (matrix supported)		A1.3	Fleshwick Unit, Lady Port Fm	Crush conglomerate	M_V (, M_I, W_U, Q_H)
$\mathbf{S_L}$	Traie Farkan (S)	45–60	< 5 (1 av)	Ripple-planar laminated	Tcde?	C2.3	Lonan Fm	Flags	S_I, M_L (, Q_H)
$\mathbf{S_I}$	Port Mooar	60–80	< 0.1–15 (7.5 av)	Quartzose, ripple-planar laminated	Tcde?	C2.3, C2.2	Lonan Fm	Flags	S_H, S_L (, W_H, Q_H)
$\mathbf{S_H}$	Port Cornaa	80–100	1-50 (10 av)	Quartzose, convolute lam. common	Tabcd	B1.1	Ny Garvain Fm (lower unit)	Flags	S_I
$\mathbf{Q_H}$	Traie Farkan (N)	80–90	3–50 (12.5 av)	Quartzite, commonly planar laminated	Tabcde	B1.2	Creg Agneash Fm	Grit	M_L (, M_P, M_I, S_L)
$\mathbf{Q_V}$	The Chasms	80–100	10–100 (40 av)	Quartzite, massive (-planar laminated)	Tabcd	B1.1	Mull Hill Fm, Keristal Mbr	Grit	M_L, S_I (, M_P)
$\mathbf{S_V}$	Purt Veg	95–100	50–300 (125 av)	Arkosic, planar-massive, coarse-gran.	Tab	A2.8	Santon Formation (rare)	Grit	S_H (, M_I ?)
$\mathbf{W_H}$	Marine Drive	80–90	3–50 (12.5 av)	Wacke, commonly planar laminated	Tbcde	B1.2	Santon Fm (Marine Drive)	Flags	S_I (, S_H)
$\mathbf{W_{HP}}$	Lag ny Keeilley	80–90	2–100 (50 av)	Wacke, conglomeratic, planar laminated	Ta	A2.5	Eary Cushlin Unit (very rare)	Unassigned	M_P, S_L, W_U
$\mathbf{W_U}$	Not defined	85–100	1–300	Commonly wacke, undifferentiated	Tabc	B1.1, C2.1	Glion Cam Unit	Grit	M_P, M_I, M_{IC}, M_H

*Pickering *et al.* (1989).

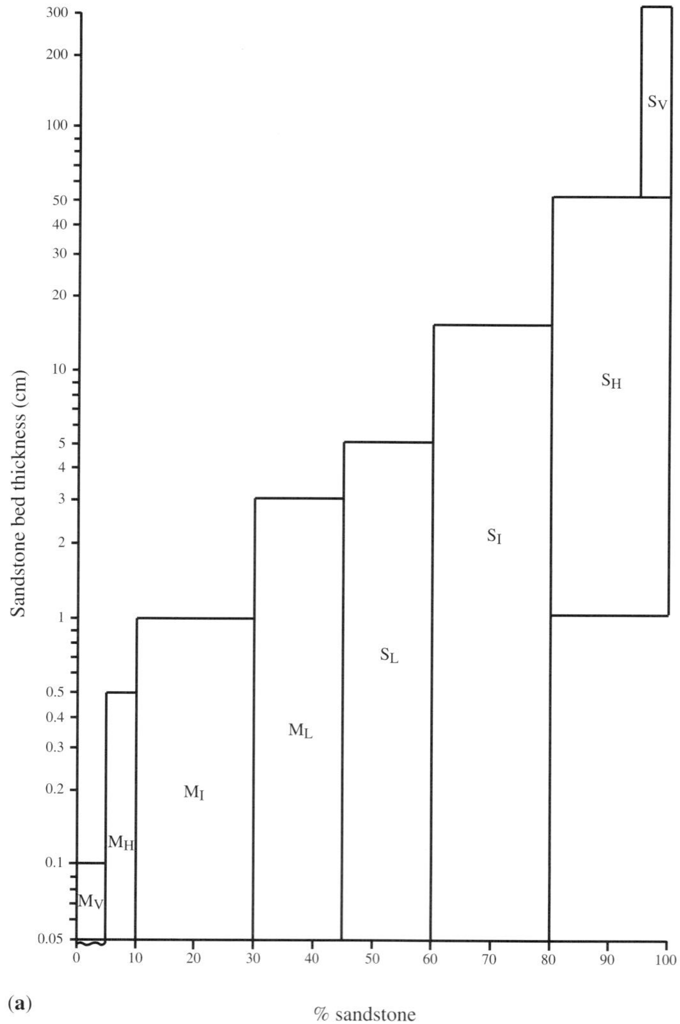

Fig. 2. (**a**) Graph comparing sandstone percentage with bed thickness for mud-rich and sand-rich lithofacies; (**b**) Photograph showing samples of lithofacies ranging from low mudstone to high sandstone percentages. The labels are 1.5 cm high. See Table 1 and Supplementary Publication No. SUP 18134 for further details.

and the observed or inferred positions of boundaries between markedly different groups of lithofacies. The interpretation was also improved by integration with the pioneering geological survey work of Lamplugh (1903) who mapped the entire island in detail (cf. Fig. 3).

The results are shown in Fig. 4 as a set of lithological columns. Each column represents a more or less continuously exposed section separated from adjacent sections by observed or inferred faults, intrusive bodies or gaps in exposure. Each lithofacies has been assigned a colour in logical progression with the next most similar lithofacies. An interval is shown as containing a single lithofacies if at least 75% of the interval conforms with the description of that lithofacies (Table 1). If the interval contains an estimated 25–75% of another lithofacies, then it is given a shared name (shown with horizontal stripes in Fig. 4). Where another lithofacies makes up only 5–25% of the interval, it is regarded as a minor constituent (shown with coloured dots in Fig. 4). Where a lithofacies forms < 5% of an interval it is generally not depicted, except where pebbly mudstone or beds of quartzite help to characterize the interval.

Correlations and vertical succession of the Manx Group

There is a virtual absence of marker beds in the Lower Palaeozoic rocks of the Isle of Man and

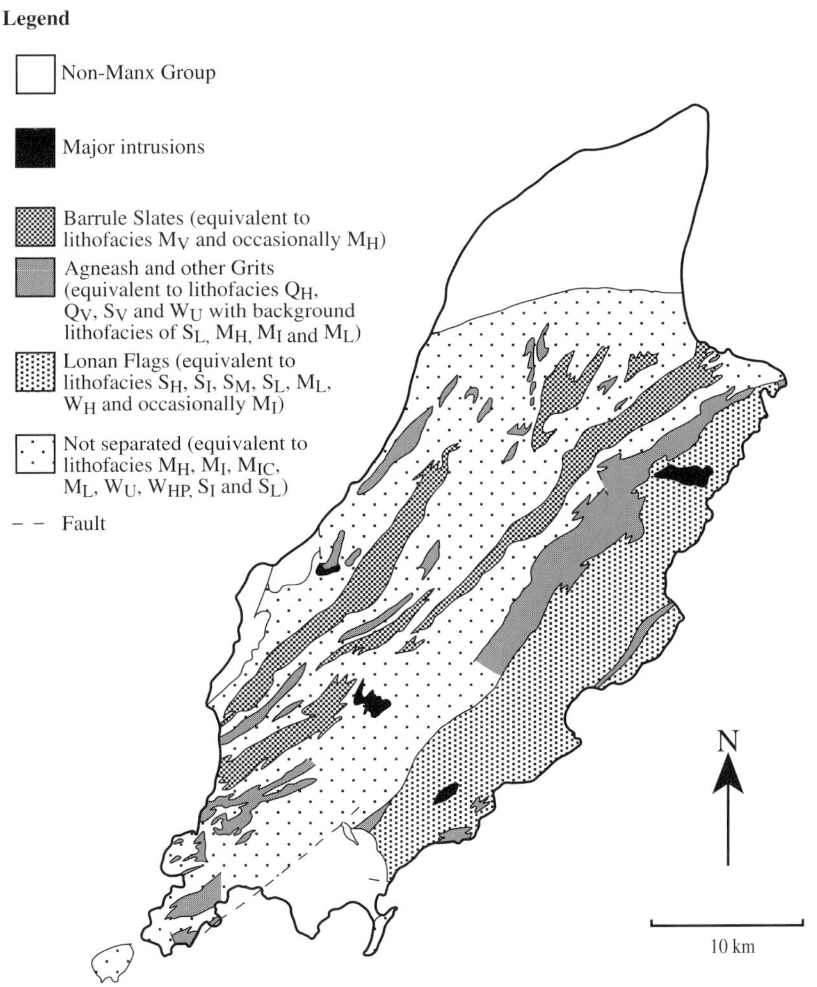

Fig. 3. Simplified map of lithotypes surveyed by Lamplugh (1903).

currently only a poor biostratigraphic framework on which to tie the stratigraphy (cf. Molyneux 1999; Orr & Howe 1999). However, certain lithofacies clearly occur grouped together in areas or tracts with a specific lithofacies signature (Fig. 4). Within each tract, correlations can be made with a fair degree of confidence but correlations between tracts are more difficult to establish. Some flexibility has been used during correlation because of inferred lateral facies variations and some imprecision in classifying certain outcrops, particularly inland where weathering and limited exposure can be a problem. Within tracts, correlations across faulted boundaries have been made using the minimum offset required to produce a good match between columns, unless there is geological evidence to suggest otherwise. Alternatively, the data in each column can be used independently of the lithostratigraphic interpretation shown on Fig. 4. Thus, in the future, the columns can be rearranged and added to as new field and laboratory data are acquired that help improve the correlations.

Probable and possible correlation surfaces are numbered upwards from oldest to youngest (Fig. 4), making the general assumptions that:

- in an overall sense, lithofacies units get younger to the northwest, except where the contrary was observed in the field;
- there are no missing sections and therefore the full stratigraphy can be reconstructed.

These assumptions are unlikely to be completely valid and therefore the correlation surfaces set a lithostratigraphic rather than chronostratigraphic framework. None the less, they have allowed units to be defined which, on the whole, conform with the formal lithostratigraphic scheme of Woodcock *et al.* (1999). However, where there is a difference in interpretation, or where new subdivisions have been made, the term 'unit' has been adopted here to distinguish it from the formations and members defined by Woodcock *et al.* (1999).

The main observations are briefly synthesized in separate sections below, for which Fig. 1 can be used as a location map, and are summarized diagrammatically in Fig. 5. This framework is then used in later sections to interpret the lateral and vertical relationships in the Manx Group with which structural cross-sections and a basin model are developed.

East coast

Access along the coast between Onchan and Maughold is generally good and has allowed confident correlations to be established within the quartzose sand-rich lithofacies (S_L, S_I and S_H) which dominate this part of the succession. Only where a probably sand-rich lithofacies (?S_H) exposed in Dhoon Bay has been strongly recrystallized due to contact metamorphism by the Dhoon Granite is there any uncertainty in the classification. In all, a total stratigraphic thickness of *c.* 1800 m is exposed but this estimate is complicated by the presence of some large folds and strike-slip faults (Fig. 4).

The key to correlating this tract is the recognition of a distinctive 500–550 m thick section (interval 20–25) of highly sand-rich lithofacies (S_H) forming the lower unit of the Ny Garvain Formation. Below and above this section the sediments are more muddy, in the Lonan Formation (interval 10–20, lithofacies M_L–S_I) and the upper unit of the Ny Garvain Formation (interval 25–30, mostly lithofacies S_L–S_I), respectively. Woodcock *et al.* (1999) have assigned the coastal section from Braggan Point to Onchan Head, north of Douglas, to the Santon Formation lying within the axis of the Douglas syncline. However, the lithofacies are identical to the lower unit of the Ny Garvain Formation north of the Dhoon Anticline (Fig. 4). The type Santon Formation (south of Douglas) is more muddy but almost certainly correlates with the lower unit of the Ny Garvain Formation. The axial trace of the Douglas Syncline appears to cut down-section from Douglas to Port Groudle (Fig. 4), suggesting that the fold axis plunges to the southwest. The top of the Lonan Formation (correlation surface 20) is marked in places by a thin package of quartzites (lithofacies Q_V), the Keristal Member of Woodcock *et al.* (1999). A faulted contact between the top of the Ny Garvain Formation and the overlying Creg Agneash Formation (correlation surface 30) is present at Maughold Lighthouse.

Northeast coast

A thickness of 1200 m of highly mud-rich lithofacies (M_V–M_H) occurs between Maughold Head and Port e Vullen. By contrast withWoodcock *et al.* (1999), these rocks are assigned to the Barrule Formation (interval 40–50) due to their close similarity in lithofacies and thickness with rocks

Fig. 4. Compilation of lithofacies mapped by authors between 1995 and 1997 presented in the form of correlation panels and simplified map.

exposed around Clagh Ouyr in the central-north area (Fig. 4). However, the possibility remains that the eastern (lower) part of the section represents a lateral variant of the upper part of the Creg Agneash Formation. The base of the Maughold Head–Port e Vullen section is exposed at [SC 497 916], where a significant part of the Creg Agneash Formation (interval 30–40) has been faulted out by a large east–west dextral fault (Quirk *et al.* 1999*b*). The top of the section is interrupted by several thick felsitic intrusions and associated shear zones around Port e Vullen and Port Lewaigue (Fig. 4). A short muddy section (lithofacies M_V, M_{IC} and M_P) at Ballure is thought to be faulted in from higher up in the succession, probably the middle part of the Injebreck Formation (interval 60–65).

Central-north area

Figure 4 shows a composite section made up of the best exposed sections along traverses between Laxey and the northern escarpment. It is estimated that the stratigraphic thickness of this part of the succession is at least 8000 m (assuming no fault repetition), dipping on average 45° to the northwest. An overall change is seen from quartzose sand-rich lithofacies in the lower (eastern) part of the succession to mud-rich lithofacies in the upper (western) part.

Probably the oldest rocks in the area, in Glen Roy, comprise low to moderately sand-rich lithofacies (S_L–S_I) correlating with the Lonan Formation (interval 10–20) on the eastern coast. Above this is the lower unit of the Ny Garvain Formation (interval 20–25), the distinctive highly sand-rich lithofacies (S_H) seen, for example, in Glen Agneash. The more muddy upper part of the Ny Garvain Formation (interval 25–30) is best seen at Creg ny Baa.

The Laxey Valley exposes a 1700 m thick unit above the Ny Garvain Formation known as the Creg Agneash Formation (interval 30–40) consisting predominantly of moderately mud-rich and quartzite-bearing lithofacies (M_I–Q_H). It appears to thin to less than half its original thickness and become slightly less sandy in a northeast direction. However, at least some of the change in thickness is due to faulting at Maughold Head (see preceding section). The Creg Agneash Formation has been divided into a lower, middle and upper unit on the basis of three apparent fining-upwards cycles (Fig. 4). At the base of the lower and middle units, the quartzites are particularly thick and well developed around Creg ny Baa and Windy Corner, reminiscent of the Mull Hill Formation in the south of the island (lithofacies Q_V). The upper unit, which is the thickest, is distinctly less sandy, comprising mostly lithofacies M_I with the occasional quartzite bed. This interval is assigned to the Maughold Formation by Woodcock *et al.* (1999) on the basis of correlations with a highly muddy section on the northeast coast, between Maughold Head and Port e Vullen (equated with the Barrule Formation here), and a pebbly mudstone-bearing interval on the south coast, between Bradda Head and Lhiattee Beinee (the Fleshwick Unit, Fig. 5). On top of the Creg Agneash Formation is the very highly mud-rich lithofacies (M_V) of the Barrule Formation, estimated to be *c.* 1200 m thick in the vicinity of Clagh Ouyr, assuming that there is no structural repetition. The Barrule Formation appears to thin rapidly between Snaefell and Beinn y Pott ([SC 397 879]–[SC 381 860]) and on the west side of North Barrule (e.g. [SC 414 900]). This thinning corresponds with the apparent truncation of units within the underlying Creg Agneash Formation and the overlying lower unit of the Injebreck Formation. This relationship, and the fact that linear aeromagnetic anomalies are observed at these boundaries, suggest that the Barrule Formation is partly fault bounded, e.g. by the North Barrule lineament at the base of the Barrule Formation (Quirk *et al.* 1999*b*). The North Barrule lineament coincides with the position of a trial mine adit at [SC 387 872], which intersected a mineralized fault of unknown orientation.

The interval between the Barrule and the Glen Rushen Formations (interval 50–70) is defined as the Injebreck Formation (Woodcock *et al.* 1999), and is here divided into three units. However, the stratigraphic picture is complicated by the inferred presence of at least one major fault (Quirk *et al.* 1999*b*). The lowest unit (interval 50–60) occurs in Glen Auldyn and Lhergyrhenny and comprises moderately mud-rich and quartzite-bearing lithofacies (M_I–Q_H) with similarities to parts of the Creg Agneash Formation. A pebbly, highly mud-rich lithofacies (M_P–M_V) in Ballakerka forms the middle part of the Injebreck Formation (interval 60–65). This Slieau Managh Unit pinches out southwards (Fig. 4). Its thickness is hard to estimate, due to probable faulting and lateral facies variations. A large east-northeast–west-southwest trending fault is interpreted to separate it from the lower unit of the Injebreck Formation. However, the extreme scarcity of pebbly mudstone (M_P) between the Barrule and Glen Rushen Formations (interval 50–70) further south suggests that some lateral facies variation may also exist, possibly due to syn-sedimentary movement on the same fault (Quirk *et al.* 1999*b*). The Slieau Managh Unit probably correlates to the east with a faulted section exposed on the coast at Ballure (Fig. 4). The base of the upper unit of the Injebreck Formation is marked by a thin package of quartzites (lithofacies Q_H) and the rest of the interval (interval 65–70) consists of a

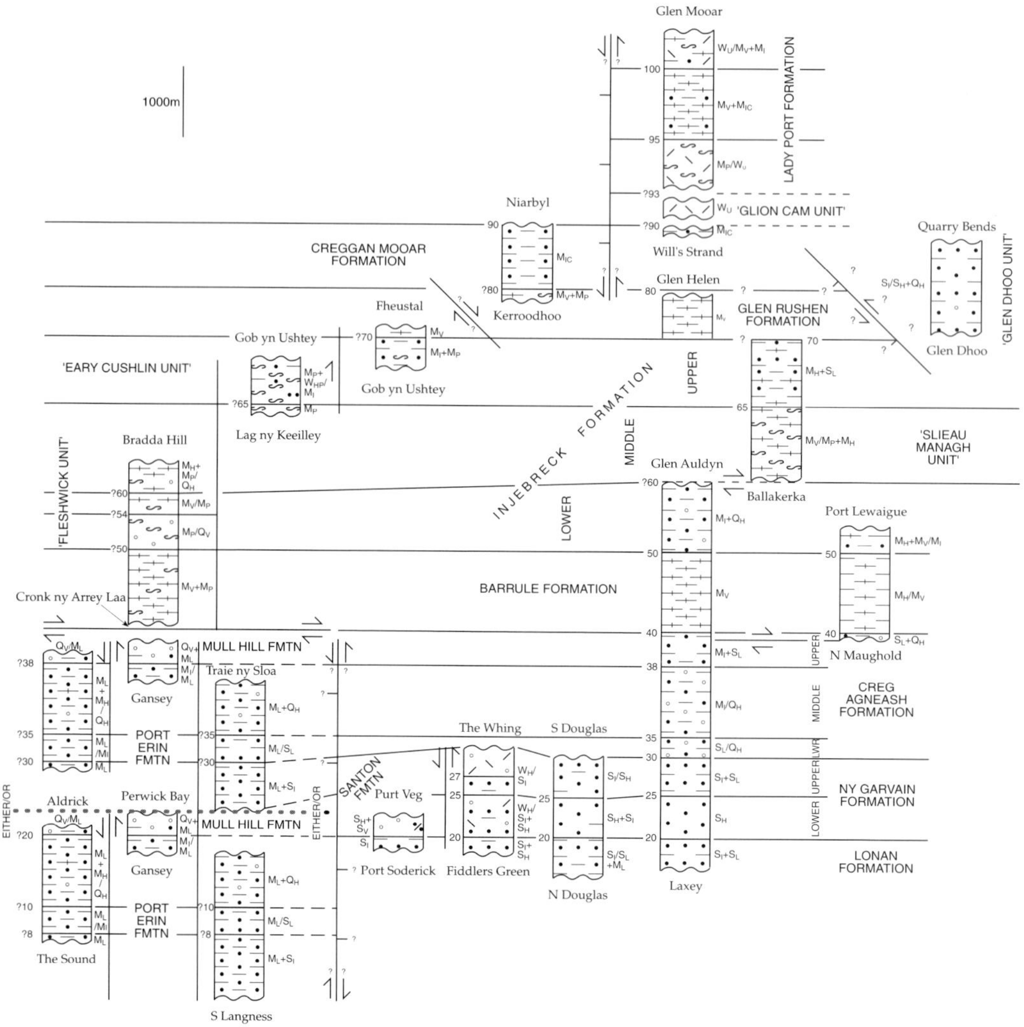

Fig. 5. Schematic lithostratigraphy of the Manx Group based on Fig. 4. The principal lithofacies in each unit are shown on the right-hand side of most columns. Note that the lithostratigraphic position of the Mull Hill and Port Erin Formations (bottom left-hand corner) is uncertain and two alternatives are given. Table 2 gives further information.

broadly mud-rich lithofacies association (M_V, M_H, M_I, M_L and S_L). It shows a higher illite crystallinity grade than the middle and lower parts of the Injebreck Formation, supporting the idea that it is bounded on its southeast side by a reverse fault (Quirk *et al.* 1999*b*). It is difficult to trace the Injebreck Formation towards the Central Valley and further south because of poor exposure. However, the Barrule Formation appears to be offset dextrally by a large east–west fault known as the Greeba Lineament, evidence for which is also seen on aeromagnetic data (Quirk *et al.* 1999*b*).

The Glen Rushen Formation occurs at the top of the succession in the central-north area (interval 70–80). Approximately 600 m of the formation is exposed in Glen Helen, where it is exclusively composed of very highly mud-rich lithofacies (M_V). However, a variable, but generally moderate to highly sand-rich lithofacies (S_I–S_H), similar to the Ny Garvain Formation, is present around Glen Dhoo and Carrick. The stratigraphic relationship of this Glen Dhoo Unit to the Glen Rushen Formation and the underlying mixed lithofacies of the Injebreck Formation is uncertain (Fig. 5). Acritarch data presented by Molyneux (1999) suggest that the Glen Dhoo Unit is early Arenig in age whereas the

Glen Rushen Formation, which occurs along-strike from it, is mid-Arenig in age (Table 2). Field mapping and aeromagnetic data (Quirk *et al.* 1999*b*) indicate that they are in faulted contact (Fig. 4), suggesting a number of possible structural models which are discussed later.

Southeast coast, including Marine Drive

Most of the succession exposed from Santon Head to Loch Promenade in Douglas is thought to correlate with the Ny Garvain Formation (interval 20–30) which is present further north (Fig. 4). However, differences in the lithofacies led Woodcock *et al.* (1999) to instead assign this succession to the Santon Formation, albeit the lateral equivalent of the Ny Garvain Formation. The area has been divided into three separate sections (Santon, Marine Drive and south Douglas) with uncertain mutual relationships.

The south Douglas section comprises *c.* 1200 m of moderate to highly sand-rich lithofacies (S_I–S_H), the base of which, on Douglas Head, may represent the top of the Lonan Formation (correlation surface 20). Marine Drive contains a highly wacke-rich lithofacies (W_H) unique to that section, which may represent an immature lateral equivalent of the highly sand-rich lithofacies (S_H) in south Douglas (Fig. 4). However, the presence of a ?northwest–southeast sinistral fault offsetting the Douglas Syncline by 1500 m at Keristal and an east–west ?dextral fault juxtaposing different lithofacies at Fiddlers Green allows for the possibility that the Marine Drive section has been faulted in from lower in the succession (Quirk *et al.* 1999*b*). Approximately 450 m of section is exposed between Keristal and Santon Head. As well as moderate to highly sand-rich lithofacies (S_I–S_H), it includes an unusually thick-bedded, very highly sand-rich lithofacies (S_V), interpreted as a turbidite channel. It appears to correlate with a moderately mud-rich lithofacies containing thin bedded quartzites ($M_I + Q_H$) (Fig. 4) which may represent a levee deposit.

South coast (Langness–Purt Veg)

A vertical northwest–southeast trending fault at Purt Veg [SC 324 703] marks an important change in lithofacies (Quirk *et al.* 1999*b*). The direction and amount of offset on this fault is uncertain and two possibilities exist, depending on how the low to moderately mud-rich lithofacies (M_L–M_I) between Purt Veg and Langness correlates with the succession to the northeast (Fig. 5). The correlation favoured by Woodcock & Barnes (1999) assigns the section to the Lonan Formation (interval 5–20) implying that it is much muddier here than, for example, around Port Jack (lithofacies S_L–S_I) further north. In this interpretation, the amount of stratigraphic offset across the fault at Purt Veg may be as much as 700 m in a dextral sense, although this estimate depends on assumptions made in reconstructing the rather faulted succession south of Purt Veg (Fig. 4). The alternative correlation is with the Creg Agneash Formation (interval 30–40) which shows a generally similar mud-rich background lithofacies (M_I) but contains significantly greater amounts of quartzite (lithofacies Q_H). Based on a tentative match with the Laxey Valley (Fig. 4), the likely stratigraphic offset in this interpretation is *c.* 3000 m by sinistral movement.

There is a marked similarity in the mud-rich lithofacies between Purt Veg and Langness and the Port Erin Formation, exposed further south (lithofacies M_I–M_L; Fig. 5), to which the section is therefore assigned. Whether the Port Erin Formation is the lateral equivalent of the Lonan Formation or the Creg Agneash Formation is yet to be established. Detailed correlations in this area are difficult to make because of structural complexity and the relationships shown on Fig. 4 remain speculative.

South coast (Port Erin–Gansey)

In the area around Port St Mary, The Sound and Port Erin, the Manx Group is affected by some large faults leading to uncertainties in correlation, particularly between The Sound and Aldrick and between Port Erin and Cregneash. Almost 2000 m of low to moderately mud-rich sediments belonging to the Port Erin Formation are overlain by the Mull Hill Formation comprising *c.* 300 m of quartzites and mudstones (mostly lithofacies Q_V and M_L). The quartzites are concentrated towards the base of the Mull Hill Formation. The contact with the Port Erin Formation (correlation surface 20–38) is best seen at The Chasms, at Port St Mary, where it is overturned, and at Aldrick, truncated by a major fault (Fig. 4). Assuming that the Port Erin Formation is not equivalent to the highly sandy parts of the Santon–Ny Garvain Formations, the Mull Hill Formation may correlate to the northeast in one of two ways; either with the upper part of the Creg Agneash Formation (interval 38–40), which contains few quartzites in the Laxey River, or with the lower parts of the Santon and Ny Garvain Formations (interval 20–25) (Fig. 5), which contain equivalent amounts of quartzose sand but in a different type of lithofacies (S_H). A second thinner package of quartzites occurs within the middle of the underlying Port Erin Formation at the north end of Port Erin. This package could correlate with the base of the middle unit of the Creg Agneash Formation (correlation surface 35) or else marks

Table 2. *Estimated thickness, sequence stratigraphic interpretation and possible correlatives for each of the main lithostratigraphic units shown in Fig. 4, north of Cronk Ny Arrey Laa*

Lithostratigraphic subdivision	Estimated thickness (m)	Sequence stratigraphic interpretation	Biostratigraphic age	Established correlative	Possible correlatives		Potential correlative
Lady Port Fm	*c*. 2100 m	Tectonically complex	Late Arenig			Creggan Mooar (±)	
Glion Cam Unit	?500–1000 m	Late highstand	?Early Arenig?		Glen Dhoo (3,4)	Ny Garvain (4)	Upper Injebreck (2)
Creggan Mooar Fm	≥900 m	Early highstand			Lower Injebreck (2, 3, 4)	Upper Injebreck (3, 4)	Lonan (4)
Glen Rushen Fm	≥600 m	Transgression	Mid-Arenig		Barrule (2, 3, 4)	Slieau Managh (3, 4)	
Glen Dhoo Unit	*c*. 1000 m	Lowstand	Early Arenig			Ny Garvain (2, 4)	Glion Cam (3, 4)
Upper Injebreck Unit	950 m	(?Late) highstand		Eary Cushlin[a] (S)	Lower Injebreck (3, 4)	Creggan Mooar (3, 4)	Lonan (4)
Slieau Managh Unit	≥1000 m	Slope instability	?Trem(-Aren)?	Upper Fleshwick[b] (S)	Barrule/Glen Rushen (3, 4)		
Lower Injebreck Unit	*c*. 1200 m	(?Early) highstand		Lower Fleshwick[c] (S)	Creggan Mooar (2, 3, 4)	Upper Injebreck (3, 4)	Lonan (4)
Barrule Fm	1200 m	Transgression			Glen Rushen (2,3,4)	Slieau Managh (3, 4)	
Creg Agneash Fm	1700 m	Late lowstand–early transgression			Port Erin/Mull Hill (S)		
Ny Garvain Fm	1050 m	Early lowstand	Early Arenig*	Santon[d] (S)	Mull Hill[e] (S)	Glen Dhoo (2, 4)	Glion Cam (4)
Lonan Fm	550 m	Late highstand			Port Erin[f] (S)	Injebreck/Creggan Mooar (4)	

Refer to Fig. 5 for an overview of the principal lithofacies present within each lithostratigraphic unit. Biostratigraphic ages are based on data in Molyneux (1999) and Orr & Howe (1999).

*based on correlation with Santon Formation; (±) and/or Glen Rushen Formation and/or Glion Cam Unit; (2), (3), (4), in structural Models 2, 3 and 4, respectively (see Fig. 7); (S) in south of Isle of Man; [a]*c*. 800 m thick; [b]*c*. 440 m thick; [c]*c*. 800 m thick; [d]*c*. 1100 m thick; [e]*c*. 300 m thick; [f]*c*. 2400 m thick.

the top of the oldest part of the Manx Group (correlation surface 10).

Southwest coast

A marked change in lithofacies occurs at the north end of Port Erin across a major east–west trending fault at [SC 193 697] (Quirk *et al.* 1999*b*). Depending on whether the Port Erin Formation on the south side of this fault is correlated with the Creg Agneash Formation or with the Lonan Formation (Fig. 5), the fault accounts for either *c.* 2800 or 5400 m of stratigraphic offset, respectively, by apparent dextral movement. In contrast to the south and east coasts, pebbly mudstones (lithofacies M_P) make up an important part of the succession north of the fault, in addition to quartzites and very highly mud-rich lithofacies (Q_H–M_V). In total, *c.* 2300 m of succession is interpreted in this southwest area. The succession is here called the Fleshwick Unit rather than the Maughold Formation (cf. Woodcock *et al.* 1999), to avoid structural inconsistencies and problems in equating the lithofacies. Roberts *et al.* (1990) show that the lower part of the Fleshwick Unit has a significantly lower illite crystallinity grade than the Port Erin Formation, supporting the idea that it is from higher in the succession. The lithofacies in the Bradda Head–Lhiattee Beinee section are most similar to those of the lower and middle units of the Injebreck Formation (interval 50–65). However, there are no pebbly mudstones in the lower part of the Injebreck Formation (interval 50–60) around Lhergyrhenny and Glen Auldyn (Fig. 4), implying that moderately mud-rich lithofacies (M_I) may pass laterally into pebbly mudstones (lithofacies M_P). The very highly mud-rich and pebbly mudstone lithofacies (M_V–M_P) on Cronk ny Arrey Laa are correlated with the underlying Barrule Formation (interval 40–50) but, similarly, there are no pebbly mudstones in the Barrule Formation in the north-east, e.g. around Clagh Ouyr (Fig. 4). However, as explained below, the pebbly Slieau Managh Unit may also correlate with the Barrule Formation because of possible fault repetition, in which case debris flows are a common feature at this level (?mid-Arenig). As in the northeast, where the base and top of the Barrule Formation are probably fault-bounded, the Cronk ny Arrey Laa section is separated from the overlying Eary Cushlin Unit and probably also from the Lhiattee Beinee section by approximately northeast–southwest trending faults (Quirk *et al.* 1999*b*).

Lag ny Keeilley

The northwest side of Cronk ny Arrey Laa is marked by a southeast dipping thrust estimated to account for *c.* 3400 m of stratigraphic offset (Quirk *et al.* 1999*b*). The section exposed on the coast north of the thrust, from Lag ny Keeilley to Gob yn Ushtey, forms part of the Eary Cushlin Unit (Fig. 5). It contains a varied lithofacies association, different from that on Cronk ny Arrey Laa, including a unique conglomeratic wacke (W_{HP}). Due to a number of faults and shear zones bounding and cutting this short section, as well as its relatively poor exposure, correlation is problematic (cf. Fitches *et al.* 1999). However, the nearest similar lithofacies association (M_V–S_L) occurs within the upper part of the Injebreck Formation (interval 65–70) around, for example, the Blaber River, which is the correlation tentatively suggested here and supported by Woodcock *et al.* (1999).

Niarbyl south

The section between Gob yn Ushtey and Niarbyl consists of a relatively poorly exposed lower part, south of Fheustal, forming the upper part of the Eary Cushlin Unit, and a well-exposed upper part, north of Fheustal (Fig. 5). On the basis of correlations with the central-north area (Fig. 4), most of the very highly mud-rich Glen Rushen Formation (interval 70–80) appears to have been faulted out at Fheustal (Quirk *et al.* 1999*b*). Consequently, the section south of here, comprising moderately mud-rich and pebbly mudstone lithofacies (M_I–M_P), is correlated with the upper unit of the Injebreck Formation (interval 65–70). The Creggan Mooar Formation (interval 80–90) lies north of Fheustal and consists mostly of moderately mud-rich lithofacies with characteristic red–brown iron–manganese carbonate bands (M_{IC}) (Kennan & Morris 1999). Inland this lithofacies is rarely exposed, but it reappears on the northwest coast within the Lady Port Formation (interval 90–100). It may, however, be indistinguishable from lithofacies M_I inland in the Injebreck Formation, as the suspicion is that the iron–manganese bands only become obvious on wave-washed outcrops. In fact, the only other place where lithofacies M_{IC} has been observed is on the coast near Ramsey at [SC 460 934], probably within the Injebreck Formation (Fig. 4), with which the Creggan Mooar Formation may correlate (see below and Table 2). Approximately 900 m of Creggan Mooar Formation is estimated to be present and it is bound to the north by a shear zone and fault separating it from the graded wackes of the Silurian Dalby Group (Morris *et al.* 1999).

Although the boundary is not well exposed, the Glion Cam Unit is thought to overlie the Creggan Mooar Formation (Woodcock *et al.* 1999). Limited outcrop suggests that it is 500–1000 m thick and consists mostly of lithofacies W_U.

Northwest coast

A highly faulted section containing diverse lithofacies (M_P, W_U, M_V and M_{IC}) is exposed between Will's Strand and Glen Mooar. This represents the Lady Port Formation (interval 90–100) which is tentatively estimated to be *c.* 2200 m thick, excluding several thick felsitic igneous bodies (Fig. 4). It is thought to represent the highest part of the succession, as supported by a late Arenig acritarch age (Molyneux 1999) and low illite crystallinity values (Roberts *et al.* 1990). The presence of lithofacies W_U, M_{IC} and M_V, often in fault-bounded packets, as well as thick intervals of pebbly mudstone (lithofacies M_P), indicate that the unit may contain slivers of Glion Cam Unit, and Creggan Mooar and Glen Rushen Formations. It is also worth noting that lithofacies M_{IC} at Gob y Deigan [SC 283 873] is remarkably similar in appearance to an interval on the coast near Ramsey at [SC 460 934] assigned to the Injebreck Formation (Fig. 4). The Lady Port Formation can only be traced for a limited distance inland where a faulted boundary is inferred with the poorly exposed Glion Cam Unit (Woodcock & Morris 1999) (Fig. 4).

Comparison with the lithostratigraphy of Woodcock *et al.* (1999)

The results of the present study have mostly been incorporated into the formal lithostratigraphy proposed by Woodcock *et al.* (1999), but there are specific differences which are briefly discussed below.

Lithostratigraphy

A number of the formations defined by Woodcock *et al.* (1999) have been subdivided here on the basis of obvious lithofacies trends. However, the present authors are more conservative in extrapolating lithostratigraphic units through areas of poor exposure and across large faults, so that few units are shown to continue uninterrupted across the island. For example the Injebreck Formation is shown confined to the north (Fig. 4) whereas Woodcock *et al.* (1999) continue it to the west coast (the Eary Cushlin Unit in this paper). The Maughold Formation of Woodcock *et al.* (1999) has been dropped due to perceived differences between the lithofacies at the southern and northern ends of the island where it is best exposed. Instead, the Maughold Formation in the south is named informally here the Fleshwick Unit, which is thought to overlie the Barrule Formation. Hence, it is tentatively correlated with the lower part of the Injebreck Formation (Fig. 5). The northern outcrop of the Maughold Formation of Woodcock *et al.* (1999), between Maughold Head and Port e Vullen, comprises lithofacies very similar to the Barrule Formation and appears to connect with it (Fig. 4). The Barrule Formation itself is cut out on the southwest side of Snaefell by the North Barrule Lineament (Fig. 6). A similarity in lithofacies between the lower and upper units of the Injebreck Formation supports the idea of possible fault repetition (see below). The rest of the area around Glen Dhoo, Cronk Sumark and Sulby Glen, a region left uninterpreted by Woodcock *et al.* (1999), has been mapped and the lithostratigraphy informally defined by the current authors (Fig. 4). Instead, however, a large area in the centre of the island, south of the Central Valley, has been left uninterpreted because of poor exposure. Unlike Woodcock *et al.* (1999), the present authors assign the rocks between Langness and Purt Veg [SC 324 703] to the Port Erin Formation rather than the Lonan Formation, again on lithofacies grounds. Likewise, the Ny Garvain Formation is correlated with the Santon Formation rather than with the Port Erin and Lonan Formations, which contain far less sandstone.

Faults

Interpretation of tectonic lineaments on the Isle of Man (Quirk & Kimbell 1997; Quirk *et al.* 1999*b*) has indicated that the Manx Group is compartmentalized into a number of fault-bound slices. The Windy Corner Fault of Woodcock & Barnes (1999) is not recognized. However, several larger faults are identified which, despite affecting the stratigraphy, have not been included in the map of Woodcock *et al.* (1999). These include an east–west mineralized fault traversing Maughold Head (e.g. [SC 497 915]), an east-northeast–west-southwest shear zone which cuts off the northern end of the Douglas Syncline (e.g. [SC 442 808]) and a number of north–south and east-northeast–west-southwest lineaments in the central-north area (Fig. 6).

Structural interpretations

The lithostratigraphic units of the Manx Group appear generally to dip and young to the northwest. Some stratigraphic repetition is likely to occur across large northwest dipping reverse faults (Quirk *et al.* 1999*b*), but its magnitude is uncertain without better biostratigraphic control. Seismic evidence suggests that at least some of the northwest dip in the Manx Group is due to post-Caledonian tilting in the footwall of the Eubonia–Lagman Faults (Quirk *et al.* 1999*b*). These offshore faults lie close to the east coast of the Isle of Man, along pre-existing Caledonian weaknesses (Quirk *et al.* 1999*a*). A

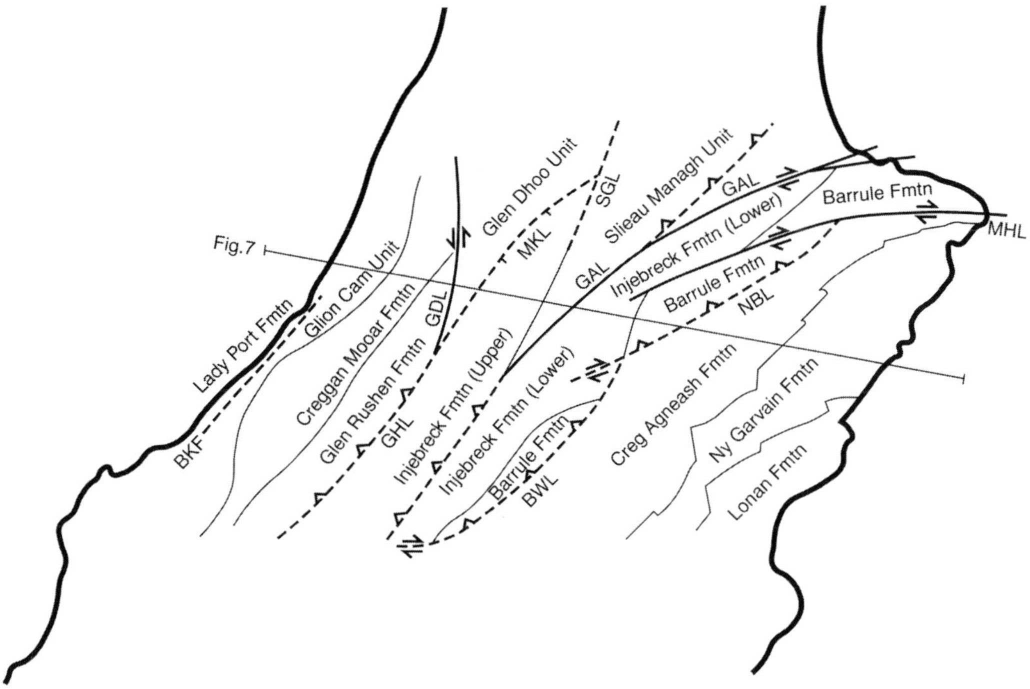

Fig. 6. Simplified lithostratigraphic map of the central-north area showing principal geological boundaries and line of section used in Fig. 7. BKF, Ballakaighin Fault; BWL, Baldwin Lineament; GAL, Glen Auldyn Lineament; GDL, Glen Dhoo Lineament, GHL, Glen Helen Lineament; MHL, Maughold Head Lineament; MKL, Mount Karrin Lineament; NBL, North Barrule Lineament; SGL, Sulby Glen Lineament.

total of 1–4 km of normal movement is recorded following extensional events in the early Carboniferous, early Permian, late Jurassic and early Tertiary (Quirk & Kimbell 1997).

On the basis of the geological boundaries mapped in Fig. 6, four alternative structural cross-sections have been constructed for the north of the island (Fig. 7). These illustrate possible strati-graphic–structural scenarios ranging from a minimum of fault repetition (Fig. 7a) to a maximum of fault repetition (Fig. 7d). The direction of dip of the main faults is generally inferred rather than observed. Lettering is used to order the succession in each model (A being the oldest, N being the youngest) and to indicate proposed correlations, such as the Ny Garvain Formation equating with the Glen Dhoo Unit in Model 2 (C_2 in Fig. 7b). Several assumptions have been made during their construction, in particular:

- different lithostratigraphic units with similar lithofacies associations may correlate (Table 2; Figs 7b–d);
- lateral facies variations are limited except in some cases where lithofacies Q_H, Q_V, M_P, M_V and M_{IC} are present (Fig. 4);
- the overall structure is not complex and is controlled by a number of observed or inferred northeast–southwest reverse or normal faults, east–west dextral faults and north–south sinistral faults (Fig. 6; Quirk *et al.* 1999*b*);
- little of the succession is missing (Fig. 5);
- early Arenig or possible Tremadoc acritarch dates from near the Peel Harbour Fault are not representative of the age of the Glion Cam Unit (cf. Molyneux 1999), except possibly in Model 4 (Fig. 7d).

The total thicknesses quoted below are based on the succession north of Lag ny Keeilley (Figs 1 and 6) and do not take into account the possible lateral equivalence of the Lonan Formation (550 m thick) with the apparently much thicker Port Erin Formation (*c.* 2400 m thick) on the south coast (Table 2). In general, thicknesses may be overestimated due to the difficulty in recognizing contractional structures in certain intervals, particularly those that are predominantly muddy such as the Barrule Formation.

Model 1

Figure 7a assumes that a continuous succession exists from the Lonan Formation to the Lady Port

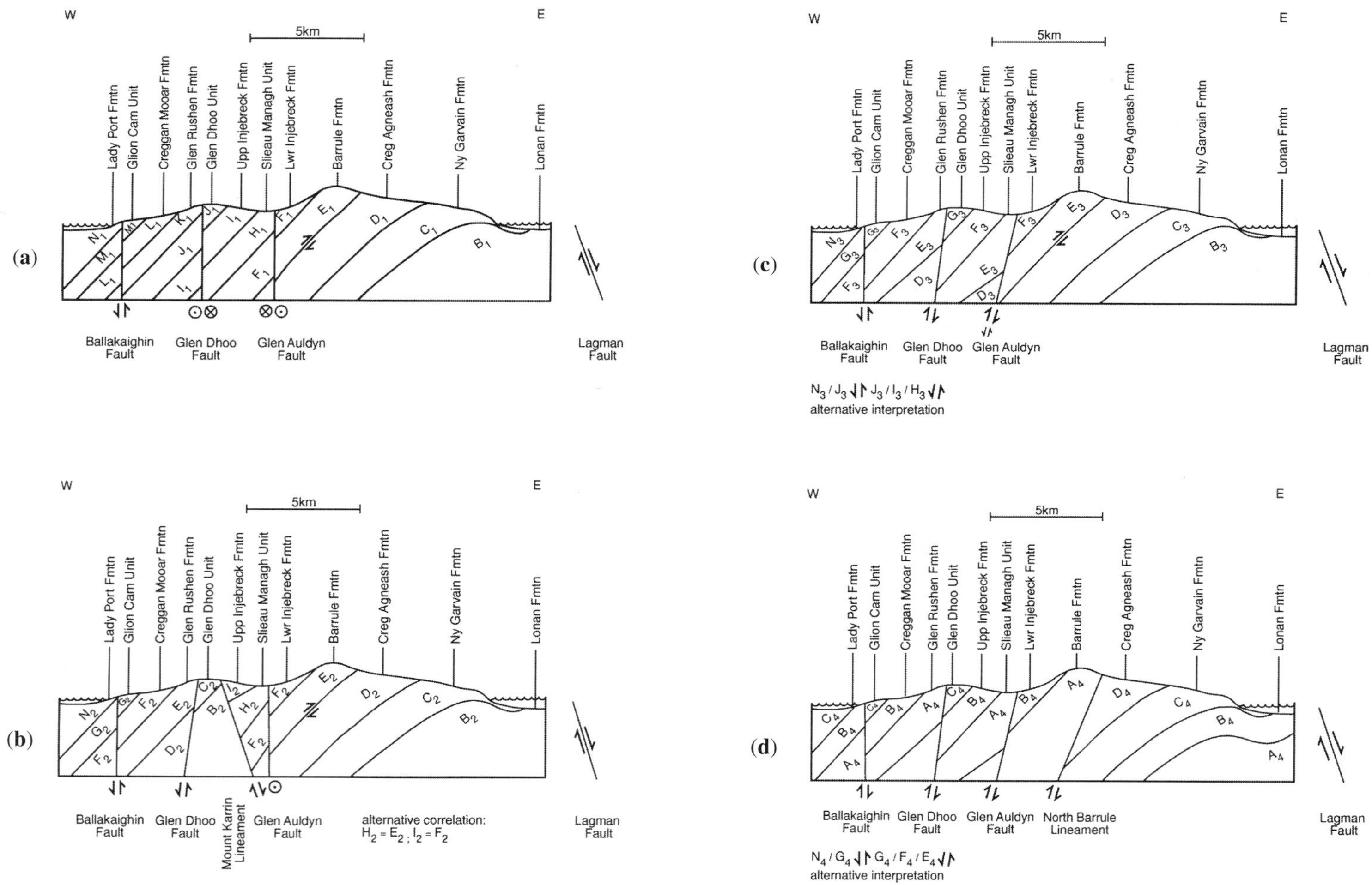

Fig. 7. Schematic cross-sections illustrating alternative structural models across the central-north area (Fig. 6). See text for details. **(a)** Model 1 (interpretation with minimum fault repetition); **(b)** Model 2 (favoured interpretation); **(c)** Model 3; **(d)** Model 4 (interpretation with maximum fault repetition).

Formation with minimal fault repetition. The Glen Auldyn, Glen Dhoo and Ballakaighin Faults throw down to the west. This interpretation implies that the Manx Group is *c.* 12 750 m thick (or 10 650 m if the tectonically complex Lady Port Formation is excluded). Based on two similar acritarch dates from the Glen Dhoo Unit and from the lateral correlative of the Ny Garvain Formation (the Santon Formation) (Table 2; Molyneux 1999), the model suggests that a probably unreasonable 6500 m of sediment was deposited during the early Arenig (*c.* 5 Ma). The youngest acritarch date is late Arenig from the Lady Port Formation at the top of the succession (Table 2).

Model 2

Figure 7b provides the best fit with the biostratigraphy of Molyneux (1999), with the geophysical interpretation of Quirk *et al.* (1999*b*) and with geologically reasonable sedimentation rates. It assumes that a continuous succession exists from the Lonan Formation to the upper unit of the Injebreck Formation, but that the early Arenig Glen Dhoo Unit is equivalent to the Ny Garvain Formation by virtue of a large normal fault (the Mount Karrin Lineament) that throws down to the east. The Slieau Managh Unit may represent a set of debris flows sourced from the fault. This unit may either be confined to the area between the Mount Karrin and Glen Auldyn Faults or it may correlate with the Barrule Formation, implying that the lower and upper units of the Injebreck Formation are lateral equivalents (Fig. 7b). Alternatively, the normal fault may be entirely post-depositional. The mid-Arenig Glen Rushen Formation is a thinner correlative of the Barrule Formation and the Creggan Mooar Formation is a thinner, more distal equivalent of the Injebreck Formation, a correlation supported by the presence of the distinctive lithofacies M_{IC} at a coastal exposure near Ramsey (Fig. 4). The Glen Auldyn, Glen Dhoo and Ballakaighin Faults throw down to the west. South of the section, the Glen Helen Lineament throws down to the east. It either represents a west dipping reverse fault reactivating the Glen Dhoo Fault or, more likely, is an extension of the east dipping Mount Karrin normal fault, implying that the Glen Dhoo Fault is also normal.

In Model 2, the Manx Group reaches a maximum thickness of 9250 m if the Slieau Managh Unit does not correlate with the Barrule Formation (or 7150 m if the Lady Port Formation is excluded). If this does correlate, the Manx Group is *c.* 7500 m thick (or 5400 m if the Lady Port Formation is excluded). Possible syn-sedimentary fault activity during deposition of the Slieau Managh Unit is consistent with a model of Arenig tectonism supported by Woodcock & Barnes (1999) and Woodcock & Morris (1999).

Model 3

Figure 7c shows a continuous succession from the Lonan Formation to the Injebreck Formation. The Glen Rushen Formation and Slieau Managh Unit are equivalent to the Barrule Formation, due to stratigraphic repetition caused by reverse offset on the Glen Dhoo and Glen Auldyn Faults. The Slieau Managh debris flows may have been sourced from a normal syn-sedimentary precursor to the Glen Auldyn Fault. Later reverse offset would be due to Caledonian reactivation. The Creggan Mooar Formation and upper Injebreck Unit are thinner, more distal correlatives of the lower Injebreck Unit, with the Glion Cam and more proximal Glen Dhoo Units interpreted to overlie these. However, biostratigraphic evidence (Molyneux 1999) does not support this interpretation, as the Glen Dhoo Unit is older than the Glen Rushen Formation. The estimated thickness of the Manx Group in this model is 7750 m (or 5650 m excluding the Lady Port Formation). However, to conform with the biostratigraphy, an alternative is also suggested on Fig. 7c where the younger Glen Rushen Formation is juxtaposed against the Glen Dhoo Unit by normal faulting along the Glen Helen Lineament. In this case, the Glen Rushen Formation is not the lateral equivalent of the Barrule Formation. This alternative interpretation adds another 2000 m to the estimated thickness of the Manx Group in Model 3.

Model 4

Figure 7d shows the most stratigraphic repetition due to reverse movement on the Glen Auldyn, Glen Dhoo and Ballakaighin Faults, and an additional fault, known as the North Barrule Lineament, which cuts out part of the Barrule Formation. The complete succession is represented by the Barrule–Slieau Managh–Glen Rushen lateral correlatives at the base, followed by the Lonan–Injebreck–Creggan Mooar correlatives, in turn overlain by the Ny Garvain–Glen Dhoo–Glion Cam correlatives and, finally, by the Creg Agneash Formation. The lateral correlatives tend to thin and fine westwards. The structural configuration shown on Fig. 7d is reminiscent of northwest dipping structures imaged on offshore seismic sections along-strike from the Glen Dhoo, Glen Auldyn and North Barrule Lineaments (Quirk *et al.* 1999*b*). None the less, as in the previous model, this stratigraphic order is not supported by acritarch dates. These indicate that the Glen Dhoo Unit is older than the underlying Glen Rushen Formation. An alternative interpretation, also shown on Fig. 7d, overcomes this problem by making the Glen Helen Lineament a normal fault juxtaposing the younger Glen Rushen Formation against the Glen Dhoo Unit. If, instead, the Glen Rushen data are ignored, then the poorly constrained biostratigraphic data from near Peel may indicate that the base of the Glion Cam Unit is pre-Arenig in age (Molyneux 1999). This interpretation implies that the Manx Group may be as little as 4500 m thick.

Summary

Although Model 2 (Fig. 7b) is favoured here, all four models have aspects to recommend them. Resulting estimates of the thickness of the Manx Group vary from 4500 to 10 650 m, with an average of 7000 m. These estimates do not include the highly faulted Lady Port Formation (*c.* 2100 m thick) at the top of the succession, nor the Port Erin Formation (*c.* 2400 m thick) in the south of the island, with which the upper part of the Lonan Formation (550 m thick) may correlate (Table 2; Fig. 5).

Sequence stratigraphic interpretation and basin model

No matter what structural interpretation is put on the Manx Group, important lithological variations are apparent within the succession (Fig. 5). On the east side of the island, the 550 m thick lower unit of the Ny Garvain Formation comprises > 80% thin–thick-bedded quartz arenites (lithofacies S_I and S_H), whereas the 1200 m thick Barrule Formation consists almost entirely of dark grey mudstone (lithofacies M_V). A similar contrast is seen between other formations, such as the Glen Dhoo Unit and the Glen Rushen Formation. Alternations on this sort of scale in deep-marine sediments are typically caused by second- (10–80 Ma) or third-order (1–10 Ma) variations in the relative height of sea level, the different units within each cycle or sequence being assigned to systems tracts (Vail *et al.* 1977). Sand bypasses the shelf during periods of falling and low relative sea level (lowstands), when it is carried down submarine canyons into the deeper parts of the basin by turbidity currents, depositing thick submarine fans in an otherwise distal environment (Galloway 1998). A subsequent rise in relative sea level causes transgression, when sands are trapped on the shelf and muds are mostly deposited further out in the basin. Thereafter, a subsequent highstand in relative sea level is associated with deltaic progradation on the shelf which may lead to oversteepening and slumping at the front of the delta causing mixed sand–mud turbidites to flow basinwards.

Therefore, a change from low to high relative sea level is likely to be reflected in the deep-marine environment as a change from clean sandstones within a turbidite fan (lowstand systems tract) to mudstones (transgressive systems tract) to mixed sandstone–mudstone turbidites with possible debris flows (highstand systems tract). This corresponds with the change seen from the base of the Santon–Ny Garvain Formations (early Arenig) to the top of the Injebreck Formation (?mid-Arenig) (Table 2). The Creg Agneash Formation, above the Ny Garvain Formation, displays a fining-upwards signature (Fig. 4) typical of the lower part of the transgressive systems tract. The Barrule Formation is thought to represent the upper part of the transgressive systems tract. There are insufficient biostratigraphic data to be sure whether there is another, younger relative sea-level cycle recorded in the Manx Group, from the base of the Glen Dhoo Unit to the top of the Glion Cam Unit (Table 2). Consequently, this part of the succession could be a partial or complete repetition of the early–?mid Arenig cycle (Fig. 7b–d).

Woodcock & Barnes (1999) record a divergence between palaeoflows recorded from flute casts (west directed) at the base of turbidite beds (typically lithofacies Q_H, Q_V, W_H and M_L) and ripple crests (north directed) at the top of other beds (typically lithofacies S_L, S_I and S_H). They propose a model for the deposition of formations exposed on the southern and eastern coast of the Isle of Man involving 2–10 km wide turbidite lobes along an actively faulted margin. The high concentration parts of the turbidity currents are shown running west, subparallel to the continental margin, due to deflection at east-northeast–west-southwest trending fault scarps, whereas the higher, low concentration parts of the currents flow northwards, undeflected by the scarps, towards the deeper basin (Woodcock & Barnes 1999). The present authors regard this model as unlikely for the following reasons:

- evidence for lobe geometries is minimal, with sand-dominated intervals easily correlated over distances of at least 10 km;
- bimodal palaeocurrent data are not usual in faulted areas where, typically, the wide variety of slopes produces a polymodal flow distribution (e.g. Boote & Gustav 1987; Prosser 1993);
- turbidites usually flow down the continental slope, not along it, whether there are faults present or not (e.g. Galloway 1998);
- evidence for thickness variations or other syn-sedimentary features indicative of active faults in the Manx Group is circumstantial.

An alternative model is developed here building on geophysical interpretations by Kimbell & Quirk

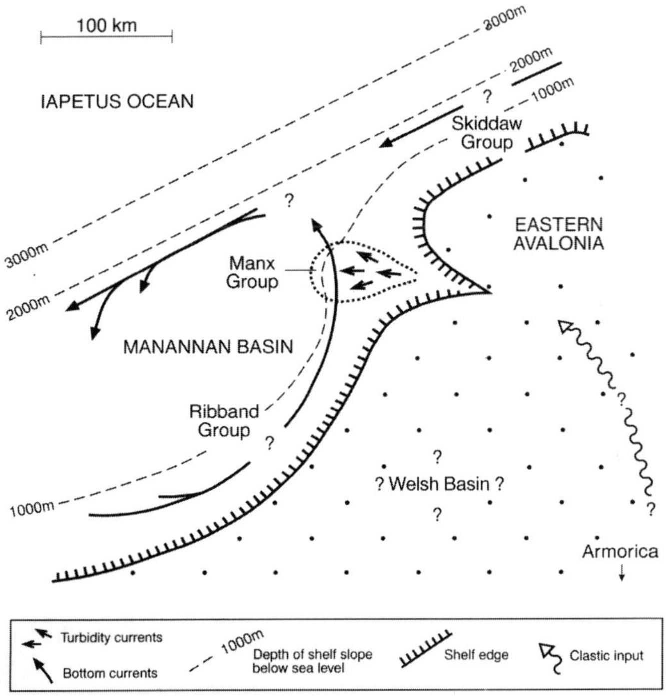

Fig. 8. Tentative palaeogeographic reconstruction of the northwest margin of Eastern Avalonia during the early Arenig showing conceptual current patterns. The relative positions of areas indicated, particularly the Welsh Basin, are speculative and differing amounts of anticlockwise rotation during the late Caledonian have not been taken into account (see Piper & Crowley 1999; Piper *et al.* 1999). See text for discussion.

(1999) and Quirk *et al.* (1999*b*). This work suggests that the Manx Group was deposited on the west dipping margin of an embayment on the northwest edge of Eastern Avalonia, termed the Manannan Basin. The thick sands of the lower part of the Santon Formation and the lower unit of the Ny Garvain Formation, its lateral equivalent, are therefore thought to represent part of an early Arenig lowstand fan, with the basal, high energy parts of turbidity currents flowing westwards down the slope of the margin (Fig. 8). However, ripples were produced by low-energy tractional currents flowing to the north, apparently perpendicular to the dip of the slope, i.e. by contour currents. These contour currents either reworked sands initially deposited as turbidites or deflected the flow of turbidity currents as they slowed down. Reworking may help to explain why the lower unit of the Ny Garvain Formation consists of a thick package of unusually clean sandstones (lithofacies S_H; Fig. 4) and also why flute casts and grading are rarely observed in lithofacies S_L and S_I in the Lonan Formation and the upper unit of the Ny Garvain Formation (Table 1). However, the northwards directed ripples are usually preserved in sandstone beds which are interbedded with thin mudstones, indicating that currents strong enough to move sand were only intermittently active. The favoured explanation is one of episodic flow due to deflection along-slope of the low-energy part of turbidity currents by north moving bottom waters (Fig. 8). What is not recorded in palaeocurrent data is the swing from downslope, westwards flow to along-slope, northwards flow as energy decreased, presumably because this was when massive and planar laminated sands were deposited (Bouma units Tab, e.g. in lithofacies S_H, Q_H and Q_V; Table 1).

A speculative lowstand flow pattern for contour currents along the northwest margin of Eastern Avalonia (Fig. 8) is one of the anticlockwise circulation system constrained only by the ripple cross-lamination palaeocurrent data from the Manx Group (Woodcock & Barnes 1999). As Eastern Avalonia lay in the southern hemisphere at this time (e.g. Noblet & Lefort 1990), this direction of circulation may have been driven directly by the Coriolis force. However, the circulation probably formed part of a larger, more complex flow pattern as is typical in present-day oceans (Stow *et al.* 1996).

A modern analogue for the Manannan Basin, albeit in the northern hemisphere, might be the Faeroe–Shetland channel where bottom waters between 500 and 1700 m below sea level flow southwest from the Norwegian Basin towards the Atlantic Ocean at rates of up to 0.5 m s^{-1} (Stoker *et al.* 1998; Masson *et al.* 1997). The Faeroe–Shetland Channel, unlike the Manannan Basin, is starved of coarse-grained turbiditic input. Nevertheless, a thin sandy contourite sheet is developed here at a water depth of 700–850 m over an area of 60×10 km^2, elongate-parallel to the shelf edge.

There are lithological similarities between the Ribband Group and mud-prone (distal) lithofacies in the Manx Group (McConnell *et al.* 1999; Morris, pers. comm.; Brück, pers. comm.), although this is the subject of ongoing research. In contrast to the Manx and Ribband Groups, a good biostratigraphic framework is available for the Skiddaw Group, although lithostratigraphic comparisons are not straightforward. Based on acritarch data (Molyneux 1999), the quartzose Santon and Ny Garvain Formations in the Manx Group seem to correlate with the mud-dominated Hope Beck Formation in the Skiddaw Group. The sandstones in the Watch Hill Formation and Loweswater Formation below and above the Hope Beck Formation are wackes rather than quartz arenites (Cooper *et al.* 1995). Even the slumps and debris flows in the upper Arenig Kirk Stile Formation are different to the pebbly mudstones of the Slieau Managh Unit and Lady Port Formation in that they represent larger scale events (olistostromes) and contain extra-formational clasts (Webb & Cooper 1988; Cooper, pers. comm.). Cooper *et al.* (1995) suggest a passive margin setting for the Skiddaw Group. Similar to most of southeast Ireland (Max *et al.* 1990), subduction-related accretionary prisms and island arcs are not evident until the Llanvirn in the Lake District, probably when Eastern Avalonia broke from Gondwana.

Kimbell & Quirk (1999) propose that the Manannan Basin was initiated by rifting, probably during the late Tremadoc. Felsitic igneous sheets occur throughout the Manx Group (Lamplugh 1903). Many of these appear to have been intruded when the sediment was still soft, perhaps indicating that a limited amount of extension was still occurring during deposition (Quirk & Kimbell 1997). However, evidence for syn-sedimentary normal faults is highly tentative (see Fig. 8 and previous discussion of Model 2, Fig. 7b). At present, the authors regard the eastern margin of the Manannan Basin as essentially passive and undergoing post-rift thermal subsidence during most of the Arenig. This is contrary to the ideas of Quirk & Kimbell (1997).

Cooper *et al.* (1995) propose that some of the more quartzose arenites in the Skiddaw Group were derived from a Gondwanan shelf where widespread shallow water and continental sandstones, known collectively as the Grès Armoricain, were deposited (Noblet & Lefort 1990). A similar link has been suggested by Woodcock & Barnes (1999) for the Manx Group, implying that a connection existed between the northwest margin of Eastern Avalonia and the coastal plain deposits of Arenig age on the Armorican Massif (Fig. 8).

The relationship of the Ribband, Manx and Skiddaw Groups with the Welsh Basin (Fig. 8) is uncertain (Kokelaar 1988). It is, however, noteworthy that Wales was the site of intracratonic rifting in the late Tremadoc, then uplift followed by transgression in the early–mid-Arenig (Kokelaar 1988; Woodcock 1990), consistent with the structural and stratigraphic evidence at the edge of the craton in the deeper-water setting of the Isle of Man.

Conclusions

The Manx Group dips generally northwest with the oldest sediments (quartzose sandstones) in the southeast and the youngest sediments (mud dominated) in the central and northwest parts of the island. The gross structure is controlled by northeast–southwest reverse faults and east–west dextral faults. With only limited biostratigraphic control, several alternative lithostratigraphic correlations are possible, depending on the amount of assumed fault repetition. The Manx Group is >4500 m thick, with distal units in the west juxtaposed by reverse faults against more proximal units in the east. A large normal fault may define the east-southeast edge of a sand-prone interval (the Glen Dhoo Unit) in the central-north area. Whether this fault was active during deposition is as yet unproven.

The lithofacies observed in the Manx Group are typical of a passive continental margin. A tentative model proposes that the succession represents the inverted, eastern side of a basin stretching at least as far as Leinster, named the Manannan Basin, which formed an embayment on the northwest edge of Eastern Avalonia. In the early Arenig, a large-scale turbidite fan developed in the Isle of Man area during a lowstand in relative sea level, when the Santon and Ny Garvain Formations were deposited, and probably also the correlative Glen Dhoo Unit. Quartzose sand bypassed the shelf and was carried by turbidity currents downslope to the west. The fan may also have been affected by north flowing bottom currents. The onset of a rise in relative sea level, probably towards the end of the early Arenig, was associated with deposition of the fining-upwards Creg Agneash Formation as the fan became inactive. A blanket of fine-grained mud was deposited during the mid-Arenig (the Glen Rushen Formation and its probable lateral equivalent, the Barrule Formation) overlain by thin-bedded turbidites and debris flows corresponding to a relative sea-level highstand (the Creggan Mooar and Injebreck Formations).

Further biostratigraphic work is now required to constrain the lithostratigraphic interpretations, particularly to test whether the Glen Rushen–Creggan Mooar Formations are distal equivalents of the Barrule–Injebreck Formations, and to assess whether the Port Erin Formation is the correlative of the Lonan or Creg Agneash Formations.

The authors are grateful for essential technical support given by Ian Pope, Kate Winder, Jon Wells, Simon Deadman, Sean Mulligan, Roger Sims, David Kelly, Richard Young, Kathleen Quirk, Lisa Hill and Graeme Foster. Many of the ideas in this paper have come from lively discussions with Nigel Woodcock, Rob Barnes, John Morris, Bill Fitches, Greg Power, Padhraig Kennan, Geoff Kimbell, Tony Cooper, Peter Brück, Brian McConnell, Paddy Orr, Mike Howe, Trevor Ford, Fred Radcliffe and Frank Cowin. The reviews of earlier manuscripts by Dick Waters, Maxine Akhurst and Nigel Woodcock proved helpful in improving the paper, although they would not endorse some parts of the final version. The work was funded by NERC research grant GR9/01834, the Isle of Man Government and Oxford Brookes University. We thank Burlington Resources, UK for sponsoring the costs of colour reproduction.

References

BOOTE, D. R. D. & GUSTAV, S. H. 1987. Evolving depositional systems within an active rift: Witch Ground Graben, North Sea. *In*: BROOKS, J. & GLENNIE, K. W. (eds) *Petroleum Geology of NW Europe*. Graham Trotman, 819–833.

COOPER, A. H., RUSHTON, A. W. A., MOLYNEUX, S. G., HUGHES, R. A., MOORE, R. M. & WEBB, B. C. 1995. The stratigraphy, correlation, provenance and palaeogeography of the Skiddaw Group (Ordovician) in the English Lake District. *Geological Magazine*, **132**, 185–211.

FITCHES, W. R., BARNES, R. P & MORRIS, J. H. 1999. Geological structure and tectonic evolution of the Lower Palaeozoic rocks of the Isle of Man. *This volume*.

FORD, T .D., WILSON, E. & BURNETT, D. J. 1999. Previous ideas and models of the stratigraphy, structure and mineral deposits of the Manx Group, Isle of Man. *This volume*.

GALLOWAY, W. E. 1998. Siliciclastic slope and base-of-slope depositional systems: component facies, stratigraphic architecture and classification. *AAPG Bulletin*, **82**, 569–595.

KENNAN, P. S. & MORRIS, J. H. 1999. Manganiferous ironstones in the early Ordovician Manx Group, Isle of Man: a protolith of coticule? *This volume*.

KIMBELL, G. S. & QUIRK, D. G. 1999. Crustal magnetic signature of the Irish Sea region: evidence for a major basement boundary beneath the Isle of Man. *This volume*.

KOKELAAR, P. 1988. Tectonic controls of Ordovician arc and marginal basin volcanism in Wales. *Journal of the Geological Society, London*, **145**, 759–775.

LAMPLUGH, G. W. 1903. *The Geology of the Isle of Man*. Memoir of the Geological Survey, United Kindom. HMSO.

MCCONNELL, B. J., MORRIS, J. H. & KENNAN, P. S. 1999. A comparison of the Ribband Group (southeastern Ireland) to the Manx Group (Isle of Man) and Skiddaw Group (northwestern England). *This volume*.

MASSON, D. G., BETT, B. J. & BIRCH, K. G. 1997. Atlantic margin environmental survey. *Sea Technology*, **10/97**, 52–59.

MAX, M. D., BARBER, A. J. & MARTINEZ, J. 1990. Terrane assemblage of the Leinster Massif, SE Ireland, during the Lower Palaeozoic. *Journal of the Geological Society, London*, **147**, 1035–1050.

MOLYNEUX, S. 1979. New evidence for the age of the Manx Group, Isle of Man. *In*: HARRIS, A. L., HOLLAND, C. H. & LEAKE, B. E. (eds) *The Caledonides of the British Isles – Reviewed*. Geological Society, London, Special Publications, 415–421.

—— 1999. A reassessment of Manx Group acritarchs, Isle of Man. *This volume*.

MORRIS, J. H., WOODCOCK, N. H. & HOWE, M. P. A. 1999. The Silurian succession of the Isle of Man: the late Wenlock Niarbyl Formation, Dalby Group. *This volume*.

NOBLET, C. & LEFORT, J. P. 1990. Sedimentological evidence for a limited separation between Armorica and Gondwana during the Early Ordovician. *Geology*, **18**, 303–306.

ORR, P. J. & HOWE, M. P. A. 1999. Macrofauna and ichnofauna of the Manx Group (early Ordovician), Isle of Man. *This volume*.

PICKERING, K. T., HISCOTT, R. N. & HEIN, F. J. 1989. *Deep Marine Environments. Clastic Sedimentation and Tectonics*. Unwin Hyman.

PIPER, J. D. A. & CROWLEY, S. F. 1999. Palaeomagnetism of (Palaeozoic) Peel Sandstone and the Langness Conglomerate Formation, Isle of Man: implications for the age and regional diagenesis of Manx red beds. *This volume*.

——, BIGGIN, A. J. & CROWLEY, S. F. 1999. Magnetic survey of the Poortown dolerite, Isle of Man. *This volume*.

PROSSER, S. D. 1993. Rift related depositional sequences and their seismic expression. *In*: WILLIAMS, G. D. & DOBB, A. (eds) *Tectonics and Seismic Stratigraphy*. Geological Society, London, Special Publications, **56**, 35–56.

QUIRK, D. G. & KIMBELL, G. S. 1997. Structural evolution of the Isle of Man and central part of the Irish Sea.

In: MEADOWS, N. S., TRUEBLOOD, S. P., HARDMAN, M. & COWAN, G. (eds) *Petroleum Geology of the Irish Sea and Adjacent Areas*. Geological Society, London, Special Publications, **124**, 135–159.

——, ROY, S., KNOTT, I. & REDFERN, J. 1999*a*. Petroleum geology and future hydrocarbon potential of the Irish Sea. *Journal of Petroleum Geology*, in press.

——, BURNETT, D. J., KIMBELL, G. S., MURPHY, C. A. & VARLEY, J. S. 1999*b*. Shallow geophysical and geological evidence for a regional-scale fault duplex in the Lower Palaeozoic of the Isle of Man. *This volume*.

ROBERTS, B., MORRISON, C. & HIRONS, S. 1990. Low grade metamorphism of the Manx Group, Isle of Man: a comparative study of white mica 'crystallinity' techniques. *Journal of the Geological Society, London*, **147**, 271–277.

SIMPSON, A. 1963. The stratigraphy and tectonics of the Manx Slate Series, Isle of Man. *Quarterly Journal of the Geological Society, London*, **119**, 367–400.

STOKER, M. S., AKHURST, M. C., HOWE, J. A. & STOW, D. A. V. 1998. Sediment drifts and contourites on the continental margin off northwest Britain. *Sedimentary Geology*, **115**, 33–51.

STOW, D. A. V., READING, H. G. & COLLINSON, J. D. 1996. Deep seas. *In*: READING, H. G. (ed.) *Sedimentary Environments: Processes, Facies and Stratigraphy*. Blackwell, 395-453.

VAIL, P. R., MITCHUM, R. M. JR, & THOMPSON, S. III 1977. Seismic stratigraphy and global changes of sea level, Part 4: global cycles of relative changes of sea level. *In*: PAYTON, C. E. (ed.) *Seismic Stratigraphy – Applications to Hydrocarbon Exploration*. American AAPG Memoir, **26**, 83–97.

WEBB, B. C. & COOPER, A. H. 1988. Slump folds and gravity slide structures in a Lower Palaeozoic marginal basin sequence (the Skiddaw Group), NW England. *Journal of Structural Geology*, **10**, 463–472.

WOODCOCK, N. H. 1990. Sequence stratigraphy of the Palaeozoic Welsh Basin. *Journal of the Geological Society, London*, **147**, 537–547.

—— & BARNES, R. P. 1999. An early Ordovician turbidite system on the Gondwana margin: the southeastern Manx Group, Isle of Man. *This volume*.

—— & MORRIS, J. H. 1999. Debris flows on the Ordovician margin of Avalonia: the Lady Port Formation, Manx Group, Isle of Man. *This volume*.

——, ——, QUIRK, D. G. *ET AL.* 1999. Revised lithostratigraphy of the Manx Group, Isle of Man. *This volume*.

An early Ordovician turbidite system on the Gondwana margin: the southeastern Manx Group, Isle of Man

N. H. WOODCOCK[1] & R. P. BARNES[2]

[1]*Department of Earth Sciences, University of Cambridge, Cambridge CB2 3EQ, UK*

[2]*British Geological Survey, Murchison House, West Mains Road, Edinburgh EH9 3LA, UK*

Abstract: The southeastern part of the Manx Group comprises sandstone-rich deep-marine turbidites of Arenig (early Ordovician) age, deposited on the Avalonian segment of the Gondwana margin. The main area comprises a 2 km thick succession from thin-bedded wacke sandstone and mudstone of the Lonan Formation up into medium-bedded wacke and arenite of the Santon Formation. This succession is interpreted as the distal part of a mixed mud–sand ramp overlain by a major distributary channel and lobe system. The Keristal Member is a unit of thick-bedded quartz arenite within the Lonan Formation recording a short-lived incised channel system with small terminal sand lobes.

Flutes and scours in the Lonan and Santon Formations show west-southwest directed transport, contrasting with north-northwest directed estimates from ripple cross-lamination. Constraint by margin topography may have caused the higher concentration components of the flows to run along-margin, whilst flow-stripping of the lower concentration flow-tops allowed them to collapse down the regional north-northwest facing palaeoslope.

The southwestern and northeastern areas contain fault-bounded successions that can be correlated only tentatively with the main area. In the southwest, thin-bedded wackes and mudstones of the Port Erin Formation pass up into the quartzose Mull Hill Formation, which shows the thickening-up motif of a fan lobe. In the northeast, the Creg Agneash Formation resembles the Mull Hill Formation, and probably overlies the quartzose wackes of the Ny Garvain Formation.

The contrast of arenite and wacke sandstones in the Manx Group may have resulted partly from the intrabasinal separation of sand and mud, e.g. by flow stripping. However, the major factor was the availability on the basin margin of both clean and muddy sediment. The quartz arenite sands were probably sourced from the widespread Armorican quartzite facies of Gondwana, constraining the rifting of Avalonia from its parent continent to mid-Arenig time or later.

The Manx Group crops out over most of the Isle of Man, overlain in the south by Carboniferous rocks (Fig. 1), in the west by the Silurian Dalby Group and the poorly dated Peel Sandstones, and in the north by a thick Quaternary succession. Limited biostratigraphical control constrains the Manx Group to the Arenig (Cooper *et al.* 1995; Orr & Howe 1999; Molyneux 1999). The group correlates broadly with the Skiddaw Group of the Lake District and with the Ribband Group of Leinster (Cooper *et al.* 1995; McConnell *et al.* 1999). These units are taken to form part of a deep-marine, predominantly turbiditic, sediment prism deposited along the margin of the Avalonian terrane, probably still attached to the Gondwana continent until late Arenig time (e.g. Pickering & Smith 1995; Prigmore *et al.* 1997).

After the century old comprehensive mapping of the Manx Group (Geological Survey 1898, Lamplugh 1903), subsequent studies have concentrated on its structure, metamorphism and igneous rocks, and its sedimentology remains poorly described. This paper focuses on a major group of sandstone-dominated units, the Port Erin, Mull Hill, Lonan, Santon, Ny Garvain and Creg Agneash Formations, cropping out along the southeastern flank of the island (Fig. 1). New data comprise mainly field-scale facies and palaeoflow observations, which allow a number of regionally important questions to be addressed. Do the six formations comprise a coherent depositional package dominated by turbidites? In particular, what is the depositional relationship of the quartzose Mull Hill and Creg Agneash Formations to the wacke-prone strata of the other formations? Do compositional and palaeoflow data support the deposition of these

From: Woodcock, N. H., Quirk, D. G., Fitches, W. R. & Barnes, R. P. (eds) 1999.
In Sight of the Suture: the Palaeozoic geology of the Isle of Man in its Iapetus Ocean context.
Geological Society, London, Special Publications, **160**, 89–107. 1-86239-046-0/99/$15.00

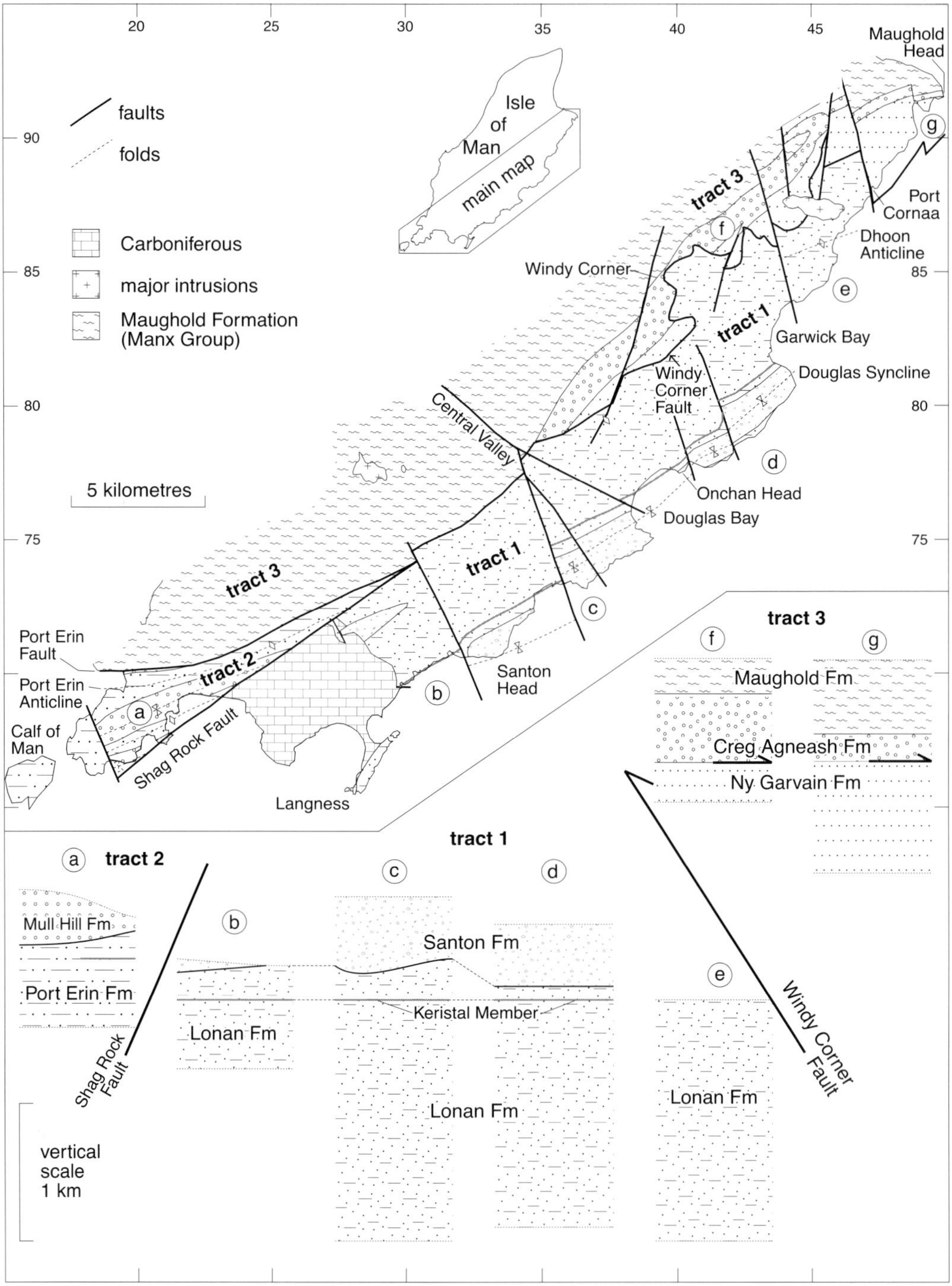

Fig. 1. Geological map showing the outcrop of the southeastern sandstone-rich formations of the Manx Group. The key is in the form of stratigraphic columns showing the relationship and approximate thickness of units along seven transects (ⓐ– ⓖ) across the area.

units on the Gondwana margin? How might the Manx Group units correlate with those of the Skiddaw and Ribband Groups?

Structural setting

Most of the Manx Group is affected by two strong phases of Caledonian deformation, the D1 and D2 events of Simpson (1963). However, the D2 deformation is weaker in the sandstone-rich formations in the southeast than it is in the mudstone units in the northwest of the island (Fitches *et al.* 1999). The southeastern sandstone units are, therefore, dominated by northeast trending F1 folds with upright or steeply southeast dipping axial surfaces. Three such major F1 folds, the Dhoon Anticline, Douglas Syncline and Port Erin Anticline, are particularly important controls on the outcrop pattern. The F1 folds are associated with an S1 cleavage, which is mainly axial-planar, but which weakly clockwise-transects the Douglas Syncline. A zone of strong cleavage transection occurs only on Langness (Fig. 1).

Northeast striking faults also play a major role in the structure of the southeastern Manx Group. Three such faults are important enough to subdivide the area into tectonostratigraphic tracts, the correlation of which is problematic. (Fig. 1; Woodcock *et al.* 1999). The steep Shag Rock Fault, the Carboniferous boundary fault of Simpson (1963), separates tracts 1 and 2. The Port Erin Fault (Quirk & Burnett 1999) separates tracts 2 and 3, and probably joins the Shag Rock Fault further northeast. The Windy Corner Fault then separates tracts 1 and 3 in the northeast of the area, and is the most important of several interpreted northwest dipping thrusts beneath and within the Ny Garvain and Creg Agneash Formations (Fitches *et al.* 1999).

Steep north-northwest striking faults complicate the structure throughout the southeast of the island. With lateral offsets, typically sinistral, hundreds of metres to several kilometres, these faults reduce the certainty of along-strike correlations. An aeromagnetic lineament following the central valley of the island has suggested to Quirk & Kimbell (1997) that a further west-northwest striking fault should underlie Douglas Bay (Fig. 1).

Despite these complications, the generally simple outcrop-scale structure and modest strain of the sandstone successions allows reliable sedimentological logging, estimation of stratigraphic thicknesses and restoration of palaeocurrent data over most of the well-exposed coastal sections.

Stratigraphy and general lithological character

The first systematic accounts of the geology of the Isle of Man (Berger 1814; Henslow 1821) recognized that its central northeast trending spine is predominantly composed of mudstone units, flanked to the northwest and southeast by sandstone-dominated units. The present paper deals with the southeastern sandstones, the lithostratigraphy of which has been revised by Woodcock *et al.* (1999) to comprise six formations; the Port Erin, Mull Hill, Lonan, Santon, Ny Garvain and Creg Agneash Formations (Fig. 1). In the previous stratigraphic schemes of Simpson (1963) and Ford (1993) these units were included either within a more widespread Lonan Flags or within the lower part of the Maughold Banded Group.

In tract 1, the central coastal area, between Langness and Port Cornaa (Fig. 1b–e), the very thin- to thin-bedded Lonan Formation (averaging *c.* 50% sandstone and 4 cm bed thickness; Fig. 2a) is overlain by the more sand-rich, thin- to medium-bedded Santon Formation (75% sandstone, 10 cm bed thickness). The Santon Formation crops out in the core of the Douglas Syncline between Santon Head and Garwick Bay. Throughout the central area, a discrete packet of thick-bedded quartzose sandstones, the Keristal Member (95% sandstone, 50 cm bed thickness; Fig. 2a), occurs within the Lonan Formation, *c.* 100–200 m below its contact with the Santon Formation (Fig. 1b–d). The upper contact to the sandstone-rich formations is not seen in tract 1, being faulted out against the Maughold Formation of tract 3.

A coarse-topped succession also occurs in tract 2 (Fig. 1a). Here, the very thin-bedded Port Erin Formation (35% sandstone, 2.5 cm average bed thickness; Fig. 2b) is overlain by the thin- to medium-bedded Mull Hill Formation (80% sandstone, 12 cm bed thickness). The top of the Mull Hill Unit is not exposed. Two discrete packets of quartzose sandstone (90% sandstone, 25 cm bed thickness) occur *c.* 150 m below the base of the Mull Hill Formation (Fig. 1a).

The lowest unit exposed in tract 3, the Ny Garvain Formation (Fig. 1f, g), is, on average, more sandstone rich and thicker bedded (80% sandstone, 11 cm bed thickness; Fig. 2b) than the Creg Agneash Formation (60% sandstone, 6 cm bed thickness), although the Ny Garvain Formation is markedly thinner bedded at the top than at the bottom. The difference between the formations is mainly due to the higher proportion of a very thin-bedded background facies in the Creg Agneash Formation. Detailed facies analysis will amplify this distinction. The Ny Garvain Formation pro-

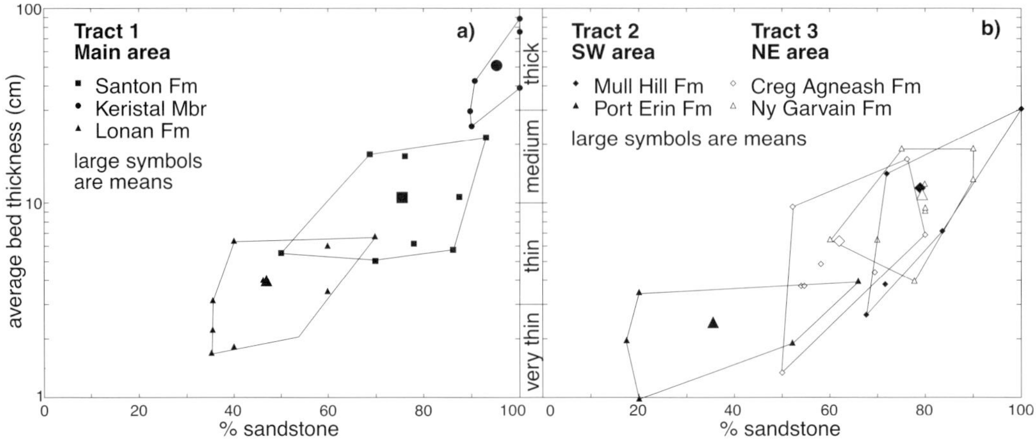

Fig. 2. Graph of percentage of sandstones v. average bed thickness for logged sections in each stratigraphic unit: (**a**) in the central tract; (**b**) in the northeast and southwest tracts. The analysis is based on 45 logged sections comprising 955 measured units and *c.* 11 000 constituent beds.

bably passes up into the Creg Agneash Formation, but this contact is within a zone of faulting where seen on the coast south of Maughold Head (Fig. 1f, g). North of Maughold Head, the sandstone-rich Creg Agneash Formation passes rapidly up into the mudstone-dominated Maughold Formation, containing more localized packets of quartz arenite.

Biostratigraphic data from this part of the Manx Group are sparse. Graptolites from the Santon Formation suggest an early Arenig (Moridunian) age (Rushton 1993). Acritarchs from the same unit, originally thought to be late Arenig (Molyneux 1979), are now interpreted to be of early Arenig age (Cooper *et al.* 1995; Molyneux 1999). These determinations place at least the Santon and Lonan Formations low down in the Manx Group (Cooper *et al.* 1995; Woodcock *et al.* 1999).

Facies classification

The facies of the southeastern Manx Group have been classified in the field using the scheme of Pickering *et al.* (1989) for deep-water clastic sediments. No evidence has been found that any of these facies were deposited above storm-wave base, and Orr & Howe (1999) report a trace fossil assemblage analogous to that from other deep-marine sequences on the Gondwana margin. The formations can be characterized (Fig. 3) using a first-order subdivision into: Class A, pebbly sandstone; Class B, sandstone; Class C, sandstone–mudstone couplets; Class D, siltstone–mudstone couplets. Class C has been further divided, into thick-bedded (C2.1), medium-bedded (C2.2) and thin- to very thin-bedded (C2.3) facies. Also distinguished is a locally significant facies, C2.0, comprising alternating intervals of parallel lamination and ripple cross-lamination, but with gradational contacts between intervals.

On this basis, the Lonan and Port Erin Formations comprise Facies C2.3 and D (Fig. 3). These units are therefore distinct from their

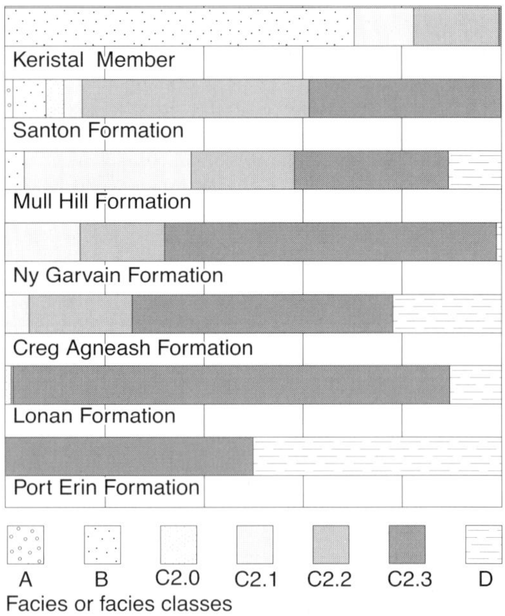

Fig. 3. Approximate proportions of lithofacies (following the scheme of Pickering *et al.* 1989) in each studied unit. The database is the same as for Fig. 2.

overlying Santon and Mull Hill Formations, respectively, which are dominated by Facies C2.2 and C2.1. The Keristal Member is conspicuous by its predominance of Facies B. The Ny Garvain Formation appears to thin upwards from Facies C2.1 and C2.2 south of the Gob ny Garvain Formation to predominanlty C2.3 near its upward transtion to the Creg Agneash Formation. The Creg Agneash Formations itself is characterized by its generally higher proportion of Facies D mudstones.

A range of bed organization has been observed within each facies class (Fig. 4). Facies D siltstone–mudstone couplets have been subdivided on the presence or absence of a cross-laminated silt interval at the base of each unit. Deposition of the couplets from low-concentration turbidity flows is preferred over a contourite origin in this study (cf. Quirk & Burnett 1999), hence the use of the Bouma notation T_{cde}. Facies C sandstone–mudstone couplets typically display sharp bed bases and the internal components of the Bouma sequence, labelled T_{a-e}. They are interpreted as the deposits of episodic turbidity currents. The presence of erosional scours and flutes on the bases of beds beginning with parallel-laminated T_b divisions suggests that these divisions were indeed deposited from turbulent flows rather than from laminar sandy flows (cf. Shanmugam 1997).

Within the Class B sandstone, some Facies B2.1 beds contain either cross-laminated T_c divisions or fluted bases to T_b divisions, and are best interpreted as deposits of turbulent flows. However, the use of the Bouma T_{ab} notation in this facies class is not meant to preclude the possibility of deposition of some of the massive or laminated sand beds from laminar sandy debris flows (Shanmugam 1997).

The facies characteristics and interrelationships are now described and interpreted for the three tracts of the southeastern Manx Group. In the following descriptions, the lower bounds of bed thickness classes are 1 (very thin), 3 (thin), 10 (medium), 30 (thick) and 100 cm (very thick). The boundary between wackes and arenites is taken at 15% matrix. All azimuths are described with respect to present-day coordinates. However, it is probable that parts of the Avalonian margin, including the Isle of Man and the Welsh Basin, were rotated anticlockwise during the Acadian Orogeny with respect to the Lake District (Piper 1997; Piper & Crowley 1999).

Facies architecture of the Lonan and Santon Formations

Lonan Formation

The Lonan Formation is dominated, apart from its included Keristal Member, by thin-bedded or very thin-bedded sandstone–mudstone couplets (Facies C2.3; Figs. 3 and 4). Each bed typically has a sharp base, sometimes with loaded scours or flutes, and grades from light grey, ripple cross-laminated, fine quartz wacke through parallel laminated silt–mudstone up to dark grey mudstone (Fig. 4). Convolute lamination is locally developed in the thicker cross-laminated divisions. Bedding surfaces commonly show straight-crested to undulatory, asymmetric current ripples. The mudstone is burrow mottled or structureless and a distinct hemipelagic interval is not apparent. Very thin-bedded couplets of siltstone–mudstone or very fine sandstone–mudstone (Facies D2.3) dominate the basal part of the Lonan succession in the core of the Dhoon Anticline (Fig. 1). Such couplets also form a significant proportion of the Lonan Formation in some other parts of the succession, e.g. in Douglas Bay.

Medium-bedded sandstone–mudstone couplets, up to 20 cm thick (Facies C2.2), are locally intercalated in the typical thin-bedded facies of the Lonan Formation. The distinctive facies C2.0 occurs between Port Soderick [SC 347 727] and Keristal Bay [SC 351 730]. These 5–20 cm alternations of parallel-laminated and ripple cross-laminated intervals have gradational rather than sharp bases, so that individual events are difficult to distinguish (Fig. 4). The beds occur in packets 0.4–4 m thick.

The typical T_{cde} and T_{de} beds of the Lonan Formation are interpreted as the product of deposition from low-concentration turbidity currents. The burrow mottling, together with the apparent biogenic homogenization of turbidite mud with any hemipelagic facies, indicates an oxygenated environment. Trace fossils comparable with those from the Skiddaw Group of the Lake District (Orr 1996; Orr & Howe 1999) suggest a deep-marine setting. The sharp bases to most turbidites suggest that the depositing flow events were discretely spaced in time. However, Facies C2.0 is interpreted as recording flows or flow pulses so closely spaced in time that each renewed pulse of sand-rich input occurred before the muddy upper part of the preceding flow had been deposited. An identical facies in the Annot Sandstones of southeast France has been similarly attributed to pulsing turbidity flow by Sinclair (1994).

Palaeoflow estimates from asymmetric ripples in the Lonan Formation show northwest directed flow swinging to west directed flow at the southwestern end of the outcrop (Fig. 5). On Langness, flutes also indicate westerly flow. The generally thinner bedded character of the Lonan Formation in the southwest suggests a proximal to distal relationship compatible with the westerly flow indicators. An alternative possibility, suggested by Quirk &

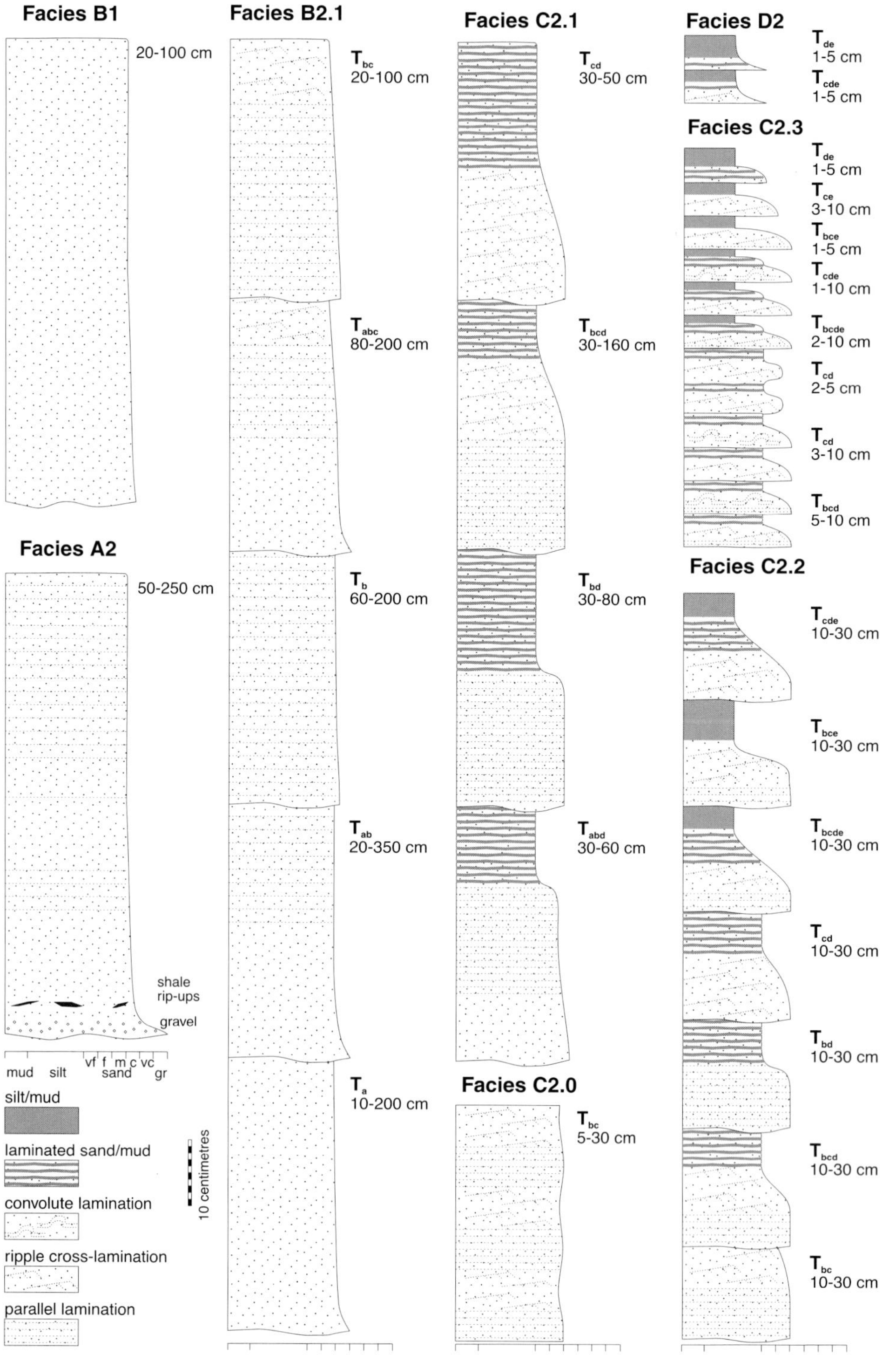

Fig. 4. Internal organization of typical beds in each common facies in the study area, with the range of bed thicknesses observed. The internal divisions are labelled following the conventions of the Bouma turbidite sequence, although doubt attaches as to whether all a and b divisions are the product of deposition from turbulent flows (see text for discussion).

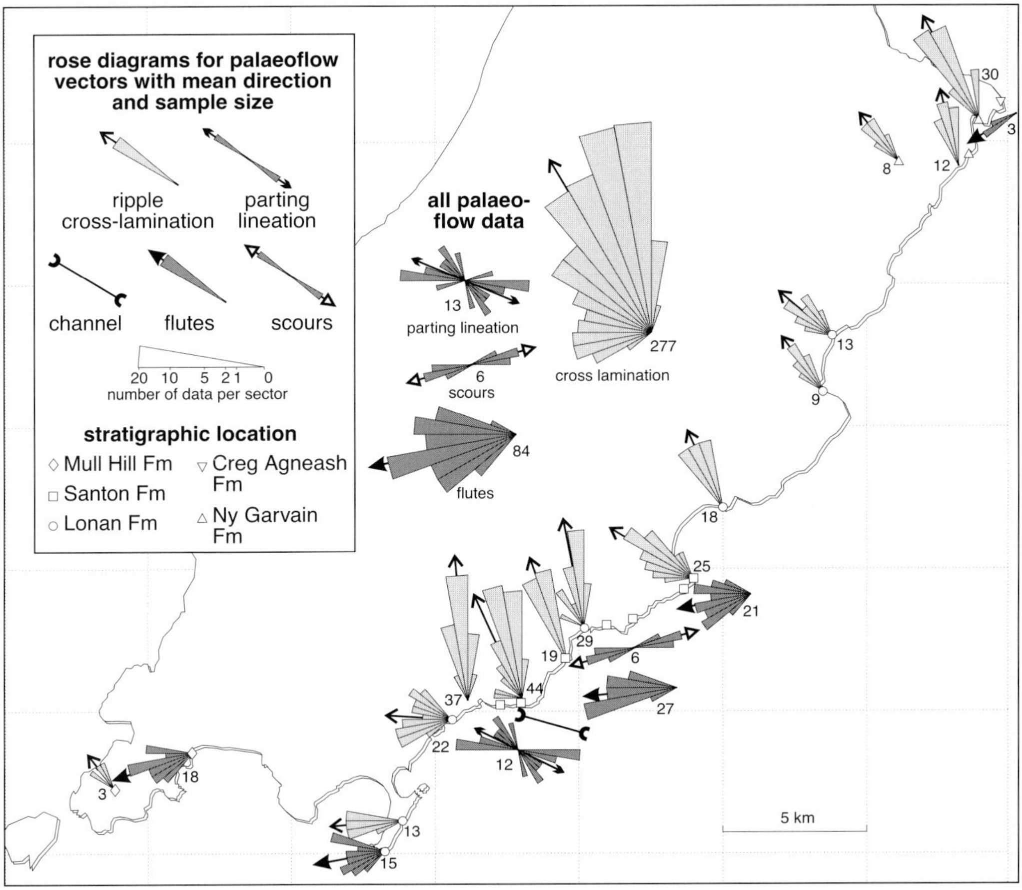

Fig. 5. Palaeoflow estimates from the southeastern Manx Group All data have been corrected for structural dip on the limbs of regionally subhorizontal folds.

Burnett (1999) but not favoured here, is that this southwestern succession is not the lateral equivalent of the Lonan Formation further northeast. In either case, indicators from the overlying Santon Formation show that ripple and flute palaeoflows are not necessarily parallel in this system, and urge caution in interpretation. Northeast of Douglas Bay, the main motif in the Lonan Formation is of overall thickening up from the very thin-bedded facies in the core of the Dhoon Anticline.

Using the classification scheme of Reading & Richards (1994), the most likely depositional setting for the Lonan Formation is on the lower part of a deep-marine fan or ramp, fed by mixed mud–sand flows from a basin margin to the east or, most likely, the southeast. The formation lacks internal bed thickness motifs that suggest strong organization into channels or lobes. The thickening-up motif in the Dhoon Anticline may be a progradational signature. However, this hypothesis, and that of a proximal to distal trend in the southwest, are difficult to test without better biostratigraphical control, and other lithostratigraphic correlations are possible (Quirk & Burnett 1999).

Keristal Member

Within the Lonan Formation, the Keristal Member comprises a discrete lenticular packet of between seven and 14 light grey to white, fine sandstone beds, totalling between 2 and 9 m in thickness. The member is typically medium to thick bedded, but includes some very thick beds: 3.5 m thick at Port Soldrick [SC 3060 6973]; 2.5 and 2.1 m at Cass ny Hawin [SC 3000 6930]; 1.8 and 1.1 m at Keristal [SC 3490 7280] (Fig. 6a–c). The sandstone is typically quartz arenite or quartz wacke with < 25% matrix and appreciably more quartzose than the

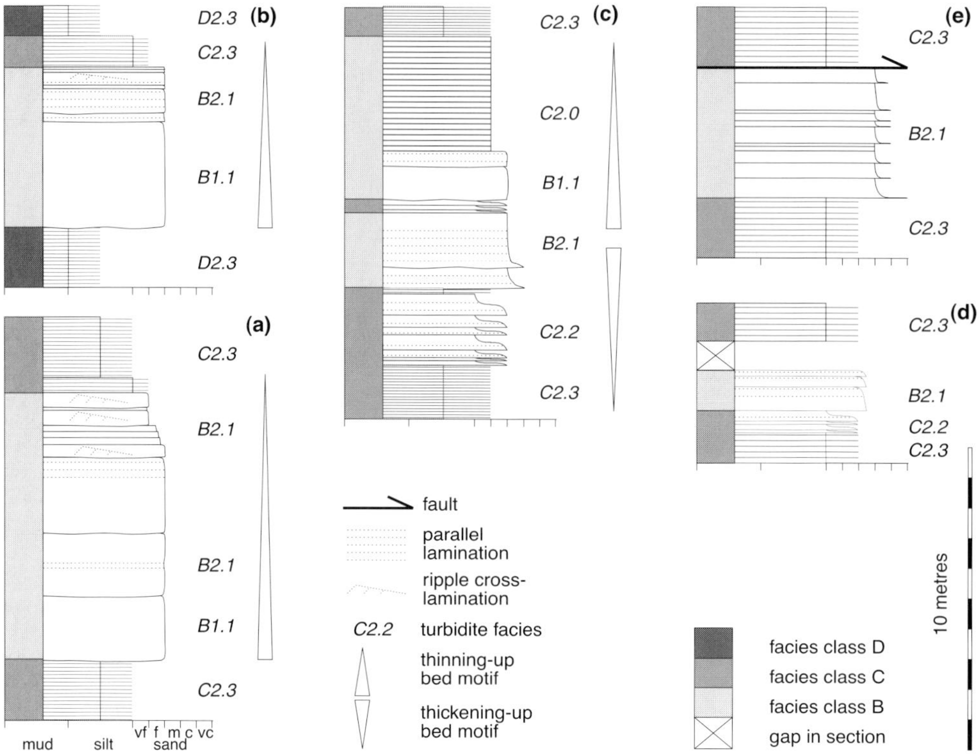

Fig. 6. Lithological logs from the Keristal Member and its possible correlatives. (**a**) Cass ny Hawin Head [SC 3000 6930]; (**b**) Port Soldrick [SC 3060 6973]; (**c**) Keristal Bay [SC 3490 7280]; (**d**) Port Jack [SC 4095 7728]; (**e**) Garwick Bay [SC 4362 8130].

wacke that typifies the bulk of the Lonan Formation. The sandstone occurs in non-graded or weakly graded beds, massive or, less commonly, parallel-laminated. Thinner beds tend to preserve ripple cross-lamination. The beds can mostly be assigned to sandstone Classes B1 or B2 of the Pickering *et al.* (1989) scheme, with subordinate Class C2 sandstone–mudstone couplets (Fig. 3). Coarsely scoured and fluted bases to the Class B sandstone beds at Cass ny Hawin suggest deposition from high-concentration turbulent flows rather than laminar debris flows. However, it is possible that these high-concentration flows could have been merely the basal parts of more muddy flows, the low-concentration tops of which have been stripped off and travelled further on into the basin (cf. Piper & Normark 1983).

The Keristal Member cannot be mapped continuously through the Lonan Formation and is demonstrably lenticular on lateral scales of the order of 100 m. The outcrop trace (Fig. 1) therefore represents the stratigraphic level of these lenticular bodies and not a continuous sheet of thick-bedded sandstone. The lenticularity is most clearly seen at Cass ny Hawin Head and Port Soldrick, where the basal units of the member are also strongly erosional into underlying thin-bedded facies. Here (Figs 6a, b and 7) a thinning-up bed motif suggests a distributary channel-fill. The basal scours imply local southwestward or northeastward flow in these channels (Figs 5 and 7). At Keristal Bay, Port Jack [SC 4095 7728] and Garwick Bay [SC 4362 8130] (Figs 6c–e and 7), and in a possible correlative at Perwick Bay [SC 2060 6715] (Fig. 9c), the quartzose packets show more symmetrical or irregular bed thickness motifs, more compatible with deposition in a sandy lobe.

The Keristal Member represents a short-lived, but profound, change in the architecture of the Lonan turbidite system. Channels were incised across the pre-existing outer fan or ramp, although perhaps not synchronously. The channels hosted sand-rich sediment gravity flows, which probably travelled southwestward. These flows either fed small terminal sand lobes, now included within the member, or they may have fed a larger quartzose

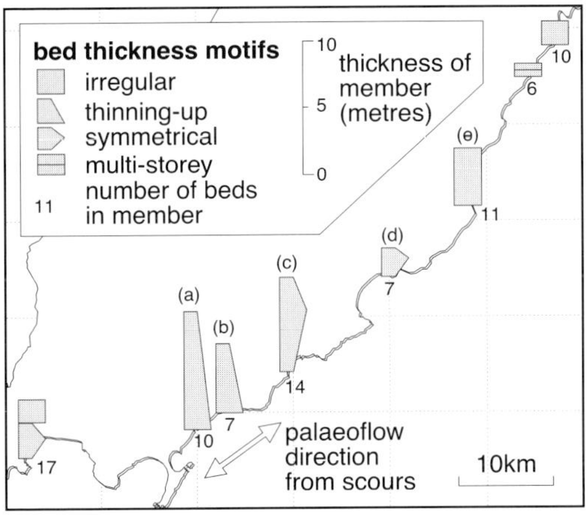

Fig. 7. Thickness and internal organization of the Keristal Member and its possible correlatives in the Port Erin and Ny Garvain Formations. Logs (a)–(e) are those detailed in Fig. 6.

turbidite lobe such as that represented by the Mull Hill Formation. The Keristal channels were progressively plugged by thick- to medium-bedded quartzose turbidites that were less muddy than were those of the enclosing Lonan Formation. The compositional difference could indicate the preferential stripping of muddy sand from the flows within the basin and/or a different source of sediment at this time.

Santon Formation

Sections in the Santon Formation are characterized by at least 25% of medium-bedded sandstone–mudstone couplets (Facies C2.2) interbedded with the thin-bedded facies (C2.3) typical of the underlying Lonan Formation (Fig. 3). The beds typically grade from light grey or light greenish grey, fine or very fine quartz wacke up to mud. Sandstone makes up 60–90% of each bed. Bed bases are sharp, and sporadically preserve flute and groove marks or horizontal burrows. Parallel and ripple cross lamination are common in the sandstone interval. The mudstone is sometimes visibly burrow mottled. These T_{cde} and T_{bcde} beds are interpreted as the product of deposition from turbidity currents in an oxygenated, deep-marine environment.

Although quartz wacke with *c.* 30–40% matrix dominates the Santon Formation, intervals of more quartzose wacke with some quartz arenite occur in places, e.g. at Port Skillion Lighthouse [SC 3900 7470] and on Little Ness [SC 3661 7293]. Another variant is seen below Wallberry Hill [SC 3700 7349] and at The Whing [SC 3603 7330], where isolated or weakly packeted medium to thick beds of quartz arenite punctuate a background of medium-bedded quartz wackes. The question arises, as in the Keristal Member, whether this cleaner sandstone records a different sediment source or forms the basal intervals of more muddy flows that have run on further down the system.

The base of the Santon Formation is typically gradational above thin-bedded strata of the Lonan Formation. One important section of the Santon Formation, on the coast at Purt Veg (Fig. 8), is unique both in having a sharp base and in being dominated by thick- or very thick-bedded sandstone. The basal 30 m of the member here comprises very thick-bedded, parallel-laminated medium-grained sandstone, with lenses of granule or coarse sand at bed bases, and is assigned to Facies A2.8 (Fig. 4; Pickering *et al.* 1989). These gravelly sands pass up, in part laterally, into 20 m of very thick- or thick-bedded medium- to fine-grained sandstone (Facies B2.1) with intercalated thin-bedded sand–mud turbidites (Facies C2.3). The upper part of this interval also contains sandstone units, of the type designated as Facies C2.0 in the lower Lonan Formation, comprising amalgamated thin- to medium-bedded couplets of parallel- and cross-lamination, but lacking any intervening mudstone. This facies, intercalated with further thick- or very thick-bedded sandstone, continues through at least a further 50 m of the succession (Fig. 8) before typical Santon Formation lithofacies (Facies C2.1, C2.2 and C2.3) predominate at Santon Head.

The Purt Veg deposits are a thinning- and fining-upward sequence, interpreted as filling a trunk distributary channel in the Santon turbidite system. The thick and very thick beds were deposited from high-concentration turbulent flows, or possibly laminar flows. The frequency of these high-concentration flows waned through the filling history of the channel, allowing thin-bedded deposits of low-concentration turbidity flows to be preserved in more quiescent intervals. Facies C2.0 probably records pulsing continuous flows, or discrete flows so closely spaced in time that each renewed event of sand-rich input occurred before the muddy upper part of the preceding flow had been deposited (cf. Sinclair 1994).

Flutes and grooves in the Santon Formation suggest that the primary flow direction through the local system was westward (Fig. 5). By contrast, the flow direction from ripple cross-lamination in the Santon Formation was between northwest and north. A strong obliquity between erosional and depositional flow indicators is common in turbidite systems (e.g. Kneller *et al.* 1991; Clayton 1993). One possible cause, primary deflection by the

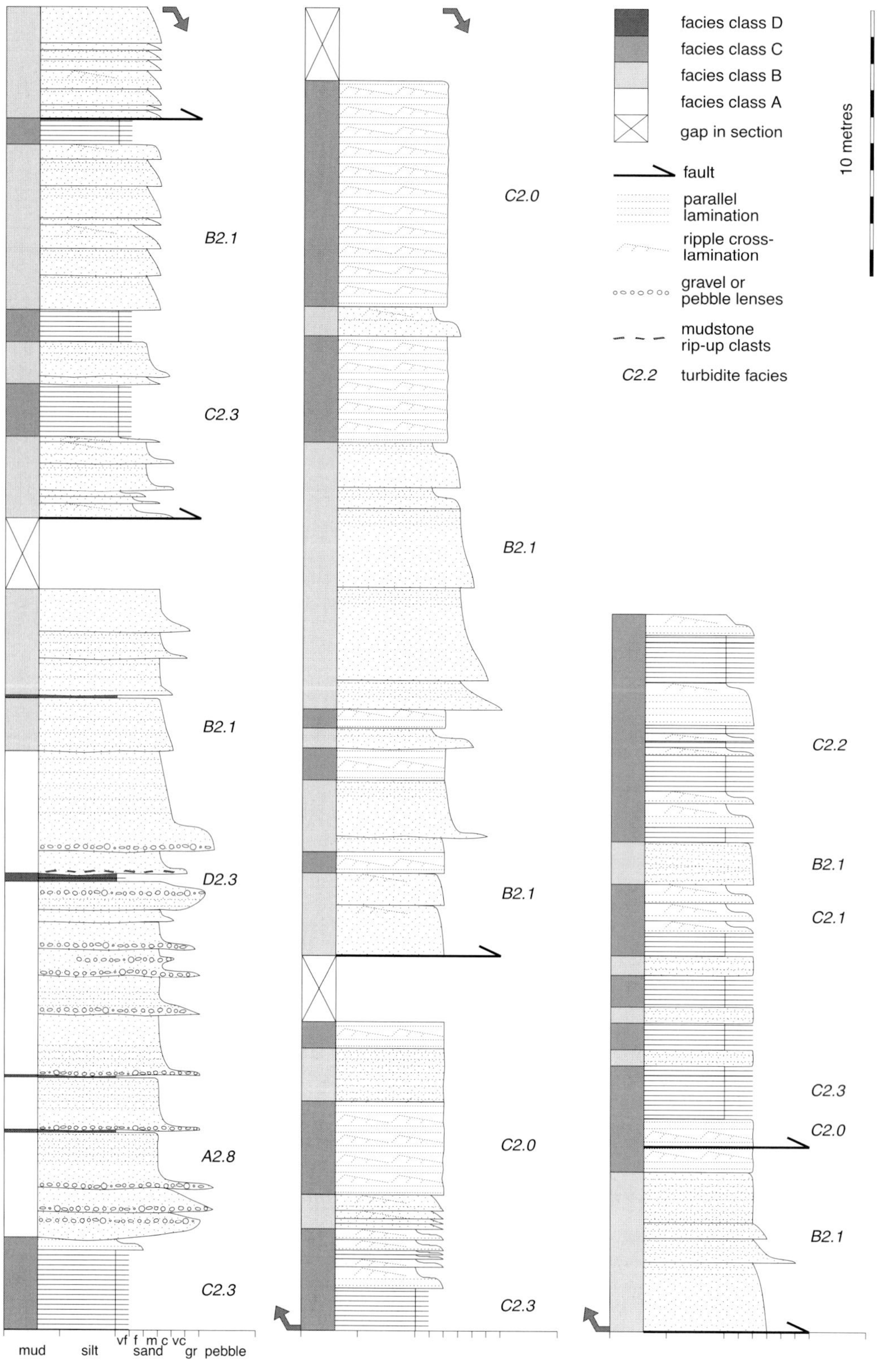
facies class D
facies class C
facies class B
facies class A
gap in section
fault
parallel lamination
ripple cross-lamination
gravel or pebble lenses
mudstone rip-up clasts
C2.2 turbidite facies
10 metres
B2.1
C2.3
B2.1
D2.3
A2.8
C2.3
C2.0
B2.1
B2.1
C2.0
C2.3
C2.2
B2.1
C2.1
C2.3
C2.0
B2.1
mud
silt
vf f m c vc
sand
gr
pebble

Coriolis force, can be ruled out in the Manx Group as being in the wrong sense for its southern-hemisphere setting in Early Ordovician time. A more local clockwise gyre remains a possible driving force. The hypothesis of reflection of internal waves off a bounding slope (Kneller *et al.* 1991) implies a discrete bimodality of palaeoflow, and is not supported in the Santon Formation by the orientation of parting lineation intermediate between that of the average flute and cross-lamination flows. Adjustment of more dilute parts of the flow to a regional palaeoslope is favoured here. The dilute flows probably overtopped the confines of the channel and were stripped from the more sandy high-concentration remnants of the flow constrained within the channel axis (cf. Piper & Normark 1983).

The thickness of the Santon Formation may reach as much as 600 m in its type area (Fig. 1) and *c.* 400 m to the northeast of Douglas Bay (e.g. Onchan Head; Fig. 8e). Mapping continuity between these two areas cannot be proved but the occurrence of the Keristal Member below the Santon Formation in both areas is a persuasive factor (Fig. 1).

In summary, the Santon Formation represents a period of more vigorous turbidity current activity than in the Lonan Formation. This model is of a major trunk distributary channel that fed flows westward across a more northwesterly dipping slope. The coarser grained sediment tended to be confined to the channel, but low-concentration flows escaped to form levees represented by thin- and very thin-bedded turbidites. On and beyond the levees, the flows adjusted to the regional palaeoslope. High-concentration quartzose flows dominated the early channel-fill. The later fill preserves more matrix-rich sands and, at times, the channel hosted pulsing long-lived flows rather than discrete episodic turbidity flows.

Facies architecture of the Port Erin and Mull Hill Formations

Port Erin Formation

The Port Erin Formation contains thin-bedded or very thin-bedded sandstone–mudstone couplets (Facies C2.3) interbedded with intervals of very thin-bedded siltstone–mudstone couplets (Facies D2.3). These two facies occur in about equal proportions, in contrast to the Lonan Formation in which Facies C2.3 predominates (Fig. 3). Both facies typically preserve the T_{cde} or T_{de} Bouma divisions and are interpreted as the product of low-concentration turbidity currents. No palaeoflow data have been collected from the Port Erin Formation. Facies comparisons suggest that it may have been deposited in a distal part of a fan or ramp.

Mull Hill Formation

The Mull Hill Formation is characterized by a high proportion of light grey to white quartzose sandstone in medium to thick beds (Fig. 9a, b and d), contrasting strongly with the thin- or very thin-bedded facies of the underlying Port Erin Formation (Fig. 9c). The contrast is also evident in the predominance of quartz arenites in the Mull Hill Formation and the high (70–100%) overall proportion of sandstone. Sandstone beds typically grade up from medium sand to very fine sand or silt and show the T_b, T_{bc} or T_{cd} divisions (Fig. 9a). Rare beds have coarse to very coarse sand bases containing mudstone rip-up clasts. In some sections (e.g. Fig. 9b and d) sandstone beds show only weak grading, are structureless or contain parallel-lamination (T_a or T_{ab}), and either pass rapidly up into a thin mudstone cap or are amalgamated with the succeeding sandstone bed. The Mull Hill sandstone beds are assigned to Facies C2.2 and C2.1 (Fig. 3), and are interpreted as the products of low- to medium-concentration turbidity flows. The thicker massive or parallel-laminated sandstones are designated as Facies B2.1 and may have been deposited from high-concentration turbulent or laminar flows, possibly the basal parts of larger flows.

Near the base (Fig. 9a), top (Fig. 9b) and postulated lateral margins (Fig. 9d) of the formation, packets of thin-bedded turbidites of Facies C2.3 are intercalated with the medium- to thick-bedded facies. There is also a wider variation in sandstone composition in these marginal zones, with quartz wacke, containing up to 30% matrix, interspersed with the quartz arenite beds. Some wacke beds have a basal interval of cleaner sand, several centimetres thick, interpreted as the deposit from a basal high-concentration carpet to an otherwise mud-rich flow (cf. Clayton 1994). This interpretation implies that arenite, at least in limited volumes, could be produced by sorting a mixed mud–sand flow within the Manx Group basin. However, the thickness of clean arenite produced is an order of magnitude less than the average beds in the Mull Hill Formation.

Fig. 8. Lithological log from channel-fill succession in the Santon Formation between Purt Veg [SC 3255 7037] and Santon Head [SC 3290 7031].

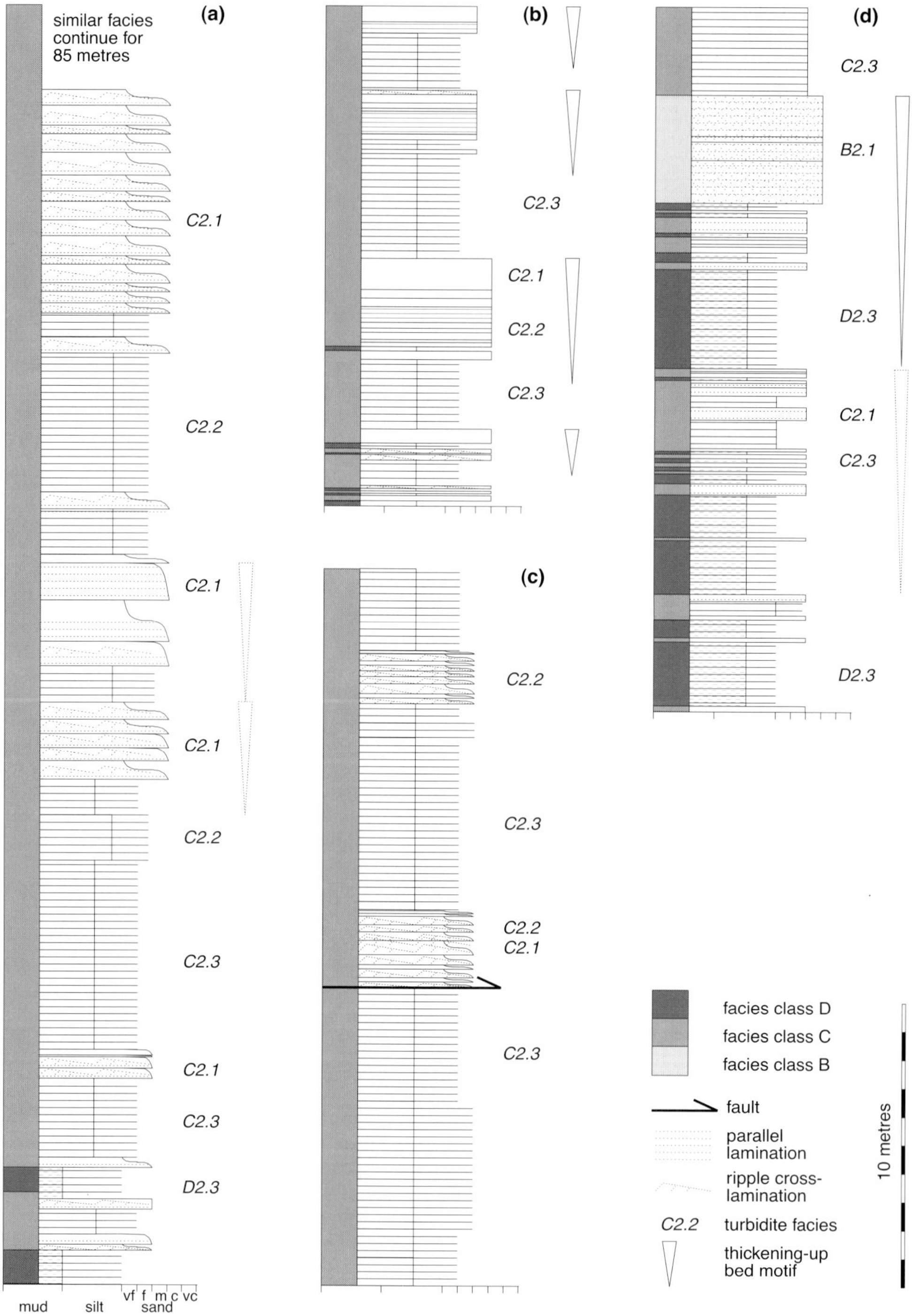

Fig. 9. Lithological logs from the Mull Hill and Port Erin Formations. (**a**) The Mull Hill Formation base at Chapel Bay [SC 2107 6792]; (**b**) near the Mull Hill Formation top at Cregneish Quarry [SC 1910 6740]; (**c**) within the uppermost Port Erin Formation at Perwick Bay [SC 2060 6715]; (**d**) the possible correlative of the Mull Hill Formation at Spaldrick [SC 1938 6952].

The Mull Hill Formation appears to be lenticular, with a maximum thickness of *c.* 400 m across Mull Hill but thinning to a preserved thickness of 40 m at Spaldrick [SC 1938 6952], a postulated correlative originally deposited *c.* 2 km away. The base of the formation is gradational over *c.* 40 m at Chapel Bay [SC 2107 6792] (Fig. 9a) and a thickening-upwards bed motif continues up through a further 100 m of section at the Chasms [SC 1936 6637]. Thickening-up motifs also occur within 1–10 m thick sandstone packets in the marginal zones of the formation (Fig. 9a, b and d). The thickening-up patterns are interpreted as prograding sandy turbidite lobes and the Mull Hill Formation as a shingled stack of such lobes, forming a small sandy fan. Flutes within the formation suggest a locally west-southwestward primary flow direction and cross-lamination in a northwestward secondary flow, a similar pattern to that in the Santon Formation (Fig. 5).

Facies architecture of the Ny Garvain and Creg Agneash Formations

Ny Garvain Formation

In the southern part of the Ny Garvain Formation, between Port Cornaa [SC 4728 8778] and Gob ny Garvain [SC 4885 8986], the succession is sand dominated. Medium to thick beds of fine- to medium-grained sandstone are separated by very thin mudstone partings. The sandstone beds are massive, parallel-laminated or ripple cross-laminated (T_{abd}, T_{bd}, T_{bcd}). Sandstone-rich packets are interspersed with intervals of thin-bedded cross-laminated sandstone–mudstone couplets (Facies C2.3). By contrast, the northern part of the Ny Garvain Formation is dominated by thin-bedded sandstone–mudstone couplets (Facies C2.3; Fig. 3), but with interspersed packets of medium-bedded sandstone–mudstone couplets (Facies C2.2). There are also intervals of very thin-bedded siltstone–mudstone couplets (Facies D2), particularly near the top of the succession. Typical C2.3 beds are sharp based and grade from green–grey ripple cross-laminated fine-grained quartz wacke through interlaminated siltstone and mudstone up to dark grey mudstone. Convolute lamination is common in the cross-laminated divisions, which show straight-crested to undulatory, asymmetric current ripples on bedding surfaces.

At Port Cornaa [SC 4728 8778] the wackes are punctuated by an interval of medium- to thick-bedded quartz arenite, similar in appearance to the Keristal Member in the Lonan Formation. Thin- to medium-bedded quartz arenite also occurs interbedded with quartzose wacke over several metres of the succession north of Gob ny Garvain [SC 4880 9015].

The Ny Garvain Formation is interpreted as the deposits of low-, medium- or, less commonly, high-concentration turbidity currents, with Bouma divisions $T_{(b)cde}$ typically preserved. Palaeoflow estimates from ripple cross-lamination show a north-northwesterly direction (Fig. 5). Flutes were not observed.

The succession within the Ny Garvain Formation is partially obscured by its structure (Fitches *et al.* 1999). It is probable that the thinner bedded and more mudstone-rich succession in the north end of the coastal outcrop overlies the units to the thicker bedded sequence to the south. The Ny Garvain Formation was probably deposited on an outer fan lobe or ramp in a mixed mud–sand system (Reading & Richards 1994).

Creg Agneash Formation

The Creg Agneash Formation, like the Mull Hill Formation, is characterized by light grey or white, quartz arenite, typically in thin to medium beds (Fig. 10a and b). Each sandstone bed grades weakly up from medium or fine sandstone to very fine sandstone, usually with a very thin mudstone parting at the top. Sandstone intervals typically have weakly defined parallel-lamination, thinning upwards within the bed, sometimes with an overlying ripple cross-laminated division (Bouma

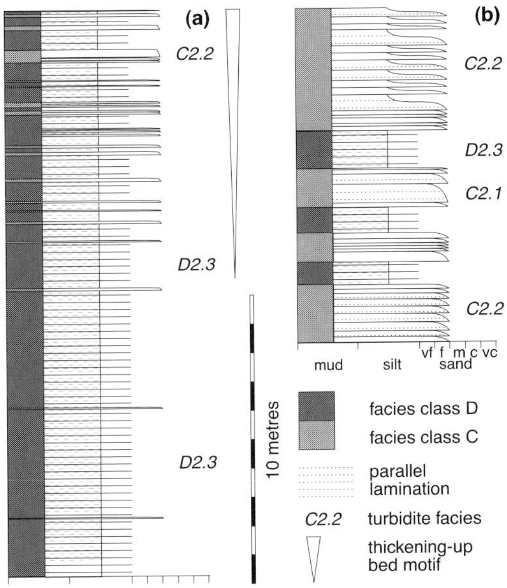

Fig. 10. Lithological logs from the Creg Agneash Formation. (**a**) The formation base in Laxey Glen [SC 4218 8618]; (**b**) near the formation top at Windy Corner Quarry [SC 3903 8400].

divisions T_{be} or T_{bce}). The sand–mud couplets correspond to Facies C2.3, C2.2 or, less commonly, C2.1 (Fig. 3), and are interpreted as the depositional products of mostly low- to medium-concentration turbidity currents. The Creg Agneash Formation differs from the Mull Hill Formation in the proportion of mudstone-rich intervals that separate packets of sandstone beds up to a few metres thick (Fig. 10). The mudstone-rich intervals comprise dark grey, very thin-bedded silt–mud couplets, often with a basal lamina or an interval of very fine sand (Facies D2.1, D2.3 or C2.3), interpreted as stacked fine-grained turbidites. Near the top and bottom of the formation, sandstone beds become isolated in the predominant thin-bedded background (Fig. 10a).

The original mapping of the 'Agneash Grit' (Geological Survey 1898; Lamplugh 1903) showed a lenticular outcrop pattern, partly because it included quartz arenite-bearing intervals of the Maughold Formation as assigned here. It is now apparent that the thinning of the Creg Agneash Formation towards the central valley of the island is structurally controlled (Fitches *et al.* 1999). Its thickest development occurs near Windy Corner [SC 3911 8450] where it may attain *c.* 1000 m depending on the extent of internal folding (for which there is little evidence). The formation thins progressively northeast from here, being *c.* 250 m thick at Maughold Head, although this may be modified by faulting (Quirk & Burnett 1999).

The base of the Creg Agneash Formation shows a gradational increase from the underlying Ny Garvain Formation in the number and thickness of quartz arenite beds within the very thin-bedded background facies. This transition is affected by faults south of Maughold Head, but is intact in the Laxey Valley. The Creg Agneash Formation therefore represents a later phase than the Ny Garvain Formation in the evolution of the turbidite system. However, no marked thickening or thinning motifs have been recognized within the Creg Agneash Formation and its assignment to a particular element of fan morphology is problematic. The formation probably represents a weakly organized stack of sandy fan lobes. Rare flutes suggest a west-southwestward primary palaeoflow direction (Fig. 5).

Correlation between Manx Group tracts

The depositional model for the southeastern Manx Group (Fig. 11) is focused on relationships in tract 1, where an early phase of moderate sand supply (Lonan Formation) is succeeded by a more sandy depositional phase (Keristal Member and Santon Formation). Integration of tracts 2 and 3 into this model is hampered by the lack of biostratigraphic control and by uncertainties over lithostratigraphical correlation of their component units with those in tract 1. The model presented here is not a unique solution – other possible correlations are discussed by Barnes *et al.* (1999) on the basis of sandstone geochemical data and by Quirk & Burnett (1999) on the basis of lithofacies mapping.

The model assumes firstly that the Port Erin Formation is approximately time-equivalent to the Lonan Formation. The Port Erin facies are a plausible, more distal, equivalent of the Lonan facies. Moreover, two medium-bedded packets of quartzose sandstone occur within the Port Erin Formation at Perwick Bay, 100 m or so below the locally faulted transition into the Mull Hill Formation (Fig. 9c). These sandstones are a possible correlative of the Keristal Member in the Lonan Formation.

Secondly, the model assumes that the Ny Garvain Formation represents the earliest of the sandy lobes to be initiated, probably during the later part of the deposition of the Port Erin and Lonan Formations further southwest (Fig. 11a). Packets of quartz arenite in the Ny Garvain Formation at Port Cornaa and Gob ny Garvain are again similar to the Keristal Member in the Lonan Formation, although the Ny Garvain Formation is, except at its top, always more sandy than the Lonan Formation.

Thirdly, the model assumes that the Mull Hill, Creg Agneash and Santon Formations represent a later phase of more sand-rich deposition (Fig. 11b), although not with the implication that these formations are precise time-equivalents. The palaeoflow evidence is consistent with the Mull Hill fan being fed by flows passing through the trunk Santon distributary channel at Purt Veg. However, this correlation is questioned by the mismatch in composition between more arenitic Mull Hill Formation and the more wacke-prone Santon Formation (Barnes *et al.* 1999). If the Mull Hill Formation was supplied through the Purt Veg channel, then the arenitic flows bypassed the channel and left little depositional record. On compositional grounds alone, it is possible that the Mull Hill fan was fed through the earlier Keristal channel system (Fig. 11a).

Depositional model and its uncertainties

The suggested correlation lead to an interpretative model (Fig. 11) in which the Lonan and Port Erin Formations form the distal part of a submarine ramp, fed by mixed mud–sand turbidity currents. These currents flowed northwestward or westward and deposited, generally, thin-bedded wacke sandstones and mudstones (Fig. 11a). Although the interval could be the distal part of a single fan,

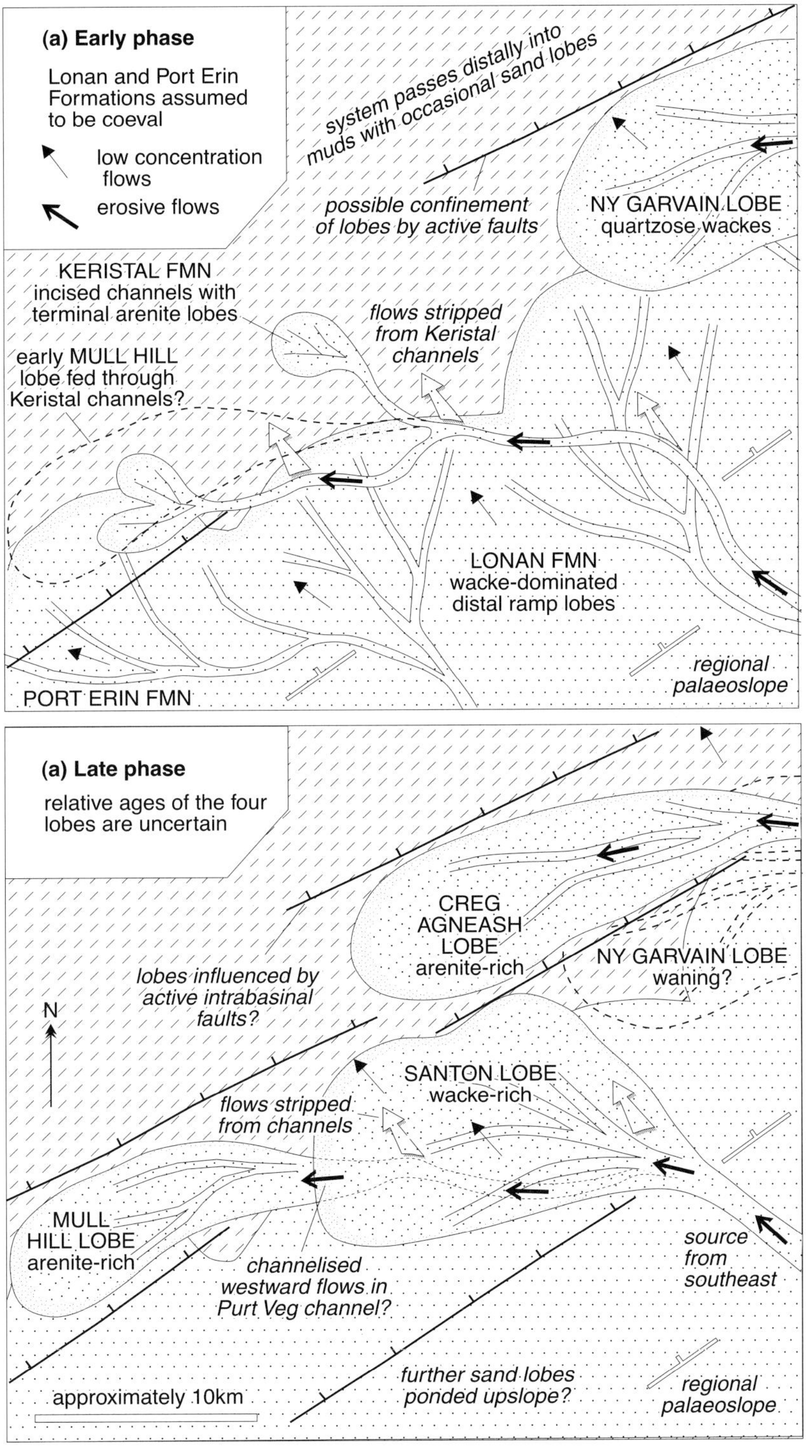

Fig. 11. Diagrammatic maps of two phases in the sedimentological evolution of the study area. The maps cover approximately the same area as the map in Fig. 1. Uncertainties over correlation of depositional events are discussed in the text. The positions and downthrow senses of faults are speculative.

subsequent events suggest that there were several point sources on the margin to the southeast.

Late in the history of the ramp, it was incised by a system of small channels localizing sand-rich flows and ending in small quartz arenitic sandstone lobes – the Keristal Member (Fig. 11a). The channels, some of which ran southwestward, were soon plugged by a limited number (< 20) of quartzose sandstone beds before less sand-rich deposition resumed. The sand-rich flows, which should have travelled less far than mixed mud–sand flows (e.g. Dade & Huppert 1994), apparently reached the distal part of the sand–mud ramp. It is possible that the sand source had effectively moved closer due to a fall in sea level. Alternatively, or additionally, the preserved sand may have been just the basal component of large mixed mud–sand flows from which the muddy fraction had been removed. This removal could have been due to flow-stripping at bends in the channels (Fig. 11a), the fine-grained fraction of the flow spilling off down the regional northwest facing palaeoslope. Whilst palaeoflow data in the Manx Group suggest that flow-stripping might have occurred, it is unlikely that this process alone could have partitioned the mud and sand fractions so effectively within a turbulent flow. Regional considerations, discussed later, make it likely that an increased supply of clean sand to the basin was the more important factor influencing preservation of the basinal arenites.

It is possible that the Mull Hill sandstone lobe was fed partly through the channels of the Keristal Formation (Fig. 11a) although, on balance, a later supply is favoured here. In any case, a lobe or small fan of quartzose wackes, represented by the Ny Garvain Formation, probably began to accumulate in the northeast of the area, coeval with the less sandy Lonan and Port Erin successions further southwest.

There was a brief resumption of wacke-rich ramp deposition following the Keristal Member. Then the whole of this part of the margin seems to have become dominated by coarser grained turbidite systems (Fig. 11b). This change may have been driven by a switch in supply routes across the basin margin or, more probably, by a relative fall in sea level. A major channel – the Purt Veg channel – incised the Lonan Formation and fed turbidity flows to the west-southwest. Here, the flows may have generated the prograding arenitic lobe of the Mull Hill Formation although, on this hypothesis, this phase left no depositional record within the channel. The Purt Veg channel was progressively plugged by quartzose gravelly sandstones and then by wacke-dominated deposits of the rest of the Santon Formation. These flows again show primary flow to the west-southwest, with postulated over-spill of finer flows to the northwest (Fig. 11b). Sporadic quartz arenite beds in the Santon Formation along Marine Drive testify to the continuing availability of clean sand as well as muddy sand during this phase of deposition.

The phase of coarser grained arenitic deposition in tract 3 is probably recorded by the Creg Agneash Formation. More limited palaeoflow data here suggest the same partitioning of flows, with more erosive components constrained to travel parallel to the margin whereas less vigorous flows spilled over down the regional palaeoslope. This pattern is compatible with a margin-parallel topography, hypothetically controlled by active faults, producing small mid-slope basins that redirected the higher concentration parts of the turbidity flows (Fig. 11b). The faults have been shown in positions that correspond to later Caledonian structures and direct evidence for their syn-depositional nature is not recorded.

Judging by relationships at the upper contact of the Creg Agneash Formation, sand-rich flows waned through a relatively short interval of time. The sand-prone fan systems were blanketed by the muds, pebbly muds and more sporadic packets of sand that dominate the overlying Maughold Formation. This event suggests a major starvation of the turbidite fans by a marine transgression over the continental margin.

Regional significance

The increased knowledge of the southeastern Manx Group allows better assessment of its original relationship to other units, first within the Manx Group and then elsewhere along the Gondwana margin.

An important correlation proposed by Lamplugh (1903), and followed by most subsequent authors, is that of the Lonan Flags – essentially the Lonan, Santon and Ny Garvain Formations here – with the Niarbyl Flags on the northwest coast of the Isle of Man [the Niarbyl Formation of Morris *et al.* (1999)]. This equivalence of sandstone-rich units was an important constraint in previous structural models, particularly the synclinorium hypothesis of Lamplugh (1903) and the Isle of Man Syncline hypothesis of Simpson (1963). Given the regional northwest facing margin deduced in the present study, the Niarbyl Formation would also give an important glimpse into a down-slope facies equivalent of the Lonan–Santon system. However, recent graptolite discoveries in the Niarbyl Formation prove its Silurian age (Howe 1999), precluding any correlation with the Manx Group (Morris *et al.* 1999). In any case, the Niarbyl Formation is distinguished sedimentologically from the Manx Group sandstone units by its anoxic hemipelagic facies, and its contrasting sandstone petrography, geochemistry and palaeocurrents (Barnes *et al.* 1999; Morris *et al.* 1999).

If down-margin equivalents of the southeastern sandstone formations do exist within the Manx Group, they are more likely to be found within the mudstone-rich units that make up the central spine of the island. E.g. the Glen Dhoo Flags of Simpson (1963) contain the same lower Arenig acritarch assemblage as the Santon Formation (Molyneux 1979, 1999, Cooper *et al.* 1995). They would be a plausible distal ramp facies to the Santon–Mull Hill turbidite systems (Quirk & Burnett 1999). However, it is likely that most of the mudstone-prone Manx Group units are of a later age than the Lonan, Creg Agneash and Mull Hill Units (Woodcock *et al.* 1999). Moreover, some of them, e.g. the Maughold, Creggan Mooar, Sulby and Lady Port Units, contain extensive pebbly mudstones, which are an improbable distal facies of the sand–mud fans to the southeast.

The general equivalence of the Manx Group to the Skiddaw Group of the Lake District is supported by the confirmation that the southeastern sandstone formations comprise deep-marine turbidites and by the diagnosis that they were derived from a continent to the southeast. Biostratigraphic data from the Santon Formation suggests an early Arenig age, but detailed lithological correlation with the Skiddaw succession is not feasible. The Skiddaw Group lacks the thick quartzose arenitic units of the Manx Group, although quartzose wackes do occur locally in the Loweswater Formation (Moore 1992; Cooper *et al.* 1995). Conversely, the Manx Group lacks clear evidence of the repetition of thick wacke-dominated successions represented by the Watch Hill and Loweswater Formations in the Skiddaw Group. Geochemical evidence (Barnes *et al.* 1999) suggests that the Lonan and Santon Formations are compositionally equivalent to the broadly contemporaneous Loweswater Formation.

The paucity of quartz arenites in the Skiddaw Group compared with the Manx Group is notable. The significance of this observation depends on which is the more important of the two hypotheses for arenite preservation given in this paper. Probably, the Skiddaw margin was more remote from a point source of quartzose sediment because the palaeocurrent data and facies analysis of Moore (1992) suggests that flow-stripping of channelized turbidity currents was also important during Skiddaw Group deposition. Arenitic sandstone bodies are again common in the correlative Ribband Group of southeastern Ireland (Shannon 1978; Max *et al.* 1990), although detailed correlation with the Manx Group is problematic (McConnell *et al.* 1999). In Ireland, the margin successions seem to be dominated by mixed quartz arenite and quartzose wacke bodies throughout Cambrian and Early Ordovician time. Indeed, one possible correlation of the Ny Garvain–Creg Agneash succession suggested by Barnes *et al.* (1999) is with these older sandstone units in southeast Ireland.

The source of the quartzose sediment on the Gondwana margin is of particular interest. It is possible that this detritus was derived from the local hinterland comprising Neoproterozoic basement with a cover of Cambrian–Tremadoc clastic sediments and a Tremadoc volcanic arc (e.g. Cope *et al.* 1992). Given this provenance, the detritus would need to have been progressively cleaned of its fine-grained fraction, by shallow-marine processes and by hydrodynamic sorting within the depositing turbidity flows. Equally likely is that quartzose sediment was being supplied to the margin from the remote interior of the large Gondwana continent. The widespread Armorican quartzite facies of Arenig age is confirmation that such sediment was at least reaching parts of the shallow-marine margin of the Gondwana continent (Noblet & Lefort 1990). The quartz arenites of the Manx and Ribband Groups could be the evidence that some of this sediment was redeposited on to the deep-water margin (Cooper *et al.* 1995). If so, the implication would be that the Avalonian fragment of Gondwana was still attached to its parent continent through much of Arenig time (Pickering & Smith 1995; Prigmore *et al.* 1997).

Any correlation of the depositional sequences in the southeastern Manx Group with global sea-level changes is necessarily tentative until better biostratigraphic control on their age is secured. The early Arenig age of the Santon Formation, derived from acritarchs (Molyneux 1979, 1999; Cooper *et al.* 1995) and graptolites (Rushton 1993), is compatible with the sandstone-rich lobes being deposited during the lowstands around the Tremadoc–Arenig boundary and into the early Arenig (Fortey 1984; Ross & Ross 1992). On this hypothesis, the mudstone-rich ramp deposits of the underlying Lonan Formation might even record higher sea levels in the late Tremadoc. Globally rising sea levels into the mid-Arenig (Fortey 1984) could have been responsible for shutting off sand supply to the margin, producing the mudstone-dominated successions of the Maughold Formation and of the rest of the central tract of the Isle of Man (Woodcock *et al.* 1999).

Dave Quirk, Dave Burnett, John Morris and Bill Fitches helped with some of the field observations in this paper, and in stimulating discussion of the results. NHW also thanks Jonathan Copus and Brian Dade for invaluable guidance on the processes of turbidite deposition. The manuscript was improved by helpful reviews from Dave Quirk, Rick Moore and Kevin Pickering. RPB publishes with the permission of the Director, BGS. This work was funded by NERC research grant GR9/01834.

References

BARNES, R. P., POWER, G. M. & COOPER, D. C. 1999. The definition of sandstone-bearing formations in the Isle of Man and correlation with adjacent areas – evidence from sandstone chemistry. *This volume.*

BERGER, J. F. 1814. Mineralogical account of the Isle of Man. *Transactions of the Geological Society, London,* **2**, 29–65.

CLAYTON, C. J. 1993. Deflection versus reflection of sediment gravity flows in the late Llandovery Rhuddnant Grits turbidite system, Welsh Basin. *Journal of the Geological Society, London,* **150**, 819–822.

—— 1994. Contrasting sediment gravity flow processes in the late Llandovery, Rhuddnant Grits turbidite system, Welsh Basin. *Geological Journal*, **29**, 167–181.

COOPER, A. H., RUSHTON, A. W. A., MOLYNEUX, S. G., HUGHES, R. A., MOORE, R. M. & WEBB, B. C. 1995. The stratigraphy, correlation, provenance and palaeogeography of the Skiddaw Group (Ordovician) in the English Lake District. *Geological Magazine*, **132**, 185–211.

COPE, J. C. W., INGHAM, J. K. & RAWSON, P. F. 1992. Atlas of palaeogeography and lithofacies. *Memoir of the Geological Society, London,* **13**, 1–153.

DADE, W. B. & HUPPERT, H. E. 1994. Predicting the geometry of channelized deep-sea turbidites. *Geology,* **22**, 645–648.

FITCHES, W. R., BARNES, R. P. & MORRIS, J. H. 1999. Geological structure and tectonic evolution of the Lower Palaeozoic rocks of the Isle of Man. *This volume.*

FORD, T. D. 1993. *The Isle of Man.* Geologists' Association, 1–94.

FORTEY, R. A. 1984. Global earlier Ordovician transgressions and regressions and their biological implications. *In:* BRUTON, D. L. (ed.) *Aspects of the Ordovician System.* Universitetsforlaget, 37–50.

GEOLOGICAL SURVEY OF UNITED KINGDOM, 1898. Isle of Man. *1:63 360 geological map.* Sheets 36, 45, 46, 56 & 57.

HENSLOW, J. S. 1821. Supplementary observations to Dr. Berger's account of the Isle of Man. *Transactions of the Geological Society, London*, **5**, 482–505.

HOWE, M. P. A. 1999. The Silurian fauna (graptolite and nautiloid) of the Niarbyl Formation, Isle of Man. *This volume.*

KNELLER, B. C., EDWARDS, D., MCCAFFREY, W. & MOORE, R. 1991. Oblique reflection of turbidity currents. *Geology*, **14**, 250–252.

LAMPLUGH, G. W. 1903. *The Geology of the Isle of Man.* Memoir of the Geological Survey, United Kingdom. HMSO.

MAX, M. D., BARBER, A. J. & MARTINEZ, J. 1990. Terrane assemblage of the Leinster Massif, SE Ireland, during the Lower Palaeozoic. *Journal of the Geological Society, London,* **147**, 1035–1050.

MCCONNELL, B. J., MORRIS, J. H. & KENNAN, P. S. 1999. A comparison of the Ribband Group (southeastern Ireland) to the Manx Group (Isle of Man) and Skiddaw Group (northeastern England). *This volume.*

MOLYNEUX, S. G. 1979. New evidence for the age of the Manx Group, Isle of Man. *In:* HARRIS, A. L., HOLLAND, C. H. & LEAKE, B. E. (eds) *Caledonides of the British Isles: Reviewed.* Geological Society, London, Special Publications, **8**, 415–421.

—— 1999. A reassessment of Manx Group acritarchs, Isle of Man. *This volume.*

MOORE, R. M. 1992. *The Skiddaw Group of Cumbria: Early Ordovician turbidite sedimentation and provenance on an evolving microcontinental margin.* PhD Thesis, University of Leeds.

MORRIS, J. H., WOODCOCK, N. H. & HOWE, M. P. A. 1999. The Silurian succession of the Isle of Man: the late Wenlock Niarbyl Formation, Dalby Group. *This volume.*

NOBLET, C. & LEFORT, L. P. 1990. Sedimentological evidence for a limited separation between Armorica and Gondwana during the Early Ordovician. *Geology*, **18**, 303–306.

ORR, P. J. 1996. The ichnofauna of the Skiddaw Group (early Ordovician) of the Lake District, England. *Geological Magazine*, **133**, 193–216.

—— & HOWE, M. P. A. 1999. Macrofauna and ichnofauna of the Manx Group (early Ordovician), Isle of Man. *This volume.*

PICKERING, K. T. & SMITH, A. G. 1995. Arcs and backarc basins in the Early Paleozoic Ocean. *The Island Arc*, **4**, 1–67.

——, HISCOTT, R. N. & HEIN, F. J. 1989. *Deep Marine Environments: Clastic Sedimentation and Tectonics.* Unwin Hyman, 1–416.

PIPER, J. D. A. 1997. Tectonic rotation within the paratectonic British Caledonides and Early Palaeozoic location of the orogen. *Journal of the Geological Society, London,* **154**, 9–14.

—— & CROWLEY, S. F. 1999. Palaeomagnetism of (Palaeozoic) Peel Sandstones and Langness Conglomerate Formation, Isle of Man: implications for the age and regional diagenesis of Manx red beds. *This volume.*

—— & NORMARK, W. R. 1983. Turbidite depositional patterns and flow characteristics, Navy submarine fan, California borderland. *Sedimentology*, **30**, 681–694.

PRIGMORE, J. K., BUTLER, A. J. & WOODCOCK, N. H. 1997. Rifting during separation of Eastern Avalonia from Gondwana: Evidence from subsidence analysis. *Geology*, **25**, 203–207.

QUIRK, D. G. & BURNETT, D. J. 1999. Lithofacies of Lower Palaeozoic deep-marine sediments in the Isle of Man: a new map and stratigraphic model of the Manx Group. *This volume.*

—— & KIMBELL, G. S. 1997. Structural evolution of the Isle of Man and central part of the Irish Sea. *In:* MEADOWS, N., TRUEBLOOD, S., COWAN, G.. & HARDMAN, M. (eds) *Petroleum Geology of the Irish*

Sea and Adjacent Areas. Geological Society, London, Special Publications, **124**, 135–160.

READING, H. G. & RICHARDS, M. 1994. Turbidite systems in deep-water basin margins classified by grain size and feeder system. *AAPG Bulletin*, **78**, 792–822.

ROSS, J. R. & ROSS, C. A. 1992. Ordovician sea-level fluctuations. *In:* WEBBY, B. & LAURIE, J. R. (eds) *Global Perspectives on Ordovician Geology*. Balkema, 327–335.

RUSHTON, A. W. A. 1993. Graptolites from the Manx Group. *Proceedings of the Yorkshire Geological Society*, **49**, 259–262.

SHANMUGAM, G. 1997. The Bouma Sequence and the turbidite mind set. *Earth Science Reviews*, **42**, 201–229.

SHANNON, P. M. 1978. The stratigraphy and sedimentology of the Lower Palaeozoic rocks of southeast Co. Wexford. *Proceedings of the Royal Irish Academy*, **78B**, 247–265.

SIMPSON, A. 1963. The stratigraphy and tectonics of the Manx Slates Series. *Quarterly Journal of the Geological Society, London*, **119**, 367–400.

SINCLAIR, H. D. 1994. The influence of lateral basin slopes on turbidite sedimentation in the Annot Sandstones of SE France. *Journal of Sedimentary Research*, **A64**, 42–54.

WOODCOCK, N. H., MORRIS, J. H., QUIRK, D. G. *ET AL*. 1999. Revised lithostratigraphy of the Manx Group, Isle of Man. *This volume*.

Manganiferous ironstones in the early Ordovician Manx Group, Isle of Man: a protolith of coticule?

P. S. KENNAN[1] & J. H. MORRIS[2]

[1]*Geology Department, University College, Belfield, Dublin 4, Ireland*

[2]*Geological Survey of Ireland, Beggars Bush, Haddington Road, Dublin 4, Ireland*

Abstract: Spessartine–garnet quartzites (coticule) are now widely recognized as extremely useful marker horizons throughout the Appalachian–Caledonian Orogen. In many instances, they appear to be of early Ordovician age. They reflect a syn-sedimentary, volcanic exhalative origin and in many areas, e.g. the Leinster region in southeast Ireland, they are intimately associated with, *inter alia*, base metal mineralization and with tourmalinite. However, though the metamorphic coticule is common, the pre-metamorphic coticule precursor has proved difficult to recognize.

Manganiferous ironstone rocks of coticule aspect are a distinctive feature in parts of the Manx Group exposed along the west coast of the Isle of Man. Though they lack any trace of the characteristic manganese garnet, perhaps reflecting a relatively low metamorphic grade, they do bear a strong morphological resemblance in outcrop to the typical coticule lithology of the Ribband Group in southeast Ireland. In both places, tourmaline-rich rocks occur nearby. On the Isle of Man, a major tourmalinite occurrence is linked to an important shear zone.

In briefly reviewing the field occurrence, petrography and chemistry of the manganiferous ironstone beds, and by comparison with coticule elsewhere, the possibility that these ironstones are a coticule precursor or a coticule facies variant is introduced. That a lithology now termed coticule might be a significant component in the Manx Group ('Slates') was noted no less than 70 years ago. Any recognition of coticule rock would have an important bearing on the understanding of the stratigraphy, metamorphism and mineralization of the Manx Group.

This paper is written for two reasons. The first of these is to redraw attention to the possible presence on the Isle of Man of a minor lithology that occurs widely in the Ordovician of the Caledonides and Appalachians. That lithology is coticule. The second reason is to briefly describe a thinly bedded manganiferous ironstone lithology that occurs as very obvious layers within pelitic and psammitic metasediments on the northwest coast of the island. These may be an associate, or a precursor, of coticule.

The significance of coticule lies in its common association elsewhere with mineralization and in its potential value as a stratigraphic marker horizon – hence the importance of any possible occurrence of this lithology on the Isle of Man. As a marker horizon, any occurrence of coticule is likely to aid correlation between the Manx Group, the Ribband Group in southest Ireland and the Skiddaw Group in the Lake District. Though coticule has not, as yet, been described from the Skiddaw Group, Harker (1950, fig. 13C) illustrates what appears to be typical coticule from metamorphosed Skiddaw Slates.

Coticule

The coticule that is typical of Caledonian–Appalachian settings is a thinly bedded, quartzose rock characterized by an abundance of small (usually < 0.2 mm) equidimensional garnets. The garnets, in the typical case, are usually described as spessartine rich. However, the garnet in coticule worldwide contains a varying, and often significant, almandine component. Garnet compositions can vary between adjacent layers on a single outcrop.

The presence of quartz makes for a poor whetstone (Latin: *cotis – coticula*). In the lithology of the type area in the Belgian Ardennes, the multitudes of tiny spessartine garnets lie in a matrix of fine sericite (Lamens *et al.* 1986). As a result, the quality of the rock as a sharpener was well known to the Romans.

In the field, coticule is usually hosted in what are otherwise unexceptional pelitic and semi-pelitic horizons and is, and has been, frequently overlooked. The initial recognition may be no more than a suspicion where thinly bedded quartzites display

From: Woodcock, N. H., Quirk, D. G., Fitches, W. R. & Barnes, R. P. (eds) 1999.
In Sight of the Suture: the Palaeozoic geology of the Isle of Man in its Iapetus Ocean context.
Geological Society, London, Special Publications, **160**, 109–119. 1-86239-046-0/99/$15.00

complex, disharmonic folds suggestive of syn-sedimentary slumping. The diagnostic garnets are usually revealed only in thin section. In this light, the comparison drawn between garnet-bearing rocks in the Manx Group, similar rocks in southeast Ireland and the coticule-bearing rocks of the Ardennes 70 years ago (Stainier 1929) was remarkable.

Coticule is but one of a group of very distinctive metamorphic lithologies that tend to occur together. An important association with tourmaline was noted by Renard (1878) when he wrote the classic description of the type coticule which is now recognized worldwide [see Slack (1996)]. Tourmalinite and coticule are but two lithologies in a recurring package (Kennan 1986) of distinctive rocks that includes, for example, rocks rich in manganiferous chloritoid (the type area for ottrelite is the same as that for coticule), other aluminous silicates, apatite, magnetite enrichments and metals. Coticule-bearing horizons in southeast Ireland, for example, are spatially associated with lead, zinc, tungsten, gold, lithium, etc., mineralization (e.g. McArdle & Kennan 1992). It is this association with mineralization, and the stratabound occurrence, that suggests a syn-sedimentary origin involving volcanic exhalation for all. What are essentially contemporaneous basic extrusive rocks are usually found nearby.

In coticule-bearing rocks, metamorphic spotting is commonplace, e.g. in the Ardennes-type area (Theunissen 1970) and in eastern Ireland (Kennan & Murphy 1993). Though a hidden granite may be the cause near Bellewstown in eastern Ireland – and on the Isle of Man (see below) – such is not the case in Belgium. There, lowermost greenschist metamorphism is adequate to promote the growth of garnet, chloritoid, andalusite and cordierite spots in the manganiferous rocks; the manganiferous chemistry of the original sediment appears to be the controlling influence.

In Belgium, the coticule-hosting sequences are of Lower Ordovician–Arenig age (Lamens *et al.* 1986). In eastern Ireland, some are also probably of the same age (Kennan & Murphy 1993). Others, including the extensive occurrences along the margin of the Leinster Granite, are poorly dated; these are, however, probably of Tremadoc or Arenig age [but see Brück et al. (1974)]. In many places, coticule has been recognized to be a field marker horizon, e.g. in southeast Ireland, where the presence of the lithology has helped in Ribband Group correlation (Brück *et al.* 1979). In Newfoundland, and elsewhere along the Caledonian–Appalachian Orogen, the lithology is contributing to the correlation of Lower–Middle Ordovician sequences over very much greater distances (e.g. Kennan & Kennedy 1983; Gardiner & Venugopal 1992; Schofield *et al.* 1998). As a rock that can be tracked in the field over hundreds of kilometres (Kennan & Kennedy 1983), coticule is clearly a superb lithostratigraphic marker. Palaeontological control is usually poor or lacking.

Though it can be difficult to do so with any certainty on outcrop, it is relatively easy to recognize coticule in thin section. However, it is not so easy to identify what the coticule protolith was prior to the growth of garnet. Earlier suggestions have included sands rich in manganese (e.g. Clifford 1960), chert (e.g. Doyle 1984), chamosite-rich layers (Brindley 1954), volcanic tuffs (Kramm 1976) and manganese ironstones (Stanton 1976). More recently, increased emphasis has been given to the possibility that coticule is essentially a replacement – during diagenesis or later – of carbonate sediment (e.g. Lamens *et al.* 1986; Bennett 1989; Jones 1994). In southeast Ireland, Shannon (1977) has observed the transition directly; within the aureoles of dolerite dykes, small carbonate nodules are gradually replaced from the outside in by ringlets of garnets. These ringlets match textures that are common to many coticule beds (Kennan 1972). Perhaps, the garnetiferous calcareous nodules mentioned by Gillott (1955, p. 146) are comparable features.

It is the recognition of a possible carbonate precursor that appears most germane on the Isle of Man. The extensive development of buff-coloured, iron- and manganese-rich carbonates in the layers of coticule aspect characterizing some horizons outcropping along the northwest coast of the island, invite examination.

Possible occurrences of coticule in the Manx Slates

When Stainier (1929) first compared rocks on the Isle of Man to then recently described garnet hornfels (now recognized as coticule) in southeast Ireland and to rocks in the coticule-type area in the Belgian Ardennes, part of the reason for the comparison was the fact that a northwest–southwest trending belt of porphyroblasts marks the central spine of the island (Lamplugh 1903; Simpson 1964; Gillott 1955). This metamorphic zone is also characterized by occurrences of tourmaline that cannot easily be related to the aureoles of the small exposed granites (Fig. 1). If there is a large granite below, it is completely hidden; the aureoles surrounding the Foxdale and Dhoon Granites are distinctively different (Lamplugh 1903; Power & Barnes 1999).

Gillott (1955) described and analysed garnets (13.1% spessartine) and a dark band of very garnetiferous rock (0.09% MnO) from a quarry –

the Bungalow Quarry [SC 4000 8652] (Fig. 1) – on the south flank of Snaefell. Siliceous beds in this quarry are thinly bedded (1 cm) but none found during this study quite compare in garnet abundance or garnet size with typical coticule. Garnets from the quarry range in size up to 1.5 mm in diameter (Fig. 2a) whereas those of the typical coticule are usually much smaller (< 0.5 mm) and far more numerous. The whole rock from this quarry analysed for this study (see Table 1) is, however, not particularly manganiferous – in agreement with the earlier finding of Gillott (1955). The cause of the garnet growth in chemically suitable layers and of the more widespread spotting, was, in Gillott's (1955, p. 152) and Stainier's (1929) views, a granite batholith underlying the axial region of the island.

At Fleshwick Cove [SC 2020 7140] (Fig. 1), small garnets are clearly preferentially developed in thin (2–5 mm) layers in spotted metasediments (Fig. 2b). These occurrences bear some textural comparison with some coticule occurrences. The rocks outcropping here belong to the same Maughold Formation that maps northeastwards to the Bungalow Quarry (above). Similar garnet occurrences were not seen in any other formation on the island; nor was any typical coticule found.

Tourmalinite in the Manx Group

Most of the tourmaline encountered during this study occurs widely as small (< 5 mm long), scattered, isolated needles on occasional bedding surfaces. Local concentrations are to be found adjacent to the exposed granites. It is no surprise that this tourmaline, and the entire northeast–southwest belt of metamorphism in which the

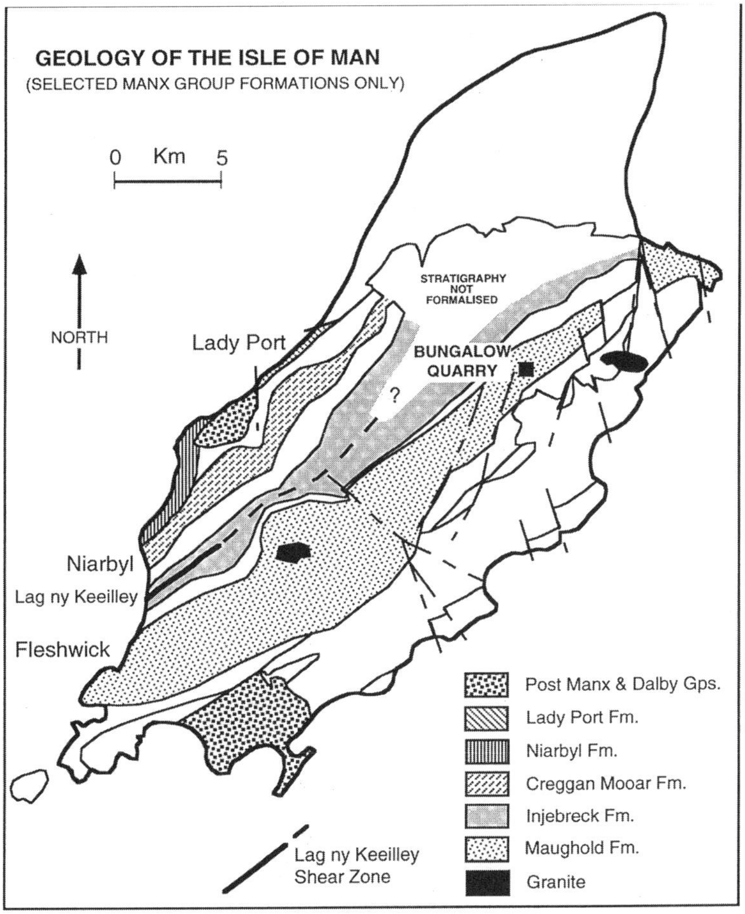

Fig. 1. Simple map outlining the field distribution of those formations that are mentioned in the text. Localities named in the text are shown.

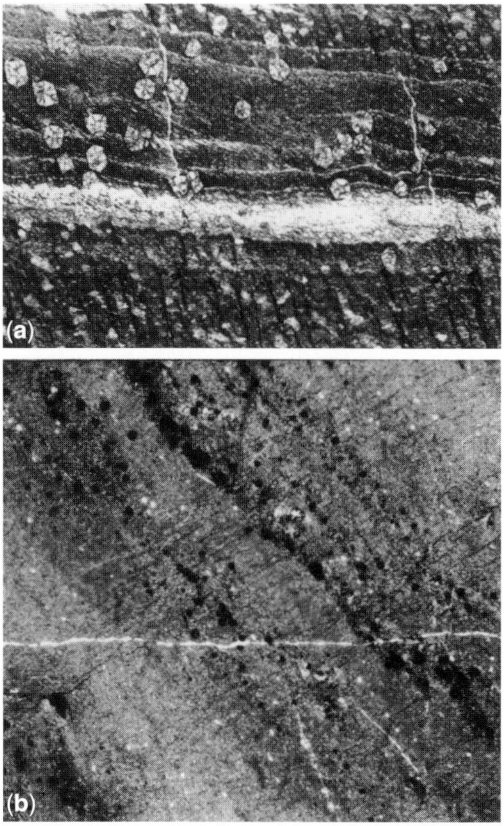

Fig. 2. (**a**) Typical garnetiferous siliceous band from the Bungalow Quarry. Metamorphic spots characterize the enclosing cleaved and crenulated pelites. Width of view, 3.5 cm. (**b**) Strings of fine garnets in banded and spotted metapelitic and metapsammitic rock from the Maughold Formation, Fleshwick Harbour. Width of view, 1.75 cm.

tourmaline is found, might be thought to reflect a granite at depth (Gillott 1955).

No tourmaline was found in any immediate or obvious association with any coticule-like lithology or with the carbonate ironstones at Niarbyl (see below). However, a short distance from Niarbyl, at Lag ny Keeilley [SC 2162 7470] on the northwest coast of the island (Fig. 1), significant tourmaline does occur. At the base of the cliffs, quartz-rich rock with copious tourmaline coincides with a significant high-strain zone (interpreted as a shear zone below). Layers of micaceous tourmalinite, which can be traced for some distance on the hill above, are closely similar to occurrences of coticule-related tourmalinite elsewhere, e.g. in southeast Ireland. Though exposure did not permit the shear zone to be mapped away from Lag ny Keeilley, it is likely to extend further northeastwards across the island.

The shear zone is perhaps best exposed on the foreshore below the hermit's chapel at Lag ny Keeilley, after which it is named here. A structural context for this shear zone is attempted in Fig. 3. The shear zone juxtaposes a north younging, quartz sandstone sequence to the south, against a *c.* 3.5 km coastal section of inverted and south younging mixed-pelite, quartzite and minor debrite sequence to the north, corresponding with the axial trace of Simpson's (1963, 1964) Isle of Man Syncline. The primary structure occurs in the lower limb of a D2 synform, the axial trace of which dips gently north. The sequences north and south of the shear zone are both assigned to the Injebreck Formation (Woodcock *et al.* 1999).

The most intense expression of the shear zone occurs over a 5 m wide zone in the immediate hanging wall of the contact, though related structures are evident for 100 m and more into both the hanging walls and footwalls. The high-strain zone is marked by a very intense fabric in which planar and down-dip folded stratiform quartz veins, 'quartz fish' and shear bands all show a prominent sinistral geometry. This zone dissipates northwards over *c.* 8 m, through a strongly foliated zone with occasional stratiform veins, and boudined and disrupted quartz sandstone beds into less disturbed bedding up to the near-vertical, faulted contact with a felsite boss. A strong fabric is evident for many tens of metres further northward up to and including the prominent tourmalinite zone.

The footwall is composed principally of quartz-veined, locally very intensely foliated, thickly bedded quartz sandstone with occasional coarse conglomerate horizons. In places, the conglomerates contain very conspicuous lenticular shear zones, up to 5 m long and 12 cm wide, defined by a very intense, steeply plunging, sinistral sigmoidally folded fabric, intensely flattened clasts, and elongated and sinistrally offset discordant quartz veins. Clasts in the adjoining wall rock are, by comparison, only mildly elongated, though many contain extensional quartz-veined fractures which are orthogonal to clast long axes, but roughly parallel to the shear zone principal compressive stress. This geometry indicates long axis extension roughly parallel to the clast axes, approximately perpendicular to the inferred principal compressive stress, although both sinistral and dextral rotations are evident. Overall, a primarily sinistral geometry is inferred for the Lag ny Keeilley Shear Zone.

The tourmaline is zoned and displays the blue and green hues of the dravitic tourmaline typical of coticule-bearing horizons in southeast Ireland (Gallagher & Kennan 1992). At Lag ny Keeilley, some of the tourmaline clearly grew during the

Table 1. Representative analyses of carbonate ironstones (Isle of Man), coticule (southeast Ireland) and questionable coticule (Isle of Man)

	1	2	3	4	5	6	7	8	9	10	11	12	13	14	15	16
[%]																
SIO_2	35.32	25.98	24.41	23.46	50.95	54.18	26.75	27.27	25.95	55.85	66.27	67.44	52.94	60.32	59.69	57.7
TIO_2	0.25	0.24	0.29	0.19	0.65	0.16	0.28	0.32	0.28	0.87	0.49	0.07	0.43	0.35	0.86	0.89
Al_2O_3	5.96	4.89	7.58	5.75	15.99	4.18	6.54	10.91	6.46	19.16	11.26	10.2	19.3	15.16	16.59	17.29
Fe_2O_3(t)	26.25	27.55	28.29	29.44	14.85	10.45	30.95	26.25	25.26	12.22	12.97	11.28	14.24	12.16	10.89	11.43
Mn_3O_4	6.29	8.95	8.76	9.25	2.17	3.63	7.37	8.54	7.53	0.74	4.37	6.93	6.31	5.82	5.67	2.97
MgO	3.46	3.45	3.65	3.67	2.26	2.69	4.27	2.92	3.21	2.78	1.84	0.75	1.45	0.98	1.41	2.52
CaO	1.96	4.38	1.83	2.57	0.98	8.41	1.83	1.44	7.31	0.69	1.26	1.96	1.54	1.65	1.06	0.59
Na_2O	0.41	0.44	0.5	0.39	0.88	0.26	0.28	0.69	0.65	0.5	0.19	0.15	2.05	2.42	0.65	1.38
K_2O	0.8	0.84	1.25	0.97	2.14	0.59	0.41	1.95	0.6	2	0.64	0.51	1.31	0.82	1.77	2.99
P_2O_5	0.78	1.66	0.52	0.8	0.64	0.13	0.98	0.67	1.62	0.38	0.25	0.86	0.05	0.38	0.13	0.12
Cr_2O_3	0.01	0.01	0.01	0	0.01	0	0	0	0	0.01	0.01	0	0.01	0.01	0.01	0.01
SrO	0.01	0.02	0.01	0.01	0.01	0.01	0	0.01	0.01	0	0	0	0.02	0.02	0.01	0.01
ZrO_2	0.01	0	0	0.01	0.01	0	0	0.01	0.01	0.02	0.01	0	0.01	0.01	0.02	0.02
BaO	0.02	0.03	0.03	0.03	0.05	0.01	0.01	0.04	0.02	0.05	0.01	0.01	0.06	0.03	0.06	0.09
LOI	18.59	21.69	22.85	23.66	8.41	15.3	20.22	18.78	21.24	4.34	0.13	–0.51	–0.06	–0.21	0.41	1.13
Total	100.12	100.13	99.98	100.2	100	100	99.89	99.8	100.15	99.61	99.7	99.65	99.64	99.92	99.23	99.14
ppm																
Sc	23	23	25	25	19	9	21	27	17	21	9	8	17	11	19	17
V	53	52	64	55	95	31	63	80	53	114	68	62	98	85	108	111
Cr	35	37	43	29	70	19	36	47	36	102	51	14	56	51	81	81
Co	< 2	< 2	< 2	< 2	30	7	< 2	< 2	< 2	20	14	< 2	< 2	< 2	< 2	21
Ni	28	17	29	22	51	11	22	34	24	42	44	19	20	9	21	57
Cu	35	47	23	35	37	12	37	23	104	14	24	11	13	8	6	5
Pb	24	26	32	6	5	7	16	11	42	15	4	6	34	37	19	26
Mo	< 1	1	2	< 1	< 1	2	< 1	3	< 1	< 1	< 1	< 1	< 1	< 1	< 1	< 1
Ag	9*	10*	10*	9*	5	6	10*	9*	11*	3	6	4	5	6	5	3
Sn	6	5	8	7	< 1	< 1	7	6	6	2	< 1	< 1	9	< 1	3	3
Sb	2	4	2	4	< 1	< 1	2	5	4	< 1	< 1	< 1	< 1	< 1	< 1	< 1
Zn	90	139	113	78	89	35	83	78	91	129	49	34	41	25	32	108
W	4	< 1	3	< 1	3	< 1	3	4	2	4	2	2	5	2	5	5
As	36	13	27	48	40	13	77	116	49	3	104	138	389	264	48	35
Rb	36	38	58	45	94	25	20	91	27	95	41	20	46	39	52	146
Sr	74	115	77	75	102	92	41	105	130	61	17	28	167	206	61	88
Ba	187	204	318	266	424	109	73	432	143	363	74	147	559	243	558	843
Y	30	60	24	33	33	11	44	25	58	49	33	26	35	29	32	35
Zr	51	33	46	36	124	20	59	57	44	141	131	21	99	83	162	165
Nb	4	3	4	3	11	3	4	5	5	15	13	2	7	6	16	18
La	27	49	30	25	31	13	34	21	38	43	23	17	25	33	39	42
Ce	84	161	77	77	82	30	88	58	116	93	58	51	72	92	91	99
Nd	34	68	23	30	37	6	23	19	40	35	15	10	13	23	26	33
Hf	< 1	< 1	< 1	2	3	< 1	< 1	< 1	< 1	4	4	< 1	2	3	4	4

1–4, carbonate ironstones, Ladyport, Isle of Man; 5–9, carbonate ironstones, Niarbyl, Isle of Man; 10, garnetiferous layer (?coticule), Bungalow Quarry, Isle of Man; 11–16, typical coticule, Leinster Granite Aureole, southeast Ireland (see text).

Major and trace element analyses (PW1480 and PW2400 X-ray Fluorescence spectrometry) by Analytical and Regional Geochemistry Group, British Geological Survey.

* Ag concentrations may be unreliable due to interference effects.

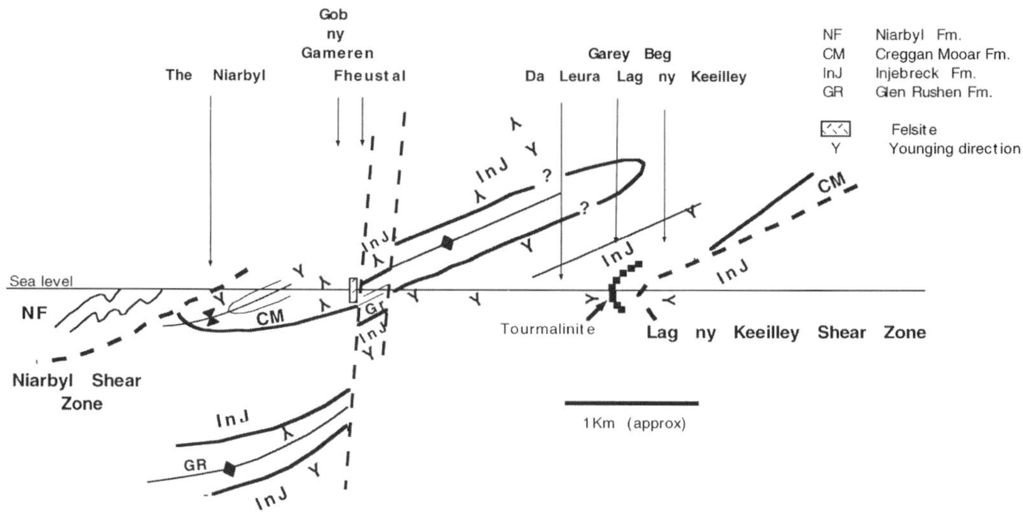

Fig. 3. A sketch structural cross-section (based largely on mapping by JHM) of the coastal section between Niarbyl and Lag ny Keeiley (see Fig. 1). The location of a major development of tourmalinite in relation to the Lag ny Keeilley Shear Zone is shown.

deformation(s) that resulted in quartz segregation and veining; coarse tourmaline developed in some of the quartz veins. Aggregates of lineated tourmaline were also rotated during the shearing schistosity (Fig. 4) and, in the coastal exposure and in the more micaceous graphitic schists on the hill above, much tourmaline (ranging up to 10% of some rock samples) is augened by the most prominent schistosity. These occurrences match many similar occurrences of tourmaline that are conspicuous around quartz segregations and in quartz veins where these occur in certain Ribband Group rocks in southeast Ireland. The segregations and the veins are typically found in horizons that are themselves rich in tourmaline – horizons that are spatially associated with occurrences of coticule.

The finding of this tourmalinite at Lag ny Keeilley introduces a number of possibilities. Firstly, it makes the shear zone, and any extension, a prospective site for mineralization. Obvious pyrite and some chalcopyrite occur within the shear zone. Secondly, if, as in many other places, the tourmaline is coticule related and reflects exhalative boron enrichment in the original sediments (Slack 1996), the fluxing qualities of the boron prior to its incorporation in tourmaline in both schists and quartz veins may have served to key the shear zone to the boron-enriched horizon(s). In the absence of clear evidence of syn-sedimentary boron enrichment on the Isle of Man, these possibilities must remain conjectural.

Small, heavily altered felsic intrusions of unknown age, but with pepperitic margins, occur close to this shear zone in rocks of the Injebreck Formation at Da Leura [SC 2172 7535]. Comparable felsitic intrusions occur in the Niarbyl

Fig. 4. Deformed tourmalinite. Coarse, aligned tourmalines in the centre are rotated and enclosed in a schistosity in which almost all the augen are tourmalines; Lag ny Keeilley. Width of view, 2.25 cm.

Shear Zone at Niarbyl (Fig. 3). Andesitic volcanics outcrop within the Creggan Mooar Formation [*c.* 1 km to the northeast at Ballaquane Farm – see Morris *et al.* (1999)]. Exhalation may have occurred in tandem with these extrusions. However, no stratigraphic, spatial association between tourmaline and volcanic or hypabyssal igneous rocks matching that in southeast Ireland (McArdle & Kennan 1992) has been recognized.

Manganese carbonate ironstones – a possible coticule precursor?

Thinly bedded, carbonate-rich, though variably siliceous, beds which are, in most instances, between 1 and 2 cm thick, outcrop extensively in two places only, south of the Niarbyl and at Lady Port to the north of Peel (Fig. 1). The term carbonate ironstone seems entirely appropriate for these layers (Jones 1994).

On outcrop, these beds are typically iron stained (in red, brown and black colours). Many are also folded in a ptygmatic manner (Fig. 5a) but many are not (Fig. 5b). Folded and unfolded layers occur on single outcrops. This disharmonic style of folding is typical of coticule elsewhere and suggests syn-sedimentary slumping. However, individual beds of these ironstones do not display the same compexity of folding that is typical of coticule in southeast Ireland. There, however, later tectonic modification obscures the early history of the folds (Brindley 1954; Kennan 1971).

Some of the individual ironstone beds at Niarbyl [SC 2118 7758] display grading upwards from a granular carbonate base to a phyllosilicate-enriched top. Evidence of way-up is also provided by some individual layers which show small-scale evidence of erosion, e.g. scoured channels filled with relatively coarse material and by occasional beds which show cross-lamination. Opaque mineral grains, typically pyrite, tend to be concentrated in some of the the carbonate layers.

These iron-rich carbonate rocks also contain chlorite, white mica, clinozoisite and secondary Fe- and Mn-oxides. Pelitic interbeds are composed of white mica and chlorite. Some of the chlorite is porphyroblastic. The enclosing pelites may also contain porphyroblasts of carbonate which appears to overgrow an earlier fabric and to either deflect a later crenulation cleavage or, in a few instances, overgrow it (Jones 1994).

The carbonate layers are, perhaps, the most distinctive individual lithological feature in the Creggan Mooar Formation and serve to distinguish the formation from all other units of the Manx Group, other than part of the Lady Port Formation. (Woodcock & Morris 1999; Woodcock *et al.* 1999). The Creggan Mooar Formation otherwise comprises thinly bedded turbidites and laminated siltstones. Most of the formation consists of thinly bedded, grey, exceptionally reddish brown, silt to very fine sand-grade turbidites, such as those which are well exposed in the cove north of Gob ny Gamera [SC 2170 7650] – a log is given for this type section in Fig. 6. The beds, prominently bioturbated on many exposures, range between 1 and 8 cm thick, typically 1–4 cm and exceptionally

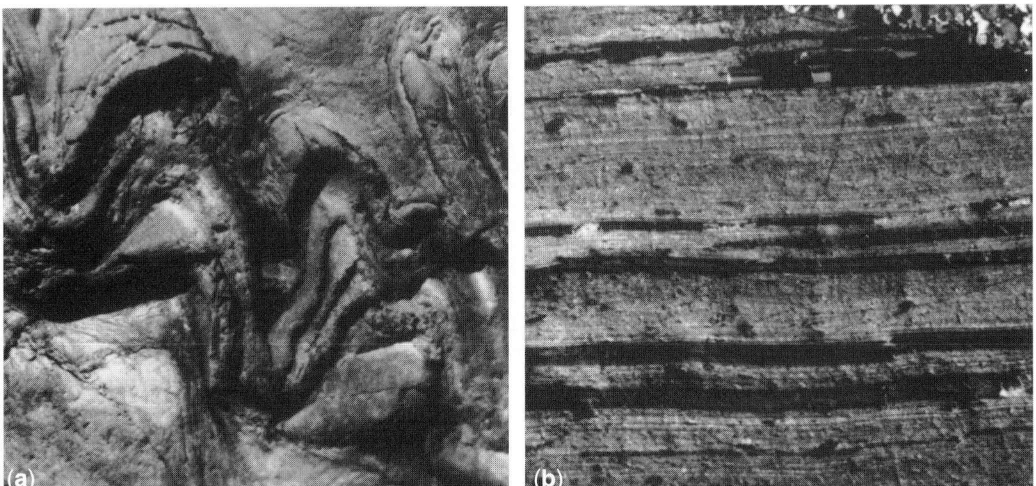

Fig. 5. (**a**) Ptygmatic style of folding typical of many carbonate ironstone layers; Niarbyl Formation. Width of view, 20 cm. (**b**) Typical aspect of relatively unfolded carbonate ironstone layers. Scale, pen.

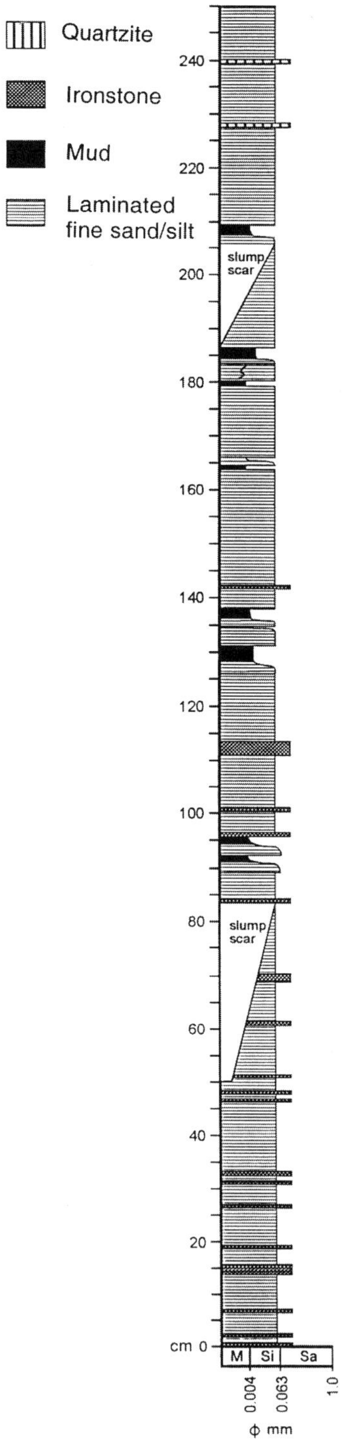

Fig. 6. Log section of interbedded manganiferous carbonate ironstone, pale grey bioturbated laminated siltstone, minor silty turbidites and quartzites in the Creggan Mooar Formation type section at Gob ny Gamera [SC 2170 7650]. Note slump scar annealed by an ironstone bed. (Fig. 5b illustrates the lowermost part of this log section.)

up to 40 cm. The proportion of sand is typically in the 60–70% range. The laminated siltstones (silt to very fine sand grade) occur in thick, monotonous, apparently non-organized, sequences with occasional very thin turbidites. Nearly perfectly spherical spots (diameter *c.* 0.2 mm), singly or in clusters, occur in silty laminae or beds.

Other lithologies include pinstripe laminated silt mudstone and thinly bedded (3–15 cm) quartzite, which is especially significant in the most northerly exposures of the formation at Niarbyl. Here, siliceous ferruginous concretions, ellipsoidal and flattened parallel to bedding, and with surfaces pockmarked by cone-in-cone structures, are also common.

Thin manganiferous–ironstone beds, in all morphological respects identical to those of the Creggan Mooar Formation, also occur in the Lady Port Formation, although restricted to the section of the formation between the prominent deformation zone at the south end of Lynague Strand [SC 2802 8704] to a point *c.* 60 m south of the Gob y Deigan Headland [SC 2837 8738] at the north end of the strand. The beds typically range between 5 and 7 mm in thickness, exceptionally reaching 20 mm, and occur either widely dispersed or, in some instances, very regularly spaced every 20–50 mm. As at Niarbyl, colours vary considerably, from dark brownish-grey, through orange or orange–brown, to red and black. The background sediment to the ironstones varies from very finely laminated pale grey and white bioturbated siltstone to thinly bedded silty turbidites, an assemblage very comparable to that in the Creggan Mooar Formation. The occurrence of the manganiferous ironstone lithofacies as a minor component in the Lady Port Formation may serve as a means of correlation between the two formations.

At Niarbyl Bay, and elsewhere, carbonate occurs with white mica, chlorite and opaque minerals in small, discrete, ovoid nodules. These nodules occur in discrete layers, or as isolated packets or lenses and, in many instances, comprise rims of radially arranged micaceous flakes which may surround a core of opaque minerals (Fig. 7). Many of these nodules are deformed in the main cleavage. In their appearance on outcrop, and in thin section, these nodules bear a close comparison with nodules in the Ribband Group in southeast Ireland in which metamorphic garnet occurs (Shannon 1977). However, in no instance was garnet recognized in either the Niarbyl or Lady Port rocks, although Ford (1993) notes the occurrence of garnet-rich rocks at the Niarbyl.

Chemistry of Isle of Man ironstone and coticule from southeast Ireland

The origin of coticule, and of associated rock types, e.g. tourmalinite, involves volcanogenic seafloor exhalation (e.g. Reinecke *et al.* 1985). In their usual continent margin setting, that exhalation interplays in a variety of ways with craton-derived sediments; e.g. manganese dissolved in sea water may precipitate as oxide in oxygenated shallow water and the precipitated oxide, in turn, adsorbs metals. Exhalated boron in sea water may adsorb on detrital micas. These are but two of a whole spectrum of possibilities for the variable mixing of material from different sources in a sedimentary package. Rocks to which the name coticule has been applied occur in sedimentary packages as different as clastic sequences on continental margins and pelagic sediments on oceanic crust (Kennan 1986) and, in addition, may or may not be associated with nearby mineralization. It is no surprise that non-metamorphosed true coticule precursors have proved difficult to identify – with a possible exception in the Ordovician Tetagouche Group in New Brunswick where red manganese siltstones–cherts seem to be a likely precursor (Gardiner & Venugopal 1992).

Chemical data relating to coticule in the literature is limited (Spry 1990). This is partly due to the fact that the lithology is often not recognized for what it is in the field; a thin section is usually required. There is, as yet, no clear picture as to what defines a coticule in chemical terms. Reported manganese contents vary from an exceptional 23% MnO (Spry & Wonder 1989) to many values of 1% or less. Even the characterizing garnet, often referred to as spessartine or spessartine-rich, is typically a spessartine–almandine solid solution.

Small representative suites of the carbonate ironstone rocks from the Creggan Mooar and Lady Port Formations, and of typical coticule from Leinster, were analysed for this study. In addition, one sample of the possible coticule from the Bungalow Quarry locality (Fig. 1) was analysed. The aim was to evaluate the possibility that the ironstones might be precursors to coticule. The data obtained are summarized in Table 1. In each instance, carbonate and coticule layers were separated as completely as possible from any enclosing rock before X-ray fluorescence spectrometry analysis. The complete data set is presented in Table 1.

One clear conclusion may be drawn. In terms of SiO_2, Al_2O_3, Fe_2O_3, CaO and P_2O_5 major element contents and of V, Cu, Zr, Cr and Ba, the carbonate rocks from Creggan Mooar and Lady Port compare closely with one another and contrast with the typical Leinster coticule. In their contents of every other element, including the rare earth elements (REEs), both the carbonate rocks and the coticule are comparable with the limited and scattered values encountered in the literature on coticule worldwide.

It is clear that the carbonate ironstones do not equate with the typical coticule from southeast Ireland; the latter are, in general, more siliceous and more aluminous. The questionable coticule from the Bungalow Quarry locality compares more closely in its chemistry with the southeast Ireland rocks but does not compare with coticule in its texture; the garnets are fewer and much larger than those of typical coticule. Compared with the type coticule rocks of the Ardennes (Lamens *et al.* 1986), all of the Isle of Man carbonate ironstones are, as might be expected, much richer in iron. They are also richer in magnesium. Some of the ironstones might compare, however, with non-metamorphic quartz–mica–rhodochrosite beds related to coticule in the Ardennes (Lamens *et al.* 1986, table 2).

In the absence of the typical coticule garnet, it is questionable if any chemical comparison between the carbonates and coticule should be driven further at this stage. There is no recognizable and definitive chemical characteristic that allows the certain conclusion that the Isle of Man rocks would, if suitably metamorphosed, become coticule in their mineralogy and texture. Better palaeontological control, or more certain structural evidence, is required before the ironstones can be related to the coticule-bearing formations in southeast Ireland. However, in outcrop aspect, in the complex folds the ironstones display and, in the case of the ironstone occurrences along the shoreline south of Niarbyl, in the nearby occurrence of copious tourmaline, it suggests that a relationship may be there for the proving.

Conclusion

There is no doubt that the manganiferous ironstones in the Creggan Mooar and Lady Port Formations have a coticule aspect on outcrop. However, there is no trace of the diagnostic coticule garnet. The metamorphic spotting, so typical of coticule-bearing sequences elsewhere and that characterizes the northeast–southwest spine of the Isle of Man, does not extend to include the carbonate rocks on the west coast. It is possible to speculate that the

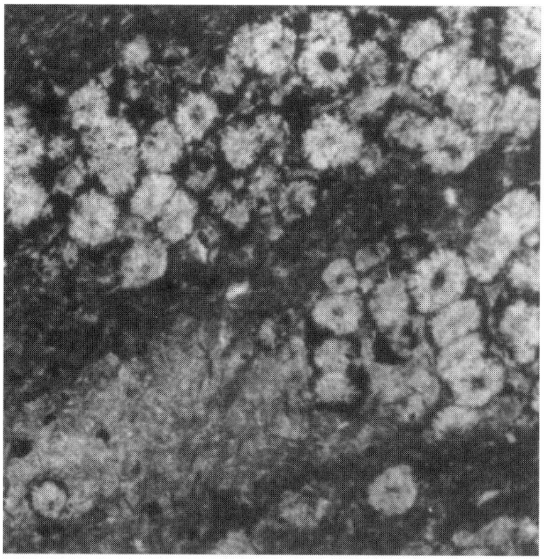

Fig. 7. Typical cluster of nodules with opaques in their cores and radiating micaceous rims in carbonate ironstone; Niarbyl. Width of view, 1.0 cm.

grade of metamorphism was simply too low or that, perhaps, the carbonate rocks are lateral equivalents of manganiferous ironstones that are now coticule elsewhere. It is possible that some of the Isle of Man rocks might bear comparison to epidote-rich rocks that occur in association with coticule, and for which the term 'epicule' was coined by Skeehan & Abu-Moustafa (1976).

The chemistry of coticule rocks, and of the constituent garnets, varies greatly and is reflected in the simple fact that these distinctive lithologies are, in many instances, recognized and mapped in the field only after they have been recognized in thin section. On the Isle of Man, it is the fact that tourmalinite has been recognized in close proximity both to the manganiferous ironstones and to those rocks long since compared with coticule in Leinster and the Ardennes, that suggests that they are both members of a recurring and complex package of exhalative sediments of which coticule is merely one. It is sound structural or fossil evidence that will define their stratigraphic relationship on the Isle of Man and in the lowermost Ordovician – and confirm, or otherwise, the original thoughts of Stainier (1929).

Grateful thanks are due to Mark Jones for his help on the rocks, to John Kennedy for his work on the illustrations, to Aileen O'Brien for her help with the script, and to N. J. Fortey and T. Young for many improving suggestions. John Morris contributes with the permission of the Director, Geological Survey of Ireland.

References

BENNETT, M. A. 1989. Quartz–spessartine metasediments (coticules) and their protoliths in North Wales. *Geological Magazine*, **126**, 435–442.

BRINDLEY, J. C. 1954. The garnetiferous beds of the Leinster Granite aureole and their small-scale structures. *Scientific Proceedings of the Royal Dublin Society*, **26**, 245–262.

BRÜCK, P. M., POTTER, T. L. & DOWNIE, C. 1974. The Lower Palaeozoic stratigraphy of the northern part of the Leinster Massif. *Proceedings of the Royal Irish Academy*, **74B**, 75–84.

——, COLTHURST, J. R. J., FEELY, M. *ET AL.* 1979. Southeast Ireland: Lower Palaeozoic stratigraphy and depositional history. *In*: HARRIS, A. L., HOLLAND, C. H. & LEAKE, B. E. (eds) *The Caledonides of the British Isles – Reviewed*. Geological Society, London, Special Publications, **8**, 533–544.

CLIFFORD, T. N. 1960. Spessartine and Mg-biotite in coticule-bearing rocks from Mill Hollow, New Hampshire, U.S.A. A contribution to the petrology of metamorphosed manganiferous sediments. *Neues Jahrbuch fuer Mineralogie. Abhandlungen*, **94**, 1369–1400.

DOYLE, E. M. 1984. *The coticule rocks of the Lower Palaeozoic Maulin Formation in Southeast Ireland*. PhD Thesis, National University of Ireland.

FORD, T. D. 1993. *The Isle of Man*. Geologists' Association Guide, 46.

GALLAGHER, V. & KENNAN, P. S. 1992. Tourmaline on the margin of the Leinster Granite, southeast Ireland: petrogenetic implications. *Irish Journal of Earth Sciences*, **11**, 131–150.

GARDINER, W. W. & VENUGOPAL, D. V. 1992. Spessartine–quartz rock (coticule) occurrences in New Brunswick, Canada, and their use in exploration for massive sulphide, tin–tungsten and gold deposits. *Transactions of the Institution of Mining and Metallurgy*, **101B**, 147–157.

GILLOTT, J. E. 1955. Metamorphism of the Manx Slates. *Geological Magazine*, **92**, 141–154.

HARKER, A. 1950. *Metamorphism*. Methuen.

JONES, M. 1994. *Coticule and related rocks: their significance in the Caledonian–Appalachian Orogen*. PhD Thesis, National University of Ireland.

KENNAN, P. S. 1971. Porphyroblast rotation and the kinematic analysis of a small fold. *Geological Magazine*, **108**, 221–228.

—— 1972. Some curious garnet clusters from the Garnetiferous Beds of the Leinster Granite Aureole. *Geological Magazine*, **109**, 165–170.

—— 1986. The coticule package: a common association of some very distinctive lithologies. *Aarkundige Mededelingen*, **3**, 139–148.

—— & KENNEDY, M. J. 1983. Coticules – a key to correlation along the Appalachian–Caledonide Orogen? *In*: SCHENK, P. E. (ed.) *Regional Trends in*

the Geology of the Appalachian–Caledonide–Hercnian–Mauritanide Orogen. Riedel, 355–361.
—— & MURPHY, F. C. 1993. Coticule in Lower Ordovician metasediments near the hidden Kentstown Granite, County Meath: a petrographic study. *Irish Journal of Earth Sciences*, **12**, 41–46.
KRAMM, U. 1976. The coticules (spessartine quartzites) of the Venn–Stavelot Massif, Ardennes, a volcano-clastic metasediment? *Contributions to Mineralogy and Petrology*, **56**, 135–155.
LAMENS, J., GEUKENS, F. & VIANNE, W. 1986. Geological setting and genesis of coticules (spessartine metapelites) in the Lower Ordovician of the Stavelot Massif, Belgium. *Journal of the Geological Society, London*, **143**, 253–258.
LAMPLUGH, G. W. 1903. *The Geology of the Isle of Man*. Memoir of the Geological Survey, UK. HMSO.
MCARDLE, P. & KENNAN, P. S. 1992. Deformational and stratigraphic influences on mineralisation in SE Ireland. *Mineralium Deposita* , **27**, 213–218.
MORRIS, J. H., WOODCOCK, N. H. & HOWE, M. 1999. The Silurian succession of the Isle of Man: the late Wenlock Niarbyl Formation, Dalby Group. *This volume.*
POWER, G. M. & BARNES, R. P. 1999. Relationships between metamorphism and structure on the northern edge of Eastern Avalonia in the Manx Group, Isle of Man. *This volume.*
REINECKE, T., OKRUSCH, M. & RICHTER, P. 1985. Geochemistry of ferromanganoan metasediments from the Island of Andros, Cycladic Blueschist Belt, Greece. *Chemical Geology* , **53**, 249–278.
RENARD, A. 1878. Sur la structure et la composition minéralogique du coticule et sur ses rapports avec le phyllade oligistfière. *Mémoires Courronés de l'Académie Royale Belge* , **41**, 1–42.
SCHOFIELD, D. I., VAN STAAL, C. R. & WINCHESTER, J. A. 1998. Tectonic setting and regional significance of the 'Port aux Basques Gneiss', SW Newfoundland. *Journal of the Geological Society, London*, **155**, 323–334.
SHANNON, P. M. 1977. Diagenetic concretions from the Ribband Group sediments of County Wexford, Ireland. *Geological Magazine*, **114**, 127–132.
SIMPSON, A. 1963. The stratigraphy and tectonics of the Manx Slate Series, Isle of Man. *Quarterly Journal of the Geologoical Society, London*, **119**, 367–400.
—— 1964. The metamorphism of the Manx Slate Series, Isle of Man. *Geological Magazine*, **101**, 20–36.
SKEEHAN, J. W. & ABU-MOUSTAFA, A. A. 1976. Stratigraphic analysis of rocks exposed in the Wachusett–Marlborough tunnel, east-central Massachusetts. *In*: PAGE, L. R. (ed.) *Contributions to the Stratigraphy of New England.* Memoir of the Geological Society of America, **148**, 217–240.
SLACK, J. F. 1996. Tourmaline associations with hydrothermal ore deposits. *In*: GREW, E. S. & ANOVITZ, L. M. (eds) *Boron: Mineralogy, Petrology and Geochemistry*. Reviews in Mineralogy, **33**, 559–643.
SPRY, P. G. 1990. Geochemistry and origin of coticules (spessartine–quartz rocks) associated with metamorphosed massive sulphide deposits. *In*: SPRY, P. G. & BRYNDZIA, L. T. (eds) *Regional Metamorphism of Ore Deposits and Genetic Implications*. Utrecht, 49–75.
—— & WONDER, J. D. 1989. Manganese-rich garnet rocks associated with the Broken Hill lead–zinc–silver deposit, New South Wales, Australia. *Canadian Mineralogist*, **27**, 275–292.
STAINIER, X. 1929. Le metamorphisme des Regions de Bastogne et de Vielsalm. Bulletin. Société Géologique de Belgique, **34**, 112–156.
STANTON, R. L. 1976. Petrochemical studies of the ore environment at Broken Hill, New South Wales, Parts 1,2,3,4. *Transactions of the Institution of Mining and Metallurgy*, **85B**, 33–46, 118–131, 132–141, 221–233.
THEUNISSEN, K. 1970. L'andalousite et ses phases de transformation dans la région de Vielsalm. *Annales de la Société Géologique de Belgique*, **73**, 363–382.
WOODCOCK, N. H. & MORRIS, J. H. 1999. Debris flows on the Ordovician margin of Avalonia: Lady Port Formation, Manx Group, Isle of Man. *This volume.*
——, ——, QUIRK, D. G. *ET AL.* 1999. Revised lithostratigraphy of the Manx Group, Isle of Man. *This volume.*

Debris flows on the Ordovician margin of Avalonia: Lady Port Formation, Manx Group, Isle of Man

N. H. WOODCOCK[1] & J. H. MORRIS[2]

[1]*Department of Earth Sciences, University of Cambridge, Cambridge CB2 3EQ, UK*

[2]*Geological Survey of Ireland, Beggars Bush, Haddington Road, Dublin 4, Ireland*

Abstract: The Lady Port Formation (Arenig, Lower Ordovician) occurs in a fault-bounded tract on the northwestern coast of the Isle of Man. It forms part of the Manx Group, deposited on the outboard edge of the Avalonian margin of Gondwana. The formation is dominated by laminated siltstones and mudstones deposited from low-concentration turbidity flows, and rare manganiferous ironstones. This facies is overlain in places by thin-bedded sand-based turbidites with conspicuously bioturbated mud tops. These sands are quartzose wackes and contrast with a third facies of quartzose arenite turbidites that punctuate the laminated mudstones. This arenite facies suggests a discrete inboard source of clean sand which was occasionally diverted into this part of a predominantly mud-prone basin.

The Lady Port Formation also features units of pebbly mudstone and associated intervals of quartz arenite or mudstone showing strongly disrupted bedding. The pebbly mudstones are the products of cohesive debris flows. Clast compositions distinguish debrites derived from the bedded facies with which they are now intercalated from debrites that contain further travelled, but still intrabasinal, clasts. The disrupted facies preserves a spectrum of soft-sediment fragmentation of bedding, culminating in bed fragments being incorporated into the base of pebbly mudstone units. The facies is interpreted as the product either of downslope slumping, the possible precursor to debris flows, or of deformation ahead of and below advancing debris flows.

The local derivation of slumps and debrites implies an actively forming topography in this part of the Manx Group basin. The slumps and debrites of the Lady Port Formation are matched in other, less well-exposed, parts of the northwestern Manx Group. The available biostratigraphic control suggests that the formation is of late Arenig (Fennian) age and contemporaneous with the Buttermere olistostrome and pebbly mudstones in the Kirk Stile Formation of the Skiddaw Group, Lake District. A widespread mass wastage event is indicated on the Gondwana margin, perhaps coeval with the rifting from Gondwana of the Avalonian microcontinent.

The outboard margin of early Ordovician Gondwana is preserved in the Ribband Group of southeast Ireland (e.g. McConnell *et al.* 1999), the Manx Group of the Isle of Man and the Skiddaw Group of the Lake District (e.g. Stone *et al.* 1999). These units comprise mostly deep-marine mudstones and turbidites, deposited on the northwestern edge of an Eastern Avalonian terrane. This terrane was probably attached to the main Gondwana continent to the south at the beginning of Arenig time but had rifted from it by the latest Arenig (e.g. Pickering & Smith 1995; Prigmore *et al.* 1997).

Distinctive components of these early Ordovician sequences are the olistostromes, pebbly mudstones and slumps that remobilize parts of the bedded sequence. These features have been described in the Ribband Group (Shannon 1978; Max *et al.* 1990) and Skiddaw (Webb & Cooper 1988; Cooper *et al.* 1995) Groups. However, it is the Manx Group that spawned the most prominent early records of pebbly mudstones and provoked debate over their origins (Lamplugh & Watts 1895; Lamplugh 1903; Blake 1905). The Manx Group offers particularly informative coastal sections of pebbly mudstones, slumps and quartzose turbidites, which have remained undescribed in detail. These exposures, of the redefined Lady Port Formation (Woodcock *et al.* 1999*b*), are documented in the present paper. The depositional processes that they reveal will allow better comparison and correlation of the Manx Group debrites with other Avalonian margin examples. Are these local phenomena along the margin, or is there a widespread mass-wastage event with a regionally important trigger?

Stratigraphy and setting

The Manx Group crops out over about three-quarters of the Isle of Man (Fig. 1), overlain tectonically by the Silurian Dalby Group and

From: WOODCOCK, N. H., QUIRK, D. G., FITCHES, W. R. & BARNES, R. P. (eds) 1999.
In Sight of the Suture: the Palaeozoic geology of the Isle of Man in its Iapetus Ocean context.
Geological Society, London, Special Publications, **160**, 121–138. 1-86239-046-0/99/$15.00

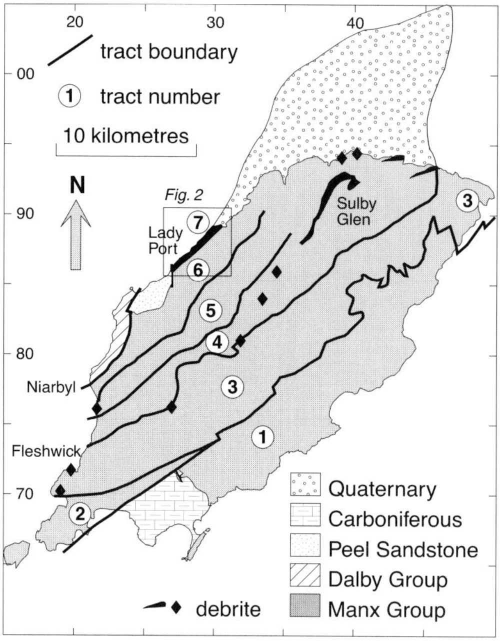

Fig. 1. Outline geological map of the Isle of Man, with locations of the Lady Port Formation and other units bearing pebbly mudstones.

Lower Devonian Peel Sandstones in the west, stratigraphically by Carboniferous rocks in the south, and a thick Quaternary sequence in the north. The lithostratigraphy is reviewed and partly redefined by Woodcock *et al.* (1999*b*). Limited biostratigraphical control allows the Manx Group to range from Tremadoc, possibly into Llanvirn time, although most of the sequence probably lies within the Arenig (Cooper *et al.* 1995; Molyneux 1999; Orr & Howe 1999).

The Lady Port Formation has been redefined by Woodcock *et al.* (1999*b*) to include all the fault-bounded tract exposed on the northwest coast of the Isle of Man from Will's Strand to Glen Mooar (Figs 1 and 2). This tract was previously assigned by Simpson (1963) to three discrete units; in upward sequence the Lady Port Banded Group, the Ballanayre Slump Breccia and the Niarbyl Flags. However, none of these divisions can yet be mapped as coherent units within or beyond the tract and their separate definition as formal lithostratigraphical units is considered premature. Units of 'slump breccia' occur at several horizons, each with distinct lithological character. The local Niarbyl Flags probably lie between slump horizons rather than at the top of the sequence (Figs 2 and 3). Moreover, these Niarbyl Flags, whilst comprising sand turbidites, are lithologically distinct from the type Niarbyl Formation southwest of Peel (Morris *et al.* 1999). The variants of the Lady Port Formation are simply described as separate lithofacies in this paper.

The redefined Lady Port Formation contrasts lithologically and structurally with the succession immediately inland, best displayed in Glion Cam (Fig. 2). A northeast–southwest fault, the Ballakaighin Fault, between the two sequences seems the most plausible structural solution, and a branch of the same fault system has been mapped on the coast at Gob ny Creggan Glassey. To the southwest, the Ballakaighin Fault may join north–south faults that eventually fault down the Peel Sandstones, or it may be truncated by those faults. Roberts *et al.* (1990, p. 276) have speculated that pebbly mudstones of the Lady Port Formation may reappear southwest of the Peel Sandstones in a poorly exposed fault-bounded tract at Peel Harbour.

The fault-bounded nature of the Lady Port Formation precludes its direct correlation with the rest of the Manx Group. The acritarch evidence of Molyneux (1979, 1999) and Cooper *et al.* (1995) suggests a late Arenig (Fennian) age, and that the formation occurs high in the Manx Group, rather than at its base as Simpson (1963) proposed. Woodcock *et al.* (1999*b*) have speculated that the black mudstones of the Lady Port Formation may correlate, at this level, with the Glen Rushen and Barrule Formations in other tracts to the southeast.

History of research

Henslow (1821) noted: 'Near Ballaneah [Ballanayre], I observed the cliffs to consist of angular fragments of clay–slate embedded in a clay–slate paste, and what is curious, these fragments are scarcely to be distinguished from the base, excepting on the surface of the rock which has been exposed to the action of the waves, where they become sufficiently apparent by the fragments assuming different tinges of colour, giving the specimen a mottled appearance.' This early record of pebbly mudstones in the Manx Group captures their essential character as matrix-supported conglomerates containing angular clasts that differ little from the surrounding bedded lithologies and, often, from the matrix itself.

Lamplugh (Lamplugh & Watts 1895; Lamplugh 1903; Geological Survey 1898) mapped these rocks as 'crush conglomerates' more extensively through the Manx Group (Fig. 1). He focused on an extensive inland outcrop centred on Sulby Glen, but recognized other examples between Fleshwick Bay and Niarbyl in the southwest, and on both the northwest and southeast flanks of the Sulby Glen

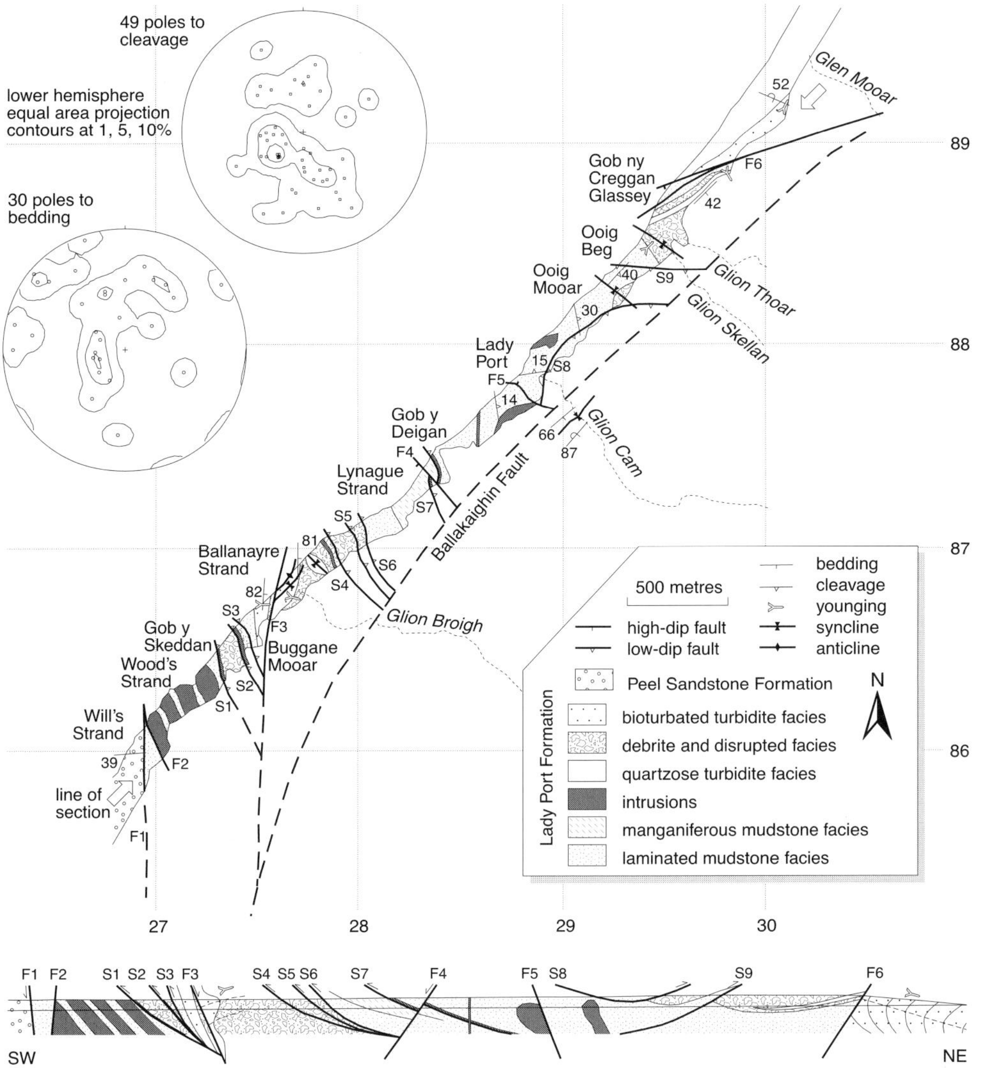

Fig. 2. Detailed map and cross-section through the Lady Port tract, with lower hemisphere equal area projections of poles to bedding and the main cleavage.

outcrop (Fig. 1). Lamplugh & Watts (1895) recognized that some crush conglomerates have a mappable continuity at particular stratigraphic levels. They also stressed the derivation of the conglomerates by progressive break-up of local bedded sequences, particularly of thin-bedded sandstones and shales. They ascribed this break-up to tectonic deformation, particularly at strength contrasts within the Manx Group sequence, so that the conglomerates were seen essentially as thick zones of fault rock. However, Lamplugh & Watts (1895) recognized that the break-up of the bedded sequence occurred before the introduction of either dykes or quartz veins, that it pre-dated the main cleavage and was not associated with an enhanced metamorphic grade.

The tectonic origin of the pebbly mudstones was doubted by Blake (1905). Referring particularly to the coast sections at Ballanayre, he noted the sharp rather than gradational nature of many contacts of the pebbly mudstone units and a degree of clast disruption and mixing incompatible with a purely

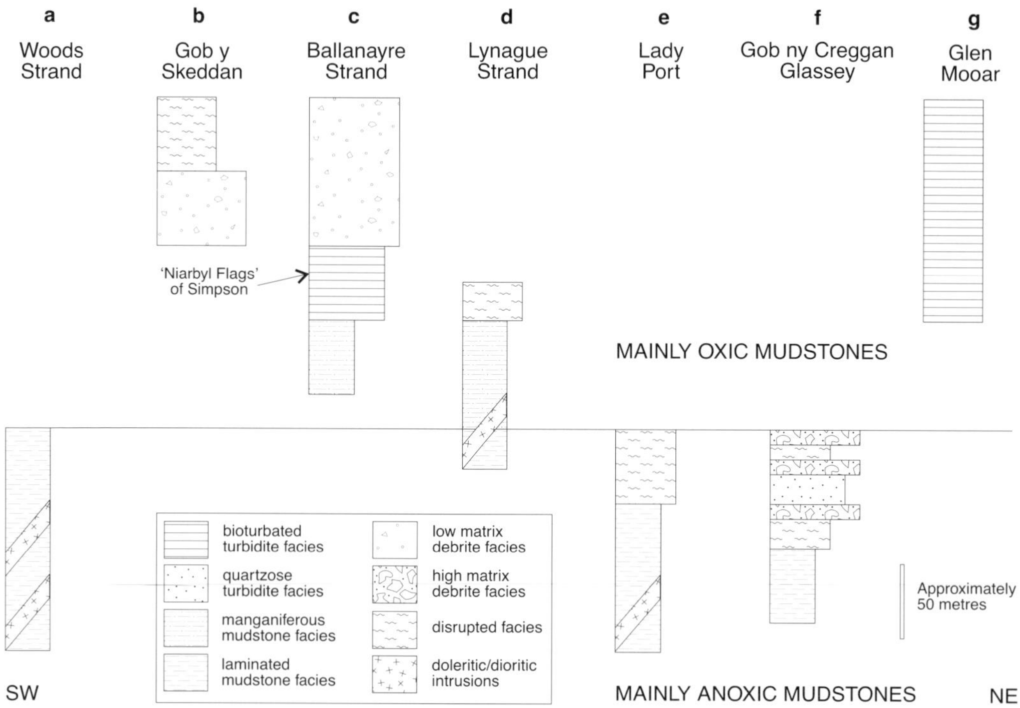

Fig. 3. Simplified summary logs through the Lady Port Formation, showing its postulated internal stratigraphic organization. Locality names are shown on Fig. 2.

cataclastic origin. Blake regarded these 'schistose breccias' as primary fragmental rocks, by implication of sedimentary origin. Gillott (1956) reinforced Blake's evidence for a primary origin, noting particularly the conformable contacts between breccias and interbedded stratified sediments, and the absence of obvious slickensided shear planes associated with the brecciation. Gillott was also able to draw on evidence from other examples of pebbly mudstones to interpret the Manx occurrences as the product of submarine sliding of partially lithified sediments.

Simpson (1963) mapped the Ballanayre and Sulby Slump Breccias as two of 11 lithological units within the Manx Group stratigraphy. His structural interpretation precluded correlation of these two major slump units. Simpson mapped out the bedded Lady Port Unit below the Ballanayre Slump Breccia. However, his structural model also required that a unit of sand turbidites, assigned to the Niarbyl Flags, overlay the slump breccia, a relationship disputed in the present study.

Since Simpson's (1963) study, Molyneux (1979) and Roberts *et al.* (1990) have reported biostratigraphical and metamorphic data, respectively, from the Lady Port Formation. However, the detailed facies relationships and their interpretation have remained undescribed.

Structure and sequence

The structure of the coastal sections of the Lady Port Formation is complex. It is described here in enough detail only for the components of the original stratigraphic sequence to be identified, and for structures and fabrics of tectonic origin to be distinguished from those due to early soft-sediment deformation. Five sedimentary lithofacies are distinguished on the geological map and section (Fig. 2), and these are subdivided further to give seven lithofacies on the tentative reconstruction of stratigraphic sequence (Fig. 3). For economy of description, the genetic terms turbidite and debrite are used for graded sandstone–mudstone couplets and for pebbly mudstone, respectively. These facies are detailed in later sections, where their origin is justified.

The structural cross-section (Fig. 2) is dominated by faults or shear zones of two types. Steep to moderately dipping faults (F1–F6) displace all other structures and contain fault gouge, suggesting late brittle displacements, probably of Mesozoic age. These faults strike mostly north or northwest, although the largest of them (F6) strikes northeast. Several (F3 and F4) show normal displacement senses. The other dominant brittle structures are gently dipping faults or shear zones (S1–S9; Fig. 2), mostly striking northwest but dipping both southwest and northeast. These zones contain quartz vein arrays that cut the main fabric in the Lady Port Formation and commonly suggest a thrust sense of displacement. The zones are often localized at the boundaries of igneous sheets (e.g. S_1, S_2 and S_7). They commonly show a late brittle phase involving rebrecciation of their component quartz vein arrays.

The main fabric in the formation is a penetrative phyllitic foliation. Its gently dipping or flat-lying attitude (stereogram; Fig. 2) suggests correlation with the S_2 fabric over the rest of the Isle of Man (Simpson 1963; Fitches *et al.* 1999). However, the Lady Port fabric only rarely displays the crenulation component that characterizes S_2 elsewhere, and is provisionally regarded here as a gently dipping variant of the S_1 fabric seen elsewhere on the island. The fabric cuts the main igneous intrusions that penetrate the Lady Port Formation and is axial planar to recumbent, or gently inclined close to tight folds. Folds facing northeast and locally downwards on this fabric can be seen in four places between Ballanayre Strand and Glen Mooar.

It is important for later diagnosis of soft-sediment deformation that, during both the ductile and brittle tectonic deformations, mechanical contrasts between the different sedimentary lithologies do not appear to have exerted a strong influence over the outcrop-scale structures and fabrics. Indeed, a prominent ductile shear zone, exposed at the south end of Lynague Strand, defines the boundary between pelite–dominant successions, debrites to the south and manganiferous ironstone bearing siltstones-pelites to the north. This contrasts with soft-sediment deformation styles, whose geometries reflect the ductility contrast of partially lithified sandstones and siltstones with enclosing unlithified mud.

An attempt has been made to correlate the lithofacies within the many fault-bounded slices of the Lady Port Formation, resulting in a bipartite division (Fig. 3). The lower division is dominated by anoxic laminated mudstones, with quartzose turbidites, low-matrix debrites and related disrupted facies. The upper division contains more oxic mudstones, bioturbated turbidites and high-matrix debrites. However, this correlation is constrained mainly by facies transitions observed in only two sections, at Ballanayre Strand (Fig. 3c) and south of Gob ny Creggan Glassey (Fig. 3f), and should be treated as provisional.

Bedded lithofacies: description and interpretation

Laminated mudstone facies

The most abundant lithofacies in the Lady Port Formation is thinly or very thinly laminated, dark grey mudstone, now metamorphosed to phyllite. The lamination is planar and typically defined by interlayering of dark grey mudstone with light grey siltstone (Fig. 4a). The siltstone laminae show either diffuse contacts or sharp bases and gradational tops. They record deposition from dilute waning flows, probably of turbiditic origin. Bioturbation is absent or inconspicuous, suggesting that dysaerobic bottom waters inhibited a burrowing benthos. In places, especially close to units of the quartzose sandstone facies, very thin graded beds of very light grey, quartzose, fine sandstone are intercalated in the mudstones. The laminated siltstone–mudstone lithology corresponds to facies D2.3 of Pickering *et al.* (1986).

The laminated mudstones crop out as screens amongst numerous intrusions between Will's Strand (Ordnance Survey Grid reference [SC 2695 8597]) and Gob y Skeddan [SC 2742 8652]. More representative exposures occur between Gob y Deigan [SC 2842 8741] and Ooig Beg [SC 2945 8843], and at Gob ny Creggan Glassey [SC 2967 8880; Fig. 5a]. At Ooig Beg and Gob ny Creggan Glassey, the mudstones border a quartzose sandstone packet and are correspondingly rich in very thin quartzose sandstone beds.

Manganiferous mudstone facies

In places, the laminated mudstone facies contains a high proportion of finely laminated pale grey–white siltstone or fine sandstone intercalated with the dark grey mudstone. These lithologies either occur in unorganized intervals up to *c.* 10 cm thick, or define 2–30 mm thick beds with parallel-laminated silt overlain by mud. The siltstone intervals are very commonly bioturbated. The sequence also contains occasional 2–3 cm thick, pale grey–white, fine-grained, convolute or cross-laminated quartz arenite beds. However, the particularly distinctive features of this facies are the 2–10 mm thick ironstone beds spaced at 1–7 cm through the sequence (Fig. 4b). These beds vary in colour from medium or buff-grey through orange to dark grey, black or patchy red and black. They are non-graded, very fine grained and invariably have sharp, non-gradational contacts with enclosing sediments. Kennan & Morris (1999) have found these ironstones to contain a high proportion of manganese carbonate. In this respect, the facies is very similar to the Cregganmoar Formation,

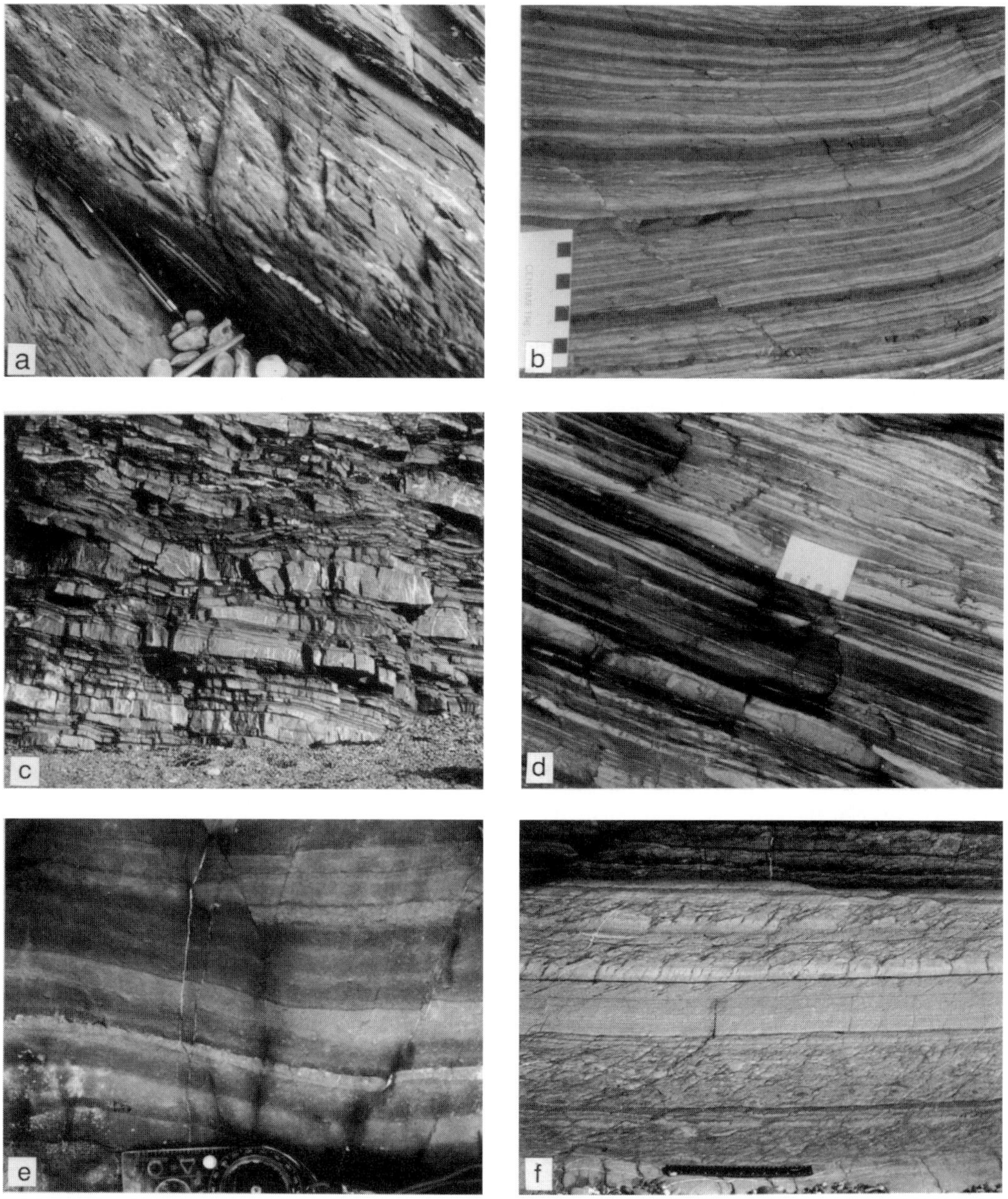

Fig. 4. Bedded facies in the Lady Port Formation. (a) Laminated siltstone–mudstone facies, here with some very thin quartzose sandstone beds (pen length, 13 cm), south of Gob ny Creggan Glassey [SC 2960 8876]. (b) Laminated very fine sandstone and mudstone with reddish brown manganiferous laminae (scale units, 1 cm), Lynague Strand [SC 2834 8733]. (c) Quartzose turbidite facies (ruler length, 31 cm), south of Gob ny Creggan Glassey [SC 2958 8868]. (d) Thin-bedded marginal zone of the quartzose turbidite facies (scale units, 1 cm), south of Gob ny Creggan Glassey [SC 2960 8876]. (e) Bioturbated graded siltstone–mudstone turbidites (compass length, 10 cm), south of Glen Mooar [SC 2983 8895]. (f) Bioturbated graded sandstone–mudstone turbidites (ruler, 31 cm), Ballanayre Strand [SC 2768 8684].

cropping out south of Niarbyl on the west coast, and suggests a possible correlation between different tracts of the Manx Group at this level (Woodcock *et al.* 1999*b*).

The increased bioturbation in the manganiferous mudstone facies compared with the laminated mudstone facies suggests bottom waters with a higher proportion of dissolved oxygen. The facies

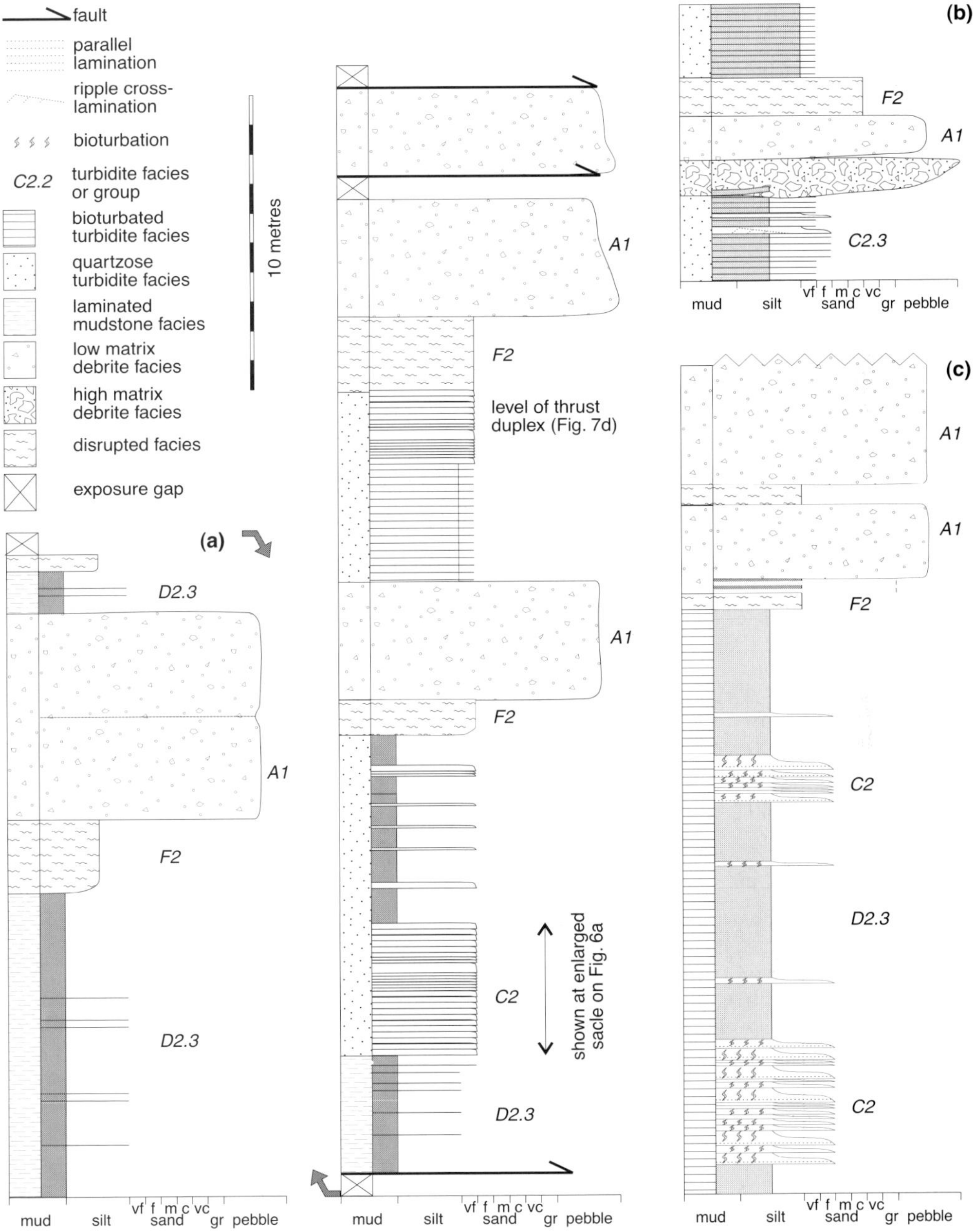

Fig. 5. Logs of debrite-bearing sequences. (a) South of Gob ny Creggan Glassey (between [SC 2958 8868] and [SC 2955 8857]). (b) Between Gob ny Creggan Glassey and Ooig Beg [SC 2955 8861]. (c) North of Ballanayre Strand [SC 2768 8684].

also records a higher input of quartzose silt and very fine sand. The manganese carbonate ironstones themselves are attributed to a high concentration of exhalative metals in the Arenig sea water. They are thought to be the protolith for the garnetiferous coticule beds widely recognized at this level in higher grade rocks of the Caledonian–Appalachian Orogen (Kennan & Kennedy 1983).

The manganiferous facies has been observed in the section between the prominent deformation zone at the south end of Lynague Strand [SC 2802 8704] and a point *c.* 60 m south [SC 2837 8738] of the Gob y Deigan headland, at the north end of the strand. However, sporadic red laminae also occur in the laminated mudstone sequence below Buggane Mooar [SC 2744 8663] and these rocks are tentatively assigned to the same facies (Fig. 3c).

Quartzose turbidite facies

This facies comprises thin, medium or thick beds of quartzose sandstone, each grading rapidly from fine or very fine, light grey sandstone up to dark grey mudstone (Fig. 4c). The sandstones are typically weakly parallel laminated (T_b division), sometimes with a T_c division showing ripple cross-lamination or tabular cross-lamination (Fig. 6a). The facies corresponds to classes C2.1, C2.2 and C2.3 of Pickering *et al.* (1986), and is interpreted as the deposits of low- to medium-concentration turbidity flows tapping a source of relatively clean quartz sand.

The quartzose sandstone facies is concentrated in one 4.5 m thick packet, mappable south of Gob ny Creggan Glassey (between [SC 2983 8887] and [SC 2955 8867]; Figs 4c and 6a). However, this packet grades upwards and downwards into laminated mudstones containing isolated thin beds of the same quartzose sandstone (Figs 4d and 5a).

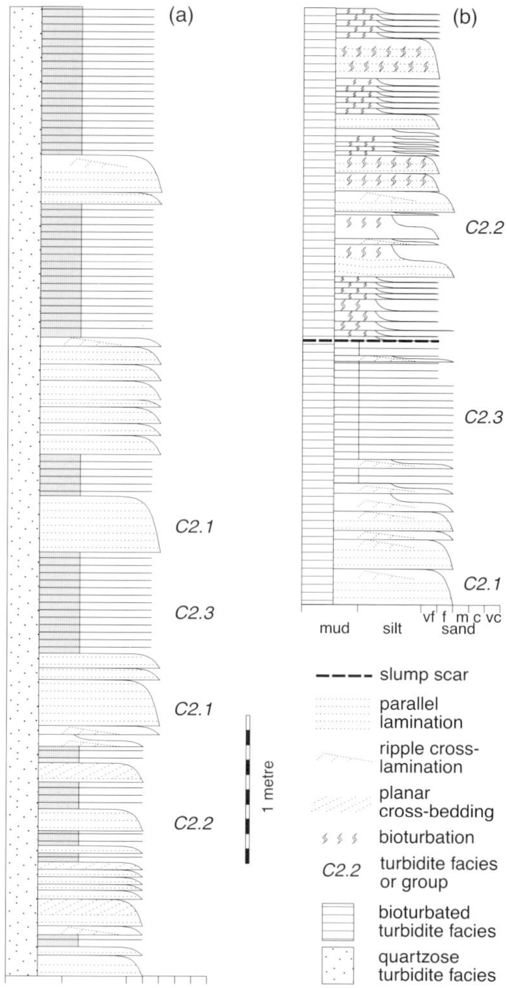

Fig. 6. Logs of contrasting turbidite sequences. (a) Quartzose turbidite facies south of Gob ny Creggan Glassey [SC 2955 8862]. (b)Bioturbated turbidite facies [SC 2772 8688].

Bioturbated turbidite facies

This facies comprises graded beds, typically with their tops conspicuously bioturbated. Most abundant are dark to light greenish grey to olive grey, thin or very thin beds grading from silt to mud (Fig. 4e). The bioturbation affects the top half of each bed, below which a horizon of small ellipsoidal phosphatic nodules is commonly seen. This siltstone turbidite subfacies dominates the sequence north of Gob ny Creggan Glassey, [SC 2973 8890] to [SC 3008 8917]. It is also exposed in the old railway cutting near Ballycarnane [SC 3010 8893], where it contains 1–2 cm thick quartzose sandstone beds at 5–30 cm intervals. Further southwest, at Ballanayre Strand [SC 2763 8678] and below Buggane Mooar [SC 2750 8668], the bioturbated graded facies is coarser and dominated by light grey to greenish grey quartz wacke sandstones, in thin, medium or thick beds (Figs 5c and 6b). Each bed grades from fine or very fine sand up to mud. Parallel and ripple cross-lamination is preserved, but both these divisions are often conspicuously bioturbated (Fig. 4f). The parallel lamination in some thick beds passes laterally into a swaley low angle cross-stratification with weak truncations of laminae. A 3D hummocky geometry is possible but cannot be verified in the 2D exposures.

The bioturbated facies corresponds to classes D2.3, C2.3 and C2.2 of Pickering *et al.* (1986). The facies was probably deposited from low- to medium-concentration turbidity flows. The bioturbation and phosphatic nodules point to better oxygenated bottom waters than in the laminated siltstone–mudstone facies, either persistently or at least in the intervals after introduction of each turbidity flow. In the absence of 3D evidence, the occasional preservation of low-angle cross-stratification is ascribed to abnormal flow conditions, perhaps confined standing waves, in a deep-water environment, and not to any shallow-marine influence.

Disrupted facies

This designation is given to intervals of the bedded facies that have apparently suffered soft-sediment deformation but not disaggregation and redeposition. Disruptive structures range from fitted-fabric fragmentation of beds (Fig. 7b), through pinch-and-swell or pull-apart of bedding, to slump folds and shear zones (Fig. 7a). With increasing deformation, the disrupted facies develops brecciated zones comprising jostled bed fragments, although retaining a ghost stratification. The various structures are almost invariably defined by the mechanically stiffer components of the bedded sequence, mainly quartz sandstones and siltstones. Mudstones behaved plastically during deformation, either passively accommodating to the shape changes in the sandstone layers, or apparently actively injecting between fragments (Fig. 7b) and intruding between layers. In one example (Figs 5b and 7c), a 9 cm thick wedge of matrix-supported debrite intrudes beneath a slab of bedded quartzose turbidite from the underlying disrupted facies sequence, raising the slab into the overlying debrite for subsequent deformation and fragmentation. A single example of a thrust duplex has been noted in medium-bedded quartzose turbidites (Fig. 7d). This structure could represent soft-sediment imbrication

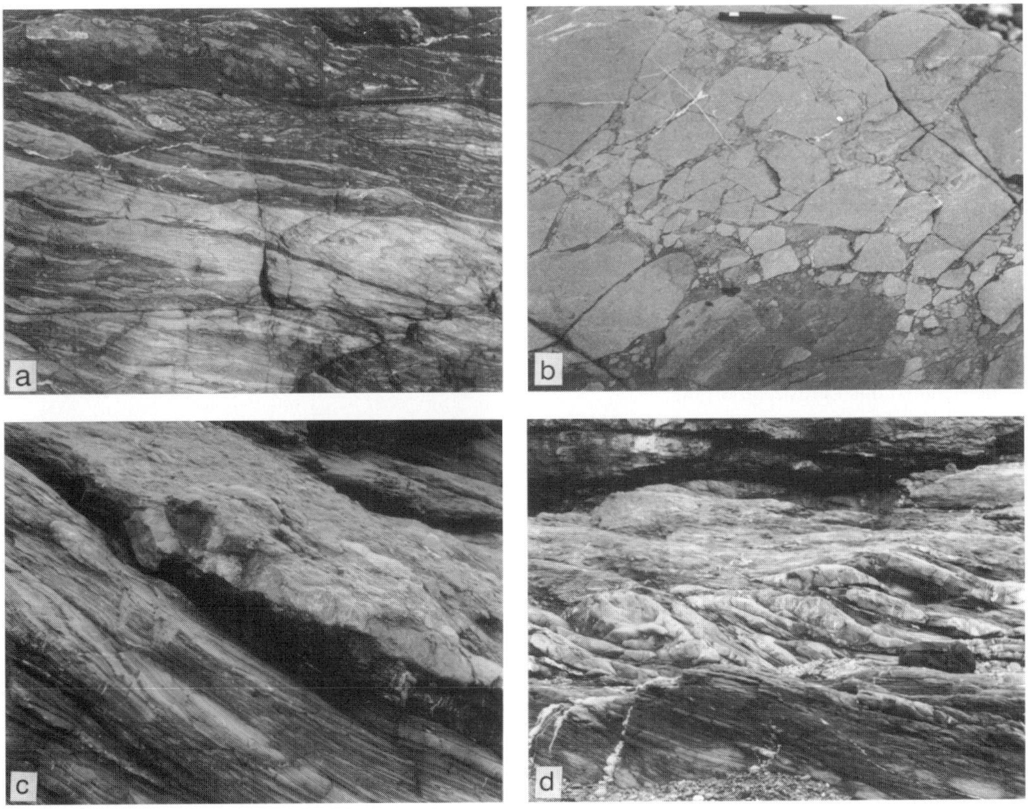

Fig. 7. Disrupted facies in the Lady Port Formation, south of Gob ny Creggan Glassey [SC 2956 8864]. (a) Sheared silts and muds (ruler length, 31 cm). (b) Fitted-fabric break-up of sandstone slab with injected mudstone (pen length, 13 cm). (c) Low-matrix proximal debrite, including a sandstone slab being detached by debrite intrusion, overlying disrupted facies (field width, 1.5 m). (d) Contractional duplex in quartzose sandstones (rucksack length, 50 cm).

within the disrupted facies, although a tectonic origin cannot be ruled out.

The best examples of the disrupted facies occur interbedded with debrites south of Gob ny Greggan Glassey, [SC 2943 8848] to [SC 2963 8873] (Fig. 5a). The facies is less common further south, although examples occur at Gob y Deigan [SC 2838 8740] and Buggane Mooar [SC 2744 8655]. Significantly, disrupted bedding is less common in association with the thick high-matrix debrites between Lynague and Ballanayre Strands, [SC 2772 8688] to [SC 2800 8704]. Localized disruption occurs at the margins of bedded fragments with the debrite, either outsize siltstone blocks or, in one case, a 30 × 10 m block of bioturbated sandstone turbidites. Minor shearing occurs at the contact of the bioturbated turbidites with overlying debrites near Ballanayre Strand (Fig. 5c) and one scar, draped by further turbidites, has been noted within the bioturbated turbidite sequence (Fig. 6b).

The detailed geometries within the disrupted facies, and its complete gradation into clast-rich debrite units, argue strongly for formation when its mud was unconsolidated and its sand partially lithified. The cross-cutting relations of cleavage show that the disruption was predominantly pre-tectonic. The disrupted facies corresponds to classes F2.1 or F2.2 of Pickering *et al.* (1986), and is interpreted as the product of gravity induced sliding or slumping of bedded sediments, or of loading by overriding sediment sheets, either further slumps or debris flows. The evidence of localized injection of matrix-supported debrite into bedded sediment raises the possibility that the entire process of disruption and fragmentation was occurring intrastratally rather than near the sediment surface. The evidence from the pebbly mudstone bodies themselves, described below, argues against this intrastratal origin.

Low-matrix debrites

This distinctive facies comprises a densely packed framework of angular clasts of pale grey quartz wacke or green–grey shale, with only a subordinate amount of dark grey mudstone matrix (Figs 5b and 8a). The clasts vary from fine gravel to cobble size, with occasional outsize boulders up to 30 cm, but are mainly coarse to very coarse gravel size (Fig. 9a). The clasts are notably angular to subangular (Fig. 9c), with a preponderance of wedge to equant shapes, and now occupy *c.* >80% of the rock volume. Spalling of angular fragments from internal sandstone blocks is common. The composition of clasts in any single unit is very restricted and closely matches its adjacent protolith. In the unit sampled in detail (Fig. 9b), green–grey shale is

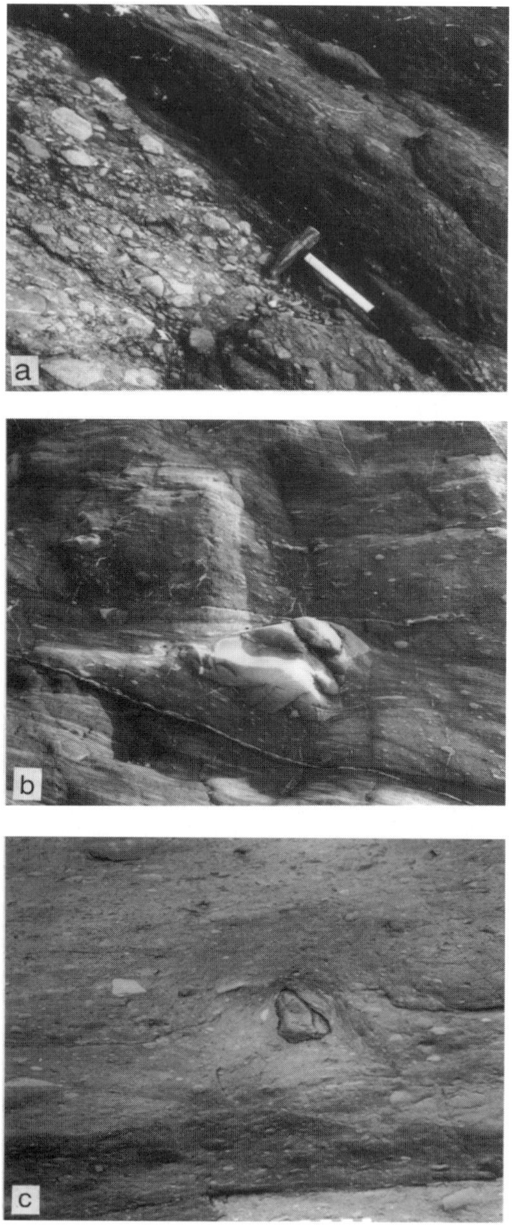

Fig. 8. Debrites in the Lady Port Formation. (a) Low-matrix proximal debrite overlain by high-matrix distal debrite (hammer, 35 cm), south of Gob ny Creggan Glassey [SC 2955 8861]. (b) High-matrix distal debrite (field width, 2 m), south of Gob ny Creggan Glassey [SC 2956 8863]. (c) High-matrix distal debrite (field width, 60 cm), south of Lynague Strand [SC 2785 8700].

common at clast sizes < 4 mm but more resistant grey quartzites dominate larger clast suites.

The 1.25 m thick unit of low-matrix debrite shown in Fig. 5b is instructive. Its upward transition from disrupted facies has already been described (Fig. 7c). Its upper contact shows the low-matrix debrite grading up into a high-matrix unit by a rapid diminution in clast size, coupled with a sharp increase in the proportion of pelite matrix (Figs 8a and 5b). Both of these units wedge out laterally from *c.* 1.5 m each to 50 cm over a distance of 6 m, the texture fining progressively into low-matrix gravel.

Low-matrix debrites are most common in the sections from Gob ny Greggan Glassey to Ooig Beg, [SC 2943 8848] to [SC 2963 8873], interstratified with disrupted facies, high-matrix debrites, quartzose turbidites and laminated mudstones. Here, the low-matrix units are interpreted as the products of surficial debris flows that have moved only a short distance from their parent strata. These units are therefore intermediate between the essentially in-place disrupted facies and the further travelled high-matrix debrites. In the best documented example (Fig. 5b), low-matrix debrite apparently formed the basal interval to an integrated debris flow that had a high-matrix upper interval. The contact above the low-matrix interval reveals a normal sedimentary grading and precludes the possibility of intra stratal injection of the low-matrix lithology. Whilst injection of fluidized sediment therefore played a role in incorporating material into the base of the Lady Port debris flows, the balance of evidence favours each flow having a free upper surface.

High-matrix debrites

Lithological character and interpretation

Although thin units of high-matrix debrite occur in the area south of Gob ny Creggan Glassey (Fig. 8b), the most extensive exposures are further south, between Ballanayre Strand [SC 2772 8688], through the sea caves, to the shear zone at the south end of Lynague Strand [SC 2800 8704]. These latter exposures, comprising an outcrop width of *c.* 300 m, are undoubtedly those first referred to by Henslow (1821).

Most of the Ballanayre debrite consists of non-graded, non-stratified gravel- and cobble-sized clasts set in a medium to dark grey fissile mudstone or mudstone–sandstone matrix (Fig. 8c). Occasional outsize boulders up to 1.2 m are also present, typically comprising pale buff, cream or medium–dark grey, fine- to coarse-grained, massive, laminated and cross-laminated quartzite, although there are also rare conglomerate clasts. Any internal bedding planes are truncated sharply at the edges of the clasts, suggesting that the resistate fragments were partly lithified prior to disruption. Also present is a single block of bioturbated turbidites, *c.* 8 m high by *c.* 30 m wide, containing a reclined syncline. In the bulk of the Ballanayre deposit, the clasts are matrix supported and are sometimes very widely dispersed.

The high-matrix units correspond to facies class A1.3 or A1.2 of Pickering *et al.* (1986) and are interpreted as the products of cohesive debris flows. Strong evidence is provided by the general absence of sorting, grading or stratification. A matrix with appreciable yield strength is indicated by the outsize metre scale blocks and the many clasts discordant to the general fabric alignment. The high-matrix debrites are seen as the most mobile end of the spectrum of fragmentation and redeposition, represented by the sequence from disrupted facies through low-matrix debrites. However, two particular questions concerning the high-matrix debrites need to be addressed; the degree of debris flow transport indicated by clast texture and compositions, and the magnitude of each event diagnosed from basal and interflow boundaries.

Clast texture and composition

Clast counts on outcrop surfaces and cut slabs reveal a modal clast size at Ballanayre in the 8–16 mm range, and at Ooig Beg in the 2–4 mm range. Both samples are skewed towards coarser clasts (Fig. 9a). The maximum clast size noted in this analysis is 50 cm, but a 1.2 m clast has been observed elsewhere and Lamplugh (1903) noted a single block measuring 4.2 m. Most clasts are subangular to subrounded (Fig. 9c). They are tabular or, less commonly, nearly equant, and are, on average, aligned parallel to the cleavage. There is, however, no evidence, other than in local shear bands, of tectonically modified shapes, and indeed many apparently unmodified clasts lie discordant to the cleavage. Only occasional clasts show any evidence of reworking by angular fracturing of previously rounded clasts. Rare well-rounded, subspherical clasts, mainly quartz arenites, are anomalous by comparison with other clasts. Lamination is notably deflected below one of these clasts.

The range of clast composition in any one unit is limited (Fig. 9b). At Ballanayre, the principal lithologies are pale buff and grey laminated quartzite, pale grey siltstone, medium–dark grey shale, rare black shale and extremely rare clasts of other types, including off-white coloured, aphanitic felsite. These lithologies can be matched with those in the Lady Port Formation in general but not with the immediately underlying bioturbated turbidites to the south or the manganiferous mudstones to the north. Shale clasts dominate only the very coarse sand grade (1–2 mm) and coarser fractions are dominated by resistate components, mainly quartzite (Fig. 9b); shale clasts > 32 mm are very rare. The bulk of the pelitic component is, of course, concentrated in the debrite matrix and therefore not

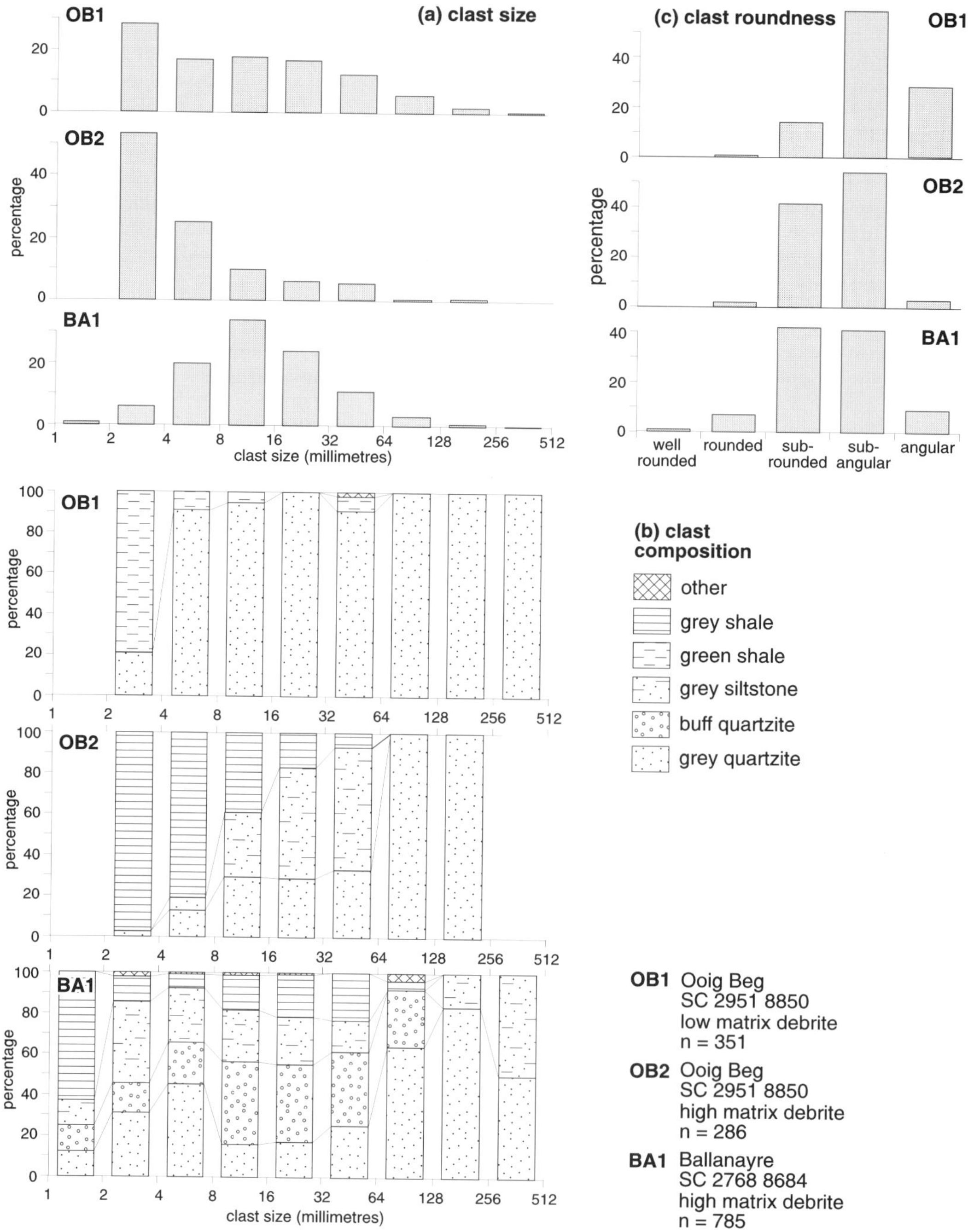

Fig. 9. Results of clast counts at one site in low-matrix debrite and two sites in high-matrix debrite. (a) Distribution of maximum dimension of clasts. (b) Proportion of clast compositions in each clast size interval. (c) Visual estimate of roundness (Pettijohn scale).

reflected in these data. At Ooig Beg (Fig. 9b) the grey shale clasts predominate in the finer clast ranges (2–8 mm) and the more resistant grey siltstone and grey sandstone clasts in the coarser range (8–256 mm).

The subangular to subrounded clast shapes, coupled with the sharp truncation of internal bedding fabrics at clast margins, strongly suggest that the quartzite, siltstone and some of the pelite were, at least partially, lithified prior to disruption of the host sequence. The observed compositional patterns are ascribed to disruption of this sequence close to its source and the progressive attrition and fragmentation of the pelitic component during transport. The clast assemblage and the clast:matrix ratio of a debrite unit can, however, be expected to reflect the composition of its source sequence. The poor match of the clasts in the high-matrix debrites at Ballanayre with the immediately surrounding bedded sequences shows that their generating debris flows travelled further from their source than the generally lower matrix flows in the assemblage south of Gob ny Creggan Glassey. A proximal to distal increase in matrix is therefore superimposed on the texture inherited from the disrupted protolith for each flow.

Stratified intervals

The basal contact of the high-matrix debrite sequence with underlying bioturbated turbidites is well exposed on the southwest side of Ballanayre Strand [SC 2759 8677] and particularly beyond the headland to the north [SC 2772 8688]. Here, the contact is marked by an abrupt, but perfectly conformable, transition from T_{ae} and T_{be} turbidites into a 55 cm thick zone of plane-parallel laminated sandstones, interstratified with gravelly horizons *c.* 10 cm thick, overlain by debrite proper. The two upper laminar zones contain isolated 'floating' clasts up to 7 cm. The lower 20 cm of the debrite is reverse graded before assuming its typical, massive non-graded aspect. This contact zone is interpreted as the product of laminar shear in the boundary zone at the base of the debris flow.

Similar stratified boundary layers might be expected at other basal flow contacts within the high-matrix debrite sequence and to provide a guide to typical flow thicknesses. Only one other such zone has been noted, *c.* 30 m north of the basal contact [SC 2773 8692]. Here, a succession of 3–5 cm, clast-supported, graded, medium gravel to coarse sand beds, are spatially associated with a laterally impersistent contact between clast-supported gravel and interbedded coarse sand horizons up to 15 cm thick. This bedded zone overlies debrite which, immediately below, contains several poorly stratified 2 cm thick fine to medium gravel bands intercalated with 1 cm thick, coarse sand horizons. The sedimentary structures in this zone clearly reflect discrete depositional events, presumably at a contact between thick debrite units. The laminated zones are not compatible with a possible injection origin for the high-matrix debrites. On this hypothesis, the boundaries of flow units should show intrusive relationships with incipient brecciation of wall rock.

With due allowance for an assumed shallow sheet dip of *c.* 30°, the thickness of the debrite between the two laminated zones is *c.* 15 m, providing the only guide to the order of flow thickness in the high-matrix debrite sequence. It is possible that the large block of bioturbated turbidites within the debrite 10 m north of the upper stratified zone [SC 2777 8693] also represents bedded sediment deposited between debris flow events. Its lower, seaward contact is transitional over 3.5 m from well-bedded, silty turbidites, through a zone of intense cleavage with wispy and transposed bedding, into intensely foliated debrite with outsize tabular cobbles and boulders of silty turbidite. A similar relationship is locally evident on the mainly fault-defined north side of the block, but neither contact precludes the possibility that the block is a raft within a flow rather than an interflow horizon.

Curious features in the bedded turbidite block are the abnormally well-rounded, gravel-sized clasts occurring either in isolation or in trains parallel to bedding. These are reminiscent of the floating clasts in the basal boundary zone of distal debrite, but cannot share the same origin. Faintly defined, heavily bioturbated bedding, notably deflected below one such clast but planar across the top of the clast, suggests that these particular clasts may even be dropstones. The occasional anomalously well-rounded clasts in the debrite itself might have a similar origin. However, there is no other indication of a syn-glacial origin and no suggestion, globally, of an Arenig glacial event (Hambrey & Harland 1981).

Processes of fragmentation and resedimentation

An integrated picture can be reconstructed from the Lady Port Formation of the progressive disruption and fragmentation of bedded sequences and their transformation into moving debris flows (Figs 10 and 11). Descriptions of debrites that can be so intimately linked with their parent sequence seem rare in the literature and new clues to debris flow behaviour can therefore be gained from the Lady Port examples.

The bedded protolith for the Lady Port debrites typically comprised laminated mudstones with

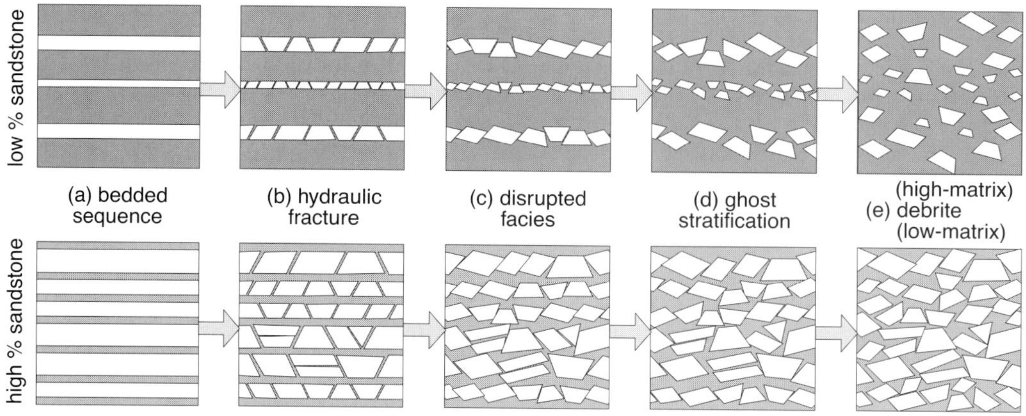

Fig. 10. The stages [(a)–(e)] in fragmentation of a bedded protolith to produce a pebbly mudstone, for two sequences with different ratios of sand:mud.

variable proportions of quartzose sandstone and siltstone beds, which the clast shape analysis suggests underwent early partial cementation. The proportion of 'brittle' sand or silt to 'plastic' mud in the protolith was particularly important in determining subsequent behaviour (Fig. 10a).

The earliest stages of disruption often involved a fitted-fabric brecciation of sandstones and siltstones, accompanied by the injection of mud between the fragments (Figs 7b and 10b). The fabrics are similar to those produced by hydraulic fracturing in fault zones, mineral veins and phreatomagmatic deposits, suggesting that the mud behaved as an overpressured fluid at this stage. Any compactional overpressure might have been enhanced by the vertically applied weight of an advancing debris flow (Fig. 11b), by the lateral flow of fluid mud in response to the horizontal pressure gradient at the head of the flow (Fig. 11a), or by collapse of a sensitive mud fabric along an incipient basal slide detachment (Fig. 11c). Overpressure might also have arisen by the pumping of

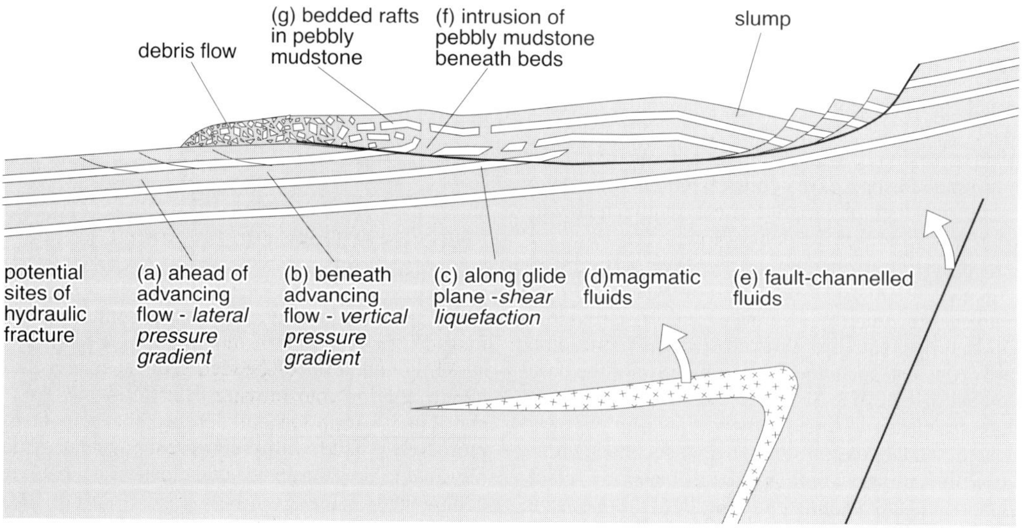

Fig. 11. The observed [(a)–(c)] and hypothetical [(d)–(e)] sites of fragmentation by hydrofracturing with the deduced mechanisms of incorporation of fractured sediment into a debris flow [(f)–(g)].

extra-formational fluids along faults, or from magmatic fluids associated with the early high level intrusions that are common in the Lady Port Formation (Fig. 11d and e). However, direct evidence for these associations is lacking. Any of these mechanisms could have been enhanced by seismic shocks.

The next stage of disruption (Fig. 10c), involving shortening or elongation of the bedding in folds, pinch-and-swell structures and plastic shears, suggests the action of deviatoric stresses rather than just elevated hydrostatic stress. Such stress states could have occurred within a coherent slide sheet or below a slide or flow (Fig. 11b and c). Side-scan sonar records of recent submarine debris flows (Prior *et al.* 1984) show that deformed 'pressure ridges' can also occur ahead of advancing flow lobes (Fig. 11a).

Deformation progressively disrupted the bedding within the host sediment until only a ghost stratification was present (Fig. 10d), typically seen as a contrast between zones of higher and lower clast concentration. Fragments of sandstone and siltstone still tended to behave in a brittle or only semi-ductile manner during this process, so that the increasingly isolated clasts which were produced had a subangular shape. Non-hydrostatic pressure gradients in the basal parts of the debrite matrix are suggested by lateral and downward injection of pebbly mudstones into bedded protolith (Figs 7c and 11f). This process aided the detachment of rafts of bedded sediment into the body of the debris flow. The rafts could have been supported by a combination of the cohesive strength of the debrite and raised fluid pressures along their base, as proposed by Leigh & Hartley (1992) for large rafts in the Pindos Basin, Greece.

Zones of sediment with a high mud:sand or mud:silt ratio would, by this stage, have been capable of independent movement as a cohesive debris flow, with clasts supported by the strength of the matrix (Lowe 1979). However, the proportion of mud in some zones, derived from the sand- or silt-rich protolith, was too low to fully support the clasts which were being generated. The clasts in these zones were in partial contact with each other, reducing the potential of the zone for plastic flow. Such zones probably moved only limited distances to yield the low-matrix debrites seen in the Lady Port Formation, unless they were isolated and transported as semi-rigid rafts within a more plastic high-matrix flow (Fig. 11g).

The high-matrix flows travelled downslope at least far enough for some of them to run out into different subenvironments of the Lady Port system. In particular, the debrites north of Ballanayre Strand were derived from the mudstone and quartzose sandstone protolith, similar to that south of Gob ny Creggan Glassey, yet were deposited at a site containing the bioturbated turbidites. The conspicuous lack of fragmentation and clast incorporation below these more distal debrites may have resulted from the lower susceptibility to hydrofracturing of the bioturbated turbidites compared with the lithologically more differentiated mudstones and quartzose sandstones. The stable planar bases to these distal flows allowed the development of stratified and inversely graded basal zones, common features of debris flows elsewhere, but still of debatable origin (e.g. Naylor 1980; Broster & Hicock 1985).

The apparent importance of hydrofracturing in generating the Lady Port debrites, together with the possibility of magmatic or fault-channelled fluid sources, leaves open the possibility that some of the 'debrites' were formed as intrastratal injections of fluidized sediment. Rare felsite clasts in the distal debrites match irregular or tabular intrusions in the Lady Port Formation, some of which display pepperitic margins suggesting essentially syn-sedimentary near-surface emplacement. Magmatic fluid pressures could have triggered fragmentation of bedded sequences that either intruded laterally as debrite sills or themselves broke surface to form unconfined flows. However, no unequivocal evidence of this genetic link has been observed in the formation. Magmatic doming accompanying hypabyssal intrusions might also have played a part in the debrite generation process.

A more conventional appeal to slope instability as the trigger for the debris flows begs the question of the origin of the necessary slopes in this supposedly distal part of the Gondwana margin. Steepening of local slopes by syn-sedimentary intrabasinal faulting is one possible cause. This tectonic trigger for the Lady Port debris flows is given further plausibility by the possibility of a synchronous mass movement event along the Gondwana margin (see below).

Regional correlations and significance

No other debrites in the Manx Group (Fig. 1) are so well exposed as those in the Lady Port Formation and none have been examined in the same detail as in this study. However, there are important similarities with the Lady Port examples. Most clasts in the Manx Group debrites appear to be intraformational and never exotic. The clasts have a wide range of size but are predominantly subangular to subrounded. There is a similar range from high- to low-matrix textures, but with the high-matrix debrites strongly predominating. Analogous fragmentation sequences, from bedded sediments to debrite, occur elsewhere in the Manx Group, particularly in the Sulby Slump Breccia of

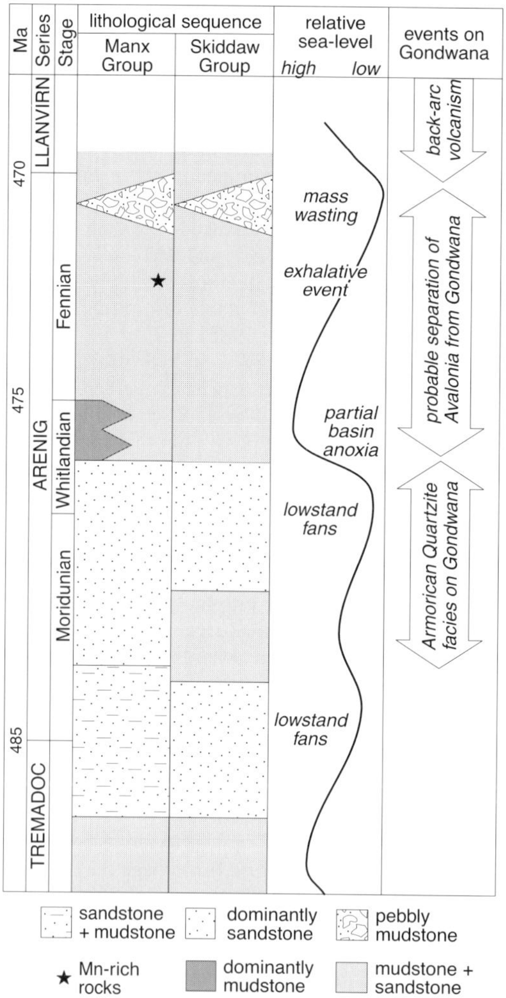

Fig. 12. Proposed correlation of Manx Group debrites, within the Upper Arenig, with the olistostromes and pebbly mudstones of the Skiddaw Group. A relative sea-level curve for the Gondwana margin is shown [after Woodcock *et al.* (1999*b*)], together with relevant regional constraints on the timing of rifting of Avalonia from Gondwana [after Prigmore *et al.* 1997].

Simpson (1963). Indeed, it was precisely these transitional zones that Lamplugh (1903) figured as evidence for the origin of the pebbly mudstones as tectonic crush breccias (e.g. Lamplugh 1903; Figs 7 and 11). These similarities encourage the interpretation of the Manx Group debrites in terms of the same range of intrabasinal fragmentation and transport mechanisms as detailed for the Lady Port examples.

The acritarchs from the Lady Port Formation, specifically from the laminated mudstone facies near Lady Port, suggest a late Arenig (Fennian) age (Molyneux 1979, 1999; Cooper *et al.* 1995). Although other debrite sequences in the Manx Group are not dated directly, the revised lithostratigraphic correlation of the Manx Group (Woodcock *et al.* 1999*b*; Fig. 9) suggests that these could correlate with the Lady Port Formation in time as well as process. In general, therefore, middle to upper Arenig mudstone-prone sequences with debrites overlie lower Arenig sandstone-prone sequences (Fig. 12).

A broadly similar trend is seen in the northern Skiddaw Group of the English Lake District (Cooper *et al.* 1995). Here, the mudstone-rich, middle to upper Arenig Kirk Stile Formation overlies a lower Arenig succession that includes two major sandstone intervals, the Loweswater and Watch Hill Formations. Although sporadic slumped units and rare pebbly mudstones occur in the lower Arenig strata, thicker examples are concentrated in the upper Arenig units. The Skiddaw Group of the Central Fells includes the Buttermere Formation, a 1500 m thick olistostrome emplaced during late Arenig time (Cooper *et al.* 1995). Both the Skiddaw and Manx Groups therefore seem to record an abundance of mass-wastage deposits formed during late Arenig time.

Woodcock *et al.* (1999*b*) support the hypothesis of Cooper *et al.* (1995) that there may have been a widespread, synchronous episode of downslope mass movement on the Gondwana margin. This instability may have been promoted by the thick mud blanket on the margin formed during the mid-Arenig transgression (Fig. 12). The succeeding latest Arenig regression might have been one specific trigger for the mass movement. However, the deduction that fault-steepened intrabasinal slopes may have driven the Lady Port debris flows also suggests a regional tectonic trigger. One possible cause of faulting is the thrusting or extensional faulting in the active fore-arc in which the Manx and Skiddaw Groups were being deposited (Moore 1992; Woodcock *et al.* 1999*a*). However, the mass movement event is also broadly coeval with the supposed time of rifting from Gondwana of the Avalonian fragment on which both the Manx and Skiddaw Groups lie (Cooper *et al.* 1995; Prigmore *et al.* 1997; Pickering & Smith 1995). In particular, a late Arenig event would just post-date the main Armorican quartzite facies that provides a sedimentary link between the Gondwana interior and its shelf, including the Midland Platform of Avalonia (Fig. 12; Noblet & Lefort 1990). In the Manx Group, quartzose debris from this possible source persists in large volumes in the Creg Agneash and Mull Hill Formations, thought to be of early or mid-Arenig age (Woodcock & Barnes 1999). Late Arenig mass movement, induced by rifting, would also just pre-date the onset of back-arc volcanism in the Welsh Basin, another indicator of active extension on Avalonia (Fig. 12).

General conclusions

This paper has detailed the origin, by debris flow, of the pebbly mudstones in the Lady Port Formation. However, several more generally applicable conclusions can be drawn about the processes of initiation and movement of these flows:

- a complete textural spectrum exists between bedded sediment protolith through disturbed facies and breccias with a ghost stratification, to either high- or low-matrix pebbly mudstones; this spectrum preserves the progressive production of debrite from its protolith;
- hydraulic fracture of early cemented sandstone or siltstone layers by overpressure in the intervening muds was probably an important mechanism in the fragmentation of the bedded sequences;
- these disruptive overpressures may have been created partly by the load of advancing debris flows; the possibility of magmatic or tectonic fluid overpressures cannot be entirely discounted;
- bedded sediments, being fragmented below an advancing debris flow, were entrained into the base of the flow, partly by injection of fluid pebbly mudstone beneath stiff-bedded fragments or rafts;
- the texture of a pebbly mudstone is strongly correlated to the sand:mud ratio of its protolith, with relatively mud-rich sequences being most capable of spawning high-matrix debris flows with the potential to travel far.

Of regional importance is the conclusion that the Lady Port debris flows may correlate with others in the Manx Group, and with even larger mass flows in the Skiddaw Group. This correlation suggests a widespread mass-wasting event on the Avalonian continental margin of Gondwana. This event may have been synchronous with the rifting of the Avalonian microcontinent from Gondwana.

This study has benefited from discussion with Padhraig Kennan on the manganiferous sediments, and with Rob Barnes and Dave Quirk on the field relationships. Ben Kneller and Martin Smith provided incisive reviews that much improved the paper. Dudley Simons is thanked for printing the photographic plates. JHM acknowledges that his contribution is published with the permission of the Director, Geological Survey of Ireland. The work was funded by NERC research grant GR9/01834.

References

BLAKE, J. F. 1905. On the order of succession of the Manx Slates in their northern half, and its bearing on the origin of the schistose breccia associated therewith. *Quarterly Journal of the Geological Society, London*, **61**, 358–373.

BROSTER, B. E. & HICOCK, S. R. 1985. Multiple flow and support mechanisms and the development of inverse grading in a subaquatic glacigenic debris flow. *Sedimentology*, **32**, 645–657.

COOPER, A. H., RUSHTON, A. W. A., MOLYNEUX, S. G., HUGHES, R. A., MOORE, R. M. & WEBB, B. C. 1995. The stratigraphy, correlation, provenance and palaeogeography of the Skiddaw Group (Ordovician) in the English Lake District. *Geological Magazine*, **132**, 185–211.

FITCHES, W. R., BARNES, R. P. & MORRIS, J. H. 1999. Geological structure and tectonic evolution of the Lower Palaeozoic rocks of the Isle of Man. *This volume*.

GEOLOGICAL SURVEY OF UNITED KINGDOM 1898. Isle of Man. *1:63 360 geological map*, Sheets 36, 45, 46, 56 and 57.

GILLOTT, J. E. 1956. Breccias in the Manx Slates: their origin and stratigraphic relationships. *Liverpool and Manchester Geological Journal*, **1**, 370–380.

HAMBREY, M. J. & HARLAND W. B. (eds) 1981. *Earth's Pre-Pleistocene Glacial Record*. Cambridge University Press.

HENSLOW, J. S. 1821. Supplementary observations to Dr. Berger's account of the Isle of Man. *Transactions of the Geological Society, London*, **5**, 482–505.

KENNAN, P. S. & KENNEDY, M. J. 1983. Coticules – a key to correlation along the Appalachian–Caledonian Orogen? *In:* SCHENK, P. E. (ed.) *Regional Trends in the Geology of the Appalachian–Caledonian–Hercynian–Mauritanide Orogen*. Reidel, 355–361.

—— & MORRIS, J. H. 1999. Manganiferous ironstones in the early Ordovician. *This volume*.

LAMPLUGH, G. W. 1903. *The Geology of the Isle of Man*. Memoir of the Geological Survey UK. HMSO.

—— & WATTS, W. W. 1895. The crush-conglomerates of the Isle of Man. *Quarterly Journal of the Geological Society, London*, **51**, 563–599.

LEIGH, S. & HARTLEY, A. J. 1992. Mega-debris flow deposits from the Oligo-Miocene Pindos foreland basin, western mainland Greece: implications for transport mechanisms in ancient deep marine basins. *Sedimentology*, **39**, 1003–1012.

LOWE, D. R. 1979. Sediment gravity flows:their classification and some problems of application to natural flows and deposits. *Special Publication of the Society of Economic Palaeontologists and Mineralogists*, **27**, 75–82.

MAX, M. D., BARBER, A. J. & MARTINEZ, J. 1990. Terrane assemblage of the Leinster Massif, SE Ireland, during the Lower Palaeozoic. *Journal of the Geological Society, London*, **147**, 1035–1050.

McCONNELL, B., MORRIS, J. H. & KENNAN, P. 1999. A comparison of the Ribband Group (southeastern Ireland) to the Manx Group (Isle of Man) and Skiddaw Group (northern England). *This volume*.

MOLYNEUX, S. G. 1979. New evidence for the age of the Manx Group, Isle of Man. *In:* HARRIS, A. L., HOLLAND, C. H. & LEAKE, B. E. (eds) *Caledonides of the British Isles: reviewed*, Special Publication of the Geological Society, London, **8**, 415–421.

—— 1999. A reassessment of Manx Group acritarchs, Isle of Man. *This volume*.

MOORE, R. M. 1992. *The Skiddaw Group of Cumbria: Early Ordovician turbidite sedimentation and provenance on an evolving microcontinental margin*. PhD Thesis, University of Leeds.

MORRIS, J. H., WOODCOCK, N. H. & HOWE, M. P. A. 1999. The Silurian succession of the Isle of Man: the late Wenlock Niarbyl Formation, Dalby Group. *This volume*.

NAYLOR, M. A. 1980. The origin of inverse grading in muddy debris flow deposits – a review. *Journal of Sedimentary Petrology*, **50**, 1111–1116.

NOBLET, C. & LEFORT, L. P. 1990. Sedimentological evidence for a limited separation between Armorica and Gondwana during the Early Ordovician. *Geology*, **18**, 303–306.

ORR, P. J. & HOWE, M. P. A. 1999. Macrofauna and ichnofauna of the Manx Group (early Ordovician), Isle of Man. *This volume*.

PICKERING, K. T. & SMITH, A. G. 1995. Arcs and backarc basins in the Early Paleozoic Ocean. *The Island Arc*, **4**, 1–67.

——, STOW, D., WATSON, M. & HISCOTT, R. N. 1986. Deep-water facies, processes and models: a review and classification scheme for modern and ancient sediments. *Earth Science Reviews*, **23**, 1–98.

PRIGMORE, J. K., BUTLER, A. J. & WOODCOCK, N. H. 1997. Rifting during separation of Eastern Avalonia from Gondwana: Evidence from subsidence analysis. *Geology*, **25**, 203–207.

PRIOR, D. B., BORNHOLD, B. D. & JOHNS, M. W. 1984. Depositional characteristics of a submarine debris flow. *Journal of Geology*, **92**, 707–727.

READING, H. G. & RICHARDS, M. 1994. Turbidite systems in deep-water basin margins classified by grain size and feeder system. *AAPG Bulletin*, **78**, 792–822.

ROBERTS, B., MORRISON, C. & HIRONS, S. 1990. Low grade metamorphism of the Manx Group, Isle of Man: a comparative study of white mica 'crystallinity' techniques. *Journal of the Geological Society, London*, **147**, 271–277.

SHANNON, P. M. 1978. The stratigraphy and sedimentology of the Lower Palaeozoic rocks of southeast Co. Wexford. *Proceedings of the Royal Irish Academy*, **78B**, 247–265.

SIMPSON, A. 1963. The stratigraphy and tectonics of the Manx Slates Series. *Quarterly Journal of the Geological Society, London*, **119**, 367–400.

STONE, P., COOPER, A. H. & EVANS, J. A. 1999. The Skiddaw Group (English Lake District) reviewed: early Palaeozoic sedimentation and tectonism at the northern margin of Avalonia. *This volume*.

WEBB, B. C. & COOPER, A. H. 1988. Slump folds and gravity slide structures in a Lower Palaeozoic marginal basin sequence (the Skiddaw Group), NW England. *Journal of Structural Geology*, **10**, 463–472.

WOODCOCK, N. H. & BARNES, R. P. 1999. An early Ordovician turbidite system on the Gondwana margin: the southeastern Manx group, Isle of Man. *This volume*.

——, QUIRK, D. G., FITCHES, W. R. & BARNES, R. P. 1999*a*. In sight of the suture: the early Palaeozoic geological history of the Isle of Man. *This volume*.

——, MORRIS, J. H., QUIRK, D. G. *ET AL.* 1999*b*. Revised lithostratigraphy of the Manx Group, Isle of Man. *This volume*.

The definition of sandstone-bearing formations in the Isle of Man and correlation with adjacent areas – evidence from sandstone chemistry

R. P. BARNES[1], G. M. POWER[2] & D. C. COOPER[3]

[1]*British Geological Survey, Murchison House, West Mains Road, Edinburgh EH9 3LA, UK*

[2]*University of Portsmouth, Department of Geology, Burnaby Road, Portsmouth PO1 3QL, UK*

[3]*British Geological Survey, Kingsley Dunham Centre, Keyworth, Nottingham NG12 5GG, UK*

Abstract: The outcrop of Lower Palaeozoic rocks on the Isle of Man is dominated by thin- to medium-bedded sand-rich turbidites over most of the southeast side of the island (Lamplugh's Lonan and Agneash Grits, comprising greywacke and quartz arenite, respectively, interbedded with mudstone) and on the northwest coast south of Peel (the Niarbyl Flags). Recent work has shown that the latter, the Niarbyl Formation, is Silurian in age and thus distinct from the otherwise early Ordovician sequences. The composition of the Niarbyl sandstone is also distinct, with comparatively low silica (SiO_2 60–66%) but elevated CaO, MgO and Cr, relative to the Ordovician sandstones.

The Ordovician sandstones fall into two compositional groups: a very mature, silica-rich (SiO_2 78–95%) quartz arenite (Agneash type) and a greywacke (Lonan type) with lower silica (SiO_2 65–78%). Most element contents vary with silica but there is a compositional hiatus. It is here inferred that the two sandstone groups represent material from separate source areas. The three tectonostratigraphical sequences distinguished in the southeast of the island all include sandstone of both compositional types in different proportions. These usually occur as units of one or other composition, but in one sequence the two are locally closely interbedded while remaining compositionally distinct. Sandstone in two possibly equivalent units may, however, vary gradationally between the two types, implying more intimate mixing. In the absence of biostratigraphical control, the geochemical data are used to constrain the various ways in which the tectonostratigraphical sequences might correlate.

The chemical signature of the Isle of Man sandstones also provides constraints on possible correlatives in adjacent areas. The Lonan type is very similar to mid-Arenig sandstone in the upper part of the Skiddaw Group of the English Lake District. The more siliceous Agneash type has no compositional comparative in the main Skiddaw Group outcrop, although sandstone comprising the enigmatic Redmain Formation is similar in composition. The Wenlock Niarbyl Formation is lithologically and chemically comparable with Wenlock turbidite sequences in the Southern Uplands terrane and in the Windermere Supergroup in the Lake District, and they may all be closely related. None provides a precise match but compositionally the Niarbyl sandstone closely resembles the sandstone of the same age in the Birk Riggs Formation of the Windermere Supergroup. Subject to a number of constraints, the chemical composition of sandstone can also provide information on the probable tectonic environment of the source rocks. On this basis, the Lonan and Niarbyl sandstones include substantial components of first- or second-cycle volcanic debris, whereas the Agneash type is dominated by mature debris reworked from older sedimentary rocks or derived from a granite/gneiss basement source.

Clastic sedimentary rocks formed at different times, or in different locations but in broadly the same depositional setting, may appear superficially similar and be difficult to distinguish. The bulk chemical composition of the rocks, particularly those of sand grain size, can be useful in this respect as it reflects, at least in part, the detrital fragments contained in them. Such data can indicate whether possible correlations are more or less likely. They also provide information on the source rocks and may help to constrain the tectonic setting, although the results need to be treated with caution as they can be misleading (e.g. Mack 1984; Floyd *et al.* 1991; McCann 1991).

The geochemical composition of a sandstone is, nevertheless, a complex interplay of a range of variables relating to provenance (debris content), transport and deposition, diagenesis/metamorphism

From: WOODCOCK, N. H., QUIRK, D. G., FITCHES, W. R. & BARNES, R. P. (eds) 1999.
In Sight of the Suture: the Palaeozoic geology of the Isle of Man in its Iapetus Ocean context.
Geological Society, London, Special Publications, **160**, 139–154. 1-86239-046-0/99/$15.00

and weathering (e.g. Bhatia 1983). Many of these variables may themselves be strongly influenced by the tectonic situation of both the source and the depositional environment. Consequently, sandstone composition has been utilized to identify tectonic setting (e.g. Pettijohn *et al.* 1972; Blatt *et al.* 1980; Roser & Korsch 1986) in a similar way to the more established geochemical investigation of igneous rocks (e.g. Rollinson 1993). Subject to a number of constraints, chemical composition also provides a convenient means of characterizing sandstones for the purposes of comparison and/or correlation, particularly as it may be less susceptible to distortion through grain size variation and weathering than petrographical methods (e.g. Barnes 1998). Such techniques may be particularly useful when, as in the Isle of Man, a combination of structural complexity, poor biostratigraphical control, limited facies variation and poor exposure restrict the application of more traditional stratigraphical methods.

This geochemical study of the sandstones which crop out in the Isle of Man was initiated in order to validate stratigraphical models being developed for the Manx Group and to constrain likely correlations with coeval sequences in adjacent areas. To be successful, comparative geochemistry requires that the sandstone composition in different formations in a sequence have defined ranges which are significantly different from one another. Previous work in the English Lake District (e.g. Cooper *et al.* 1988; see also below) has established the compositional ranges of formations within the Skiddaw Group and associated rocks and these can be used as a basis for comparison with the Manx Group (Cooper *et al.* 1995).

Lithological framework

Sandstone-dominated sequences in the Ordovician Manx Group crop out over much of the southeastern and southern parts of the Isle of Man. Due to the structural complexity of the outcrop and the nature of the exposure there is considerable uncertainty over possible correlations between different parts of the island. Therefore, lithostratigraphical successions (Woodcock *et al.* 1999) have been defined in three separate tectonostratigraphical tracts (Fitches *et al.* 1999) southeast of the outcrop of the Barrule Formation (Fig. 1):

- tract 1 – Lonan–Santon Formations along much of the southeast side of the island;
- tract 2 – Port Erin–Mull Hill Formations in the south of the island;
- tract 3 – Ny Garvain–Creg Agneash–Maughold Formations in the northeast of the island.

All three sequences contain a laminated silty mudstone background which becomes dominant in the Maughold Formation in the upper part of the Agneash sequence. In the Lonan and Port Erin Formations in the lower parts of the Lonan and Mull Hill sequences, and in parts of the Ny Garvain Formation in the Agneash sequence, the silty mudstone includes a variable proportion of very thinly bedded to laminated, very fine-grained sandstone (Bouma Tc). The remainder of the Ny Garvain Formation, and the Santon, Mull Hill and Creg Agneash Formations, are dominated by sequences with a large proportion of thin–medium-(locally thick-) bedded, fine- to medium-grained sandstone. These relatively thickly bedded units include two distinct sandstone types as recognized by Lamplugh (Lamplugh 1903; Geological Survey 1898):

- relatively impure, matrix-rich 'greywacke' termed the Lonan Flags by Lamplugh, now separated into the Lonan, Santon, Port Erin and Ny Garvain Formations;
- better sorted quartz-rich arenite termed Agneash Grit by Lamplugh, including the Mull Hill and Creg Agneash Formations, and sandstone beds and packages within the Maughold Formation and locally interbedded in the Lonan Formation [including the Keristal Member (Woodcock & Barnes 1999)].

Another sandstone sequence, now known to be of Silurian age, from a newly discovered graptolite fauna (Howe 1999) and termed the Niarbyl Formation (Morris *et al.* 1999), crops out in the northwest of the island south of Peel. It is dominated by medium- to thick-bedded sandstone which was formerly regarded (Lamplugh 1903; Simpson 1963) as equivalent to the relatively thickly bedded sequence (Santon Formation) south of Douglas, although recent work has shown that it may be distinguished by the occurrence of interbedded, carbonaceous, laminated siltstone.

Geochemical characterization of Isle of Man sandstones

Whole rock X-ray fluorescence analyses of a set of 40 samples from the Ordovician rocks of the Manx Group and five samples from the Silurian Niarbyl Formation (Fig. 1) are used to characterize the sandstone geochemistry. The full analytical and sample details are presented along with a wide range of bivariate plots in Barnes *et al.* (1998). Most of the samples were taken from well-characterized sections in the different formations and are used to assess possible correlations. Samples from quartz-rich material interbedded within the Lonan sequence and the relatively dirty

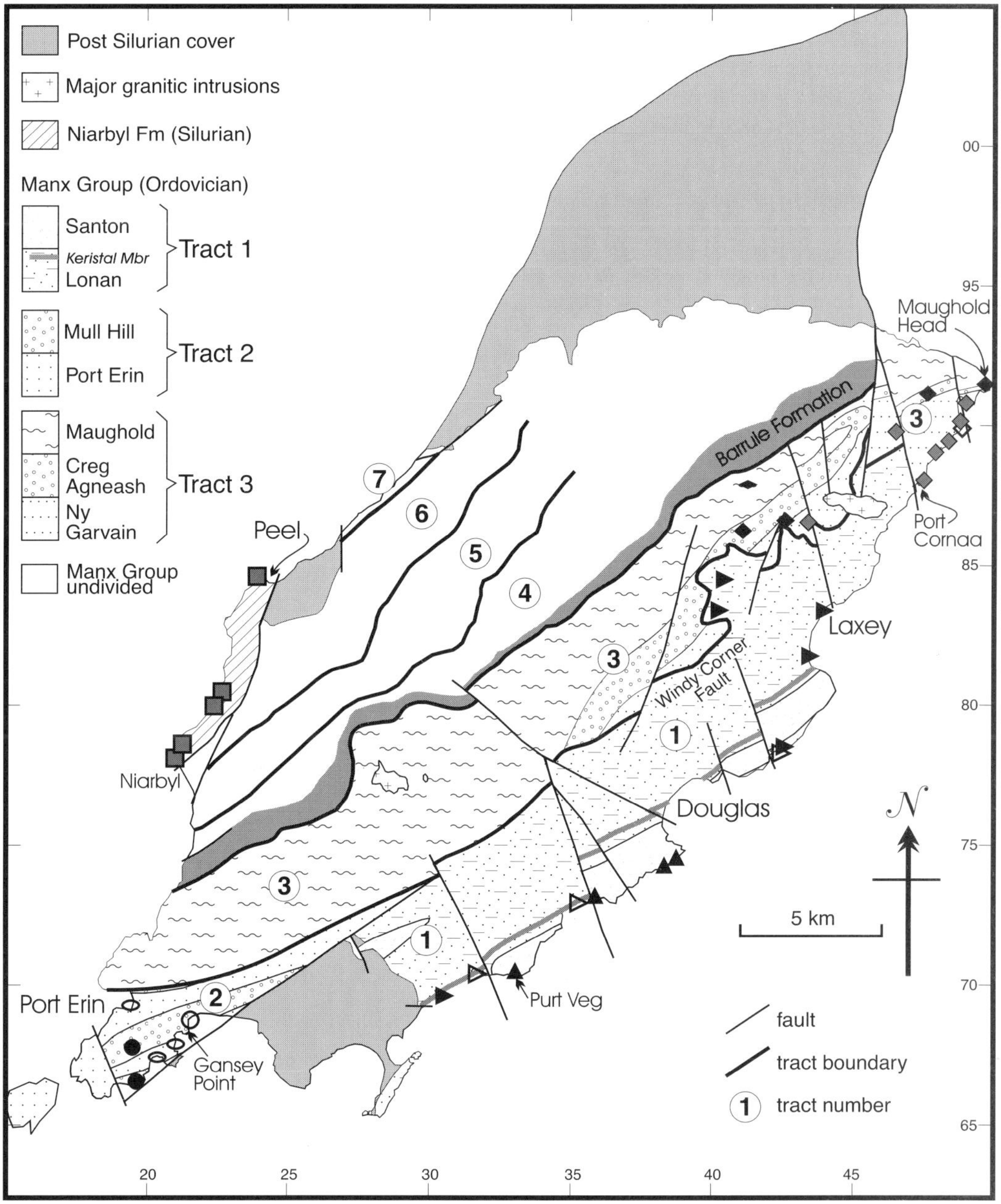

Fig. 1. Geological map of the Isle of Man based on Woodcock *et al.* (1999) and showing sample locations (symbols as in key to Fig. 2).

material within the quartz arenite sequence of the Mull Hill Formation at Gansey Point are used to assess the affinity of these materials to the host formations. The small number of samples available from the different units discussed necessarily limits the confidence that can be placed on the analysis of these data.

The samples from stratigraphically well-characterized sections fall into three compositional groups (Figs 2–4):

- Agneash type – characterized by the distinctive quartz arenite of the Creg Agneash and Mull Hill Formations; high (> 78%) SiO_2 and relatively low concentrations of most other elements, usually varying with SiO_2;

- Lonan type – characterized by the Lonan–Santon Formations; SiO_2 contents 64–77% and proportionately higher concentrations of other elements forming an extension of the variation trend from the Agneash type;
- Niarbyl type – five samples from the Niarbyl Formation have a very limited compositional range; the SiO_2 content (60–66%) lies close to the lower end of the range of the Lonan type but the Niarbyl Formation is distinguished by relatively high CaO, MgO and Na_2O, and relatively low concentrations of other major elements.

Trace element contents generally vary with TiO_2, with, in most cases, all three sandstone types lying on well-defined linear trends (e.g. Fig. 4a). Chromium (Fig. 4b) and, to a lesser extent, nickel and strontium contents are distinctive in the Niarbyl sandstone, while a plot of Cr v. TiO_2 (Fig. 4b) appears to define completely separate compositional fields for the three types.

Two types of relatively quartzose material of uncertain affinity are interbedded with the greywacke of tract 1 (Woodcock & Barnes 1999):

- thin beds of distinctive quartz arenite occur in the upper part of the Lonan Formation and the base of the Santon Formation, along with a more thickly bedded, sometimes channelized, sequence in the former (the Keristal Member);
- apparently quartz-rich sandstone also occurs in the thickly bedded fill to a large channel exposed in the base of the Santon Formation at Purt Veg [SC 326 703].

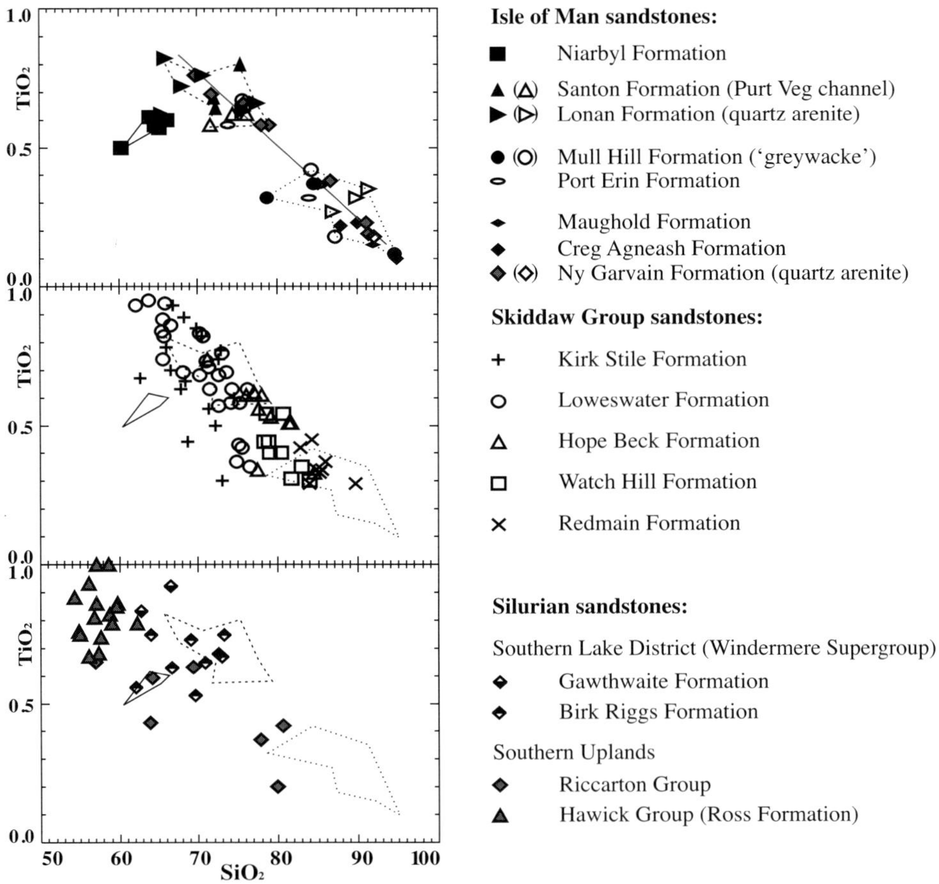

Fig. 2. Bivariate plot of TiO_2 v. SiO_2 (wt%); Isle of Man data (top) and as fields (——, Niarbyl Formation; - - - - - , Lonan type; . . . . , Agneash type) compared with data from Ordovician sandstones in the Skiddaw Group (middle) and Wenlock sandstones in the Southern Uplands and Windermere Supergroup (bottom).

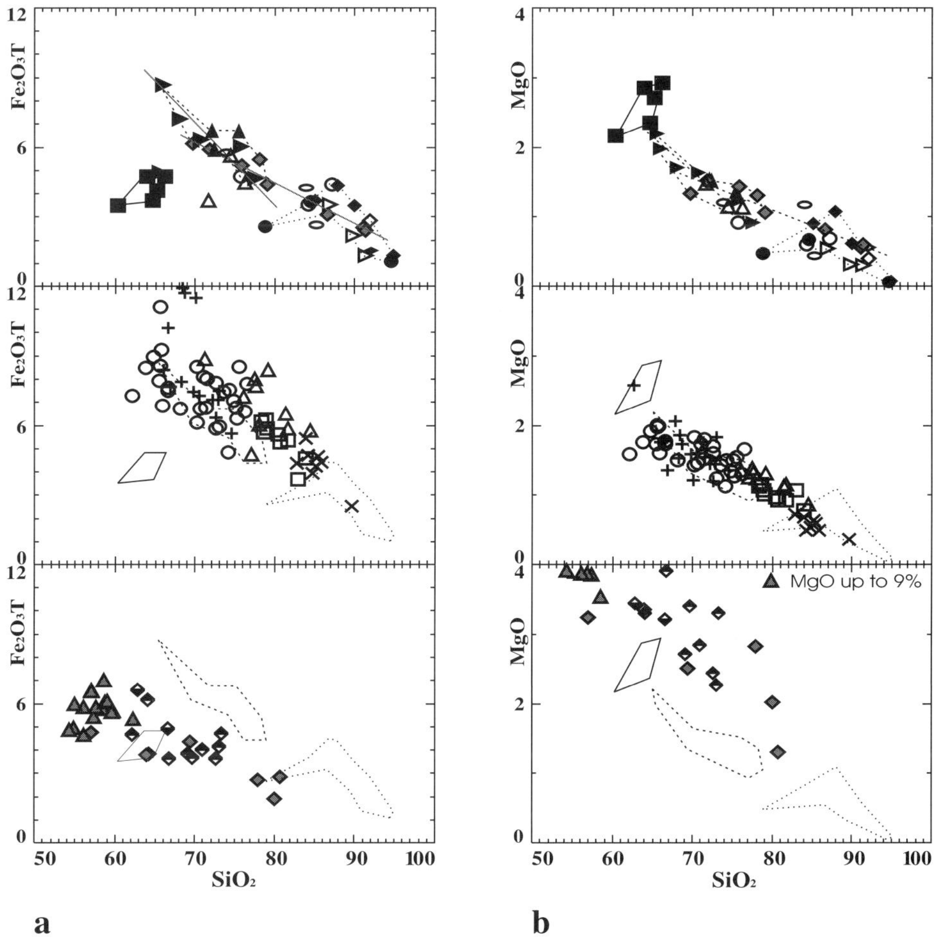

Fig. 3. Bivariate plots of Fe_2O_3T v. SiO_2 (a) and MgO v. SiO_2 (b): key to symbols as in Fig. 2.

On the basis of its composition, the interbedded quartz arenite is clearly of Agneash type, demonstrating contemporaneous availability of both sandstone types at some time in the early Arenig. On the other hand, the material in the Purt Veg Channel is of Lonan composition and this channel could thus have been a feeder to a lobe of the Santon Formation (see also below).

The converse situation, of greywacke interbedded within the dominantly quartz arenite sequences is most obvious in the transitional base of the Creg Agneash Formation where the quartz arenite beds occur within a background of mudstone and very thin greywacke beds typical of the upper part of the Ny Garvain Formation (see below). The same relationship occurs at the base of the Mull Hill Formation in the south of the island. At Gansey Point (Fig. 1) the latter includes thicker beds of greywacke, closely associated with quartz arenite (interlayered within single beds in some cases), which have the same compositional character as the underlying Port Erin Formation (see below).

Discrimination of the Port Erin and Ny Garvain formations

These two formations are lithologically similar to other units which crop out on the Isle of Man [e.g. Quirk & Burnett (1999) and Woodcock & Barnes (1999)] but biostratigraphically unconstrained and hence possible correlations are subject to considerable uncertainty. Sandstone samples were collected from superficially uniform 'wacke', as distinct from associated quartz arenite (Woodcock

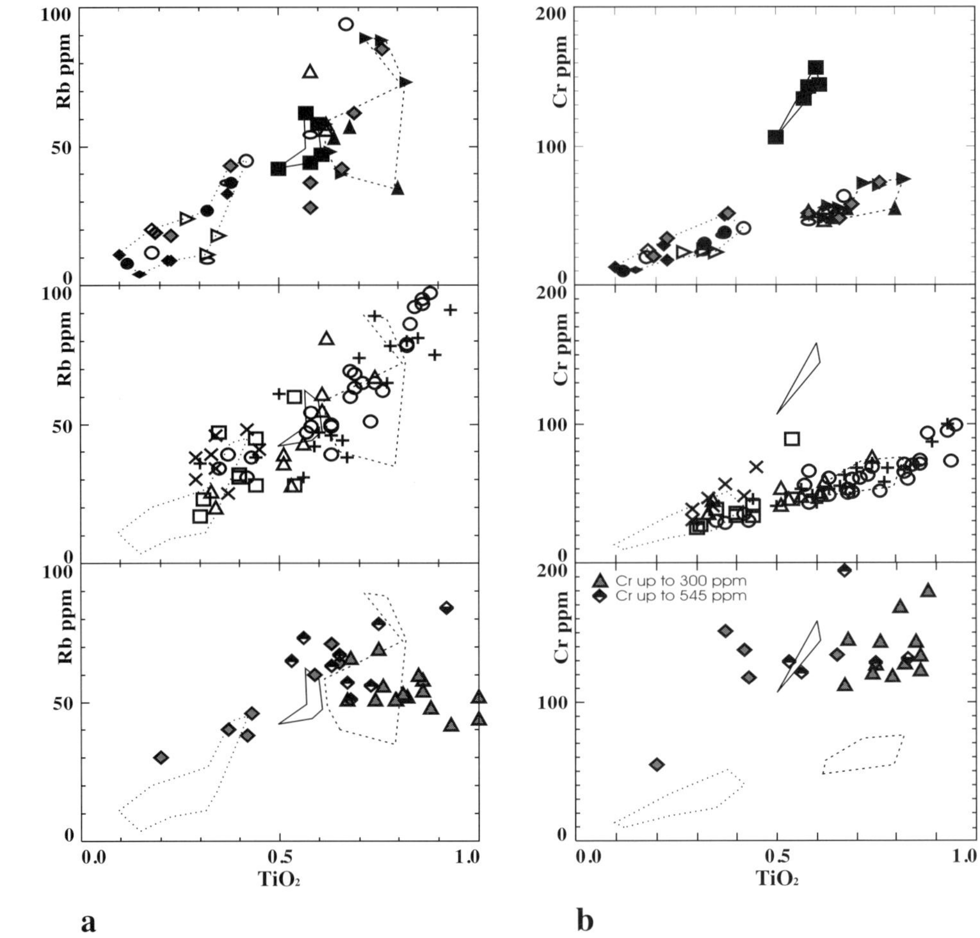

Fig. 4. Bivariate plots of Rb v. TiO_2 (a) and Cr v. TiO_2 (b); key to symbols as in Fig. 2.

& Barnes 1999), from the Port Erin and Ny Garvain Formations in order to address this problem. Compositionally, they span the range of both of the Agneash and Lonan types. Six samples from the Port Erin Formation and the base of the Mull Hill Formation at Gansey Point plot around the compositional break between the Agneash and Lonan types (Figs 2–4). Samples from the Ny Garvain Formation, mainly from the coastal section between Port Cornaa and Port Mooar (Fig. 1), have a wider range which, at the SiO_2-rich end of the spectrum, is close to the composition of thin interbeds of distinct quartz arenite exposed at Gob ny Garvain.

On the bivariate plots of major elements against silica, the material from the Ny Garvain and Port Erin Formations, and the greywacke from the base of the Mull Hill Formation, define a marked linear trend (e.g. Fig. 3) which links the Lonan and Agneash fields. This may be due to more intimate mixing of the two end-member types than is apparent in, for example, the Lonan Formation where the two types are interbedded but compositionally distinct. The data which define the Lonan and Agneash fields are generally relatively scattered, but MgO and Fe_2O_3 contents relative to SiO_2 in the Lonan material (Fig. 3) form a steeper trend than that of the Ny Garvain and Port Erin material. This suggests that sandstone of an Agneash-type composition could not simply be derived from a Lonan-type end member.

Simple bivariate plots of wt% of elements, particularly using SiO_2, are subject to the problem of closure (e.g. Chayes 1971; Le Maitre 1982). One way to attempt to avoid this effect, and that of any resulting interelement correlations, is to use a principal component analysis. Principal components analysis reduces the overall dimensionality of the data by the formation of a new set of variables, or components, which are not themselves correlated. The first component has the greatest proportion of the total variance, the second a lesser proportion and so on. For the principal components analysis carried out here only the major elements and loss-on-ignition values were used, and a logarithmic (base ten) transform was applied. A covariance matrix (Le Maitre 1982) was used for the calculation of the principal components and principal component scores for each rock in the data set from the Isle of Man sandstones. The statistical package *Minitab* was used for these calculations [further details are given in Barnes *et al.* (1998)]. A plot of the first two principal component scores (Fig. 5) shows four distinct groups of data, on one side of the plot a close cluster of the Niarbyl samples and on the other an elongate group comprising the Agneash-type quartz arenite, whilst in between these two is an elongated steep (parallel to the second-component axis) trend comprising the Lonan-type sandstone samples with a nearby subsidiary cluster of some Ny Garvain and Mull Hill greywacke samples. Thus, as in the bivariate analysis, the Ny Garvain material partly overlaps with the Lonan and Agneash fields but, in contrast to the steep Lonan trend, forms a trend near-parallel to the first-principal component axis, again suggesting that it may represent a different source.

It must be concluded, therefore, that the Port Erin and Ny Garvain Formations comprise sandstone of a range of compositions linking the end-member Lonan and Agneash sandstone types. This seems likely to be due to relatively intimate mixing of the two components which otherwise form separate units, even where locally closely interbedded. However, the data from the Port Erin and Ny Garvain sandstones still preserve some suggestion of a compositional gap (e.g. Fig. 2) suggesting that mixing was either incomplete or that two degrees of mixing are present, although additional samples may close this gap. Further investigation is therefore required to define the extent and relationships of the sandstones of different compositions within these formations.

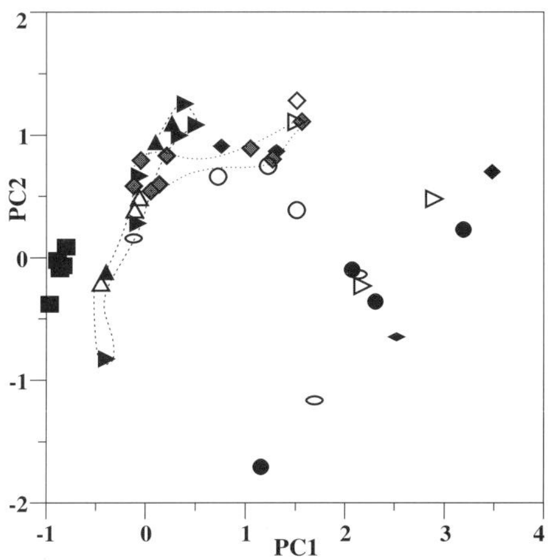

Fig. 5. Principal component analysis using a covariance matrix: plot of first-principal component (PC1) v. second-principal component (PC2) for major element data; key to symbols as in Fig. 2.

Correlation of the sandstone-bearing sequences within the Isle of Man

Evidence of the biostratigraphical age(s) of the sandstone-bearing sequences in the eastern part of the Manx Group outcrop (Fig. 6) is restricted to the Santon Formation in tract 1, in which graptolites (Rushton 1993) and microflora are probably of an early to mid-Arenig age (e.g. Molyneux 1999). The sequences in tracts 2 and 3 are biostratigraphically unconstrained and thus possible correlations between them, and with the tract 1 sequence, may only be argued on the basis of their lithostratigraphy and sandstone composition.

The tract 2 and 3 sequences

These sequences are dominated by the Mull Hill and Creg Agneash Formations which, although structurally separate, occur along-strike from one another (Fig. 1). They are composed of compositionally similar quartz arenite, although it is more thickly bedded in the Mull Hill Formation. Both of these formations overlie very thinly bedded turbidite sequences comprising sandstone–mudstone couplets (the Port Erin Formation and the upper part of the Ny Garvain Formation) with some interbedding of sandstone of Lonan and/or mixed Lonan–Agneash composition near the base. Otherwise, the sandstone from the Port Erin and the Ny Garvain Formations varies compositionally across the range of the Lonan and Agneash types. It seems likely, therefore, that the tract 2 and 3 sequences can be simply correlated in this way (Fig. 6), although the more thickly bedded sandstone seen in the lower part of the Ny Garvain Formation (Woodcock & Barnes 1999) does not occur in that part of the Port Erin Formation which is exposed.

The tract 3 sequence extends above the quartz arenite-dominated package into the Maughold Formation, which is seen to be stratigraphically

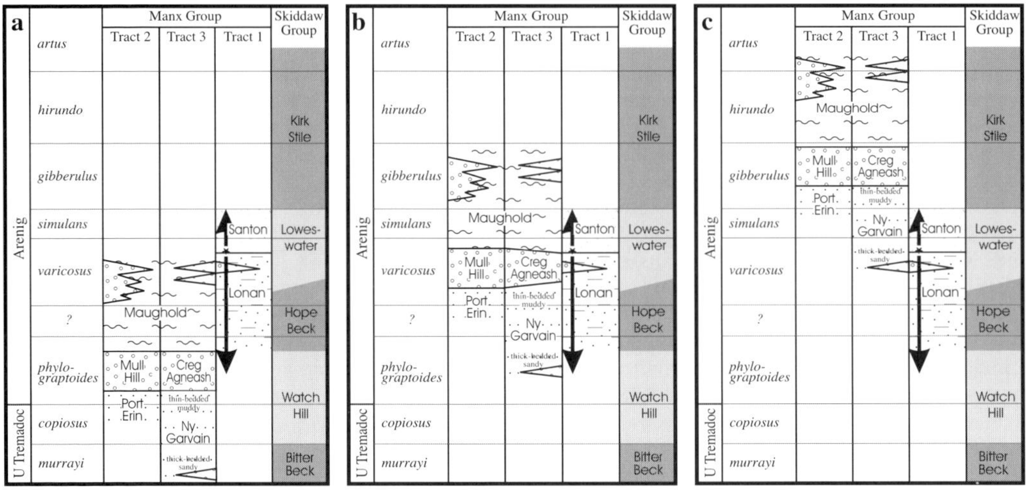

Fig. 6. Alternative correlations between tracts 1–3 in the Isle of Man (Fig. 1) based on different possible correlations of the Keristal Member in tract 1 with different quartz arenite packages in tracts 2–3 (assuming a simple correlation between tracts 2 and 3 as shown). The Skiddaw Group is shown for comparison from Cooper *et al.* (1995). Note that the only biostratigraphical control in the three tracts is the graptolite and acritarch faunas from the Santon Formation, the uncertainty of which is indicated by the arrow. There is no control over the biostratigraphical position of the tract 2–3 sequence(s) nor the biostratigraphical extent of any of the formations shown.

continuous with the Creg Agneash Formation in the northeast of the island. The Maughold Formation is dominated by laminated silty mudstone but includes a variable proportion of interbedded quartz arenite of Agneash type, either as dispersed beds or as packets several tens of metres in thickness comprising medium- to thickly bedded sequences. The outcrop of the Maughold Formation extends along the length of the island (Fig. 1); it occurs adjacent to the Mull Hill sequence in the southwest but with a faulted contact, although this may be a relatively minor modification of an originally stratigraphical boundary as is preserved further northeast.

The tract 1 and 2–3 sequences

The greywacke in the Lonan–Santon Formations is consistently of Lonan composition throughout and is compositionally distinct from the quartz arenite of the Mull Hill–Agneash Formations. The Lonan Formation, a thinly bedded turbidite sequence with a variable proportion of mudstone, is, in parts, not dissimilar to the Port Erin Formation or the upper part of the Ny Garvain Formation. However, the samples analysed from the Lonan Formation show none of the tendency towards the more silica-rich sandstones seen in the latter. Quartz arenite does occur near the top of the Lonan Formation in the Keristal Member, a few metres of thickly bedded, sometimes channelized, sandstone. The overlying, dominantly medium-bedded turbidite sequence which comprises the Santon Formation also includes sporadic, locally abundant, interbeds of quartz arenite (e.g. the Whing and Douglas Head; Woodcock & Barnes 1999).

If the appearance of the quartz arenite within the sequence is taken as the primary means of correlation, a range of correlations is possible (Fig. 6) between the 1 and 2–3 sequences:

- the Keristal Member equates with quartz arenite high in the Maughold Formation, implying that this formation is, at least in part, equivalent to the Lonan Formation. The Ny Garvain and Creg Agneash Formations could then be older than anything seen in tract 1, with the more thickly bedded part of the Ny Garvain Formation possibly equivalent to the possibly Cambrian Bray Group in southeast Ireland (Brück & Reeves 1976), as suggested by its sedimentological characteristics (Brück & Kennan, pers. comm);
- the Keristal Member is a lateral equivalent of part, or all, of the Creg Agneash Formation, allowing correlation of the Lonan Formation with the superficially similar Port Erin Formation and the upper part of the Ny Garvain Formation. If equivalent to the basal part only,

the bulk of the Agneash and Mull Hill Formations may be equivalent to the Santon Formation. This is preferred by Woodcock *et al.* (1999, fig. 9) who correlate these relatively thickly bedded units, irrespective of composition, as representative of a single phase of sea-level lowstand. Woodcock & Barnes (1999); on the other hand, consider the possibility that the Keristal Member preserves the channels across the Lonan slope apron that fed the Agneash and/or Mull Hill fan lobes. This correlation suggests that the sand-dominated Santon Formation is in part equivalent to the mud-dominated Maughold Formation;

- the Keristal Member equates with a quartz arenite package of similar character which occurs in the Ny Garvain Formation exposed in Port Cornaa (Woodcock & Barnes 1999). This allows correlation of the relatively thickly bedded lower part of the Ny Garvain Formation and the Santon Formation, but suggests that the tract 1 and 2–3 sequences do not otherwise overlap.

Such correlations are somewhat tenuous and remain unconstrained without independent age data, serving only to illustrate the current level of uncertainty in the Manx Group succession as a whole. Quartz arenite occurs throughout the upper part of the Agneash sequence and also in sequences which crop out further west in the Isle of Man, leading to a variety of other possible correlations with the Keristal Member.

Niarbyl Formation

Although relatively few sandstone analyses are available from the Niarbyl Formation, the samples spread along the length of the coastal outcrop have remarkably consistent compositions. From these data it is clear that the Niarbyl Formation is compositionally distinct from the other sandstone-bearing sequences sampled in the Isle of Man, consistent with its younger age.

Correlation with other Caledonian sandstone-bearing sequences

The Manx Group outcrop on the Isle of Man lies along-strike from, and has long been regarded as equivalent to, the Skiddaw Group. This is well exposed in the Lake District Inlier in northwest England and is of comparable age and lithofacies (e.g. Cooper *et al.* 1995), although detailed correlation has never been achieved. In the opposite direction, rocks of similar age and lithology crop out in southeast Ireland as the Ribband Group (Brück *et al.* 1979). The latter is spatially associated with the mid-Cambrian Bray Group (Brück & Reeves 1976), although the precise relationships between the two are unknown because the contacts are faulted everywhere. The Wenlock Niarbyl Formation on the Isle of Man is of comparable age and lithological character to turbidite sequences of the Hawick and Riccarton Groups which crop out in the Southern Uplands terrane to the north (e.g. Barnes & Stone 1999; Lintern & Floyd 1999) and to parts of the Windermere Supergroup in southern parts of the Lake District (Kneller *et al.* 1994; Johnson *et al.* 1999). Possible correlations with these areas are investigated below using available sandstone compositional data.

Northwest England: Manx–Skiddaw Group correlation

The most complete Skiddaw Group sequence, exposed in the northern part of the Lake District, contains alternating mudstone-dominated and sandstone-rich turbidite formations ranging from upper Tremadoc to early Llanvirn in age (Cooper *et al.* 1995; Stone *et al.* 1999). In ascending order these are:

- Bitter Beck Formation – dominantly dark grey mudstone–siltstone with minor thin- to medium-bedded fine-grained greywacke best developed in the lowest exposed part of the formation;
- Watch Hill Formation – thin to thickly bedded, fine- to coarse-grained lithic greywacke and subordinate lithic arenite with palaeocurrent directions from the east, interbedded with siltstone and mudstone;
- Hope Beck Formation – dominated by laminated turbidite siltstone and mudstone but containing distinct thin to medium beds of medium- to coarse-grained, quartz-rich lithic greywacke;
- Loweswater Formation – quartz-rich greywacke, dominantly fine- to medium-grained, thinly bedded at the top and bottom of the formation but thickening to *c.* 1 m beds in the middle, with palaeocurrent directions from the southeast; interbedded with up to 50% siltstone and mudstone;
- Kirk Stile Formation – dominated by laminated to very thinly bedded turbidite siltstone and mudstone but with two *c.* 100 m thick units including 20–30% very thin- to thin-bedded lithic sandstone; thick slumped units occur in the upper part.

A separate sandstone-dominated sequence exposed in a stream section near Cockermouth, termed the Redmain Formation by Allen & Cooper (1986), has been described as part of the Skiddaw Group but cannot be related to the main sequence.

Extensive geochemical characterization, carried out as part of British Geological Survey work, in the Lake District over the last 15 years (e.g. Allen & Cooper 1986; Cooper *et al.* 1988, 1995, 1999), provides a substantial database of stratigraphically well-constrained sandstone analyses from all but the Bitter Beck Formation. These data, from samples collected and analysed as described in Cooper *et al.* (1988), are illustrated for comparison with the Isle of Man data by Barnes *et al.* (1998) and in Figs 2–4.

Irrespective of formation, most major elements in the Skiddaw Group sandstone data form a strong negative linear trend against SiO_2 (e.g. Figs 2 and 3). Data from the Redmain Formation cluster at the SiO_2-rich end of this trend, adjacent to but largely separate from the overlapping clusters formed by the data from the Watch Hill and Hope Beck Formations. The data from the younger Loweswater and Kirk Stile Formations overlap and are more scattered along the SiO_2-poor end of the trend. There is a remarkably sharp cut-off in the SiO_2 content of the sandstone in the Loweswater and Kirk Stile Formations compared with that of the older formations. Trace element data are more scattered but generally lie along a positive linear trend against TiO_2, within which they define the formations in the same way as the major elements.

The Manx Group (Lonan- and Agneash-type) data correspond closely with the major and trace element trends of the Skiddaw Group data, but the Agneash-type compositions extend the SiO_2-rich end of the trend and overlap significantly only with samples from the enigmatic Redmain Formation. The spread of the major and trace element data from the Lonan sandstone type fits well with the Loweswater and Kirk Stile Formations (e.g. Fig. 3), with the same lower cut-off for SiO_2 [although TiO_2 (Fig. 2) has a more restricted range], and is clearly distinct from the older formations. Correlation of the Lonan and Loweswater Formations is further supported by the Nd isotope data reported by Stone & Evans (1995). The composition of the Niarbyl Formation is distinct from any part of the Skiddaw Group (e.g. Figs 2 and 3).

Southeast Ireland: Manx–Ribband–Bray Groups correlation

The Ribband Group may have correlatives in the pelitic parts of the Manx Group sequence which crop out in the western part of the Isle of Man (e.g. McConnell *et al.* 1999), both being of Arenig age (Brück *et al.* 1979; Molyneux 1999), and including manganese- and boron-rich coticule-bearing horizons (Kennan & Morris 1999). The Ribband Group contains little sandstone but the associated Bray Group, considered to be mid-Cambrian in age, is dominated by greywacke with interbedded quartz arenite (Brück & Reeves 1976). The greywacke in the Bray Group sequence is commonly very thickly bedded, but the lithological character of the more thinly bedded upper part of the Bray Head Formation is similar to that of the relatively thickly bedded, lower part of the undated Ny Garvain Formation exposed south of Gob ny Garvain (Brück & Kennan, pers comm.). Unfortunately, no sandstone geochemistry is available from the Bray Group, but possible correlation with the Ny Garvain Formation in the Isle of Man and with Lake District rocks is under investigation.

Possible Silurian correlatives

The Niarbyl Formation is shown to be of Wenlock age (Howe 1999) by sparse graptolites from the distinctive laminated hemipelagite which is interbedded with the otherwise sand-dominated turbidite sequence. Two sequences, similar in lithology and age, which crop out in the southern part of the Lake District and in southern Scotland are possible correlatives. These are briefly described then compared compositionally with the Niarbyl Formation.

Lake District: the Windermere Supergroup. The Windermere Supergroup in the southern Lake District (Kneller *et al.* 1994; Johnson *et al.* 1999; Millward *et al.* 1999) comprises a marine sequence which records almost continuous deposition from late Ordovician to late Silurian times:

- Dent Group – the Caradoc–Ashgill part of the sequence composed of variable, largely shallow marine deposits and volcanic rocks;
- Stockdale Group – a condensed Llandovery sequence of graptolitic black shale and calcareous siltstone (the Skelgill Formation) followed by pale green siltstone with red beds in the upper part (the Browgill Formation);
- unnamed group(s) of Wenlock strata, dominated by distinctive hemipelagite with a well-developed millimetre scale varve-like lamination of alternating siltstone–mudstone couplets. Up to 300 m of almost pure hemipelagite of the Brathay Formation includes interbedded sandstone in the *C. lundgreni* Biozone (Birk Riggs Formation) which, reaching a thickness of 380 m locally, marks the onset of a rapid expansion of the sequence by turbidites. The overlying Coldwell Formation is defined by two *c.* 15 m units of calcareous siltstone separated by hemipelagite;
- Coniston Group – of Ludlow age, comprising three sandstone turbidite-dominated units (Gawthwaite Formation, 0–520 m; Poolscar Formation, 430–700 m; Yewbank Formation, *c.*

750 m) interbedded with, and separated by, sequences dominated by hemipelagite;
- unnamed group(s) of late Ludlow–Pridoli mudstone, siltstone and fine-grained sandstone deposited in shallow-marine and fluviatile environments.

The sand-dominated turbidites of the Birk Riggs Formation and the Coniston Group are characterized by thin- to medium-bedded, fine- to medium-grained sandstone in packeted sequences interbedded with hemipelagite. Palaeocurrent directions vary from axial (dominantly southwest directed) to north or northwest derivation.

Southern Uplands terrane: the Hawick and Riccarton Groups. The Lower Palaeozoic rocks of southern Scotland and northeastern Ireland (Fig. 1) are dominated by a Caradoc–Wenlock sandstone-rich turbidite sequence overlying a condensed sequence of black mudstone and chert which ranges from Caradoc to mid-Llandovery in age. The sedimentary rocks are steeply dipping, contained in a series of northeast trending, strike-parallel fault-bounded tracts in which the mudstone passes northwards into sandstone which becomes progressively younger southwards (e.g. Leggett *et al.* 1979).

The Hawick Group, ranging from late Llandovery to early Wenlock in age (White *et al.* 1991), crops out over a large area in several tectonostratigraphic tracts in the south of the terrane. It is generally of extremely uniform lithological character, usually comprising classical turbidites with fine- to medium-grained calcareous greywacke in medium- to thick-bedded packets, several tens of metres in thickness, separated by thin-bedded sandstone and silty mudstone in packets a few metres thick. However, the occurrence of two additional lithologies interbedded in parts of the Hawick Group sequence, although partly reflecting regional trends (Kemp 1991), may suggest the first sedimentary links with the Lake District succession (Lintern *et al.* 1992). Red mudstone, locally well developed within the late Llandovery part of the Hawick Group, corresponds with that in the Browgill Formation. The youngest part of the Hawick Group (the early Wenlock Ross Formation, *murchisoni–antennularius* Biozones) includes small amounts of laminated hemipelagite, in beds from a few centimetres to 3 m thick, closely comparable with the Brathay Flag lithology.

In the southernmost tracts of the Southern Uplands terrane, the Wenlock (*rigidus–lundgreni* Biozones) Riccarton Group sequence comprises more varied turbidite facies than the Hawick Group and, although also rotated to steep dip, it is significantly less deformed (Kemp 1986; Lintern & Floyd 1999). Hemipelagite is more common than in the Ross Formation and in lower parts of the sequence it may occur in continuous units up to 25 m thick.

Silurian sandstone chemistry compared. A large quantity of sandstone geochemistry data is available from the Southern Uplands (Duller & Floyd 1995; Barnes 1998), including 17 analyses from the Ross Formation and seven from the Riccarton Group in southwest Scotland. Sandstone analyses from the Windermere Supergroup are provided by McCaffrey & Kneller (1996), although data are mainly from the Birk Riggs Formation (nine analyses) with only one or two from each of the higher units. Data from the Ross and Birk Riggs Formations and the Riccarton Group are compared with the Isle of Man data on Figs 2–4.

The major element data from the Ross Formation form a cluster with the lowest SiO_2 content (< 64%), representative of the whole of the Hawick Group which is characterized by remarkable uniformity compared with other Southern Uplands sandstones with little or no difference between formations (Barnes 1998). Data from the Riccarton Group, on the other hand, show considerable variability, ranging from 57 to 81% SiO_2. The Birk Riggs Formation forms a separate loose cluster with SiO_2 ranging from 62 to 73%. This seems to be generally comparable, compositionally, to the other sandstone-bearing formations in the Windermere Supergroup on the basis of the few analyses available (McCaffrey & Kneller 1996).

Overall, most of the major elements from the Silurian formations lie on a common trend against SiO_2 (e.g. Figs 2 and 3) from the late Llandovery–early Wenlock Hawick Group material at low SiO_2 content, through the Birk Riggs Formation into the more siliceous sandstone samples of the Riccarton Group of the same age. This trend, best seen in Fig. 3a, is distinct from that seen in the Skiddaw and Manx Groups. A similar situation is apparent for several trace elements when plotted against TiO_2, the Hawick Group data generally having higher values passing to progressively lower values in the Windermere Supergroup and the more siliceous samples from the Riccarton Group. However, the Y and Zr, and to some extent Rb, contents of the Hawick Group (e.g. Fig. 4a) are noticeably low, lying beneath the trend of the other Silurian sandstones.

Due to the widely variable CaO and MgO contents (at least partly representing carbonate as seen in high loss on ignition values) of the Silurian sandstones it is difficult to compare the simple bivariate plots of major element data. Removing these components from the data, and recalculating the remainder to 100%, focuses the trend in the

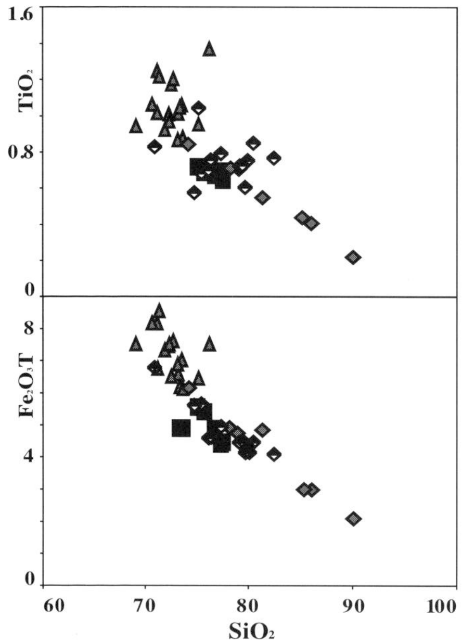

Fig. 7. Data from the Silurian sandstones recalculated to 100% without CaO, MgO and loss on ignition; key to symbols as in Fig. 2.

remaining major elements against SiO_2 (e.g. Fig. 7). On this basis, the Niarbyl Formation sandstone is most similar to that of the Birk Riggs Formation. The data from the Ross Formation (Hawick Group) are largely separate. The Riccarton Group, in contrast with the relatively restricted compositional range of the other formations (particularly the Niarbyl Formation on the basis of the five samples analysed), is characterized by widely variable sandstone composition and is, in this respect, distinctive. However, data for all of the formations lie close to the trend of the Riccarton Group material, suggesting that all are parts of a single system. This is consistent with indications of lithological correlation between the Southern Uplands terrane and the Lake District (e.g. Barnes *et al.* 1989; Lintern *et al.* 1992). Although analyses for all parts of the Windermere Supergroup are not included, specific correlation of the Niarbyl Formation with the Birk Riggs Formation in the Lake District appears reasonable and is consistent with their age (*lundgreni* Biozone).

Geotectonic setting of the source(s) of clastic debris

Determination of the tectonic setting of the source of detritus from sandstone geochemistry can be used to good effect in juvenile deposits, particularly where compositional differences result from contemporaneous volcanism (e.g. Bhatia 1983; Van de Kamp & Leake 1985). However, insufficient data, reworking of older sedimentary deposits generating inherited signatures and/or degradation of the relatively unstable components during weathering, transport and diagenesis, resulting in more mature compositions, may mask the original signature and cause misleading conclusions (e.g. Mack 1984; Haughton *et al.* 1991).

Bhatia (1983) presented discriminant diagrams for sandstones in simplified plate tectonic settings as follows:

- oceanic island arc – basins adjacent to, and dominated by, sediments from a contemporaneous basic volcanic arc;
- continental island arc – inter-arc, back-arc and fore-arc basins in which sediment is mainly derived from felsic volcanic rocks;
- active continental margin – basins on, or adjacent to, thick continental crust with sediment derived from granite–gneiss and siliceous volcanic rocks of the uplifted basement;
- passive margin – highly mature sediments derived by recycling of older sedimentary and metamorphic rocks, basins may include intracratonic and rift-bounded graben.

The trends of the Lower Palaeozoic sandstones lie close to the fields defined by Bhatia for TiO_2 (Fig. 8a). They fall below the fields for SiO_2 (Fig. 8b), apparently due to a relative lack of Al_2O_3-bearing phases (possibly mainly feldspar) compared with the limited suite of samples used by Bhatia. The fields of Niarbyl and Lonan types, overlapping the Hawick–Windermere and Loweswater–Kirk Stile material, respectively, suggest significant components of volcanic material. The signature of increasing volcanic components in the Skiddaw Group sequence (cf. Moore 1992) and Lonan sandstone, as apparent in Fig. 8, may reflect precursor volcanicity to the Eycott–Borrowdale Volcanic Group or progressive unroofing of an older volcanic sequence in the source terrane. The latter was preferred by Moore (1992), except in the upper part of the sequence (the Tarn Moor Formation, Cooper *et al.* 1995) where primary volcanic material is recognized.

McCaffrey & Kneller (1996), however, suggested that in the upper part of the Windermere Supergroup this signature may be inherited, the volcanic material forming a component of debris reworked from the accretionary margins of Laurentia. This scenario may be extended to include the Niarbyl Formation, although in the latter the debris is derived from further west (Morris *et al.* 1999). A volcanic-rich source may

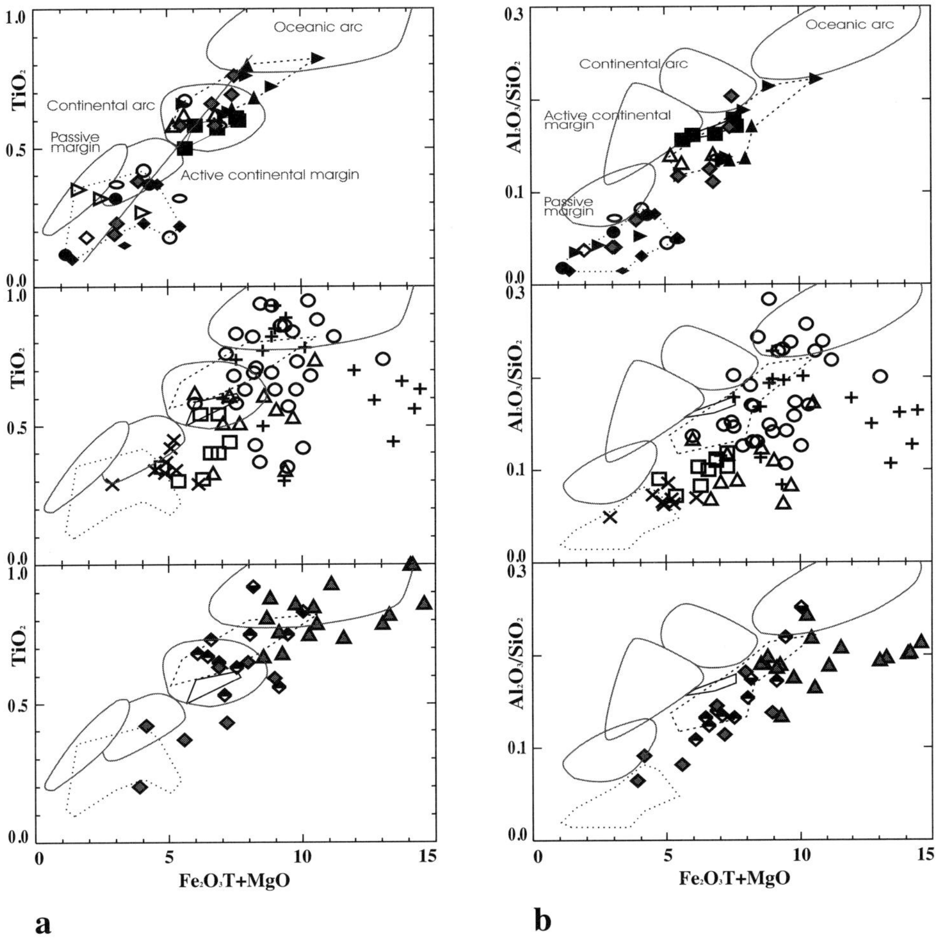

Fig. 8. Sandstone environment discrimination diagrams for simplified plate tectonic settings from Bhatia (1983); key to symbols as in Fig. 2.

also be suggested for the Hawick Group. In this case much of the material may have been reworked from volcanic-rich sandstone in older parts of the Southern Uplands thrust stack (e.g. Stone *et al.* 1987), although additional basic volcanic and bioclastic carbonate components are evident petrographically (Kemp 1985; Barnes 1999). The Riccarton Group sandstone, with a relatively wide compositional spread, probably also represents reworked debris, but from more varied source areas.

The very mature Agneash sandstone type, classified as having a passive margin source on both diagrams, probably represents a very mature source of cratonic material and may be comparable with the extensive, early Ordovician Grès Armorican quartz arenite (cf. Woodcock & Barnes 1999).

An alternative discrimination diagram (Fig. 9; Roser & Korsch 1986) neatly separates the three sandstone types identified in the Isle of Man into three fields, defined in a similar way to those of Bhatia (1983). The Lonan sandstone, closely comparable with the Loweswater Formation material, again falls into an active continental margin setting, as does the younger material from the Windermere Supergroup. The location of the Niarbyl and Hawick Group material in the island arc field is at least in part due to the dilution of SiO_2 by carbonate, although relatively high TiO_2 would partially support such an assignment for the Hawick Group (Fig. 9). The Agneash-type material

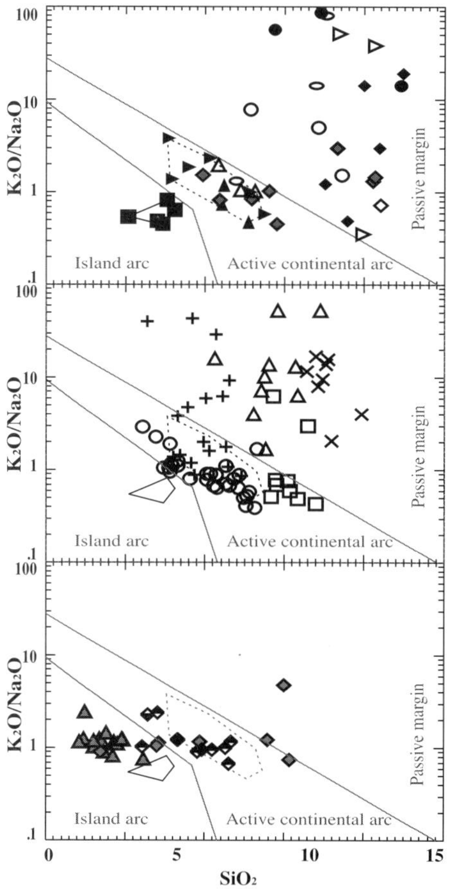

Fig. 9. Sandstone environment discrimination diagram from Roser & Korsch (1986); key to symbols as in Fig. 2.

is again classified as representing a passive margin setting.

Conclusions

Sandstone geochemistry can be an appropriate tool for characterization of sedimentary sequences, particularly when, as in the Isle of Man, a combination of structural complexity, poor biostratigraphical control, limited facies variation and poor exposure restrict the application of more traditional stratigraphical methods. These data may be used to establish possible correlations both locally and regionally, and provide information relating to the tectonic environment of sedimentation, although more data are necessary.

Sandstones from parts of the Ordovician Manx Group fall into two compositional groups: silica-rich (SiO_2 78–95%) quartz arenite (Agneash type) and greywacke (Lonan type) with lower silica (SiO_2 65–78%). Other element contents vary with silica, but some show a compositional hiatus suggesting that the two sandstone groups are distinct, with the possible inference that they represent material from separate source areas. In the three tectonostratigraphical sequences defined in the southeast of the island, sandstone of both types occurs in varying proportions as separate units at various scales, either as formations, members or locally more closely interbedded. In the Ny Garvain and Port Erin Formations, however, the sandstone composition is more variable between the two end-member types. In the absence of biostratigraphical control, these data are used to suggest various ways in which the sequences might correlate.

Regionally, the Lonan sandstone type is compositionally distinct from the early Arenig sandstones but very similar to mid- to upper Arenig sandstones which occur in the Skiddaw Group (the Loweswater and Kirk Stile Formations) of the Lake District in northwest England. There is no obvious equivalent of the Agneash quartz arenite sandstone type in this area [see also discussion in Woodcock & Barnes (1999)], with the possible exception of the Redmain Sandstone (Allen & Cooper 1986). Major incursions of quartz arenite do, however, occur in the mid-Cambrian Bray Group in southeast Ireland which has some sedimentological similarity to the undated Ny Garvain Formation in the Isle of Man.

The Silurian sandstone of the Niarbyl Formation is distinct from anything else in the Isle of Man but is compositionally similar to sedimentologically similar sequences which occur in the Hawick Group in the Southern Uplands of Scotland and in the Windermere Supergroup of the southern Lake District. The general similarity of these sequences suggests that they are parts of a single depositional system developed during the final stages of crustal shortening following closure of the Iapetus Ocean. The available data indicate that the closest compositional comparisons occur with material in the Birk Riggs Formation of the Windermere Supergroup.

Use of the sandstone geochemistry to constrain the tectonic environment from which the sandstone debris was derived gives consistent results on discriminant diagrams. These suggest that the Lonan sandstone type and the Silurian sandstones include significant volcanic components. The Lonan sandstone forms part of a trend of increasing volcanic material apparent in the Skiddaw Group, thought to represent primary volcanic material. This may have been principally derived from an older volcanic sequence but also includes input from contemporaneous volcanicity, at least in the younger part of the sequence. Following McCaffrey & Kneller (1996), the volcanic material in the Silurian sandstones, including the Niarbyl Formation, may be second-cycle debris reworked from the accretionary margins of Laurentia. The silica-rich Agneash sandstone type is classified as a passive margin deposit, inferred to contain very mature material derived from reworked older sedimentary and/or granite–gneiss basement sources.

The new analyses reported herein were made at the University of Portsmouth, University of Nottingham and BGS Keyworth, and the assistance of the staff of the analytical facilities is gratefully acknowledged. Colleagues in the Isle of Man research group, particularly David Burnett, David Quirk and Nigel Woodcock, are thanked for guidance in the field and discussion of the analytical results in the context of the lithostratigraphy. We are grateful to Maria Mange, Bill McCaffrey and Phil Stone for constructive reviews which significantly improved this contribution. Field work was funded by NERC grant no. GR9/01834. RPB and DCC publish with the permission of the Director, British Geological Survey (NERC).

References

ALLEN, P. M. & COOPER, D. C. 1986. The stratigraphy and composition of the Latterbarrow and Redmain sandstones, Lake District, England. *Geological Journal*, **21**, 59–76.

BARNES, R. P. 1998. *Graphical presentation of element abundances in Ordovician and Silurian sandstone formations in the Southern Uplands of Scotland.* British Geological Survey Technical Report.

—— 1999. *The geology of the Whithorn, Kirkcowan and Wigtown districts*. Memoir of the British Geological Survey. Sheets 2, 4W and 4E (Scotland).

—— & STONE, P. 1999. Trans-Iapetus contrasts in the geological development of southern Scotland (Laurentia) and the Lakesman Terrane (Avalonia). *This volume.*

——, LINTERN, B. C. & STONE, P. 1989. Timing and regional implications of deformation in the Southern Uplands of Scotland. *Journal of the Geological Society, London*, **146**, 905–908.

——, POWER, G. M. & COOPER, D. C. 1998. *New geochemical data from Ordovician and Silurian sedimentary rocks in the Isle of Man.* British Geological Survey Technical Report.

BHATIA, M. R. 1983. Plate tectonics and the geochemical composition of sandstones. *Journal of Geology*, **91**, 611–627.

BLATT, H., MIDDLETON G. V. & MURRAY R. C. 1980. *Origin of Sedimentary Rocks.* Prentice Hall.

BRÜCK, P. M. & REEVES, T. J. 1976. Stratigraphy, sedimentology and structure of the Bray Group in County Wicklow and south County Dublin. *Proceedings of the Royal Irish Academy*, **76**, 53–77.

——, COLTHRUST, J. R. J., FEELY, M. *ET AL.* 1979. South-east Ireland: Lower Palaeozoic stratigraphy and depositional history. *In*: HARRIS, A. L., HOLLAND, C. H. & LEAKE, B. E. (eds) *The Caledonides of the British Isles – Reviewed.* Geological Society, London, Special Publications, **8**, 533–544.

CHAYES, F. 1971. *Ratio Correlation.* Chicago University Press.

COOPER, A. H., FORTEY, N. J., MOLYNEUX, S. G., RUSHTON, A. W. A. & STONE, P. 1999. *The Geology of the Skiddaw Group, English Lake District.* Memoir of the British Geological Survey, UK .

——, RUSHTON, A. W. A., MOLYNEUX, S. G., HUGHES, R. A., MOORE, R. M. & WEBB, B. C. 1995. The stratigraphy, correlation, provenance and palaeogeography of the Skiddaw Group (Ordovician) in the English Lake District. *Geological Magazine*, **132**, 185–211.

——, LEE, M. K., FORTEY, N. J., COOPER, A. H., RUNDLE, C. C., WEBB, B. C. & ALLEN, P. M. 1988. The Crummock Water aureole: a zone of metasomatism and source of ore metals in the English Lake District. *Journal of the Geological Society, London*, **145**, 523–540.

DULLER, P. R., & FLOYD, J. D. 1995. Turbidite geochemistry and provenance studies in the Southern Uplands of Scotland. *Geological Magazine,* **132**, 557–569.

FITCHES, W. R., BARNES, R. P. & MORRIS, J. H. 1999. Geological structure and tectonic evolution of the Lower Palaeozoic rocks of the Isle of Man. *This volume.*

FLOYD, P. A., SHAIL, R., LEVERIDGE, B. E. & FRANKE, W. 1991. Geochemistry and provenance of Rhenohercynian synorogenic sandstones: implications for tectonic environment discrimination. *In:* MORTON, A. C., TODD, S. P. & HAUGHTON, P. D. W. (eds) *Developments in Sedimentary Provenance Studies.* Geological Society, London, Special Publications, **57**, 174–188.

GEOLOGICAL SURVEY. 1898. *Isle of Man. Solid and Drift Geology. 1:63 360 scale.* Reprinted at 1:50 000 scale by Institute of Geological Sciences 1975. Ordnance Survey.

HAUGHTON, P. D. W., TODD, S. P. & MORTON, A. C. 1991. Sedimentary provenance studies. *In*: MORTON, A. C., TODD, S. P. & HAUGHTON, P. D. W. (eds) *Developments in Sedimentary Provenance Studies.* Geological Society, London, Special Publications, **57**, 1–11.

HOWE, M. P. A. 1999. The Silurian fauna (graptolite and nautiloid) of the Niarbyl Formation, Isle of Man. *This volume.*

JOHNSON, E. W., SOPER, N. J., BURGESS, I. C. *ET AL.* 1999. *The Geology of the Country Around Ulverston.* Memoir of the British Geological Survey, Sheet 48 (England & Wales).

KEMP, A. E. S. 1985. *The later (Silurian) sedimentary and tectonic evolution of the Southern Uplands accretionary terrane.* PhD Thesis, University of Edinburgh.

—— 1986. Tectonostratigraphy of the Southern Belt of

the Southern Uplands. *Scottish Journal of Geology*, **22**, 241–256.

—— 1991. Discussion on Silurian collision and sediment dispersal patterns in Southern Britain. *Geological Magazine*, **128**, 673.

KENNAN, P. S. & MORRIS, J. H. 1999. Manganiferous ironstones in the early Ordovician Manx Group, Isle of Man: a protolith of coticule? *This volume.*

KNELLER, B. C., SCOTT, R. W., SOPER, N. J., JOHNSON, E. W. & ALLEN, P. M. 1994. Lithostratigraphy of the Windermere Supergroup, Northern England. *Geological Journal*, **29**, 219–240.

LAMPLUGH, G. W. 1903. *The Geology of the Isle of Man.* Memoir of the Geological Survey, UK. HMSO.

LEGGETT, J. K., MCKERROW, W. S. & EALES, M. H. 1979. The Southern Uplands of Scotland: a Lower Palaeozoic accretionary prism. *Journal of the Geological Society of London*, **136**, 755–770.

LE MAITRE, R. W. 1982. *Numerical Petrology: Statistical Interpretation of Geochemical Data.* Elsevier.

LINTERN, B. C. & FLOYD, J. D. 1999. The Kirkcudbright–Dalbeattie district – a concise account of the geology. *Memoir of the British Geological Survey*, Sheets 5W, 5E and part of 6W (Scotland).

——, BARNES, R. P. & STONE, P. 1992. Discussion on Silurian and early Devonian sinistral deformation of the Ratagain Granite, Scotland: constraints on the age of Caledonian movements on the Great Glen system. *Journal of the Geological Society, London*, **149**, 858.

MCCAFFREY, W. D. & KNELLER, B. C. 1996. Silurian turbidite provenance on the northern Avalonian margin. *Journal of the Geological Society, London*, **153**, 437–450.

MCCANN, T. 1991. Petrological and geochemical determination of provenance in the southern Welsh Basin. *In*: MORTON, A. C., TODD, S. P. & HAUGHTON, P. D. W. (eds) *Developments in Sedimentary Provenance Studies*. Geological Society, London, Special Publications, **57**, 215–230.

MCCONNELL, B. J., MORRIS, J. H. & KENNAN, P. S. 1999. A comparison of the Ribband Group (southeastern Ireland) to the Manx Group (Isle of Man) and Skiddaw Group (northwestern England). *This volume.*

MACK, G. H. 1984. Exceptions to the relationship between plate tectonics and sandstone composition. *Journal of Sedimentary Petrology*, **54**, 212–220.

MILLWARD, D., JOHNSON, E. W., BEDDOE-STEPHENS, B. & YOUNG, B. 1999. *The Geology of the Ambleside District.* Memoir of the British Geological Survey, Sheet 38 (England & Wales).

MOLYNEUX, S. 1999. A reassessment of Manx Group acritarchs. *This volume.*

MOORE, R. M. 1992. *The Skiddaw Group of Cumbria: early Ordovician turbidite sedimentation and provenance on an evolving microcontinental margin.* PhD Thesis, University of Leeds.

MORRIS, J. H., WOODCOCK, N. H. & HOWE, M. P. A. 1999. The Silurian succession of the Isle of Man: the late Silurian Niarbryl Formation, Dalby Group. *This volume.*

PETTIJOHN, F. J., POTTER, P. E. & SIEVER, R. 1972. *Sand and Sandstones.* Springer Verlag.

QUIRK, D. G. & BURNETT, D. J. 1999. Lithofacies of Lower Palaeozoic deep marine sediments in the Isle of Man: a new map and stratigraphic model for the Manx Group. *This volume.*

ROBERTS, B., MORRISON, C. & HIRONS, S. 1990. Low grade metamorphism of the Manx Group, Isle of Man: a comparative study of white mica 'crystallinity' techniques. *Journal of the Geological Society, London*, **147**, 271–277.

ROLLINSON, H. R. 1993. *Using Geochemical Data: Evaluation, Presentation, Interpretation.* Longman.

ROSER, B. P. & KORSCH, R. J. 1986. Determination of tectonic setting of sandstone-mudstone suites using SiO_2 content and K_2O/Na_2O ratio. *Journal of Geology*, **94**, 635–650.

RUSHTON, A. W. A. 1993. Graptolites from the Manx Group. *Proceedings of the Yorkshire Geological Society*, **49**, 259–262.

SIMPSON, A. 1963. The stratigraphy and tectonics of the Manx Slates Series. *Quarterly Journal of the Geological Society, London*, **119**, 367–400.

STONE, P. & EVANS, J. A. 1995. Nd isotope study of provenance patterns across the British sector of the Iapetus suture. *Geological Magazine*, **132**, 571–580.

——, COOPER, A. H. & EVANS, J. A. 1999. The Skiddaw Group (English Lake District) reviewed: early Palaeozoic sedimentation and tectonism at the northern margin of Avalonia. *This volume.*

——., FLOYD, J. D., BARNES, R. P. & LINTERN, B. C. 1987. A sequential back-arc and foreland basin thrust duplex model for the Southern Uplands of Scotland. *Journal of the Geological Society, London*, **144**, 753–764.

VAN DE KAMP, P. C. & LEAKE, B. E. 1985. Petrography and geochemistry of feldspathic and mafic sediments in the northeastern Pacific margin. *Transactions of the Royal Society of Edinburgh: Earth Sciences*, **76**, 411–450.

WHITE, D. E., BARRON, H. F., BARNES, R. P. & LINTERN, B. C. 1991. Biostratigraphy of late Llandovery (Telychian) and Wenlock turbiditic sequences in the SW Southern Uplands, Scotland. *Transactions of the Royal Society of Edinburgh: Earth Sciences*, **82**, 297–322.

WOODCOCK, N. H. & BARNES, R. P. 1999. An early Ordovician turbidite system on the Gondwana margin: the southeastern Manx group, Isle of Man. *This volume.*

——, MORRIS J. H., QUIRK, D. G. *ET AL.* 1999. Revised lithostratigraphy of the Manx Group, Isle of Man. *This volume.*

Magnetic survey of the Poortown Dolerite, Isle of Man

J. D. A. PIPER[1], A. J. BIGGIN[2] & S. F. CROWLEY[1]

[1]*Department of Earth Sciences, University of Liverpool, Liverpool L69 3BX, UK*

[2]*School of Geological Sciences, University of Kingston, Penrhyn Road, Kingston upon Thames, Surrey KT1 2EE, UK*

Abstract: A ground magnetic survey of a 1.5 × 1 km^2 area surrounding the Poortown Dolerite emplaced into Early Ordovician Manx Group metasediments has identified two regions of high-amplitude and short-wavelength anomalies; the first region extends for 300 m north of the quarry outcrop and the second lies between 100 and 500 m to the east. Linear anomalies have trends ranging from east–west to east-northwest–west-southwest. Palaeomagnetic and rock magnetic studies of the exposed intrusion show that the magnetization resides in Ti-poor titanomagnetite, which is predominantly multidomained. The natural remanent magnetization is dominated by a viscous remanence in the present geomagnetic field and the magnetic anomalies can therefore be modelled using the ambient field direction. A smaller high-blocking temperature component has a westerly direction of negative inclination and is compatible with a normal magnetization acquired during Late Ordovician times. 2D Geometrical models are developed to fit the observed magnetic anomalies. They show that the igneous complex comprises one or two north dipping sheets at its eastern and western peripheries, which expand into a set of multiple sheets separated by screens of Manx Group country rock in the area immediately north of the present quarry outcrop. The sheets may be components of a smaller number of sill-like intrusions repeated by faulting. This general model is supported by results from six boreholes.

The igneous body exposed at Poortown Quarry 3 km east of Peel (National Grid reference [SC 269 832]) is the largest of a suite of basic intrusions emplaced into Lower Palaeozoic rocks of the Isle of Man (Lamplugh 1903). They occur along a north-northeast–south-southwest trend located in the western and southern parts of the island (Fig. 1).

Owing to the poor quality of exposure, the architecture of the Poortown Intrusion is unknown. Lampugh (1903) referred to it as a 'diabase' and proposed that it has the form of a 'lens-like sheet inclining northwestwards', probably on the basis of the prominent ridge-like topographic feature with a southeast facing escarpment which rises above surrounding, relatively flat, cultivated farm land. Exploration of potential quarry reserves based on four boreholes (Holmes Grace Consulting Engineers Ltd 1992) led to the suggestion that a more complex intrusive geometry comprising a series of 'pod-like bodies' exists beyond the confines of the existing quarry.

Further exploration of reserves during 1995 (Davies *et al.* 1995) resulted in aquisition of substantial additional borehole data. Simplified logs of the boreholes are given in Fig. 2 and their locations are shown in Fig. 3. Whilst up to 50 m of

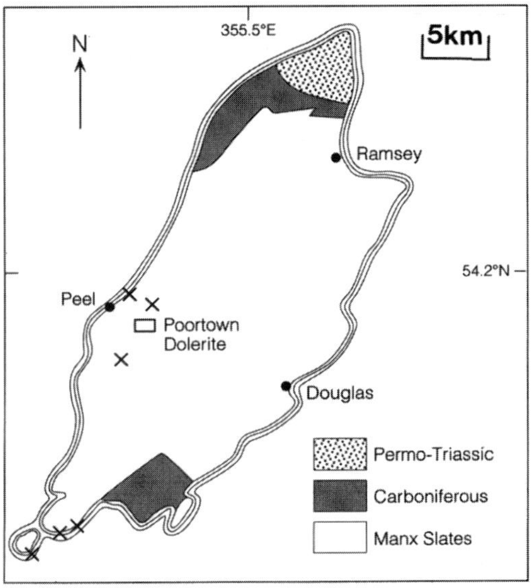

Fig. 1. Regional location of the Poortown Dolerite. The crosses show the distribution of the basic intrusive suite described by Lamplugh (1903).

From: WOODCOCK, N. H., QUIRK, D. G., FITCHES, W. R. & BARNES, R. P. (eds) 1999.
In Sight of the Suture: the Palaeozoic geology of the Isle of Man in its Iapetus Ocean context.
Geological Society, London, Special Publications, **160**, 155–163. 1-86239-046-0/99/$15.00

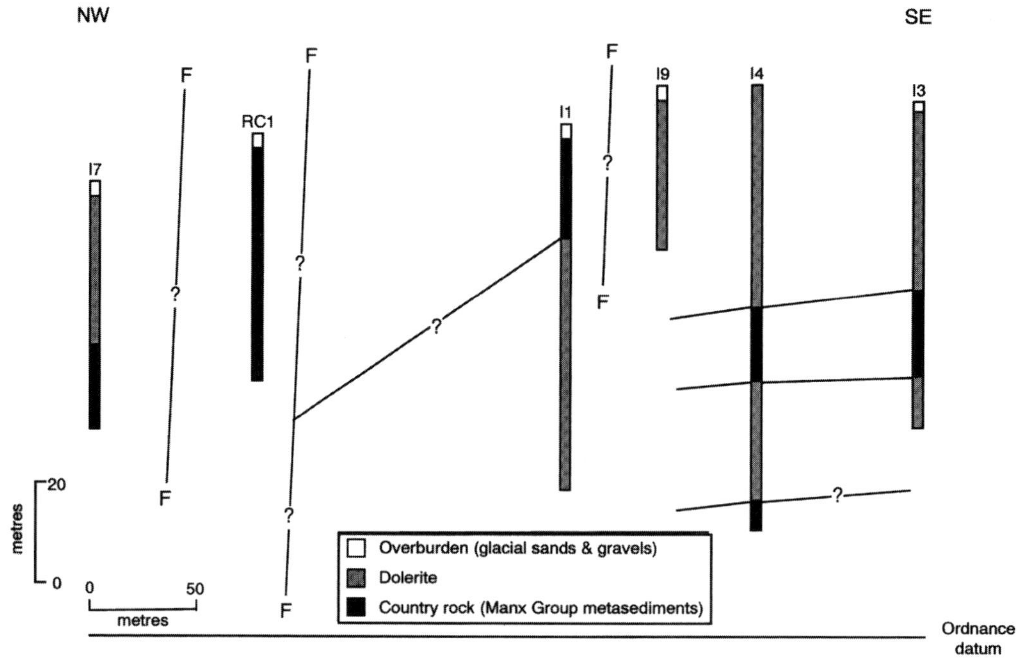

Fig. 2. Distribution of dolerite and Manx Group rocks logged in boreholes in the vicinity of the Poortown Quarry. Locations of the boreholes are shown in Fig. 3.

continuous intrusive body are present in some logs, they show that significant lateral variations in both the distribution and apparent thickness occur over short distances (Fig. 2). The intrusion is evidently not a single unit but consists of at least two sheets separated by Manx Group low-grade metasediments (see boreholes I3 and I4 in Fig. 3).

Interest in the form of the Poortown Intrusion has increased in recent years because it represents the only economic source of roadstone on the Isle of Man. It is also of general interest because improved understanding of the geometry should help to resolve the origin of the dolerite. As a relatively strongly magnetized basic body emplaced into weakly magnetized metasediments, in which paramagnetic chlorite and pyrite appear to be the main inducing phases, a magnetic survey provides an effective geophysical method for establishing the wider extent of the body and forms the main topic of this paper.

Geological background

Although parochially referred to as 'gabbro' (e.g. Ford 1993), the quarried body at Poortown appears to form part of a rapidly cooled, high-level intrusion composed of variable amounts of clinopyroxene and plagioclase in a matrix dominated by chlorite, epidote and carbonate minerals (calcite, Fe-rich dolomite). Despite extensive alteration during low-grade metamorphism, geochemical analyses of least-altered samples from the quarry indicate equivalent primary rock compositions ranging from high-Mg tholeiitic to calc-alkaline basalt and basaltic andesite of possible arc-related origin (Power & Crowley 1999). Within this overall textural and geochemical context the Poortown Intrusion is more correctly described as a dolerite.

Examination of the quarry exposure identifies sharp, sill-like contacts which are conformable with the bedding in Manx Group country rock (Fig. 4). The chilled margin is thin (*c.* 5 cm) and grades from a microcrystalline lithology containing sparse millimetre scale pyroxene phenocrysts into a more coarse-grained porphyritic rock which forms an outer zone to the equicrystalline centre of the intrusion. Although the country rock (particularly the sand-dominated quartzitic beds) is locally silicified within a few metres of the contact, there is no evidence of significant growth of contact metamorphic aluminosilicate minerals.

Deformation in the form of repeated faulting of the intrusion is evident from: the occurrence of

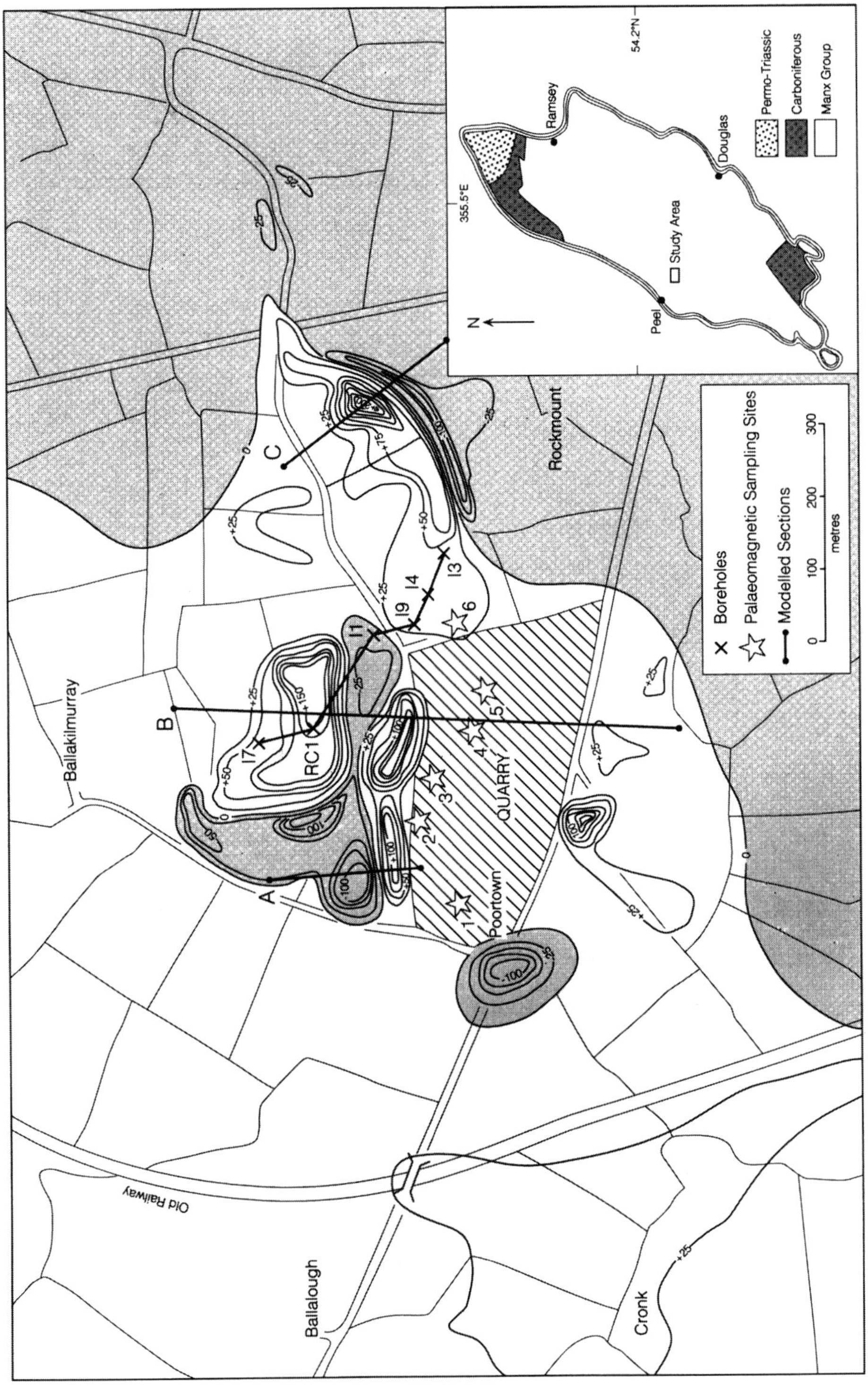

Fig. 3. Regional location and magnetic anomaly map of the Poortown district, Isle of Man. The smoothed total field magnetic anomalies are plotted with a contour interval of 25 n T relative to a base value of 48 881 nT. The three-lettered lines are the profiles fitted by the modelling as described in the text. Parameters for the magnetic modelling are derived from palaeomagnetic samples drilled in the quarry exposure at locations shown by the stars. Borehole locations are also indicated and refer to the logs shown in Fig. 2.

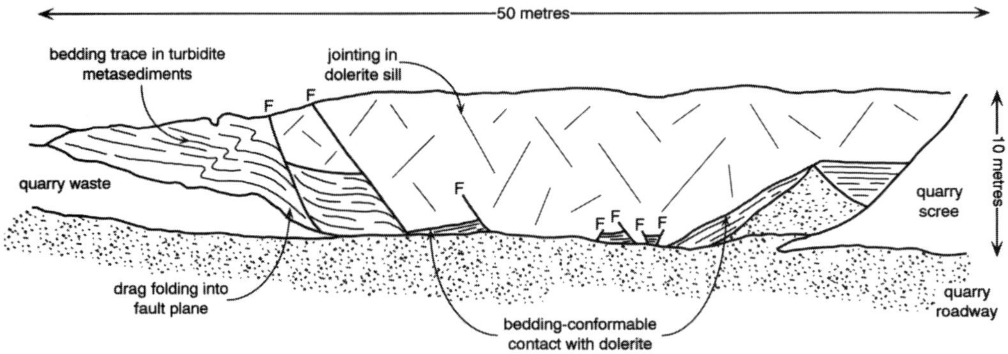

Fig. 4. Composite field drawing showing relationships between dolerite and Manx Group country rocks in the north face of the Poortown Quarry

steep, scarp surfaces in fields immediately east of the quarry; sharp changes in the dip of bedding and drag folding of Manx Group country rock (Fig. 4); numerous well-developed slickensided surfaces mineralized with hematite and calcite exposed on quarried blocks of the dolerite; and the apparent offset of intrusive bodies between adjoining borehole records (cf. boreholes I4 and I1, and RC1 and I7 in Fig. 2).

Although displacements on specific faults which juxtapose dolerite and Manx Group are obvious (Fig. 4), many *in situ* fault planes are difficult to identify in the quarry walls. However, sufficient evidence is available to indicate a near-vertical, orthogonal fault set, with metre scale displacements, oriented approximately north–south and east-northeast–west-southwest. Close to some faults, phenocrysts present in porphyritic dolerite exhibit textures consistent with the development of ductile shear bands in which stretched and flattened phenocrysts are drawn into fault plane alignment. Shear bands of this type occur frequently in cores recovered from a horizontal borehole (referenced RC4) drilled into the north face of the quarry near the northeast corner; these are accompanied by extensive alteration and replacement of primary dolerite mineralogy by carbonates. A further important observation from dolerite cores is the apparent occurrence of chilled margin contacts within the body of the intrusion, suggesting that it may consist of multiple intrusive sheets as opposed to simple sill-like units.

There is no direct evidence for the age of the Poortown Dolerite. Although not pervasively foliated, a fabric defined by shape orientation of chlorite is approximately parallel to a cleavage in the adjoining Manx Group (Power & Crowley 1999) and suggests that the dolerite was emplaced early in the history of the group.

The magnetic survey

The regional magnetic survey was conducted by pacing long traverse lines designed to establish the extent of the dolerite outwards from the quarried outcrop. To provide a basis for modelling the geometry of the intrusion, the survey was extended well into adjoining regions underlain by Manx Group country rocks. A proton precession magnetometer reading the magnitude of the ambient total magnetic field directly in nanotesla (nT) to an accuracy of 0.1 nT was used for the survey. Traverse lines were surveyed by pacing lines between reference points on the 1:10 560 scale topographic map. Fields were mostly surveyed between their corners and along trajectories following their perimeters at a fixed distance away, usually 50 m, to avoid magnetic interference from fences and field boundaries. The magnetic field was recorded at intervals of 10 paces, reduced to 5, 2 or 1 paces in regions of steep gradients. A base station was visited at intervals of 1–2 h during the survey period to identify variations of the geomagnetic field during the survey intervals.

The survey lines were extended into regions where the magnetic field was flat or comprised small random variations which identified the probable presence of underlying Manx Group rocks. Regions of high-amplitude and short-wavelength anomalies were recognized as being due to artificial noise from the quarry, roads, the derelict railway line, wire fences and some sections of the bridle tracks. These sections of the survey lines were excluded from compilation of the magnetic base map. Other magnetic anomalies due to non-geological sources correlated with the overhead powerlines and underground cables related to quarry activities crossing the region;

these were found to influence the survey for up to 8 m on either side and comprise narrow bands of excluded information on the base map. Because some fields west of the track leading to Ballakilmurray (see Fig. 3) were in-crop at the time of the survey, this area was only partially surveyed. Accordingly, it has not been contoured, although no magnetic anomalies of > 25 nT were recognized by the limited surveying in this region.

Over 1800 stations were recorded during the survey. The resultant data were processed in three steps: (1) removal of temporal variations of the geomagnetic field; (2) filtering of 'noisy' segments of the profiles to suppress short-wavelength anomalies; (3) subtraction of the regional background field.

Diurnal changes in the ambient magnetic field were recognized from the base station measurements recorded at intervals during the field survey – corrections of up to ±20 nT were required to cancel effects of this variation. Short-period fluctuations which could not be excluded on the grounds of an obvious artificial source were smoothed over 50 m data sets using a five-point filter. Since the survey covered a relatively small area, the regional background was taken to be the average field value recorded over the Manx Group country rocks and determined to be 48 881 nT. This value is subtracted from the field values to derive the magnetic anomalies.

Reduced values were plotted on an expanded copy of the 1:10 560 map and sampled for two levels of interpretation: (1) a contour map was produced joining values of equal field strength, at intervals of 25 nT, to permit a qualitative regional interpretation; (2) the three profiles located in Fig. 3 were compiled to run approximately orthogonal to the trend of the anomalies and provide a basis for magnetic modelling.

Regional extent of the Poortown Complex

The regional survey shows two areas of strong near-surface magnetization (Fig. 3): the first extends for 300 m north of the present quarry and the second lies between 100 and 500 m to the east-northeast. These zones of high-amplitude and short-wavelength anomalies are surrounded by magnetically flat ground which is interpreted to be underlain by Manx Group metasediments only. The eastern anomaly is a curvilinear feature and changes from +220 to –80 nT in just 40 m; it is located across a prominent topographic feature where strongly magnetized dolerite is probably upfaulted to the north against weakly magnetized slates to the south. The anomaly to the north of the quarry comprises several positive and negative features, probably attributable to interleaving of intrusions and country rock (cf. Fig. 2). With the exception of two small non-linear features, there are no anomalies south of the quarry and road, and any extension of the intrusion in this direction must lie at depth.

Magnetic properties of the dolerite

A palaeomagnetic and rock magnetic study of the Poortown Dolerite has been conducted on 62 cores drilled at six sites in the vicinity of the quarry (see locations in Fig. 3). The rock magnetic experiments show that the remanence carrier is Ti-poor titanomagnetite. The magnetic structure of this mineral is predominantly multidomained but significant fractions of single domains are also present. The ubiquitous presence of magnetite in the dolerite is confirmed by petrographic study (Power & Crowley 1999). The total natural remanent magnetizations (NRM) show considerable scatter which, in this instance, may include viscous remanent magnetizations (VRM) acquired during four months of laboratory storage (Fig. 5). However, NRM directions tend to have positive inclinations with northerly to westerly declination. The mean direction (D/I = 335/48°; cone of 95% confidence, α_{95} = 10°; precision parameter, k = 4.4) has a shallower inclination and more westerly declination than the present geomagnetic field; this implies that remanence recorded by the NRM is typically the vector resultant of a VRM in the present field and a smaller high-blocking temperature component.

Progressive thermal demagnetization (Fig. 6) demonstrates that NRM are composite and comprise of one or more low-blocking temperature components plus a high-blocking temperature component of probable ancient, and possible primary, origin. This is recognized in typical component structures (Fig. 6) which show a northerly positive magnetization unblocked over a broad range of temperatures up to 400°C to isolate a negative westerly component over a much narrower temperature range (typically 520–560°C). Directions of the subtracted low blocking temperature components are seldom precisely in the present Earth's field direction (Fig. 5), but collectively they yield a mean (D/I = 355/62°, α_{95} = 6, k = 9.6°) with identical declination to the present field and an inclination which is only marginally shallower. This is better defined than the mean of the NRM because it excludes the high-unblocking temperature components.

The high-blocking temperature components yield a mean direction of D/I = 266/–48° (α_{95} = 5°, k = 18). It does not correlate with any post-Lower Palaeozoic field direction from the British Isles and, since Britain lay in the southern hemisphere

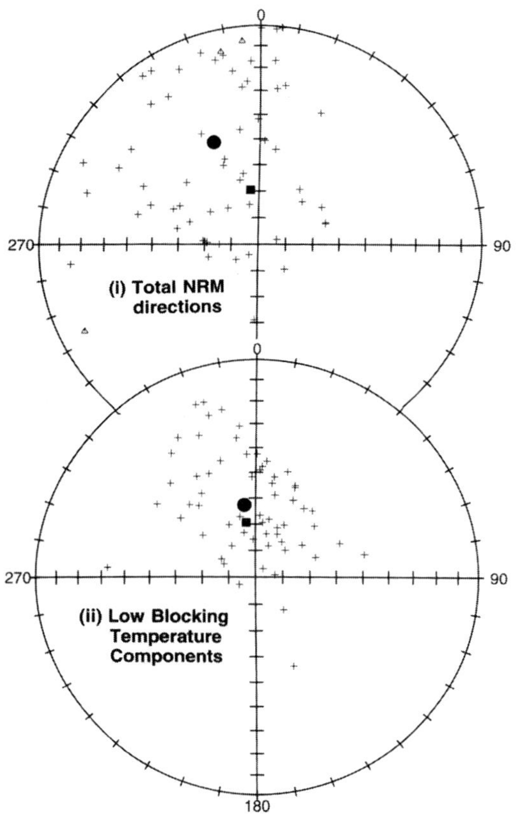

Fig. 5. Distributions of total NRM directions of magnetization in the Poortown Gabbro and subtracted low-blocking temperature components. +, Plots on the lower hemisphere; △, plots on the upper hemisphere; ●, the mean directions of the distributions; ■, direction of the present geomagnetic field in the study region.

prior to Carboniferous times, it is equivalent to a normal polarity. The polarity and palaeolatitude (29° S *in situ*) are compatible with acquisition during Late Ordovician times (cf. Piper *et al.* 1997), although, in view of the alteration and deformation of the dolerite, this is likely to be of secondary origin. This conclusion relates magnetization, and possibly basic magmatism, on the Isle of Man to Caledonian tectonomagmatic activity of this age in North Wales and the Lake District. Emplacement at a late stage of the closure of Iapetus would be compatible with the arc-related chemistry (Power & Crowley 1999). Unfortunately, the palaeomagnetic solution remains uncertain because possible post-magnetization tilting of the complex is unknown; the direction of this ancient component of magnetization cannot therefore be defined unambiguously.

Although a high-blocking temperature remanence divergent from the present field direction is present in the quarried dolerite, it is the smaller fraction (typically one-quarter to one-third) of the total NRM. Because the induced magnetization is also in the present field direction, and typically larger than the remanent contribution, for purposes of modelling the magnetization of the body it may be taken to be in the direction of the ambient geomagnetic field. The volume susceptibilities of the cores measured using a Bartington Bridge range from 0.02 to 1.00×10^{-3} SI units. Since this range includes three orders of magnitude, the results are best summarized by a log 10 normalized distribution (Irving *et al.* 1966), which yields a mean and standard deviation of $0.54 \pm 0.26 \times 10^{-3}$ SI units. The intensities of magnetization before treatment show a corresponding variation in the 0.05–4.20×10^{-2} A m^{-1} range; the log mean value is $1.9 \pm 0.52 \times 10^{-2}$ A m^{-1}.

Models for the magnetic anomalies

Of the three profiles used for modelling (see locations in Fig. 3), A is interpolated from the contour map, and lines B and C are compiled directly from reduced field profiles with suitable orientations. B is derived by aligning two separated traverses (the central part could not be surveyed because the quarry intervenes) in an attempt to model the full width of the intrusive complex. Profile C is used to model the eastern anomaly at Rockmount. These profiles comprise input to the *GRAVMAG* program and the model is developed as a series of polygons to match the observed anomalies. The model is described as 2.5D because a half-strike is entered for the whole model and the polygons are constrained to continue on either side of the 2D profile for this distance. All polygons were assigned the same magnetic properties using values derived from the field and laboratory study: geomagnetic field strength 48 880 nT; magnetic susceptibility 0.54×10^{-3} SI units; intensity of remanent magnetization 1.9×10^{-2} A m^{-1}; direction of remanence, D/I = 353/70°. A half-strike of 300 m was used, although the selected value was found to have little influence on the models provided that it exceeded a few tens of metres.

Profile A (Fig. 7) requires at least two near-surface intrusions dipping to the north to match the observed profile. Polygon 1 may imply a faulted south margin. Although it appears unrealistic for a geological feature, it stresses that the magnetized volume diminishes rapidly to the north at this point and may be faulted out altogether. Profile B crosses the most complex part of the anomaly (note that the short-wavelength features are smoothed on the

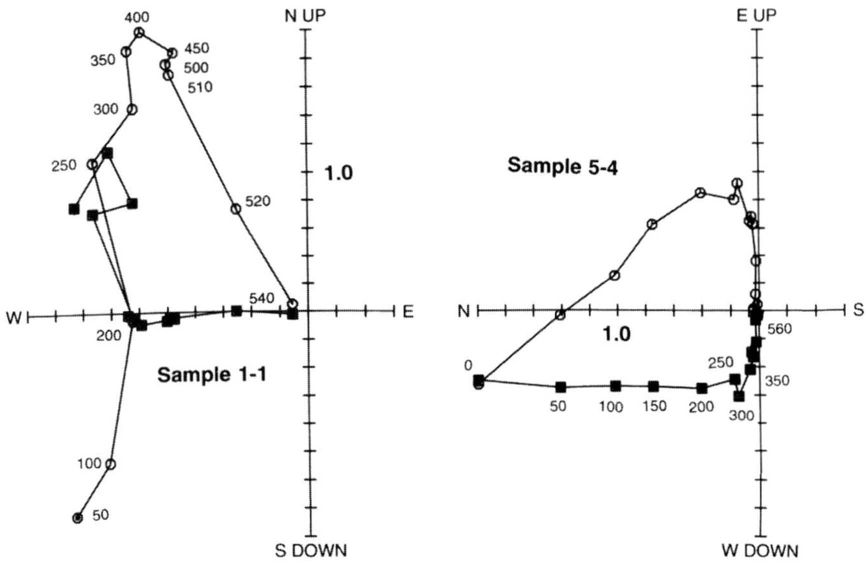

Fig. 6. Thermal demagnetization results from two samples of the Poortown Gabbro plotted as orthogonal plots with the magnetization vector projected on to the horizontal (■) and the vertical (○) planes. Demagnetization steps are indicated in °C and axes are calibrated in units of $\times 10^{-5}$ A m^2 kg^{-1}. Note that the northerly positive magnetization acquired in the present geomagnetic field is subtracted by treatment to 300–400°C to recover a high-unblocking temperature convergent component with westerly direction and negative inclination.

regional map of Fig. 4). Added complexities to the modelling procedure come from the space occupied by the quarry and the fall in topography from north to south; the latter point could not be accommodated by the modelling and level ground is assumed for the profiles shown in Fig. 7. This limitation does not affect the main features of the model which requires a series of northerly dipping sheets, each a few tens of metres in thickness, separated by screens of non-magnetic country rock to explain the oscillations in the profile. Because a number of variables is involved, these models are inevitably a compromise and should be interpreted in general, rather than specific, ways. Thus, the presence of dolerite across the floor of the quarry has not been accommodated; attempts to link the bodies with horizontal polygons at this level produced a deterioration in the fit, although the geometry of the sheets was only affected in detail. It is concluded that the sequence of sheets required to explain the magnetic anomalies are either: (1) connected to a single body at a depth greater than the modelling in Fig. 7; or (2) parts of one or more original sill-like bodies which have been repeated by faulting.

A model fit to line C is shown in Fig. 7 where polygons 1 and 2 suggest a possible sill and feeder. Whilst a third body is required to the south to produce the fit shown, this is excluded from the figure because it is an artefact of the arbitrary background and is not supported by the flat magnetic field around Rockmount (Fig. 3).

Conclusions

The presence of a strongly magnetised basic intrusion at Poortown, emplaced into essentially non-magnetic Manx Group country rock, makes magnetic survey the most suitable method for mapping the intrusion in unexposed terrain. The survey identifies a continuation of the intrusion to the north and east of the quarry outcrop and shows that extraction can potentially extend into this ground. However, high-amplitude and short-wavelength anomalies show that the unexposed complex comprises a number of dipping sheets; extraction will therefore need to contend with screens of country rock. The presence of a dominant viscous magnetization in the dolerite, acquired in the present geomagnetic field and compounded with an induced magnetization, permits the form of the intrusive complex to be modelled. The modelling shows that a simple form at the east and west peripheries, probably comprising two north dipping bodies, exapands into a sequence of multiple sheets in the region immediately north of the present quarry. The sheets

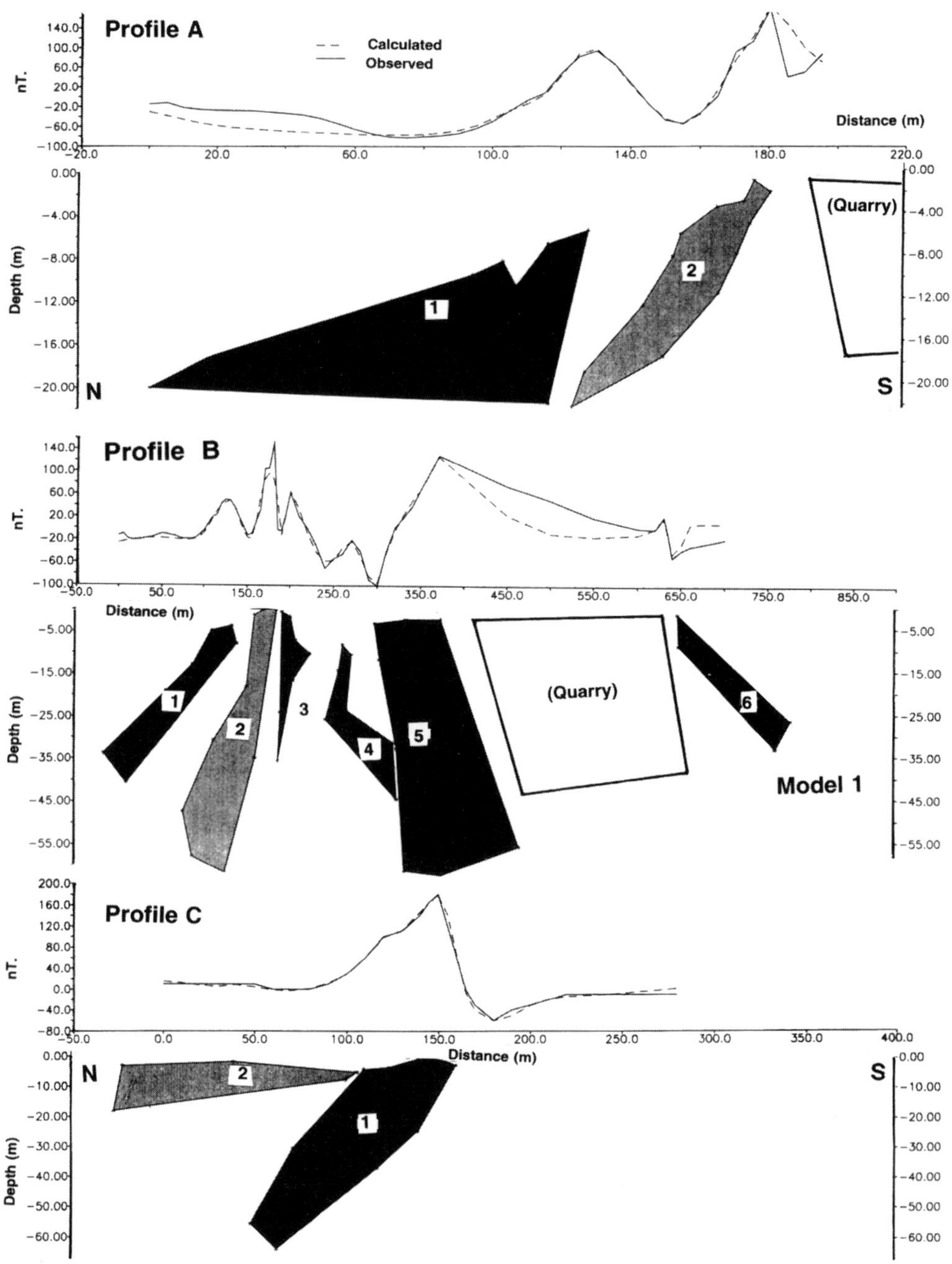

Fig. 7. Magnetic profiles A–C across the Poortown Complex (see Fig. 3 for locations) with 2D polygonal models derived to fit the observed anomalies. Magnetic parameters assigned to the polygons are given in the text and the vertical exaggeration is 5.3.

are typically a few tens of metres in thickness and attain a maximum thickness of *c.* 80 m; variations in intensity and susceptibility within the bodies would, of course, influence the thicknesses derived from the modelling.

We are grateful to local landowners in the Poortown district for allowing us to conduct the magnetic survey on their land, the Department of Highways, Properties and Ports for allowing us to collect samples in the quarry, and to G. S. Kimbell and an anonymous reviewer for their valued criticisms of the manuscript.

References

DAVIES, M., GUARD, J. & WRIGHT, A. 1995. *Poortown Quarry, Isle of Man. Geological Interpretive Report.* CSA-RDL report no. CSA 95.95.

FORD, T. D. 1993. *The Isle of Man.* Geological Association Guide, **46**.

HOLMES GRACE CONSULTING ENGINEERS LTD. 1992. *A geological investigation of the Poortown Quarry.* Report for the DHPP.

IRVING, E., MOYNEUX, L. & RUNCORN, S. K. 1966. The analysis of remanent magnetisation intensities and susceptibilities of rocks. *Geophysical Journal of the Royal Astronomical Society*, **10**, 451–464.

LAMPLUGH, G. W. 1903. *The Geology of the Isle of Man.* Memoir of the Geological Survey, UK. HMSO.

PIPER, J. D. A., STEPHEN, J. C. & BRANNEY, M. J. 1997. Palaeomagnetism of the Borrowdale and Eycott volcanic groups, English Lake District: primary and secondary magnetisation during a single late Ordovician polarity chron. *Geological Magazine*, **134**, 481–506.

POWER, G. M. & CROWLEY, S. F. 1999. Petrological and geochemical evidence for the tectonic affinity of the (?) Ordovician Poortwon Basic Intrusive Complex, Isle of Man. *This volume*.

WARDELL ARMSTRONG CONSULTANTS LTD. 1994. *Isle of Man mineral resources plan.* Vols. 1–4, prepared for the Dol, March 1994.

Petrological and geochemical evidence for the tectonic affinity of the (?)Ordovician Poortown Basic Intrusive Complex, Isle of Man

G. M. POWER[1] & S. F. CROWLEY[2]

[1]*School of Earth, Environmental and Physical Sciences, University of Portsmouth, Burnaby Road, Portsmouth PO1 3QL, UK*

[2]*Department of Earth Sciences, The Jane Herdman Laboratories, University of Liverpool, Brownlow Street, Liverpool L69 3BX, UK*

Abstract: The rocks of the Poortown Quarry, 3 km east of Peel, Isle of Man, are shown to comprise a complex series of sills of pyroxene-rich dolerite, plagioclase-rich dolerite and plagioclase–phyric andesite intruded into Manx Group deep-water marine sedimentary rocks of Arenig age. They have suffered early Devonian deformation and greenschist facies metamorphism, together with later alteration and faulting. The pyroxene-rich dolerite has the composition of a Mg-rich basalt relatively enriched in Fe, Cr and Ni. It contains up to 60% augite and is likely to have been produced by fractionation in a high-level magma chamber before intrusion into its present position. Some of the pyroxene grains have more primitive (higher Mg, Cr and lower Ti), partly resorbed cores which supports a multi-stage history for this magma.

The sills cover a range of compositions from Mg-rich basalt to calc-alkaline basaltic andesite and the geochemistry of the more immobile elements suggests a calc-alkaline volcanic arc origin in an active continental margin environment. Although the age of the Poortown Complex is poorly constrained, a tentative comparison is made with the lower part of the Borrowdale Volcanic Group of the English Lake District.

The Lower Palaeozoic succession of the Isle of Man, together with that of the Lake District and southeastern Ireland, forms part of the northeast–southwest trending Lake District–Wexford Terrane (Hutton 1987), the exposed remnants of which record the progressive Ordovician closure of the Iapetus Ocean. This closure culminated in the oblique collision of the northern margin of the East Avalonian microcontinent with Laurentia during the Silurian, and the subsequent development of a Silurian foreland basin (McKerrow *et al.* 1991; Soper *et al.* 1992). The Ordovician history of the Lake District–Wexford Terrane is distinguished by the occurrence of voluminous subduction-related extrusive volcanism of Llanvirn–Ashgill age (Stillman 1988). This is represented in the Lake District by the Eycott and Borrowdale Volcanic Groups (Branney & Soper 1988; Beddoe-Stephens *et al.* 1995), and in southeast Ireland by numerous volcanic successions (Stillman & Williams 1978; Stillman 1988). However, no record of similar volcanic activity is preserved on the Isle of Man because any Ordovician rocks later than earliest Llanvirn in age that may have existed have been removed by erosion (Woodcock *et al.* 1999). Consequently, if centres of arc volcanism did exist within the portion of the Lake District–Wexford Terrane crust represented by the Isle of Man, then this magmatic episode could be reflected by subvolcanic intrusions within pre-Silurian Manx crust. The identification of such intrusions offers the potential for inferring a subduction-related volcanic component to the crustal evolution of the Isle of Man.

The Lower Ordovician Manx Group (Tremadoc–Llanvirn age) of the Isle of Man comprises a succession of deep-marine clastic turbidite sedimentary rocks (Woodcock *et al.* 1999) which, it has been proposed (Fitches *et al.* 1999), lie in a series of northeast striking tectonostratigraphical tracts. Extrusive igneous activity of Ordovician age in the Isle of Man is known only from the Peel Volcanic Group, an extremely poorly exposed series of andesitic volcaniclastic deposits, considered to be of early Arenig age (Molyneux 1999). The Peel Volcanic Group is, therefore, possibly equivalent to the earliest stages of destructive margin volcanism in the Lake District–Wexford Terrane represented by the Dowery Hill Member of the Ribband Group, southeast Ireland (McConnell & Morris 1997). Intrusive igneous rocks are more common and Lamplugh (1903) described many thin basic dykes

From: Woodcock, N. H., Quirk, D. G., Fitches, W. R. & Barnes, R. P. (eds) 1999.
In Sight of the Suture: the Palaeozoic geology of the Isle of Man in its Iapetus Ocean context.
Geological Society, London, Special Publications, **160**, 165–175. 1-86239-046-0/99/$15.00

of pre-Carboniferous age emplaced into the Manx Group. They are usually deformed and extremely altered. The largest of these, generally poorly exposed, basic intrusions is the Poortown Basic Intrusive Complex which was intruded into Arenig turbidites of the Cregganmoar Formation (Woodcock *et al.* 1999). Although no radiometric age is available for any of these basic intrusions, a recent palaeomagnetic investigation of the Poortown Complex (Piper *et al.* 1999) reveals remenance signatures consistent with an Upper Ordovician age for Poortown magmatism. Given the geological evidence on a regional scale for the former existence of an Ordovician magmatic arc across the Lake District–Wexford Terrane, the occurrence of high-level basic intrusions within the Lower Ordovician Manx Group may provide a proxy record of Ordovician volcanism on the Isle of Man for which no other information exists. As a consequence, geochemical signatures preserved in the Poortown intrusives (and other basic intrusions within the Manx Group) may provide valuable information regarding the occurrence of potential Ordovician destructive margin magmatic activity within the Isle of Man. This paper reports the results of a petrographic and geochemical investigation of the basic intrusive complex exposed at Poortown with a view to examining evidence for a Manx arc volcanic centre. The data obtained are: (1) used to infer the magmatic and tectonic affinities of Poortown magmatism; (2) tentatively assessed in terms of evidence for an episode of Ordovician arc magmatism in the Isle of Man.

Geological relationships of the Poortown Basic Intrusive Complex

The main exposure of the Poortown Basic Intrusive Complex occurs in the Poortown Quarry [SC 269 832], 3 km east of Peel (Fig. 1). There is very little exposure outside the quarry and geophysical methods have been used (Piper *et al.* 1999) to attempt to define the limits of the complex. The results indicate that the complex underlies an area of *c.* $1 \times 0.5\ km^2$ extending north and east from the present quarry. Furthermore, modelling of the data from the magnetic surveys strongly suggests that the complex is made up, not of a single body, but of a series of sheet-like bodies interspersed with metasedimentary layers, all dipping gently towards the north.

Poortown Quarry has been an important source of road stone for the Isle of Man for many years and a series of exploratory boreholes have been drilled in and around the present quarry to evaluate possible reserves. The positions of boreholes referred to in this paper are shown in Fig. 1 and a summary of the logs constructed by the authors from examination of cores is given in Table 1. The geophysical model for the form of the intrusions (Piper *et al.* 1999) is supported by the boreholes which show a variety of igneous rock types with a sill-like relationship to intervening layers of sandstone and mudstone. Both the upper and lower contacts of some of these igneous bodies have been recorded and their fine-grained chilled contacts have been confirmed in thin section. There can be

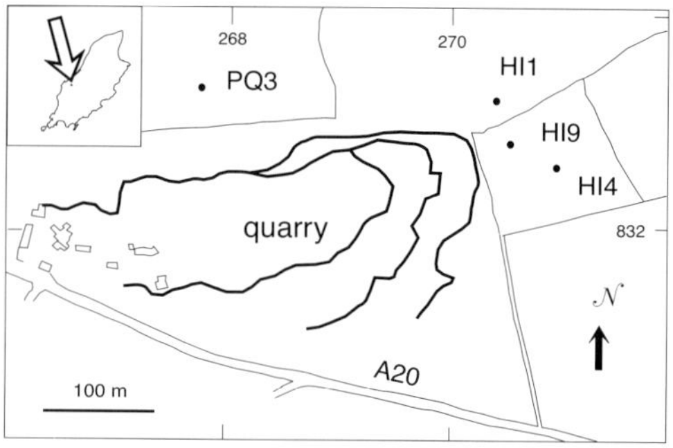

Fig. 1. Location of boreholes PQ3, HI1, HI4 and HI9 around the Poortown Quarry, Peel, Isle of Man. Strong lines define the levels of the working quarry; thin lines indicate field boundaries.

Table 1. *Summary of selected borehole logs, Poortown Quarry, Isle of Man*

PQ3	[SC 2677 8322]
Field north of main quarry	
10 m	Overburden and boulders
10 m	Plagioclase-rich dolerite
15 m	Pyroxene-rich dolerite
(base not reached)	
HI1	[SC 2704 8331]
Northeast of quarry	
13 m	Sandstones and mudstones with thin basic veins
9 m	Foliated dolerite
41 m	Plagioclase-rich dolerite
7 m	Altered dolerite
HI4	[SC 2709 8325]
Middle of first field east of quarry	
6 m	Overburden
44 m	Altered plagioclase-rich dolerite
10 m	Fine sandstones cut by thin dolerite veins
20 m	Plagioclase-phyric andesite
6m	Siltstones
HI9	[SC 2705 8328]
Northwest corner first field east of quarry	
6 m	Overburden
29 m	Olivine dolerite

little doubt that the main igneous bodies are sills. Again, within the quarry, although some of the contacts have undergone later tectonic modification, it is clear that the igneous rocks were intrusive in origin. The igneous sheets have very sharp lower contacts that are sub parallel to bedding in the Cregganmoar Formation, although locally contacts transgress bedding and thin dolerite veins cut across bedding at a high angle. No evidence has been found that the sills were intruded into soft sediments. The sharp cross-cutting contacts of the veins and sills suggest that the Manx Group was lithified at the time of emplacement. There is no indication of the development of contact metamorphic minerals in the Manx Group but bleaching and silicification does occur. The sills must have been emplaced at a fairly high level in the crust as the rocks into which they were intruded have not been metamorphosed above middle greenschist facies grade (Power & Barnes 1999).

The sills are not pervasively foliated but, in places, particularly in the more basic units, a fabric is defined by the alignment of metamorphic chlorite flakes. This fabric dips at *c.* 40° towards the north, similar in orientation to the first cleavage in the Manx Group. Fitches *et al.* (1999) argue that the first deformation of the Manx Group probably took place in the early Devonian, as it affects rocks of Wenlock age. Thus, the age of emplacement of the Poortown Complex is poorly constrained between Arenig and early Devonian times.

Two main sets of faults cut the sills, one set trending north–south and the other set close to east–west. Shear zones and the development of localized fabrics are associated with each of these fault sets. Fault surfaces are slickensided and mineralized with quartz, hematite and chlorite. Because of the rarity, or absence, of markers it is difficult to deduce the overall displacements on these faults and this makes correlations between boreholes almost impossible.

General compositional variations of the Poortown Complex

Lamplugh (1903, p. 156) referred to the 'Poortown diabase' as 'a handsome dark green porphyritic rock crowded with augite crystals'. Petrographic examination of 80 thin sections, together with chemical analyses of 27 selected rock samples, reveals that a considerably greater range of rock compositions is present than suggested by Lamplugh's concise description. A brief general overview of this variation will be given as an introduction and then two examples will be considered in more detail.

All of the rocks examined are holocrystalline and the majority are medium grained. The main original igneous minerals were augite, plagioclase, olivine and magnetite, present in varying proportions. Unaltered olivine is extremely rare and its former presence may usually only be inferred from characteristic euhedral pseudomorphs. Rock types range from pyroxene-rich dolerite, with pyroxene as a phenocryst phase (Fig. 2), to plagioclase–phyric andesite (Fig. 3). There is some systematic compositional variation within sills but there is more variation between sills.

It should be emphasized that all the rocks examined have been extensively altered from their original igneous mineralogy and should be considered as metamorphic, or possibly metasomatic, rocks. The plagioclase is now usually albite in composition. The dark green colour that Lamplugh observed is the result of the considerable

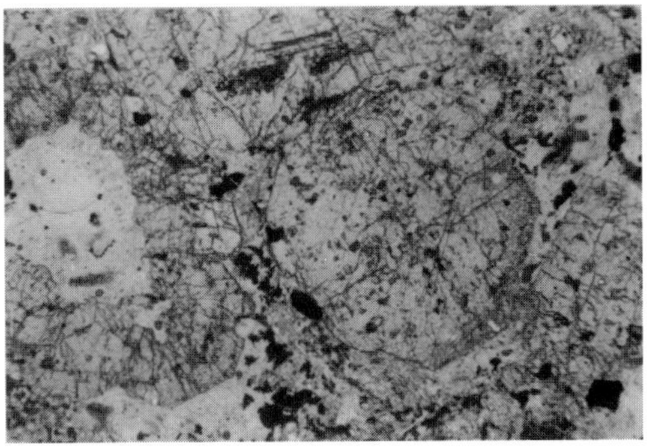

Fig. 2. Photomicrograph of pyroxene-rich dolerite, PQ3/19 from 28.3 m depth. Width of field, 5.5 mm. Pale subhedral area of chlorite, middle left, is probably a pseudomorph after olivine.

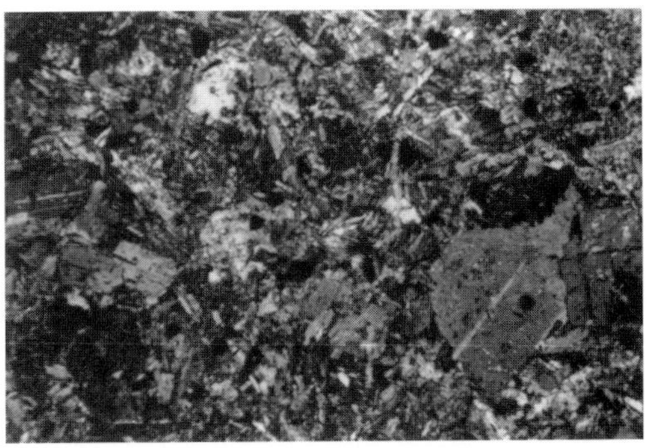

Fig. 3. Photomicrograph of plagioclase–phyric andesite from the lower sill of HI4. Width of field, 5.5 mm, crossed polars. Fine-grained groundmass mainly of plagioclase with some plagioclase phenocrysts (e.g. lower right).

new growth of chlorite together with minor amounts of epidote. Metamorphic amphibole is rare but one example of actinolitic hornblende has been recorded. Secondary calcite occurs as veins, along grain boundaries and, in some rocks, replacing pyroxene. Other secondary minerals include rare pumpellyite and also K-feldspar, but these are confined to zones of higher permeability such as well-foliated areas or contacts between contrasting rock types

Representative chemical analyses are given in Table 2 and a complete data set may be obtained from the authors. The chemical compositional range of the Poortown rocks is well displayed (Fig. 4) by the (FeO^* + TiO_2)–Al_2O_3–MgO diagram of Jensen (1976). The pyroxene-rich dolerite of borehole PQ3 has an abnormal composition and plots in the basaltic komatiite field. There is a range of compositions across the diagram to the plagioclase–phyric andesite of the lower sill of borehole HI4 which plots in the calc-alkaline andesite field.

Plagioclase–phyric andesite

Borehole HI4 is sited in the middle of the field immediately to the east of the quarry. It passes through two igneous bodies separated by 10 m of fine-grained sandstone which is cut by thin dolerite veins.

The lower body, a 20 m thick sill, is of particular interest because it is the most andesitic of all the analysed rocks. This fine-grained, holocrystalline, plagioclase-rich rock is composed of plagioclase, chlorite pseudomorphs and magnetite, with calcite (5–20%) present as a secondary mineral. Plagioclase, now all albite in composition, occurs as subhedral phenocrysts (15–20%) up to 1 mm in size in a finer grained groundmass composed of small (2–300 µm) prismatic plagioclase grains (40–60%) and chlorite. Fine-grained chlorite forms the pseudomorphs which are commonly 3–500 µm in size, have euhedral outlines and, from their shapes, may be after pyroxene. They poikilitically enclose some of the small plagioclase grains. The lowest part of the sill contains up to 5% quartz as individual grains (250 µm), intergrown with the groundmass plagioclase. As this quartz is only present near the contact of the sill, it is likely that it is evidence of contamination of the andesitic magma by assimilation of country rocks. The andesite immediately adjacent to the lower contact is fine-grained with euhedral plagioclase phenocrysts (up to 1 mm) in a groundmass composed of plagioclase crystallites (50 µm), sometimes exhibiting a trachytic texture that wraps around the phenocrysts, together with indeterminate very fine-grained brown material, possibly altered glass.

The upper body is 44 m thick and is a medium-grained plagioclase–phyric dolerite. It is extensively altered with much of the plagioclase replaced by colourless mica and with secondary calcite

Table 2. *Representative analyses of Poortown basic rocks*

	PQ3 7	PQ3 13	PQ3 23	HI1 3005	HI4 6025
SiO_2	50.26	45.76	46.18	46.24	52.93
Al_2O_3	15.28	9.84	9.33	14.48	16.18
Fe_2O_3	11.03	13.89	13.79	11.04	9.08
MgO	7.47	12.17	13.53	7.31	3.46
CaO	7.15	11.15	11.15	5.23	4.88
Na_2O	3.64	1.90	0.98	4.26	0.77
K_2O	0.84	0.16	0.16	0.43	2.04
TiO_2	0.97	1.06	0.88	1.16	1.18
MnO	0.17	0.22	0.24	0.18	0.20
P_2O_5	0.21	0.16	0.12	0.23	0.22
Total	97.02	96.31	96.36	90.56	90.94
Ni	74	122	142	62	14
Cu	131	100	78	111	23
Zn	91	103	99	94	112
Zr	117	73	62	91	146
Sr	657	394	258	462	96
Rb	36	5	9	18	95
Cr	249	488	558	186	54
V	264	298	264	321	292
Ba	488	112	92	207	390
Sc	32	42	48	33	35
Y	24.58	20.27	16.28	20.55	24.63
Nb	6.789	4.164	3.305	5.614	12.43
La	17.95	10.99	7.983	14.77	17.76
Ce	38.99	27.32	19.45	36.19	45.03
Pr	4.608	3.349	2.457	4.282	5.414
Nd	19.18	15.01	11.65	19.54	23.08
Sm	4.394	3.582	2.770	3.957	4.544
Eu	1.236	1.004	0.801	1.157	1.094
Gd	4.126	3.769	3.064	4.126	4.704
Tb	0.656	0.626	0.519	0.650	0.755
Dy	3.975	3.502	2.898	3.687	4.260
Ho	0.85	0.746	0.609	0.789	0.923
Er	2.373	2.024	1.651	2.159	2.532
Tm	0.322	0.270	0.219	0.285	0.338
Yb	2.061	1.766	1.493	1.893	2.255
Lu	0.339	0.279	0.230	0.301	0.355
Hf	2.897	2.024	1.594	2.431	3.673
Ta	0.633	0.470	0.385	0.580	1.145
Th	4.100	2.246	1.654	3.030	5.019
U	1.401	0.785	0.559	1.025	1.225

Samples were analysed by X-ray fluorescence spectroscopy at Portsmouth using lithium metacarbonate fusion disks for major elements and pressed powder pellets for a range of trace elements. Major oxides in wt% (total Fe as Fe_2O_3) and some trace elements in ppm. A subset of the samples was analysed for rare earth elements and Y, Nb, Hf, Ta and Th by inductively coupled plasma source mass spectrometry following hydrofluoric-perchloric acid digestion in pressurized PTFE vessels at the Department of Geology, University of Southampton.

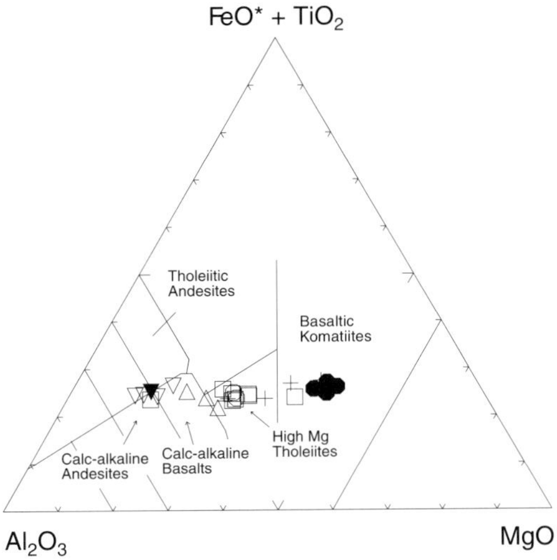

Fig. 4. Plot of (FeO^* + TiO_2)–Al_2O_3–MgO [after Jensen (1976)] showing compositional range of the Poortown rocks. ●, Pyroxene-rich dolerite, lower PQ3; +, plagioclase-rich dolerite, upper PQ3; □, HI1; ○, HI9; △, HI4 upper body; ▽, plagioclase–phyric andesite, HI4 lower body; ▼, base of HI4 lower body.

common throughout. Pyroxene (5–25%), now entirely replaced by carbonate, formed isolated euhedral phenocrysts.

Pyroxene-rich dolerite

Borehole PQ3 is sited in the field immediately north of the west part of the quarry and the rocks recovered are very similar to those worked in the northwest part of the quarry. They correspond most closely to the rock described by Lamplugh (1903). They are of particular interest because the pyroxene-rich dolerite comprising the lower 15 m of the borehole has an unusually high pyroxene content and is likely to have required special conditions for its formation. It is medium grained and composed of 50–60% subhedral augite (4–5 mm), 5–20% plagioclase laths (2–300 μm), 10–30% chlorite and 5% magnetite. Pyroxene forms such a major part of the rock that grains may be in direct contact with each other with only relatively minor interstitial plagioclase and chlorite. The pyroxene shows growth zones marked by opaque dust.

The upper 10 m of core recovered from PQ3, with between 40 and 60% plagioclase (1 mm), is up to three times as plagioclase rich as the lower 15 m. Pyroxene forms widely separated subhedral phenocrysts up to 4 mm in size comprising 15–30% of the rock. Chlorite pseudomorphs (1 mm) with regular crystal outlines occur within the interlocking plagioclase laths and some have bipyramidal forms similar to those of olivine. Similarly shaped pseudomorphs, with some relict kernels of olivine in samples from borehole HI9, lend support to this interpretation.

Chemical variation with depth

The distinct differences in mineralogy between the upper and lower parts of core PQ3 are reflected in the chemistry. Ten samples from different depths have been chemically analysed and results for nine elements are plotted v. depth in Fig. 5. Generally,

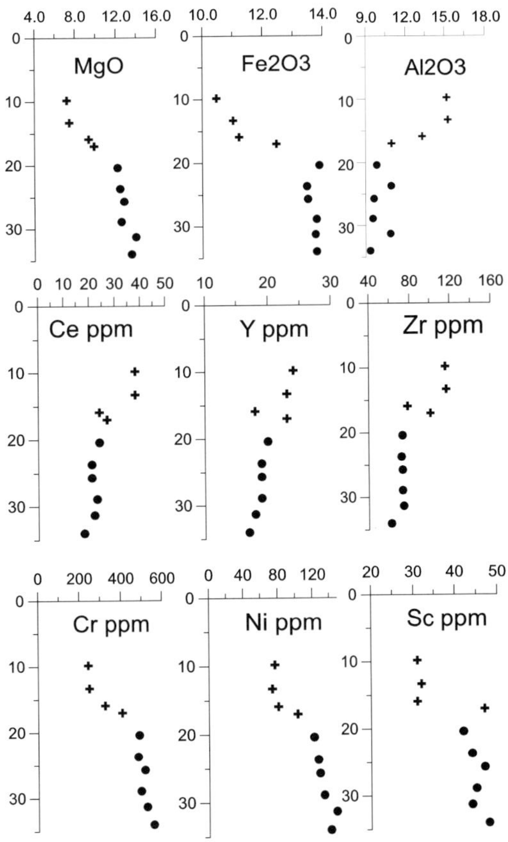

Fig. 5. Plots of MgO, total iron as Fe_2O_3 and Al_2O_3 (all in wt% oxides), and Ce, Y, Zr, Cr, Ni and Sc (ppm) v. depth (in m) down borehole PQ3. +, plagioclase-rich dolerites of the upper section; ●, pyroxene-rich dolerites of the lower section. There is some systematic change in composition with depth for the plagioclase-rich dolerite but very little change for the pyroxene-rich dolerites.

the plagioclase-rich dolerite shows some progressive changes in composition with depth, whilst the pyroxene-rich dolerite shows little change. There is a distinct overall change in chemistry between the upper and lower parts of the core: MgO, Fe_2O_3, Cr, Ni and Sc are greater, and Al_2O_3, Ce, Y and Zr are lower in the pyroxene-rich dolerite of the lower part of the core

Pyroxene chemistry

The chemistry of pyroxenes has been used as an indicator of the origins of basalt, even for rocks which have undergone low-grade metamorphism (Leterrier *et al.* 1982). In the light of the altered state of the Poortown rocks, clinopyroxenes from one plagioclase-rich and three pyroxene-rich samples from core PQ3 were analysed by electron microprobe at the University of Manchester as an independent procedure for investigating the origin of the Poortown magma. A total of 250 individual points were analysed from 11 cored and eight uncored pyroxene grains from the four rocks. Zoning visible in thin section tends to be more obvious in the outer parts of grains and there is a small range in composition across these zones. However, other grains, obviously zoned or not, have a much less uniform chemical composition. Chemical analyses on two traverses across one of these grains are shown in Fig. 6. The inner, irregular parts of the grain display a darker shade on the back-scatter electron image in the lower

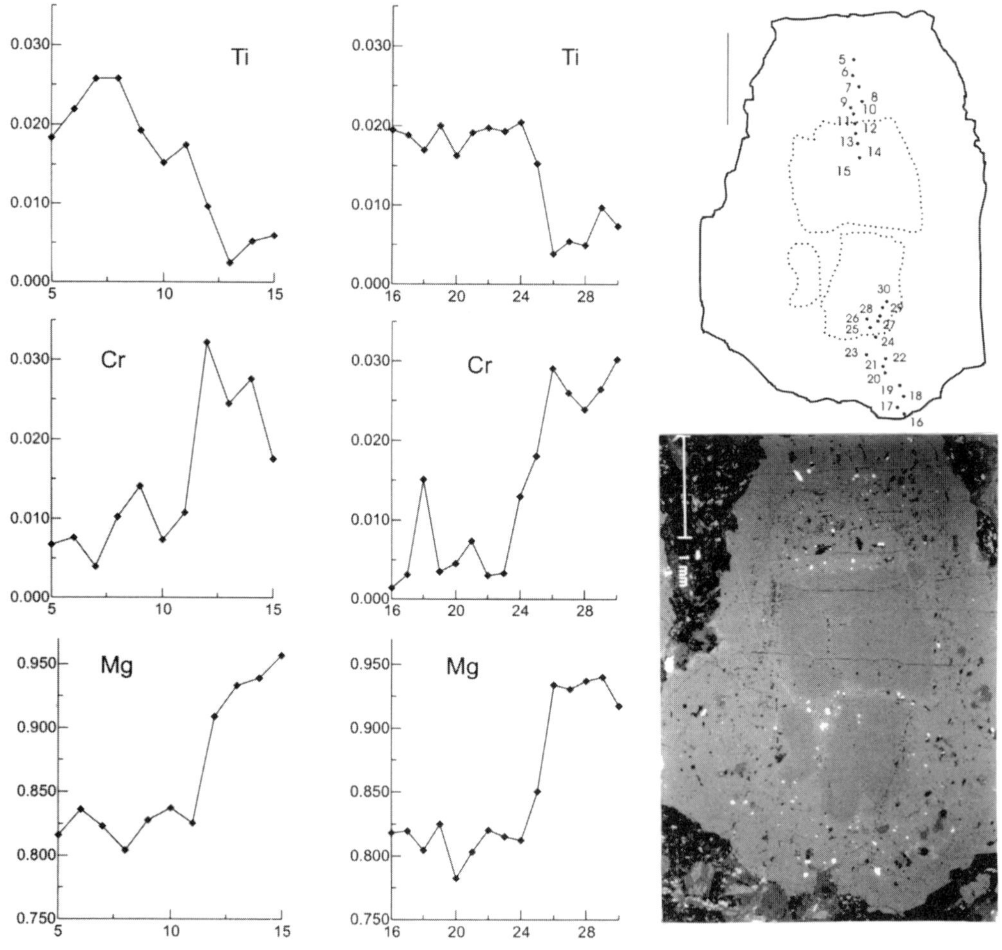

Fig. 6. Plots of point analyses in cation wt% for a pyroxene grain from PQ3 borehole. Lower right is a back-scatter electron image showing the patchy areas within the grain which are different in chemical composition from the rest of the grain.

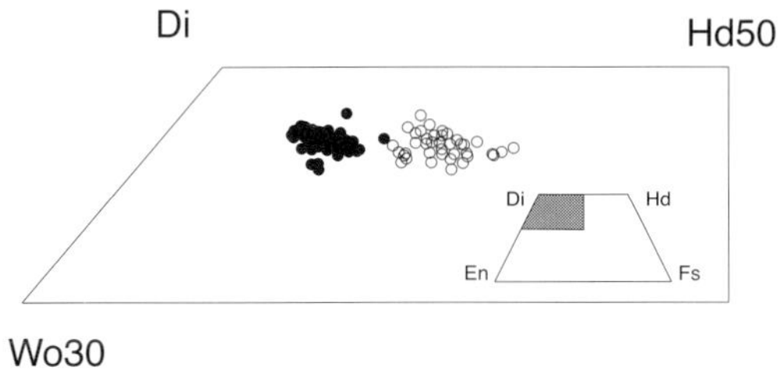

Fig. 7. Plot of part of the pyroxene quadrilateral showing the difference in composition of the relict cores (●) and outer areas (○) of PQ3 pyroxenes.

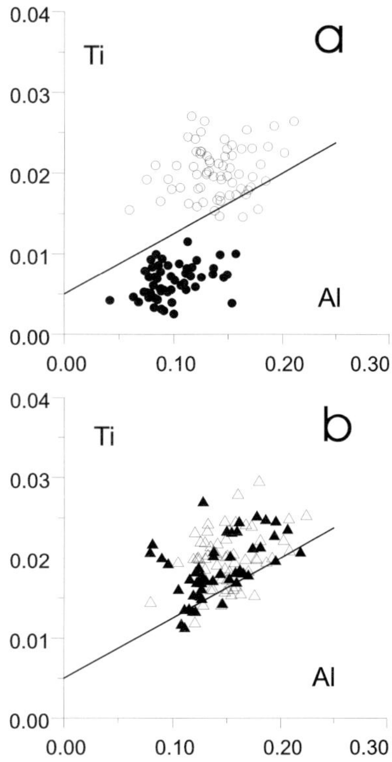

Fig. 8. (**a**) Relict cores (●) and outer areas of PQ3 (○) pyroxenes form two distinct clusters when plotted on the Ti v. Al discrimination diagram of Leterrier *et al.* (1982). (**b**) PQ3 pyroxenes without relict cores, inner (▲) and outer (△) parts of grains, plotted on the Ti v. Al discrimination diagram of Leterrier *et al.* (1982). Only a single cluster is apparent.

right of the diagram. The plots show that these cores have more primitive chemical compositions, having higher Mg and Cr, and lower Ti, than the outer parts of the grains.

Figure 7 shows some of the pyroxene analyses plotted on to part of the pyroxene quadrilateral. They all fall into the augite compositional field but with a range of Mg–Fe content. Two separate compositional groups are apparent. The cluster of filled circles represents points from the cores mentioned above and the open circles represent points from the outer parts of the same grains. The patchy cores do not have euhedral boundaries, some boundaries are embayed and a complex form in 3D is likely. Pyroxene grains with no evidence of cores occur randomly in the same rock, and no obvious differences in size and physical characteristics between cored and uncored grains have been detected. Differences might be expected if the absence of cores was simply an artifact introduced by a cut through a random array of cored grains. The relict cores are interpreted as partially resorbed, more primitive pyroxene which has been partly or completely replaced by more evolved compositions.

Leterrier *et al.* (1982) proposed various discrimination diagrams for basaltic rocks, constructed using the compositions of pyroxenes of known origins. They concluded that it was possible to use this sort of diagram to assign pyroxenes to either a tholeiitic or calc-alkaline parental magmatic origin. For their Ti v. Al discrimination diagram (see Fig. 8), they claimed a high probability that populations plotting above the dividing line would prove to be calc-alkaline in origin and those plotting below the line, tholeiitic. However, their original data for populations of known origin showed considerable overlap across the line and any interpretation should be treated with caution. The same data used in Fig. 7 is shown in Fig. 8a; it forms two clear clusters, one for the cores to grains and the other for the outer parts of grains enclosing the cores. Regardless of the exact significance, the clusters plot either side of the dividing line and the diagram thus discriminates between the two populations. Analyses for grains without any cores are plotted on Fig. 8b; in this case most of the analyses plot above the line in the calc-alkaline field, matching the outer parts of cored grains. The relict cores, i.e. the earliest pyroxene to crystallize, have more 'tholeiitic' characteristics and were overgrown by pyroxene with more 'calc-alkaline' characteristics.

No systematic difference in composition was detected between the pyroxene from the plagioclase-rich dolerite and that from the pyroxene-rich dolerite. The sample from the plagioclase-rich dolerite included some grains with relict cores indicating that the cored pyroxenes are not a unique feature of the pyroxene-rich dolerite.

Origin of the compositional division of core PQ3

No internal igneous contacts were apparent in examination of core PQ3. However, the abrupt changes in mineralogy and chemistry that have been demonstrated are unlikely to have been produced by *in situ* crystal settling of pyroxene to the lower part of a single body. In very simplistic terms, to give the observed distribution of *c.* 55% pyroxene in the lower 15 m and 20% pyroxene in the upper 10 m of the core would require an initial magma with a uniform composition of *c.* 40% pyroxene and an extremly efficient mechanism capable of transferring half the pyroxene from the upper 10 m to the lower part of the body. A more likely explanation for the differences is the existence of two intersecting sills formed by two different pulses of magma. Concentration of pyroxene would have taken place in a high-level magma chamber and expulsion of increments of different composition would have been triggered by replenishment of the chamber. The evidence of the relict pyroxene cores suggests that there may have been several episodes of replenishment.

Tectonic discrimination

It is apparent from petrographic study that the Poortown basaltic andesites are often extremely altered, which limits the amount of information that chemical composition may yield regarding the tectonic environment at the time of their formation. Although some comments will be made about the distribution of the more mobile elements, only those elements generally considered to be immobile will be used for discrimination purposes.

Rare earth elements

Rocks from both the pyroxene-rich and plagioclase-rich dolerites of borehole PQ3 (Fig. 9a), and the plagioclase–phyric andesite of borehole HI4 (Fig. 9b), have rare earth element distributions typical of volcanic arc basaltic rocks, i.e. moderate enrichment of light rare earth relative to heavy rare earth elements. The relative enrichment, as measured by the normalized La/Yb ratio, increases from *c.* 4 in the pyroxene-rich dolerite of borehole PQ3 to *c.* 6 in the plagioclase-rich rocks of borehole HI4. The total rare earth element content is also higher in the more plagioclase-rich rocks. These

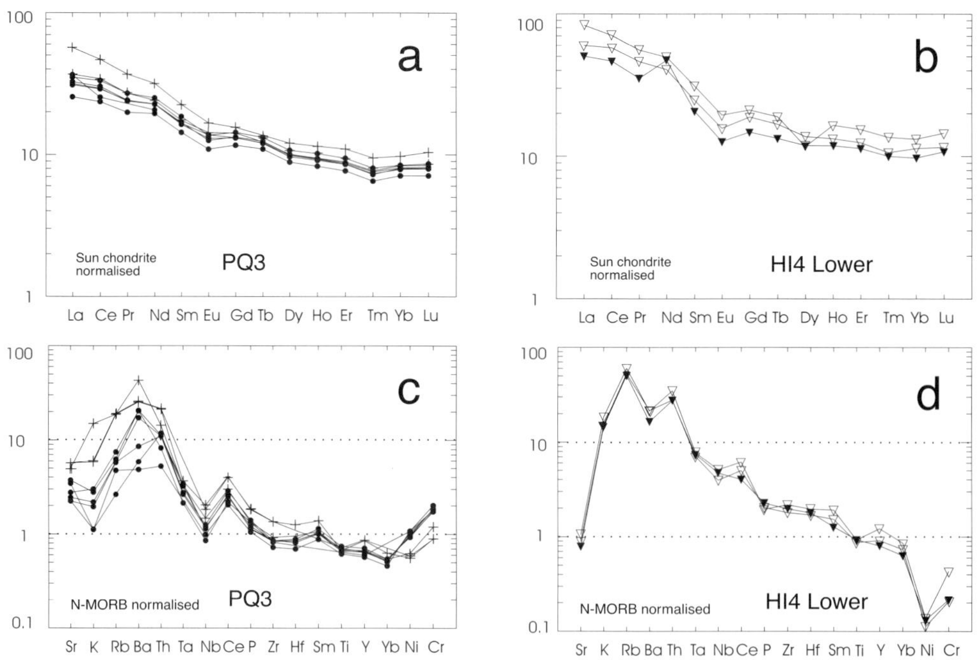

Fig. 9. Rare earth element and multi-element plots (symbols as for Fig. 4). Chondrite normalized rare earth element plots [Sun (1982) values]: (**a**) borehole PQ3; (**b**) lower body in borehole HI4. N-MORB normalized plots of selected elements [after Pearce (1983)]: (**c**) borehole PQ3; (**d**) lower body in borehole HI4.

trends indicate that relative and absolute rare earth distributions are influenced by crystal fractionation.

A slight depletion in europium, compared with the elements either side, is visible on all the rare earth element plots. Negative europium anomalies may result from many influences: e.g. changes in oxidation potential during early formation of magnetite; separation of plagioclase incorporating europium; assimilation of crustal material by the magma; possible modification by various alteration processes. The negative europium anomaly of the plagioclase-rich andesites of borehole HI4 is of similar size to that of the pyroxene-rich dolerites of borehole PQ3 and they are, therefore, unlikely to result from plagioclase separation. Hence, an important influence on the observed distribution of europium in the Poortown basaltic andesites is more likely to be the effects of low-grade metamorphism (Sun & Nesbitt 1978).

Multi-element plots

Relative enrichment of large-ion lithophile elements is apparent on the multi-element plots (Fig. 9). Because of the later alteration, the levels of these more mobile elements cannot be regarded as necessarily indicative of the original composition, however, elevated abundances of these elements and Th are typical of a subduction zone component (Pearce 1983). The samples from core PQ3 have quite scattered large-ion lithophile distributions (Fig. 9c), suggesting variable amounts of alteration. Those from core HI4 (Fig. 9d), on the other hand, are all very similar, suggesting either more complete change, as indicated by the petrography, or, less likely, more limited mobility.

Negative Nb anomalies, a feature typical of subduction-related rocks (Wilson 1989), are displayed for all the samples. The Nb anomaly is much more pronounced for the pyroxene-rich dolerites (Fig. 9c) and is clearly influenced by the concentration of pyroxene in these rocks. The relatively elevated value of Cr in the pyroxene-rich dolerite, suggesting some form of differentiation, is also apparent from the multi-element plots.

The multi-element plots for the Poortown basalts are remarkably similar to those for the lower Borrowdale Volcanic Group basalts presented by Beddoe-Stephens *et al.* (1995). They conclude that subduction-modified, enriched lithospheric mantle most likely acted as a source for, or interacted with, ascending magmas.

Other chemical characteristics

Nb/Th ratios of < 5 in the Poortown basalts suggest a volcanic arc origin (Jenner *et al.* 1991). On the Th–Hf/3-Ta discrimination diagram of Wood (1980) (Fig. 10) the Poortown samples all fall in the arc-related field; together with Hf/Th ratios of < 3 a calc-alkaline arc origin is inferred. The Poortown samples all plot well within the 'active continental margin' field on the Th/Yb v. Ta/Yb diagram of Pearce (1983) (Fig. 11). The Zr/Y v. Zr diagram of Pearce & Norry (1979) is shown in Fig. 12. The Poortown samples all plot in the 'within plate' field, having Zr/Y values > 3. However, Pearce (1983) showed that continental arc basalt also has this type of Zr/Y ratio which, taken in the context of the other evidence, would suggest that a continental arc origin is the most likely.

In conclusion, the geochemical evidence suggests that the Poortown basalt originated as a calc-alkaline volcanic-arc basalt in an active continental margin environment.

Discussion

There is abundant evidence for widespread volcanic activity in the Ordovician of the Lake District (Branney & Soper 1988), southeastern Ireland (Stillman & Williams 1978; Stillman 1988) and Wales (Kokelaar 1988). From its relationships, the Poortown Complex must belong to this general period of activity attributable to the closure of the Iapetus Ocean. However, in the absence of a more precise age for the emplacement of the Poortown Complex, it appears that there are two possibilities – an Arenig or Caradocian age.

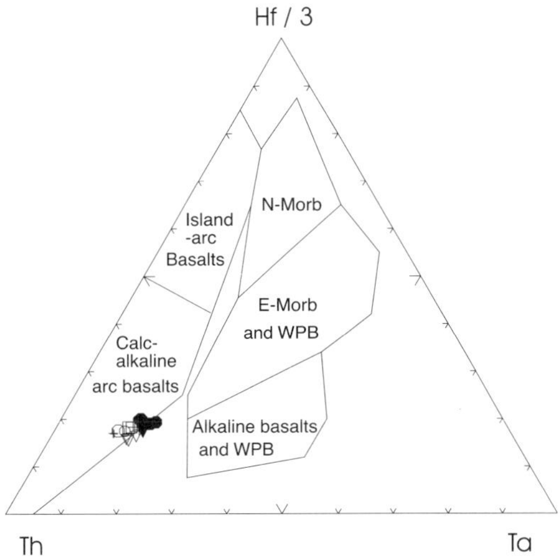

Fig. 10. Th–Hf–Ta discrimination diagram (Wood 1980) showing Poortown rocks plotting in the calc-alkaline arc basalt field. Symbols as for Fig. 4.

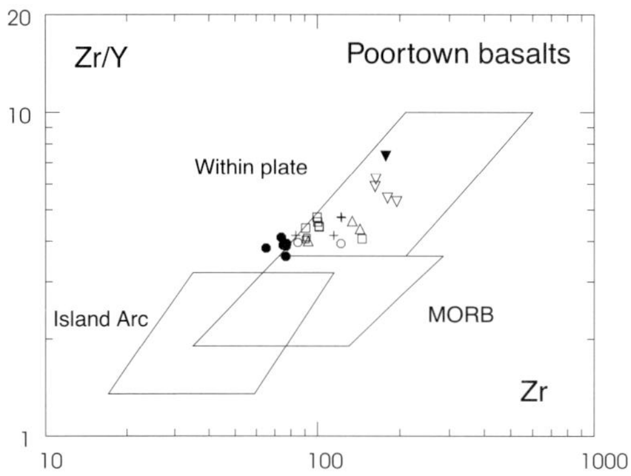

Fig. 11. Plot of Zr/Y v. Zr (Pearce & Norry, 1979). Poortown rocks fall in the within plate field. Symbols as for Fig. 4.

Initiation of subduction under Avalonia has been suggested to have taken place during the Tremadoc in Wales (Rhobell volcanics; Kokelaar 1988) and Arenig in southeastern Ireland, (Dowery Hill volcanics, McConnell & Morris 1997). The Peel Volcanic Group in the Isle of Man has been assigned an Arenig age (Woodcock *et al.* 1999). The Poortown Complex could be the subsurface expression of the subaerial Peel volcaniclastic deposits, as they both have broadly similar basaltic andesitic compositions (Power, unpublished analyses).

Another possible correlation, developing stratigraphic comparisons between the Lake District and Isle of Man stratigraphy (Cooper *et al.* 1995; Woodcock *et al.* 1999), is with the Eycott or Borrowdale Volcanic Groups. The Eycott Volcanic Group has recently been re-examined by Millward & Molyneux (1992). They reinterpret andesite sheets near the base of the group as sills rather than lava flows, and suggest that, as they are unable to confirm a Llanvirn age for the Group, they could be penecontemporaneous with the Llandeilo–Caradoc Borrowdale Volcanic Group episode. Fitton *et al.* (1982) state that the Eycott Volcanic Group is mostly composed of basalt and basaltic andesite, with no andesite, insignificant volumes of acid rocks and only a small proportion of pyroclastic rocks. The Borrowdale Volcanic Group, on the other hand, has a higher proportion of andesites in the lower part and pyroclastic rocks are dominant in the upper part. Beddoe-Stephens *et al.* (1995) give a detailed account of the geochemical variation in the lower Borrowdale Volcanic Group, concluding that it represents a calc-alkaline, plateau andesite pile of continental-arc affinity. The Borrowdale Volcanic Group includes many more acidic components than have been found at Poortown but there are similarities between the two groups for the restricted compositional range represented at Poortown. In particular, both include examples of primitive (high Mg, Ni and Cr) basaltic lavas. At present there seems to be no evidence that precludes the Poortown Basic Igneous Complex from representing a fragment of one of the volcanic centres that made up the Borrowdale Volcanic Field.

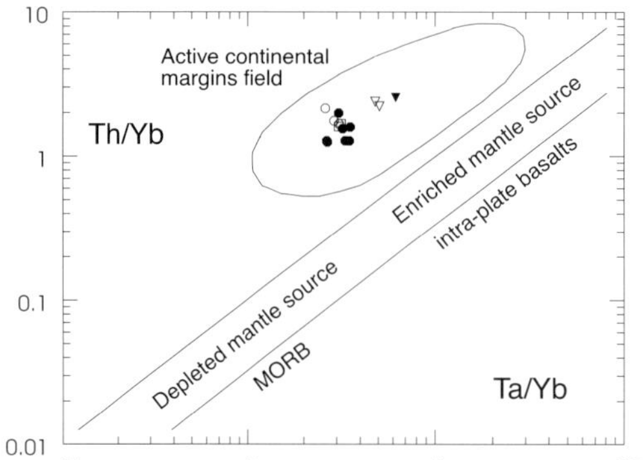

Fig. 12. Plot of Th/Yb v. Ta/Yb (Pearce 1983). Poortown rocks fall in the active continental margin field. Symbols as for Fig. 4.

Conclusions

The Poortown Basic Igneous Complex is made up of a series of gently dipping sills of a range of compositions from Mg-rich basalt and basaltic andesite to andesite. They were intruded into the deep-water marine Cregganmoar Formation of probable Arenig age and underwent deformation and low-grade metamorphism during the early Devonian.

The pyroxene-rich dolerite contains up to 60% augite and has relatively high levels of Mg, Fe, Cr and Ni. It is likely to have been produced by fractionation in a high-level magma chamber before emplacement of the pyroxene–phyric magma into its present position. Partly resorbed cores of more primitive pyroxene (higher Mg and Cr and lower Ti) composition mantled with pyroxene of more evolved composition support a multi-stage history for this magma.

The geochemistry of the more immobile elements supports a calc-alkaline volcanic arc origin in an active continental margin environment. The Poortown Basic Igneous Complex could be a centre related to the Borrowdale Volcanic Field.

We are very grateful to Mr Kevin Brookes (Isle of Man, Department of Highways, Ports and Properties) for access to the quarry and for permission to sample the core material. We would like to thank Derek Weights, University of Portsmouth, for the XRF analyses, Andy Milton, University of Southampton, for the ICP-MS data and Dave Plant, University of Manchester, for his assistance during microprobe analysis. Brett Beddoe-Stephens and an anonymous referee are thanked for their considerable contributions. The authors would also like to acknowledge funding received from NERC Small Grant number GR9/01834 (GMP) and the Stable Isotope Laboratory, University of Liverpool (SFC) to cover fieldwork expenses and other costs.

References

BEDDOE-STEPHENS, B., PETTERSON, M. G., MILLWARD, D. & MARRINER, G. F. 1995. Geochemical variation and magmatic cyclicity within an Ordovician continental-arc volcanic field: the lower Borrowdale Volcanic Group, English Lake District. *Journal of Volcanology and Geothermal Research*, **65**, 81–110.

BRANNEY, M. J. & SOPER, N. J. 1988. Ordovician volcano-tectonics in the English Lake District. *Journal of the Geological Society, London,* **145**, 367–376.

COOPER, A. H., RUSHTON, A. W. A., MOLYNEUX, S. G., HUGHES, R. A., MOORE, R. M. & WEBB, B. C. 1995. The stratigraphy, correlation, provenance and palaeogeography of the Skiddaw Group (Ordovician) in the English Lake District. *Geological Magazine,* **132**, 185–211.

FITCHES, W. R., BARNES, R. P. & MORRIS, J. H. 1999. Geological structure and tectonic evolution of the Lower Palaeozoic rocks of the Isle of Man. *This volume.*

FITTON, J. G., THIRLWALL, M. F. & HUGHES, D. J. 1982. Volcanism in the Caledonian orogenic belt of Britain. *In:* THORPE, R. S. (ed.) *Andesites: Orogenic Andesites and Related Rocks.* Wiley, 611–636.

HUTTON, D. H. W. 1987. Strike-slip terranes and a model for the evolution of the British and Irish Caledonides. *Geological Magazine*, **124**, 405–425.

JENNER, G. A., DUNNING, G. R., MALPAS, J. & BRACE, T. 1991. Bay of Islands and Little Port complexes, revisited: Age, geochemical and isotopic evidence confirm suprasubduction-zone origin. *Canadian Journal of Earth Sciences,* **28**, 1635–1652.

JENSEN, L. S. 1976. *A new cation plot for classifying subalkalic volcanic rocks.* Ontario Division of Mines Miscellaneous Paper 66.

KOKELAAR, P. 1988. Tectonic controls of Ordovician arc and marginal basin volcanism in Wales. *Journal of the Geological Society, London,* **145**, 759–775.

LAMPLUGH, G. W. 1903. *The Geology of the Isle of Man.* Memoir of the Geological Survey, UK. HMSO.

LETERRIER, J., MAURY, R. C., THONON, P., GIRARD, D. & MARCHAL, M. 1982. Clinopyroxene composition as a method of identification of the magmatic affinities of paleo-volcanic series. *Earth and Planetary Science Letters*, **59**, 139–154.

MCCONNELL, B. & MORRIS, J. 1997. Initiation of Iapetus subduction under Irish Avalonia. *Geological Magazine,* **134**, 213–218.

MCKERROW, W. S., DEWEY, J. F. & SCOTESE, C. R. 1991. The Ordovician and Silurian development of the Iapetus Ocean. *Special Papers in Palaeontology,* **44**, 165–178.

MILLWARD, D. & MOLYNEUX, S. G. 1992. Field and biostratigraphic evidence for an unconformity at the base of the Eycott volcanic group in the English Lake District. *Geological Magazine,* **129**, 77–92.

MOLYNEUX, S. G. 1999. A reassessment of Manx group acritarchs, Isle of Man. *This volume.*

PEARCE, J. A. 1983. Role of the sub-continental lithosphere in magma genesis at active continental margins. *In:* HAWKESWORTH C. J. & NORRY, M. J. (eds) *Continental Basalts and Mantle Xenoliths.* Shiva, 230–249.

—— & NORRY, M. J. 1979. Petrogenetic implications of Ti, Zr, Y and Nb variations in volcanic rocks. *Contributions to Mineralogy and Petrology*, **69**, 33–47.

PIPER, J. D. A., BIGGIN, A. J. & CROWLEY, S. F. 1999. Magnetic survey of the Poortown Dolerite, Isle of Man. *This volume.*

POWER, G. M. & BARNES, R. P. 1999. Relationships between metamorphism and structure on the northern edge of Eastern Avalonia in the Manx Group, Isle of Man. *This volume.*

SOPER, N. J., STRACHAN, R. A., HOLDSWORTH, R. E., GAYER, R. A. & GREILING, R. O. 1992. Sinistral transpression and the Silurian closure of Iapetus. *Journal of the Geological Society, London*, **149**, 871–880.

STILLMAN, C. J. 1988. Ordovician to Silurian volcanism in the Appalachian-Caledonian orogen. *In:* HARRIS, A. L. & FETTES, D. J. (eds) *The Caledonian–Appalachian Orogen.* Geological Society, London, Special Publications, **38**, 275–290.

—— & WILLIAMS, C. T. 1978. Geochemistry and tectonic setting of some Upper Ordovician volcanic rocks in east and southeast Ireland. *Earth and Planetary Science Letters*, **41**, 288–310.

SUN, S. S. 1982. Chemical composition and origin of the Earth's primitive mantle. *Geochimica et Cosmochimica Acta*, **46**, 179–192.

—— & NESBITT, R. W. 1978. Petrogenesis of Archean ultrabasic and basic volcanics: evidence from rare earth elements. *Contributions to Mineralogy and Petrology,* **65**, 301–325.

WILSON, M. 1989. *Igneous Petrogenesis.* Chapman & Hall, 1–466.

WOOD, D. A. 1980. The application of a Th–Hf–Ta diagram to problems of tectonomagmatic classification and to establishing the nature of crustal contamination of basaltic lavas of the British Tertiary volcanic province. *Earth and Planetary Science Letters,* **50**, 11–30.

WOODCOCK, N. H., MORRIS, J. H., QUIRK, D. G. *ET AL.* 1999. Revised lithostratigraphy of the Manx Group, Isle of Man. *This volume.*

The Silurian fauna (graptolite and nautiloid) of the Niarbyl Formation, Isle of Man

M. P. A. HOWE
Department of Geology, University of Leicester, University Road, Leicester LE1 7RH, UK

Abstract: The discovery of a mid–late Wenlock (Silurian) graptolite and orthoconic nautiloid fauna at Traie Dullish Quarry, Peel Hill, in the Niarbyl Formation (Dalby Group) of the Isle of Man, disproves all earlier correlations between the Niarbyl and Lonan Flags on the west and east coasts of the island, respectively. All previous structural hypotheses require re-examination.The graptolites comprise *Cyrtograptus* cf. *lundgreni, Monograptus flemingii* cf. *warreni* and *Monograptus* ex gr. *flemingii*. They suggest, but do not prove, a *lundgreni* Biozone age, thus indicating a possible correlation between the Niarbyl Formation and the Birk Riggs Formation of the English Lake District, and the Denhamstown Formation of the Balbriggan Inlier, Southern Ireland.

Previous workers on the Isle of Man, such as Lamplugh (1903) and Simpson (1963), recognized sequences of 'Flags' outcropping on the west coast (the Niarbyl Flags) and the east coast (the Lonan Flags), and have suggested that they might be equivalent and repeated on opposite limbs of a synclinorium (Lamplugh 1903; Simpson 1963) or anticlinorium (e.g. Harkness & Nicholson 1866). For a fuller discussion see Morris *et al.* (1999) and Woodcock *et al.* (1999).

Work by Molyneux (1979) on the acritarch faunas, and the discovery of graptolites at Baltic Rock (Rushton 1993), suggested an Arenig age for the Lonan Flags [the Lonan and Santon Formations *sensu* Woodcock *et al.* (1999)], and this date was broadly assumed to also apply to the Niarbyl Flags [the Niarbyl Formation *sensu* Morris *et al.* (1999)]. A late Tremadoc or early Arenig age acritarch fauna from the presumed Niarbyl Formation of the Glenfaba Brickworks (Molyneux 1979; Cooper *et al.* 1995) was taken as supporting this. A graptolite, collected by Trevor Ford, from the Niarbyl Formation in the northernmost quarry on Peel Hill, was tentatively identified as a didymograptid of Arenig age (A. Rushton, pers. comm.).

The present study has discovered a mixed graptolite and nautiloid fauna of Wenlock age from the Niarbyl Formation at Traie Dullish Quarry on Peel Hill, demonstrating that the Niarbyl Formation cannot be considered part of the early Ordovician Manx Group [see Morris *et al.* (1999) for the formal definition of the Niarbyl Formation and Woodcock *et al.* (1999) for a discussion of the Manx Group]. The flags on the east and west coasts of the Isle of Man cannot therefore be correlated, thus questioning the fundamental structural idea of the Isle of Man Synclinorium.

The Niarbyl Formation fauna

Background to the discovery

During recent mapping of the Peel to Niarbyl Point section of the coast, John Morris observed blocks of hemipelagite in the walls of Peel Castle. In view of the common association of hemipelagite with graptolites, he searched for the lithology in nearby quarries, locating a considerable thickness in Traie Dullish Quarry and subsequently at a number of other localities within the Niarbyl Formation. For further details see Morris *et al.* (1999).

Localities

The key localities discussed below are shown on the sketch map (Fig. 1). Grid references are as follows: Traie Dullish Quarry, Peel Hill [SC 2370 8401]; 'the northernmost quarry on Peel Hill' [SC 2390 8417]; Glenfaba Brickworks [SC 241 828]

Museum depositories

All the material figured or described herein is in the collections of the Isle of Man Manx Museum, Douglas (IOMMM) or the British Geological Survey at Keyworth, Nottinghamshire (BGS).

From: Woodcock, N. H., Quirk, D. G., Fitches, W. R. & Barnes, R. P. (eds) 1999.
In Sight of the Suture: the Palaeozoic geology of the Isle of Man in its Iapetus Ocean context.
Geological Society, London, Special Publications, **160**, 177–187. 1-86239-046-0/99/$15.00

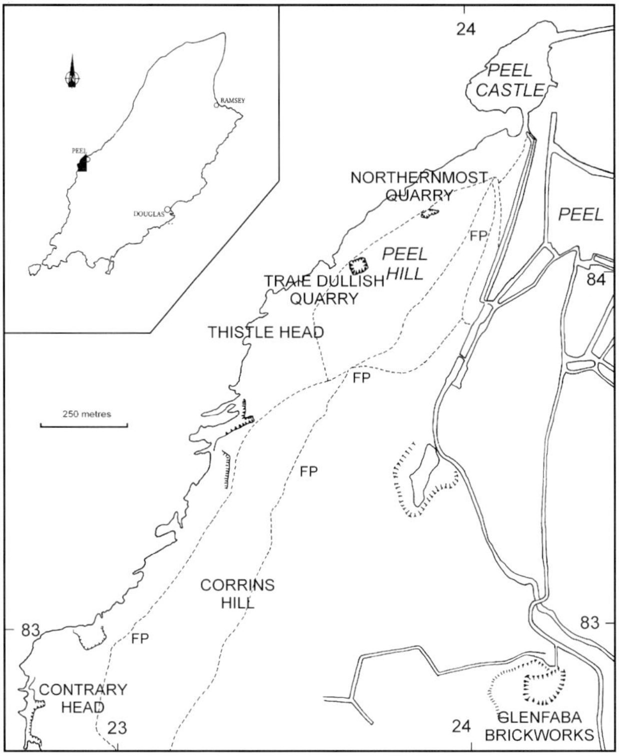

Fig. 1. Sketch map showing location of key localities in the Peel area. FP, footpath.

The Peel Hill quarries

Traie Dullish Quarry is a disused quarry exposing *c.* 10 m of the Niarbyl Formation in a parasitic syncline. A detailed section measured by John Morris is given in Fig. 2. A large prominent bedding plane, near the bottom of the outcrop (0.0 m on the measured section), is exposed in the northern corner of the quarry. Examination by the author revealed numerous poorly preserved nautiloids and a few graptolite fragments on the surface. Further collecting revealed graptolites and nautiloids from *in situ* hemipelagite, *c.* 0.6 m above the prominent bedding plane, and additional material was collected from loose spoil within the quarry. The following fauna was identified: *Cyrtograptus* cf. *lundgreni*; *Monograptus flemingii* cf. *warreni*; *Monograptus* ex gr. *flemingii*; orthoceratid nautiloids.

The northernmost quarry, where Trevor Ford had collected the graptolite previously identified as a 'didymograptid' (BGS Zx 295), was also studied, but no further specimens could be located. The original specimen has been re-examined by the author. Careful preparation has revealed hooked thecae, justifying reidentification as *Monograptus* ex gr. *flemingii* (Fig. 4c).

Assessment of tectonic deformation

The importance of attempting to quantify the tectonic deformation of fossils has only been fully appreciated over the past few years, as taxonomic descriptions have become more precise. There are now numerous studies that use computers, variable *XY*-zoom photocopiers, or similar methods, to restore deformed fossils to their original appearance. An appropriate example is provided by Rushton (1993), who studied two dendroid specimens from Cronk Sumark, Isle of Man, originally figured by Bolton (1899, plate 1, figs 1 and 2) as *Dictyonema sociale* and *Dendrograptus flexuosus*. By calculating the deformation, Rushton was able to show that they were both the same species, the different appearances being due to their different orientations relative to the direction of maximum strain.

It is normally possible to calculate the relative strains within a bedding plane if a number of

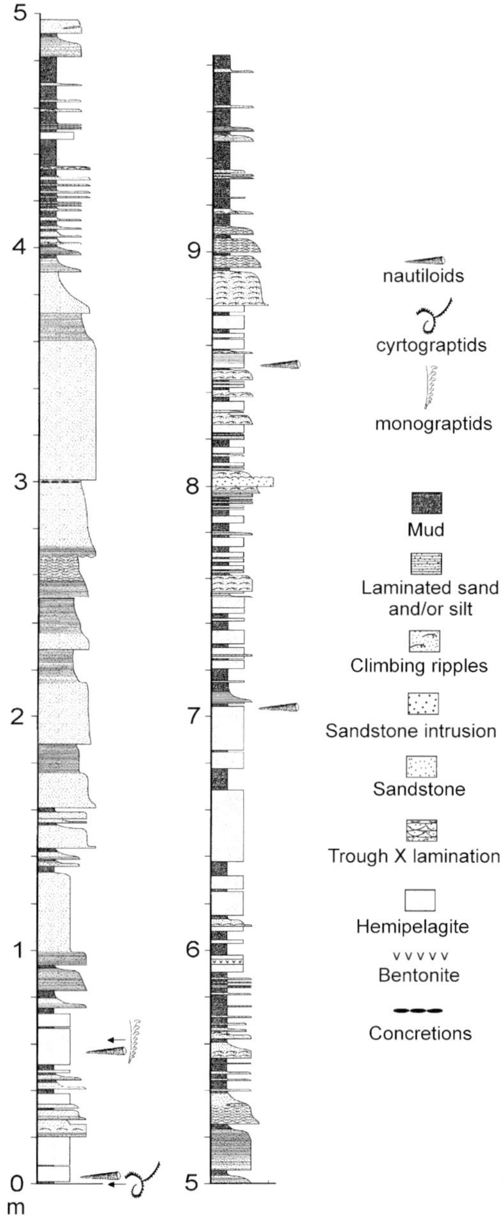

Fig. 2. Sedimentological log of part of the Niarbyl Formation exposed in Traie Dullish Quarry, showing major fossiliferous horizons. Modified from Morris *et al.* (1999, fig. 4).

specimens of the same species are present and if they are preserved at different orientations to the maximum strain direction, which is frequently visible as a lineation on the bedding surface. To calculate the absolute strains requires either the absolute measurement of one of the strains or the making of certain assumptions, e.g. the conservation of dimensions along-strike.

It is not considered possible to calculate the deformation from the various specimens of *Monograptus flemingii* s.l. because they are from at least three different horizons within two different localities and there is good reason to believe that at least two different subspecies are present. In addition, some specimens are from the steep limb of the syncline and others from the shallow limb; one might reasonably expect the strains to vary between the two limbs.

The specimen of *Cyrtograptus* cf. *lundgreni* (IOMMM 98-140) is more useful in estimating the deformation. It must be stressed, however, that because of the contrast in physical properties between the variably pyritized graptolite and the surrounding sediment, the strain is likely to be heterogeneous. Examination of the presumed second cladium shows that the pyrite has fractured and pulled apart along several planes perpendicular to the stretching lineation on the bedding plane. By carefully measuring these fractures and comparing their sum to the total length of the cladium, the ratio of deformed length : original length can be calculated as *c.* 1.05. The main stipe of the cyrtograptid has thecae (th) both perpendicular and parallel to the stretching lineation. The 2TRD at th8 (perpendicular to lineation) is 1.4 mm and distally (parallel to lineation) is 2.5 mm. [For the formal definition of 2TRD see Howe (1983).] Allowing for the observed extension parallel to the lineation gives an undeformed distal 2TRD of 2.38 mm. Assuming that the 2TRD will not vary substantially between th8 and th-distal, the ratio of the original length:deformed length perpendicular to the lineation is *c.* 0.59. Using a computer to remove the deformation from the drawing of the actual specimen (Fig. 3b) produces an estimate of the appearance of the original graptolite prior to deformation (Fig. 3a). The result is extremely close to relatively undeformed material of *Cyrtograptus lundgreni* collected from the *lundgreni* Biozone of the River Irthon, Builth Wells, e.g. specimen BGS MWL98. The only significant difference is that the two cladia on the computer restoration appear to have greater widths. This is probably because, being strongly pyritized, they were not compressed to the predicted extent by the original tectonic deformation.

It is likely that the approximate strain ratios calculated above apply only in the part of Traie Dullish Quarry where the cyrtograptids were collected. To use them in other parts of the same quarry, or in adjacent quarries, would be much more speculative.

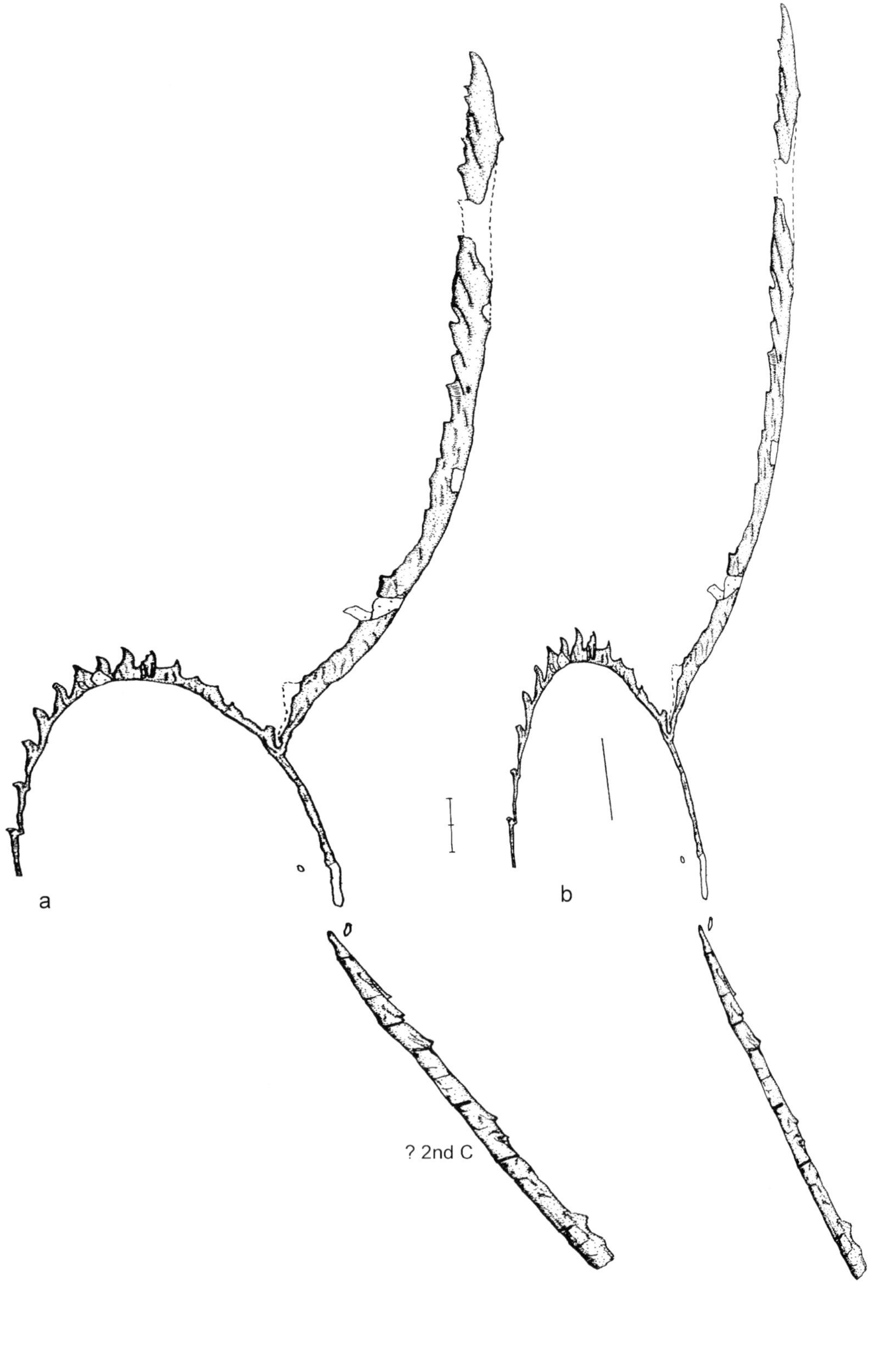
a
b
? 2nd C

Graptolite taxonomic notes

Cyrtograptus *cf.* lundgreni *Tullberg, 1883* *(Fig. 3a and b)*

Material. Two specimens, both from the hemipelagite bedding plane at 0.0 m on the measured section (Fig. 2) at Traie Dullish Quarry. Both specimens are preserved in medium to full relief in pyrite. One is largely complete (IOMMM 98-140; Fig. 3b) and the other fragmentary (IOMMM 98-141).

Description. The fragmentary specimen (IOMMM 98-141) is 2.5 cm long and is a distal fragment of either a main stipe or a cladium. The thecae are hooked, of typical cyrtograptid appearance and are orientated with apertures into the sediment. The (lateral) width is 0.8 mm and the 2TRD is 2.0 mm (ten thecae/10 mm). The specimen exhibits a sharp bend, being abruptly deflected through *c.* 100°. This type of preservation is commonly seen in some of the slender cyrtograptids of the Builth Wells district (Williams & Zalasiewicz, pers. comm.). The specimen is best considered as ?*Cyrtograptus* sp.

The largely complete specimen (IOMMM 98-140; Fig. 3b) was collected by the author during the Southern Uplands Workshop Field Meeting, April 1998. The primary stipe is 2.0 cm long and the first cladium 5.5 cm in length (a 3.0 cm fragment on a piece of rock that broke away from the main specimen was not included in the drawing). A probable second cladium is 1.4 cm long.

The sicula is missing, although there is a possible fragment attached to the most proximal thecae which most likely is th1. The latter is highly elongate, being 1.8 mm in length and 0.25 mm in dorsoventral width; the metatheca is hooked. Th2 (presumed) has a dorso-ventral width of 0.3 mm and a 2TRD (th1–3) of 3.3 mm (six thecae/10 mm). Both of these thecae are long-axis parallel to the pronounced lineation in the rock and have probably been considerably deformed.

By th5 (presumed), the dorso-ventral width is 0.75 mm and the 2TRD (th4–6) measures 2.0 mm. At th10, the dorso-ventral width has reached 1.1 mm and the 2TRD (th9–11) is 1.2 mm (16.6 thecae/10 mm), but these thecae are perpendicular to the deformation lineation. Distally on both the primary stipe and the cladium, the dorso-ventral width measures 0.8 mm and the 2TRD is 2.5 mm (eight thecae/10 mm).

The first cladium appears to originate at about the level of th15. Beyond this, the primary stipe is diffusely pyritized and preserved with thecal hooks pointing into the sediment. The distal fragment is more strongly pyritized, it is preserved in full relief and it shows cracking where tectonic extension has occurred. It is interpreted as a second cladium, although it has not been possible to prove the continuation of the primary stipe beneath it: preparing away the matrix would destroy the specimen. It is therefore possible that it is a continuation of the primary stipe, but this is considered unlikely in view of its shape and relative orientation to the preceeding section of the primary stipe. It must be remembered that the tectonic deformation has increased the angle of separation between the two cladia. As described above, an attempt has been made to remove the estimated tectonic deformation (see Fig. 3a).

Discussion. If the presence of a second cladium is accepted, the general dimensions of the rhabdosome suggest referral to *Cyrtograptus lundgreni* Tullberg 1883. There is some evidence that the Swedish type and topotype material is slightly more robust (Williams & Zalasiewicz, pers. comm.), but the Isle of Man specimen agrees well with other material described from the British Isles (Elles 1900; Elles & Wood 1914; Cope 1954) and elsewhere (e.g. Štorch 1994; Lenz 1988).

Schauer (1968) described a new gracile species, *C. pseudolundgreni*, to which he referred Elles' (1900, fig. 1a) Builth Wells specimen. However, his figures do not show sufficient detail for an accurate comparison to be made and Teller (1976) considered *C. pseudolundgreni* as a junior synonym of *C. lundgreni*. A detailed biometric study is required to ascertain whether there is a single variable species or two separate species.

There are a number of other slender cyrtograptid taxa that show some similarities with the Isle of Man specimen. *Cyrtograptus rigidus rigidus* Tullberg 1883 and *Cyrtograptus rigidus cautleyensis* Rickards 1967 may be excluded because of their greater maximum dorso-ventral widths (1.5–1.6 mm and 1.2–1.3 mm, respectively) and their cladia originating at th5–7 and th7–8. (Williams & Zalasiewicz, pers. comm.; Tullberg 1883; Rickards 1967). *Cyrtograptus linnarssoni*

Fig. 3. *Cyrtograptus* cf. *lundgreni* Tullberg 1883. Specimen IOMMM 98-140 from 0.0 m level in Traie Dullish Quarry. Magnification approximately ×4.25. Scale-bar represents 2 mm. (**a**) Drawing of specimen after computer processing to remove tectonic deformation. (**b**) Drawing of actual specimen prior to removal of deformation. Direction of tectonic lineation denoted by heavy line.

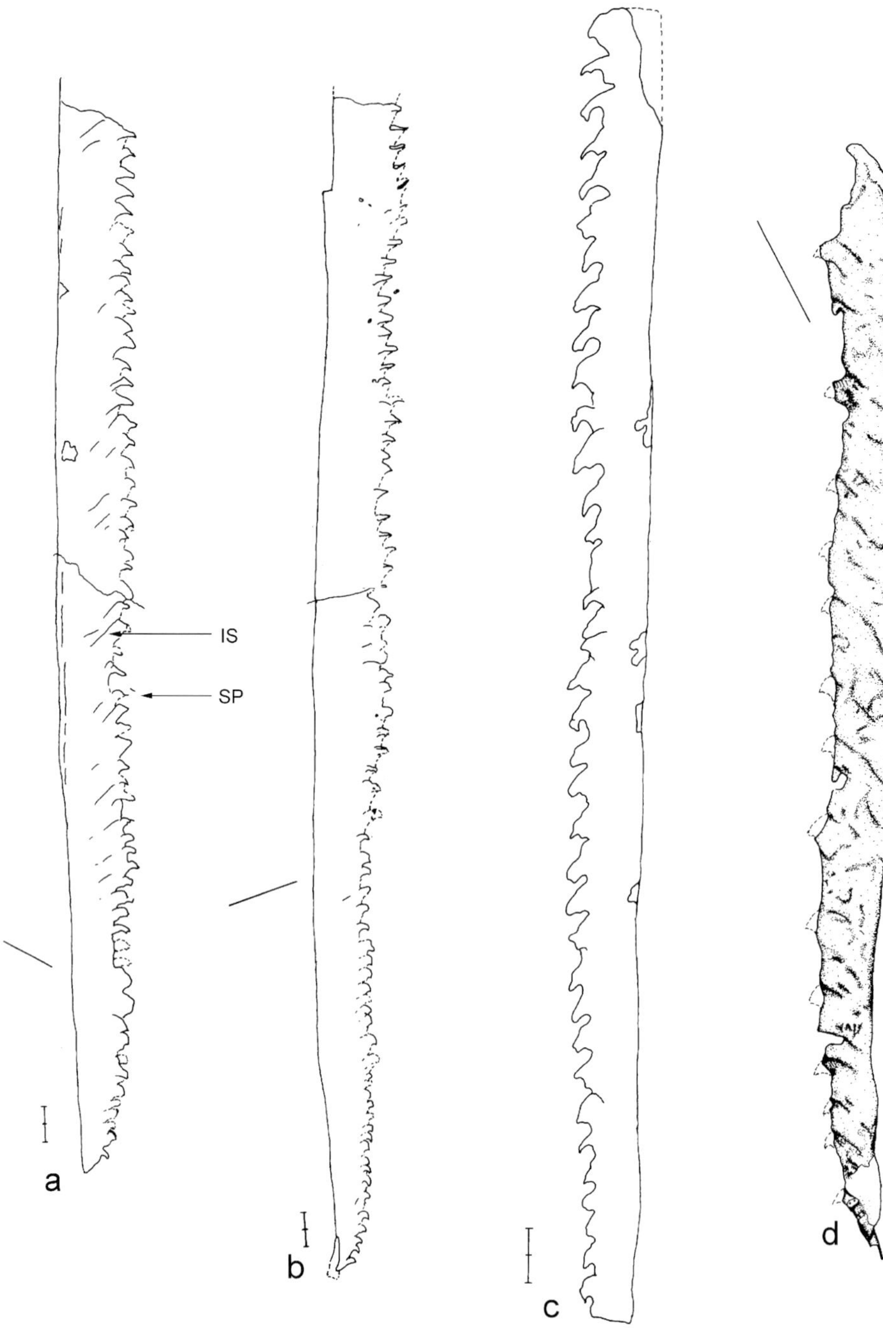

Fig. 4. (**a**) *Monograptus flemingii* cf. *warreni* Burns & Rickards 1993. Specimen IOMMM 98-142. From scree in Traie Dullish Quarry; ×2.25. IS, Interthecal septum; SP, thecal spine. (**b**) *Monograptus flemingii* cf. *warreni*. Specimen IOMMM 98-143. From scree in Traie Dullish Quarry, ×2.25. (**c**) *Monograptus* ex gr. *flemingii* (Salter). Specimen BGS Zx 295 . Collected by Trevor Ford from scree in the northernmost quarry on Peel Hill; ×3.6. (**d**) *Monograptus* ex gr. *flemingii*. Specimen IOMMM 98-144/1. From 0.6 m level in Traie Dullish Quarry; ×6.75. In each case, the scale-bar represents 2 mm and the direction of tectonic lineation (where visible) is denoted by the heavy line. Quoted magnifications are approximate.

Lapworth 1880 is excluded because the first cladium appears *c.* th5–7 (see Elles & Wood 1914, plate 51, fig. 4). *Cyrtograptus perneri* Bouček 1933 may be discounted because it also has fewer precladial thecae (eight in the Builth Wells district (Williams & Zalasiewicz, pers. comm.), its proximal thecae are highly triangulate and the main stipe curves through up to 270°. In addition, *C. perneri* possesses only a single cladium.

If the possible second cladium of the Isle of Man specimen were a deformed piece of main stipe, then comparison could be made with *Cyrtograptus ellesae* Gortani 1923. *C. ellesae* was introduced by Gortani for material from the Builth Wells district figured by Elles (1900) and Elles & Wood (1914) as *Cyrtograptus rigidus*. New material from the Builth Wells district shows that *C. ellesae* has 16–19 precladial thecae, narrow proximal thecae with short metathecal hooks and the main stipe curves through *c.* 160° (Williams & Zalasiewicz, pers. comm.).

Because, to date, only a single sizeable specimen has been recovered from the Isle of Man, and in view of the extent of the tectonic deformation and the uncertainty in the variability of *C. lundgreni*, specimen IOMMM 98-141 is best considered as *Cyrtograptus* cf. *lundgreni*.

Monograptus *ex gr.* flemingii *(Salter) (Fig. 4c and d)*

Material. Two specimens. One, collected by Trevor Ford from the northernmost quarry on Peel Hill (Fig. 4c; BGS Zx 295), is in very low relief with fragments of carbonized periderm remaining. The second specimen is a proximal fragment (Fig. 4d; IOMMM 98-144/1 and /2, part and counterpart) with possible sicula, and is from *c.* 0.6 m on the measured section (Fig. 2) at Traie Dullish Quarry. This specimen is preserved in moderate relief with a pyrite infill.

Description. Specimen BGS Zx 295 is a mesial fragment with a dorso-ventral width increasing from 1.9 mm proximally to 2.75 mm distally. The thecal spacing varies from a 2TRD of 2.5 mm (eight thecae/10 mm) proximally to a 2TRD of 3 mm (6.7 thecae/10 mm) distally. Specimen IOMMM 98-144/1 is a proximal fragment 21 mm long, with the possible trace of a sicula. It is a 'subscalariform' preservation, the thecal hooks being largely hidden in the sediment. The dorso-ventral width increases from 0.8 mm at th1, to 1.0 mm at th5 and 1.5 mm at th10. Thecal spacing varies between a 2TRD of 2.0 mm (ten thecae/10 mm) at th2 (th1–th3) and a 2TRD of 2.6 mm (7.7 thecae/10 mm) distally.

Discussion. The dimensions fall well within limits recorded for the species group, but in view of the rarity and deformation of the material, and the lack of published detailed biometric studies, a more precise taxonomic assignment is not attempted.

Monograptus flemingii *cf.* warreni *Burns & Rickards 1993 Figs 4a and b.*

Material. Two specimens [IOMMM 98-142 (Fig. 4a) and IOMMM 98-143 (Fig. 4b)] on the bases of mud turbidites from loose blocks in Traie Dullish Quarry. Collected by members of the Southern Uplands Field Workshop, April 1998. Horizon uncertain.

Description. Both specimens are very poorly preserved in low relief with a thin, variable film of pyrite between strongly carbonized and fractured periderm. Both specimens are long-axis subperpendicular to the deformation lineation. Specimen IOMMM 98-142 (Fig. 4a) is 67 mm long. The sicula and extreme proximal end are missing but the dorso-ventral width starts at 1.95 mm and reaches 4.1 mm distally. The thecal spacing varies from a 2TRD of 1.5 mm (13.3 thecae/10 mm) proximally to a TRD of 2.1mm (9.5 thecae/10 mm) distally. The other specimen (IOMMM 98-143; Fig. 4b) is 125 mm long and possesses what appears to be the trace of a sicula. The dorso-ventral width increases from 1.25 (th1) to 1.8 mm (th5), 3.5 (th40) and 4.1 mm (distally). The thecal spacing varies from a 2TRD of 1.3 mm (15.4 thecae/10 mm) at th2 (th1–th3) to a 2TRD of 1.7 mm (11.8 thecae/10 mm) at th40 (th39–41), and 2TRD of 1.9 mm (10.5 thecae/10 mm) distally. Details of thecal apertures are poorly preserved but traces of hooks, and even occasional spines, are visible. One specimen (IOMMM 98-142) clearly shows traces of interthecal septa.

Attempting to correct for the tectonic deformation using the strains calculated above would predict maximum dorsoventral widths slightly reduced to 3.9 mm, but distal 2TRD values of 3.2–3.5 mm (5.7–6.2 thecae/10 mm). Whilst the dorso-ventral width seems reasonable, the 2TRD appears too large, indicating that the tectonic deformation does indeed vary across the quarry. It is not unexpected that specimens preserved on hard sandstone beds have been deformed less than specimens in relatively softer hemipelagite.

Discussion. This subspecies was described by Burns & Rickards (1993), based on two specimens from Denbigh, North Wales, originally identified by Bulman (1965) as *Monograptus* cf. *flemingii* (Salter), and on a collection of 11 specimens from

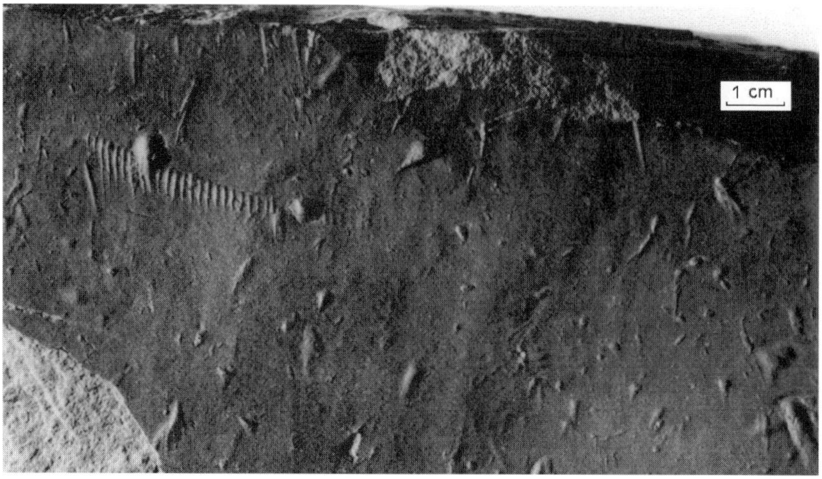

Fig. 5. Graptolite tool mark, probably made by *Monograptus* ex gr. *flemingii*. Specimen IOMMM 7366, from scree in Traie Dullish Quarry (but see text). After BGS, photograph F1728, ©NERC, all rights reserved.

Gormanston Beach, Balbriggan Inlier, Southern Ireland. Burns & Rickards (1993) quote a dorso-ventral width of 2.0 mm at 20 mm from the sicula, 3.0 mm by 100 mm and distally 3.4–3.8 mm in the Welsh specimens, and occasionally up to 4.4 mm in the Irish specimens (although just < 4.0 mm is more common). Thecal spacings vary from 12 thecae/10 mm proximally to ten thecae/10 mm distally, occasionally reaching seven to eight thecae/10 mm in the very widest fragments. They also mention the prominent interthecal septa and a distal thecal overlap reaching three-quarters. The Isle of Man specimens agree well with this description, possibly reaching close to the maximum width more proximally. However, it is not possible to compare thecal spacings proximally as Burns & Rickards (1993) do not quote the 2TRD values.

There is no doubt that the Isle of Man material has been tectonically broadened: small pull-apart fractures can be observed in the periderm. However, these cracks only add a small percentage to the width. Given the good agreement between both dorso-ventral widths and thecal spacings, identification with the subspecies would probably be correct, but in view of the limited deformed material, and the current impossibility of a statistical biometric comparison, the Isle of Man material is only compared to the subspecies.

Graptolite tool mark

Whilst searching the Isle of Man collections at the British Geological Survey, Keyworth, the author discovered a photograph (BGS F1728) of a slab of rock showing a variety of tool and prod marks, including one trace with at least 24 (and possibly as many as 33) short subparallel striations (see Fig. 5). A note by C. J. Stubblefield reads "Photograph of slab submitted by A. M. Cubbon of Manx Museum 1952 of a specimen from the Niarbyl Flags 'loose in the scree of a small disused quarry, the most northerly of a series of such quarries on the western side of Peel Hill, Isle of Man'. It resembles specimen (BGS) 88823 from the Ingletonian (photograph enclosed)." The latter specimen is the one figured by Rayner (1957, plate 1, fig. 1).

Enquiries with K. M. Hawkins (Manx Museum Assistant Keeper: Natural History) revealed that the slab was registered on 21st January 1952 as IOMMM 7366, from 'scree in the northeast corner of the second most northerly quarry on the west side of Peel Hill'. K. M. Hawkins also confirmed that the scale of the photograph was almost exactly 1:1. A letter to A. M. Cubbon, dated 6th February 1952, from R. M. C. Eagar of the Manchester Museum (to whom the slab had been sent initially for identification), implied that A. M. Cubbon had suggested that '... the markings could be the remains of graptolites'. After being examined by several authorities of the day, a letter from R. M. C. Eagar to A. M. Cubbon, dated 6th January 1953, stated '... I am sorry to say he [C. J. Stubblefield] is not yet convinced of its organic origin and nothing definite can be said.'

There is a discrepancy in the locality information, but the Manx Museum record (i.e. Traie Dullish Quarry) is considered most likely. The author has examined and measured the photograph in detail, and compared it to other similar figures, such as Smith (1957, plate XIV), Rayner (1957, plate 1, fig. 1), Trewin (1979, plates 1 and 2) and Palmer & Rickards (1991, plate 136). On specimen IOMMM 7366, the longest striation was just over 11 mm, although most were only 3–5 mm long. The spacing of the striae varied between 1.25 and 1.4 mm. By comparison, the thecal spacing on specimen (BGS Zx 295) of *M.* ex gr. *flemingii* varies between 1.2 and 1.5 mm (a 2TRD of 2.4–3.0 mm). The match is so close that it must be concluded that the tool mark was probably made by a rhabdosome of *M.* ex gr. *flemingii* scraping across the sediment–water interface whilst being transported by a turbidity current.

Such traces are important because they record the precise movements of graptolite rhabdosomes as they are transported by turbidity currents over sediment surfaces. They are one possible way of testing some of the predictions from recent fluid dynamics studies (e.g. Rickards *et al.* 1998).

Orthocone nautiloids

Material. Currently all material is from the Traie Dullish Quarry. One large slab (IOMMM 98-19) with three specimens was found in scree. One specimen (IOMMM 98-144/1 and /2), with part and counterpart, was collected from close to 0.6 m on the measured section (Fig. 2). The bedding plane at 0.0 m on the measured section contains numerous current orientated specimens (see Morris *et al.* 1999). Nautiloids were also observed at two more horizons within the quarry (see Fig. 2).

Discussion. Material collected during the 1996 field season, together with photographs of specimens on the 0.0 m bedding plane were kindly examined by D. H. Evans of English Nature, Peterborough, who reported (Evans, pers. comm.) that they were slender, slowly expanding orthocones, almost certainly orthocerids. However, the preservation was such that only the gross shape of the shell remained and there was therefore no taxonomic evidence as to the age of the material, although the gross morphology indicated a middle Ordovician or younger age (most probably Ashgill or younger).

Despite the poor preservation, D. H. Evans believed that they were all the same species. The monotypic nature of the fauna and its preservation (imploded and completely infilled with sediment) indicated a deep-water environment. Similar material could be seen in the Brathay Formation of the English Lake District (Loubere 1977) and its equivalents in Ireland from County Louth to Tipperary.

The association of graptolites and orthocone nautiloids is common within much of the deeper water facies of the Wenlock of the British Isles. In the Builth Wells district the orthocone nautiloids outnumber the graptolites at some horizons (Williams, pers. comm.). In the Denbigh area of North Wales, the Lower Nantglyn Flags Group contain graptolites and orthocone nautiloids at many localities (Warren *et al.* 1984, p. 59–106) and in the Wenlock Riccarton Group of the Langholm area of the Southern Uplands, '... the commonest fossils are cephalopods, crustaceans and graptolites' (Smith in Lumsden *et al.* 1967, p. 17).

Biostratigraphical implications

Cyrtograptus lundgreni is most abundant within the *lundgreni* Biozone, although Rickards (1976) recorded it extending down into the top of his underlying *ellesae* Biozone. *Monograptus flemingii warreni* was recorded by Burns & Rickards (1993) from the upper *lundgreni* Biozone in Wales, but they could not be more precise than *lundgreni* Biozone in Ireland. The *Monograptus flemingii* plexus is much longer ranging, occurring throughout the middle Wenlock, the *lundgreni* Biozone, and overlapping with *Gothograptus nassa* (Rickards 1976).

If future finds were to indicate that the cyrtograptid material was closer to *Cyrtograptus ellesae* than to *Cyrtograptus lundgreni*, a *lundgreni* Biozone age would still be possible. Rickards (1976) records *Cyrtograptus ellesae* from the underlying *ellesae* Biozone, but also extending up into the *lundgreni* Biozone. Recent detailed collecting in the Builth Wells district suggests that the biostratigraphical range of *Cyrtograptus ellesae* may lie entirely within the range of *Cyrtograptus lundgreni* (Williams & Zalasiewicz, pers. comm.) . In view of the limited amount of Isle of Man material available, and its deformation, a probable, but not proven, *lundgreni* Biozone age is suggested. Confirmation and refinement of this will require the collection of additional, better preserved material, and the development of a consensus of opinion over the precise ranges of cyrtograptids within the middle and upper Wenlock.

Implications for regional correlation

The probable *lundgreni* Biozone age for the Niarbyl Formation suggests a correlation with the Birk Riggs Formation of the English Lake District (Kneller *et al.* 1994) and the Denhamstown Formation of the Irish Balbriggan Inlier (Burns & Rickards 1993). This is discussed in more detail by Morris *et al.* (1999).

Discussion of anomalous records

Glenfaba Brickworks

An acritarch fauna from Glenfaba Brickworks indicates a very late Tremadoc or early Arenig age (Molyneux 1979; Cooper *et al.* 1995), placing the horizon near the base of the Manx Group. The brickworks was presumed to lie within the Niarbyl Formation, but detailed mapping by John Morris [for details see Morris *et al.* (1999)] suggests that it lies to the east of the Niarbyl Formation contact and within the Manx Group.

Peel Volcanic Formation

Molyneux (1979) recorded an abundant acritarch fauna from the Peel Volcanic Formation – sediments interbedded with volcanic rocks in a quarry on the southwest outskirts of Peel - [SC 2386 8334] (but incorrectly quoted in Molyneux (1979)). This was reinterpreted by Cooper *et al.* (1995) as very late Tremadoc or early Arenig in age and assigned to the Niarbyl Formation. Similar detailed mapping by John Morris (Morris *et al.* 1999) suggests that this also lies within the Manx Group to the east of the Niarbyl Formation.

Taylor's (1864) record of Orthoceras

Both Bolton (1899) and Lamplugh (1903) refer to a paper read by J. E. Taylor before the Manchester Geological Society (Taylor 1864), in which he stated 'In the slate quarry of Messrs Ashe at Mount Craig I have found what I esteem the section of an orthoceratite – the specimen showing the chambers and also the gradual tapering of the body of the shell. The outline, as well as the transverse divisions of the septa, are composed of carbonate of lime.'

Bolton (1899) dismissed this: 'In the absence of either the specimen or any figure, neither of which seem to have been exhibited, very little reliance can be placed upon the description. It is not unlikely that it refers to some partially weathered out worm casting, or to thin intersecting mineral veins, which in these rocks often simulate chambered organisms.'

Mount Craig is not a place name in common use today. Taylor (1864, p. 73) described the locality as 'on a mountain opposite to Greeba, and to the east of it, the Craig, we find the beds dipping at an angle of about 45°, but these are finely laminated blue slates, quite sonorous and hard, and fit for building purposes.' This implies that 'the Craig' is the same as The Creg on modern Ordnance Survey maps [SC 345 831]. Lamplugh (1903, p. 157) mentions 'The Creg, 400 yards SW of the old slate quarries ...', which would place Taylor's original locality at about [SC 348 833], within the Barrule Formation of the Manx Group of Woodcock *et al.* (1999). This locality lies within the more strongly metamorphic part of the island (Simpson 1964) and agrees well with lithological description quoted above.

Taylor's record of shell and septa preserved in calcium carbonate seems most improbable because of the much higher metamorphic grade of his locality when compared to the Niarbyl Formation, where the nautiloids are poorly preserved without any of the original shell material, and where they do not show septa. In the absence of the original specimen, and in view of the lack to date of any molluscan fauna from the Manx Group, Taylor's record should probably be discounted.

Conclusions

Re-examination of a graptolite collected by Trevor Ford from the most northerly quarry on Peel Hill and the discovery by the author of a mixed graptolite–nautiloid fauna in Traie Dullish Quarry, proves that the Niarbyl Formation, at least at its northern end, is of Wenlock age and cannot therefore be correlated with the Lonan Formation of Arenig age.

The graptolite fauna is identified as *Cyrtograptus* cf. *lundgreni*, *Monograptus flemingii* cf. *warreni* and *Monograptus* ex. gr. *flemingii*. It suggests, but does not prove, a *lundgreni* Biozone age. This would support correlation with the Birk Riggs Formation of the Lake District.

The members of BIGG (British & Irish Graptolite Group), especially Adrian Rushton, Mark Williams, Barrie Rickards and Jan Zalasiewicz, provided invaluable advice. (MW and JZ made freely available unpublished results from their detailed biostratigraphical study of the Wenlock of the Builth Wells district.) Nigel Woodcock, John Morris and David Quirk provided encouragement throughout, and Kate Hawkins assisted with information on the Manx Museum collections. Trevor Ford and Lady Wilson helped with historical and locality details. Receipt of funds from NERC Small Grant GR9/01834 assisted with fieldwork.

References

BOLTON, H. 1899. The palaeontology of the Manx Slates of the Isle of Man. *Memoirs and Proceedings of the Manchester Literary and Philosophical Society*, **43**, 1–15.

BOUČEK, B. 1933. Monographie der obersilurischen Graptoliten aus der Familie Cyrtograptidae. *Práce geologicko-palaeontologického ústavu Karlovy university v Praze*, **1**, 1–84.

BULMAN, O. M. B. 1965. Giant rhabdosome of *Monograptus* cf. *flemingii* (Salter). *Proceedings of the Geological Society, London*, **1624**, 5.

BURNS, V. & RICKARDS, R. B. 1993. Silurian graptolite faunas of the Balbriggan Inlier, counties Dublin and Meath, and their evolutionary, stratigraphical and structural significance. *Proceedings of the Yorkshire Geological Society*, **49**, 283–291.

COOPER, A. H., RUSHTON, A. W. A., MOLYNEUX, S. G., HUGHES, R. A. & WEBB, B. C. 1995. The stratigraphy, correlation, provenance and palaeogeography of the Skiddaw Group (Ordovician) in the English Lake District. *Geological Magazine*, **132**, 185–211.

COPE, R. N. 1954. Cyrtograptids and retiolitids from County Tipperary. *Geological Magazine*, **91**, 319–324.

ELLES, G. L. 1900. The zonal classification of the Wenlock Shales of the Welsh Borderland. *Quarterly Journal of the Geological Society, London*, **53**, 370–414.

—— & WOOD, E. M. R. 1901–1918. A monograph of British graptolites. *Monograph of the Palaeontographical Society*.

GORTANI, M. 1923. Faune Paleozoiche della Sardegna. Part 1. Le graptoliti di Goni. *Palaeontographica Italica*, **28**, (for 1922), 41–67.

HARKNESS, R. & NICHOLSON, H. 1866. On the Lower Silurian Rocks of the Isle of Man. *Quarterly Journal of the Geological Society, London*, **22**, 488–491.

HOWE, M. P. A. 1983. Measurement of thecal spacing in graptolites. *Geological Magazine*, **120**, 635–638.

KNELLER, B. C., SCOTT, R. W., SOPER, N. J., JOHNSON, E. W. & ALLEN, P. M. 1994. Lithostratigraphy of the Windermere Supergroup, Northern England. *Geological Journal*, **29**, 219–240.

LAMPLUGH, G. W. 1903. *The Geology of the Isle of Man*. Memoir of the Geological Survey, United Kingdom. HMSO.

LAPWORTH, C. 1880. On new British graptolites. *Annals and Magazine of Natural History* (*Series 5*), **5**, 149–177.

LENZ, A.C. 1988. Upper Llandovery and Wenlock graptolites from Prairie Creek, southern Mackenzie Mountains, Northwest Territories. *Canadian Journal of Earth Sciences*, **25**, 1955–1971.

LOUBERE, P. 1977. Orientation of Orthcones in the English Lake District based on field observations and experimental work in a flume. *Journal of Sedimentary Petrology*, **47**, 419–427.

LUMSDEN, G. I., TULLOCH, W., HOWELL, M. F. & DAVIES, A. 1967. *The Geology of the Neighbourhood of Langholm*. Memoir of the Geological Survey of Scotland. HMSO.

MOLYNEUX, S. G. 1979. New evidence for the age of the Manx Group, Isle of Man. *In*: HARRIS, A. L., HOLLAND, C. H. & LEAKE, B. E. (eds) *The Caledonides of the British Isles – Reviewed*. Geological Society, London, Special Publications, **8**, 415–421.

MORRIS, J. H., WOODCOCK, N. H. & HOWE, M. P. A. 1999. The Silurian succession of the Isle of Man: the late Wenlock Niarbyl Formation, Dalby Group. *This volume*.

PALMER, D. & RICKARDS, R. B. (eds) 1991. *Graptolites. Writing in the Rocks*. The Boydell Press.

RAYNER, D. H. 1957. A problematical structure from the Ingletonian rocks, Yorkshire. *Transactions of the Leeds Geological Association*, **7**, 34–43.

RICKARDS, R. B. 1967. The Wenlock and Ludlow succession in the Howgill Fells (north-west Yorkshire and Westmoreland). *Quarterly Journal of the Geological Society, London*, **123**, 215–249.

—— 1976. The sequence of Silurian graptolite Biozones in the British Isles. *Geological Journal*, **11**, 153–187.

——, RIGBY, S., RICKARDS, J. & SWALES, C. 1998. Fluid dynamics of the graptolite rhabdosome recorded by laser Doppler anemometry. *Palaeontology*, **41**, 737–752.

RUSHTON, A. W. A. 1993. Graptolites from the Manx Group. *Proceedings of the Yorkshire Geological Society*, **49**, 259–262.

SALTER, J. W. 1852. Description of some graptolites from the South of Scotland. *Quarterly Journal of the Geological Society, London*, **8**, 388–392.

SCHAUER, M. 1968. Zur Taxonomie und Stratigraphie der Gattung *Cyrtograptus* (Graptolithina). *Freiberger Forschungshefte*, **C.221** (Paläontologie), 32–41.

SIMPSON, A. 1963. The stratigraphy and tectonics of the Manx Slate Series, Isle of Man. *Quarterly Journal of the Geological Society, London*, **119**, 367–400.

—— 1964. The metamorphism of the Manx Slate Series. *Geological Magazine*, **101**, 20–36.

SMITH, J. D. D. 1957. Graptolites with associated sedimentary grooving. *Geological Magazine*, **94**, 425–428

ŠTORCH, P. 1994. Graptolite biostratigraphy of the Lower Silurian (Llandovery and Wenlock) of Bohemia. *Geological Journal*, **29**, 137–165.

TAYLOR, J. E. 1864. The Cambrian strata of the Isle of Man. *Transactions of the Manchester Geological Society*, **4**, 70–84.

TELLER, L. 1976. Morphology of some Upper Wenlock Cyrtograptinae from Zawada 1 profile (NE Poland). *Acta Geologica Polonica*, **26**, 469–484.

TREWIN, N. H. 1979. Transported graptolites and associated tool marks from Grieston Quarry, Innerleithen, Peeblesshire. *Scottish Journal of Geology*, **15**, 287–292.

TULLBERG, S. A. 1883. Skånes graptoliter, II. Graptolitfaunorna i Cardiolaskiffern och Cyrtograptusskiffrarne. *Sveriges Geologiska Undersöknung*, Series C, **55**, 1–43.

WARREN, P. T., PRICE, D., NUTT, M.J.C. & SMITH, E.G. 1984. *Geology of the Country Around Rhyl and Denbigh*. British Geological Survey, England and Wales, HMSO, i–x, 1–217.

WOODCOCK, N. H., MORRIS, J. H., QUIRK, D. G. *ET AL.* 1999. Revised lithostratigraphy of the Manx Group, Isle of Man. *This volume*.

The Silurian succession of the Isle of Man: the late Wenlock Niarbyl Formation, Dalby Group

J. H. MORRIS,[1] N. H. WOODCOCK[2] & M. P. A. HOWE[3]

[1]*Geological Survey of Ireland, Beggars Bush, Haddington Road, Dublin 4, Ireland*

[2]*Department of Earth Sciences, University of Cambridge, Downing Street, Cambridge CB2 3EQ, UK*

[3]*Department of Geology, University of Leicester, University Road, Leicester LE1 7RH, UK*

Abstract: Interpretation of the stratigraphy and structure of the Lower Palaeozoic Manx Group has been strongly influenced by the assumed equivalence of distinct sandstone sequences, the Niarbyl and Lonan Flags, exposed along the west and east coasts of the Isle of Man, respectively. However, new palaeontological evidence confirms an Arenig age for the Lonan Flags and indicates a mid-Silurian age for the Niarbyl Flags, thereby necessitating a complete revision of previous interpretations. The Niarbyl Flags are formally defined here as the Niarbyl Formation and are assigned to a new group, the Dalby Group. A previously unrecognized lithofacies in the formation, laminated hemipelagite, is also distinguished and described and this indicator of anoxic deposition is compared and contrasted with the Manx Group and other Lower Palaeozoic sequences. Combined turbidite facies, palaeocurrent and petrofacies analysis of the turbidite suite further serves to reinforce the distinction with the Manx Group, in particular palaeocurrent and petrofacies data that indicates provenance from a magmatic-arc source to the west-northwest. The turbidite sequence is interpreted as a mid–lower fan, slope-apron sequence of interlobe and sandy lobe packets. Finally, the regional correlation and significance of the formation in its Iapetus Ocean setting, as well as the significance of the Niarbyl Shear Zone that defines its lower contact with the Manx Group, is considered and compared.

Lamplugh (1903) and Simpson (1963) recognized distinct sandy sequences of 'Flags' outcropping on opposite coasts of the Isle of Man, the Lonan Flags along the east coast and the Niarbyl Flags along the west coast. Lamplugh (1903) postulated that the sequences were equivalent and repeated in the opposite limbs of a synclinorium extending along the northeast–southwest axis of the island, rather than an anticlinorium as proposed in earlier interpretations (Cumming 1846; Harkness & Nicholson 1866). Lamplugh did, however, express two reservations about this proposed correlation: the Niarbyl Flags, unlike the Lonan Flags, were not seen to be overlain by the Agneash Grits (the Creg Agneash Formation, Woodcock *et al.* 1999); and a possible 'overthrust fault' was noted at the boundary between the Niarbyl Flags and underlying slates (Lamplugh 1903).

New palaeontological evidence (Howe 1999) demonstrates that the Niarbyl Flags are mid-Silurian–late Wenlock in age, not Cambrian or early Ordovician as previously supposed. This discovery has major implications for the understanding of the significance and position of the Niarbyl Flags in the Lower Palaeozoic evolution of the Isle of Man. Accordingly, a comprehensive reassessment of the stratigraphy, sedimentology and regional context of the Niarbyl Flags is presented here. Only a brief discussion of the structural geology is presented here as it is considered in detail in Fitches *et al.* (1999).

Stratigraphy

All previous studies have, explicitly or implicitly, included the Niarbyl Flags with the other rocks loosely termed the Manx Slate Series by Lamplugh (1903) and Simpson (1963) or, more formally, the Manx Group by Simpson (1968, p. 135). The age of the group has been the subject of much debate because of the paucity and unreliability of palaeontological data. Cumming (1848) and Harkness & Nicholson (1866) proposed that the Manx Group was Lower Silurian (= Ordovician) in age, whereas Taylor (1864), Lamplugh (1903) and Simpson (1963) suggested a Cambrian age.

More recent studies, particularly of acritarch faunas by Downie & Ford (1966) and Molyneux (1979), of the first undisputed *in situ* graptolites by Rushton (1993), and a summary by Cooper *et al.*

From: Woodcock, N. H., Quirk, D. G., Fitches, W. R. & Barnes, R. P. (eds) 1999.
In Sight of the Suture: the Palaeozoic geology of the Isle of Man in its Iapetus Ocean context.
Geological Society, London, Special Publications, **160**, 189–211. 1-86239-046-0/99/$15.00

(1995), indicate Tremadoc or Arenig ages for parts of the group. An acritarch fauna recovered from the Glenfaba Brickworks, near Peel [SC 241 828], was presumed to lie within the Niarbyl Formation. It indicates a very late Tremadoc or early Arenig age, which places the unit almost at the base of the Manx Group (Molyneux 1979; Cooper *et al.* 1995).

In contrast to any of these earlier studies, Howe (1999) reports an *in situ*, mid-Silurian–Wenlock age *lundgreni* Biozone mixed graptolite–orthocone fauna in the Niarbyl Formation. This discovery renders redundant a number of previous suppositions: the Cambrian or Ordovician age of the formation; its inclusion in the Manx Group; the presumed equivalence of the formation with the Arenig age Lonan Flags (Molyneux 1979; Rushton 1993; the Lonan and Santon Formations of Woodcock *et al.* 1999); and the inferred fold repetition of the Lonan and Niarbyl Flags.

Name and equivalence

The mid-Silurian age Niarbyl Formation takes its name from The Niarbyl [SC 2150 7764] at the north end of Niarbyl Bay and supercedes the previous informal term of the Niarbyl Flags (Lamplugh 1903; Simpson 1963). The outcrop of the formation is restricted to the area originally defined by Lamplugh (1903). This comprises the outcrop north from Niarbyl to Peel Hill [SC 2380 8395] and St Patrick's Isle [SC 241 845], and inland only as far as the ridge west and south of Knockaloe [SC 2370 8225], through Glenmaye [SC 2365 7990] and south to Dalby, roughly along the line of the Glenmaye–Dalby main road (Fig. 1b). The Niarbyl Formation specifically excludes the northward extension suggested by Simpson (1963), to the east and north from Peel to Kirk Michael and Ballaugh. Although greywackes occur in that zone, they lack all the features diagnostic of the Niarbyl Formation. All previous correlations with the early Ordovician age Lonan and Santon Formations on the east coast of the Isle of Man are discounted on the grounds of contrasting age, as well as sedimentary characteristics.

Formal lithostratigraphic definitions: type area and subsidiary localities

For consistency with established nomenclature, the formation type section is taken as the low, east facing cliff section [SC 2107 7758] of the island opposite the end of the slipway at Niarbyl, accessible across a rocky causeway other than at high tide (Fig. 3a). The formation is equally well exposed in the continuous outcrop along the easily accessible coast section north from Niarbyl to the mouth of the Glen Mooar River, at the seaward end of the Glen Maye ravine [SC 2250 7995]. Subsidiary localities are provided by three disused slate quarries north of Contrary Head. These comprise the larger of two disused quarries on the west side of Peel Hill, immediately above Traie Dullish [SC 2370 8401], termed the Traie Dullish Quarry, a cliff-top quarry on the north edge of Contrary Head [SC 2293 8297], and a narrow cliff-top quarry just south of Thistle Head [SC 2339 8357]. The formation is also well exposed on St Patrick's Isle.

In the light of the new palaeontological evidence, it is proposed that the Niarbyl Formation be removed from the early Ordovician Manx Group and assigned instead to the new Dalby Group. The group is named after the excellent coastal exposures of the Niarbyl Formation north and south of Dalby Point [SC 2125 7875]. The group contains only one formation, although this does not preclude revision by future studies.

Lithological characteristics

The Niarbyl Formation is composed principally of turbidites, infrequent, though characteristic, hemipelagites and rare metabentonites. Lamplugh (1903) and Simpson (1963) also noted two occurrences of volcanic rocks which they included in the formation but which are excluded here for reasons outlined later.

Niarbyl Formation sandstones [Ta–Tc divisions of the Ta–Td turbidite division nomenclature defined by Pickering *et al.* (1986)] are distinctly rusty brown, buff or green–buff coloured on weathered surfaces, the weathering colour reflecting substantial carbonate cement content. They are pale grey on fresh surfaces. Pelitic material, invariably turbiditic silt and pelite (Td, where discernible, and Te), is mainly grey–green coloured. Most turbidite beds range between 1 and 40 cm in thickness, with infrequent very thick or massive beds ranging from 1 to 4 m. Sand-grain size, even in the thickest beds, rarely exceeds medium grade, although some beds may contain dispersed fine granules in 'coarse tail' layers (*sensu* Middleton & Hampton 1976).

Laminated hemipelagite is a volumetrically minor lithofacies characteristic of this formation. It is not observed in the Manx Group *sensu stricto* and it is not comparable to the clay-rich hemipelagite noted in some Manx Group formations by Woodcock *et al.* (1999). The facies is particularly well exposed in the Traie Dullish Quarry on Peel Hill [SC 2370 8401], although it occurs intermittently throughout the formation outcrop. It consists of very thinly interlaminated siltstone and carbonaceous pelite, grey in fresh outcrops but

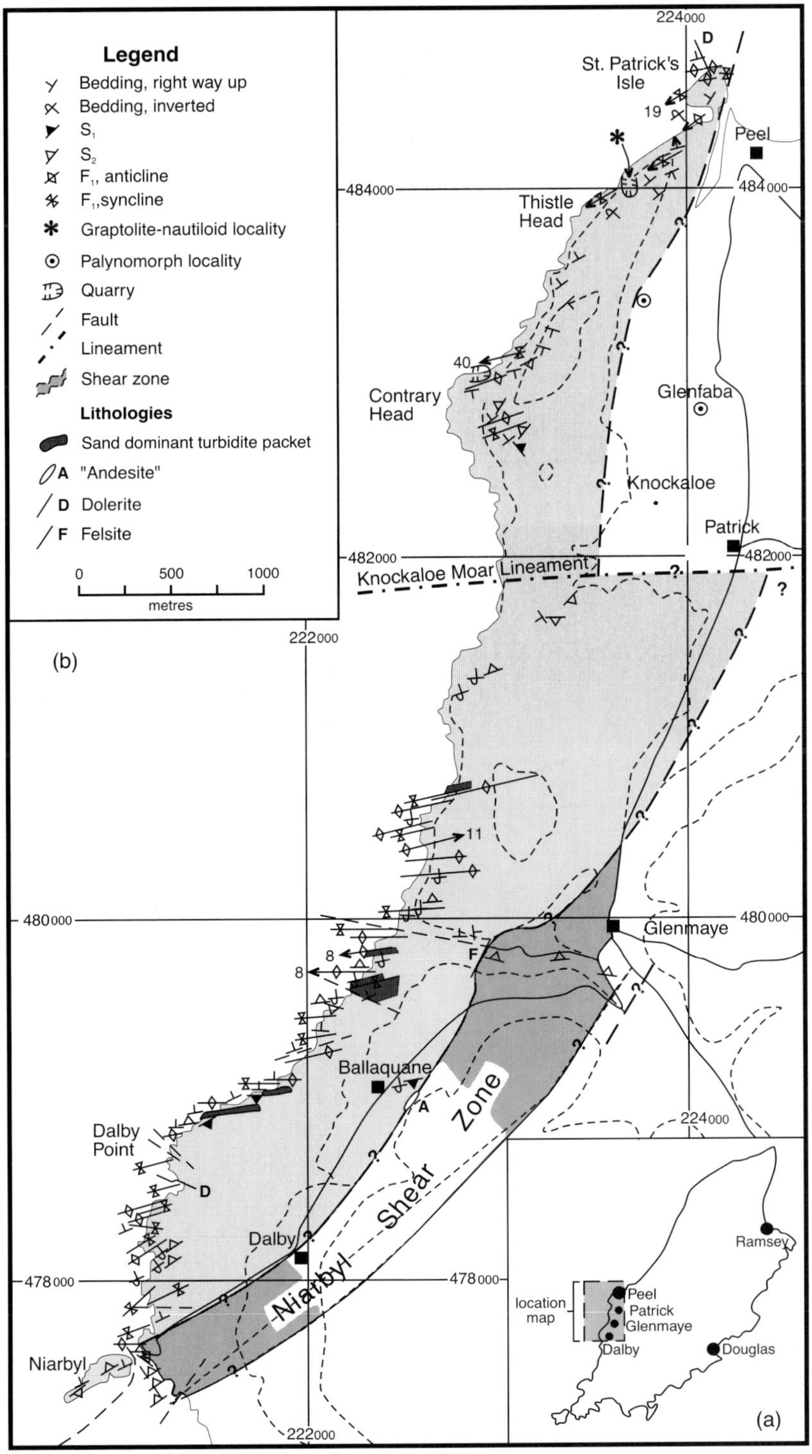

Fig. 1. (**a**) The Isle of Man, showing the location of the study area. (The location of the Isle of Man, relative to Britain and Ireland, is shown in Fig. 7.) (**b**) Geological map of the Niarbyl Formation. Fold traces and structural readings greatly simplified. UK zone SC grid coordinates, 200 (60 m) and 400 (121 m) ft contour intervals.

(a)

(b)

weathering to pale brown or buff (Fig. 2a). The lamination is defined by submillimetric alternations of carbonaceous laminae, hosting rare graptolites and orthocones, with silty mud laminae to form silt–carbon couplets. Whilst the lamination appears quite regular in outcrop, the carbon laminae are more wispy and discontinuous in thin section. However, counts of 297 couplets in three thin sections reveal an average frequency of *c.* 30 couplets/cm, remarkably constant throughout the sequence. Intervals of hemipelagite, termed 'packets' here, typically range in thickness from a few millimetres to 10 cm, with occasional thicker packets up to 30 cm and exceptionally up to 2 m (Fig. 3d). The facies is most easily visible when intercalated with the mudstone turbidite facies, although occasional packets or individual laminae occur in more sandy sequences. The sparse, mixed graptolite–orthocone fauna described by Howe (1999) was recovered from several hemipelagite horizons in the Traie Dullish Quarry, as well as loose scree, and from another small quarry just to the north. The facies, its significance and correlation are described in detail below.

Metabentonite beds have been observed in three instances, two near Dalby Point and the other in the Traie Dullish Quarry. The beds are typically pale green–buff, notably sericitic and may contain thin epidote–chlorite veinlets or quartz microphenocrysts, as at Boirane [SC 2150 7890]. A single metabentonite bed is exposed in the Traie Dullish Quarry: this bed averages *c.* 3 cm thick but is locally thickened to 30 cm by thrusting. Lamplugh (1903, p. 148) also notes the very rare occurrence of 'thin, less than 4″' (10 cm) thick horizons of a 'curious pale-yellowish green steatic material' of possible volcanic origin on the north side of Contrary Head, and on the coast in the vicinity of

(c)

Fig. 2. (**a**) Hemipelagite intervals (a) interbedded with turbidite mud (b) and sets of climbing ripples (c). Note overturning of ripple foresets in downcurrent direction and loaded bases. Traie Dullish Quarry [SC 2370 8401]. Scale in cm. (**b**) Intercalated, non-cyclical sandy lobe (b) and interlobe (a) turbidite packets. Hammer, 14 cm. (**c**) Fining-up 'compensation cycles' [(a) and (c)] and neutral [(b)] turbidite packet. (**d**) Channel, non-erosional base, Elby Point [SC 2114 7777]. Hammer, 14 cm.

(d)

Dalby. There is little doubt that these occurrences are also metabentonite.

The volcanic rocks noted by Lamplugh (1903), Simpson (1963), Molyneux (1979) and Cooper *et al.* (1995) occur in two disused quarries, which are isolated from known outcrops of the Niarbyl Formation: one just southwest of Peel (at [SC 2386 8334], not to be confused with the Glenfaba Brickworks Quarry site at [SC 241 828]); and another *c.* 250 m east-sotheast of Ballaquane Farm at [SC 2257 7900]. Molyneux (1979) and Cooper *et al.* (1995) noted the recovery of very abundant, well-preserved, (?)early Arenig age acritarch faunas from the quarry at [SC 2386 8334], at first assigning the sequence to a new unit, the Peel Volcanic Formation (Molyneux 1979), but later incorporating it into the Niarbyl Formation (Cooper *et al.* 1995). The quarry near Ballaquane Farm was revisited in this study, confirming the presence of a pale grey–green to green, aphanitic, net-veined igneous rock not unreasonably termed 'andesite'. A distinct tuffaceous texture is locally evident, defined by ovoid clasts ranging from a few millimetres up to 9 cm in size, and a distinct particulate texture in devitrified glass confirms the volcanic origin of this rock (McConnell, pers. comm.).

The previous assignment of both these occurrences of volcanic rocks to the Niarbyl Formation, is doubtful, as is the affinity of the Glenfaba Brickworks locality. Similarly, Lamplugh (1903) expressed uncertainty over the relationship of the Ballaquane Farm occurrence to surrounding rocks. Apart from metabentonite, no volcanic rocks have been observed or previously recorded elsewhere in the Niarbyl Formation. It is therefore considered more likely that both the outcrops of volcanic rocks and the Glenfaba Brickworks site lie within the Manx Group. This stratigraphic position would place the volcanic rocks in the regional context of widespread, early Ordovician age volcanism in Ireland and Britain.

Thickness and its relationship with adjoining units

The Niarbyl Formation is exposed in an 8 km long coastal section which extends, at maximum, *c.* 1 km inland. Estimates of the true thickness of the formation are severely hampered by the fold structure and by the absence of marker horizons. With due regard for structural complexities, a thickness of *c.* 1250 m is estimated here, compared with Simpson's (1963) estimate of 600–3000 m for the undifferentiated Lonan and Niarbyl Flags. North of Knockaloe Moar, a minimum thickness of 350 m is estimated in the northern limb of the Peel Hill anticline, and a further 200 m in the poorly exposed ground between Contrary Head and Knockaloe Moar. To the south of Knockaloe Moar, the F1 fold train involves a stratal thickness in the order of 500–700 m. Combining the northern and southern figures yields an overall formational thickness in the order of 1250 m.

The lower contact of the Niarbyl Formation with the (?)mid-Arenig (Molyneux 1999) Creggan Mooar Formation (*sensu* Woodcock *et al.* 1999) of the Manx Group is defined primarily by a compound zone of sinistral shear and southeast directed thrusting. The fault zone is best exposed at The Niarbyl, although the phyllonitic fault zone fabric outcrops extensively in the Glen Maye ravine. The zone is cut and displaced by late brittle faults which locally define the boundary, e.g. at The Niarbyl, in the Glen Maye ravine and along the east flank of Peel Hill.

Structural geology

A detailed account of the structural geology of the Niarbyl Formation is presented by Fitches *et al.* (1999) and only a brief summary is provided here. In common with Simpson (1963), two major deformation episodes in the Niarbyl Formation are recognized, denoted D1 and D2, as well as a variety of other structures. These include the Niarbyl Shear Zone, which defines the lower contact of the formation, and several low-angle, brittle thrust fault–quartz vein complexes.

The formation is dominated by a continuous train of tight to very tight, upward-facing F1 folds, with limb widths normally < 60 m but ranging up to 150 m. A prominent axial-planar to locally transecting S1 cleavage is associated with these folds. The cleavage morphology varies from penetrative in pelitic lithologies to a spaced foliation in sandy units and locally to a pressure solution cleavage.

North of Knockaloe Moar, F1 folds are upright to very steeply inclined and generally plunge gently southwest. In contrast, to the south of Knockaloe Moar, asymmetric, generally southeast verging, gently northeast plunging F1 folds have axial surfaces dipping on average *c.* 50° NW. The sense of vergence is consistent with thrust vectors defined in the Niarbyl Shear Zone and in minor brittle thrust systems in this and the Creggan Mooar Formation.

D2 structures are relatively minor, represented mainly by mesoscopic F2 folds associated with a fanned or axial-planar S2 crenulation cleavage. The folds occur sporadically throughout the formation as asymmetrical, generally upward-facing, kinks with wavelengths and amplitudes up to a few metres. Examples occur in the limbs of the F1 synclines exposed in the Traie Dullish Quarry [SC 2370 8401] and another quarry *c.* 200 m to the north. Larger, isolated F2 folds occur rarely: these have more rounded hinges and maximum exposed limb widths up to *c.* 15 m.

Quartz veins are associated with many F2 folds, either stratiform in the limbs or folded about the hinges. This spatial association is also evident in several prominent zones of brittle thrusts, folds, fractures and en echelon and pinnate quartz veins. A good example of such a zone is exposed *c.* 90 m north of the Niarbyl Shear Zone at [SC 2117 7780]. The gross morphology of this system and, in particular, the orientation of en echelon and pinnate veins, indicates a top-to-the-southeast sense of shear, as do all other such systems in the Niarbyl Formation. Identical systems are also evident in the adjoining Creggan Mooar Formation up to 800 m south of the Niarbyl Shear Zone, e.g. at [SC 2157 7694]. These systems attest to late-stage, primarily brittle, thrusting towards the southeast in both hanging wall and footwall sequences of the Niarbyl Shear Zone.

The Niarbyl Shear Zone is best exposed at The Niarbyl, although its phyllonitic fault zone fabric is also extensively exposed in the Glen Maye ravine. Lamplugh (1903), Simpson (1963) and Morrison (1989) all draw attention to the shallow, north, northeast to northwest dipping, ductile–brittle fault zone exposed at the base of the Niarbyl Formation at The Niarbyl [SC 2150 7764], generally describing it as an overthrust. Simpson (1963) observes that it is cross-cut by S2 and S3 cleavages, while Morrison (1989) notes that the dominantly sinistral mylonitic fault zone fabric overprints regional S1 and is, in turn, overprinted by fabrics related to southeast directed thrusting.

Our observations, detailed by Fitches *et al.* (1999), are in general agreement with those of the above authors. However, it is noted that, while the phyllonitic fault zone fabric is developed in the Creggan Mooar Formation in Glen Maye, it is primarily developed in the Niarbyl Formation at The Niarbyl. This observation contrasts with earlier studies which assigned the shear zone protolith at The Niarbyl to the Creggan Mooar Formation (*sensu* Woodcock *et al.* 1999). The shear zone is a compound structure, consisting principally of a heterogenous ductile shear zone, brittle thrust structures and other faults. The ductile shear zone dips moderately west or northwest, defining an inferred listric fault zone beneath the outcrop of the Niarbyl Formation.

At The Niarbyl, the most intense expression of the ductile fault zone outcrops west of the slipway and at the southeast end of the nearby island. Here, a pervasive east-southeast trending fabric comprises boudinaged to totally disrupted bedding; layer-parallel to totally disrupted quartz segregation veins and asymmetric 'quartz fish'; shear bands; and folded quartz veins and fabric. Very heavily altered and augened intrusions and post-fabric, downward-facing folds occur in the highest strain parts of the Creggan Mooar Formation *c.* 20–40 m south of the slipway. Kinematic indicators in the high-strain zone record an overwhelming sinistral sense of shear. These fabrics are overprinted, apparently transitionally at one location, by top-to-the-southeast brittle thrusts, thrust duplexes and ramp-flat thrusts (Fitches *et al.* 1999). The inferred sense of thrusting is compatible with that of the thrust–quartz vein complexes described above and also with the general southeast sense of vergence of the F1 fold train south of Knockaloe Moar. Overall, the fault zone is interpreted as a compound ductile sinistral shear zone with a top-to-the-southeast brittle thrust component.

The ductile shear zone is cut and dissected by numerous brittle faults, which not only mark the contact between the (?)mid-Arenig Creggan Mooar Formation and the mid–late Wenlock Niarbyl Formation, but also create an impression of one sided ductile shear zones at both The Niarbyl and in the Glen Maye ravine. However, the combined evidence from both locations clearly indicates that ductile high-strain zones are equally well developed, and therefore bilaterally symmetrical in both formations.

The interpretation of the Niarbyl Shear Zone as a compound sinistral ductile shear zone and thrust fault paradoxically implies southeast directed thrusting of younger rocks over older. This configuration might reflect one of three situations: parautochthonous overthrusting of a previously deformed part of the sequence by a younger part of the same sequence; overthrusting by an allochthonous unit; or simply a thrust-modified unconformity. It is implicit in the first scenario, and virtually implicit in the third, that at least D1 in sequences above and below the contact cannot be contemporaneous. Morrison (1989) notes that the phyllonitic fabric overprints S1 (in both the Creggan Mooar and Niarbyl Formations as defined here) but that it is, in turn, overprinted by late thrust-related fabrics. It is further noted here that regional S2 crenulation cleavage apparently 'stitches' across the boundary, as it is coplanar in both the Creggan Mooar and Niarbyl Formations.

The age of the Niarbyl Formation implies that the shear zone, together with D2 and D3 as defined by Simpson (1963), are all younger than Wenlock, and most probably reflect early Devonian Acadian deformation. The relative age of D1 in the Dalby and Manx Groups is, however, more debatable. Geometric evidence suggests that D1 in the Niarbyl Formation is related to thrusting associated with the sole Niarbyl Shear Zone and that it too is, therefore, post-Wenlock in age. In contrast, D1 in the Manx Group can only be constrained to be post-Arenig in age. Murphy & Hutton (1986), Hutton & Murphy (1987), Gallagher *et al.* (1994) and Tietzsch-Tyler (1996) all note evidence of Ordovician deformation in Cambro-Ordovician sequences along-strike to the southwest in Ireland. It is therefore conceivable that D1 in the Manx Group may be of Ordovician age and entirely unrelated to D1 in the Dalby Group. In this case, the thrust relationship between the two groups could reflect overthrusting across an already deformed sequence.

Sedimentology

Turbidite association facies

Five distinct, though generally intergradational, turbidite facies are recognized within the Niarbyl Formation – summarized in Table 1 – as well as the

Table 1. *Summary descriptions of the principal turbidite lithofacies in the Niarbyl Formation*

Facies name	Pickering facies*	Bouma division(s)	Thickness (cm)	Description
Massive beds	B1.1		to 400	Beds sand throughout, amalgamation common, particularly of thinner bed examples and in channel margins. Maximum grain size to only medium grain size, though may possess coarse tail layers to a few centimetres thick of quartz, feldspar, ?felsics, and black and grey shale flakes to 6 cm
Thick-bedded turbidites	C2.1	ab(d)e, abc(e)	40–100, rare to 200	Generally middle absent sequences, where Te < 5%. Ta to medium-grained sand. Some Tab amalgamated beds, also Tb alone
Medium-bedded turbidites	C2.2	ae, abe, abce, ab(d)e, be	10/15–40	Generally middle absent sequences. Ta, Tb, Tc divisions generally > 90%, mainly fine sand grain size, though medium grain size not uncommon and may occur even in quite thin beds. Thin beds may contain shale flakes to *c*.5 cm
Thin-bedded turbidites, sandy	C2.3s	abe, be, ce, bce, b(d)e, bd	< 1–15, av. 1–3	Te < 5–20%. Tb, Tc divisions generally fine sand to silt grade. One instance of two *c*. 15 cm amalgamated Tab–Tabe beds
Thin-bedded turbidites, pelitic	C2.3p	b(d)e, c(d)e, bc(d)e, ce.	<1–5, rare to 20, av. 1–3	Generally base absent sequences, T(d)e dominant, range 50–80%. Tb, Tc invariably fine-grained sand or silt
Hemipelagite	E2.2			See detailed description above

* Designations after the scheme of Pickering *et al.* (1986). Note that the classification system is based upon both sand content and bed thickness.

highly distinctive hemipelagite facies which has been described above. Various internal structures are evident in the turbidites: Tc division ripple lamination, concretions and sand intrusions. Ripple lamination is common as: trough cross-lamination; straight to slightly curvilinear asymmetric sets on bedding surfaces; and particularly in outcrops north of Contrary Head, climbing ripples. The climbing ripples, best exposed in the Traie Dullish Quarry, are defined by very low amplitude, relatively long wavelength ripples in which the angle of climb, relative to bedding, averages *c.* 5° (Fig. 2a). Climbing ripple bases are commonly loaded and foresets occasionally show incipient to well-developed overturning in the downcurrent direction (Fig. 2a). Trains of climbing ripples commonly fade downcurrent within individual beds and, in all such instances, the current azimuth progressively swings down bed dip from a 'lateral' (from 323°) to an 'axial' (from 015° to 048°) orientation, with respect to the Caledonoid basin trend. (This pattern is considered further under Palaeocurrents.)

Partially weathered-out carbonate concretions, up to 20 cm long and 7 cm thick, occur commonly flattened parallel to bedding. Two examples of soft-sediment intrusions are noted, both consisting of discordantly coarse-grained sand in pelite. One example, at [SC 2212 7954], consists of a 10 cm thick complex of anastomosing semiconcordant sills projecting downward from a bed base into underlying pelite. The other example, exposed in the Traie Dullish Quarry, is defined by a 1.5 cm wide, pale grey–white medium-grained quartz sandstone dyke that wedges out across a 0.5 m section (Fig. 4).

Turbidite facies architecture

Assessment of the overall facies organization of the Niarbyl Formation is complicated by restricted outcrop extent, effectively limited to an 8 km long coastal section, intense folding resulting in homoclinal sections < 150 m thick and lack of definitive marker horizons to assess lateral facies variations. These constraints not only severely restrict architecture interpretation to an essentially vertical facies analysis but they also render as tentative any assessment of the turbidite system as part of a submarine fan, ramp or slope apron (*sensu* Stow *et al.* 1996).

The formation is composed essentially of a non-channelized sequence of predominantly parallel-sided, thin- to thick-bedded turbidites of various facies types (Table 1). The entire sequence is notably 'packeted'. Randomly intercalated, non-cyclical (neutral) packets of thin-bedded facies and thicker bedded, sandy turbidites are interpreted as interlobe and sandy lobe deposits, respectively (Figs 2b, 3a and b; see Interpretation). Thin-bedded interlobe packets may range up to 10 m or more in thickness, whereas sandy lobe packets are normally < 2 m in thickness and consist of < 12 beds, usually less than five beds (Figs 3a, b and d). Exceptionally thick sandy packets, up to 40 m, are locally evident in the central part of the formation (Figs 1b and 3c). Internal contacts within and between packets, and between individual beds, are generally planar and non-erosive. Erosional contacts do occur locally, however, as evidenced by: sole structures (Figs 3b, d and 5); low-angle erosional bases to sand packets (Fig. 3d); and two channels which outcrop near the base of the formation (Figs 2d and 5).

Sandy lobe packets are evident in all log sections (Figs 3a, b, d and 4: sections marked a–b, etc. on each log). These generally range in thickness from 0.5 to 12.25 m, but reach > 15 m at [SC 2220 7957] (Fig. 3c); the average thickness is 3.5 m. The sand lobe packets are most commonly defined by sequences of seven to 15 principal sand deposition events (e.g. Fig. 3b, a–b, c–d, e–f) and up to 35 or so events in thicker packets (e.g. Fig. 3d, a–b). The thinner packets are dominated by fine–medium sand grade Ta-, Tb- or Tc-based beds of the C2.2 facies classes and Tb- or Tc-based beds of the C2.3s facies class (Table 1; e.g. Fig. 3a and b). The thicker packets are dominated by similar grain size Ta based C2.1 and C2.2 beds (Figs 3d and 4). Exceptionally thick packets are dominated by medium–coarse sand grain size B1.1, C2.1 and C2.2 bed sequences, which are very commonly amalgamated (Fig. 3c).

Thinning- and thickening-up sequences are evident within many sand lobe packets. In some instances these are arranged as successive minor thickening-up cycles in the lower part of the packet, overlain by one or more thinning-up cycles in the upper part of the packet (e.g. Figs 3a, c–e and 4, a–c). This type of arrangement may reasonably be interpreted to reflect initial progradation of the sand lobe, culminating with lobe-abandonment cycles. In other instances, thinning-up cycles dominate either entire packets or subordinate cycles within packets (e.g. Figs 3a, a–b, d, a–b and 4, f); or the packets may define neutral trends (e.g. Fig. 3b, a–b, e–f). These types of trends are best defined as abandonment or failed lobe-development cycles although, in the case of Fig. 3d (a–b), the base of the packet is noticeably erosional as the prominently flute-marked base erodes across underlying beds at an angle of *c.* 12°.

There are two distinct types of interlobe packets: turbidite dominated and turbiditic mud–hemipelagite-dominated sequences. The turbidite-dominated type characterizes the formation south of the Knockaloe Moar Lineament. The packets range in thickness, in logged sequences, from 0.6 to 6.6 m (e.g. Figs 3a, b–c and 3b, b–c, d–e) and up to 10 m or more in unlogged sections. The packets are composed primarily of C2.3s and C2.3p facies beds, normally arranged in symmetric (neutral) successions (Fig. 2b), although locally arranged in minor thinning- or thickening-up sequences, interpreted as 'compensation cycles' (Mutti & Normark 1987: Figs 2c, 3a and b). These packets frequently contain isolated or minor packets of C2.2 facies beds, presumably reflecting the extreme fringes of sand lobes (e.g. Figs 3a, b–c and 3b, d–e).

Turbidite mud–hemipelagite dominant assemblages characterize interlobe packets north of the Knockaloe Moar Lineament. The packets vary from very thin C2.3p silt–mud turbidites with interspersed hemipelagite horizons up to 2 m thick (Fig. 3d, base–a, b–c), to mud-dominant, medium-bedded C2.3p beds with abundant hemipelagite intercalations (Fig. 4, base–a, e–f). The proportion of hemipelagite in such sequences can vary from *c.* 70% (Fig. 4, base–a) to *c.* 40–60% (Fig. 4, e–f). Hemipelagite packets vary in thickness from sub-millimetre single couplets up to *c.* 35 cm. A noticeable feature of these sections, which cannot be represented graphically, is the frequent, very delicate, plane-parallel interlamination of extremely thin hemipelagite in mud or mud in hemipelagite.

Interpretation: turbidite facies

With due regard for the limited outcrop extent of the formation, its sedimentological characteristics suggest a mid–lower, non-channelized, submarine-fan environment. The lower part of the formation represents the sandy or sandy–silty distal lobe facies association (Stow 1985) and the upper portion the even more distal, perhaps lower, fan silt–mud–hemipelagite lobe association (Stow *et al.* 1996). Thin bedded packets are interpreted as interlobe deposits and the sandy packets as lobe deposits. The intermittent character of the sandy lobes could reflect various control mechanisms, such as episodic avulsion from distributary channels or depositional constraint by subtle relief features on the fan.

The formation is sandier at its inferred base at Niarbyl than in its northern outcrop, north of the Knockaloe Moar Lineament, suggesting a stratigraphically upward fining trend. The less abundant sand input into the upper part of the depositional system could, for example, reflect slowly rising sea levels, thereby increasing the width of shelf sediment entrapment areas and reducing sand transport to the shelf–slope break. Such patterns are defined in other submarine fan associations, e.g. the Navy Fan in California (Piper & Normark 1983), and match the more mud- and silt-rich sequences in late Wenlock sequences in adjoining regions in Scotland and Ireland (Kemp 1987; Murphy 1984).

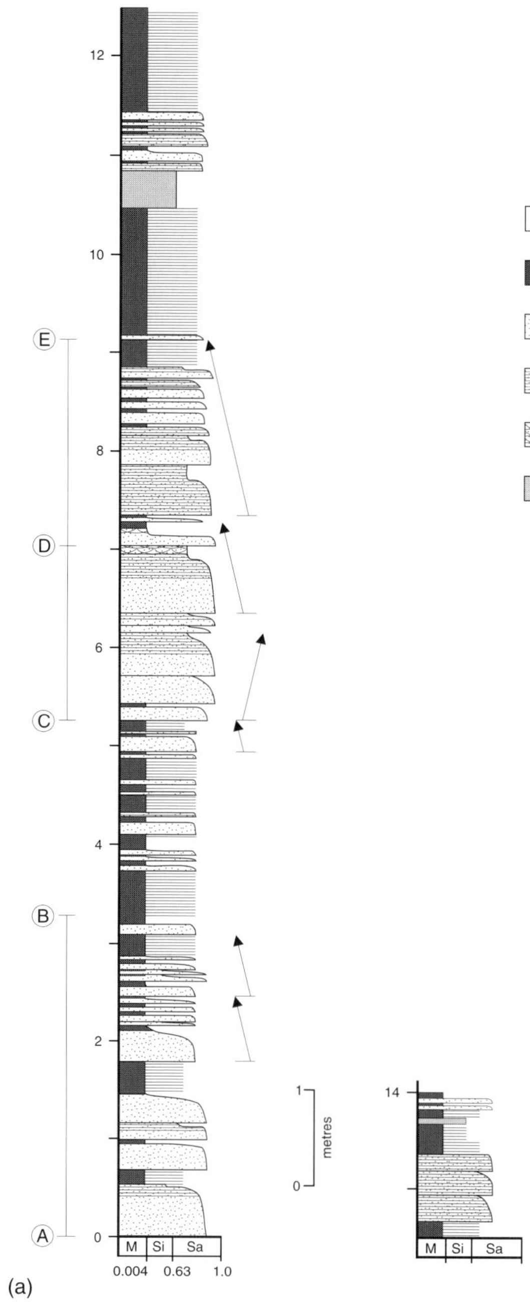

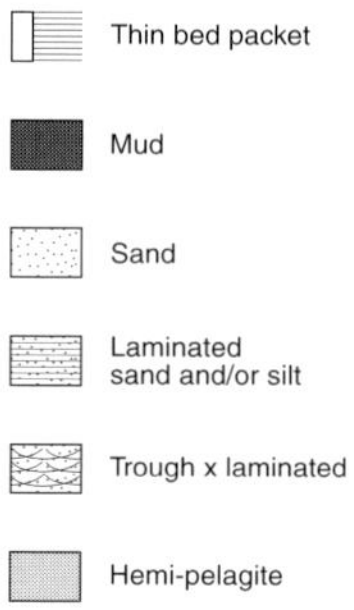

Fig. 3. Sedimentological logs representative of different turbidite facies associations in the Niarbyl Formation. Arrows indicate generalized thinning or thickening trends; palaeocurrent rose diagrams show type, mean flow vector and number of readings at each sample point; M, Si and Sa at the base of each log section denotes mud, silt and sand grade fractions, respectively. Location of each log section shown in Fig. 5.(**a**) Type section of the Niarbyl Formation, east face of the island at the The Niarbyl [SC 2107 7758]. (**b**) Rock platform *c.* 300 m northeast of Boirane [SC 2191 7916]. (**c**) Adjacent to the mouth of the Lag ny Troagh Stream [SC 2220 7957]. (**d**) Disused slate quarry, Thistle Head [SC 2339 8357].

Interpretation: hemipelagite

The laminated hemipelagite facies has been described extensively from other mid-Silurian (Wenlock or lower Ludlow) sequences around the former Iapetus Ocean (Kemp 1991) and its origin has been much debated. Earlier views on the examples in the English Lake District are summarized by Rickards (1965), who developed the idea of Marr (1927) that each silt–carbon laminae couplet may represent an annual cycle of deposition. Rickards envisaged a constant rain of algal organic carbon deposited in anaerobic bottom waters, interrupted periodically, though not necessarily annually, by silt deposition from low concentration turbidity currents. Alternatively, Kemp (1991) suggested that a number of carbon and silty–mud laminae could be deposited during one turbidity flow event. The carbonaceous laminae would represent discontinuous films of algal organic material deposited in intimate association with clay and fine silt, their apparent continuity being enhanced by later compaction. In support of the first hypothesis of discrete lamina-by-lamina deposition, Dimberline *et al.* (1990) drew analogies between the Wenlock hemipelagite in Wales and recent sediments in basins on the California Borderland (Thornton 1984). There the lamination represents an annual climatic cycle of high warm-season productivity with high wet-season sediment run-off. The silt component of the lamination may be sedimented by vertical fall-out from nepheloid suspensions rather than directly from turbidity flows. On either hypothesis, the

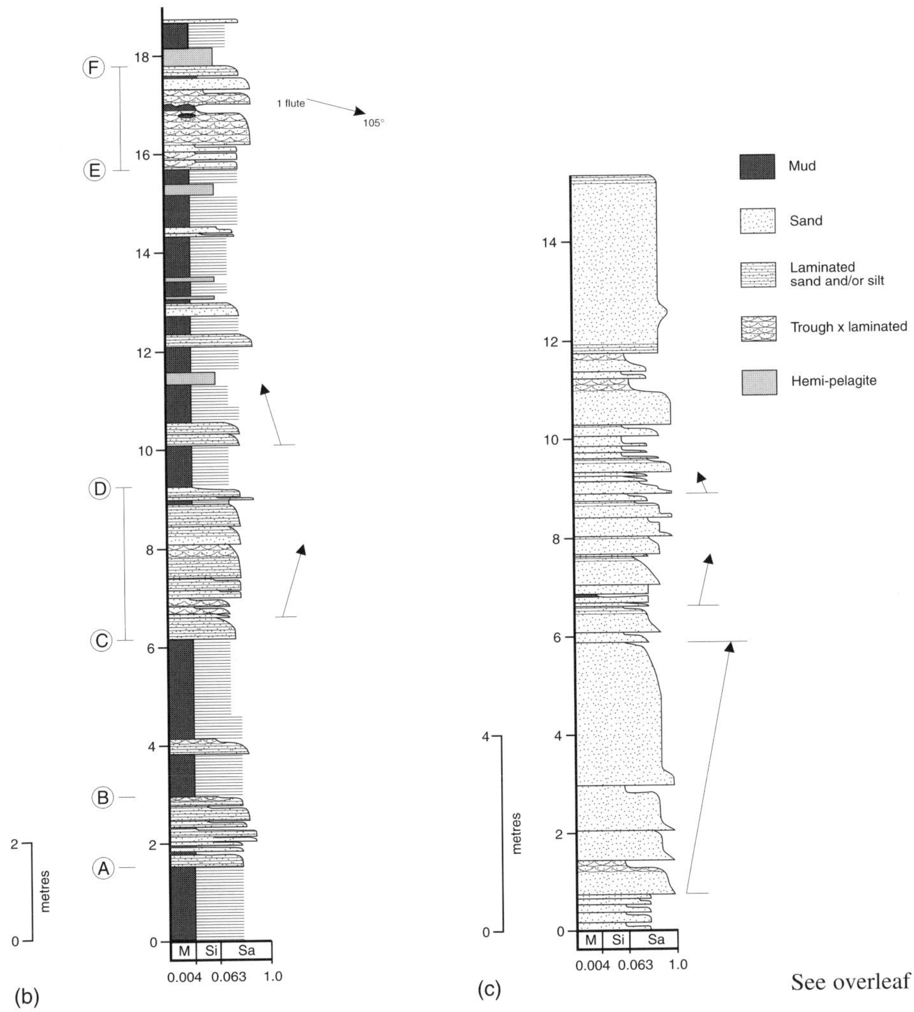

See overleaf for Fig. 3d

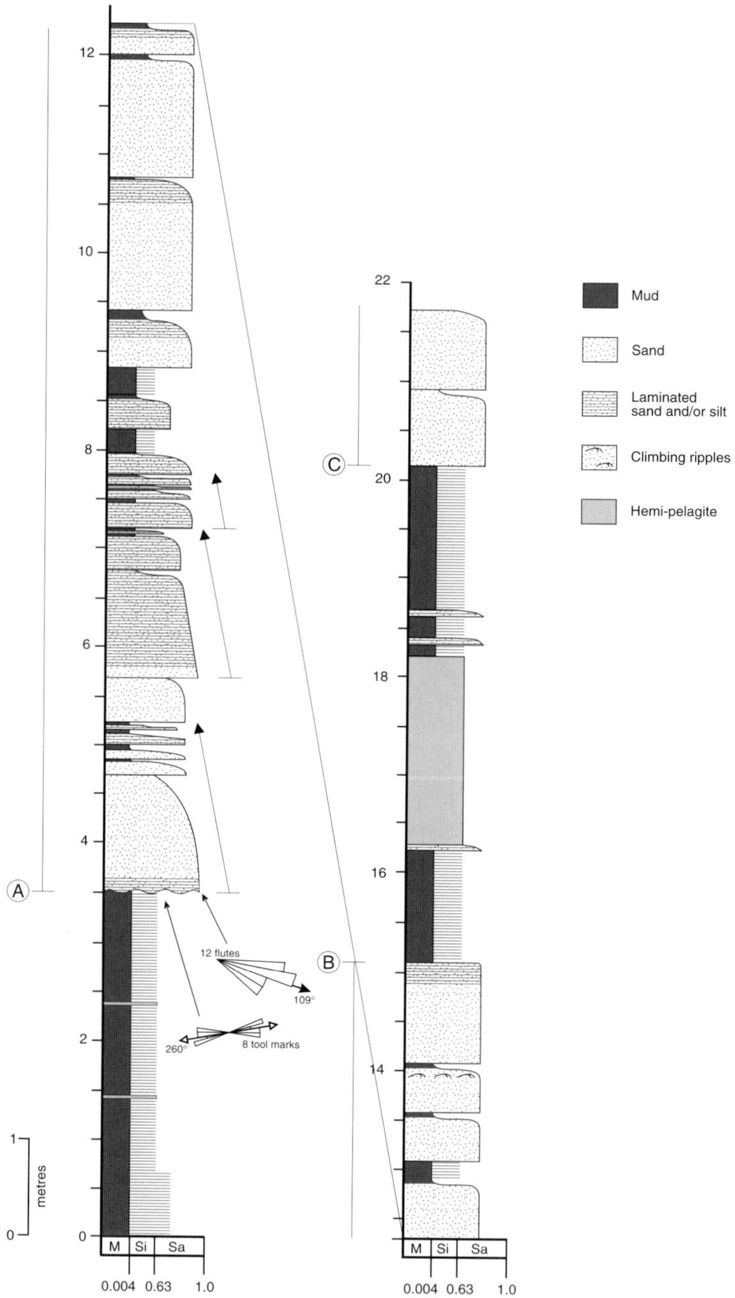

(d)

paucity of bioturbation in the laminated mudstone implies a sparse benthonic fauna with no burrowers and probable dysaerobic bottom waters.

Estimates of silt–carbon couplet frequency in the Brathay Formation of the Lake District, using graptolite zonation and its chronometric calibration (King 1992), suggest a periodicity of three or four years. A near-annual driving influence is not precluded by this observation, given the potential for erosion and non-preservation of laminae. This

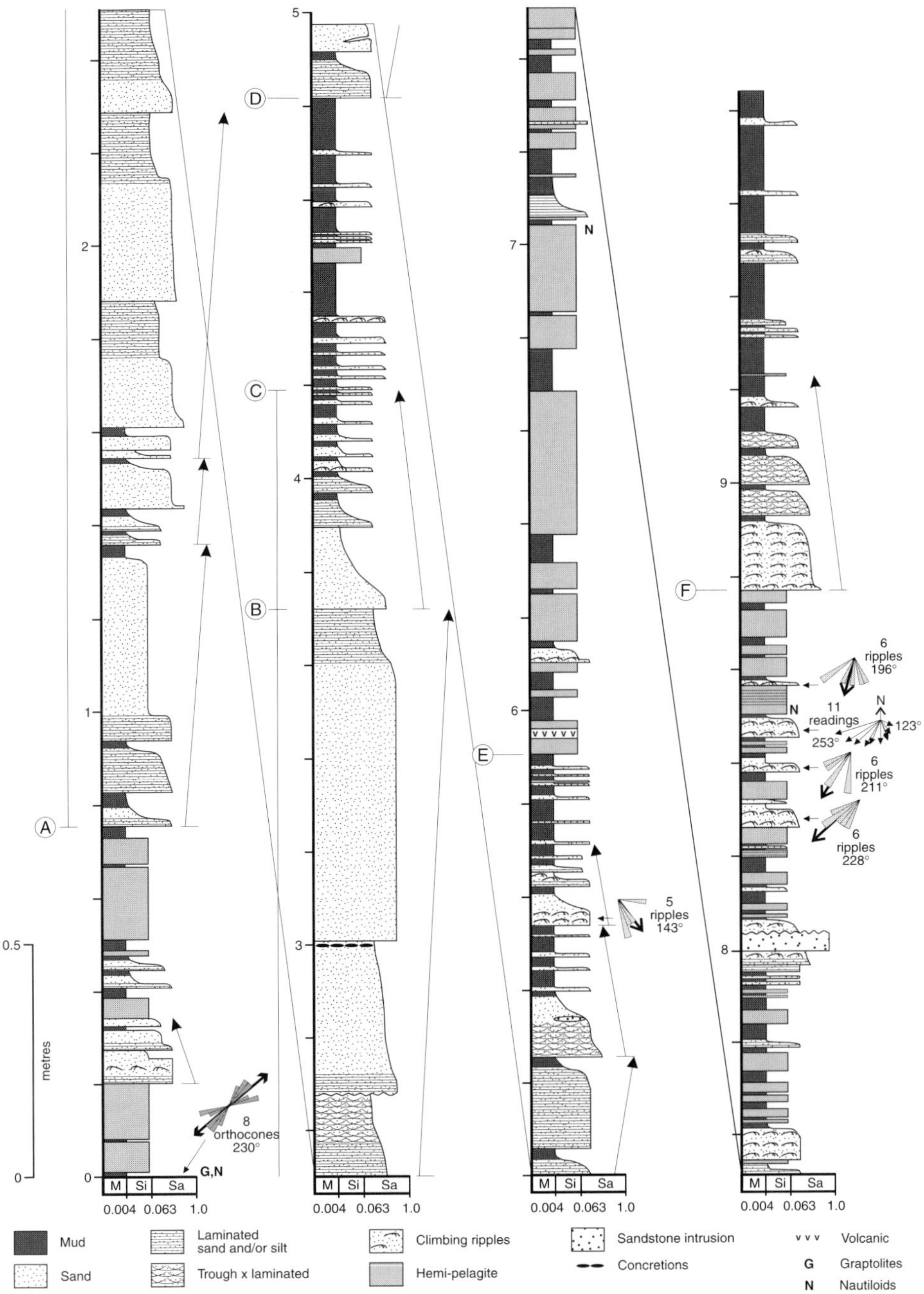

Fig. 4. Sedimentological log, Traie Dullish Quarry, Peel Hill [SC 2370 8401]. Arrows indicate generalized thinning or thickening trends; palaeocurrent rose diagrams show type, mean flow vector and number of readings at each sample point; M, Si and Sa at the base of each log section denotes mud, silt and sand grade fractions, respectively. Location of log section shown in Fig. 5. This sequence, along with Fig. 3d, is representative of the more mud-silt-dominant upper part of the Niarbyl Formation.

assumed time calibration has several implications for the sedimentology of the Niarbyl Formation. First, the typical time interval between deposition of sand turbidites was relatively short, less than that represented by the thickness of hemipelagite that could be eroded by one flow. At 30 couplets/cm and *c.* 3.33 years/couplet, 1 cm of hemipelagite represents 100 years. Erosion of a few centimetres of hemipelagite would be compatible with repeat times of several hundred years. Secondly, however, there was at least one period during Niarbyl Formation deposition, represented in the Traie Dullish Quarry, when sandy turbidity flows shut down for long enough to allow preservation of mud turbidites and intervening hemipelagites. The average thickness of hemipelagite packages during this period was *c.* 4 cm, representing an average frequency of muddy turbidity flows of the order of 400 years. The period of quiet deposition lasted for at least 50 000 years, based on a sequence thickness of 25 m and a hemipelagite proportion of 20%.

Palaeocurrents

Palaeocurrent trends have been deduced from uni- and bidirectional sole marks – flutes, tool marks and longitudinal ridges and grooves – from asymmetric ripple trains and climbing ripple sets, and from rare channels (Fig. 5). The morphology of ripple marks is described above. Flutes vary from bulbous to linguiform, relatively deeply incised marks up to 10 cm long and 2 cm deep (e.g. Thistle Head Quarry; Fig. 3d) to very shallow linguiform types, up to 5 cm long and < 0.5 cm in depth. Flutes of the latter type are well exposed on several bedding surfaces in the Contrary Head Quarry (Fig. 5), where longitudinal ridge and groove patterns, tool marks ranging up to 10 cm in length and 0.5 cm deep, prod and chevron marks and cauliflower structures are also exposed.

Only two channel-like structures are noted in the whole formation outcrop, both near its southern contact (Fig. 5), and only a single markedly erosional base to a sandy lobe packet, in the Thistle Head Quarry (Figs 3d and 5). One margin of a prominent *c.* 2 m thick channel-like structure is well exposed over *c.* 12 m in the hinge of an F1 syncline in a sea stack on the north side of Elby Point (Figs 2d and 5: [SC 2114 7777]), while its opposite margin is exposed *c.* 30 m across-strike further north [SC 2117 7784]. Though channel-like in form, the base of the roughly 075° trending structure is not noticeably erosional (Fig. 2d). Instead, individual facies C2.2–C2.1 beds amalgamate laterally towards the margin, where they are overstepped by evenly bedded similar facies turbidites. A dip-slope bed surface within the 'channel', close to its northern margin, is covered by an array of slightly undulose asymmetric ripples, which indicate a palaeoflow transverse to its orientation (Fig. 5). Some 300 m north, just south of Creg Mooar, a *c.* 080° trending channel is exposed (Fig. 5). In this instance, a *c.* 1m thick packet of facies C2.3 beds is cut out by the erosional base of a 4 m thick thinning- and fining-up cycle defined by facies C2.1 and C2.2 beds.

The unidirectional current marks and channels define two broad trends which are paralleled by bidirectional indicators. Palaeoflows from the northwest quadrant represent lateral input with respect to the northeast–southwest Caledonian trend, and flows from both the northeast and southwest represent axial input (Fig. 5). Palaeoflows from the southeast quadrant are conspicuously absent. The northwest lateral vector, which is taken to include individual data points ranging from west to north, is defined by mainly west-northwest orientated flutes and north-northwest orientated ripples alike, and is very pronounced throughout the formation (Fig. 5). Flutes, ripples, tool marks, ridges and grooves, and nautiloid orthocone orientations define axial trends (Fig. 5). Both vector sets display some degree of variability. In the Contrary Head Quarry, flutes and tool marks may diverge by up to 20° on a single bed base, this presumably reflects fluctuating flow directions in a single current.

In the Traie Dullish Quarry, climbing ripple sets in successive beds show either progressive or fluctuating deflection of palaeoflow vectors (Figs 4 and 5). The deflection pattern in these ripples, from initial southeast or south directed flows, swinging towards south, southwest or west-southwest directed flows (trends within individual beds display flow orientation deflections ranging from *c.* 105 to 165°, 205–245°, 181–247° and 123–253°, Figs 4 and 5), might be interpreted in several ways: as steering by the Coriolis force; as interference between turbidity flows and ocean bottom currents; as constraint of flows by topographic relief. The clockwise deflection pattern is opposite to the Coriolis effect expected in the southern hemisphere position of Avalonia in the mid-Silurian (Torsvik *et al.* 1993). Persistent thermohaline bottom currents would be unlikely in the small restricted marine basins of the closing Iapetus Ocean.

Topographically controlled reorientation is more likely, either by subtle or more substantial relief features. While the former possibility cannot be entirely ruled out, there is little evidence at outcrop scale of any relief controlled features affecting the intercalated hemipelagites (Fig. 4). In contrast, the presence of axially orientated channels suggests that the turbidity flows were being deflected, either towards the northeast or the southwest, by substantial seafloor topographic features. This axial

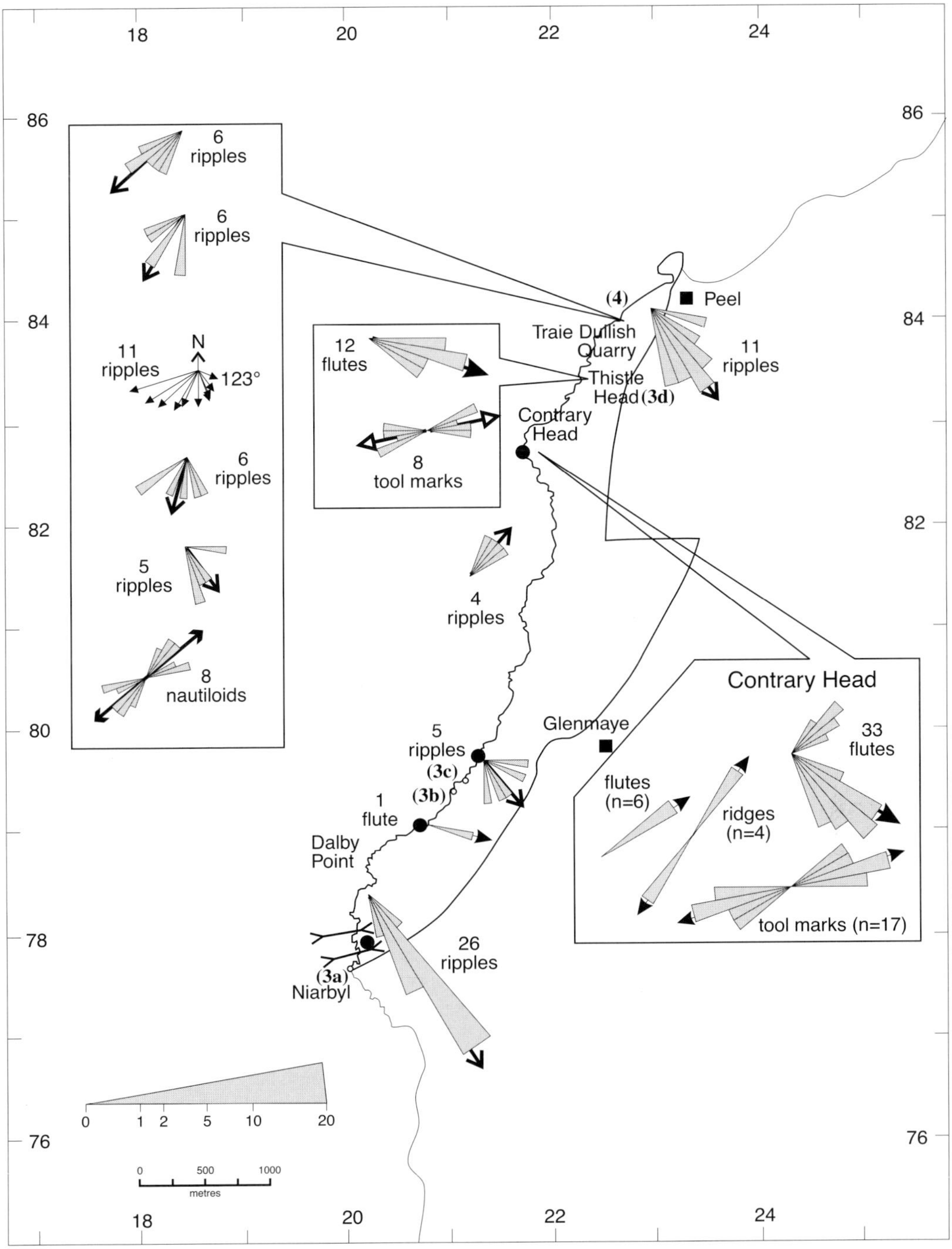

Fig. 5. Palaeocurrent trends in the Niarbyl Formation, also showing the location of sedimentological logs presented in Figs 3 and 4. Orientations reconstructed by rotation about-strike only as fold plunges are essentially horizontal, other than at Contrary Head. Here, correction first for fold plunge (40°) and then rotation about-strike, results in a clockwise rotation of all readings by only 15° and has consequently been ignored. Channels >—< .

deflection is not just a local effect. Kemp (1987) notes a dominant southwest directed flow vector in the Southern Uplands, while the influence of inherited Ordovician fabrics on Wenlock sedimentation is implicit in Hutton & Murphy's (1987) observation that an axially orientated Ordovician terrane-bounding fault is stitched across by Wenlock sediments in the Balbriggan Inlier in Ireland. It is conceivable that minor structures of this type influenced the development of the channel-like structure at Elby Point (Fig. 2d).

Turbidite sandstone petrography

Niarbyl Formation turbidite sandstone is, with only a single exception, characterized by heterolithic, angular to subrounded, quartz–felsic mineral and lithic clast assemblages. A sample from a single turbidite bed, though identical in outcrop appearance to any other sandstone, is essentially composed of a monolithic assemblage of mono- and polycrystalline quartz.

In the heterolithic sandstones, the most abundant clasts are non- to weakly undulose monocrystalline quartz (including exceptionally rare, very well-rounded clasts), undulose polycrystalline quartz and microcrystalline felsite. There are lesser proportions of fine-grained equigranular, fasicular and spherulitic textured rhyolite, microgranite, white mica, plagioclase and K-feldspar; and relatively minor amounts of shale, siltstone, carbonaceous shale, mica schist (sometimes crenulated), granophyre, quartz porphyry, spilite, myrmekite, trachyte, green tourmaline, zircon, orthite, pyrite, and quartz that bears rutile and vermicular chlorite. The polycrystalline quartz class includes polygonized quartz, microcrystalline quartz of uncertain affinity (chert, fine-grained quartzite or felsite), as well as equidimensional and ribbon textured cataclasite. The matrix consists of finely comminuted clastic detritus, a very prominent S1 white mica foliation, which is particularly well developed as pressure fringes at clast edges, and abundant carbonate. The carbonate occurs both aligned in S1 pressure fringes (MS1), and randomly overgrowing and replacing all types of clasts and interstitial matrix components.

Ingersoll *et al.* (1993) succinctly demonstrate that the composition of turbidite sands is a good predictor of the plate tectonic setting from which the sands have been derived. For the Niarbyl Formation, a petrofacies analysis is presented based upon the modal analysis of 14 samples of generally medium- but occasionally coarse-grained sandstones. Point counting followed the Gazzi–Dickinson convention (Ingersoll *et al.* 1984), with the addition of carbonate as a separate class distinct from matrix. Each analysis is based upon a count of 500 points per sample, each stained for K-feldspar. Distinction of microcrystalline quartz was the most problematic operation (cf. Dickinson 1970). The many examples containing quartz or sericitized feldspar microphenocrysts are assigned to class Lv; those with mica to Lm; and those composed of apparently monocrystalline quartz to Qp (a table of complete modal analyses is available from the senior author if required).

The results are summarized in four provenance-source discrimination diagrams (Fig. 6): QFL and QmFLt diagrams proposed by Dickinson *et al.* (1983), and QpLvmLsm and LvLsLm diagrams proposed by Ingersoll & Suczek (1979) – see also Suczek & Ingersoll 1985. Dickinson *et al.* (1983) note that their two diagrams are useful only for 'provisional classification' of likely source regions, while Ingersoll *et al.* (1985) strongly discourage use of the QFL plot in isolation and note that the QpLvmLsm and LvLsLm plots, as well as QmFLt, are more useful discriminators. Analyses of six samples of sandstone collected from the base of the exceptionally thick-bedded, and coarse-grained, Purt Veg distributary channel sequence in the Santon Formation of the Manx Group (Woodcock *et al.* 1999) are included in these diagrams for comparative purposes.

With the exception of the single quartz-rich sample noted above, the Niarbyl Formation sandstones cluster in both the QFL and QmFLt diagrams, confirming both their compositional uniformity and relative maturity. All, again bar the quartz-rich sample, plot in the '(transitional) recycled-orogen' provenance-source fields. The Santon Formation (Manx Group) samples also plot in the same fields, although they are notably less well clustered, spreading into the 'lithic recycled-orogen' field (Fig. 6a, b), suggesting a higher degree of compositional variability. This distinction is also provided by the relative proportions of secondary carbonate. In the Niarbyl Formation, the proportion ranges between 0.4 and 33%, and averages 13.34%. In the Santon Formation sample set, three samples contain no carbonate and the other three range between 0.6 and 14%, and averages 3.23%. Matrix contents in the two formations are, however, reasonably comparable: in the Niarbyl Formation, the proportion ranges between 24.2 and 38.2%, and averages 33.93%, while the comparable figures for the Santon Formation sample set are 19.6–40.6%, and averages 26.76%.

Dickinson *et al.* (1983) note that recycled-orogen sources are dominated by the sedimentary and volcanic rocks (whether metamorphosed or not) that characterize thin-skinned foreland fold–thrust belts, subduction complexes and collision-orogen suture belts. The derived sands are poor in feldspar,

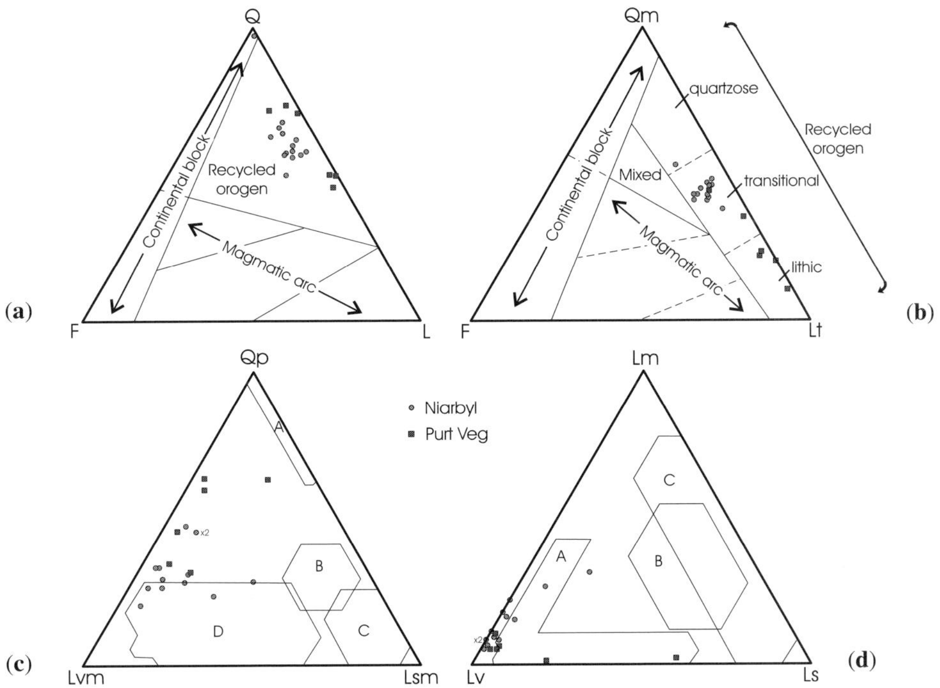

Fig. 6. Petrofacies discrimination diagrams. (**a**) and (**b**) After Dickinson *et al.* (1983); (**c**) and (**d**) after Ingersoll & Suczek (1979). Fields in (c): A, continental margin; B, subduction complex and magmatic arc; C, suture belt; D, magmatic arc. Fields in (d): A, magmatic arc; B, subduction complex and magmatic arc; C, suture belt.

since plutonic rocks are not a primary source in such regions, in contrast to active and dissected magmatic arcs. An arc affinity is indicated by the qualitative sandstone composition noted above, rather than a recycled-orogen source, an apparent discrepancy resolved by reference to the QpLvmLsm and LvLsLm plots (Fig. 6c, d).

On both diagrams, the Niarbyl Formation samples plot in or near the magmatic-arc provenance fields, whereas the Santon Formation samples are more widely scattered and generally plot outside any of the diagnostic fields. The contrasts between the two Niarbyl plots largely reflects the relative proportions of feldspar as a mineral clast and microcrystalline felsite as a lithic clast type. These are the principal determinants of magmatic-arc affinity. The Niarbyl sandstones contain a relatively low volume of feldspar (although note a spread of samples towards the magmatic-arc fields in both QFL and QmFLt diagrams) and a relatively high proportion of felsite. This composition suggests a primary source from a felsic igneous volcanic or hypabyssal intrusive complex, rather than a coarser grained plutonic zone, and implies a juvenile to intermediate stage in the unroofing of the magmatic-arc source zone. In both the Niarbyl and, particularly, the Santon Formation sample sets, there is a considerable degree of compositional variability. This variation will influence the chemical composition of the sandstones and therefore their plot affinities in chemical discrimination diagrams, as well as apparent contrasts between lithostratigraphic units.

Discussion: the regional context of the Niarbyl Formation

Consideration of the regional context of the Silurian succession of the Isle of Man is constrained by several factors: its late Wenlock, probably *lundgreni* Biozone, age; the presence of the distinctive hemipelagite lithofacies; and by combined palaeocurrent and turbidite compositional evidence. Each of these issues is considered below.

Age

Equivalent age sequences occur in several areas of the surrounding regions of Britain and Ireland: the

Brathay, Birk Riggs and Coldwell Formations in the Lake District (Kneller *et al.* 1994); the Riccarton Group in the Southern Uplands (*sensu* Barnes *et al.* 1989); and in various inliers in Ireland, most proximally in the Skerries, Denhamstown and Clatterstown Formations in the Balbriggan Inlier (Murphy 1984; Fig. 7). Comparison with each of these successions is considered in more detail after consideration of the regional position of hemipelagite.

Hemipelagite

The laminated hemipelagite is a helpful lithofacies for regional correlation purposes. The closest match amongst northern margin sequences is with the Riccarton Group of the Scottish Southern Uplands, where laminated hemipelagite is interbedded with very fine sand or silt–mud turbidites throughout all but the top two graptolite zones (*nassa* and *ludensis* Biozones) of the Wenlock (Kemp 1991). Analogous beds are said to be exposed on the southern side of Dundalk Bay in eastern Ireland (Kemp 1991). On the southern margin, the Wenlock sequence is predominantly composed of laminated hemipelagite. Examples occur in the Lake District (Brathay Formation; Kneller *et al.* 1994), the lower parts of sequences in the Craven Inliers in northern England (Kneller *et al.* 1994) and the Balbriggan sequence of eastern Ireland (Rickards *et al.* 1973; Murphy 1984). However, sand-rich turbidites occur in the higher parts of all these sequences, the Birk Riggs Formation by the *nassa* Biozone in the Lake District, the Austwick Formation by *rigidus* time at Craven, and the Skerries Formation first in *linnarssoni* time and then persistently from *lundgreni* time at Balbriggan. These sandy Wenlock sequences, or their continuation into the lower Ludlow, offer the closest lithological matches with the Niarbyl Formation. However, the Niarbyl Formation is unlikely to include the latest Wenlock *nassa* and *ludensis* Biozones, which contain prominent bioturbated intervals in most southern margin sequences (Kemp 1991), representing periods of oxygenated bottom water.

Comparison with Silurian successions in the regions surrounding the Isle of Man

Hutton & Murphy (1987) provide a compelling argument that Silurian successions in the Lake District, Southern Uplands and various inliers in Ireland represent geographically isolated deposits in a single basin. This 'successor basin' sequence infilled the suture zone between the Avalonian and Laurentian continental margins after their docking between late Ordovician and early Wenlock time. Combined palaeocurrent and compositional evidence indicates a bilaterally symmetrical basin infilling from magmatic-arc sources to the northwest and southeast during the Llandovery, switching to asymmetrical, progradational filling from a magmatic-arc source to the northwest from the Wenlock onward. The regional synthesis of Hutton & Murphy (1987) provides a useful framework for making an initial assessment of the Niarbyl Formation in its regional context.

The Balbriggan Inlier in Ireland provides the nearest equivalent part of the 'successor-basin' sequence (Fig. 7a). Murphy (1984) identifies several mid–late Wenlock age formations in the inlier, which form part of two distinct terrane sequences separated by the Lowther Lodge Fault (Murphy *et al.* 1991). The Skerries Formation occurs in the Leinster Terrane south of the fault, and the Denhamstown and Clatterstown Formations in the Bellewstown Terrane to the north. The Skerries and Denhamstown Formations span the early–late Wenlock, *murchisoni–lundgreni* Biozones interval and overlap in time across the terrane boundary (Fig. 7b). Hemipelagite dominates the basal sections of both formations, whereas the mid–upper parts of both are composed primarily of blue–grey, quartz-rich, calcareous greywackes and siltstones, arranged in two distinct, *riccartonensis–linnarssoni* and *lundgreni* Biozones, sandy packets. The packets are interpreted as essentially non-channelized, progradational mid–lower fan association sequences, derived principally from sources to the southeast, northeast and southwest. In distinct contrast, the succeeding, *ludensis* Biozone Clatterstown Formation is dominated by thin-bedded silty turbidites, which are punctuated intermittently by channelized sandy packets derived from the north and northwest. Murphy (1984) notes that not only does this mark the first pulse of northwest derived sediments, which persisted here and elsewhere into the Ludlow, but also a regionally widespread development of mud-dominated turbidite facies in the late Wenlock.

The entirely Wenlock age Riccarton Group in the Southern Uplands (Fig. 7a and b) consists of a single formation, the Raeberry Castle Formation as redefined by Barnes *et al.* (1989) – see also Kemp (1986) for an alternative interpretation. Kemp (1987) divides the formation into three distinct tectonostratigraphic units, only the youngest of which, the mid–late Wenlock age, *linnarssoni–lundgreni* Biozones, Mulloch Bay Unit is temporally equivalent to the Niarbyl Formation (Fig. 7b). The unit contains two members, a 228 m thick lower sequence of generally thin-bedded, fine-grained and sporadic coarse-grained facies, interpreted as a channel–overbank complex; and a

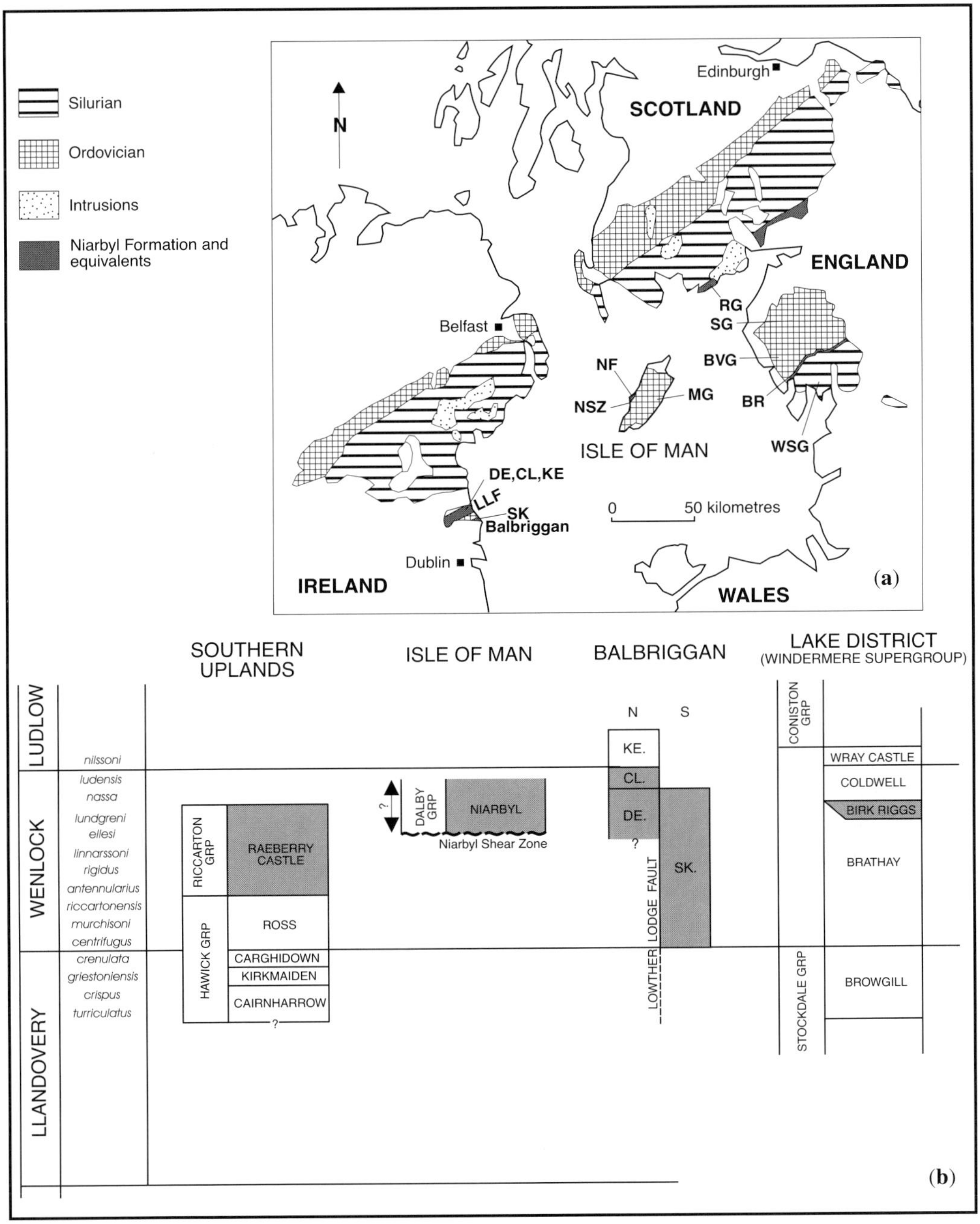

Fig. 7. The regional context of the Niarbyl Formation. (**a**) Map showing the location of late Wenlock sequences in adjoining parts of Scotland, England and Ireland. BR, Birk Riggs Formation; BVG, Borrowdale Volcanic Group; CL, Clatterstown Formation; DE, Denhamstown Formation; KE, Kentstown Formation; LLF, Lowther Lodge Fault; MG, Manx Group; NF, Niarbyl Formation; NSZ, Niarbyl Shear Zone; RG, Riccarton Group; SG, Skiddaw Group; SK, Skerries Formation; WSG, Windermere Supergroup. (**b**) Stratigraphic correlation diagram.

346 m thick upper sequence of packeted, cyclical intercalations of thin, fine-grained to medium-bedded sandy facies, interpreted as a succession of progradational mid–outer fan depositional lobes. This upper part of the Mulloch Bay Unit bears a very strong resemblance to the Niarbyl Formation, in general bed morphology as well as turbidite facies architecture and depositional system interpretation: cf. Kemp's (1987) figs 25–27 with Figs 3a, b and 2b respectively. Palaeocurrent trends are dominantly axial, from the southwest or northeast, with some from the west or west-northwest, in contrast to underlying units in which lateral trends, from the northwest and west, are more prevalent. Kemp (1987) also notes variable palaeocurrent trends within stratigraphically comparable intervals at widely separated points along-strike. He suggests that this may reflect multiple sediment inputs into the depositional system, a configuration to be expected in a slope apron rather than a single-point source submarine fan.

Finally, comparison may also be considered with various Wenlock age formations in the Lake District (Fig. 7a and b). The Niarbyl Formation does not match the major part of the Wenlock sequence of the Lake District, represented by the Brathay Formation (*centrifugus–lundgreni* Biozones), which is composed almost exclusively of laminated hemipelagite. However, the overlying Birk Riggs Formation (*lundgreni* Biozone) comprises sandstone–turbidite packets intercalated with units of hemipelagite that contain variable amounts of thin-bedded mud turbidites (Kneller *et al.* 1994). The sandstone-rich parts of the Birk Riggs Formation resemble the Niarbyl Formation south of the Knockaloe Moar Lineament, particularly as they are generally calcareous (Kneller 1990). The hemipelagite-rich parts of the Birk Riggs Formation, which may be up to 200 m thick, resemble the Niarbyl Formation north of the same lineament. The Austwick Formation is a possible, though longer ranging, equivalent in the Craven Inliers (Kneller *et al.* 1994). Another possible match for the northern part of the Niarbyl sequence is with the lowest Ludlow (*nilssoni* Biozone) Wray Castle Formation, also comprising laminated hemipelagite with silt–mud turbidites. However, this correlation is made less likely by the apparent absence in the Niarbyl Formation of an equivalent of the distinctive Coldwell Formation of the Lake District (uppermost Ludlow, *nassa–ludensis* Biozones). This heavily bioturbated, locally shelly, calcareous siltstone intervenes between the Birk Riggs and Wray Castle Formations, and its absence in the Niarbyl Formation suggests that levels higher than the *lundgreni* Biozone are not preserved there.

There are a number of specific, as well general, similarities between all of these sequences and that of the Niarbyl Formation, in particular: mixed hemipelagite–sandy facies assemblages; turbidite systems architecture; a trend towards more mud–silt dominant facies in the upper part of all sequences and an indicated dissected magmatic-arc source for Balbriggan and Niarbyl turbidite sands.

Such similarities are entirely consistent with Hutton & Murphy's (1987) successor-basin hypothesis, as each of the successions are merely geographically isolated outcrops of what was once a single, contiguous system. Favouring comparison with one area rather than another is therefore rather irrelevant, although gross patterns obviously contribute to a better understanding of the architecture of the basin system. In that context, it may be noted that the indicated sinistral shear vector on the sole Niarbyl Shear Zone indicates a post-Wenlock displacement of the Niarbyl Formation from a position to the northeast of its present outcrop, a direction closer to the current location of the Riccarton Group in the Southern Uplands (Fig. 7a). The magnitude of that displacement is presently unknown, as is the corollary displacement away from the present position of the Balbriggan sequence along strike to the southwest.

The relationship with the Lake District and its equivalent successions is more complex, as the entire bulk of the Ordovician age, Avalonian margin magmatic-arc complex, and fringing sedimentary sequences (the Borrowdale, Skiddaw and Manx Groups), now lie between the Niarbyl and Birk Riggs Formations (Fig. 7a). It is unlikely that the present orthogonal relationships reflect those that existed during the Wenlock as terrane-bounding, as well as sinistral transpressive pre-Wenlock and Acadian, strike faults are well documented in the region (Hutton & Murphy 1987; Murphy 1984; Murphy *et al.* 1991). Comparable faults are now recognized on the Isle of Man, including the Niarbyl Shear Zone and the newly discovered Lag ny Keeilley Shear Zone (Fitches *et al.* 1999). It is consequently prudent to assume that there has been some degree of strike displacements, almost certainly sinistral given general regional patterns (e.g. Hutton & Murphy 1987).

A further complication is provided by the late stage, top-to-the southeast thrust vector on the Niarbyl Shear Zone. While the age of this thrusting is uncertain, either Acadian or possibly even Variscan (Fitches *et al.* 1999), it does imply southeast directed tectonic foreshortening, across the orogen, perhaps as an imbricate thrust stack, which has displaced the Niarbyl Formation towards the present location of the both the Manx Group and the Lake District (Fig. 7a). It also implies that the Niarbyl and Birk Riggs Formations are now closer to each other than they were during the Wenlock.

There are some differences between the Wenlock successor basin sequences, particularly contrasting palaeocurrent trends at equivalent stratigraphic levels. A pronounced northwest–west-northwest source during the *lundgreni* Biozone interval in the Niarbyl Formation contrasts with a dominant southwest directed vector in the Birk Riggs Formation (Kneller *et al.* 1994), and axial or southeast vectors during the same time interval in Balbriggan and the Southern Uplands. Northwest source vectors in the latter areas either occur immediately after the *lundgreni* Biozone interval, as at Balbriggan, or before it, as in the Southern Uplands. This regional pattern may be interpreted in several ways. It may reflect establishment of the increasingly dominant northwest source region at slightly different Wenlock times alongstrike, as well as varying influence of supply input points along the northwest and southeast margin slope apron depositional systems. Kemp (1987) specifically applies this interpretation to the Riccarton Group, noting that the variable palaeoflow vectors primarily reflect the influence of lateral supply points and orientation of specific distributary systems. It is considered here that the Dalby Group represents an isolated and areally restricted portion of that system.

Conclusions

There are a number of significant conclusions arising from this study:

- the inferred minimum 1250 m thick Niarbyl Formation, as formally defined here, is of Silurian, mid–late Wenlock age (Howe 1999), rather than early Ordovician. All previous stratigraphic correlations with the Manx Group may therefore be discounted, as may reliance on the correlation of the Niarbyl and Lonan Flags to define the Isle of Man Synclinorium;
- the Niarbyl Flags are formally redefined as the Niarbyl Formation, removing them from the otherwise entirely early Ordovician Manx Group and assigning them to a new group – the Dalby Group;
- laminated hemipelagite, characteristic of Wenlock and early Ludlow basinal successions elsewhere in the Caledonides is described. This facies contains a sparse graptolite–orthocone fauna, described in detail by Howe (1999). Quartzites, debrites and manganiferous ironstones, which characterize various Manx Group formations, are conspicuously absent;
- the turbidite-dominated sequence is interpreted as a mid–lower fan succession composed of mainly non-channelized interlobe and sandy lobe packets, which, on regional criteria, probably forms an isolated segment of a slope-apron system;
- the very mature, quartz–felsic igneous clast-rich turbidite sandstones were derived from magmatic-arc and recycled-orogen sources to the northwest of the present formation outcrop. Here, and elsewhere in the region, these systems become more mud- and silt-rich in the upper Wenlock, possibly reflecting rising sea levels and sediment entrapment on shelf margins;
- the Dalby Group is readily compared with other isolated remnants of the Silurian successor-basin sequence established between the Laurentian and Avalonian continental margins after docking at the end Ordovician: the Raeberry Castle Formation (Riccarton Group), Southern Uplands; the Birk Riggs Formation (Windermere Supergroup), Lake District; the Skerries and Denhamstown Formations, Balbriggan, Ireland;
- the base of the Niarbyl Formation is defined by a compound sinistral shear zone – southeast directed thrust with an inferred listric geometry. The shear zone, and D1 and D2, in the Niarbyl Formation are post-Wenlock in age and probably reflect early Devonian Acadian deformation. There is no evidence to indicate whether or not D1 in the Manx Group is the same age as D1 in the Dalby Group.

We wish to acknowledge the constructive and good humoured discussion with many individuals during this study funded by NERC Research Grant GR9/01834: Trevor Ford, in particular, for stimulating the research initiative; Dave Quirk for his enthusiasm – and for daring to go places where we would hesitate; and to all other members of the Isle of Man research group, Bill Fitches, Greg Power, Rob Barnes, Padhraig Kennan, Sean Mulligan and Dave Burnett. JHM acknowledges that his contribution is published with the permission of the Director, Geological Survey of Ireland, and we thank Charles Morrison and Barry Murphy for permission to cite from their unpublished PhD theses.

References

BARNES, R. P., LINTERN, B. C. & STONE , P. 1989. Timing and regional implications of deformation in the Southern Uplands of Scotland. *Journal of the Geological Society, London*, **146**, 905–908.

COOPER, A. H., RUSHTON, A. W. A., MOLYNEUX, S. G., HUGHES, R. A., MOORE, R. M. & WEBB, B. C. 1995. The stratigraphy, correlation, provenance and palaeogeography of the Skiddaw Group (Ordovician) in the English Lake District. *Geological Magazine*, **132**, 185–211.

CUMMING, J. G. 1846. On the geology of the Isle of Man. *Quarterly Journal of the Geological Society, London*, **2**, 317–348.

—— 1848. *The Isle of Man; Its History, Physical, Ecclesiastical, Civil and Legendary.* Van Voorst.

DICKINSON, W. R. 1970. Interpreting detrital modes of graywacke and arkose. *Journal of Sedimentary Geology*, **40**, 695–707.

——, BEARD, L. S., BRAKENRIDGE, G. R., ET AL. 1983. Provenance of North American Phanerozoic sandstones in relation to tectonic setting. *Geological Society of America Bulletin*, **94**, 222–235.

DIMBERLINE, A. J., BELL, A. & WOODCOCK, N. H. 1990. A laminated hemipelagic facies from the Wenlock and Ludlow of the Welsh Basin. *Journal of the Geological Society London*, **147**, 693–701.

DOWNIE, C. & FORD, T. D. 1966. Microfossils from the Manx Slate Series. *Proceedings of the Yorkshire Geological Society*, **35**, 307–322.

FITCHES, W. R., BARNES, R. P. & MORRIS, J. H. 1999. Geological structure and tectonic evolution of the Lower Palaeozoic rocks of the Isle of Man. *This volume.*

GALLAGHER, V., O'CONNOR, P. J. & AFTALION, M. 1994. Intra-Ordovician deformation in SE Ireland: Evidence from the geological setting, geochemical affinities and U–Pb zircon age of the Croghan Kinshelagh granite. *Geological Magazine*, **131**, 669–684.

HARKNESS, R. & NICHOLSON, H. 1866. On the Lower Silurian rocks of the Isle of Man. *Quarterly Journal of the Geological Society, London*, **22**, 488–491.

HOWE, M. P. A. 1999. The Silurian fauna (graptolite and nautiloid) of the Niarbyl Formation, Isle of Man. *This volume.*

HUTTON, D. W. & MURPHY, F. C. 1987. The Silurian of the Southern Uplands and Ireland as a successor basin to the end-Ordovician closure of Iapetus. *Journal of the Geological Society, London*, **144**, 765–772.

INGERSOLL, R. V. & SUCZEK, C. A. 1979. Petrology and provenance of Neogene sand from Nicobar and Bengal fans, DSDP sites 211 and 218. *Journal of Sedimentary Petrology*, **49**, 1217–1228.

——, KRETCHMER, A. G. & VALLES, P. K. 1993. The effect of sampling scale on actualistic sandstone petrofacies. *Sedimentology*, **40**, 937–953.

——, BULLARD, T. F., FORD, R. L. & PICKLE, J. D. 1985. The effect of grain size on detrital modes: a test of the Gazzi–Dickinson point-counting method – reply to discussion of Lee J. Suttner and Abhijit Basu. *Journal of Sedimentary Petrology*, **55**, 617–618.

——, ——, ——, GRIMM, J. P., PICKLE, J. D. & SARES, S. W. 1984. The effect of grain size on detrital modes: a test of the Gazzi-Dickinson point-counting method. *Journal of Sedimentary Petrology*, **54**, 103–116.

KEMP, A. E. S. 1986. Tectonstratigraphy of the Southern Belt of the Southern Uplands. *Scottish Journal of Geology*, **22**, 241–256.

—— 1987. Evolution of Silurian depositional systems in the Southern Uplands, Scotland. *In*: LEGGETT, J. K & ZUFFA, G. G. (eds) *Marine Clastic Sedimentology: Concepts and Case Studies.* Graham & Trotman, 124–155.

—— 1991. Mid Silurian pelagic and hemipelagic sedimentation and palaeoceanography. *Special Papers in Palaeontology London*, **44**, 261–299.

KING, L. M. 1992. *A basin study of the early Palaeozoic Windermere Group, NW England.* PhD Thesis, University of Cambridge.

KNELLER, B. C. 1990. *The Wenlock rocks of Sheet 38 (Ambleside).* British Geological Survey Technical Report, WA/90/64.

——, SCOTT, R. W., SOPER, N. J., JOHNSON, E. W. & ALLEN, P. M. 1994. Lithostratigraphy of the Windermere Supergroup, Northern England. *Geological Journal*, **29**, 219–240.

LAMPLUGH, G. W. 1903. *The Geology of the Isle of Man.* Memoir of the Geological Survey, United Kingdom.

MARR, J. E. 1927. The deposition of the later Silurian rocks of the Lake District. *Geological Magazine*, **64**, 494–500.

MIDDLETON, G. V. & HAMPTON, M. A. 1976. Subaqeous sediment transport and deposition by sediment gravity flows. *In*: STANLEY, D. J. & SWIFT, D. J. P. (eds) *Marine Sediment Transport and Environmental Management.* Wiley, 197–218.

MOLYNEUX, S. G. 1979. New evidence for the age of the Manx Group, Isle of Man. *In*: HARRIS, A. L., HOLLAND, C. H. & LEAKE, B. E. (eds) *The Caledonides of the British Isles – Reviewed.* Geological Society, London, Special Publications, **8**, 415–421.

—— 1999. A reassessment of Manx Group acritarchs, Isle of Man. *This volume.*

MORRISON, C. W. K. 1989. *A study of the Anchizone–Epizone metamorphic transition on the Isle of Man.* PhD Thesis, University of St Andrew's.

MURPHY, F. C. 1984. *The Lower Palaeozoic stratigraphy and structural geology of the Balbriggan Inlier, Counties Meath and Dublin.* PhD Thesis, University of Dublin.

—— & HUTTON, D. W. 1986. Is the Southern Uplands of Scotland really an accretionary prism? *Geology*, **14**, 354–357.

——, ANDERSON, T. B., DALY, J. S. ET AL. 1991. An appraisal of Caledonian suspect terranes in Ireland. *Irish Journal of Earth Sciences*, **11**, 11–41.

MUTTI, E. & NORMARK, W. R. 1987. Comparing examples of modern and ancient turbidite systems: Problems and concepts. *In*: LEGGETT, J. K. & ZUFFA, G. G. (eds) *Marine Clastic Sedimentology: Concepts and Case Studies.* Graham & Trotman, 1–38.

PICKERING, K. T., STOW, D. A. V., WATSON, M. & HISCOTT, R. N. 1986. Deep-water facies, processes and models: a review and classification scheme for modern and ancient sediments. *Earth Science Reviews*, **22**, 75–174.

PIPER, D. J. W. & NORMARK, W. R. 1983. Turbidite depositional patterns and flow characteristics, Navy submarine fan, California borderland. *Sedimentology*, **30**, 681–694.

RICKARDS, R. B. 1965. The graptolitic mudstone and associated facies in the Silurian strata of the Howgill Fells. *Geological Magazine*, **101**, 435–451.

——, BURNS, V. & ARCHER, J. 1973. The Silurian sequence at Balbriggan, Co. Dublin. *Proceedings of the Royal Irish Academy*, **73**, 303–316.

RUSHTON, A. W. A. 1993. Graptolites from the Manx Group. *Proceedings of the Yorkshire Geological Society*, **49**, 259–262.

SIMPSON, A. 1963. The stratigraphy and tectonics of the Manx Slate Series, Isle of Man. *Quarterly Journal of the Geological Society, London*, **119**, 367–400.

—— 1968. The Caledonian history of the north-eastern Irish Sea region and its relation to surrounding areas. *Scottish Journal of Geology*, **4**, 135–163.

STOW, D. A. V. 1985. Fine-grained sediments in deep-water – an overview of processes and facies models. *Geo-marine Letters*, **5**, 17–23.

——., READING, H. G. & COLLINSON, J. D. 1996. Deep Seas. *In*: READING H. G. (ed.) *Sedimentary Environments: Processes, Facies and Stratigraphy*. Blackwell.

SUCZEK, C. A. & INGERSOLL, R. V. 1985. Petrology and provenance of Cenozoic sand from the Indus Cone and the Arabian Basin, DSDP sites 221, 222 and 224. *Journal of Sedimentary Petrology*, **55**, 340–346.

TAYLOR, J. 1864. The Cambrian strata of the Isle of Man. *Transactions of the Manchester Geological Society*, **4**, 70–78.

THORNTON, S. E. 1984. Basin model of hemipelagite sedimentation in a tectonically active continental margin. *In*: STOW, D. A. V. & PIPER, D. J. W. (eds) *Fine-grained Sediments: Deep-water Processes and Facies*. Geological Society, London, Special Publications, **15**, 377–394.

TIETZSCH-TYLER, D. 1996. Precambrian and early Caledonian orogeny in south-east Ireland. *Irish Journal of Earth Sciences*, **15**, 19–39.

TORSVIK, T. H., TRENCH, A., SVENSSON, I & WALDERHAUG, H. 1993. Mid Silurian palaeomagnetic results from the East Mendips Inlier, Southern Britain: Palaeogeographic significance and major revision of the apparent polar wander path for Eastern Avalonia. *Geophysical Journal International*, **113**, 651–668.

WOODCOCK, N. H., MORRIS, J. H., QUIRK, D. G. *ET AL.* 1999. Revised lithostratigraphy of the Manx Group, Isle of Man. *This volume.*

Palaeomagnetism of (Palaeozoic) Peel Sandstones and Langness Conglomerate Formation, Isle of Man: implications for the age and regional diagenesis of Manx red beds

J. D. A. PIPER & S. F. CROWLEY

Department of Earth Sciences, University of Liverpool, Liverpool L69 3BX, UK

Abstract: The Peel Sandstones are a sequence of fault-bounded fluvially deposited red beds of uncertain age exposed on the western coast of the Isle of Man. Palaeomagnetic study identifies partial preservation of an A component pre-dating deformation at four of eight sampled sites; the presence of both polarities in this remanence defines a positive class C reversal test. This component adjusted for tilt is *D*/*I* = 356.9/–47.1° (48 samples, α_{95} = 5.6°) and is interpreted to be a post-depositional detrital remanence. The palaeolatitude (29° S) corresponds to the position of Britain during Late Silurian–Early Devonian times and implies that these sediments are correlative with Lower Old Red Sandstone molasse from elsewhere in the British Caledonides; a post-mid-Carboniferous age is excluded. Palaeomagnetic and geological evidence are used to constrain an age of *c*. 410–400 Ma for the Peel Sandstones, whilst a declination contrast of 40–50° is interpreted to reflect either: (a) tectonic emplacement of a unit including the Peel Sandstone outcrop; or (b) wider post-Early Devonian and pre-Late Carboniferous counterclockwise rotation between the Isle of Man and the mainland during final shaping of the Caledonides. A post-folding B component of uniform reversed polarity (*D*/*I* = 195.6/–30.8°, 45 samples, α_{95} = 4.8°) is present at all sites and was acquired during diagenesis at *c*. 250 Ma.

Red sandstones within the Langness Conglomerate Formation, near the base of the Lower Carboniferous (Chadian–Arundian) succession exposed in the southeast of the Isle of Man, have a uniform reversed magnetization (*D*/*I* = 209.7/–24.2°, 20 samples, α_{95} = 6.8°). Like the B magnetization in the Peel Sandstones, this was acquired by diagenesis in Early Permian times (*c*. 250 Ma) and during the Permo-Carboniferous Reversed Superchron. It is temporally related to early extensional tectonism in the Irish Sea region.

The magnetic remanence preserved in red sediments is often complex and usually comprises a mixture of components acquired at different stages of the geological history. These may include a record of deposition and initial compaction preserved as a depositional remanent magnetization (DRM) or post-depositional detrital magnetization (PDRM), as well as a record imparted by diagnetic events which can be much later and record chemical remanent magnetizations (CRM). As a consequence, palaeomagnetism is finding an expanding role in the study of sedimentary rocks with potential applications to all stages of their geological evolution (Turner & Turner 1995).

This paper reports a palaeomagnetic investigation of two Palaeozoic red bed sequences on the Isle of Man. In the first example, the Peel Sandstones, the age of the sediments is only loosely defined by the occurrence of derived Ordovician and Silurian (Wenlock) fossils in conglomeratic facies (Gill 1903; Lewis 1934); hence, this study was designed to improve the constraints placed on the depositional age by the very limited lithostratigraphic and palaeontological evidence (Allen & Crowley 1983). In the second example, the Langness Conglomerate Formation, the depositional age is well constrained within the Early Carboniferous [Chadian–Arundian; see Dickson *et al.* (1987)] and the palaeomagnetic investigation has therefore focused on dating the diagenetic event responsible for the red pigment and effecting a comparison with the Peel Sandstones.

In each case, samples were drilled with a portable motorized coring device and oriented by sun and magnetic compasses. Sample locations are summarized in Table 1. The sample suite at each site typically comprised 10–15 cores spanning several metres of exposure. In the laboratory, cores were sliced into 2.2 cm long cylinders and their magnetizations measured using a SQUID magnetometer. Each core was subjected to stepwise thermal demagnetization. Steps employed were 100–500°C and were decreased to 30 or 15° between 500 and *c*. 680°C, the latter temperature being above the Curie

From: WOODCOCK, N. H., QUIRK, D. G., FITCHES, W. R. & BARNES, R. P. (eds) 1999.
In Sight of the Suture: the Palaeozoic geology of the Isle of Man in its Iapetus Ocean context.
Geological Society, London, Special Publications, **160**, 213–225. 1-86239-046-0/99/$15.00

Table 1. *Summary of palaeomagnetic sampling sites*

Site No.	Location [UK grid reference]	Description
1 and 2	Langness [SC 2811 6543]	Massive lenticular red litharenite
3 and 4	Traie Fogog Bay [SC 2523 8466]	Parallel laminated sublitharenite
5 and 6	Cain's Strand [SC 2626 8527]	Parallel laminated sublitharenite
7 and 8	White Strand [SC 2646 8541]	Parallel laminated sublitharenite
9	Creg Malin [SC 2502 8444]	Parallel laminated sublitharenite
10	Traie Fogog Bay [SC 2516 8463]	Intraformational mudstone clasts overlying scoured erosional surface

point of magnetite and close to the Curie point of hematite. All demagnetization was undertaken within large sets of Rubens coils in which the ambient field was continuously nullified to minimize secondary remanence acquisition. Component structures of the magnetizations were identified by study of orthogonal projections of the remanence vector; the separate components were then isolated and their directions calculated by principal component analysis (Sherwood 1988).

The Peel Sandstones

Geological framework

The Peel Sandstones comprise a succession of fluvial (braided river and distal/fan alluvial plain depositional environments), predominantly coarse-grained (gravel and sand facies) red sediments which crop out in a series of approximate dip sections along the coast northeast of Peel (Fig. 1a). The succession dips predominantly to the north at angles of 10–50° and is repeatedly dissected by normal faults with displacements ranging from a few centimetres to (?)tens of metres. Evidence for folding is restricted to a single open anticlinal structure developed at Cain's Strand (Fig. 1a), drag folds related to motion on steep normal faults and to shortening structures associated with local thrusts. Although high-strain zones with a localized weak cleavage occur in association with the thrust structures, there is no evidence for systematic development of cleavage in these sediments.

Contact relationships between the Peel Sandstones and the surrounding Lower Palaeozoic turbidites (Early Ordovician Manx Group and Silurian Dalby Group) are restricted to a single normal-faulted boundary at the northeastern limit of the exposure (Will's Strand; Fig. 1a). A further fault contact (possibly a transfer fault related to the Central Valley Lineament on the Isle of Man) is inferred to follow along the Neb River Valley (Fig. 1a) on the basis of published geophysical evidence (Quirk & Kimbell 1998). At present, it is assumed that the Peel Sandstones are unconformable on Manx or Dalby Groups' sediments, although no direct evidence is available to substantiate this. Alternatively, the occurrence of thrust structures within the Peel Sandstones may indicate that these sediments have been tectonically emplaced and that all contacts between the Peel Sandstones and Lower Palaeozoic rocks are fault controlled.

No formal stratigraphic nomenclature has been defined for the Peel Sandstones but, for convenience, the succession has been divided into four broad lithofacies associations (Fig. 1a) which constitute potentially mappable lithostratigraphic units. Due to a lack of any evidence for sedimentary relationships with the Lower Palaeozoic turbidites and an absence of geological contacts with younger rocks, there is no reliable estimate for the original thickness of the Peel Sandstones. Estimates based on the exposed lithostratigraphy (Fig. 1a) indicate a total exposed thickness of *c.* 350 m.

Although identification of a depositional age is crucial if the sedimentology of the Peel Sandstones is to be interpreted within an appropriate geotectonic setting, the age of sediment deposition remains uncertain despite a prolonged history of investigation. In the absence of an indigenous fauna or flora [although see Ford (1971)] it has proved impossible to define the age of the Peel Sandstones using conventional palaeontological methods. Previous suggestions for the age and/or affinity of these sediments, based on a variety of palaeontological, lithostratigraphic and structural criteria, include Lower Old Red Sandstone (Allen & Crowley 1983), Old Red Sandstone (Cumming 1846), Lower Carboniferous (Lamplugh 1903; Ford 1971), late Lower or (possibly) Upper Carboniferous (Lewis 1930), Late Carboniferous (Quirk & Kimbell 1998) and Permian (Boyd Dawkins 1902). At present, the only reliable evidence for the maximum age of the Peel Sandstones is provided by the occurrence of a diverse, derived marine fauna in clasts (both fossiliferous limestones and isolated fossils) recovered from gravelly sandstone and conglomerate facies (Facies

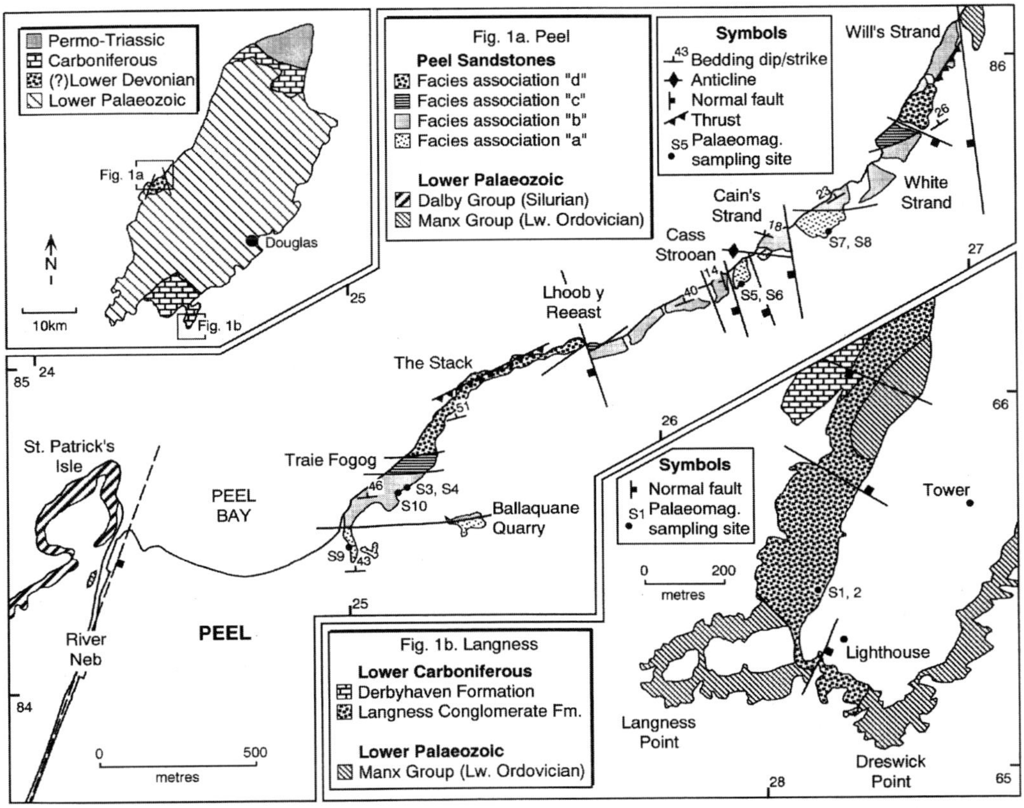

Fig. 1. Geological maps showing: (**a**) the outcrop of the Peel Sandstones; (**b**) the Langness Conglomerate Formation [after Dickson *et al.* (1987)]. The facies associations are: 'a', sandy braided river; 'b', ephemeral sheet flood; 'c', alluvial plain; 'd', distal alluvial fan. The palaeomagnetic sampling sites of this study are indicated as S1–S10. The inset shows the regional location.

Association d; Fig. 1a) exposed at The Stack and Whitestrand (Lewis 1934). Although subject to some dispute (Boyd Dawkins 1902; Gill 1903; Lamplugh 1903), this derived fauna is dominated by corals, bryozoans and brachiopods of Wenlock age (Lewis 1934). A Wenlock age for these clasts has subsequently been confirmed (Scrutton, pers. comm.) by examination of a small number of samples from a collection of fossiliferous material from The Stack. Palaeontologic and lithologic similarities between these clasts and the Wenlock (Homerian Stage) limestones of the Welsh Borderlands indicate the former existence of reefal carbonates of late Wenlock age in the central Irish Sea region. No evidence is currently available to permit a minimum age to be defined for the Peel Sandstones, but it is geologically reasonable to assume that they are pre-Jurassic given the absence of post-Triassic continental red bed sequences with well-developed calcrete profiles from the lithostratigraphy of the British Isles.

Palaeomagnetic study

Demagnetization results

A substantial part of the remanence in the Peel Sandstones has blocking temperatures above the Curie point of magnetite. A significant high-blocking temperature fraction is therefore resident in hematite grains of either detrital or authigenic specularite. The lower blocking temperature components are of variable magnitude (cf. samples 3-10, 4-1, 8-12 and 10-7 in Fig. 2) and tend to be removed by *c.* 550°C. Component declinations are southerly and inclinations are negative (Fig. 3); these directions of reversed magnetization are referred to as the B population. They do not converge to the origin of the projection and a higher blocking temperature component is also clearly present. However, in most samples at sites 4, 7, 8 and 10 trajectories tend to become erratic at higher temperatures (580–680°C) probably because

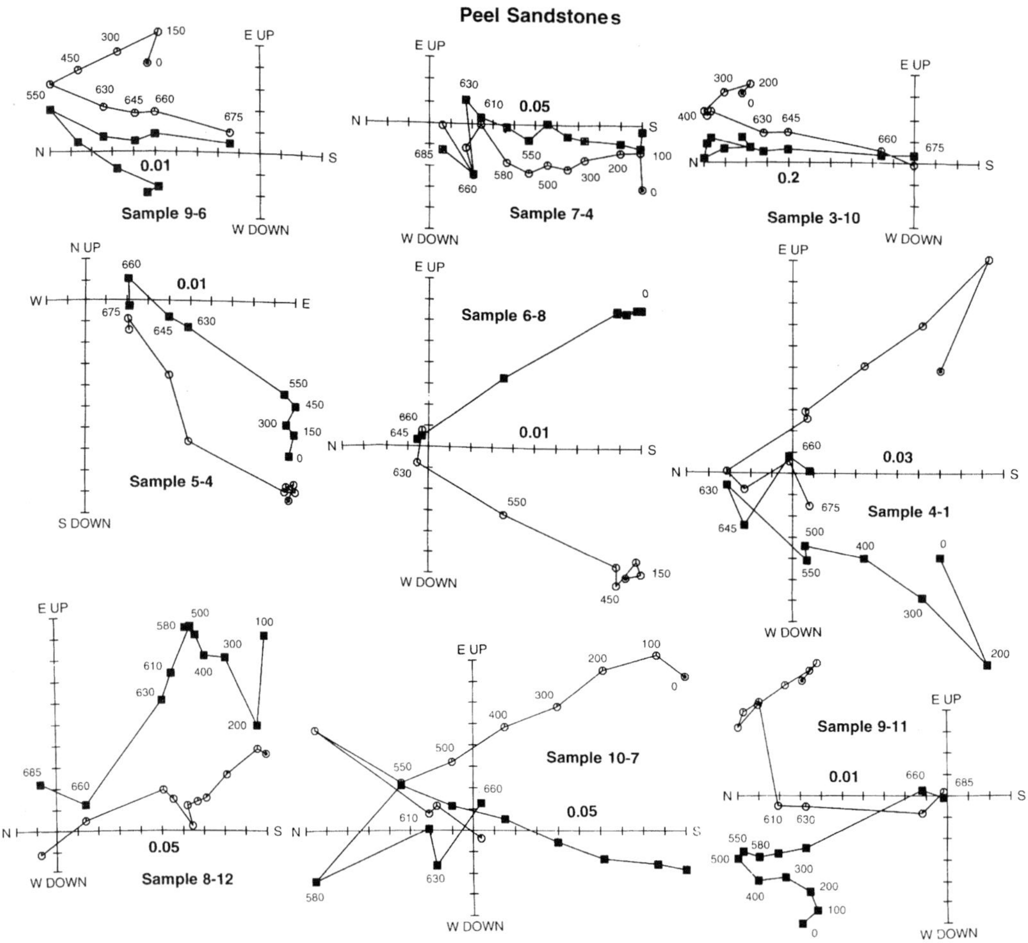

Fig. 2. Orthogonal projections showing representative palaeomagnetic results (*in situ*) derived from thermal demagnetization of Peel Sandstones. Palaeomagnetic vectors are projected on to the horizontal (■) and vertical (○) planes. Intensity values are × 10^{-5} A m^2 kg^{-1}.

magnetomineralogic changes are induced by the thermal treatment; the examples of samples 4-1, 7-4, 8-12 and 10-7 in Fig. 2 are illustrative of the poor, or unacceptable, resolution of components above 550–580°C.

The high-blocking temperature components are successfully isolated by the thermal demagnetization between 610 and 680°C in most cores from sites 3, 5, 6 and 9 (Table 2). At sites 3 and 9 the B component is relatively small and the high-blocking temperature component is of northerly shallow direction (samples 3-10, 9-6 and 9-11 in Fig. 2). Analogous components, referred to as the A component, are present as a smaller fraction of the remanence at sites 4 and 10, although their definition is poor (samples 4-1 and 10-7 in Fig. 2).

Sites 5 and 6 are dominated by single components with southeasterly positive directions which are only fully unblocked close to the Curie point of hematite (samples 5-4 and 6-8 in Fig. 2). These appear to be of opposite polarity to the components resolved at sites 3 and 10, a point which is more evident when the directions are adjusted for tilt (*c.* 44° NW at sites 3 and 10 cf. with *c.* 17° WNW at sites 5 and 6). Samples at site 7 experienced chemical change above 500°C and only the B component could be isolated. Site 10 comprises cores from seven intraformational clasts of red siltstone; unfortunately all clasts experienced some chemical change above 550°C following recovery of the B component. Hence, B is identified as a coherent chemical overprinted magnetization

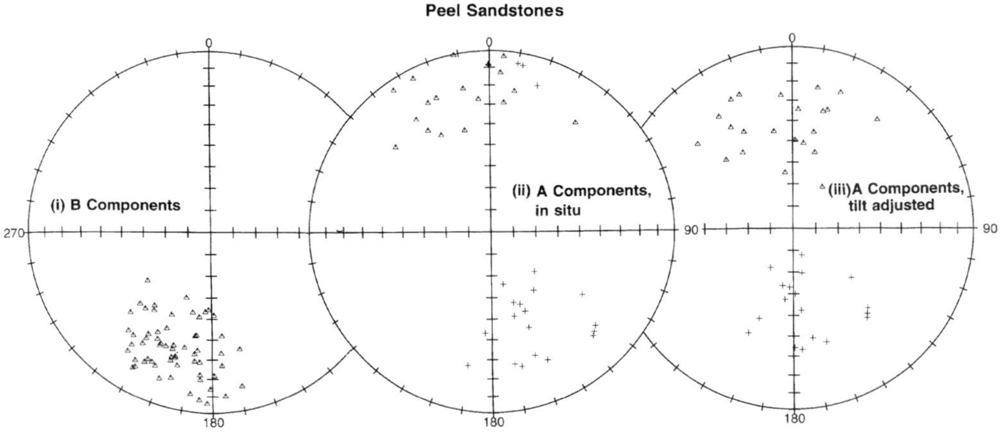

Fig. 3. Equal-area projections showing directions of A and B components derived from the Peel Sandstones. Plots are on the lower hemisphere (+) and projections on to the upper hemisphere(△).

whereas the origin of A is not resolved by this test.

Fold and reversal tests

A palaeomagnetic fold test applied to the A remanence, according to the method of McFadden (1990), is summarized in Table 3. The test statistic (ξ) for the sample grouping in the *in situ* position (32.95) is much higher than its 95% confidence value (8.06), whereas for the unfolded position it is about the same. This suggests that there is correlation between the magnetic directions and the bedding orientation in the in situ position, but the effect of tilt adjustment is unclear. Since all but four of the A components come from sites 3, 5, 6 and 9, the fold test may be conducted satisfactorily on the mean directions of these four sites. In this case, the test statistic for the *in situ* position (3.82) is higher than its critical value (2.34) but the value for the unfolded position (0.85) is much less. This confirms correlation between the magnetization directions and the bedding orientation in the *in situ* position, but identifies no significant correlation in the unfolded position. Hence it can confidently be concluded that the A magnetization in the Peel Sandstones pre-dates the deformation. Furthermore, since precision is only optimized by complete unfolding, there is no palaeomagnetic case for remanence acquisition during the deformation episode.

The angle between the normally magnetized direction (sites 3 and 9) and the reversely magnetized direction (sites 5 and 6) is 7.5°. This is less than the critical angle of 12.1° and yields a positive reversal test of classification C according to the scheme of McFadden & McElhinny (1990). The fold and reversal tests and high thermal stability of the A remanence imply residence in specular hematite grains largely, or wholly, of detrital origin. If this is the case, then the palaeomagnetic direction will be representative of the time of deposition of these sediments.

In contrast, the grouping of the B magnetization deteriorates with tilt adjustment (α_{95} increases from 4.8 to 9.6°). A fold test is not diagnostic in this case (Table 3) but provides no case for rejecting the conclusion that this remanence is post-folding. The *in situ* direction may therefore be interpreted in the context of the palaeofield migration path for Britain.

Interpretation of the A magnetization

In Fig. 4 the palaeolatitude from the Peel A magnetization is compared with the palaeolatitude of Britain during Palaeozoic times as resolved from the modern data set (Piper 1997*a*). It is observed that the palaeolatitude of 29° S is too high to accord with Early Silurian or later Lower Devonian results, as well as with the palaeolatitude in subsequent times. It is therefore most compatible with a Late Silurian–Early Devonian age, a correlation which implies that the dominant red clastic fluviatile facies in the Peel Sandstones merit comparison with the Lower Old Red Sandstone facies elsewhere in the British Isles. The second possibility is that the Peel Sandstones correlate with the excursion of Britain back into higher southerly palaeolatitudes during Early Carboniferous times

Table 2. *Site and group mean palaeomagnetic results from the Peel Sandstones, Isle of Man*

Site/component		N	R	k	α_{95}	D (*In situ*)	I (*In situ*)	Tilt	D (Tilt adjusted)	I (Tilt adjusted)
3	B	6	5.60	12.5	19.7	202.0	–40.9	44/347		
	A	12	11.39	18.1	10.5	357.6	–7.0	44/347	3.5	–49.9
4	B	7	6.93	85.0	6.6	207.7	–34.4	44/347		
5	A	10	9.57	20.8	10.9	159.9	48.4	17/291	179.3	57.4
6	B	8	7.72	24.9	11.3	181.3	–21.1	17/291		
	A	9	8.54	17.3	12.7	155.8	34.6	17/291	166.9	45.5
7	B	7	6.76	25.5	12.2	198.4	–28.0	14/354		
8	B	7	6.76	24.9	12.3	190.7	–27.0	14/355		
9	B	13	12.52	25.2	8.4	201.7	–32.5	44/355		
	A	13	12.29	17.0	10.4	348.9	3.2	44/355	347.0	–40.5
10	B	18	17.38	27.6	6.7	200.4	–33.0	47/346		

Group mean calculations

B components, *in situ*

	N	R	k	α_{95}	D	I
66 samples	66	63.02	21.8	3.8	197.7	–31.4

Palaeomagnetic pole: 148.2° E, 50.3° N dp/dm = 2.4/4.3°

A components, tilt adjusted*

	N	R	k	α_{95}	D	I
48 samples	48	44.76	14.5	5.6	356.9	–47.1

Palaeomagnetic pole: 358.0° E, 7.6° S *dp*/*dm* = 4.7/7.2°

4 sites 3.96 78.8 10.4 353.5 –48.6

Palaeomagnetic pole: 1.0° E, 6.2° S *dp*/*dm* = 9.0/13.7°

D and *I*, mean declination and inclination, respectively, derived from characteristic remanent magnetizations isolated by thermal cleaning in *N* samples; α_{95} radius of the cone of 95% confidence about the mean direction; *k*, estimate of the Fisher precision parameter [= $(N-1)/(N-R)$]; *dp* and *dm*, radii of the oval of confidence about the derived pole position in the co-latitude direction and at right angles to it, respectively.

*Cores from sites 3, 5, 6 and 9 plus four isolated A magnetizations from the remaining sites.

(Piper *et al.* 1991; Fig. 4). This possibility is less likely for two reasons: firstly, the declination of the field axis in northern England was then 25/205° E; the difference of *c.* 30° from the A magnetization could only be explained by postulating large post-Lower Carboniferous differential rotations across northern Britain. Secondly, the maximum latitude recorded in Early Carboniferous rocks is 20° S, which is appreciably less than that indicated by the A magnetization in the Peel Sandstones.

Although palaeolatitudes of *c.* 30°S of Ordovician–early Silurian age are observed in both the Orthotectonic and Paratectonic Caledonides (Piper 1998), the absence of a regional cleavage development in the Peel Sandstones precludes a comparison with rocks of pre-Late Silurian age. The climactic cleavage-producing deformation in northwest England is older than the Shap and Skiddaw Granites dated at 399–394 Ma (Soper *et al.* 1987); these rocks yield palaeolatitudes of

Table 3. *Fold test statistical parameters for A and B magnetizations in the Peel Sandstones*

N		ξ_{95}	ξ_b	ξ_a	ξ_{max}	X%	ξ_{min}	Y%
A magnetizations								
48	samples	8.059	32.954	8.671	8.671	100	0.606	100
4	sites	2.335	3.827	0.902	0.663	100	0.505	100
B magnetizations								
7	sites	3.086	5.212	5.933	0.413	38	0.016	72

N, number of samples or sites used for the test; ξ_{95}, critical value of the test statistic ξ, $\xi_{b(a)}$, value of the test statistic *in situ* (after 100% unfolding); ξ_{max}, value of the statistic at the maximum precision (*k*) position; *X*%, percentage of complete unfolding required for maximum *k*, ξ_{min}, minimum ξ statistic; *Y*%, percentage of complete unfolding required for minimum ξ statistic.

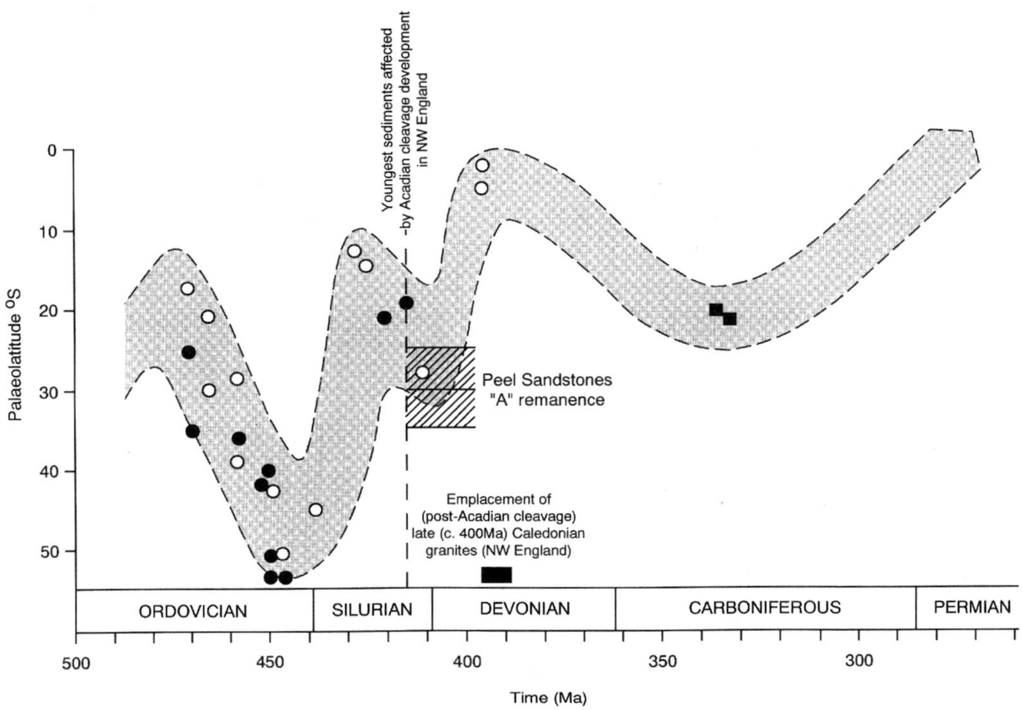

Fig. 4. Palaeolatitude path of Britain derived from a modern data set summarized in Piper (1997*a*) and comprising results from Lake District (●) and Welsh Basin (○) sectors of the paratectonic Caledonides plus two post-Caledonian results (■); specific mid Carboniferous and younger results are not shown. The palaeolatitude derived from the Peel Sandstone A magnetization with 95% confidence limits is indicated.

< 10°, showing that Britain had moved into lower latitudes by late Early Devonian times (Piper 1997*b*). Further north, in the Southern Uplands, major deformation had ceased by about mid-Wenlock times and Barnes *et al.* (1989) propose that deformation was diachronous across Britain, taking place progressively later to the south as successive terrane elements were accreted. According to this model a likely maximum age of *c.* 415 Ma for end-Caledonian deformation in Northern England would also be a maximum possible age for the Peel Sandstones (Fig. 4). Since the Lower Palaeozoic Manx and Dalby Groups are both affected by Acadian cleavage (Woodcock *et al.* 1999; Fitches & Barnes 1999), rapid uplift and erosion is implied before the unconformable or tectonic emplacement of the Peel Sandstones.

On the other hand movement of Britain into low latitudes by *c.* 395 Ma (Fig. 4) would appear to limit this entire cycle to a period of 10–15 Ma. In the Lake District this interval includes detachment of the Windermere Supergroup basin from its basement in late Ludlow times and subsequent inversion (Kneller *et al.* 1993). The basin migrated southwards on to the Avalonian foreland through early Devonian times ahead of an advancing orogen, with motion finally terminating in Emsian times. This stage is considered to define the end of Acadian deformation in Northern England (Soper *et al.* 1987) and commenced at *c.* 400 Ma (Tucker & McKerrow 1995).

Palaeomagnetic results from Old Red Sandstone rocks elsewhere in Britain define a palaeofield axis of northeast–southwest orientation (Tarling 1985; Piper 1988), contrasting with the north–south declination of the Peel Sandstone A remanence. Modern results, based on detailed thermal demagnetization and principal component analysis include a mean reversed magnetization direction of *D*/*I* = 232/32° from Lower Old Red Sandstones of the Anglo-Welsh cuvette (Channell *et al.* 1992), which pre-dates Acadian deformation in this region. The equivalent palaeolatitude (λ) is 17° S and compares with a further result from the same region which yielded a mean *D*/*I* = 246/38° (λ = 21° S) (Setiabudidaya *et al.* 1994). Lower Old Red

Sandstone lavas from the Midland Valley of Scotland yield a mean of D/I = 225/46° (λ = 27.5° S) Torsvik 1985). The stratigraphic placement of these results assigned to the Lower Old Red Sandstone has been considered to be close to the Silurian–Devonian boundary. In contrast, early Silurian sediments of the the Windermere Supergroup yield a shallower inclination of D/I = 223/24° and λ = 12° S) (Channell *et al.* 1993.

Thus, there is a declination contrast of 50–60° between the Peel Sandstone A remanence and Lower Old Red Sandstone magnetizations on mainland Britain, implying that either this correlation is invalid or that the Isle of Man has been rotated counterclockwise by this amount since Early Devonian times. The first possibility seems unlikely because there is no correlation between the A remanence declination and other post-Early Silurian data until the Late Carboniferous, by which time Britain was crossing the equator into the northern hemisphere (Fig. 5). Thus, the authors prefer to regard the declination of the A remanence as a record of later rotation. Although palaeomagnetic inclinations are accordant across the British Isles from Early Silurian times, and record closure of Iapetus by these times (Soper & Hutton 1984), more subtle differences in declination persist until at least Devonian times (Piper 1997*b*) and appear to record, at least in part, the effects of final shaping of the orogen. Specifically, it is noted that late Ordovician palaeomagnetic directions from the Welsh Caledonides are rotated anticlockwise by 55° from contemporaneous directions in the Lake District (Piper 1997*a*), whilst the Poortown Dolerite on the Isle of Man, also of probable late Ordovician age, is rotated counterclockwise by *c.* 40° from the late Ordovician vector in North Wales (Piper *et al.* 1999). Thus, it is possible that the Late Ordovician anomaly persisted into Early Devonian times and records a later regional rotation of the Isle of Man completed by Late Carboniferous times. Since contact relationships between the Peel Sandstones and the Manx and Dalby Groups are unexposed, and remain enigmatic, it is also possible that the outcrop is part of an allochthonous block which was rotated during emplacment.

In Fig. 5 the pole position derived from the A magnetization is plotted before and after clockwise rotation of the mean remanence direction by 40 and 90°. It is observed that the pole position before rotation coincides only with Late Ordovician poles; rotation by 40° moves the pole close to Old Red Sandstone poles from the Welsh Borderlands. Rotation by 90° moves the declination towards late Early Devonian results, although the pole is by then removed from other data because the inclination is too steep and therefore the palaeolatitude estimate is too high (Fig. 4). Consequently, it is concluded that the A remanence is older than late Lower Devonian in age and was rotated counterclockwise prior to Late Carboniferous times.

Although red beds of Permo-Triassic age crop out on the adjoining British mainland (St Bees and Penrith Sandstone Formations) and are present offshore, the presence of the A magnetization excludes the possibility that the Peel Sandstones are as young as this. Furthermore, a Permo-Triassic age would require magnetic inclinations of ≥ –40° (southerly) or ≥ +40° (northerly), which is observed in only a small minority of cores (Fig. 3). The B magnetization in the Peel Sandstone is interpreted below, together with a comparable remanence identified in the Langness Conglomerate Formation.

The Langness Conglomerate Formation

Geological framework

This Langness Conglomerate Formation [see Dickson *et al.* (1987) for details] comprises 30 m of coarse, predominantly conglomeratic, clastic sediments in the basal unit of a gently dipping Lower Carboniferous succession exposed along the southeast coast of the Isle of Man (Fig. 1). It underlies the Derbyhaven Formation of Arundian age and is therefore of early Dinantian, possibly Chadian, age (Dickson *et al.* 1987). The palaeomagnetic sample comprised 28 cores from two sites (Table 1) in lenticular red litharenite bodies between conglomerates.

Red coloration in the lower part of the conglomerate is due to hematite in the matrix and hematite-stained greywacke and pelite clasts derived locally from Manx Group metasediments. There are two possible origins for the pigment:

- it may be penecontemporaneous because grey coloured rudites, which occur higher in the Langness Conglomerate Formation, contain dolomite and pyrite but no hematite. The mineralogical change in the Lower Carboniferous succession would then reflect either a change in redox potential of the depositional environment occurring concurrently with the change in textural and compositional maturity, or represent later formation of dolomite and pyrite when pores of the younger conglomerates were flooded by the transgressive sea in which the overlying marine carbonates were deposited (Dickson *et al.* 1987).
- alternatively, the Langness Conglomerate Formation may have provided a conduit for flushing by fluids either during the Variscan orogenic episode (e.g. McCabe & Elmore 1989)

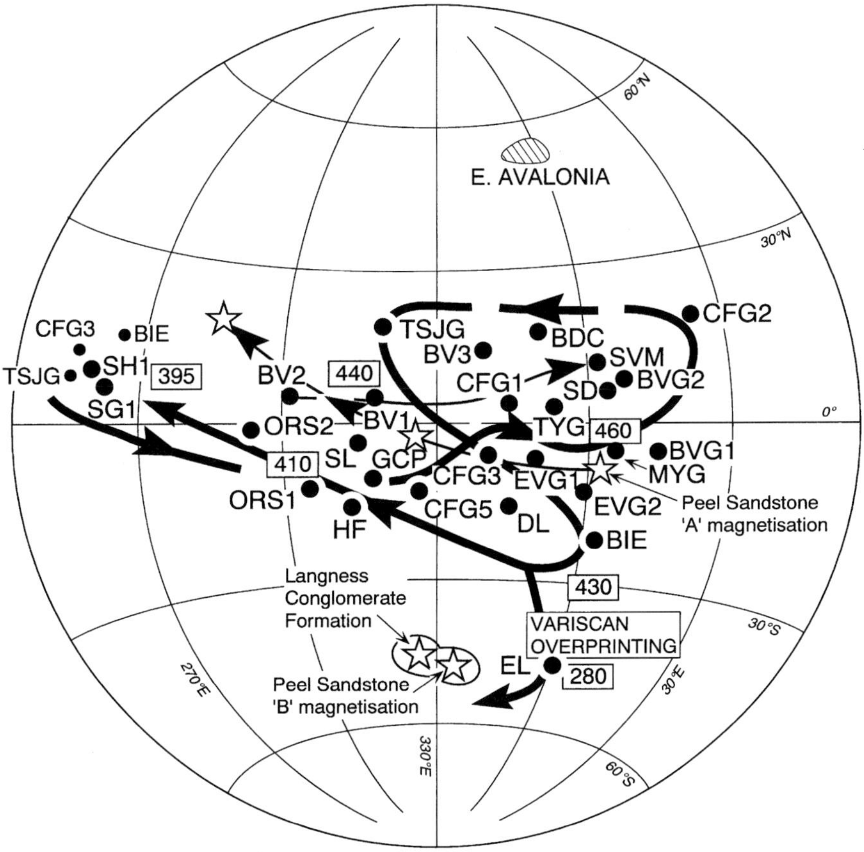

Fig. 5. Late Ordovician–mid Carboniferous apparent polar wander (APW) path (south pole) from Britain with pole positions from this study (☆). The pole from the Peel Sandstone A magnetization is also shown, following 40° and 90°clockwise rotation of the declination. Pole positions (●) with ages in bracket are coded: Lake District Block: EVG1, Binsey Volcanic Group (Late Ordovician); EVG2, High Ireby Volcanic Group (Late Ordovician); GCP, Great Cockup Picrite (457 Ma); BVG1, Borrowdale Volcanic Group (Late Ordovician, 458 Ma); CFG1, Carrock Fell Gabbro Series (Late Ordovician); BVG2, Borrowdale Volcanic Group overprint (Upper Ordovician); CFG2, Carrock granophyre; TSJG, Threlkeld–St Johns Granite (Late Ordovician, 438 Ma plus late Lower Devonian regional overprint in small symbol); CFG3, Round Knott Dolerite; HF, Howgill Mudstones (Lower Silurian); BIE, Eskdale Granite overprint; CFG3, regional late Lower Devonian overprint in Carrock Fell Complex; CFG5, Carrock district dyke swarm; SH1, Skiddaw hornfels (399 Ma); Shap adamellite (394 Ma). Welsh Basin results (with directions adjusted for 55° of counterclockwise rotation) are: BV1 and BV2, Builth Volcanic Series southern outcrop (mid-Ordovician); BV3 Builth Volcanic Group northern outcrop (mid–Late Ordovician); MYG, Moel-y-Golfa andesite (Late Ordovician); SVM, Stapely Volcanic Member overprint (Late Ordovician); SD, Shelve Dolerites (Late Ordovician); TYG, Tan y Grisiau Granite (Late Ordovician); DDC, Breidden Dolerites (Late Ordovician). Other poles shown are: SL, Strathmore Lavas (Late Silurian); ORS1 and ORS2, 'Old Red Sandstones' of the Anglo-Welsh cuvette); DL, Derbyshire Lavas (Early Carboniferous); EL, Exeter Lavas (Early Permian).

or the succeeding extensional phase of basin evolution responsible for the development of the East Irish Sea Basin (Quirk & Kimbell 1998). This explanation would imply a diagenetic origin for the remanence much later than the time of deposition. In this case the age of magnetization would serve to date the diagenesis.

Palaeomagnetic results

These sediments typically have broad unblocking temperature spectra isolated between 200 and 600°C. In some samples a single component remains following removal of a low-unblocking temperature remanence below 200°C (samples 1-3

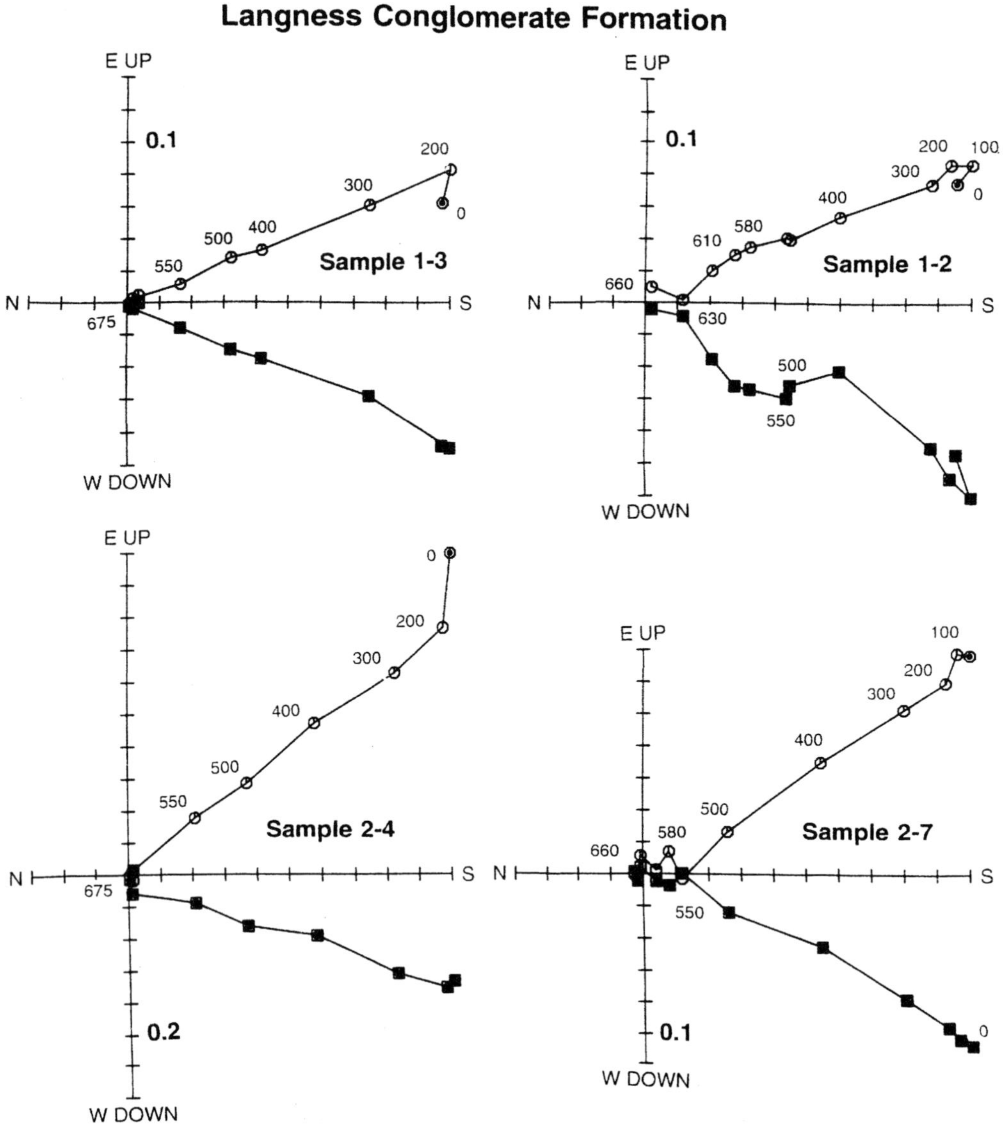

Fig. 6. Orthogonal projections showing palaeomagnetic results from massive litharenites in the Langness Conglomerate Formation. Symbols are as in Fig. 2.

and 2-4 in Fig. 6). These CRMs all have southerly declinations with shallow to negative inclinations. In a few samples, the component structures are more complex (e.g. sample 1-2) and comprise two or three components following removal of a viscous remanence, or have a small residual component with high-unblocking temperatures (580–680°C) resolved in hematite (sample 2-7 in Fig. 6). However, in all cases these components retain the general southerly negative direction and may therefore record different stages of diagenetic mineral growth.

All components have a common reversed polarity and, with the exception of two samples with steeper negative inclinations near –57°, are distributed between $I = -1$ and –41°, comparable to the distribution of declinations (Fig. 7a). The mean of this population of 20 component directions is D/I = 210/–24°. Tilts are low throughout the outcrop and average 13° in a northwesterly direction

Table 4. *Palaeomagnetic results from the Langness Conglomerate Formation, Isle of Man*

N	*R*	*D* (*In situ*)	*I* (*In situ*)	*D* (Tilt adjusted)	*I* (Tilt adjusted)	α_{95}	*k*
20	19.22	209.7	−24.4	203.8	−23.2	6.8	24.3
Palaeomagnetic pole: 142.0° E, 44.0° N, *dp*/*dm* = 3.8/7.9°							

Symbols are as for Table 1. Site coordinates used for pole calculation are 355.3° N, 54.1° N.

(300° E). The mean remanence is moved only marginally to *D*/*I* = 204/−23° following adjustment for this tilt.

Interpretation

In Fig. 7b the mean palaeofield direction at the study location is calculated from the palaeopole positions for Eurasia back to 200 Ma (Besse & Courtillot 1991), for Europe from 200 to 300 Ma (Piper 1988) and for Britain back to 330 Ma (Piper *et al.* 1991). The latter age is close to the Dinantian depositional age of these rocks. There is no evidence for a remanence dating from deposition in the magnetization record of these rocks. This would require a southerly direction of intermediate positive inclination (+30–40°). Instead, the mean remanence direction correlates with the regional field inclination at *c.* 250 Ma. Acquisition of magnetization during the Permo-Carboniferous Reversed Superchron (*c.* 320–250 Ma) is also suggested by the common reversed polarity of all samples.

The remanence in this formation is therefore of diagenetic origin and was acquired during mid-Permian time. This is too late to attribute the magnetization to fluid flow by orogenic loading during the Variscan orogenic episode (e.g. Oliver 1986; Miller & Kent 1988; and cf. the palaeofield direction at *c.* 300 Ma in Fig. 7b). By Early Permian

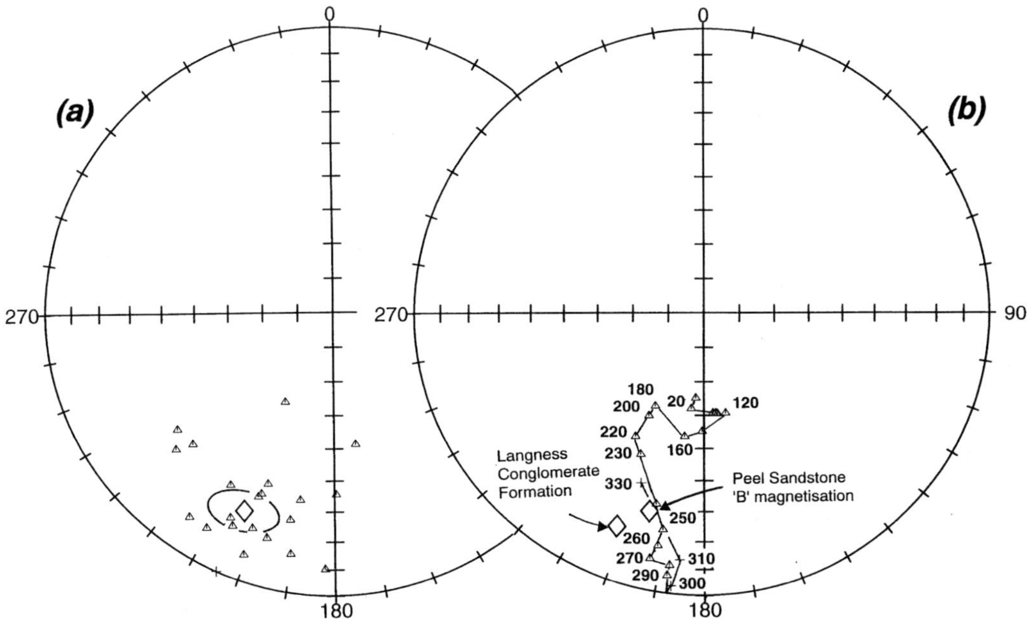

Fig. 7. Equal-area projections showing: (**a**) directions of characteristic remanent magnetization components (*in situ*) derived from the Langness Conglomerate Formation; (**b**) palaeofield migration path of the reversed palaeomagnetic field from 330 Ma to the present day calculated from regional apparent polar wander paths for the location of the Isle of Man; ages of palaeofield directions are in Ma. ◇, Results from the present study; other symbols as in Fig. 3.

times this region was sited in an extensional tectonic regime responsible for initiating the Irish Sea Basin (Quirk & Kimbell 1998). Hence, local mobilization of iron derived from unstable iron-bearing silicates of detrital origin by the convection of basinal fluids is likely to have been responsible for precipitation of authigenic hematite.

This interpretation is not affected by the tilt adjustment, although correction for dip shifts the mean slightly closer to the predicted palaeofield path (Fig. 7b), suggesting that the tilt is a consequence of post-Permian uplift and basinal development in the Irish Sea region. Both *in situ-* and tilt-adjusted means are rotated by *c.* 15° clockwise from the Eurasian palaeofield path, identifying a possible small post-Permian component of regional rotation. Pole positions derived from the Langness remanence and the B remanence in the Peel Sandstones are plotted on Fig. 5.

Summary

Palaeomagnetic study of the Peel Sandstones identifies the partial preservation of dual polarity pre-folding magnetization which correlates with no post-Lower Carboniferous field direction for the British Isles. It is interpreted as a Lower Devonian remanence of probable PDRM origin which has since been rotated counterclockwise. A ubiquitous overprinted magnetization is post-folding in origin and similar to a secondary remanence in the Langness Conglomerate Formation in the southeast Isle of Man. This regional diagenetic magnetization is mid-Permian in age (*c.* 260–250 Ma) and linked to hematite deposition during early phases of extensional tectonism in the Irish Sea region.

We are grateful to Mrs D. Rowlands for undertaking the palaeomagnetic measurements for this study, Kay Lancaster for drafting some of the figures, and G. J. Potts, G. K. Taylor and P. Turner for critical reviews of the manuscript.

References

ALLEN, J. R. L. & CROWLEY, S. F. 1983. Lower Old Red Sandstone fluvial dispersal systems in the British Isles. *Transactions of the Royal Society of Edinburgh: Earth Sciences*, **74**, 61–68.

BARNES, R. P., LINTERN, B. C. & STONE, P. 1989. Timing and regional implications of deformation in the Southern Uplands of Scotland. *Journal of the Geological Society, London*, **146**, 905–908.

BESSE, J. & COURTILLOT, V. 1991. Revised and synthetic apparent polar wander paths of the African, Eurasian, North American and Indian Plates, and true polar wander since 200 Ma. *Journal of Geophysical Research*, **96**, 4029–4050.

BOYD DAWKINS, W. 1902. The red sandstone rocks of Peel (Isle of Man). *Quarterly Journal of the Geological Society of London*, **58**, 633–646.

CHANNELL, J. E. T., MCCABE, C. & WOODCOCK, N. H. 1992. Early Devonian (pre-Acadian) magnetization directions in Lower Old Red Sandstone of South Wales (UK). *Geophysical Journal International*, **108**, 883–894.

——, —— & —— 1993. Palaeomagnetic study of Llandovery (Lower Silurian) red beds in north-west England. *Geophysical Journal International*, **115**, 1085–1094.

CUMMING, J. G. 1847. On the geology of the Isle of Man. *Quarterly Journal of the Geological Society of London*, **2**, 317–348.

DICKSON, J. A. D., FORD, T. D. & SWIFT, A. 1987. The stratigraphy of the Carboniferous rocks around Castletown, Isle of Man. *Proceedings of the Yorkshire Geological Society*, **46**, 203–229.

FITCHES, W. R., BARNES, R. P., MORRIS, J. H. 1999. Geological structure and rectonic evolution. *This volume.*

FORD, T. D. 1971. Slump structures in the Peel Sandstone Series, Isle of Man. *Isle of Man Natural History and Antiquities Journal*, **7**, 440–448.

GILL, E. L. 1903. Keisley Limestone pebbles from the Isle of Man. *Quarterly Journal of the Geological Society of London*, **59**, 307–310.

KNELLER, B. C., KING, L. M. & BELL, A. M. 1993. Foreland basin development and tectonics on the northwest margin of eastern Avalonia. *Geological Magazine*, **130**, 691–697.

LAMPLUGH, G. W. 1903. *The Geology of the Isle of Man.* Memoir of the Geological Survey, United Kingdom. HMSO.

LEWIS, H. P. 1930. The Avonian succession in the south of the Isle of Man. *Quarterly Journal of the Geological Society of London*, **86**, 234–290.

—— 1934. The occurrence of fossiliferous pebbles of Salopian age in the Peel Sandstones (Isle of Man). *Summary of Progress of the Geological Survey for 1933*, **part II**, 91–108.

MCCABE, C. & ELMORE, R. D. 1989. The occurence and origin of Late Palaeozoic remagnetization in the the sedimentary rocks of North America. *Reviews of Geophysics*, **27**, 471–494.

MCFADDEN, P. L. 1990. A new fold test for palaeomagnetic studies. *Geophysical Journal International*, **103**, 163–169.

—— & MCELHINNY, M. W. 1990. Classification of the reversal test in palaeomagnetism. *Geophysical Journal International*, **103**, 725–729.

MILLER, J. D. & KENT, D. V. 1988. Regional trends in the timing of Alleghenian remagnetization in the Appalachians. *Geology*, **16**, 588–591.

OLIVER, J. 1986. Fluids expelled tectonically from orogenic belts: their role in hydrocarbon migration and other geological phenomena. *Geology*, **14**, 99–102.

PIPER, J. D. A. 1988. *Palaeomagnetic Database.* Wiley.

—— 1997*a*. Tectonic rotation within the British paratectonic Caledonides and Early Palaeozoic location of the orogen. *Journal of the Geological Society, London*, **154**, 9–13.

—— 1997*b*. Palaeomagnetism of igneous rocks of the Lake District terrane, northern England: Lower Palaeozoic motions and deformation at a leading edge of the Avalonian Plate. *Geological Journal*, **32**, 211–246.

—— 1998. Palaeomagnetic study of the Ross of Mull granite Complex, western Scotland: Lower Palaeozoic apparent polar wander of the orthotectonic Caledonides. *Geophysical Journal International*, **132**, 133–148.

——, BIGGIN, A. J. & CROWLEY, S. F. 1999. Magnetic Survey of the Poortown Dolerite, Isle of Man. *This volume*.

——, ATKINSON, D., NORRIS, S. & THOMAS, S. 1991. Palaeomagnetic study of the Derbyshire lavas and intrusions central England: definition of Carboniferous apparent polar wander. *Physics of the Earth and Planetary Interiors*, **69**, 37–55.

QUIRK, D. G. & KIMBELL, G. S. 1998. Structural evolution of the Isle of Man and central part of the Irish Sea. *In*: MEADOWS, N., TRUEBLOOD, S., COWAN, G. & HARDMAN, M. (eds) *Petroleum Geology of the Irish Sea and Adjacent Areas*. Geological Society, London, Special Publications, **124**, 135–160.

SETIABUDIDAYA, D., PIPER, J. D. A. & SHAW, J. 1994. Palaeomagnetism of the (Early Devonian) Lower Old Red Sandstones of South Wales: implications to Variscan overprinting and differential regional rotations. *Tectonophysics*, **231**, 257–280.

SHERWOOD, G. J. 1988. MATZJI-a basic program to determine palaeomagnetic remanence directions using principal component analysis. *Computers in Geoscience*, **15**, 1173–1183.

SOPER, N. J. & HUTTON, D. H. W. 1984. Late Caledonian sinistral displacements in Britain: implications for a three-plate collision model. *Tectonics*, **3**, 781–794.

——, WEBB, B. C. & WOODCOCK, N. H. 1987. Late Caledonian (Acadian) transpression in north-west England: times, geometry and geotectonic significance. *Proceedings of the Yorkshire Geological Society*, **46**, 175–192.

TARLING, D. H. 1985. Palaeomagnetic studies of the Orcadian Basin. *Scottish Journal of Geology*, **21**, 261–273

TORSVIK, T. H. 1985. Magnetic properties of the Lower Old Red Sandstone lavas in the Midland Valley, Scotland; palaeomagnetic and tectonic considerations. *Physics of the Earth and Planetary Interiors*, **39**, 194–207.

TUCKER, R. D. & MCKERROW, W. S. 1995. Early Palaeozoic chronology: a review in light of new U–Pb zircon ages from Newfoundland and Britain. *Canadian Journal of Earth Sciences*, **32**, 368–379.

TURNER, P. & TURNER, A. (eds) 1995. *Palaeomagnetic Applications in Hydrocarbon Exploration and Production*. Geological Society, London, Special Publications, **98**.

WOODCOCK, N. H., QUIRK, D. G., FITCHES, W. R. & BARNES, R. P. 1999. In sight of the suture: The Early Palaeozoic geological history of the Isle of Man. *This volume*.

Crustal magnetic structure of the Irish Sea region: evidence for a major basement boundary beneath the Isle of Man

G. S. KIMBELL[1] & D. G. QUIRK[2]

[1]*British Geological Survey, Keyworth, Nottingham NG12 5GG, UK*

[2]*Oxford Brookes University, Gipsy Lane, Oxford OX3 0BP, UK*

Abstract: Imaging and modelling of regional aeromagnetic data indicate a significant change in crustal magnetization across a boundary which extends northeast–southwest across the Irish Sea. A concealed magnetic basement is interpreted to underlie the area to the southeast of the boundary at mid-crustal levels, whereas relatively non-magnetic rocks lie at these levels to the northwest. The boundary does not coincide with the Iapetus Suture as defined by seismic evidence in the Irish Sea and palaeontological evidence on the east coast of Ireland. Instead, it appears to extend towards the northern margin of the Precambrian rocks of southeast Ireland. Its orientation swings sharply in the vicinity of the Isle of Man from north-northeast–south-southwest in the south to east-northeast–west-southwest in the northeast. There is an apparent correlation between the magnetic basement structure and an overlying, shallow, anomalous zone along the axis of the island which is revealed by gravity and high-resolution aeromagnetic data. Modelling indicates that shallow and deep structures may be related if both are assumed to dip to the northwest, a dip direction supported by seismic reflection data acquired immediately to the south of the island. The preferred interpretation is that the mid-crustal magnetic block is of Precambrian age and represents the northern part of an Avalonian basement which was assembled in late Precambrian–early Cambrian times. The geometry of the boundary beneath the Isle of Man may have been inherited from the pre-existing Avalonian basement architecture. Planes of weakness could have been reactivated as major, early Palaeozoic extensional structures and subsequently been active under a compressional regime during the closure of the Iapetus Ocean and the Acadian Orogeny.

The Irish Sea region spans terranes derived from the northern (Laurentian) and southern (Avalonian) sides of the ancient Iapetus Ocean. The boundary between them – the Iapetus Suture – has been interpreted from a combination of seismic reflection and faunal evidence to be a north dipping structure which projects to the surface of the Caledonian basement immediately to the north of the Isle of Man (Soper *et al.* 1992; England & Soper 1997). Long-wavelength magnetic field variations have been interpreted in terms of mid-crustal magnetization variations in the vicinity of the suture (Kimbell & Stone 1995; Morris & Max 1995). The aim of this paper is to extend the analysis of the regional magnetic anomaly pattern to the south of the suture and relate this to a model for the concealed structure beneath the critical Caledonian basement outcrop on the Isle of Man. In Fig. 1, a position is shown for the Iapetus Suture that is offset 20–30 km to the north of its inferred trace at the top of the Caledonian basement (Soper *et al.* 1992; British Geological Survey 1996); this provides an indication of its approximate location at the depths at which the principal magnetic anomalies are generated.

The regional magnetic anomaly pattern

Aeromagnetic data from a broad region centred on the Irish Sea are displayed in Fig. 2: Fig. 2a shows reduced-to-pole anomalies, upward continued by 2 km to emphasize longer wavelength effects; Fig. 2b shows the horizontal gradient of pseudogravity (Baranov 1957), a transform which associates maxima with the edges of deep-seated magnetic bodies. Images of the horizontal gradient of pseudogravity have proved valuable in the identification of major boundaries from regional aeromagnetic data (e.g. Cordell & Grauch 1985; Kimbell & Stone 1995).

Magnetic anomalies to the north of the inferred Iapetus Suture

The Galloway magnetic anomaly (G in Fig. 2a; Powell 1970) has been interpreted by Kimbell &

From: Woodcock, N. H., Quirk, D. G., Fitches, W. R. & Barnes, R. P. (eds) 1999.
In Sight of the Suture: the Palaeozoic geology of the Isle of Man in its Iapetus Ocean context.
Geological Society, London, Special Publications, **160**, 227–238. 1-86239-046-0/99/$15.00

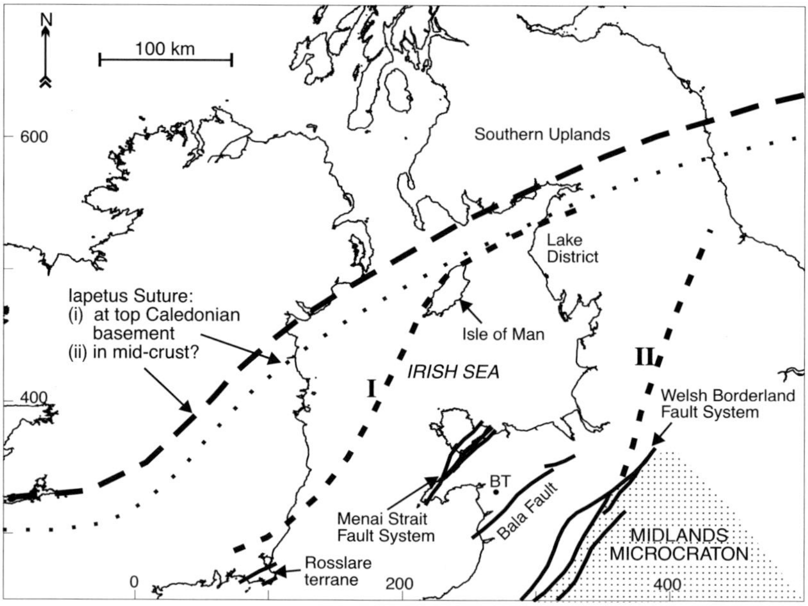

Fig. 1. Map of the region around the Irish Sea showing the approximate location of the north dipping Iapetus Suture at the top of the Caledonian basement [after Soper *et al.* (1992) and British Geological Survey (1996)] and in the mid-crust. Short dashes indicate locations of north-northeast trending magnetic lineaments (I and II). Location of Midlands Microcraton after Pharaoh *et al.* (1987); Welsh fault systems after Woodcock & Gibbons (1988). BT, Bryn-teg Borehole.

Stone (1995) as the effect of a block of magnetic mid-crustal rocks lying in the hanging wall of the Iapetus Suture; they suggest that the source is associated with Precambrian crystalline basement and/or subduction-related magmatic rocks of Ordovician age. The magnetic body may represent a distinct microterrane which originally rifted from the Avalonian continent during its northwards drift and accreted against the Laurentian margin prior to Wenlock time (Kimbell and Stone 1995; Stone *et al.* 1997). An Avalonian origin could explain the similarities in isotopic composition between granitoids subsequently intruded through this crust and those of the Lake District (Thirlwall *et al.* 1989). The position of the Virginia and Nenagh magnetic anomalies invites comparison with the Galloway feature, and the interpretation of these features by Morris & Max (1995) involves a similar juxtaposition of magnetic crustal blocks to the north of the Iapetus Suture and a non-magnetic zone to the south.

Magnetic anomalies to the south of the inferred Iapetus Suture

A conspicuous magnetic low lies immediately to the south of the Iapetus Suture in southern Ireland and extends across the Irish Sea to the Solway Firth. To the southeast of this feature are a series of prominent magnetic anomalies with a variety of trends. Magnetic anomalies immediately to the south and east of the Isle of Man (IoM in Fig. 2a) have been interpreted as evidence of a concealed magnetic basement underlying this area which could be a Precambrian crystalline basement, Ordovician magnetic intrusive and/or sedimentary rocks, or a combination of these (Lee 1989; Kimbell & Stone 1995).

A distinct magnetic anomaly can be traced between southeast Ireland and northwest Wales (RNW in Fig. 2a). In the west, the anomaly extends onshore in the vicinity of the Rosslare Terrane (Fig. 1), a metamorphic complex that may have originated in Palaeoproterozoic times, although the earliest reliable dates relate to a *c.* 620 Ma retrogressive event (Max *et al.* 1990; Winchester *et al.* 1990; Murphy *et al.* 1991; Gibbons *et al.* 1994). The magnetic anomaly appears to extend to the north of the Ballycogly Mylonite Zone, which forms the northern margin of the Rosslare Terrane, over the adjacent ?Precambrian–Cambrian metasedimentary rocks of the Cullenstown Formation. Max *et al.* (1983) identified a magnetic lineament (the Wexford Boundary Linear) corresponding to

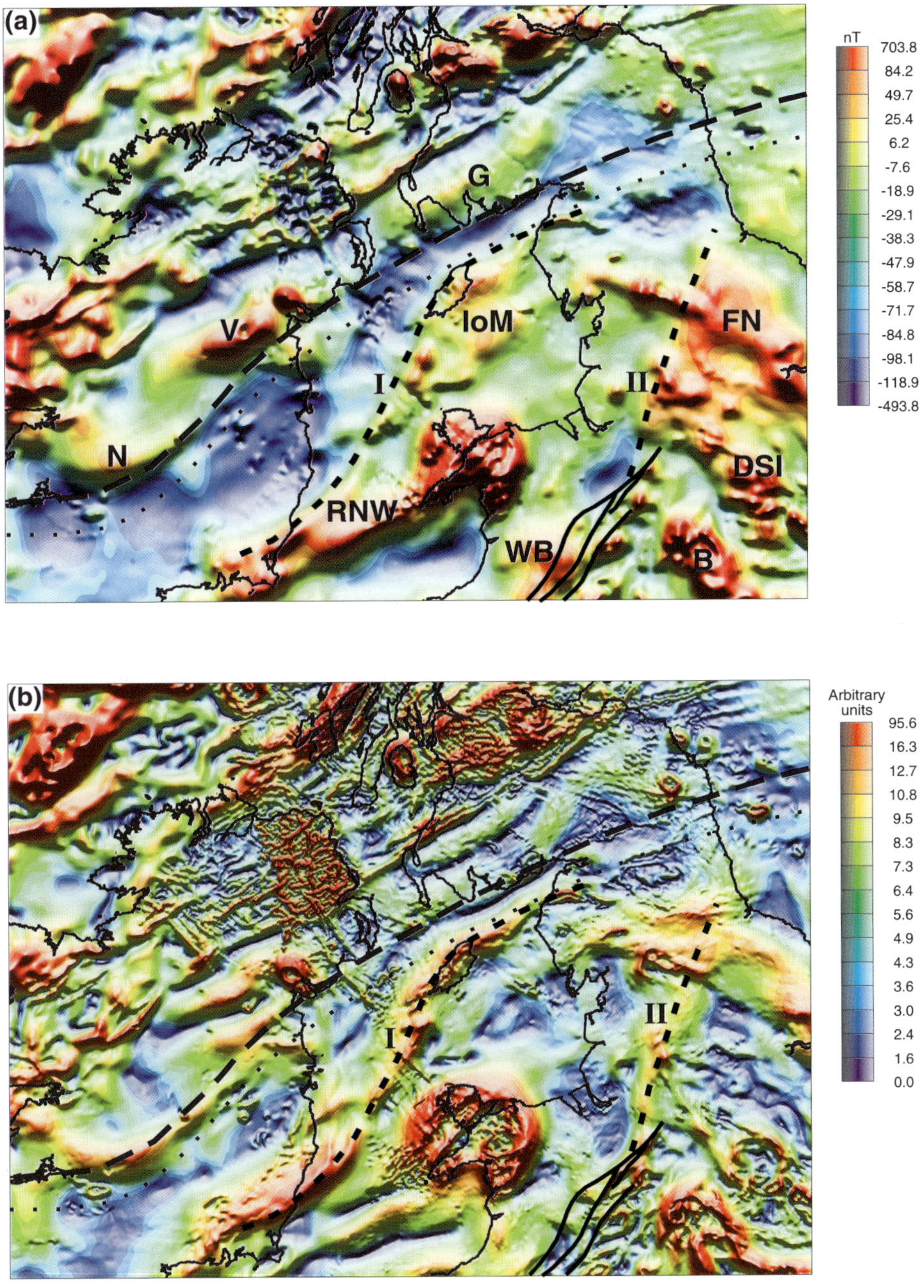

Fig. 2. Aeromagnetic images derived from the gridded data set of British Geological Survey (1998). Colour shaded-relief displays employing equal colour area and illumination from the north. Long dashes and dotted lines indicate the approximate position of the Iapetus Suture (cf. Fig. 1); short dashes indicate north-northeast trending magnetic lineaments (I and II). (**a**) Reduced-to-pole aeromagnetic field, upward-continued by 2 km. Labelled magnetic anomalies: B, Birmingham; DSI, Derby–St. Ives; FN, Furness–Norfolk; G, Galloway; IoM, Isle of Man; N, Nenagh; RNW, Rosslare–North Wales; V, Virginia; WB, Welsh Basin. (**b**) Horizontal gradient of pseudogravity.

the northern edge of the anomaly and suggested that it marks the northern limit of rocks deformed during the Cadomian Orogeny. Where the anomalous zone extends into North Wales, short-wavelength magnetic disturbances can be correlated with pyrrhotite/magnetite-bearing Lower Palaeozoic sedimentary and volcanic rocks (Allen *et al.* 1979; Evans & Greenwood 1988), but an additional, deeper source is required to explain the observed broad, high amplitude anomaly (Allen *et al.* 1979; Howells & Smith 1997). This source underlies volcanic rocks of probable Neoproterozoic age intersected in the Bryn-teg Borehole (Fig. 1; Allen & Jackson 1978), and is therefore most probably Precambrian in age, although a later (Ordovician?) intrusive origin cannot be ruled out.

Several conspicuous, northwest trending magnetic anomaly belts occur in central and northern England. The anomalies extending southeastwards from the Birmingham area (B in Fig. 2a) have been interpreted as the magnetic signature of the magmatic core of a late Precambrian (Charnian) volcanic arc embedded within the Midlands Microcraton (Pharaoh *et al.* 1991; Busby *et al.* 1993). The Derby–St Ives magnetic anomaly (DSI in Fig. 2a) has been ascribed to a belt of Ordovician magnetic intrusive rocks such as the Mountsorrel Granodiorite (Allsop 1987). Pharaoh *et al.* (1993, 1995) suggest that this, and the Furness–Norfolk feature (FN in Fig. 2a) relate to Ordovician (*c.* 450 Ma) arc magmatism associated with southwestward subduction of the Tornquist Sea beneath the Midlands Microcraton. Other explanations for the Furness–Norfolk magnetic anomaly include: shallow Proterozoic magnetic rocks (Wills 1978) and early Ordovician magnetic sedimentary rocks (as intersected in the Beckermonds Scar Borehole; Wilson & Cornwell 1982).

North-northeast trending regional magnetic lineaments south of the Iapetus Suture

Two north-northeast trending lines have been identified from the magnetic anomaly pattern and are shown in Figs 1 and 2. Line I marks the boundary between relatively magnetic crust in the southeastern part of the Irish Sea and the less magnetic zone to the northwest. This feature is particularly well imaged using the horizontal gradient of pseudogravity (Fig. 2b). It extends in a north-northeast direction between southeast Ireland and the Isle of Man, where it swings sharply to an east-northeast trend, parallel to the inferred Iapetus Suture.

Line II (Figs 1 and 2) lies on the projection of the Welsh Borderland Fault System (Woodcock & Gibbons 1988) and marks the western edge of a zone of higher anomaly values upon which the (shallower) Derby–St Ives and Furness–Norfolk features are superimposed. The latter cross-cuts this line, extending northwestward to the southern Lake District. A small dextral offset in the Furness–Norfolk axis and a change in the apparent depth to its source (deeper to the east) occurs in the vicinity of its intersection with Line II, but correlates more closely with Carboniferous displacements along the east-northeast trending North Craven Fault (between the Askrigg Block and Harrogate Basin; Kirby *et al.* 1999) than with the deeper feature. Where it is exposed, the Welsh Borderland Fault System is characterized by local magnetic anomalies associated with shallow, Precambrian magnetic rocks in structurally controlled basement highs; the Birmingham anomaly lies to the east of this fault system but further to the south the clearest evidence for deeper seated magnetic basement lies on its western side beneath the Welsh Basin (WB in Fig. 2a). Woodcock & Gibbons (1988) present evidence that admits, but does not demonstrate, major Ashgill or earlier transcurrent movements on the Welsh Borderland Fault System (in particular, on the Pontesford and Tywi Lineaments). The pattern of magnetic anomalies associated with this fault system and its northward projection are more easily explained if early tectonic movements (predating the source of the Furness–Norfolk magnetic anomaly) are invoked.

3D modelling of magnetic basement

A model has been derived which can account, in a quantitative fashion, for the broad features of the magnetic anomaly pattern in the region around the Isle of Man. The modelling process involves a number of major simplifying assumptions and these must be borne in mind when assessing the significance of the results.

Modelling procedure and assumptions

Magnetic modelling is non-unique, as it is possible to generate the same anomaly field by a variety of model geometries and magnetic property distributions. A model derived by an inversion of the observed anomalies is therefore dependent on the modelling assumptions, which inevitably represent a substantial simplification when compared with the likely complexity of the structures involved. In the example illustrated here it was assumed that the long-wavelength component of the observed magnetic field was generated by variations in the depth to the top of a 'magnetic basement' with uniform magnetization. On the basis of the arguments presented by Kimbell & Stone (1995), it has been assumed that the magnetization is in the direction of the Earth's present magnetic field and the base of the layer giving rise to the magnetic anomalies lies at a depth of 20 km. The longer wavelength component of the magnetic field was separated from anomalies due to shallow sources by applying wavelength filtering as part of the model optimization process; the model illustrated employed a low-pass filter with a ramp between 27 and 56 km. A magnetization of 1.5 A m^{-1} was assumed; this value is clearly not well constrained, but it becomes difficult to model all the long-wavelength magnetic field variations observed across the region if the level is significantly lower than this, and much higher average magnetizations would be difficult to justify in terms of the likely source rocks.

The model was generated using the *Bmod* program (Z. K. Dabek, British Geological Survey), which employs wavenumber domain algorithms based on those of Parker (1972) and Parker & Huestis (1974). In order to minimize edge effects, the modelled area (460 × 380 km) was significantly larger than that shown. The inputs were regular grids with a 2 km node spacing representing the observed total magnetic field, topography and the observation surface (305 m above topography in the onshore area and a similar height above sea level in the offshore area). A generalized starting model was created by suitable shifting and scaling of a smoothed version of the pseudogravity field; the geometry of the upper surface of the magnetic layer was then optimized by an iterative process. An initial assumption was made that the average observed field over the full model area was zero; this involved applying a shift of +30 nT to the observed total field values.

Some of the shortcomings of this approach are:

- long-wavelength magnetic anomalies may be generated by lateral variations in crustal magnetization rather than topography on a uniformly magnetized basement;
- the optimization process generates a simple envelope for the magnetic source. The methods used could not, for example, generate a model incorporating dipping slabs with contrasting magnetic properties (cf. Kimbell & Stone 1995);
- separation of the observed field into 'shallow' and 'deep' components is far from straightforward. Even if there are such separate groups of sources, there is likely to be spectral overlap between the anomalies they generate, such that a 'deep' anomaly may be distorted by removal of its shorter wavelength components or superimposition of the longer wavelength components of shallow sources. The filter applied represents a compromise and, in places, suppresses features which may be part of the 'magnetic basement' (e.g. the northern extension of the Furness–Norfolk anomaly and some of the Irish Sea features). Attempts to extend the model to incorporate shallow sources proved unsuccessful because the optimization became unstable.

In view of these limitations, the derived model does not necessarily represent a 'real' surface but is rather a way of assessing the scale of the magnetization contrasts required to explain the regional long-wavelength magnetic anomaly pattern. It provides a valuable aid in identifying the location of major crustal boundaries (large offsets in a magnetic interface and/or lateral changes in magnetization).

Features of the 3D model

The model is displayed in Fig. 3a, while Fig. 3b and c show the observed and computed fields, respectively; both fields have been upward-continued by 2 km to allow easier comparison of long-wavelength features.

The model indicates changes in crustal magnetization across the region equivalent to the assumed basement magnetization extending over a vertical interval of *c.* 15 km (approximately half the crustal thickness). Given that this magnetization (equivalent to an induced magnetic susceptibility of 0.038 SI units) would be considered high even for a local, near-surface source, the inferred geometries imply substantial changes in the nature of the crust.

One of the most striking features of the model is the major reduction in crustal magnetization required to explain the low magnetic field values to the north and west of the Isle of Man. Although significant post-Lower Palaeozoic subsidence has occurred within the Solway and Peel Basins, which lie along the axis of this magnetic low, Kimbell & Stone (1995) argue that it is primarily due to pre-existing magnetization structures rather than post-collisional subsidence. Evidence to support this assertion is provided by the apparent continuity of the magnetic low beyond the limits of these basins; it crosses the southern part of Ireland (Fig. 2a) and perhaps is related to a similar magnetic feature which characterizes the Gander Terrane in Newfoundland (Jacobi & Kristoffersen 1981).

A substantial change in crustal magnetization is required to explain the broad magnetic gradient zone which crosses the west side of the Isle of Man and forms part of Line I identified from regional magnetic images (Figs 1 and 2; same label used in Fig. 3a). The 3D model confirms the change in the orientation of this magnetic boundary from north-northeast–south-southwest to the south of the island to east-northeast–west-southwest to the northeast.

By contrast, the model shows that long-wavelength magnetic field variations observed for some distance to the east-southeast of the island can be explained by relatively modest variations in the magnetic basement. However, a significant change in the depth and/or magnetization of the magnetic basement is required to explain the change in magnetic field level across Line II (Figs 1, 2 and 3a).

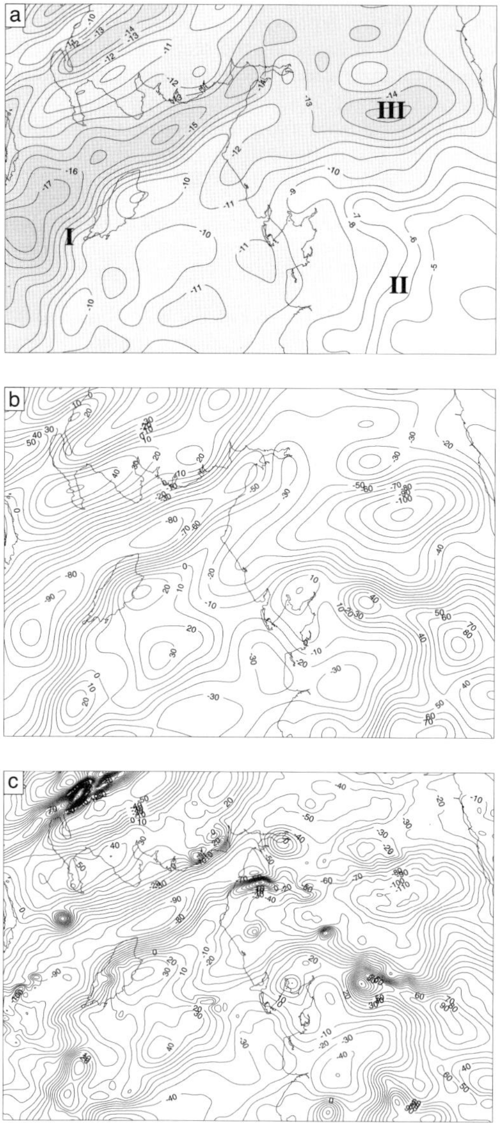

Fig. 3. Results of 3D magnetic basement modelling. (**a**) Apparent depth to magnetic basement (in km below sea level). Basement magnetization is 1.5 A m^{-1} and depth-to-base is 20 km. I–III are features discussed in the text (I and II correspond to lines with the same labels in Figs 1 and 2). (**b**) Total magnetic field (nT) computed using the model shown in (a). (**c**) Observed total magnetic field (nT) shifted by +30 nT.

A further significant perturbation in the magnetic basement in northeast England (feature III in Fig. 3a) correlates spatially with the concealed Weardale Granite (Bott 1967). This may be because of the magnetization contrast between the deep-seated granite and surrounding basement (Bott & Masson Smith 1957), although detailed magnetic modelling suggests that the magnetization contrast may extend to greater depths than those required by gravity modelling of the granite. This feature lies on an axis extending southwestwards across the Lake District (where a granite batholith also exists) and into the offshore area, where it projects towards the relative low between the two magnetic highs south-southeast of the Isle of Man (where a major granite batholith is unlikely). It is possible that, in addition to the local influence of the granites, regional east-northeast trending structures may influence the geometry of the magnetic basement in this area. Structures with this trend were active during the closure of the Iapetus Ocean and subsequent Acadian deformation (cf. Kneller & Bell 1993; Kneller *et al.* 1993; England & Soper 1997), although an earlier (Avalonian) influence is possible.

The 3D model does not allow detailed resolution of the magnetic basement structures. Its approximate nature is illustrated by the 20–30 nT difference between the maximum anomaly amplitude in the upward-continued observed and calculated fields in the vicinity of the Isle of Man (Fig. 3). In order to study the magnetic boundary further, the following sections briefly assess complementary geophysical and geological evidence from the island and present more detailed, local modelling.

Other geophysical and geological evidence

Figure 4 is a Bouguer gravity anomaly map of the Isle of Man and environs based on land stations by Cornwell (1972) and offshore data acquired by the British Geological Survey and Western Geophysical. The offshore area is dominated by the effects of variations in the thickness of the relatively low density cover sequence; major Bouguer anomaly lows are associated with the Peel Basin to the west of the island, the Solway Basin to the north and the Onchan Depression to the east (Quirk & Kimbell 1997). Onshore, Bouguer gravity anomaly lows in the vicinity of the granite exposures at Foxdale and Dhoon have been interpreted by Cornwell (1972) as the expression of voluminous granite bodies underlying these relatively restricted outcrops. In the zone between

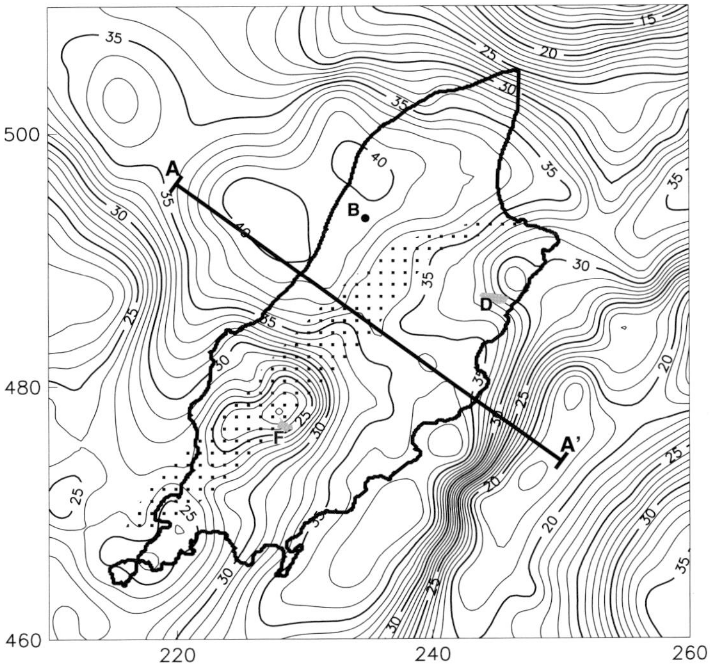

Fig. 4. Bouguer gravity anomaly map of the Isle of Man and surrounding waters. Based on surveys by Cornwell (1972), British Geological Survey and Western Geophysical. Contour interval, 1 mGal (10^{-5} m s^{-2}); reduction densities, 2.70 (onshore) and 2.20 Mg m^{-3} (offshore). Coarse stipple indicates the area of magnetic disturbance revealed by high-resolution airborne survey (Quirk & Kimbell 1997; Quirk *et al.* 1999). AA′, Location of model profile; B; Ballaugh. Granite outcrops (grey): F, Foxdale; D, Dhoon.

the two granite-related features, Bouguer anomaly values increase from east to west across the central part of the island, and in particular across a local gravity gradient zone that extends north-northeastward from the Foxdale anomaly to the vicinity of Ballaugh, then swings to an east-northeast trend (Fig. 4). Cornwell (1972) concluded that this gradient zone was unlikely to be due to concealed granite and preferred an explanation involving either a lateral density contrast within the Lower Palaeozoic rocks or the presence of denser, shallow basement underlying the exposed Lower Palaeozoic rocks to the northwest.

A high-resolution aeromagnetic survey has revealed a zone of magnetic disturbance extending in a northeast–southwest direction along the Isle of Man (Quirk & Kimbell 1997; Quirk *et al.* 1999; approximate extent indicated by stippled zone in Fig. 4). A detailed discussion of this zone is beyond the scope of this paper, but it is noted that the short wavelength of the magnetic disturbances indicates that they are due to near-surface magnetization contrasts and are thus not directly related to the underlying deep magnetic basement. Although the anomalies may relate to the original magnetization of the Lower Palaeozoic rocks, an alternative possibility is that secondary magnetic minerals have been concentrated as a result of later processes, perhaps associated with the higher metamorphism along this axis (Simpson 1964; Power & Barnes 1999). A candidate magnetic mineral is ilmenite, which developed in this zone during D_2 deformation (Power & Barnes 1999). Figure 4 illustrates that there is an approximate correspondence between the zone of magnetic disturbance and the gravity gradient zone, with the former lying immediately to the east of the latter. The magnetic data thus provide an indication of the continuity of the feature into the area where the gravity signature is masked by the anomaly due to the Foxdale Granite.

The gravity and detailed aeromagnetic data therefore indicate an anomalous, near-surface zone crossing the Isle of Man which parallels the underlying magnetic basement boundary and exhibits a similar change in trend across the island. Quirk *et al.* (1999) suggest that individual magnetic lineaments within this zone correlate with major geological boundaries, mostly faults, which separate areas of contrasting lithologies within the

Manx Group. For example, a 20 km segment of the southeast margin of the anomalous zone coincides with the faulted contact between the sand-prone Creg Agneash Formation (typical of the eastern side of the island) and the mudstone-dominated Barrule Formation, which is confined to the central axis of the island (Quirk *et al.* 1999). The centre and northwest edge of the anomalous zone coincides with a set of mostly northwest dipping reverse faults, ductile shear zones and intrusions. The current interpretation is that the anomalous zone represents a fault duplex formed during northwest–southeast compression in late Caledonian times (Fitches *et al.* 1999; Quirk *et al.* 1999).

The British Institutions Seismic Reflection Profiling Syndicate (BIRPS) WINCH-2 profile detected a north dipping reflectivity boundary to the northwest of the Isle of Man which has been interpreted as the seismic signature of the Iapetus Suture (Brewer *et al.* 1983; Hall *et al.* 1984; Soper *et al.* 1992). This feature has been correlated with the northern edge of the non-magnetic zone beneath the Solway Basin by Kimbell & Stone (1995) and does not appear to coincide with the major magnetic boundary beneath the Isle of Man, which forms the southern margin of the non-magnetic zone. The only candidate on WINCH-2 for this boundary is a set of north dipping reflections which occur towards its southern end (shot points 17 400 to 17 600) at relatively shallow crustal levels [1–3 s two-way travel time (TWTT)] within the inferred Caledonian basement (e.g. Hall *et al.* 1984, fig. 7). Recent seismic surveys by JEBCO Seismic Ltd have provided clear evidence of a northwest dipping reflective zone at relatively shallow depth [< 1.5 s TWTT (*c.* 4 km)] within the basement to the southeast of the island (Quirk *et al.* 1999, figs 3 and 5). Quirk *et al.* (1999) correlate this zone with the northeast trending belt of shallow magnetic anomalies discussed above. At deeper levels (> 5 s TWTT) two southward dipping reflector packages have been identified to the south of the island by England & Soper (1997), who infer that these are part of a conjugate set of structures developed during Acadian compression.

A possible correlation between deep and shallow structures

Given the similarities in location and trend between the deep magnetic basement boundary and shallower gravity and magnetic features, a close association is possible. This could arise, for example, because units at different crustal levels have been offset by the same, deep-seated structure, or because reactivation of an early, deep structure has influenced the subsequent evolution of the overlying rocks. The possibility of a relationship between the various anomalous bodies has been explored using 2D modelling methods along profile AA′ (Fig. 5; location shown in Fig. 4).

The gravity modelling assumes a highly simplified density structure, but none the less provides some corroboration for the conclusion of Cornwell (1972) that it is not necessary for there to be a substantial connection between the Foxdale and Dhoon Plutons in order to explain the observed gravity variations across the central part of the

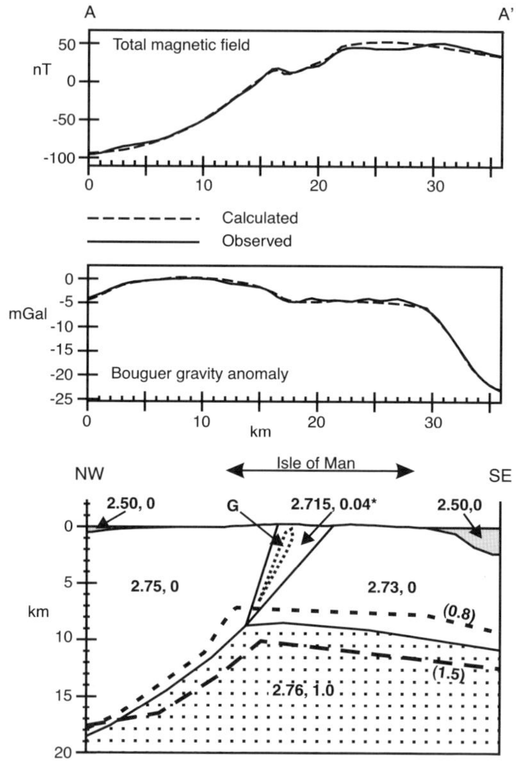

Fig. 5. 2D magnetic and gravity model for profile AA′. Stippled unit, ?Avalonian magnetic basement; no ornament, Lower Palaeozoic sequence; shaded, post Lower Palaeozoic cover. Numbers indicate density (Mg m^{-3}) and magnetization (A m^{-1}). Deep unit is magnetized in the direction of the Earth's present field while shallow unit (*) is magnetized in the opposite direction. Dashed lines show the modelled geometry of the deep magnetic source if the alternative magnetizations shown in parentheses are adopted. Dotted line shows the geometry of a granite body (G; density, 2.62 Mg m^{-3}) which could be accommodated in the model if a density of 2.73 Mg m^{-3} is assumed for the unit containing it. Assumed background fields: magnetic, –30 nT; gravity, +40 mGal.

island. These variations can be replicated by assuming relatively modest density contrasts between different Lower Palaeozoic units. However, the presence of granitic rocks cannot be ruled out altogether as the gravity effect of the central, low density zone in the current model could, for example, be generated by a concealed, sheet or lens of granite with a width of *c.* 0.5 km and a similar northwestward dip (indicated by the dotted line in Fig. 4).

A local depression in the magnetic field has been replicated by assuming a zone with weak reversed magnetization along the axis of the island. The observed magnetic profile is based on the regional data rather than the results of the high-resolution aeromagnetic survey, which indicate that the response will vary markedly depending on where the profile is drawn (Quirk *et al.* 1999). None the less, the model provides a schematic view of the spatial relationship between the zone of anomalous magnetization and the source of the local gravity gradient. More detailed investigations are required to determine whether there really is a direct correlation between anomalous magnetization and relatively low density in this zone.

The long-wavelength magnetic gradient across the island has been modelled as the effect of the northwestward truncation of a mid-crustal magnetic slab. The model includes mid-crustal magnetization variations well beyond the ends of the profile shown, because distant structures have a significant influence on such long-wavelength effects. To illustrate the range of solutions possible, derived geometries assuming three different basement magnetizations are shown. The results indicate that it is feasible to propose a link between the structures responsible for shallow geophysical anomalies across the island and the underlying northwest edge of a major magnetic basement block, providing the structures dip towards the northwest. Such a dip direction is compatible with the geometry of shallow basement reflections in the offshore area (Quirk *et al.* 1999, fig. 6). These reflections, however, have only been observed at shallower depth than the inferred truncation of the magnetic basement, which appears to coincide with a relatively 'blank' zone in deep seismic profiles (cf. England & Soper 1997, fig. 5).

Discussion

The magnetic anomaly pattern across the Irish Sea and central and southern Ireland (Fig. 2a) comprises a northeast trending magnetic low, flanked to north and south by highs characteristic of major magnetic units at mid-crustal depths. The non-magnetic zone narrows in a northeastward direction and cannot be traced along this strike beyond the UK mainland. Kimbell & Stone (1995) correlated northward dipping reflections, previously identified as the seismic expression of the Iapetus Suture, with the boundary between a northern magnetic unit and a non-magnetic zone that they inferred to be a deep wedge of partially subducted sedimentary strata. This is broadly compatible with the interpretation by Morris & Max (1995) of analogous features in Ireland. The focus of the present investigation is the southern margin of the zone of low magnetization which lies beneath the Isle of Man.

Hypotheses for the nature of the magnetic basement underlying the region to the southeast of the Isle of Man are based on an assessment of the regional anomaly pattern extending from southeast Ireland across north and central England. From this, the most likely sources of the observed long-wavelength magnetic anomalies are either parts of a Precambrian (Avalonian) basement or igneous rocks associated with Ordovician arc magmatism. Our preferred interpretation is that the magnetic basement in the vicinity of the island is principally Precambrian in age. This is because the inferred margin of the magnetic basement extends in a south-southeast direction towards the probable edge of the Precambrian basement in southeast Ireland, rather than towards any of the known Lower Palaeozoic volcanic outcrops on the east coast of Ireland. A likely contributor to the higher magnetization is the presence of magnetic Neoproterozoic igneous rocks characteristic of Avalonian basement. The magnetic signature of these rocks has been observed in the Avalon zone of Newfoundland and traced into adjacent offshore areas (Haworth & Lefort 1979). It is possible that older Precambrian magnetic units contribute to the observed anomalies, although direct evidence for such rocks within the area of interest is limited to the Rosslare Terrane, where they have not been reliably dated. Busby *et al.* (1993) suggest the possibility of ancient (pre–late Proterozoic) magnetic basement to the south of the present study area beneath the London Platform.

The above interpretation, which places Avalonian basement on either side of the Menai Strait Fault System (Fig. 1), appears difficult to reconcile with the identification of this fault system as a terrane boundary (Gibbons 1987). However, our interpretation is compatible with the model of Horák *et al.* (1996) in which the Precambrian units juxtaposed by these faults come from the same Avalonian arc system, which was dismembered by transcurrent faulting in late Precambrian–early Cambrian times (Dallmeyer & Gibbons 1987) after the Precambrian magmatism.

Therefore, it is proposed that the Isle of Man overlies the northern edge of the Precambrian

(Avalonian) magnetic, crystalline basement. The non-magnetic unit to the north and west of this boundary is inferred to be a thick succession of predominantly sedimentary Lower Palaeozoic rocks, derived from the northern margin of the Avalonian continent and stacked and thickened during the final closure of the Iapetus Ocean. The basement boundary beneath the island may have been a major extensional structure in early Palaeozoic times, perhaps associated with the rifting of Avalonian fragments such as those postulated by Kimbell & Stone (1995). The north-northeast trending segment of this boundary marks the inferred faulted margin of the Manannan Basin of Quirk & Burnett (1999) and Quirk *et al.* (1999). Rifting may have occurred along planes of weakness within the pre-existing basement and the influence of such structures could provide an explanation for the sharp change in strike observed in the vicinity of the island. Pre-existing basement structures are likely to have formed during the late Precambrian–early Cambrian assembly of Avalonia. The north-northeast–south-southwest orientation of the segment of the basement margin lying between the Isle of Man and southeast Ireland (Line I in Figs 1–3) parallels that of the Welsh Borderland Fault System (Line II), which could have been initiated at a similar time (Kokelaar 1988; Woodcock & Gibbons 1988).

The crustal boundary formed by the edge of the Avalonian basement is likely to have influenced the subsequent geological evolution of the region. In particular, it could have played an important role during compressional episodes relating to the closure of the Iapetus Ocean and subsequent Acadian deformation. The ongoing research on the Isle of Man will provide key evidence to test whether that influence can be recognized; e.g. could changes in deformation style between the east and west side of the island be related to the nature of the underlying basement? It appears likely from the form and location of the Foxdale and Dhoon Granites that the basement structure has exercised control over their emplacement, together with the processes that led to the observed zones of anomalous metamorphism and near-surface magnetization along the axis of the island.

Part of the work reported here was funded by NERC research grant No. GR9/01834. This paper is published with the permission of the Director, British Geological Survey (NERC).

References

ALLEN, P. M. & JACKSON, A. A. 1978. Bryn-teg Borehole, North Wales. *Bulletin of the Geological Survey of Great Britain*, **61**.

——, COOPER, D. C. & SMITH, I. F. 1979. *Mineral exploration in the Harlech Dome, North Wales.* Mineral Reconnaissance Programme Report, Institute of Geological Sciences, **29**.

ALLSOP, J. M. 1987. Patterns of late Caledonian intrusive activity in eastern and southern England from geophysics, radiometric dating and basement geology. *Proceedings of the Yorkshire Geological Society*, **46**, 335–353.

BARANOV, V. 1957. A new method for the interpretation of aeromagnetic maps: pseudogravimetric anomalies. *Geophysics*, **22**, 359–383.

BOTT, M. H. P. 1967. Geophysical investigations of northern Pennine basement rocks. *Proceedings of the Yorkshire Geological Society*, **36**, 139–168.

—— & MASSON SMITH, D. 1957. Interpretation of a vertical field magnetic survey in north-east England. *Quarterly Journal of the Geological Society, London*, **113**, 119–136.

BREWER, J. A., MATTHEWS, D. H., WARNER, M. R., HALL, J., SMYTHE, D. K. & WHITTINGTON, R. J. 1983. BIRPS deep seismic reflection studies of the British Caledonides – the WINCH profile. *Nature*, **305**, 206–210.

BRITISH GEOLOGICAL SURVEY. 1996. *Tectonic Map of Britain, Ireland and Adjacent Areas (1:1 500 000).* PHARAOH, T. C., MORRIS, J. H., LONG, C. B. & RYAN, P. D. (compilers). British Geological Survey, Keyworth, Nottingham.

—— 1998. *Magnetic Anomaly map of Britain, Ireland and Adjacent Areas (1:1 500 000).* ROYLES, C. P. & SMITH, I. F. (compilers). British Geological Survey, Keyworth, Nottingham.

BUSBY, J. P., KIMBELL, G. S. & PHARAOH, T. C. 1993. Integrated geophysical/geological modelling of the Caledonian and Precambrian basement of southern Britain. *Geological Magazine*, **130**, 593–604.

CORDELL, L. E. & GRAUCH, V. J. S. 1985. Mapping basement magnetization zones from aeromagnetic data in the San Juan basin, New Mexico. *In*: HINZMAN, W. J. (ed.) *The Utility of Regional Gravity and Magnetic Anomaly Maps.* Society of Exploration Geophysicists, Tulsa, 181–197.

CORNWELL, J. D. 1972. A gravity survey of the Isle of Man. *Proceedings of the Yorkshire Geological Society*, **39**, 93–106.

DALLMEYER, R. D. & GIBBONS, W. 1987. The age of blueschist metamorphism in Anglesey, north Wales: evidence from $^{40}Ar/^{39}Ar$ mineral dates of the Penmynydd Schists. *Journal of the Geological Society, London*, **144**, 843–850.

ENGLAND, R. W. & SOPER, N. J. 1997. Lower crustal structure of the East Irish Sea from deep seismic reflection data. *In:* MEADOWS, N. S., TRUEBLOOD, S. P., HARDMAN, M. & COWAN, G. (eds) *Petroleum Geology of the Irish Sea and Adjacent Areas.* Geological Society, London, Special Publications, **124**, 61–72.

EVANS, R. B. & GREENWOOD, P. G. 1988. Magnetic susceptibility measurements as a means of differentiating different rock types and their mineralisation. *Proceedings of the Asian Mining '88 Conference, Kuala Lumpa, Malaysia* (Institution of Mining and Metallurgy, London), 45–57.

FITCHES, W. R., BARNES, R. P. & MORRIS, J. H. 1999. Geological structure and tectonic evolution of the Lower Palaeozoic rocks of the Isle of Man. *This volume.*

GIBBONS, W. 1987. The Menai Strait Fault System: an early Caledonian terrane boundary in North Wales. *Geology*, **15**, 744–747.

——, TIEZCH-TYLER, D., HORÁK, J. M. & MURPHY, F. C. 1994. Precambrian rocks in Anglesey, southwest Llyn and southeast Ireland. *In*: GIBBONS, W. & HARRIS, A. L. (eds) *A Revised Correlation of Precambrian Rocks in the British Isles*. Geological Society, London, Special Reports, **22**, 73–83.

HALL, J., BREWER, J. A., MATTHEWS, D. H. & WARNER, M. R. 1984. Crustal structure across the Caledonides from the 'WINCH' seismic reflection profile: influences on the evolution of the Midland Valley of Scotland. *Transactions of the Royal Society of Edinburgh*, **75**, 97–109.

HAWORTH, R. T. & LEFORT, J. P. 1979. Geophysical evidence for the extent of the Avalon zone in Atlantic Canada. *Canadian Journal of Earth Sciences*, **16**, 552–567.

HORÁK, J. M., DOIG, R., EVANS, J. A. & GIBBONS, W. 1996. Avalonian magmatism and terrane linkage: new isotopic data from the Precambrian of North Wales. *Journal of the Geological Society, London*, **153**, 91–99.

HOWELLS , M. F. & SMITH, M. 1997. The geology of the country around Snowdon. *Memoir of the British Geological Survey*, **Sheet 119** (England and Wales).

JACOBI, R. D. & KRISTOFFERSEN, Y. 1981. Transatlantic correlations of geophysical anomalies on Newfoundland, British Isles, France and adjacent continental shelves. *In:* KERR, J. W. M. & FERGUSSON, A. J. *Geology of the North Atlantic Borderlands*. Canadian Society of Petroleum Geologists Memoir, **7**, 197–229.

KIMBELL, G. S. & STONE, P. 1995. Crustal magnetization variations across the Iapetus Suture Zone. *Geological Magazine*, **132**, 599–609.

KIRBY, G. A., AITKENHEAD, N., ALLSOP, J. M. *ET AL.* 1999. The structure and evolution of the Craven Basin and adjacent areas. *Subsurface Memoir of the British Geological Survey*, in press.

KNELLER, B. C. & BELL, A. M. 1993. An Acadian mountain front in the English Lake District: the Westmorland Monocline. *Geological Magazine*, **130**, 203–213.

——, KING, L. M. & BELL, A. M. 1993 Foreland basin development and tectonics on the northwest margin of eastern Avalonia. *Geological Magazine*, **130**, 691–697.

KOKELAAR, P. 1988. Tectonic controls of Ordovician arc and marginal basin volcanism in Wales. *Journal of the Geological Society, London*, **145**, 759–775.

LEE, M. K. 1989. *Upper crustal structure of the Lake District from modelling and image processing of potential field data.* British Geological Survey Technical Report WK/89/1.

MAX, M. D., BARBER, A. J. & MARTINEZ, J. 1990. Terrane assemblage of the Leinster Massif, SE Ireland during the Lower Palaeozoic. *Journal of the Geological Society, London*, **147**, 1035–1050.

——, RYAN, P. D. & INAMDAR, D. D. 1983. A magnetic deep structural interpretation of Ireland. *Tectonics*, **2**, 431–451.

MORRIS, P. & MAX, M. D. 1995. Magnetic crustal character in Central Ireland. *Geological Journal*, **30**, 49–67.

MURPHY, F. C., ANDERSON, T. B., DALY, J-S *ET AL.* 1991. An appraisal of Caledonian suspect terranes in Ireland. *Irish Journal of Earth Sciences*, **11**, 11–41.

PARKER, R. L. 1972. The rapid calculation of potential anomalies. *Geophysical Journal of the Royal Astronomical Society*, **31**, 447–455.

—— & HUESTIS, S. P. 1974. The inversion of magnetic anomalies in the presence of topography. *Journal of Geophysical Research*, **79**, 1587–1593.

PHARAOH, T. C., BREWER, T. S. & WEBB, P. C. 1993. Subduction-related magmatism of late Ordovician age in eastern England. *Geological Magazine*, **130**, 647–656.

——, ENGLAND, R. W. & LEE, M. K. 1995. The concealed Caledonide basement of eastern England and the southern North Sea – a review. *Studia geophysica et geodetica*, **39**, 330–346.

——, LEE, M. K., EVANS, C. J., BREWER, T. S. & WEBB, P. C. 1991. A cryptic late Proterozoic island arc and marginal basin complex in the heart of England. *Terra abstracts*, **3**, 58.

——, MERRIMAN, R. J., WEBB, P. C. & BECKINSALE, R. D. 1987. The concealed Caledonides of eastern England: preliminary results of a multidisciplinary study. *Proceedings of the Yorkshire Geological Society*, **46**, 355–369.

POWELL, D. W. 1970. Magnetised rocks within the Lewisian of Western Scotland and under the Southern Uplands. *Scottish Journal of Geology*, **6**, 353–369.

POWER, G. & BARNES, R. P. 1999. Relationships between metamorphism and structure on the northern edge of Eastern Avalonia: the Manx Group, Isle of Man. *This volume.*

QUIRK, D. G. & BURNETT, D. J. 1999. Lithofacies of Lower Palaeozoic deep-marine sediments in the Isle of Man: a new map and stratigraphic model for the Manx Group. *This volume.*

—— & KIMBELL, G. S. 1997. Structural evolution of the Isle of Man and central part of the Irish Sea. *In:* MEADOWS, N. S., TRUEBLOOD, S. P., HARDMAN, M. & COWAN, G. (eds) *Petroleum Geology of the Irish Sea and Adjacent Areas*. Geological Society, London, Special Publications, **124**, 135–159.

——, BURNETT, D. J., KIMBELL, G. S., MURPHY, C. A. & VARLEY, J. S. 1999. Shallow geophysical and geological evidence for a regional-scale fault duplex in the Lower Palaeozoic of the Isle of Man. *This volume.*

SIMPSON, A. 1964. The metamorphism of the Manx Slate Series, Isle of Man. *Geological Magazine*, **101**, 20–36.

SOPER, N. J., ENGLAND, R. W., SNYDER, D. B. & RYAN, P. D. 1992. The Iapetus suture zone in England, Scotland and eastern Ireland: a reconciliation of geological and deep seismic data. *Journal of the Geological Society, London*, **149**, 697–700.

STONE, P., KIMBELL, G. S. & HENNEY, P. J. 1997. Basement control on the location of strike-slip shear in the Southern Uplands of Scotland. *Journal of the Geological Society, London*, **154**, 141–144.

THIRWALL, M. F., MAYNARD, J., STEPHENS, W. E. & SHAND, P. 1989. Calc-alkaline magmagenesis in the Scottish Southern Uplands forearc. *Terra abstracts*, **1**, 178.

WILLS, L. J. 1978. A palaeogeographical map of the Lower Palaeozoic floor below the cover of Upper Devonian. *Memoir of the Geological Society of London*, **8**.

WILSON, A. A. & CORNWELL, J. D. 1982. Institute of Geological Sciences borehole at Beckermonds Scar, North Yorkshire. *Proceedings of the Yorkshire Geological Society*, **44**, 59–88.

WINCHESTER, J. A., MAX, M. D. & MURPHY, F. C. 1990. The Rosslare Complex: a displaced terrane in southeast Ireland*In:* STRACHAN, R. A. & TAYLOR, G. K. (eds) *Avalonian and Cadomian Geology of the North Atlantic*. Blackie, 49–64.

WOODCOCK, N. J. & GIBBONS, W. 1988. Is the Welsh Borderland Fault System a terrane boundary? *Journal of the Geological Society, London*. **145**, 915–923.

Shallow geophysical and geological evidence for a regional-scale fault duplex in the Lower Palaeozoic of the Isle of Man

D. G. QUIRK[1,2], D. J. BURNETT[1], G. S. KIMBELL[3], C. A. MURPHY[4] & J. S. VARLEY[5]

[1]*Oxford Brookes University, Gipsy Lane, Oxford OX3 0BP, UK*

[2]*Present address: Burlington Resources (Irish Sea) Ltd, 1 Canada Square, Canary Wharf, London E14 5AA, UK*

[3]*British Geological Survey, Keyworth, Nottingham NG12 5GG, UK*

[4]*World Geoscience (UK) Ltd, 3 Walnut Tree Park, Walnut Tree Close, Guildford GU1 4TR, UK*

[5]*JEBCO Seismic Ltd, 1st Floor, St George's House, Station Approach, Cheam, Surrey SM2 7AT, UK*

Abstract: A distinctive set of linear anomalies is seen on potential field data crossing the Isle of Man in a northeast–southwest belt, 5–6 km wide. The lineaments occur in an imbricate pattern with three constituent trends: northeast–southwest, east–west (to east-northeast–west-southwest) and north–south. The belt can be traced into the offshore where it ties with a northwest dipping set of anomalous high-amplitude seismic reflections interpreted as fluid-filled fractures or intrusions along fractures. In the field, the lineaments coincide with northeast–southwest reverse faults, east–west dextral strike slip faults and north–south sinistral strike slip faults interpreted to have formed during northwest–southeast compression in the late Caledonian. Several of the strike-slip faults were later sites of mineralization. In addition, there is limited kinematic evidence for an earlier period of sinistral transpression on east–west ductile shear zones. A tentative model is proposed where the Manx Group is located on the eastern side of an inverted Lower Palaeozoic basin (the Manannan Basin) forming an embayment on the northwest margin of Eastern Avalonia. During closure of Iapetus, the direction of maximum principal stress (σ_1) rotated from north-northeast–south-southwest to northwest–southeast as Eastern Avalonia docked and then locked against Laurentia. The imbricate belt developed as a duplex at the eastern edge of the basin during the later stages of contraction. The implications of the model is that the stratigraphy of the Manx Group is telescoped.

Quirk & Kimbell (1997) presented work carried out in 1994 showing that a prominent set of shallow-sourced northeast–southwest to east–west linear geophysical anomalies run along the central axis of the Isle of Man, where Lower Palaeozoic rocks assigned to the Manx Group crop out. These lineaments occupy a 5 km wide belt orientated approximately northeast–southwest with a rhomboid shape, similar in some ways to the outline of the island itself. Due to the linked en echelon and lens-shaped arrangement of the lineaments it was originally termed 'imbricate zone' but is renamed the Manx Imbricate Belt in this paper. Few of the implied faults were recognized in earlier structural studies, although Blake (1905) introduced the possibility of stratigraphic repetition by thrusts in the centre of the island. It is, however, worth noting that the central part of the imbricate belt coincides with a change in fold vergence within the Manx Group interpreted by Lamplugh (1903) and Simpson (1963) to represent the trace of a major synclinorium (cf. Fitches *et al.* 1999).

Between 1995 and 1997 fieldwork was carried out on the Isle of Man in order to partially resurvey the Lower Palaeozoic Manx Group (Woodcock *et al.* 1999). By 1996 researchers in the group carrying out this work began to interpret, often independently, the presence of major northeast–southwest, east–west and north–south faults (e.g. Fig. 1). These faults are rarely observed due to poor exposure but are necessary to explain the apparent juxtaposition of different lithostratigraphic and structural domains (Fitches *et al.* 1999; Quirk & Burnett 1999). It was after recognizing the differences between these domains that the idea of

From: Woodcock, N. H., Quirk, D. G., Fitches, W. R. & Barnes, R. P. (eds) 1999.
In Sight of the Suture: the Palaeozoic geology of the Isle of Man in its Iapetus Ocean context.
Geological Society, London, Special Publications, **160**, 239–257. 1-86239-046-0/99/$15.00

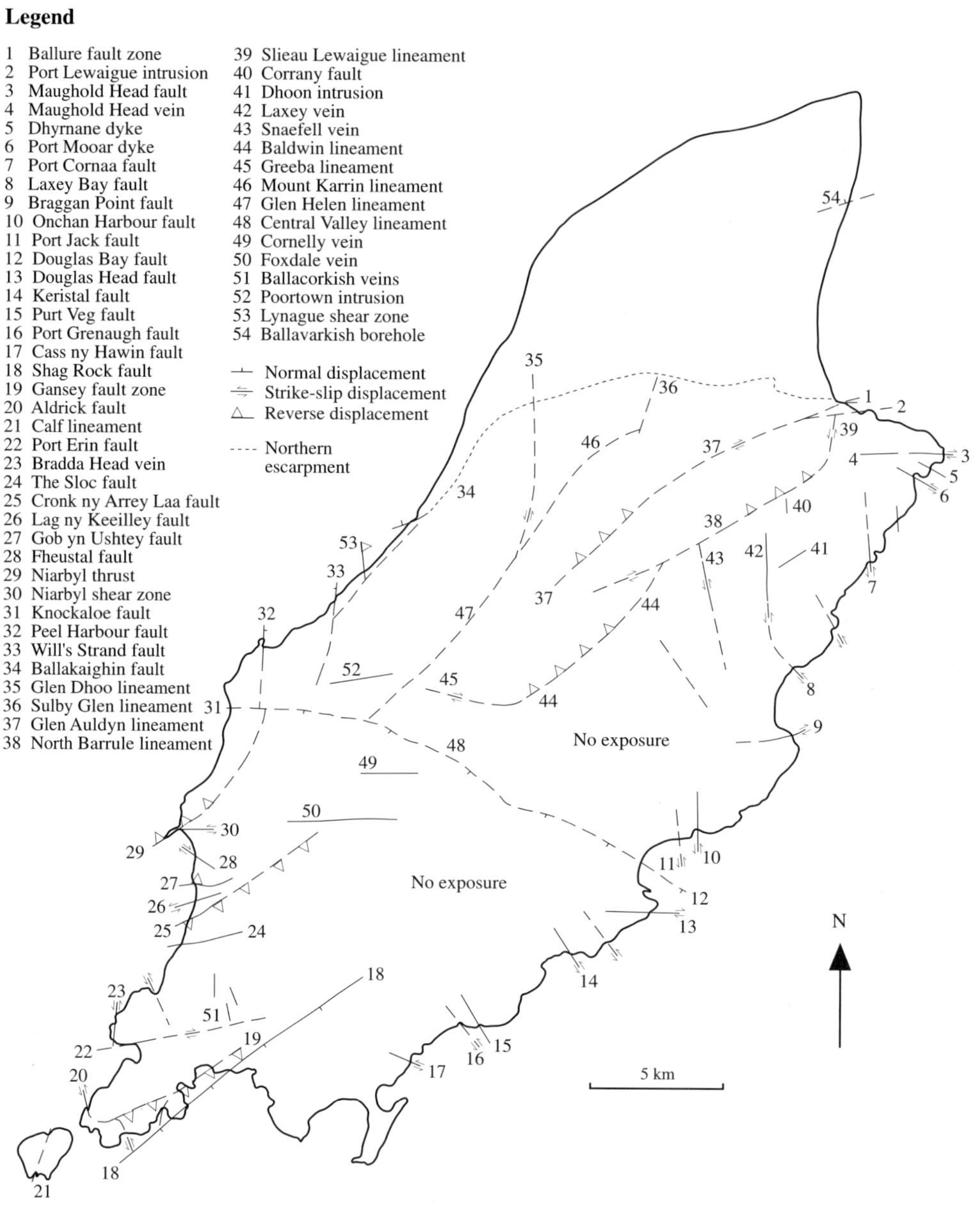

Fig. 1. Map showing observed and inferred faults affecting Palaeozoic rocks of the Isle of Man, based mostly on field work reported in Quirk & Burnett (1999).

dividing the Manx Group into a number of geological tracts was divised (Woodcock *et al.* 1999). However, it was only in 1998, when a lithofacies map had been compiled by Quirk & Burnett (1999), was it realized that there was an apparent fit between some of the important geological boundaries mapped in the field (faulted or otherwise) and the shallow geophysical lineaments originally reported by Quirk & Kimbell (1997). The objective of this paper is to compare these features in order to erect a tentative tectonic model. Detailed work on the kinematic history is required before a more rigorous structural interpretation can be developed although a start has already been made by Fitches *et al.* (1999).

Available data

The data used in this study comprise high resolution aeromagnetic images owned by World Geoscience Corporation, Bouguer gravity anomaly data owned by the British Geological Survey and Western Geophysical, marine 2D seismic data owned by JEBCO, new 1:10 000 geological field maps of the Manx Group (Quirk & Burnett 1999), published and unpublished mine data (Lamplugh 1903; Mackay & Schnellman 1963; Ford 1993; Cowin, pers. comm.), proprietary aerial photographs loaned by the Isle of Man Department of Local Government and the Environment (Kelly, pers. comm.) and sparse borehole information (Young, pers. comm.; Quirk& Kimbell 1997).

The aeromagnetic data were acquired at a height of 80 m with a 400 m traverse line spacing in a 020° direction with tie lines every 1200 m perpendicular to the traverse. The data were processed by application of standard corrections (i.e. International Geomagnetic Reference Field correction, diurnal variation removal, etc.) and by application of microlevelling and culture (human-made objects) suppression techniques proprietary to World Geoscience Corporation. The resultant total magnetic intensity anomaly data were then reduced to the magnetic pole with the aim of centring anomalies over their causative bodies. Subsequent frequency filtering operations allowed edges to be identified and mapped and semi-quantitative depth estimates of subtle magnetic anomaly variations in the subsurface geology to be made. Of particular use in this were derivative (or gradient) images and pseudodepth (or wavelength separation) slices. All aeromagnetic data were displayed using World Geoscience Corporation proprietary software making use of different colour schemes and directions of false illumination, as appropriate (e.g. Fig. 2). The onshore Bouguer gravity anomaly data were derived from 360 stations over a total area of 572 km^2 and displayed as horizontal gradient to highlight edges (e.g. Fig. 3). In addition, *c.* 200 km of migrated 2D seismic data were used to map reflective structures in the Lower Palaeozoic basement offshore to the southwest of the island.

Figure 1 is a map of observed and inferred faults and mineralized fractures based on field mapping made with reference to mine plans, unpublished aerial photographs and limited borehole data. The sense of offset shown on many of the faults was determined mostly by whether rocks on one side of the fault are thought to be older or younger than those on the other side, according to lithostratigraphic correlations presented in detail by Quirk & Burnett (1999). However, it is acknowledged that a degree of interpretation is involved in this. Whether a fault is shown as having dip-slip or strike-slip movement was based simply on its apparent dip. An arbitary value of 80° of dip was taken to discriminate between faults interpreted as normal or reverse (< 80°) and those interpreted as dextral or sinistral (> 80°). In only a few shear zones have kinematic indicators been observed (e.g. Quirk & Kimbell 1997; Fitches *et al.* 1999; Holdsworth, pers. comm.). Faults shown on Fig. 1 with no sense of movement indicated are those where correlation across the fault is equivocal. Most of the large faults are named using nomenclature from Quirk & Burnett (1999) and Woodcock *et al.* (1999), which utilize the nearest obvious place name.

Linear anomalies observed in potential field data

The magnetic response of sedimentary and metasedimentary rocks is determined mostly by the amount of iron minerals such as magnetite, ilmenite and pyrrhotite they contain. It is measured as magnetic susceptibility or how easily a rock becomes magnetized within the Earth's magnetic field. With the use of a hand-held magnetic susceptibility meter during field work, a small but significant difference has been found in the magnetic response of mudstone-rich lithofacies compared with sandstone-rich lithofacies in the Manx Group. For example, rocks containing > 90% mudstone have an average magnetic susceptibility of 0.44×10^{-3} SI (based on 73 readings, with a range of $0.2–2.1 \times 10^{-3}$ SI) whereas sandstone containing < 40% mudstone, has an average magnetic susceptibility of 0.26×10^{-3} SI (60 readings, with a range of $0.0–0.4 \times 10^{-3}$ SI). The relatively high magnetic susceptibility of mudstones in the Manx Group is probably due to the presence of significant amounts of iron, particularly ilmenite which is a common metamorphic phase in the Manx Group (Power & Barnes 1999). A difference in magnetic response in the order of a few nanoteslas is likely to occur across any steeply dipping geological boundary separating Manx Group units with contrasting amounts of mudstone. The most obvious of these types of boundaries are faults (McIntyre 1980) which are likely to appear as linear edge-type anomalies similar to those observed on aeromagnetic data from the Isle of Man (Quirk & Kimbell 1997). In addition, if the fault has been subject to fluid flow, particularly hydrothermal fluid, then it may itself produce an anomalous

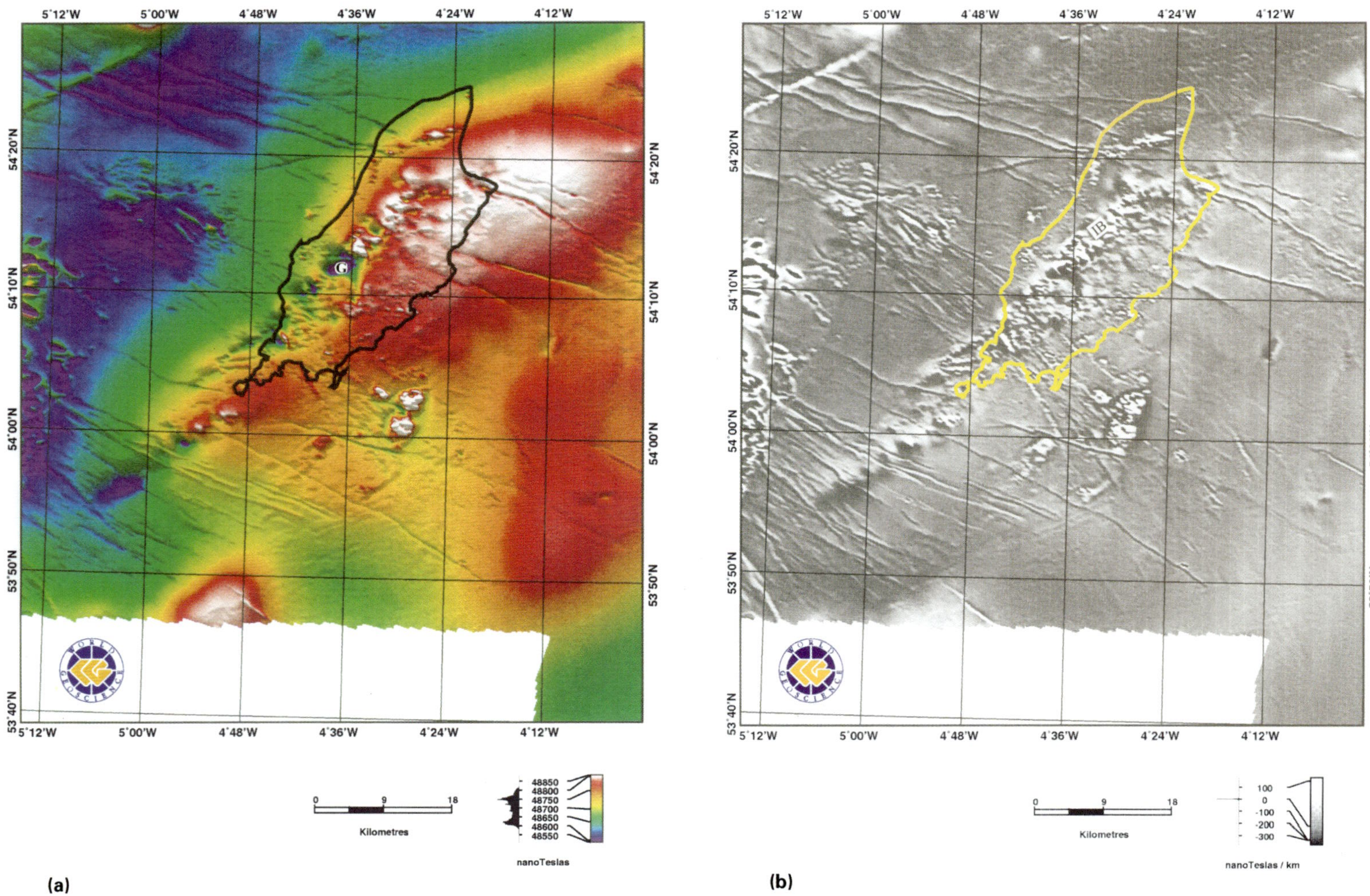

Fig. 2. (**a**) Culture-suppressed total magnetic intensity image (reduced to pole) of the Isle of Man illuminated from the north-northwest (courtesy of World Geoscience). Grid cell size, 130 m; G, Greeba Magnetic Low. (**b**) Grey scale image of first vertical derivative of total magnetic intensity image shown in (a). IB, Manx Imbricate Belt.

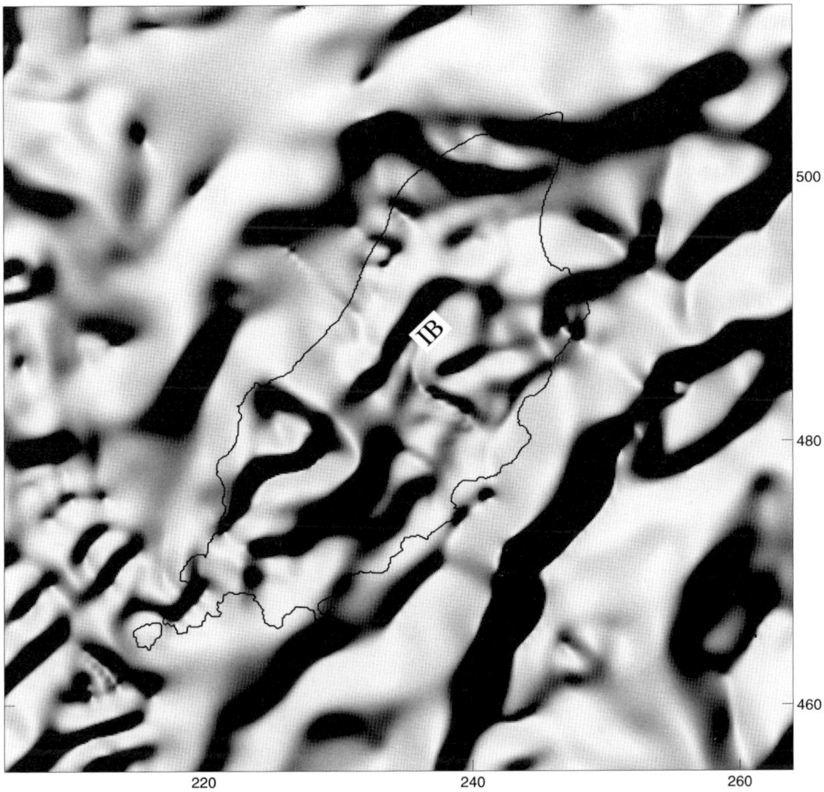

Fig. 3. Shaded relief image of the horizontal gradient of the Bouguer gravity anomaly. The original data is displayed for a larger area in Kimbell & Quirk (1999, fig. 4). Illumination is from 330°. This display highlights local features, particularly edges, but note that some apparent changes in character can relate to differences in data density and quality. IB, Manx Imbricate Belt.

response because of the effects of mineralization and wall-rock alteration (McIntyre 1980). This is likely to be the case in the Isle of Man where large east–west and north–south lead–zinc–copper–quartz veins are present (Ford 1993).

Due to the close spacing of the flight lines, horizontal resolution of the aeromagnetic data is already as high as a few hundred metres, and vertical and horizontal derivatives of the total magnetic field (e.g. Fig. 2b) help to further enhance the definition of edges. In contrast, the resolution of the Bouguer gravity data over the Isle of Man is almost an order of magnitude lower but, here again, with the careful use of derivative images (e.g. Fig. 3), it is still possible to pick out steep, near surface boundaries between rocks of contrasting density (cf. Cornwell 1972). In fact, the two potential field data sets complement each other so well that often where a lineament appears to die out on one it becomes prominent on the other (Quirk & Kimbell 1997).

The most obvious features on the aeromagnetic data are:

1. short wavelength (shallow) lineaments forming the imbricate belt that runs along the northeast–southwest axis of the island (e.g. Fig. 2b);
2. a long wavelength change from high total magnetic intensity east of the island to low total magnetic intensity west of the island at a deep boundary lying subparallel to the imbricate belt (e.g. Fig. 2a);
3. narrow, elongate anomalies trending west-northwest–east-southeast that correspond to dolerite dykes ('Tertiary-type' lineaments).

Features 2 and 3 are discussed in detail by Kimbell & Quirk (1999) and Horak *et al.* (1999), respectively, and hence this paper will concentrate on the imbricate belt and related onshore structures (1).

Manx Imbricate Belt

The short-wavelength nature of the lineaments within the imbricate belt indicates that the causes of the aeromagnetic anomalies lie probably within the upper 1 km of the crust and may therefore intersect the surface. Three trends exist: east–west (to east-northeast–west-southwest), northeast–southwest and approximately north–south (Fig. 4). In addition, west-northwest–east-southeast (to northwest–southeast) trending lineaments cross the imbricate belt (Fig. 2), corresponding to Tertiary dykes. These can lead to aliasing (false alignment) effects during interpretation because of interference with other trends. The margins of the imbricate belt are also clearly expressed on Bouguer gravity data but the internal structure is only poorly resolved (Fig. 3).

The imbricate belt consists of two segments separated by the Central Valley Lineament stretching from Peel to Douglas (Fig. 1; Quirk & Kimbell 1997). To the southwest of the Central Valley Lineament is a bow-shaped segment, 6 km wide and 20 km long. This has a straight northwest side which extends from the west coast to St Johns in the Central Valley and a curved southeast side

Fig. 4. Interpretation of potential field lineaments observed on Figs 2 and 3 made without reference to surface geology. Thick lines indicate the position of strong lineaments visible on high-resolution aeromagnetic data or where lineaments coincide on more than one data set; thin lines indicate the position of other lineaments visible on aeromagnetic data, Bouguer gravity anomaly derivatives and aerial photographs. Localities referred to in the text are also shown: V, Port e Vullen; M, Glen Mooar; B, Beary Mountain; J, St Johns; G, Greeba; C, Crosby; F, Foxdale Granite; A, Cronk ny Arrey Laa; S, St Marks; O, Oatlands Intrusion; E, Port Erin.

running from Port Erin to Crosby (Fig. 4). East–west lineaments within this segment have a left-stepping en echelon arrangement. A similar shaped, but less clearly defined, group of lineaments lies to the southwest of Port Erin, located mostly in the offshore but also including an area between Port St Mary and the Calf of Man. The imbricate belt is offset slightly to the left by the Central Valley Lineament (Fig. 2b). Northeast of here, the imbricate belt comprises a rhomboid-shaped segment, 5 km wide and 20 km long, defined by northeast–southwest to east–west lineaments. It seems to pinch out between Ramsey and Maughold (Fig. 4). In addition, the aeromagnetic data show a number of east–west lineaments branching off from the main imbricate belt which intersect the eastern coast of the Isle of Man.

Areas of anomalous magnetization occur within the imbricate belt, the most prominent of these

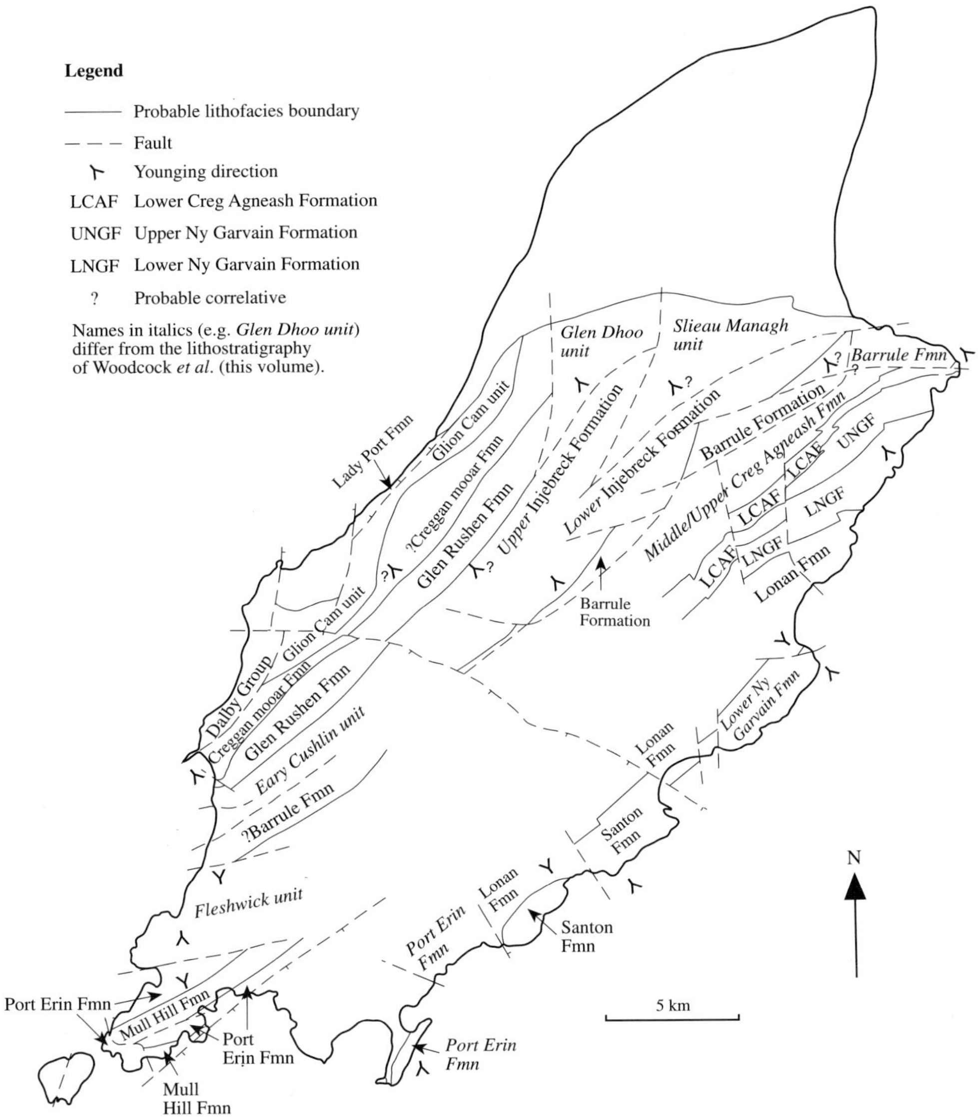

Fig. 5. Map of main lithostratigraphic units within the Lower Palaeozoic of the Isle of Man. The principal faults are also shown on Fig. 1.

being a marked low covering an area of at least 10 km^2 centred around Greeba in the Central Valley (Fig. 2a). This low is juxtaposed on its northeast side by an equally well-defined area of anomalously high magnetization around Beary Mountain (Fig. 4). A similar but smaller pair of low–high anomalies occurs close to mineral veins at Ballacorkish Vein (Fig. 1) in the south of the island lining up with a west-northwest–east-southeast Tertiary type lineament on the southeast side of the imbricate belt (Fig. 2a). Field mapping has provided no lithological explanation for these areas of anomalous magnetization but Quirk & Kimbell (1997) suggested that they may have formed during metasomatism associated with a period of hydrothermal fluid activity recorded in the Irish Sea during the early Tertiary (e.g. Green *et al.* 1997).

Coincidence of geophysical lineaments with geological boundaries

Despite the fact that the potential field lineaments shown in Fig. 4 were interpreted without reference to faults mapped in the field (Fig. 1), there is more than a passing similarity between the two maps. In fact, every strong lineament seen to intersect the coast where exposure is good seems to coincide with a faulted geological boundary and, conversely, only a few of the major faults in the Manx Group do not have a clear geophysical expression. In addition, offshore seismic data, particularly to the southwest of the island, provide clear evidence for the 3D nature of the imbricate belt (see later). The coincidence between potential field anomalies and surface geology is discussed below, separated into geographical areas for which Figs 1 and 4–6 can be used for reference.

Port Erin–Port St Mary

One of the most obvious aeromagnetic features observed on the Isle of Man is an east-northeast–west-southwest lineament intersecting the southwest coast at the north end of Port Erin (Fig. 2a). Here, a steeply north dipping fault or shear zone at [SC 193 697] is inferred from geological mapping and aerial photographs (the Port Erin Fault; Fig. 1). The fault itself cannot be observed due to a 15 m wide gap in exposure but it marks the boundary between quartzites belonging to the Port Erin Formation to the south (Fig. 5) and pebbly mudstones of the Fleshwick Unit (or Maughold Formation of Woodcock *et al.* 1999). The Fleshwick Unit has significantly lower illite crystallinity grades than the Port Erin Formation on the other side of the fault (Roberts *et al.* 1990). Depending on how the correlations are made, it accounts for between 2.5 and 5.5 km of stratigraphic offset by apparent dextral movement cutting out the Barrule Formation, at least part of the Creg Agneash Formation and possibly the Santon–Ny Garvain formations (Fig. 5; Quirk & Burnett 1999). Major folds verge towards the Port Erin fault on both sides (Fitches *et al.* 1999) and deformed pebbles within the northern wall display a gently west dipping linear fabric parallel to the inferred direction of shear on the fault.

A few kilometres east of here, between Port St Mary and Gansey, a set of southeast dipping thrusts occur in association with a major recumbent fold (the Gansey Fault Zone; Fig. 1). One of these faults extends southwest to Cregneash where it is responsible for thrust repetition of a thick quartzite interval (the Mull Hill Formation; Fig. 5). Although the continuation of this thrust into the offshore is complicated by the presence of a major north-northwest–south-southeast striking sinistral cross-fault at the coast (the Aldrick Fault; Fig. 1), it probably links up with a strong northeast–southwest trending lineament on the north side of the Calf of Man (Fig. 4).

A short north–south aeromagnetic lineament on Bradda Head, just north of the Port Erin Lineament, coincides with the Bradda Head Vein (Fig. 4). This comprises a vertical, quartz-filled and metasomatized fault which accounts for a few hundred metres of sinistral offset based on lithostratigraphic correlations presented in Quirk & Burnett (1999). A similar but north-northeast–south-southwest trending structure is observed on aerial photographs crossing the Calf of Man which is reported to be associated with quartz mineralization (Fitches, pers. comm.).

Offshore extension of the imbricate belt

The imbricate belt continues offshore to the southwest of the Isle of Man where it is crossed by several 2D seismic lines (Fig. 6a). One line in particular (Fig. 6b and c) images an unusual set of northwest dipping, high-amplitude reflections within the basement coinciding at the sea bed with the central part of the imbricate belt observed on aeromagnetic data (Fig. 2). Near the surface, the reflections occupy a zone *c.* 2.5 km wide which dip between 25 and 35°. They converge downwards and become difficult to resolve below 1.5 s two-way time travel (*c.* 4 km depth). The strength of these reflections is unusual within basement rocks, particularly in view of their relatively steep dip, and the most likely interpretation is that they image either fluid-filled fracture zones or intrusions with high acoustic impedance. Even in the case of intrusions, by analogy with the onshore, they probably occur in association with major faults or shear zones. Although not as clearly expressed on other seismic lines, the dipping reflections can be mapped for a distance of almost 25 km in a northeast–southwest (strike) direction (Fig. 6a).

The reflections on the southeast side of the dipping structure are truncated near surface by a major northeast–southwest trending normal fault which is the offshore extension of the Shag Rock Fault (Fig. 1). Onshore, this fault delimits the northwest edge of the lower Carboniferous Castletown Group cropping out between Port St Mary and Langness. Where this fault has been traced in the field using geological evidence alone (e.g. Lamplugh 1903) it shows little, if any, potential field expression. However, it coincides with the projected southeast edge of the imbricate belt linking the onshore with the offshore extension (Fig. 7) and the fault has probably exploited this zone of crustal weakening (cf. Quirk & Kimbell 1997).

A structural extension of the Isle of Man known as the Ramsey–Whitehaven Ridge occurs in the offshore, northeast of the island (Quirk & Kimbell 1997). This represents a basement block underlying a thin Carboniferous cover that is tilted by 10–15° to the northwest as a result of uplift in the footwall of the northeast–southwest Lagman Fault (Quirk *et al.* 1999). The Lagman Fault and its southern extension (the Eubonia Fault) run close and parallel to the east coast of the Isle of Man and, judging from their northeast–southwest trend, similar to the Shag Rock Fault, they may have exploited a pre-existing Caledonian structural grain (Jackson & Mullholland 1993). Both faults are normal with 1–3 km of Carboniferous and Permo-Triassic strata preserved in their southeast hanging walls. Analogous faults are not observed on the western side of the Isle of Man and therefore the Manx Group is thought to have been tilted to the northwest after the Caledonian by an amount similar to that recorded on the Ramsey–Whitehaven Ridge (10–15°).

Cronk ny Arrey Laa

A set of northeast–southwest and east–west aeromagnetic and Bouguer gravity lineaments intersect the west coast of the Isle of Man between The Sloc and Gob yn Ushtey, with the hill of Cronk ny Arrey Laa in the centre (Fig. 4). This is a steep section of coastline but, where access to the shore is possible, e.g. below Lag ny Keeilley [SC 215 745] and at Gob yn Ushtey [SC 216 756], a number of shear zones and brittle faults are exposed together with ?pre-kinematic felsitic intrusions and brecciated quartz veins (Fig. 1). At Gob yn Ushtey a 50 m wide fault zone is present, the northern wall of which trends 060°/70° NW. An 8 m wide, steeply north dipping ductile shear zone with sinistral kinematic indicators is exposed at high watermark to the northwest of Lag ny Keeilley [SC 216 747]. A short distance north of here a set of quartz veins and metasomatic tourmaline occurs in a 100 m wide fracture zone (Morris, pers. comm.) and includes a large reverse fault displaying slickensides pitching 70° SW on 045°/45° NW.

Where the coast is not accessible immediately south of Lag ny Keeilley a strong topographic feature has been used to infer the position of a major south-southeast dipping fault running from Stroin Vuigh [SC 213 742] to the north side of Cronk ny Arrey Laa, which marks the northwest boundary of the Barrule Formation (the Cronk ny Arrey Laa Fault; Fig. 1). Tentative lithostratigraphic correlations suggest that this fault accounts for almost 3.5 km of stratigraphic offset cutting out the Fleshwick Unit, the probable lateral equivalent of the lower part of the Injebreck Formation (Fig. 5; Quirk & Burnett 1999). At The Sloc, a similar change in topography and an observation by Lamplugh (1903) of quartz veining and breccia in a mine trial at The Stacks [SC 211 735] suggests that another east-northeast–west-southwest trending fault may be present at the mapped southeast boundary of the Barrule Formation (the Sloc Fault; Fig. 1).

Overall, the area between Gob yn Ushtey and The Sloc is thought to represent a zone of extensive heterogeneous deformation with both ductile and brittle movement, and evidence of intrusions and hydrothermal mineralization. It corresponds approximately with the position of Lamplugh's (1903) and Simpson's (1963) synclinorium axis to which folds verge on both sides (Fitches *et al.* 1999). It is thought likely that a similar sort of complex structure has produced the zone of dipping reflections imaged on seismic in the offshore extension of the imbricate belt (Fig. 6).

Niarbyl

Beyond the apparent northern limit of the imbricate belt at Gob yn Ushtey, an important ductile shear zone and major brittle fault is exposed at Niarbyl [SC 211 776] (Fig. 1). The fault marks the contact between the lower Ordovician Manx Group (Creggan Mooar Formation) in the footwall and the mid-Silurian Dalby Group in the hanging wall (Fig. 5), but is none the less interpreted as a thrust because of the presence of minor compressional structures (Quirk & Kimbell 1997; Morris *et al.* 1999). This faulted boundary has been mapped northwards for 7.5 km to Peel. Here it joins the Central Valley Lineament to become the Peel Harbour Fault (Fig. 1), where it also defines the western edge of the ?Devonian Peel Group.

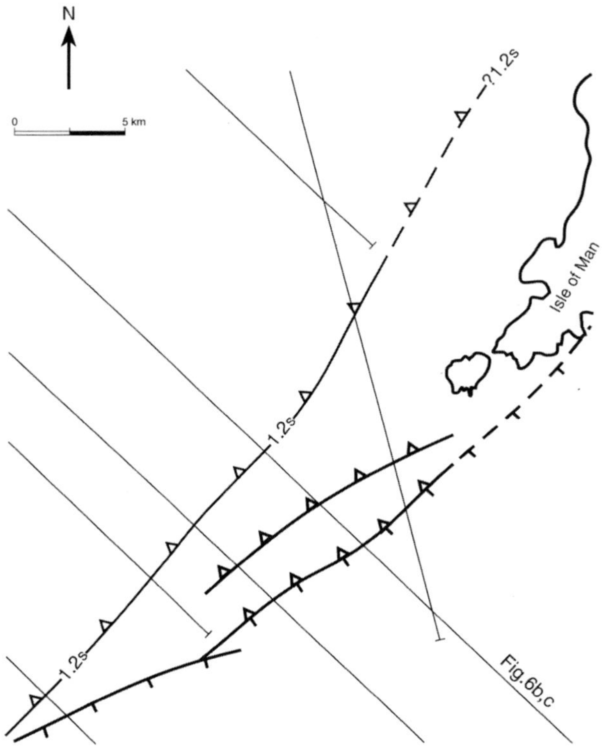

(a)

Fig. 6. (**a**) Map showing interpreted position of northwest dipping basement reflections near the sea bed (thick line ornamented with triangles) and at 1.2 s two-way time travel (equivalent to a depth of *c.* 3.5 km). Also shown are the seismic lines used (thin, straight lines) and a Carboniferous normal fault (thick line ornamented with ticks). (**b**) Northwest-southeast stacked 2D seismic line located *c.* 10 km southwest of the Isle of Man (courtesy of JEBCO). Note the prominent northwest dipping reflections within Lower Palaeozoic basement. The vertical scale is in s two-way time travel. (**c**) Geoseismic section of migrated version of the 2D seismic line shown in (b), showing prominent northwest dipping reflections within the Lower Palaeozoic basement. The vertical scale is in s two-way time travel.

Although it is poorly expressed on aeromagnetic data, a clear edge is visible on Bouguer gravity data (Fig. 2) and Roberts *et al.* (1990) recognized the presence of anomalously low illite crystallinity values around Niarbyl probably due to strain-induced metamorphic regression. At this locality the fault trends between 070°/15° NW and 000°/60° W, and truncates the ductile shear zone occupying the footwall. The shear zone itself is several tens of metres wide and contains highly deformed sediments, disrupted felsitic intrusions and disaggregated quartz veins with a strong phyllonitic fabric orientated 105°/75° N. Kinematic indicators described by Fitches *et al.* (1999) suggest overall sinistral displacement.

Northwest coast

A ductile shear zone, smaller than the one exposed at Niarbyl, occurs on the northwest coast at Lynague [SC 281 871] (Fig. 1). In contrast to Niarbyl, the Lynague Shear Zone is orientated north-northwest–south-southeast, displaying a planar phyllonitic fabric which trends between 160°/65° NE and 125°/50° NE with a reverse (–?dextral) sense of shear based mostly on folded and boudinaged quartz veins.

Elsewhere along the coast between Will's Strand and Glen Mooar the structure of this part of the Manx Group (the Lady Port Formation; Fig. 5) is complicated by younger faults but also includes numerous northwest dipping thrusts. A north–south vertical fault forms the boundary between the Manx Group and the ?Devonian Peel Group at Will's Strand, whereas the northern limit of the Manx Group is marked by an east-northeast–west-southwest vertical fault near Glen Mooar. The Lady Port Formation is interpreted to be in faulted contact with the Glion Cam Unit a few hundred metres inland. This putative fault (the Ballakaighin Fault; Fig. 1) is thought to run parallel to the coast (Woodcock *et al.* 1999). None of the faults or shear zones in this area are clearly imaged on potential field data.

The orientation of bedding and cleavage in the Lady Port Formation is also unusual in that instead of striking northeast–southwest to east-northeast–west-southwest, as is the case for the majority of the Manx Group, north–south trends are common. Inland of Glen Mooar, around Glen Dhoo and

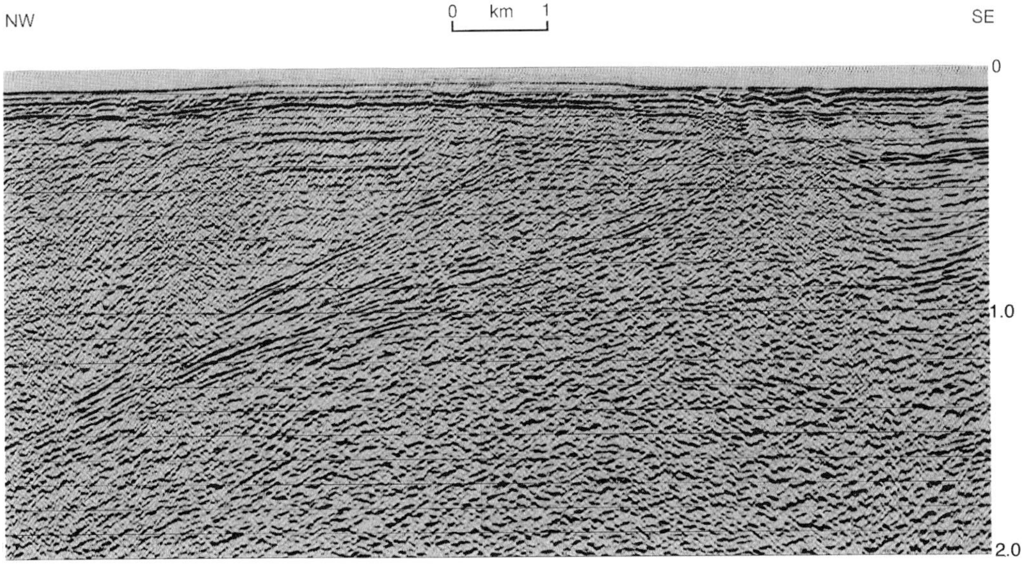

(b)

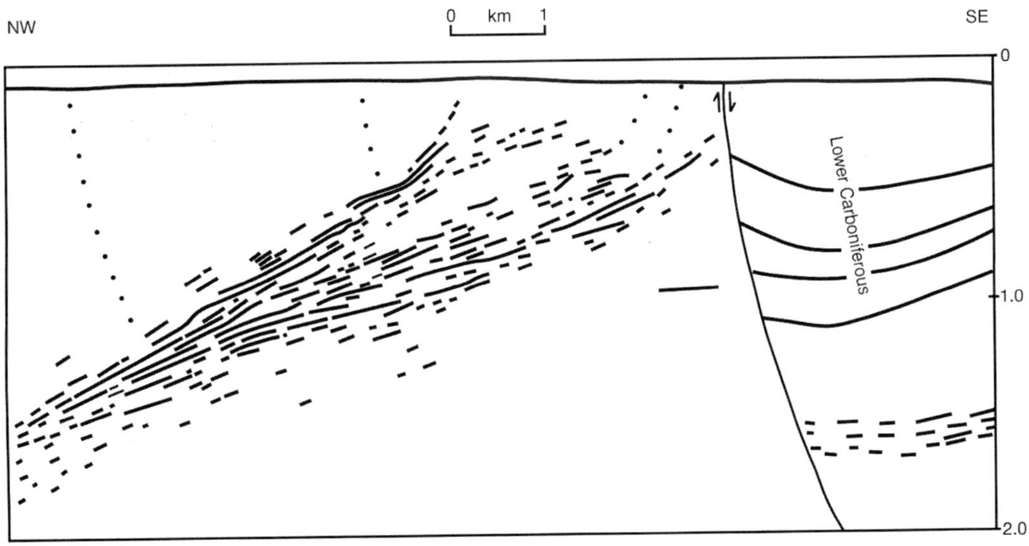

(c)

Sulby Glen, two approximately north–south trending faults are interpreted (Fig. 1) bounding the western and eastern sides of an unusual sandstone-dominated interval known as the Glen Dhoo Unit (Quirk & Burnett 1999). The mapped position of the fault bounding the western side of the Glen Dhoo Unit is supported by the presence of a mine waste heap at [SC 340 890] consisting, to a large degree, of fault breccia. It also coincides with the approximate location of a north–south trending aeromagnetic lineament (Fig. 4). The southern boundary of the Glen Dhoo Unit is probably defined by a northeast–southwest trending normal fault (the Mount Karrin lineament; Fig. 1).

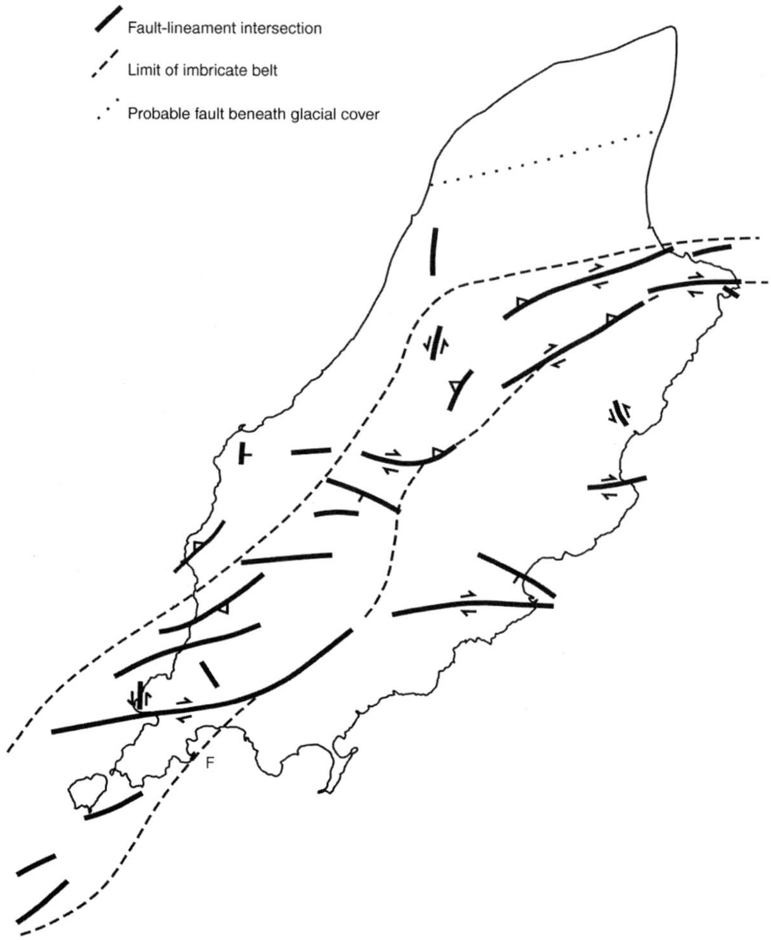

Fig. 7. Map made by integrating Figs 1 and 4 showing coincidence between faults mapped in the field and prominent potential field lineaments.

Central Valley

Aeromagnetic data show that the imbricate belt is slightly offset to the left by a west-northwest–east-southeast trending lineament running along the Central Valley between Peel and Douglas (Figs 2b and 4). On Bouguer gravity data (e.g. Fig. 3), the western half of the lineament marks the northeast edge of a gravity low, at least in part, related to the Foxdale Granite (Cornwell 1972). Field mapping and aerial photographs suggest that the Central Valley Lineament bifurcates near Peel, one strand of which coincides with an east–west fault near Knockaloe, offsetting the basal contact of the Dalby Group (Figs 1 and 5; Morris *et al.* 1999). Geophysical data described by Quirk *et al.* (1999) and Quirk & Burnett (1999) show that the lineament extends into the offshore as a set of normal faults in younger strata defining, for example, the northeast margin of the Permo-Triassic Peel Basin. Although onshore it has a marked topographic expression in that the Central Valley cuts straight through the uplands at < 50 m above sea level, it is nowhere exposed. However, field mapping of lithofacies around Douglas suggests that it corresponds to a linear feature on the shore at [SC 386 762], described as a 'pre-glacial valley' by Lamplugh (1903) but here interpreted as a fault throwing down to the southwest with the Lonan Formation on the northeast side juxtaposed against the younger Santon Formation on the southwest side (Fig. 5). The left lateral apparent sense offset of the imbricate belt is consistent with this sense of throw provided that the imbricate belt dips to the northwest as observed on seismic data (Fig. 6) and in the field at Gob yn Ushtey and near Ramsey (see below). Quirk & Kimbell (1997) suggested that the Central Valley Lineament was active during extension in the early Permian and early Tertiary but may represent an older pre-Caledonian trend.

A northern strand of the Central Valley Lineament is seen on aeromagnetic data to extend from Peel to the northeast edge of the Greeba Magnetic Low (Figs 2a and 4) where it coincides with an east–west fault (the Greeba Lineament; Fig. 1), inferred from field mapping to have cut out part of the Injebreck Formation (Fig. 5; Quirk & Burnett 1999). The lineament passes close to the Poortown mafic intrusion at the limit of the magnetic low where geophysical and borehole data show that this igneous body is broken up by east–west shear zones (Piper *et al.* 1999; Power and Crowley 1999). The Greeba Lineament is linked further east to another (probably faulted) boundary forming the southeast edge of the Barrule Formation which it appears to partly cut out (Baldwin Lineament, Fig. 5).

Maughold

The imbricate belt is seen to feather out at the northern end of Maughold (Fig. 4), directly east of the escarpment marking the northern limit of the Manx Group (Fig. 1). The aeromagnetic lineaments in the Maughold area have a dominant east-northeast–west-southwest trend, although there is interference between these and a strong west-northwest–east-southeast Tertiary-type anomaly (Fig. 4).

At Maughold Head [SC 497 915] a minimum of 250 m dextral offset is evident across a vertical or steeply north dipping east–west fault which juxtaposes younger muddy facies to the north against quartzites to the south (Fig. 1). Lithostratigraphic correlations suggest that there may be as much as 2 km dextral offset across the fault cutting out almost all of the Creg Agneash Formation (Quirk & Burnett 1999) (Fig. 5). It lines up with a major east–west vein which was mined underground near Maughold Head for copper (Cowin pers. comm.). Further to the west, the structure may join an east-northeast–west-southwest fault which is interpreted to form the southeast boundary of the Barrule Formation (the North Barrule Lineament; Figs 1 and 5). Just south of Maughold Head, at [SC 498 917], the faulted base of the Creg Agneash Formation is exposed. The contact consists of a tilted ramp-flat structure trending between 100°/70° N and 090°/20° N, which appears to be linked to the Maughold Head Fault. It has accommodated at least 50 m of shortening during the development of chevron folds in the hanging wall (Creg Agneash Formation) whereas the underlying Ny Garvain Formation is undeformed. It probably represents a minor accommodation structure developed in association with movement on the Maughold Head fault.

Between Port e Vullen and Ramsey numerous east–west and northeast–southwest shear zones, faults and brecciated quartz veins are exposed which coincide with a set of similarly orientated lineaments visible on aeromagnetic and Bouguer gravity data (Figs 1–4). The shear zones are mostly steeply north dipping associated with several pre-kinematic felsitic intrusions (Quirk & Burnett 1999) whereas the brittle faults tend to dip to the northwest and are associated with quartz veins. In addition, there are several north–south trending faults and quartz veins, one of which accounts for *c.* 50 m of sinistral offset in Port Lewaigue [SC 469 930], others having been mined for lead and copper (Lamplugh 1903; Ford 1993).

Between Maughold Head and Port Cornaa a number of northwest–southeast Tertiary dykes and north–south faults are exposed (Fig. 1). The Tertiary dykes and associated haematite mineralization are described by Quirk & Kimbell (1997) and the north–south faults by Woodcock & Barnes (1999). Corresponding northwest–southeast trending linear anomalies are clearly visible on the aeromagnetic data but north–south features are more subtle (Figs 2 and 4).

Northern inland area

Quirk & Burnett (1999) have used field evidence to suggest that a number of northeast–southwest to east-northeast–west-southwest trending thrusts and dextral strike-slip faults are present in the northern uplands of the Isle of Man (e.g. the Glen Auldyn Lineament, the North Barrule Lineament and the Baldwin Lineament; Fig. 1). These faults are responsible for cutting out parts of the Barrule and Injebreck Formations (Fig. 5). Aeromagnetic data (Fig. 2) and, to a lesser extent, Bouguer gravity data (Fig. 3) show anomalies supporting the existence of these lineaments (Fig. 4). Physical evidence for the existence of the Glen Auldyn Lineament occurs at its western end where a set of faults and shear zones were encountered in boreholes drilled in the vicinity of [SC 372 891] for the Tholt y Will reservoir (Young pers. comm.). Significantly higher illite crystallinity grades occur on the northwest side of the lineament (Roberts *et al.* 1990), supporting the interpretation that it represents a northwest dipping reverse fault (Fig. 1). Quirk & Burnett (1999) note that an unusual thickness of pebbly mudstone (> 500 m) occurs north of the Glen Auldyn Lineament, perhaps indicating that it, or a related structure, was active during sedimentation. The east-northeast–west-southwest trending Causey Pike Thrust in the Lake District is thought to have a similar origin (Webb & Cooper 1988; Hughes *et al.* 1993). The North Barrule Lineament is probably also a thrust; it coincides with the position of a trial mine adit at [SC 387 872], suggesting that it is partly mineralized. By extrapolation from offshore seismic data, Quirk & Kimbell (1997) also suggest that the steep escarpment forming the northern edge of the uplands (Fig. 1) is defined by a set of east–west faults. A corresponding lineament is clearly imaged on Bouguer gravity data and lines up with a vertical fault at Glen Mooar marking the northern limit of the Manx Group (Figs 1, 3 and 4).

In other areas inland where there is little evidence of missing stratigraphy, there is none the less some coincidence between the mapped bases and tops of thick mud-rich intervals (the Barrule and Glen Rushen Formations; Fig. 5) and the position of apparent linear aeromagnetic anomalies (Figs 2 and 4). Whether these anomalies are due only to the high magnetic susceptibility of these sediments relative to more sandstone-rich intervals (see above) or whether they show that faults have tended to concentrate at these boundaries cannot be proven because of limited exposure.

Clay Head

An east–west aeromagnetic lineament extends from near Greeba in the Central Valley to Braggan Point on Clay Head on the eastern coast of the Isle of Man (Fig. 4). Independently, field mapping has identified a 5 m wide, steeply south-southeast dipping shear zone at [SC 442 808], in the position of the lineament, which truncates the axis of the Douglas Syncline (the Braggan Point Fault; Fig. 1). It consists of foliated and disaggregated sediments and quartz mineralization, although closer examination for kinematic indicators has so far proved impossible as the shear zone occupies a precipitous gully. However, from lithostratigraphic correlations, it is calculated that *c.* 1 km of Ny Garvain Formation [equating with the Santon Formation of Woodcock *et al.* (1999)] has been cut out by apparent dextral offset (Fig. 5). Adjacent to the fault, where it is accessible, the beds in the southeast wall are overturned whereas a gentle antiform is present to the northwest. The antiform is bounded some 400 m away on its western side by a brittle thrust trending 055°/60° SE which, on the basis of drag folds in both walls, is interpreted to have dextral–reverse offset. A 2 m wide felsitic dyke with a strong foliation lies within a few metres of the fault.

Marine Drive

An east–west aeromagnetic lineament links Douglas Head at the northern end of Marine Drive to a bend in the imbricate belt close to St Marks (Fig. 4). At the coast it coincides with a 3 m wide, subvertical brecciated fault zone near Douglas Head at [SC 387 745] (Fig. 1). Based on rather tentative lithostratigraphic correlations, the fault may account for 100–200 m of apparent dextral offset (Quirk & Burnett 1999).

The fault approximately marks the northern limit of a wacke-rich lithofacies type in the Manx Group unique to Marine Drive (Quirk & Burnett 1999). This interval is in turn bounded to the south by a northwest-southeast fault at Keristal ([SC 357 732]) which offsets the Santon Formation in a sinistral sense by *c.* 1.5 km (Fig. 5). Three interpretations are possible:

- that the change in lithofacies is coincidental;
- that the Marine Drive interval occupies a fault block that has moved west, juxtaposing it against younger strata at its southern and northern ends (Fig. 1);
- that the east–west fault is an old trend controlling sedimentation of the wacke-rich lithofacies (Quirk & Burnett 1999).

Purt Veg–Cass ny Hawin

A major geological boundary lies close to Purt Veg at [SC 324 703] where thick-bedded sandstones to the east belonging to the Santon Formation are juxtaposed against thin-bedded mudstones belonging to the Port Erin Formation across a fault trending 140°/85° NE (Fig. 5). A 2 m wide fault breccia is present here with a thin Tertiary dyke occupying the northeast wall. Depending on how lithostratigraphic correlations are made, this fault may have cut out *c.* 3 km of succession by apparent sinistral movement or 0.7 km of stratigraphy by apparent dextral movement (Quirk & Burnett 1999). By analogy with the Keristal Fault, sinistral displacement is perhaps most likely, implying that the mudstones to the west of the fault are the lateral equivalent of the Creg Agneash Formation further north (Fig. 5). Alternatively, on lithofacies grounds, the mudstones show similarities with the Lonan Formation implying a dextral sense of displacement (Quirk & Burnett 1999; Woodcock *et al.* 1999). A faint aeromagnetic lineament can be traced from Purt Veg to the bend in the imbricate belt at St Marks but a more obvious northwest-southeast Tertiary-type anomaly runs through Port Grenaugh at [SC 316 705] where another sinistral fault is inferred to exist (Fig. 1).

Further to the southwest at Cass ny Hawin ([SC 298 692]), the northern boundary fault to the

Castletown Group trends 095°/75° N with sub-horizontal slickensides interpreted to have formed in the Carboniferous during northwest-southeast extension (Quirk & Kimbell 1997). However, similar to the boundary fault at Port St Mary, little evidence for the fault is seen on potential field data.

Lineaments associated with mineral veins

A number of large east–west and north–south quartz veins have been exploited on the Isle of Man for galena, sphalerite and chalcopyrite during last century and the early part of this century (Lamplugh 1903; Mackay & Schnellman 1963;Ford 1993). These are all approximately vertical, as are similar trending faults exposed on the coast. Field evidence (Fig. 5; Quirk & Burnett 1999) and mine data (e.g. Lamplugh 1903; Mackay & Schnellman 1963; Jespersen 1970; Ford 1993) suggest that they represent brittle strike-slip faults with east–west faults displaying predominantly dextral offset and north–south faults displaying sinistral offset (Fig. 1). The east–west Foxdale Vein is the longest of the exploited mineral veins and at its eastern end it forms the northern margin of the Foxdale Granite (Fig. 4). The vein corresponds with a clear east–west lineament on aeromagnetic data on which other smaller east–west veins, such as at Cornelly lead mine and Maughold Head copper mine are also visible (Figs 1, 2 and 4). North–south lineaments associated with veins such as Laxey and Snaefell mines are less clearly expressed than their east–west counterparts, possibly due to interference with other trends; e.g. with a west-northwest–east-southeast Tertiary-type anomaly at Laxey (Fig. 4). Lithostratigraphic correlations suggest that the fractures hosting the Laxey and Snaefell Veins may have accommodated somewhere in the order of 500–1000 m of apparent sinistral displacement (Fig. 5). The southern end of the Laxey Vein appears to swing to the southeast so that it joins the Laxey Bay Fault. The presence of this fault is inferred on the basis of an apparent (left lateral) mismatch between lithofacies within the Lonan Formation north of the bay and those south of the bay (Quirk & Burnett 1999)

Major intrusions

The four main igneous bodies exposed in the Isle of Man are associated with lineaments observed on aeromagnetic and Bouguer gravity data. The Poortown mafic intrusion lies close to an east–west lineament at the northwest corner of the magnetic low near Greeba (Figs 1 and 2). The Foxdale Granite occurs at the eastern end of the Foxdale Vein in the centre of a gravity low (Cornwell 1972). The older Dhoon Granite (Fig. 1) is found at the intersection of a northeast–southwest lineament and an east–west lineament which coincide with a fault and shear zone in the field (Lamplugh 1903; Mulligan, pers. comm.). It too is associated with a gravity low. The Oatlands granite–diorite complex is marked by a minor northeast–southwest lineament seen on both aeromagnetic and Bouguer gravity data (Fig. 4). These lineaments represent proven or speculative faults which are likely to have accommodated emplacement or uplift of the igneous bodies. The ages of the intrusions are poorly constrained but the Poortown body is probably late Ordovician (Piper *et al.* 1999), the Dhoon Granite is probably early Caledonian (?late Silurian), equivalent to syn-D1 (Mulligan, pers. comm.), and the Foxdale Granite is thought to be late Caledonian (early–mid Devonian; Brown *et al.* 1968), probably syn-D2 (Simpson 1965). The Oatlands intrusion is no longer exposed and its age is unconstrained. However, it is also worth noting that large and small felsitic intrusions of probable pre-kinematic origin are found within or adjacent to many of the major shear zones and faults described above, implying that they are long-lived tectonic structures.

Tectonic interpretation

By integrating the evidence for faults and shear zones observed or inferred in the field (Fig. 1) with lineaments identified on potential field data (Fig. 4), it appears that the Manx Group is traversed by a set of major faults forming the imbricate belt first identified by Quirk & Kimbell (1997). The most common fault trend is east-northeast–west-southwest (Fig. 7). Based mostly on the shape of the lineaments observed on aeromagnetic data, Quirk & Kimbell (1997) suggested that the Manx Imbricate Belt represents a fault duplex formed by sinistral transpression during closure of the Iapetus Ocean in the Silurian. Recent field mapping generally does not support this kinematic interpretation in that: east–west to east-northeast–west-southwest faults are usually steep or vertical with evidence of apparent dextral offset; northeast–southwest faults seem to represent thrusts; and west-northwest–east-southeast faults are typically normal. Only steep or vertical north–south to north-northwest–south-southeast faults show evidence of sinistral offset (Fig. 1). The implication is that the imbricate belt is a duplex formed instead by northwest-southeast contraction (Fig. 8). However, the offset recorded on these faults may only represent that of the latest stage of movement which, in most cases where the faults are exposed, is brittle in nature and post-dates major Caledonian structures such as east-northeast–west-southwest

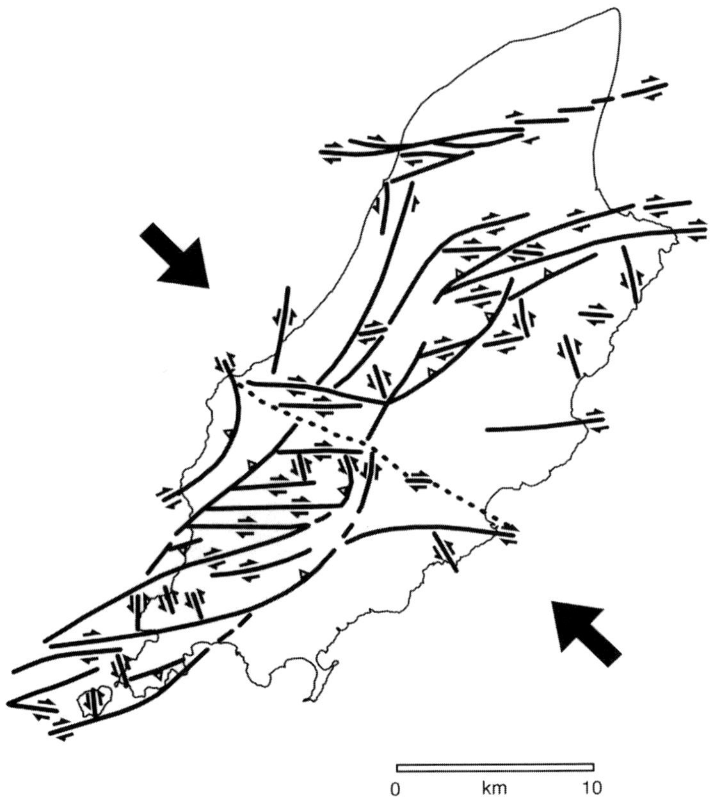

Fig. 8. Conceptual interpretation of fault lineaments active in the Isle of Man as a result of northwest-southeast compression during the late Caledonian. The imbricate belt trending northeast–southwest along the axis of the island is interpreted as a contractional duplex.

trending D1 folds and cleavage (cf. Fitches *et al.* 1999). In contrast to the faults, limited kinematic evidence on older ductile shear zones indicates that east–west lineaments, such as the Niarbyl Shear Zone, were mostly subject to sinistral movement and north-northwest–south-southeast shear zones, such as the Lynague Shear Zone to dextral movement. This suggests that there was an earlier phase of north-northeast–south-southwest directed compression, possibly associated with sinistral transpression within the imbricate belt. Although evidence for the timing of shearing relative to cleavage is not clear-cut, it is assumed here that at least some of this movement pre-dates D1 structures.

Two explanations for these observations are possible:

- that the brittle structures were formed in a tectonic event separate to that responsible for the ductile structures, e.g. in the Variscan rather than the Caledonian Orogeny;
- that the brittle structures were only the latest stage of an evolving Caledonian collisional event.

Quirk & Kimbell (1997) have already described north-northeast–south-southwest orientated Variscan reverse faults in Carboniferous strata imaged on marine seismic data close to the Central Valley Lineament. However, the offshore extension of the imbricate belt, either the Ramsey–Whitehaven Ridge to the northeast of the island or the Shag Rock fault to the southwest, seems unaffected by Variscan compression (Quirk *et al.* 1999). Therefore, the second explanation is favoured here with an early Caledonian (?late Silurian) period of ductile shearing associated with sinistral transpression and a late Caledonian or Acadian phase (?early Devonian) of reverse and dextral strike-slip faulting due to northwest–southeast compression. A possibly analogous change from sinistral transpression in the late Silurian to dextral transpression in the early Devonian is recorded

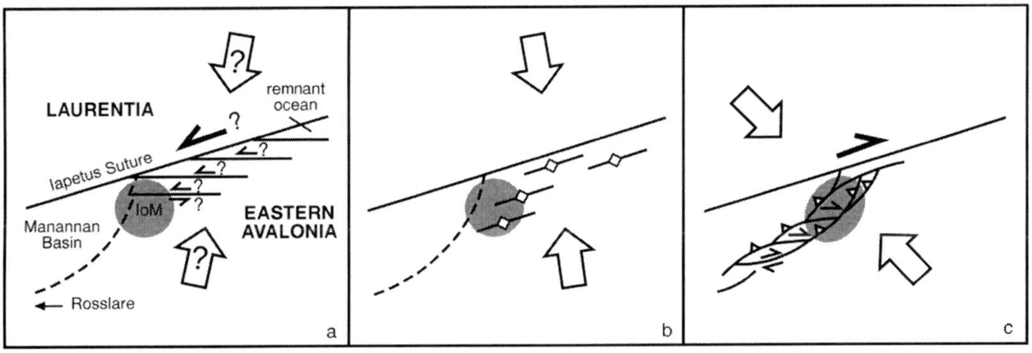

Fig. 9. Simplified model of possible plate interactions of Eastern Avalonia with Laurentia during the Lower Palaeozoic based on the orientation and kinematics of structures interpreted on the Isle of Man (stylized in grey). See text for discussion. (**a**) Possible sinistral transpression and ductile shear (convergence oblique to the Iapetus Suture); (**b**) D1 deformation (early orogenic shortening perpendicular to the Iapetus Suture); (**c**) brittle contraction (final orogenic shortening perpendicular to eastern margin of the Manannan Basin).

further west on the southeast side of the Iapetus Suture in Newfoundland (D'Lemos *et al.* 1997).

As shown in Fig. 9, the imbricate belt in the Isle of Man overlies a deep magnetic boundary, thought by Kimbell & Quirk (1999) to represent the eastern edge of a thick succession of poorly magnetic Lower Palaeozoic sediments occupying the newly named Manannan Basin. This basin is interpreted to lie south of the Iapetus Suture as an embayment on the northwest side of Eastern Avalonia and stretches between the Isle of Man and Ireland, north of Rosslare (Fig. 9a). It was probably formed by rifting during the Tremadoc. Arenig-age sediments of the Manx Group in the Isle of Man and the Ribband Group in Ireland (McConnell *et al.* 1999) represent exposed parts of the basin. The preceding discussion indicates that the northeast tip of the Manannan Basin was subject to an anticlockwise rotation of stress as Eastern Avalonia docked with Laurentia in a manner similar to regional models proposed by, for example, Soper *et al.* (1992) and Piper (1997). Initial closure of Iapetus is interpreted to have been oblique in a north-northeast direction such that some of the movement was taken up by sinistral strike-slip along east–west shear zones (Fig. 9a). As the remaining oceanic crust was eventually consumed, the two continents locked up, causing D1 folds and cleavage to develop parallel to the suture as σ_1 rotated anticlockwise (Fig. 9b). Finally, deformation was accommodated by contraction in a direction approximately perpendicular to the eastern edge of the Manannan Basin forming a thrust duplex with associated dextral strike-slip faults (Fig. 9c). The timing of D3 structures is uncertain (Fitches *et al.* 1999) but flat-lying D2 cleavage and folds may have formed as a result of thrust stacking during the late stages of movement in the duplex.

Three important implications follow on from this model. Firstly, the stratigraphy of the Manx Group is telescoped (cf. Quirk & Burnett 1999); secondly, correlations with the Ribband Group are justifiable (cf. McConnell *et al.* 1999); finally, rather than a large batholith underlying the whole of the Isle of Man, granite intrusions form discrete plutons, probably emplaced by late-stage movement on faults within the imbricate belt [cf. Cornwell (1972), Crowley & Power (1999) and Kimbell & Quirk (1999)].

Conclusions

Three Lower Palaeozoic trends identified on high-resolution aeromagnetic data and in the field they represent northeast–southwest thrusts, east–west to east-northeast–west-southwest dextral strike-slip faults and north–south sinistral strike-slip faults. These were active during northwest-southeast compression in the late Caledonian when a northwest dipping contractional duplex is thought to have formed at the eastern margin of a Lower Palaeozoic basin developed on the northwest side of Eastern Avalonia. Earlier tectonic movement in the opposite sense is recorded on ductile shear zones, associated with disruption of pre- or syn-kinematic felsitic intrusions and disaggregation of quartz veins. This may reflect an episode of sinistral transpression during closure of Iapetus before the Laurentia–Eastern Avalonia plate boundary became fully locked. D1 structures, such as folds and cleavage, are thought to have formed during an intermediate stage of north-northwest–south-southeast compression. The Isle of Man was tilted to the northwest during post-Caledonian tectonic events. However, except for the west-northwest–east-southeast trending Central Valley Lineament, and similarly orientated Tertiary dykes, younger structures are rarely imaged onshore with potential field data.

The authors wish to thank Andy Bell, Doug Fettes and Rob Barnes for essential suggestions on how to improve an earlier version of this paper. Stimulating discussions with John Morris, Nigel Woodcock, Bill Fitches, Greg Power, Bob Holdsworth and Jack Soper helped formulate many of the ideas reported here. Dave Kelly, Karen Braithwaite, Richard Young, Fred Radcliffe, Frank Cowin and Kathleen Quirk provided valuable assistance during the research, and Graeme Foster and Lisa Hill draughted most of the diagrams. The work was funded by NERC research grant GR9/01834, Oxford Brookes University, the Isle of Man Government and BG Exploration and Production Ltd. GSK publishes with permission of the Director, British Geological Survey (NERC).

References

BLAKE, J. F. 1905. On the order of succession of the Manx Slates. *Quarterly Journal of the Geological Society of London*, **61**, 358–373.

BROWN, P. E., MILLER, J. A. & GRASTY, R. L. 1968. Isotopic ages of late Caledonian granitic intrusions in the British Isles. *Proceedings of the Yorkshire Geological Society*, **36**, 251–276.

CORNWELL, J. D. 1972. A gravity survey of the Isle of Man. *Proceedings of the Yorkshire Geological Society*, **39**, 93–106.

FITCHES, W. R., BARNES, R. P. & MORRIS, J. H. 1999. Geological structure and tectonic evolution of the Lower Palaeozoic rocks of the Isle of Man. *This volume.*

FORD, T. D. 1993. *The Isle of Man*. Geologists' Association Guide **46**.

GREEN, P. F., DUDDY, I. R. & BRAY, R. J. 1997. Variation in thermal history styles around the Irish Sea and adjacent areas: implications for hydrocarbon occurrence and tectonic evolution. *In:* MEADOWS, N. S., TRUEBLOOD, S. P., HARDMAN, M. & COWAN, G. (eds) *Petroleum Geology of the Irish Sea and Adjacent Areas*. Geological Society, London, Special Publications, **124**, 73–93.

HORAK, J. M., BEVINS, R. E. & LEES, G. J. 1999. Palaeogene magmatism in the Isle of Man and Irish Sea region. *Journal of Petroleum Geology*, in press.

HUGHES, R. A., COOPER, A. H. & STONE, P. 1993. Structural evolution of the Skiddaw Group (English Lake District) on the northern margin of eastern Avalonia. *Geological Magazine*, **130**, 621–629.

JACKSON, D. I. & MULLHOLLAND, P. 1993. Tectonic and stratigraphic aspects of the East Irish Sea Basin and adjacent areas: contrasts in their post-Carboniferous structural styles. *In*: PARKER, J. R. (ed.) *Petroleum Geology of Northwest Europe: Proceedings of the 4th Conference*. Geological Society, London, 791–808.

JESPERSEN, A. 1970. *The Lady Isabella Waterwheel of the Great Laxey Mining Company, Isle of Man, 1854–1954*. 3rd Revised edition, Viborg, Denmark.

KIMBELL, G. S. & QUIRK, D. G. 1999. Crustal magnetic structure of the Irish Sea region: evidence for a major basement boundary beneath the Isle of Man. *This volume.*

—— & STONE, P. 1995. Crustal magnetization variations across the Iapetus Suture Zone. *Geological Magazine*, **132**, 599–609.

LAMPLUGH, G. W. 1903. *The Geology of the Isle of Man*. Memoir of the Geological Survey, United Kingdom, HMSO.

D'LEMOS, R. S., SCHOFIELD, D. I., HOLDSWORTH, R. E. & KING, T. R. 1997. Deep crustal and local rheological controls on the siting and reactivation of fault and shear zones, northeastern Newfoundland. *Journal of the Geological Society, London*, **154**, 117–121.

MCCONNELL, B. J., MORRIS, J. H. & KENNAN, P. S. 1999. A comparison of the Ribband Group (southeastern Ireland) to the Manx Group (Isle of Man) and Skiddaw Group (northwestern England). *This volume.*

MCINTYRE, J. I. 1980. Geological significance of magnetic patterns related to magnetite in sediments and metasediments – a review. *Bulletin of the Australian Society for Exploration Geophysics*, **11**, 19–33.

MACKAY, R. A. & SCHNELLMAN, G. A. 1963. *The mines and minerals of the Isle of Man*. Report to Industrial Officer, Isle of Man Government (through Brian Colquhon and Partners, London).

MORRIS, J. H., WOODCOCK, N. H. & HOWE, M. P. A. 1999. The Silurian succession of the Isle of Man: the late Wenlock Niarbyl Formation, Dalby Group. *This volume.*

PIPER, J. D. A. 1997. Tectonic rotation within the British paratectonic Caledonides and Early Palaeozoic location of the orogen. *Journal of the Geological Society, London*, **154**, 9–13.

——, BIGGIN, A. J. & CROWLEY, S. F. 1999. Magnetic survey of the Poortown Dolerite, Isle of Man. *This volume.*

POWER, G. M. & BARNES, R. P. 1999. Relationships between metamorphism and structure on the northern edge of Eastern Avalonia: in the Manx Group, Isle of Man. *This volume.*

—— & CROWLEY, S. F. 1999. Petrological and geochemical evidence for the tectonic affinity of the (?) Ordovician Poortown basic intrusive complex, Isle of Man. *This volume.*

QUIRK, D. G. & BURNETT, D. J. 1999. Lithofacies of Lower Palaeozoic deep-marine sediments in the Isle of Man: a new map and stratigraphic model for the Manx Group. *This volume.*

—— & KIMBELL, G. S. 1997. Structural evolution of the Isle of Man and central part of the Irish Sea. *In:* MEADOWS, N. S., TRUEBLOOD, S. P., HARDMAN, M. & COWAN, G. (eds) *Petroleum Geology of the Irish*

Sea and Adjacent Areas. Geological Society, London, Special Publications, **124**, 135–159.

——, ROY, S., KNOTT, I. & REDFERN, J. 1999. Petroleum geology and future hydrocarbon potential of the Irish Sea. *Journal of Petroleum Geology*, in press.

ROBERTS, B., MORRISON, C. & HIRONS, S. 1990. Low grade metamorphism of the Manx Group, Isle of Man: a comparative study of white mica 'crystallinity' techniques. *Journal of the Geological Society, London*, **147**, 271–277.

SIMPSON, A. 1963. The stratigraphy and tectonics of the Manx Slate Series, Isle of Man. *Quarterly Journal of the Geological Society of London*, **119**, 367–400.

—— 1965. The syn-tectonic Foxdale–Archallagan granite and its metamorphic aureole, Isle of Man. *Geological Journal*, **4**, 415–434.

SOPER, N. J., STRACHAN, R. A., HOLDSWORTH, R. E., GAYER, R. A. & GREILING, R. O. 1992. Sinistral transpression and the Silurian closure of Iapetus. *Journal of the Geological Society, London*, **149**, 871–880.

WEBB, B. C. & COOPER, A. H. 1988. Slump folds and gravity slide structures in a Lower Palaeozoic marginal basin sequence (the Skiddaw Group), northwest England. *Journal of Structural Geology*, **10**, 463–472.

WOODCOCK, N. H. & BARNES, R. P. 1999. An early Ordovician turbidite system on the Gondwana margin: the southeastern Manx Group, Isle of Man. *This volume*.

——, MORRIS, J. H., QUIRK, D. G. ET AL. 1999. Revised lithostratigraphy of the Manx Group, Isle of Man. *This volume*.

Geological structure and tectonic evolution of the Lower Palaeozoic rocks of the Isle of Man

W .R. FITCHES,[1] R. P. BARNES[2] & J. H. MORRIS[3]
[1]*Robertson Research International Ltd, Llandudno LL30 1SA, UK*
[2]*British Geological Survey, Murchison House, West Mains Road, Edinburgh EH9 3LA, UK*
[3]*Geological Survey of Ireland, Beggars Bush, Haddington Road, Dublin 4, Ireland*

Abstract: The Lower Palaeozoic sedimentary rocks of the Isle of Man, deposited near the margin of the Avalonian plate, were folded and cleaved during continental closure and collision. Although folds on a scale of several kilometres can be inferred, the large-scale structure of the Isle of Man remains uncertain in the absence of detailed and widespread biostratigraphical controls on the stratigraphy and other difficulties. A working model suggests that the island is composed of several northeast trending, strike-parallel 'tracts' separated by vertical or steeply northwest dipping faults. Stratigraphic sequences may be apparent within tracts but cannot easily be correlated between tracts. The within-tract structure is the result of two main stages of deformation (D1 and D2) and one or more localized later events (D3). F1 folds have axial surfaces which dip steeply northwest or southeast, near-horizontal hinges, and wavelengths ranging up to several kilometres. The F2 folds, seen mostly on a small scale, are reclined to recumbent and coaxial with F1 folds. The D1 and D2 structures were imposed on the Wenlock rocks of the Niarbyl Formation but are absent from the Peel Sandstone (late Silurian–early Devonian), constraining the main deformation to the Caledonian (Acadian) orogeny.

The D1 structures are the products of collisional tectonics but the origin of the flat-lying D2 structures is unclear, although vertical flattening is more likely than thrusting. Most of the inferred tract-bounding structures cannot be characterized in the absence of exposure or precise stratigraphical controls. Two of them, however, are inferred to be northwest dipping faults. One of these faults cuts D1 structures whilst the other appears to be pinned by a syn-D1 intrusion, suggesting broadly syn-D1 age. Localized, steep northeast striking zones of late to post-D1 high strain, which are exposed on the west coast of the island, in part reflect partitioning of coaxial or transpressive deformation into the boundaries between rocks of contrasted ductility. The Niarbyl Fault Zone ('shear zone') is shown to be a belt of ductile and brittle deformation whose tectonic history and regional significance have yet to be fully resolved. Brittle thrust faults are exposed in several parts of the Isle of Man and include the Niarbyl Thrust, which, in one interpretation, is taken to separate the Niarbyl high-strain phyllonite belt from the overlying Niarbyl Formation. Most thrusts are north to northwest dipping and displacements are typically on a metre scale. Many of them may be attributed to the latest stages of the Caledonian collision processes on the basis that similar structures also deform the late Silurian–early Devonian Peel Sandstone of the Isle of Man and rocks of similar age elsewhere in the region.

Lower Palaeozoic sedimentary rocks (Woodcock *et al.* 1999; Quirk & Burnett 1999) crop out over a large area on the Isle of Man (Fig. 1), but are generally only well exposed in coastal sections. Together with faulted boundaries and little biostratigraphical control, the lack of critical exposure has prevented recognition of an unequivocal lithostratigraphy. Consequently, although the small-scale structure has been appreciated since the primary geological survey by Lamplugh (1903 [see also Geological Survey (1898)], interpretation of the regional structure has largely been a matter of conjecture. Lamplugh proposed that the two parallel outcrops of black mudstone along the spine of the island (Barrule and Glen Rushen Formations; Fig. 1) and the sandstone-dominated rocks which crop out to the northwest and southeast are linked around a D1 anticlinorium. In his seminal studies, Simpson (1963) developed this theme, although he preferred a major D1 syncline along the axis of the island and D2 refolding to explain the outcrop pattern. Simpson ascribed the regional, steep northwest or southeast dips to several large D1 folds with gently plunging hinges and a pervasive, axial-planar cleavage. Common small-scale, recumbent, open

From: Woodcock, N. H., Quirk, D. G., Fitches, W. R. & Barnes, R. P. (eds) 1999.
In Sight of the Suture: the Palaeozoic geology of the Isle of Man in its Iapetus Ocean context.
Geological Society, London, Special Publications, **160**, 259–287. 1-86239-046-0/99/$15.00

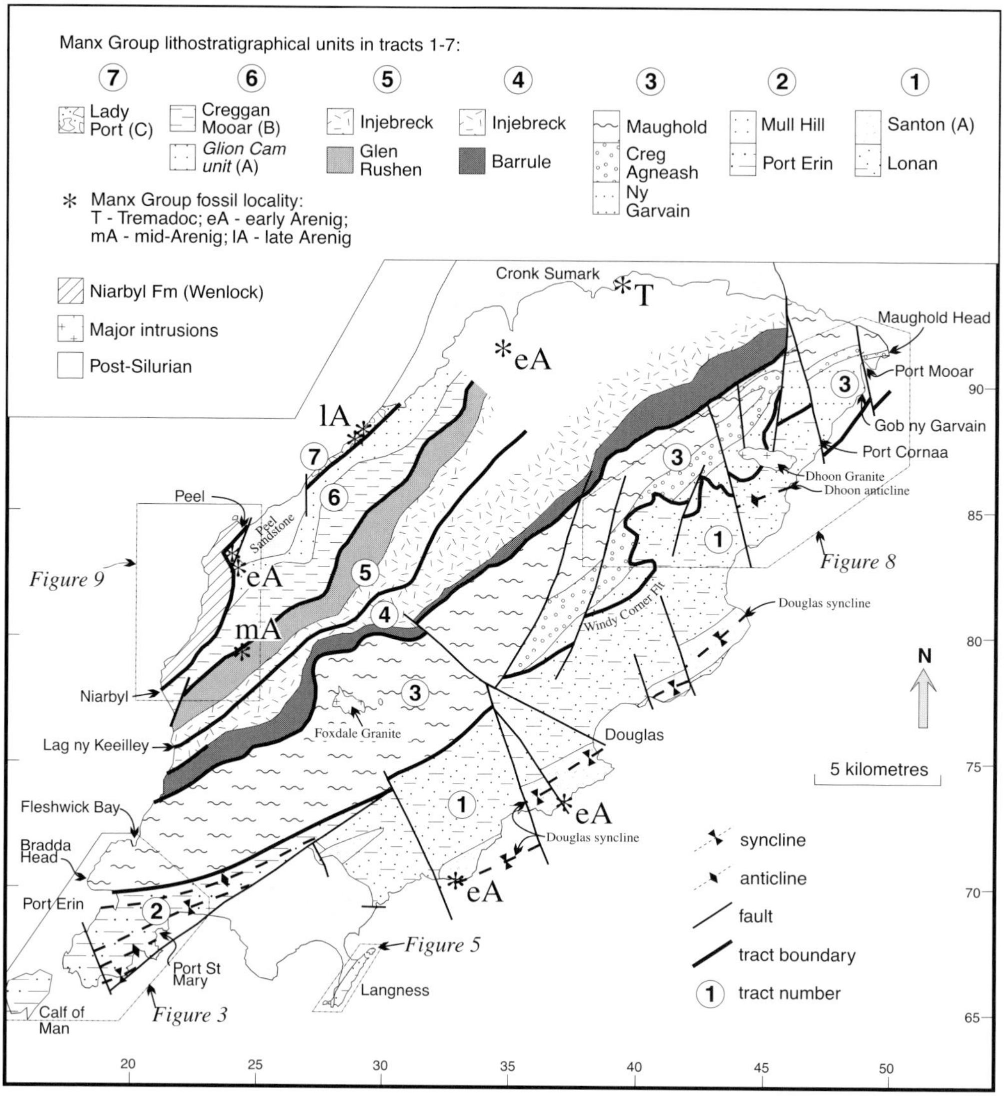

Fig. 1. Tectonostratigraphic map of the Isle of Man with biostratigraphical control points, showing localities discussed in text.

folds with an axial-planar crenulation cleavage were assigned to D2, but Simpson also interpreted some large-scale D2 structures. A third event (D3) was also recognized locally, with steep, northwest trending axial surfaces and a steep crenulation cleavage. Low-grade regional metamorphism occurred during this deformation sequence (Gillott 1955; Simpson 1964 & Morrison 1989; Power & Barnes 1999).

Acritarch faunas from a few locations, reported by Molyneux (1979, 1999), revealed that Simpson's regional model was untenable, although all of the rocks were still considered to be of late Cambrian or Arenig age. It has now been established that the sandstone sequence in the northwest of the island, the Niarbyl Formation, is Wenlock in age (Howe 1999), removing any possibility of correlation with the Arenig sequence in the southeast of the island, although uncertainty remains over the possible equivalence of the black mudstone units (see Woodcock *et al.* 1999 and below).

In general terms, the three-fold structural sequence described by Simpson (1963) are corroborated and some of the major D1 structures identified by him confirmed. However, detailed mapping in several parts of the Isle of Man provides a new insight into the nature of the intermediate and large-scale D1 structure, as described below. It is also suggested that major strike-parallel faults may play a significant role in the regional structure of the island, a theme further developed in consideration of high-strain zones which are exposed on the west coast. Separate consideration of the structure of the Silurian Niarbyl Formation allows comparison with the structure of the early Ordovician rocks and provides constraints on the timing of the deformation history. This contribution concludes by placing the Isle of Man in its regional tectonic context by comparison with the other, nearby, parts of the southern Caledonides.

Large-scale structure of the Isle of Man: fault-bounded 'tracts'

Definition of the large-scale structure of the Lower Palaeozoic rocks of the Isle of Man is still significantly hampered by lack of an overall stratigraphy. Lithostratigraphical units can be defined (e.g. Woodcock *et al.* 1999; Quirk & Burnett 1999) but the contacts critical to understanding the succession are commonly faulted or not exposed. Biostratigraphical data, discussed in detail by Molyneux (1999) and Orr & Howe (1999), provide age constraints at only seven locations (Fig. 1). Several units, notably the black mudstone of the Barrule and Glen Rushen Formations, can be traced along-strike through most of the outcrop. However, these and intervening units cannot be unequivocally correlated, or in many cases ordered into an across-strike sequence. Even the overall direction of younging is difficult to determine in mudstone-rich formations. Where there is no evidence of a stratigraphical contact between adjacent lithostratigraphical units, a fault is possible and the boundary is considered suspect. In this way, the island has been divided into strike-parallel tracts (Fig. 1) within which there is some evidence of lithostratigraphical continuity, as described by Woodcock *et al.* (1999), but between which tectonostratigraphical relationships are uncertain. In some cases, a faulted tract boundary can be demonstrated from detailed mapping work or consideration of the available age data, as described below from southeast to northwest.

The best preserved successions occur in the southeast of the island (Woodcock & Barnes 1999). Around Douglas, a thick turbidite sequence is dominated by the Lonan Formation, the lowest part of which occurs in the core of the Dhoon Anticline (Fig. 1), passing up eastwards into the Santon Formation in the Douglas Syncline. Early Arenig acritarchs and graptolites occur in the Santon Formation (Rushton 1993; Molyneux 1999). In the northeastern part of the island, turbidites of the Ny Garvain Formation (Fig. 1) pass gradationally up into the Creg Agneash Formation, dominated by thin- to medium-bedded quartz arenite. The latter is seen to be stratigraphically overlain by mudstone-rich Maughold Formation at Maughold Head. The boundary between the two sequences is not exposed, but mapping in the northeast of the island (see below) suggests that it is a northwest dipping, syn-D1 fault. Consequently, they are separated as tracts 1 and 3 (Fig. 1). There are no fossils from the Ny Garvain–Maughold sequence which would allow assessment of its age relative to the Lonan–Santon sequence, but possible correlations are discussed by Barnes *et al.* (1999) and Woodcock & Barnes (1999).

In the south of the island, quartz arenite is prominent in the Mull Hill Formation, overlying thinly bedded turbidites of the Port Erin Formation (Woodcock *et al.* 1999). The contact between this sequence and that in the adjacent tract 1 is obscured by the Carboniferous cover. To the north, the junction with the Maughold Formation in the north of Port Erin Bay is also not exposed, but a major topographic hollow suggests a fault. The Mull Hill Formation may be the direct equivalent of the Creg Agneash Formation exposed along-strike to the northeast (e.g. Barnes *et al.* 1999). However, other correlations are possible (e.g. Woodcock & Barnes 1999) and the outcrop of the Port Erin–Mull Hill Formations is separated as tract 2 (Fig.1).

Further northeast, the outcrop of massive black mudstone which forms the Barrule Formation is one of the most consistently mapped units on the Isle of Man (e.g. Lamplugh 1903; Simpson 1963). The boundary between the Maughold Formation and the Barrule Formation is nowhere exposed but, north of the Dhoon Intrusion (Fig.1), it cuts across a large-scale fold structure in tract 3 as described below. Traced southwestwards, the contact continues to be discordant with the sequence to the east, with the outcrop of the Maughold Formation progressively widening westwards. On the west coast, north of Fleshwick Bay, the contact lies within an inaccessible cliff section, but a distinct topographic lineament is provisionally interpreted to reflect a tectonic feature. Consequently, the Barrule Formation is separated into tract 4 (Fig. 1).

Further tectonostratigraphical subdivision of the Ordovician rocks northwest of the outcrop of the Barrule Formation is hampered by inaccessible coastal sections. The contact between the Barrule and Injebreck Formations follows a topographic lineament along the edge of the ridge from Cronk ny Arrey Laa [SC 224 747] to Burro Meanagh [SC 217 739]. This separates southeast dipping Barrule Formation above from the northwest dipping Injebreck Formation below and is interpreted as a gentle southeast dipping fault. However, this fault is unlikely to form the Barrule–Injebreck junction along most of its length because it maps as a steeply dipping contact. Based on the long-standing assumption that the Barrule and Glen Rushen Formations are equivalent, this junction was taken as a stratigraphical contact by Woodcock *et al.* (1999). The Glen Rushen Formation, of mid-Arenig age (Molyneux 1999), passes stratigraphically up into the Injebreck Formation in tract 5. Tracts 4 and 5 are separated at the west coast at the Lag ny Keeilley high-strain zone, described in more detail below, although this is not strictly a boundary of the type discussed above because it occurs within a single lithostratigraphical unit.

The contact between tracts 5 and 6 is offset by a late fault at the coast and is not exposed inland. In the southeastern part of tract 6, the Creggan Mooar Formation is characterized by centimetre scale manganiferous–ironstone beds (Kennan & Morris 1999), which also occur locally within the Lady Port Formation (Woodcock & Morris 1999). Correlation on this basis suggests that the Creggan Mooar Formation may be of late Arenig age (Woodcock *et al.* (1999, fig. 9), in which case the tract 5–6 junction is likely to be faulted. The northwestern part of tract 6 is occupied by a poorly exposed sequence, informally termed the Glion Cam Unit by Woodcock *et al.* (1999), which has yielded Lower Arenig acritarchs (Molyneux 1999). However, its tectonostratigraphical relationship to the Creggan Mooar Formation is unknown. A fault must be present at the northeastern boundary of the Glion Cam Unit where it is juxtaposed against the northwest striking, late Arenig Lady Port Formation (Molyneux 1999; Woodcock & Morris 1999).

The lithostratigraphy of the northern part of the outcrop of the Lower Palaeozoic rocks on the Isle of Man is left unresolved by Woodcock *et al.* (1999), although Quirk & Burnett (1999) suggest some lithological subdivisions. Tremadoc graptolites (Rushton 1993) from Cronk Sumark (Fig. 1) indicate the oldest rocks yet proved on the island. To the west, mudstone at Glen Dhoo has yielded acritarchs of early Arenig age (Molyneux 1999). These suggest a northwest younging sequence, albeit of uncertain continuity and probably faulted against the Barrule or overlying Injebreck Formations east of Cronk Sumark.

The Silurian Niarbyl Formation rests structurally above the Arenig components of tract 6 in the west of the Isle of Man and forms an additional tectonostratigraphical unit. At the coast, the boundary between the Silurian and Arenig rocks coincides with the Niarbyl Fault Zone, the significance of which is discussed fully below.

In summary, the distribution of rock assemblages and the few biostratigraphical control points currently available in the outcrop of the Ordovician Manx Group requires that at least two or three major strike-parallel tectonic boundaries are present within the Isle of Man (Fig. 1). Consideration of the uncertain nature of junctions with other undated units allows the possibility that additional tectonic boundaries may be present. The tract system, proposed in this study and adopted by Woodcock *et al.* (1999), is at one level a means of expressing this uncertainty. At another level it provides a large-scale structural model of the Isle of Man which can be tested by future work.

Structural character of the Ordovician Manx Group

The Manx Group is generally composed of sandstone, siltstone and mudstone, interbedded on a scale varying from millimetres to metres, which behaved as multilayers with different rigidities during deformation. Thick packets of slate derived from homogeneous or only faintly laminated mudrock, occurring mainly in the Barrule, Glen Rushen and Maughold Formations and locally in the Lonan and Port Erin Formations, are the most ductile components of the stratigraphy. At the other extreme, the closely bedded quartz arenite of the Mull Hill and Creg Agneash Formations was much less ductile and behaved more as a single, thick massive unit, controlling the morphology of several major folds. Widespread igneous dykes and sills, typically up to a few metres in thickness, behaved in various ways during the deformation depending on their composition and relative timing of emplacement.

D1 structures

Bedding has a regional east-northeast to northeast strike (Fig. 2) and is generally moderate to steeply dipping in abundant, open to isoclinal F1 folds with wavelengths varying from a few millimetres to several kilometres. Axial surfaces are steeply northwest or southeast dipping and hinges are typically gently plunging. Major F1 hinge zones are rarely exposed, although in tract 1 the Douglas Syncline is seen at several localities (e.g. Onchan Harbour [SC 4056 7762] and in the tramway cutting at Bank Howe [SC 4185 7824]) and the Dhoon Anticline is exposed in the road section above Bulgham Bay [SC 455 868]. In tract 3 the crest of the Port Erin Anticline crops out as an area

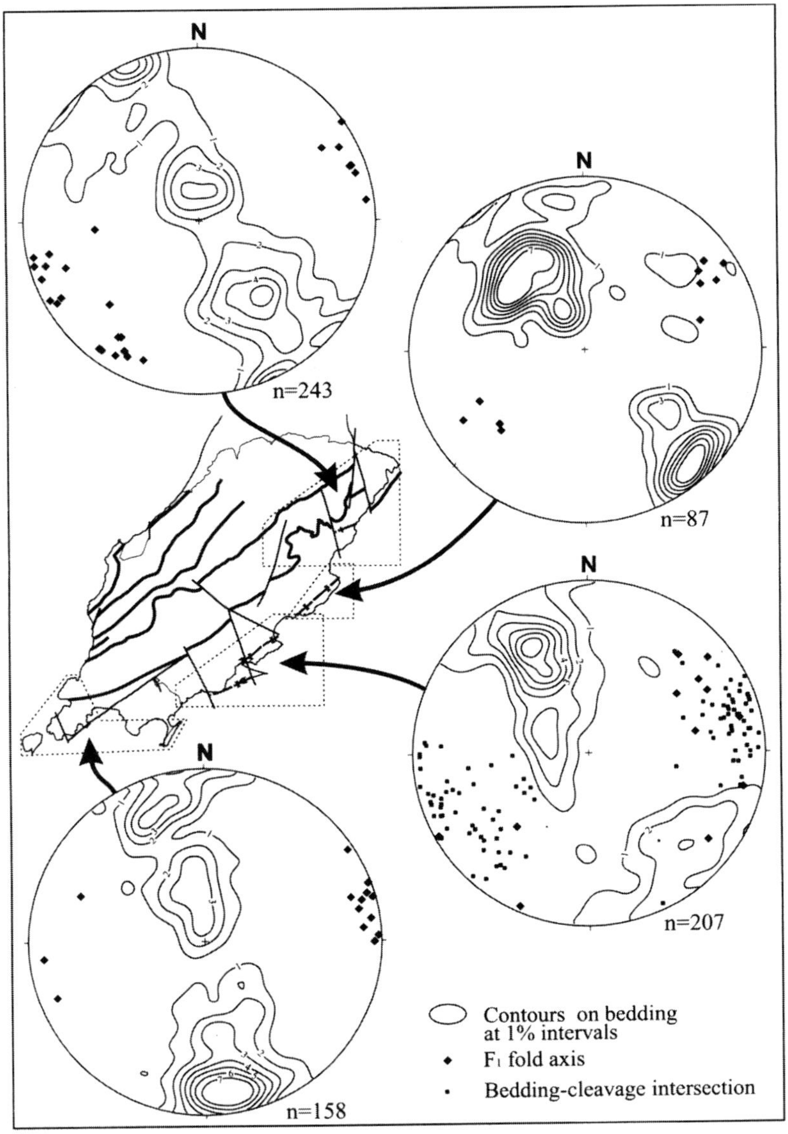

Fig. 2. Orientations of bedding and D1 structural elements from: the areas in Figs 3, 5 and 8; southeast coastal strip between Garwick Bay [SC 434 814] and Port Soldrick [SC 303 696] (data collected by N. H. Woodcock); and between Port e Vullen [SC 474 928] and Laxey Bay [SC 435 825] (data collected by D. G. Quirk). Equal area, lower hemisphere projections.

of flat-lying rocks near the Marine Biological Station (Fig. 1). Otherwise, the larger structures are identified mainly from changes in younging direction, bedding–cleavage relationships and parasitic fold vergence.

S1 cleavage is the dominant tectonic fabric in most parts of the Manx Group, although locally, particularly in mud-rich layers or units, it is often overprinted by S2. In pelitic rocks it is defined mainly by aligned flakes of white mica and chlorite (Power & Barnes 1999) and may locally become phyllitic. Pressure-solution striping in S1 occurs in several places, for example on St Michael's Island [SC 293 673] and Langness [SC 287 657]. In sandstone, S1 is commonly a weak, spaced (few millimetres), pressure-solution fabric, but in places it comprises aligned, flattened detrital grains. In many areas S1 is axial-planar or fans and refracts from bed to bed through D1 folds. There are numerous examples, however, of S1 cleavage lying virtually parallel with bedding in F1 hinge zones, especially in thin pelitic layers intercalated with medium- to thick-bedded sandstone. In some parts of the Isle of Man, the S1 cleavage transects the F1 folds by a few degrees, usually clockwise but locally anticlockwise.

Boudinage, generally of sandstone beds, early veins and igneous sheets, is commonly associated with the D1 deformation, producing boudins of various forms and orientations. In many places, the boudins are square-ended to barrel-shaped. Boudin necks, commonly marked by quartz segregations, usually plunge in one direction parallel to the local F1 fold hinges, implying stretching in the limbs and hinges. Less commonly, boudin axes plunge steeply at a high angle to F1 fold hinges, as on the coast below Lag ny Keeilley [SC 215 745]. In other places, e.g. near Milner's Tower [SC 184 699], boudins of steeply dipping quartz arenite beds are chocolate tablet type, with some axes plunging steeply and others near horizontal, implying uniaxial strain. Many of the boudins have the inverse form produced as a consequence of initial brittle behaviour (square ends due to high-strain rate and/or large ductility contrast with host rocks), then ductile necking within the boudins as strain rate and or ductility contrasts declined. The later part of this two-stage process might be attributed in some instances to the D2 deformation. However, most examples appear to have been produced during D1 as rheological and strain conditions changed, because they occur even in areas where D2 is weak or absent.

D2 structures

F2 folds occur throughout the outcrop of the Ordovician rocks, though they are best developed in tracts 2–5 and are rare in most of tract 1. F2 folds visible in exposure are generally small, with wavelengths less than a few metres and commonly only a few tens of centimetres, and a close to tight chevron profile, although rounded hinges are also common. Axial surfaces are gently northwest or southeast dipping and hinges are gently northeast or southwest plunging. Some larger F2 folds are inferred from small-scale structures and variation in the dip of bedding and F1 axial surfaces.

F2 folds are usually clearly distinguishable from the co-axial but much steeper F1 folds. Where the two sets of folds are superimposed, Type III interference hook folds are produced. Particularly clear examples are exposed in Port Erin Bay [SC 194 694]. Where D2 deformation was particularly intense, F1 axial surfaces and S1 have been rotated into alignment with F2 axial surfaces. The two generations of structures are then almost indistinguishable.

A gently dipping S2 crenulation cleavage, best developed in finer grained rocks, occurs widely in the Manx Group with the exception of tract 1, where it is only developed very locally in association with rare minor folds. It is zonal in some places, discrete in others (*sensu* Powell 1979); pressure-solution striping is developed along S2 in some pelitic rocks, e.g. on Langness [SC 288 660]. S1 and S2 are easily distinguished in most places, however, where the D2 vertical flattening was particularly intense and S1 rotated to a near-horizontal attitude, a combined S1–S2 fabric has been formed. Similarly, in the outer arcs of large F1 folds, where the original attitude of S1 was flat lying and bedding parallel, the D2 deformation has further developed S1 without generating a new cleavage.

D3 structures

D1 and D2 structures and fabrics in the Manx Group are locally deformed by upright folds associated with a broadly north striking crenulation cleavage. According to Simpson (1963), these 'F3' structures have a dominant northwest strike. However, in southwestern parts of the Isle of Man similar structures strike nearly north–south or northeast–southwest, but are highly variable. It is likely that the various directions denote the presence of conjugate sets of structures, but it is also possible that these structures were produced by more than one stage of deformation. The term D3 and allied labels are thus used here to refer collectively to late structures which may be of several generations. Most F3 folds are open, with rounded hinges and wavelengths in the 10 cm to 10 m range, although chevron and kink folds also occur. Simpson (1963) inferred kilometre scale F3 folds, mainly in northwestern parts of the Manx Group outcrop, but none has been confirmed during the present study.

Detailed structure – southwest Isle of Man

This region (Figs 3 and 4) is dominated by almost continuous coastal exposure in the Maughold, Port Erin and Mull Hill Formations (Woodcock *et al.* 1999). The boundary between tracts 2 and 3 is provisionally placed in the north of Port Erin Bay, in unexposed ground which may mark a fault; Quirk *et al.* (1999) also infer a fault zone there on geophysical grounds. A separate small area of the Lonan Formation (tract 1) on the Langness peninsula (Fig. 5) is also considered because its structure includes some characteristics otherwise unusual on the Isle of Man. The orientation of the

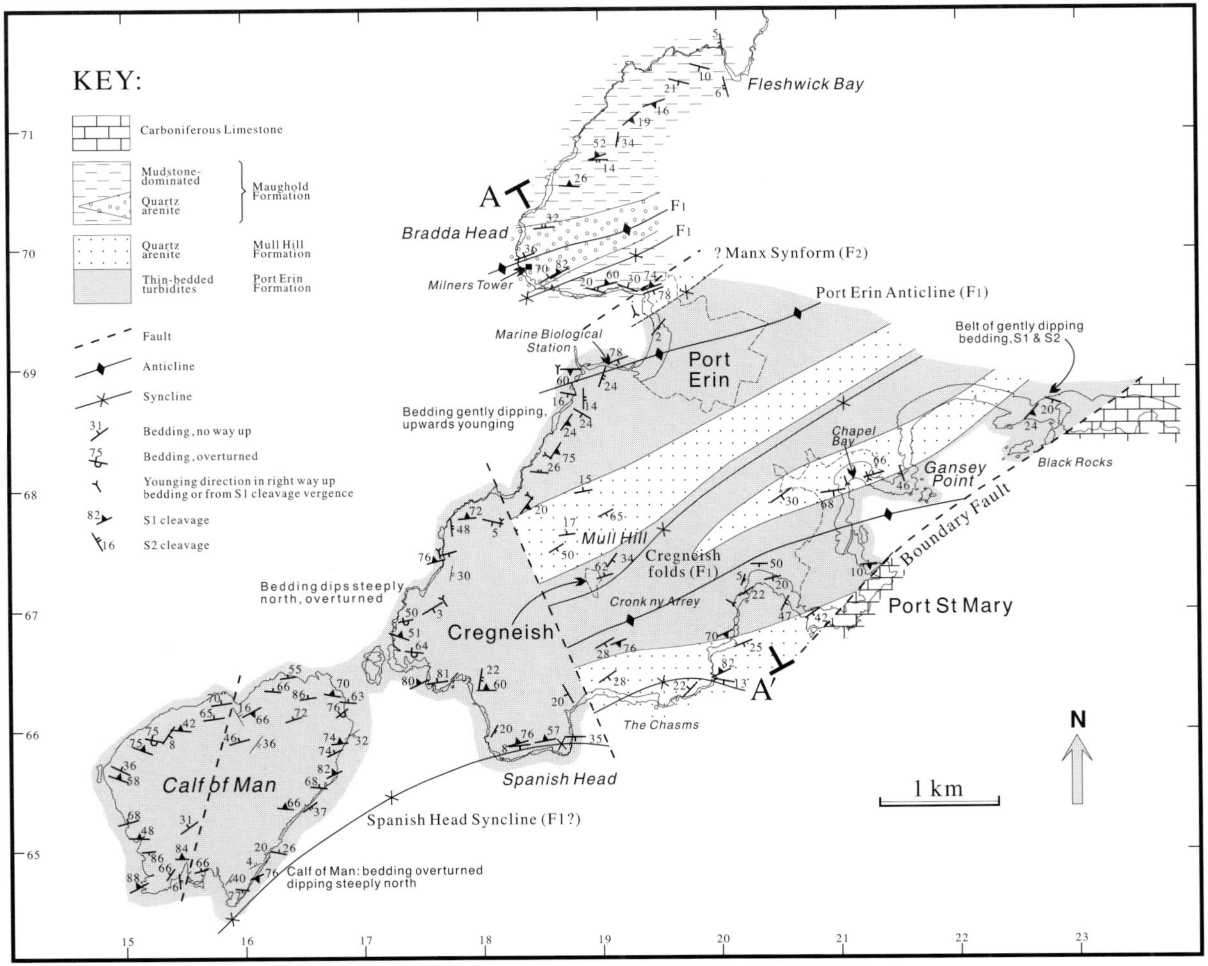

Fig. 3. Structural map of the southwest part of the Isle of Man. Location shown on Fig. 1.

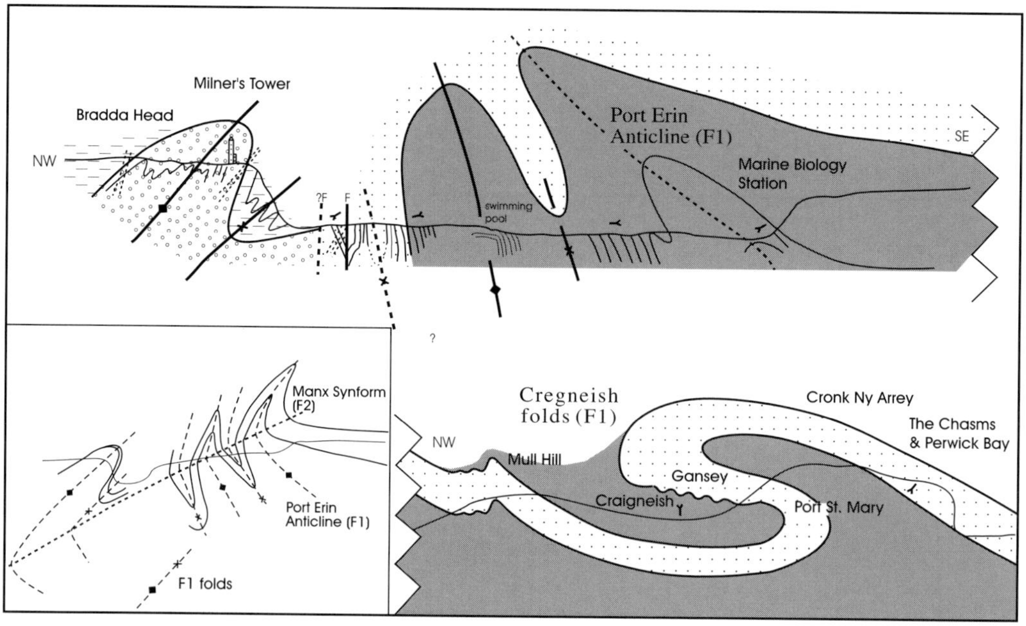

Fig. 4. Schematic cross-sections through the southwest peninsula of the Isle of Man. Composite section from the Bradda Head area to The Chasms area (A–A′ on Fig. 3). Projected on the line of section is information from the areas of Perwick Bay [SC 204 672], Port St Mary [SC 211 676] and Gansey Point [SC 215 684]. Inset: an interpretation of the effect of the major F2 Manx Synform on D1 structures in the Port Erin Bay area. Grey = thin bedded turbidites; other ornaments as on Fig. 3.

principal structural elements in these areas is represented in Fig. 6, whilst Fig. 7 illustrates several of the structures at outcrop.

D1 deformation. The area around Port Erin and Port St Mary includes several large-scale folds which repeat the outcrop of the Mull Hill Formation and adjacent units (Figs 3 and 4). Some of these structures were recognized by Simpson (1963), but he did not recognize a major F1 structure, the Cregneish Fold Pair, which occupies much of the ground in the Cregneish–Port St Mary region (Figs 3 and 4).

The long limbs of the Cregneish Fold Pair are gently southeast dipping, as seen in the lower limb in Cregneish Quarry [SC 191 674] and in the upper limb at The Chasms [SC 194 663]. The steep part of the syncline hinge zone crops out on the coast in Chapel Bay at Port St Mary. The common limb of the fold pair is exposed on the eastern foreshore of Gansey Point, where beds are gentle to moderately southeast dipping and overturned. Here, several gently plunging, parasitic F1 folds are sideways closing to downward facing, suggesting that their axial surfaces having been rotated horizontally. This, together with the unusually low dips throughout the structure, is probably a result of rotation during D2.

The structure and stratigraphy are truncated west of Cregneish by a north-northwest striking fault of undetermined throw. West of this fault, strata assigned to the Port Erin Formation mainly dip moderately to steeply northwards, but are overturned over a strike width of almost 2 km (Fig. 3). This arrangement was interpreted as the northern limb of a major F2 syncline by Simpson (1963) as his Spanish Head Syncline, but is more in keeping with a major southeast vergent F1 syncline.

Another large area of overturned strata occurs on the Langness peninsula, where bedding is moderately to steeply south dipping (Fig. 5) and some F1 folds have unusually steep plunges. For example, on St Michael's Island, folds are typically moderate east-northeast plunging but locally their plunge is steep to vertical. Systematic changes of plunge, as in periclines, sheath folds or Type 1 (basin-and-dome) refolds, have not been detected and reasons for the variation are unclear. On the east coast of Langness, there are few exposure-scale F1 folds. S1–bedding intersections are commonly moderate to steep east-northeast to west-southwest plunging, implying that F1 axes have these attitudes if S1 is axial-planar, or S1 transection of the folds. Clockwise transection is observed on the northwest coast of St Michael's

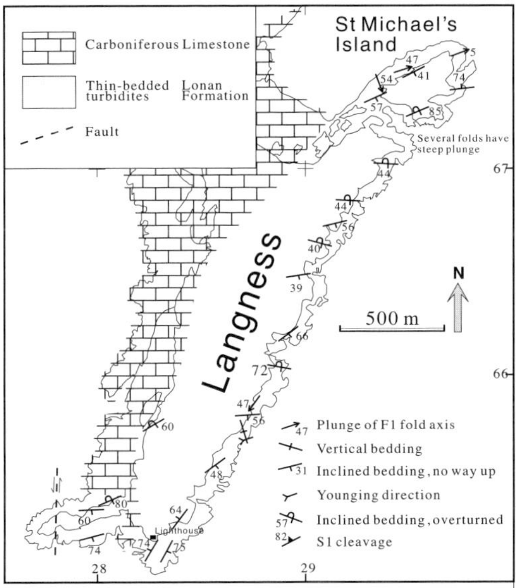

Fig. 5. Structural map of Langness. Location shown on Fig. 1.

Island [SC 297 676], representing the general pattern in tract 1 (Woodcock, pers. comm.). However, near The Chasms [SC 194 664], cleavage dominantly transects small F1 folds in an anticlockwise sense (Fig. 7b).

In the south of the Isle of Man, S1 cleavage is generally well developed at an angle to bedding and provides a reliable indication of the vergence. However, locally the S1 cleavage may lie near-parallel to bedding, even within large-scale folds. In some instances, this arrangement is due to strong outer arc extension in the crests of large folds, as in the Port Erin Anticline near the Marine Biological Station at Port Erin (Fig. 3). In the vicinity of The Chasms [SC 194 664], and locally elsewhere, strain seems to have partitioned into thin ductile pelites between thicker sandstone layers during bedding slip which accompanied folding.

Early veins. Veins are widespread in this part of the island, mostly a few centimetres wide and composed of quartz with minor chlorite and carbonate. Some have formed at the necks of square-ended D1 boudins in quartz arenite layers or igneous sheets, and record *in situ*, syn-tectonic segregation processes. Other veins, formed by early brittle deformation, have been folded and/or boudinaged during D1 depending on their orientation with respect to strain axes (Fig. 7c).

D2 deformation. Recumbent, open, small-scale F2 folds are common, especially in the thinly bedded Port Erin Formation (e.g. Fig 7d). Their vergence changes depending on their position with respect to the dip of bedding in the limbs of F1 folds, with important implications for the deformation mechanism as discussed below.

A large-scale F2 structure, corresponding with the Manx Synform of Simpson (1963), is recognized in the southwest of the Isle of Man, although no evidence has been found for its continuation throughout the island as Simpson suggested. The gently northwest dipping axial surface of the fold, situated in Port Erin Bay, is marked by a change in the regional dip of bedding and F1 axial surfaces (inset Fig. 4). Above the F2 axial surface, F1 folds, such as the large-scale Bradda Anticline, are inclined steeply northwest. Below the F2 axial surface, F1 folds have moderately to gently southeast dipping axial surfaces. The reclined mesoscopic F1 folds at Gansey Point, within the Cregneish D1 Fold Pair, are in this situation.

The S2 cleavage occurs widely in the Manx Group of the southwest of the island and in places it is the dominant tectonic fabric. Where S2 is at a high angle to S1 or bedding lamination a pencil cleavage has commonly formed along the intersection, typically gently northeast or southwest plunging; examples occur on Spanish Head [SC 181 659] and in Port Erin Bay [SC 189 697].

Ductile shear bands. Shear bands are locally abundant on Langness, notably near the lighthouse [SC 282 652] (Fig. 7e) and occur sporadically on the Calf of Man (e.g. [SC 157 662]), forming centimetre to metre wide, steeply southeast dipping structures which deflect S1 and bedding in a sinistral sense. They also occur on the southeast flank of Snaefell [SC 401 879] but have not been noted elsewhere on the Isle of Man. These structures are younger than S1 but are deformed by S3 crenulations on the Calf of Man [SC 150 652]. Their relationships with the D2 deformation have not been determined. Consequently, it is not known whether they record a minor local effect developed late in the deformation sequence (possibly 'D3') or if they represent a regionally significant, sinistral shear event which took place during or shortly after D1. These structures offer scope for further investigation.

D3 deformation. F3 folds, as described above, occur locally in southern parts of the Isle of Man. In some places they are abundant, especially where pelitic rocks contain a well-developed, steeply dipping fabric, as in parts of Port St Mary Bay [SC 21 67]. At Black Rocks [SC 224 687] and Perwick Bay [SC 203 673], F3 folds occur in belts 100 m or

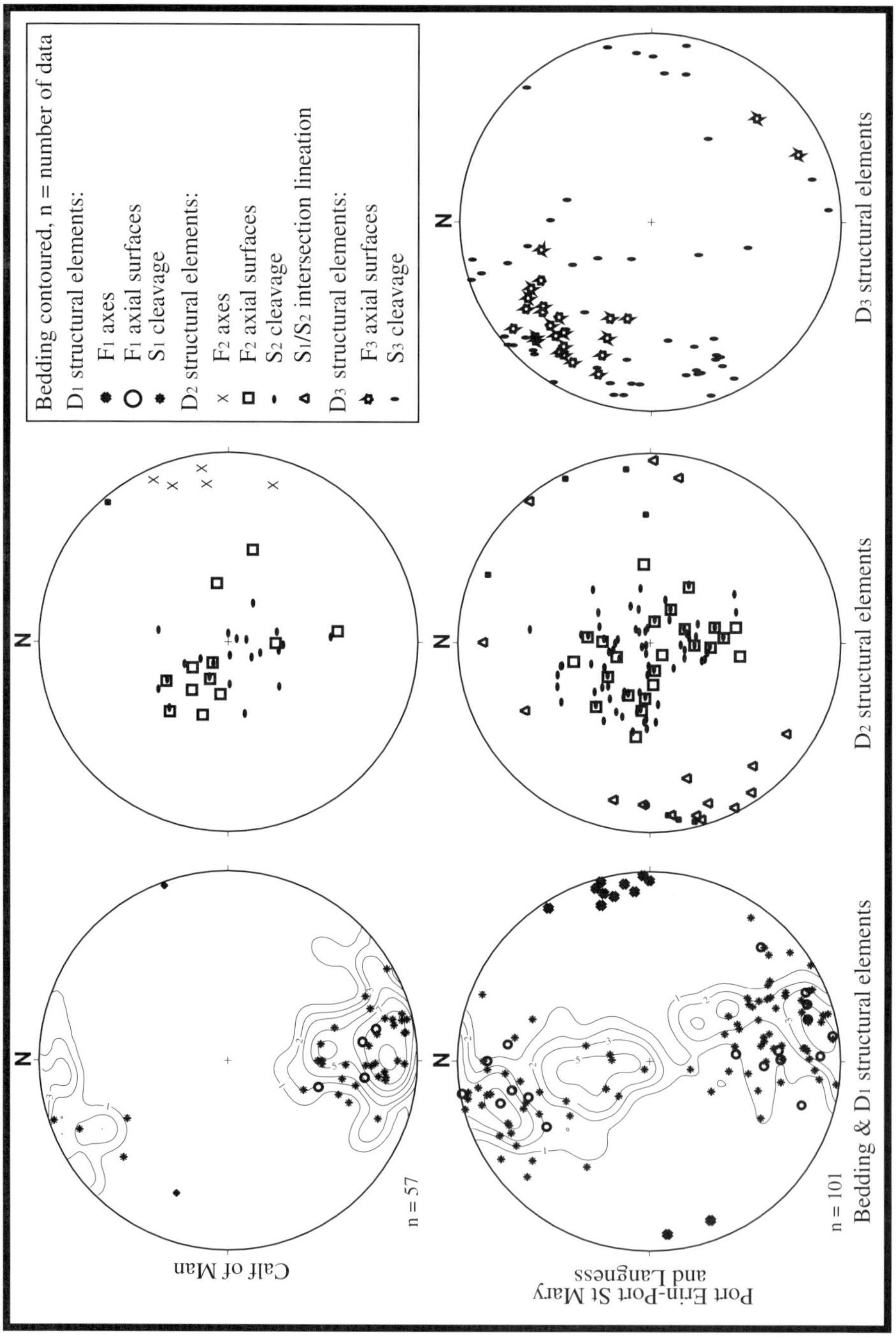

Fig. 6. Orientation data for all structural elements in the southwest of the Isle of Man. Equal area, lower hemisphere projections.

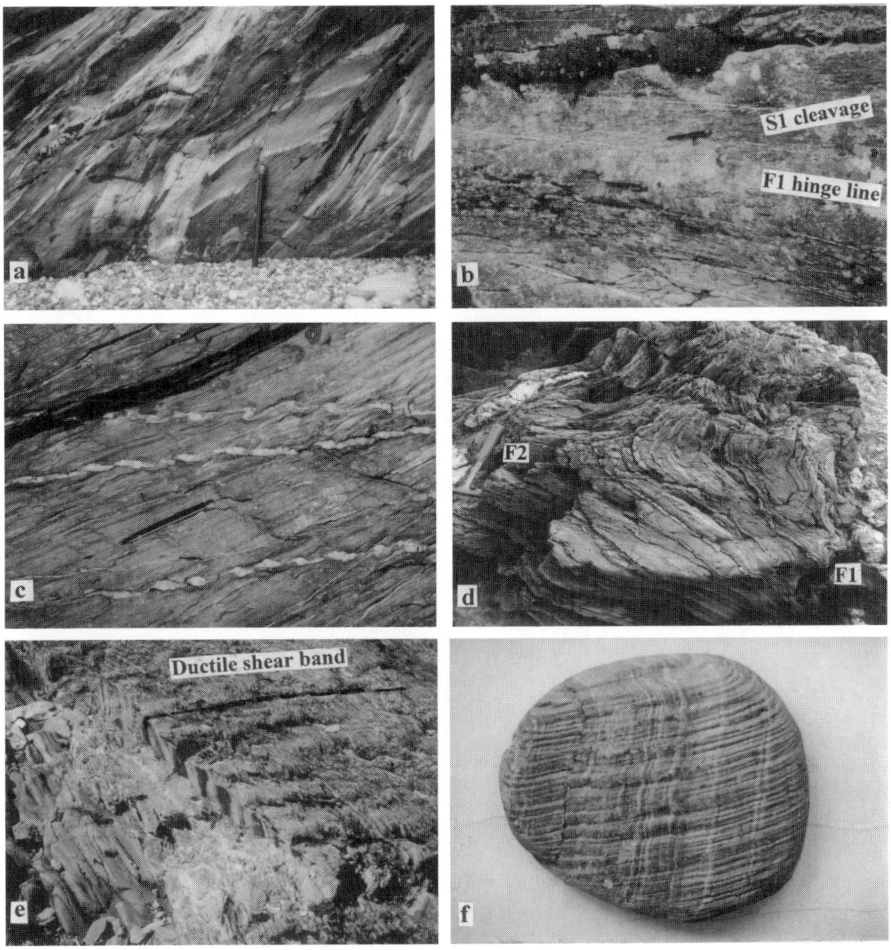

Fig. 7. Illustrations of structures in the Ordovician Manx Group, southwest Isle of Man. (**a**) F1 folds of felsite dyke (left of pen) and bedding (right of pen). Port Erin Bay [SC 195 693]. (**b**) F1 fold hinge transected anticlockwise by S1 cleavage. East end of The Chasms [SC 197 663]. (**c**) Early quartz veins deformed by D1: boudinage or folding took place, depending on orientations of veins with respect to principal stresses. Calf of Man [SC 153 651]. (**d**) Type III interference between recumbent F2 refolding upright F1 folds. Port Erin Bay [SC 195 693]. (**e**) Steep sinistral shear bands deforming S1 cleavage. Langness [SC 282 652]. (**f**) S3 crenulation cleavage at high angle to bedding. Pebble sample, Gansey Point [SC 215 684].

more in width in which parasitic chevron- or kink-style folds, 1–5 cm in wavelength, are accompanied by an axial-planar S3 crenulation cleavage (Fig. 7f). Locally, these folds represent up to 30% horizontal, northeast–southwest shortening.

Detailed structure – northeast Isle of Man

This region (Fig. 8) includes continuous exposure in the coastal section and a large area of variable inland exposure extending southwest on to the upland ridge along the axis of the outcrop of Lower Palaeozoic rocks. In this area, the Ny Garvain, Creg Agneash and Maughold Formations are considered to form a conformable sequence (Woodcock & Barnes 1999). Minor pre- or syn-tectonic intrusions occur throughout the section. The outcrop of these formations is distinguished as tract 3 (Fig. 1) because relationships with the Lonan Formation to the south and the Barrule Formation to the northwest are uncertain.

The tract 3 sequence is juxtaposed against the Lonan Formation by a late north-northwest trending fault at the coast. To the southwest, however, in hilly terrain with deeply incised valleys, the boundary between the quartz arenite of the Creg

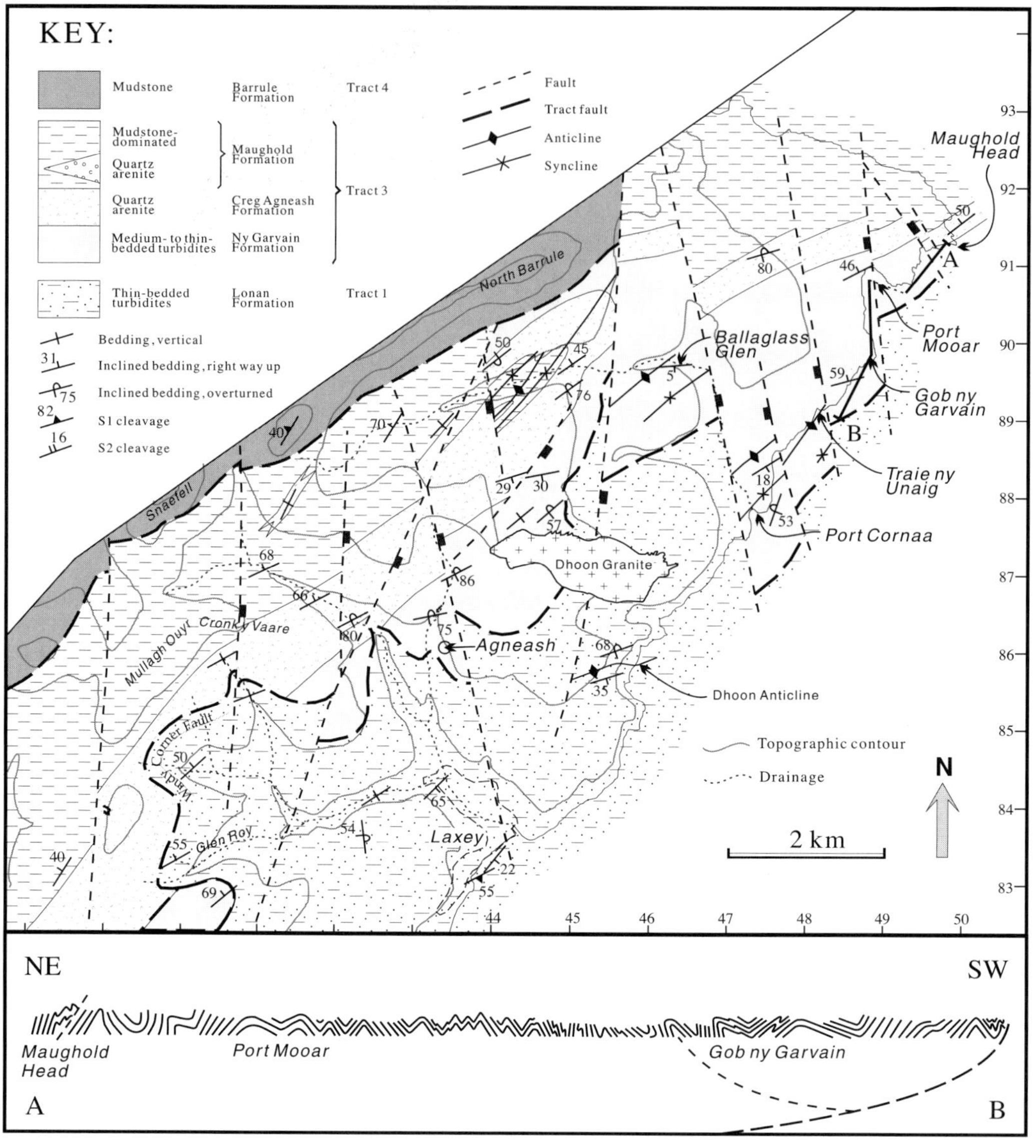

Fig. 8. Structural map of the northeastern part of the Manx Group outcrop on the Isle of Man, with structural section between Maughold Head [SC 498 915] and Gob ny Garvain [SC 489 899] illustrating the variable attitudes of F1 folds and their relationships with large faults.

Agneash Formation and the thin-bedded muddy turbidites attributed to the Lonan Formation, maps as a gentle northwest dipping surface. This is oblique to the steeply dipping to vertical bedding above and below. Lamplugh (1903) recognized that in this area the Agneash Grit forms the hill tops with Lonan Flags exposed in intervening valleys. He explained this as a stratigraphical contact folded by upright mesoscopic folds, causing an overall gentle sheet dip. However, there is little evidence of such folding, the exposed rocks being consistently northwest younging, leaving two other possible interpretations: sedimentary intercalation of the two sequences or a faulted junction. Sedimentary intercalation across the boundary should be apparent from interbedding of the two sandstone types, for which there is no evidence in this area. The contact, at least locally, is thus inferred to be a fault separating tract 3 from tract 1 (Fig. 1), termed here the 'Windy Corner Fault' (Fig. 8).

A structural boundary also appears likely northwest of Douglas. Here, the Creg Agneash Formation strikes north-northwest, its outcrop oblique to and truncated by the boundary with the Lonan Formation. Whilst this may reflect original sedimentary geometry, it is more probable that the Creg Agneash Formation is cut out by a broadly strike-parallel fault climbing-up sequence to the southeast. The fault may be pinned by the syn-D1 (Power & Barnes 1999) Dhoon Granite (Fig. 8), constraining the timing of the main displacement, although the poor exposure does not preclude minor later movement.

North of the Dhoon Intrusion, repetition of the outcrop of the Creg Agneash Formation with opposed younging direction suggests a large-scale, southwest plunging, F1 syncline with a steeply northwest dipping axial surface. This structure does not, however, appear to affect the southern boundary of the Barrule Formation immediately to the north, which maps as a planar, moderate to steep northwest dipping surface broadly parallel with the regional dip. This suggests that the northwestern boundary of tract 3 is also faulted, at least in the northeast of the island.

D1 deformation. Gently plunging D1 folds are abundant at a range of scales throughout the section, although their style and the dip of their axial surfaces vary widely (Fig. 8). The largest structures, exposed in the section from Port Cornaa to Traie ny Unaig, are two southeast inclined, southeast verging fold pairs, with steep east-southeast dipping, overturned long limbs and gentle south-southeast dipping short limbs. Gently southwest plunging minor folds are well developed in one of the anticlinal hinge zones exposed just south of Traie ny Unaig. Northwards, bedding in the long limb is typically moderate north-northwest dipping over *c*. 250 m across-strike, although open folds with subhorizontal short limbs include a *c*.100 m wide zone south of Gob ny Garvain. A zone of mesoscale folding at Gob ny Garvain is succeeded northwards by several folds with axial surfaces spaced 50–80 m apart. These larger folds are steeply northwest inclined, with moderate northwest dipping bedding in long limbs. To the north they rotate through vertical to steep southeast inclined as bedding in the long limbs becomes near vertical, locally southeast dipping and overturned. This change in style and attitude of the D1 folds from Traie ny Unaig to north of Gob ny Garvain suggests that the large-scale structure may be a pop-up between a pair of northwest and southeast dipping thrusts (Fig. 8). The northwest dipping structure is plausibly the Windy Corner Fault inferred above to underlie this sequence and which here crops out offshore.

Intense folding on a small to intermediate scale, mostly with neutral vergence at the largest scale (tens of metres half-wavelength), is characteristic of the entire section from Gob ny Garvain to Port Mooar and also the north side of Port Mooar Bay. These folds are generally upright, tight in the south but more open to the north. North of Port Mooar, several large close folds have gently dipping short limbs with southeast dipping axial surfaces. Bedding around Maughold Head is moderate to steep northwest dipping in right-way-up strata. A steep, north dipping reverse fault exposed in the south of the headland is associated with disharmonic folding of the thin-bedded quartz arenite in the hanging wall. However, as quartz arenite is also present in its footwall, the fault is not considered to have major displacement, although it may locally form the boundary between the Creg Agneash and Ny Garvain Formations.

D2 deformation. The gently dipping S2 cleavage becomes well developed north of Gob ny Garvain and in Port Mooar, and to the north it is the dominant cleavage, lying at 90° to the near-vertical F1 axial surfaces. The cleavage is axial planar to F2 folds which are sporadically developed as gently plunging, close structures with short limbs generally only a few centimetres in length, although more open flexures with a wavelength up to several metres may be associated. The small folds are best developed in moderate to steep northwest dipping strata but they also occur in southeast dipping beds where they have the opposite sense of asymmetry, always verging down-dip. Larger flexures tend to occur in very steeply dipping to vertical strata where they have neutral vergence.

Minor intrusions. Numerous dykes, up to 10 m thick and with basic to intermediate compositions, occur in this coastal section, usually steeply dipping and from northeast to southeast trending. The dykes have various ages with respect to the tectonic sequence. One suite was emplaced along (?late) northwest trending faults. Some dykes of this suite show evidence of shearing as a result of continued movement after emplacement. However, many dykes were emplaced early in the tectonic development and were cleaved by S1 and S2. Some were emplaced along fold axial surfaces.

Summary structure – northwest Isle of Man (Tracts 4–7)

This section summarizes the structural features evident along the west coast of the island, from

Fleshwick Bay in the south to Kirk Michael in the north. The sequence immediately north of Fleshwick Bay, within tract 3, is affected by a series of large, upward-facing, gently southwest plunging, monoclinal F2 flexures with gently northwest or southeast dipping axial surfaces. The overall fold envelope descends towards the southeast (southeast verging). An associated S2 crenulation cleavage is superimposed upon the penetrative S1 foliation. A similar suite of minor southeast verging F2 folds is present in the southern part of tract 4, south of a steep south dipping fault near Burroo Sodjey [SC 221 743]. In northwest dipping bedding north of the fault, minor F2 folds are northwest verging.

The gently dipping boundary between the Barrule and Injebreck Formations within tract 4, with bedding dipping southeast and northwest above and below, respectively, was interpreted as part of the D1 Isle of Man Synclinorium by Simpson (1963) but is probably a southeast dipping fault. To the north, the Lag ny Keeilley high-strain zone within the Injebreck Formation, described below, juxtaposes northwest dipping sequences with opposed younging directions. The northern, overturned section has been interpreted Kennan & Morris (1999) as the short limb of a complementary 'anticlinorium', with the Glen Rushen Formation in its core. Recumbent, northwest verging F2 folds, plunging gently northeast, are well exposed at Da Leura [SC 2172 7535].

Tract 6 is dominated by trains of tight, metre wide F1 folds associated with a penetrative, commonly transpositional, S1 foliation. The folds generally plunge gently northeast or southwest, but steepen to *c.* 65° northeast towards the southern tract boundary. Way-up is commonly difficult to discern and hence the facing direction of F1 folds is not always clear; however, apparent downward-facing relationships in several places suggest that some folds may pre-date D1. At a larger scale, Kennan & Morris (1999) interpret a major reclined D1 syncline, the right-way-up limb occupying the southern part of the tract and the overturned, northern limb forming the northern part. D2 is again represented by minor asymmetric folds, many with associated quartz vein arrays. A prominent, gently northwest dipping, brittle D2 thrust fault–quartz vein array occurs just north of Gob ny Gamera [SC 2158 7690]. Further north, from the north end of Traie Vrish [SC 2130 7743], a series of late northeast striking brittle faults, cutting alternating zones of relatively coherent to markedly disrupted bedding, arguably form part of the outer deformation envelope of the Niarbyl Fault Zone.

Deformation in tract 7 (Woodcock & Morris 1999) is dominated by faults and shear zones of two types: an array of northwest striking, mainly brittle shear zones, gently southwest and northeast dipping, define a flower structure pattern; late, steeply dipping, northwest to northeast striking brittle faults, possibly of Mesozoic age. A shallow dipping, penetrative phyllitic foliation is present throughout, even in igneous intrusions, associated with mesoscopic gently inclined, close to tight folds. Both the folds and fabric are assigned to D1, despite the unusually shallow dips and the even more unusual northwest strike, an orientation evident again in part of the Niarbyl Fault Zone, described below.

Structural character of the Silurian Niarbyl Formation

The Niarbyl Formation is well exposed in an almost continuous coastal section from Peel to Niarbyl (Fig. 9). In common with Simpson (1963), two distinct deformation episodes are recognized, designated D1 and D2, similar in character to D1 and D2 in the Ordovician rocks. A variety of other structures occur within the formation, including several gently inclined brittle thrust fault–quartz vein complexes.

D1 deformation

The structure of the Niarbyl Formation varies either side of an east trending feature termed the Knockaloe Moar Lineament (Fig. 10). To the north, the structure is dominated by the upright F1 Peel Hill Anticline (Simpson 1963); parasitic F1 folds are upright to steeply inclined and generally plunge gently southwest (Fig. 10), although a fold pair at Contrary Head plunges *c.* 40° southwest. South of the lineament, a continuous train of F1 folds generally have wavelengths between 1.5 to 50 m, exceptionally up to 150 m, such that homoclinal sections are normally < 60 m in width. The folds, ranging from open, round-hinged structures, through tight, angular folds to almost isoclinal, are asymmetric, southeast verging and gently northeast plunging (Figs 10 and 11a, c). Axial surfaces generally dip *c.* 50° northwest, although some examples are almost recumbent. Fold amplitude generally exceeds the 5–35 m height of cliff sections.

S1 cleavage associated with the folds varies from penetrative in pelite to a spaced foliation in sandstone where it is locally enhanced by pressure solution, forming lithons up to 1 cm wide (e.g. [SC 2237 7983]). The cleavage fans around many F1 fold hinges, up to 55° in one instance, and refracts up to 30° between sandstone and pelite beds. S1 cleavage and F1 folds are everywhere upward facing. South of the Knockaloe Moar Lineament,

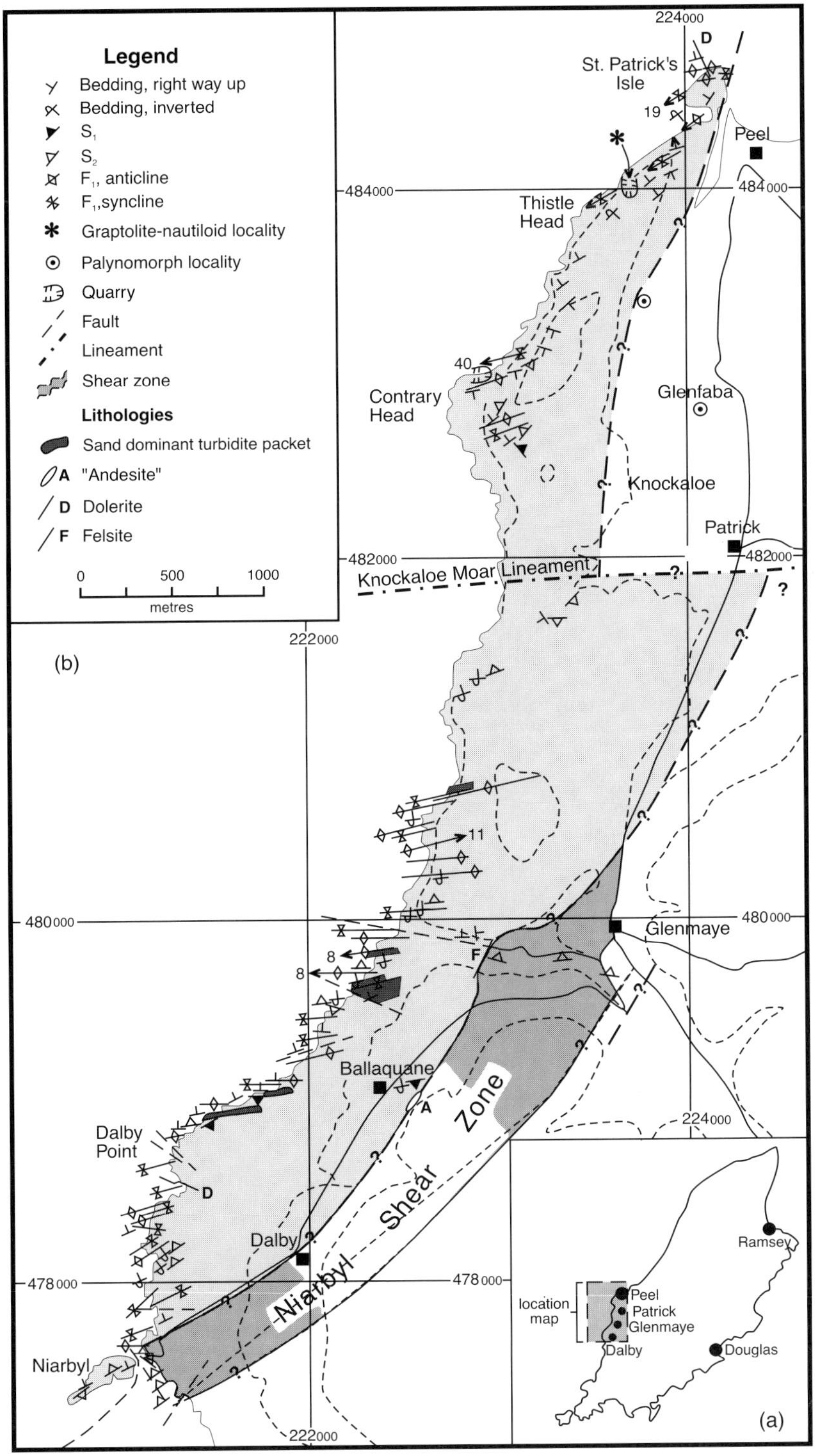

Fig. 9. Structural map of the Niarbyl Formation outcrop (from Morris *et al.* 1999).

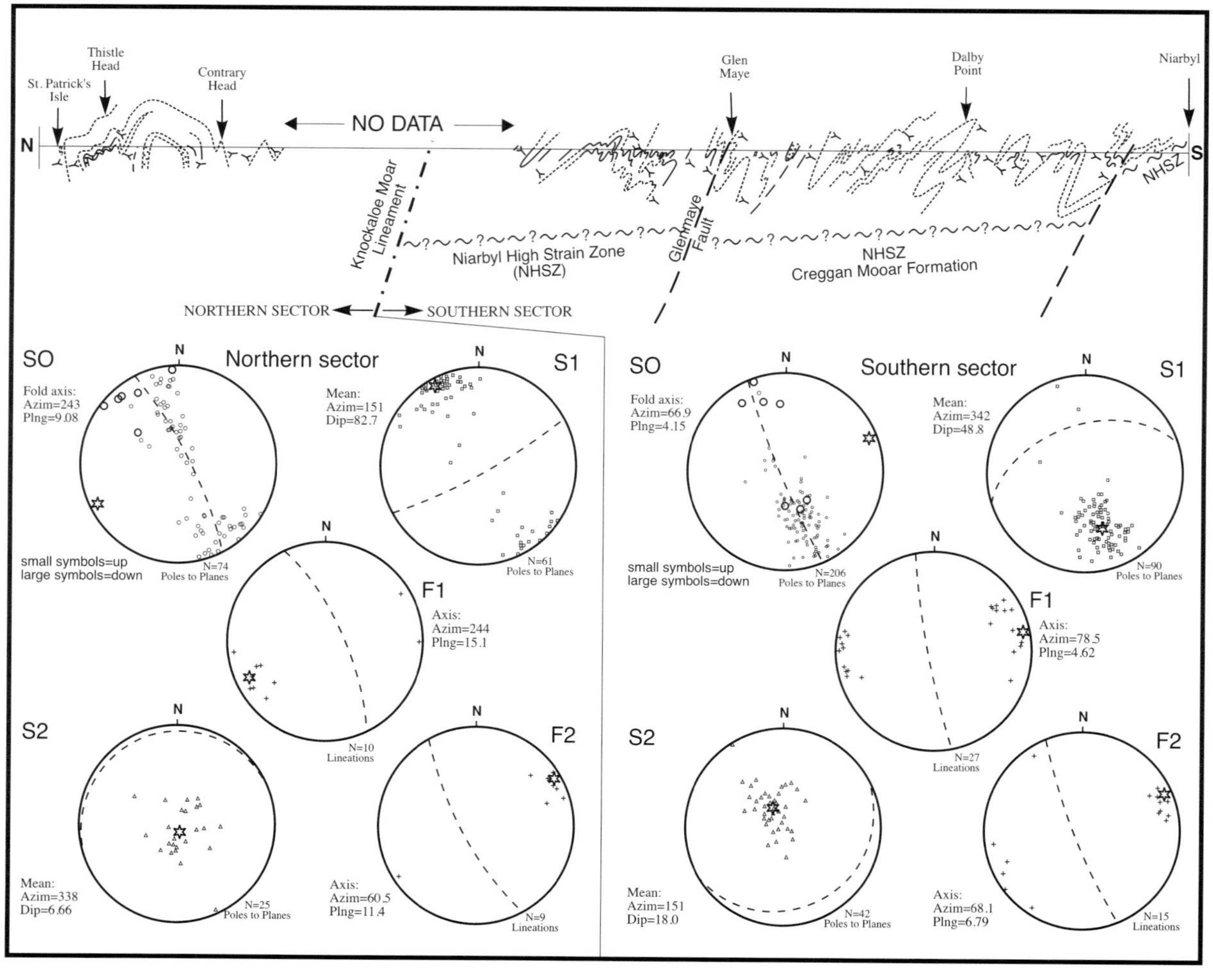

Fig. 10. Cross-section and orientations of structural elements in the Niarbyl Formation, showing principal structural elements north and south of the Knockaloe Moar Lineament. Lower hemisphere, equal area projections.

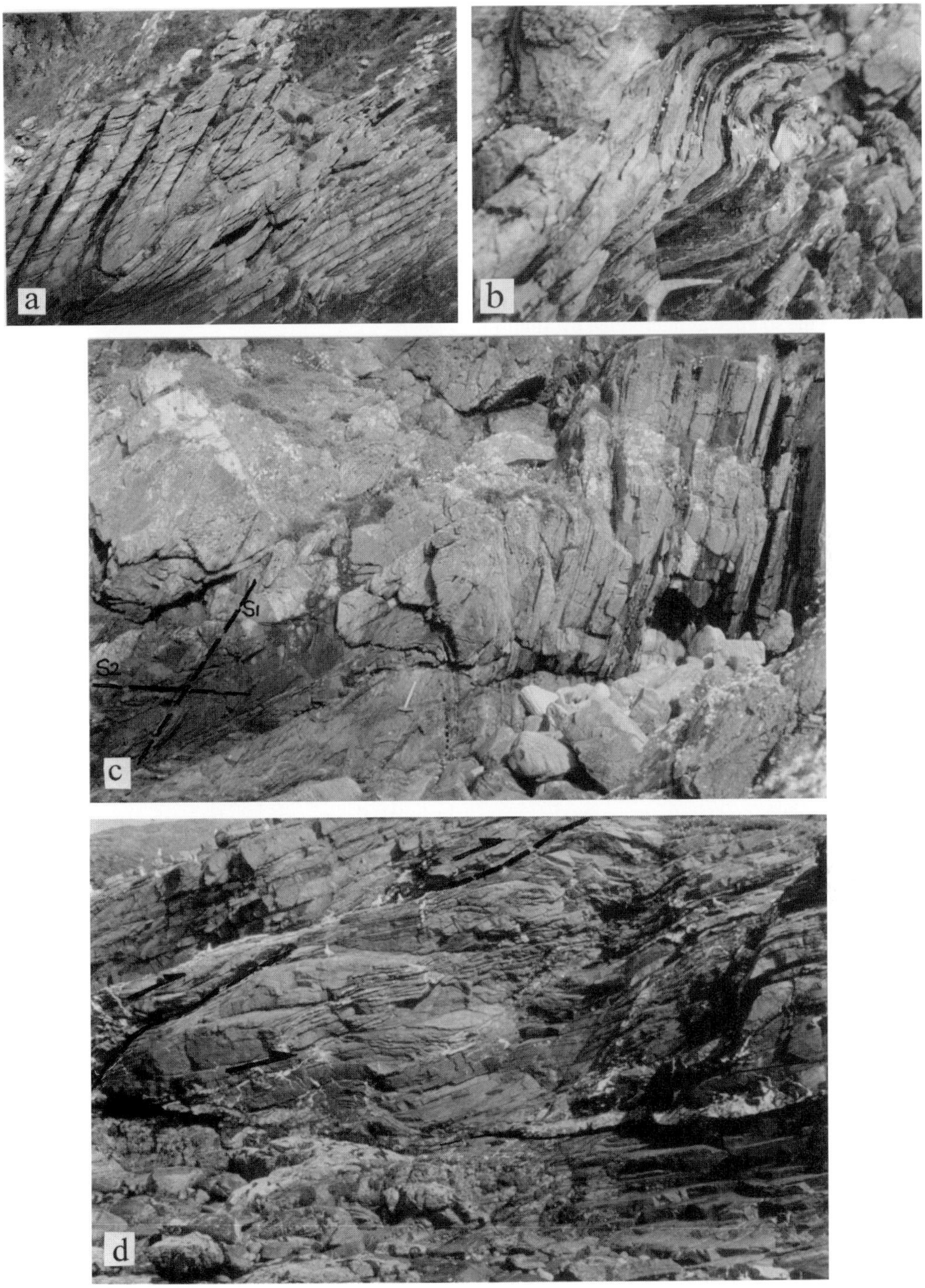

Fig. 11. Illustrations of structures in the Silurian Niarbyl Formation. (**a**) F1 folds in medium-bedded turbidites, south of the mouth of Glen Maye [SC 2237 7983]. (**b**) F2 fold, Niarbyl Bay. (**c**) F1 anticline in thinly bedded turbidites showing prominent, gently dipping S2 in thin muddy layers superimposed on S1, still visible in sandstone beds, causing a herring-bone effect, Elby Point [SC 2112 7775]. (**d**) Brittle thrust/fold/fracture and quartz vein complex, 90 m north of The Niarbyl [SC 2117 7780].

clockwise transection up to 5° is common, whilst north of the lineament anticlockwise (up to 15°) and clockwise (up to 10°) patterns are evident. Although the Knockaloe Moar Lineament appears to mark a change in the sense of cleavage transection, a rigorous assessment of this pattern has not been undertaken.

D2 deformation

The variation either side of Knockaloe Moar Lineament, from upright F1 structures to the north to inclined to the south, was interpreted by Simpson (1963) to reflect refolding about a northeast trending, large-scale D2 fold, the Peel–Ballaugh Antiform. The lineament is defined by a prominent step in the topography aligned with the Central Valley Lineament of Quirk *et al.* (1999), hence, it could represent a fault. However, as the cliffs in the vicinity of Knockaloe Moar are inaccessible the precise nature and significance of this structure remains to be determined.

F2 folds of relatively small size, with a locally developed S2 crenulation cleavage, occur sporadically throughout the formation (e.g. Fig. 11b). In almost all instances they occur as markedly asymmetrical, kink-shaped folds with wavelengths and amplitudes of a few metres at most. Good examples are exposed in the limbs of the F1 synclines exposed in the Traie Dullish Quarry [SC 2370 8401]. Larger, isolated folds occur very rarely and usually have more rounded hinges and maximum exposed limb-to-limb widths up to *c.* 15 m. A number of these folds, downward facing in inverted bedding, are evident in the rock platform at [SC 2113 7806]; one is associated with intense, fanned S2 cleavage. Most F2 folds occur in isolation; they plunge gently northeast with gently dipping axial surfaces (Fig. 10), although rarely they define minor conjugate systems or box folds.

The S2 crenulation cleavage, gently northwest or southeast dipping and either axial planar to the folds or fanned, is widely, but not universally, developed in the Niarbyl Formation. It is particularly intense in thin pelitic beds in the southern part of the outcrop (Fig. 11d).

Brittle thrust systems

Quartz veins are associated with many F2 folds, either stratiform in the limbs or folded about the hinges. This spatial association is also evident in several prominent brittle thrust–fold fracture and en echelon and pinnate quartz vein complexes, a good example of which is exposed *c.* 90 m north of The Niarbyl [SC 2117 7780] (Fig. 11d). Here, a 15 m thick zone of braided fractures and gently northwest dipping, brittle faults is associated with F2 folds and quartz veins up to 50 cm thick. Many of the veins are stratiform, others are arranged in en echelon arrays; sigmoidally folded pinnate veins occur along the hanging wall of one particularly thick stratiform vein. The gross morphology and the orientation of the veins in this system, consistent with others in the Niarbyl Formation, indicates a top-to-the-southeast sense of shear. Similar late brittle fracture systems are also evident in the Creggan Mooar Formation to the south (e.g. [SC 2157 7694]) (Kennan & Morris 1999).

High-strain zones

The mudrock and siltstone of the Manx Group behaved in a relatively ductile manner during deformation compared with interbedded packets of sandstone and early, bedding-parallel intrusions which were more rigid. During the D1 deformation, strain was commonly focused into the ductile layers that border the rigid packets. The pattern is, however, not uniform or consistent. For example, the prominent high-strain zone in the Lady Port Formation at the south end of Lynague Strand [SC 2801 8705] lies between two pelitic sequences, both of which contain early rigid intrusions. Conversely, many contacts between quartz arenite packets and pelitic sequences, e.g. along the summit of Lhiattee ny Beinee [SC 2125 7290], are not zones of high strain. Two high-strain zones exposed at Lag Ny Keeilley and Niarbyl (Fig. 1), described below, exemplify the complexities and uncertainties of such zones.

Lag Ny Keeilley

A zone of high strain exposed on the foreshore below the hermit's chapel at Lag ny Keeilley [SC 215 745] lies within northwest dipping strata assigned to the Injebreck Formation (Woodcock *et al.* 1999). It separates a north younging quartz arenite sequence to the south and an inverted sequence of pelite, quartz arenite and minor pebbly mudstone to the north. Simpson (1963) interpreted his F1 'Isle of Man Syncline' at this location and continued it across the island to explain the gross outcrop pattern of the Manx Group. It is possible that such a fold may exist along the line of the boundary between tracts 4 and 5 (Fig. 1), repeating the black mudstone of the Barrule and Glen Rushen Formations (see above). In contrast, Kennan & Morris (1999) infer that the Creggan Mooar Formation is cut out across this zone.

The most intensely deformed rocks occur in the hanging wall of the zone in a *c.* 5 m wide belt in pelites abutting the north younging quartz arenite

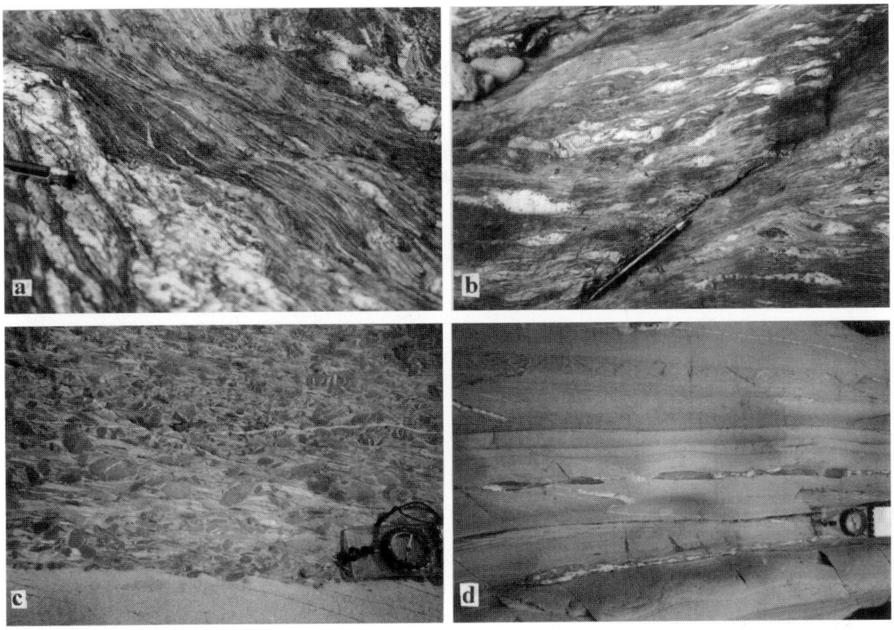

Fig. 12. Structures in the Lag ny Keeilley high-strain zone [SC 215 745]. (**a**) and (**b**) Sinistral riedel shears or extensional crenulation fabric deforming S1 cleavage in mudstone. (**c**) Conglomerate with sandstone intraclasts deformed by early ductile strain and later brittle boudinage during D1 deformation. (**d**) Mudstone layer in quartz arenite boudinaged during D1 deformation (see text for interpretation).

sequence. It is marked by an intense fabric with 'quartz fish' deformed by variably orientated shear bands, both showing a predominantly sinistral geometry (Fig. 12a). Some stratiform quartz veins up to 8 cm thick are deformed by small F1 folds. Northwards, over *c.* 8 m, this zone dissipates into a strongly foliated zone with occasional stratiform veins, and boudinaged and disrupted quartz arenite beds. This passes into less disturbed bedding adjacent to a felsite sill which contains the S1 cleavage and is also boudinaged. To the north, pelites contain a phyllitic S1 fabric. The sill and pelites are deformed by F2 folds and contain a patchily developed S2 cleavage.

The footwall is composed principally of thick-bedded quartz arenite with occasional lenses of conglomerate which are locally strongly deformed (Fig. 12c). There are few steep exposure surfaces, so only 2D strain in horizontal surfaces is readily assessed in the field. Ductile flattening of clasts in S1 produced axial ratios of *c.* 3:1. Many clasts were then broken into square-ended boudins with quartz-filled necks, recording a further 25% extension. Near the top of the quartz arenite sequence, mudrock layers are stretched by at least 50% about vertical boudin axes (Fig. 12d). This is unusual because the mudrock was probably the more ductile lithology, but the mudrock boudins now seen may occupy spaces left when early brittle boudins in the adjacent rocks underwent ductile necking as they deformed into inverse boudins. Discordant quartz veins are offset sinistrally in the fabric. Although dextral rotations are evident locally, kinematic indications are primarily sinistral.

The Lag ny Keeilley High-Strain Zone, situated at the junction between two thick sequences in which younging is opposed but also between two rock units with a marked competence contrast, is open to various interpretations between two end members:

- a major strike-slip shear zone;
- a localized belt of D1 strain partitioning at the boundary between rheologically different rock packets without regional significance.

The Lag Ny Keeilley High-Strain Zone lies within a single lithostratigraphical unit and thus does not have the same status as the tract boundaries between potentially unrelated sequences. The opposed younging directions either side suggest that the zone at least records deformation within a major F1 fold.

All indications are that the most intensely deformed zone developed during the D1 deformation in a package of ductile pelites. It's location between rigid igneous and coarse clastic

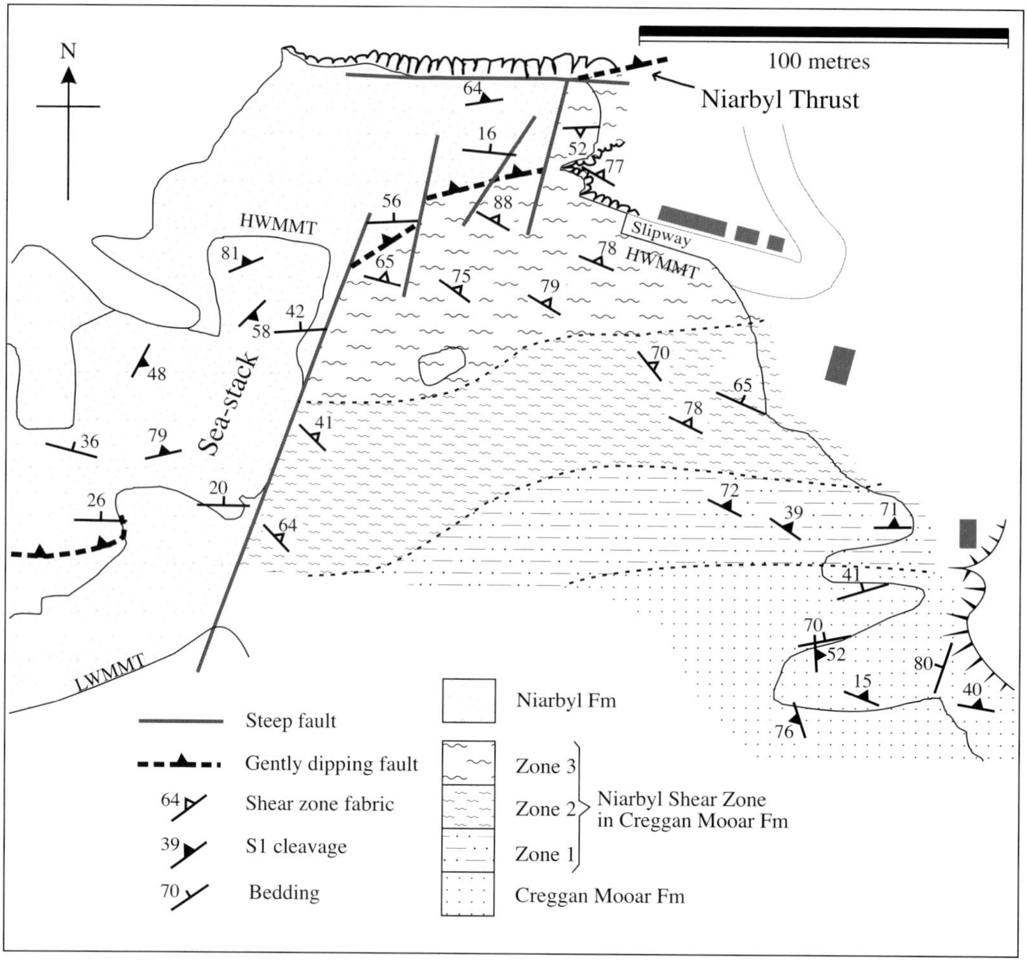

Fig. 13. The Niarbyl High-Strain Zone at The Niarbyl [SC 210 776 area], showing textural zones in the phyllonitic mylonite (Morrison 1989, Fig. 3.2) and some of the later steep and gently inclined brittle faults. Intrusions are omitted.

sedimentary rock units was an ideal location for strain partitioning. However, interpretation of the mode of D1 deformation remains unresolved. One interpretation is that it is a zone of sinistral ductile shear in which the sinistral shear bands which might represent Riedel shears. However, there is no evidence for a horizontal linear fabric component in S1. The alternative is that it is a zone of strong flattening normal to the lithological boundary, with little or no simple shear along it. In that case, the shear bands could be an extensional crenulation fabric of the type found in zones of intense flattening (Platt & Vissers 1980). The shear bands may also have been formed during D2 and later stages of deformation, and hence may not be specific to the high-strain zone.

The Niarbyl Fault Zone

Lamplugh (1903), Simpson (1963), Morrison (1989) and Roberts *et al.* (1990) all noted the conspicuous zone of ductile deformation exposed across strike for 150 m on the wave-cut platform at The Niarbyl [SC 215 776]. Simpson (1963) considered the Niarbyl Slide to be the most pronounced dislocation seen in the island, regarding it as a vertical early structure rotated into a moderate north-northwest dip by D2. He recorded it as a belt of strongly sheared, laminated and boudinaged rocks, and noted that the schistosity is cut by S2 and S3 cleavages, and deformed by small F2 and F3 folds.

Morrison (1989) made a detailed structural map

of the zone, studied its microfabrics, determined a sequence of tectonic events and drew conclusions which are markedly different from those of Simpson. Roberts *et al.* (1990) and Merriman *et al.* (1995) summarized their views in the context of metamorphic processes. In particular, Morrison showed that the Niarbyl area comprises two structural elements, both of which are imposed on D1 structures. The earlier structure is a steep, west-northwest trending belt of phyllonitic mylonite recording sinistral ductile deformation, in which deformation intensity increases northwards (Fig. 13). The younger, east-northeast striking Niarbyl Thrust truncates the phyllonite in the north and carries rocks of the Niarbyl Formation southwards over the phyllonite; this is a southward directed, compressional brittle structure, which is strongly oblique to the older belt.

Controversy persists about the nature and importance of the ductile and brittle components of this fault zone, and even the protolith in which it was developed. Even it's extent is unclear, one interpretation given below being that the zone does not extend significantly inland, the other is that it extends at least as far as Glen Maye (Fig. 9). In the following sections the fault zone is first described then interpreted in various ways. The continuing uncertainty highlights the need for more detailed work than has been possible during the course of this study.

The 'Niarbyl High-Strain Zone' is defined as being the ductile shear zone exposed in Niarbyl Bay, equivalent to: the Niarbyl Slide of Simpson (1963); the phyllonitic fault zone of Morrison (1989); the Niarbyl Shear Zone of Morris *et al.* (1999). The Niarbyl Thrust, one of a plexus of brittle faults and related minor structures which cut the Niarbyl High-Strain Zone, is taken as defined by Simpson. The term Niarbyl Fault Zone is used here to describe the compound, ductile–brittle structural complex represented, *inter alia*, by the combination of the Niarbyl High-Strain Zone and the Niarbyl Thrust and related structures (Figs 13 and 14).

The fault zone affects a varied suite of rocks. Large parts of Niarbyl Bay are occupied by pelitic rocks with thin siltstone and sandstone intercalations and distinctive ironstone laminae attributed to the early Ordovician Creggan Mooar Formation (Kennan & Morris 1999). To the north is the Silurian, sandstone-dominated Niarbyl Formation. Heavily altered igneous sheets are also present, strongly disrupted by the deformation. One of the current uncertainties regarding the Niarbyl Fault Zone is to what extent the Silurian rocks are involved in the ductile deformation (Figs 13 and 14). One interpretation (Fig. 13) is that the Niarbyl High-Strain Zone is entirely restricted to the Creggan Mooar Formation. Figure 13, which largely accords with Morrison (1989), records the continuity and trend of the phyllonite texture zones and the northward intensification of strain from zone 1 to zone 3; it indicates that the boundary between the Silurian Niarbyl Formation and the Niarbyl High-Strain Zone is the gently dipping, brittle Niarbyl Thrust, offset by later steep faults. The alternative interpretation (Fig. 14), suggests that the Niarbyl High-Strain Zone affected both formations; that the boundary between the formations is primarily defined by a set of brittle faults which is distinct, apart from area A, from those interpreted as the Niarbyl Thrust; and that distinct high-strain zones do not fully accord with the texture zones identified by Morrison (1989). These contrasts have major interpretation implications which are considered below.

The Niarbyl Fault Zone at Niarbyl. The first deformation in the high strain zone produced steep F1 folds and S1 cleavage, which are best preserved in southern, least deformed parts of the zone (zone 1 in Figs 13 and 14). These early structures are essentially similar to D1 structures in rocks of the Niarbyl and Creggan Mooar Formations out with the high-strain zone. Within the high-strain zone, however, steeply dipping, mainly west-northwest striking, S- and C-fabrics were imposed on the D1 structures during ductile shearing which converted pelitic rocks in the belt to phyllonite. The intensity of this shear deformation overprint increases northwards. In southern parts (zone 1), bedding is still visible but is dismembered into lenses by C-planes with mainly sinistral sense which are oblique to S1. Partitioned strain zones, marked by wide belts of coherent bedded rocks alternating with narrow zones of disrupted and transposed bedding, are exposed for several hundred metres in the Creggan Mooar Formation further to the south. To the north, in zone 2, disruption of bedding is more intense, C-planes are more common, quartz veins are deformed, and quartz and carbonates are segregated to form augen.

The most intensely sheared part of the high-strain zone comprises phyllonite with abundant quartz and carbonate augen, tightly folded quartz veins and well-developed S–C fabrics. Pull-apart zones and Riedel shears have caused further disaggregation. Morrison (1989) noted that the S–C fabrics and asymmetry of augen point to both sinistral and dextral shear (Fig. 15a). We, in common with Morrison, note a preponderance of sinistral shear indicators. Morrison was unable to determine unequivocally the timing of the shear deformation with respect to the regional deformation sequence, but suspected that movement began shortly after the D1 deformation. Here, it is

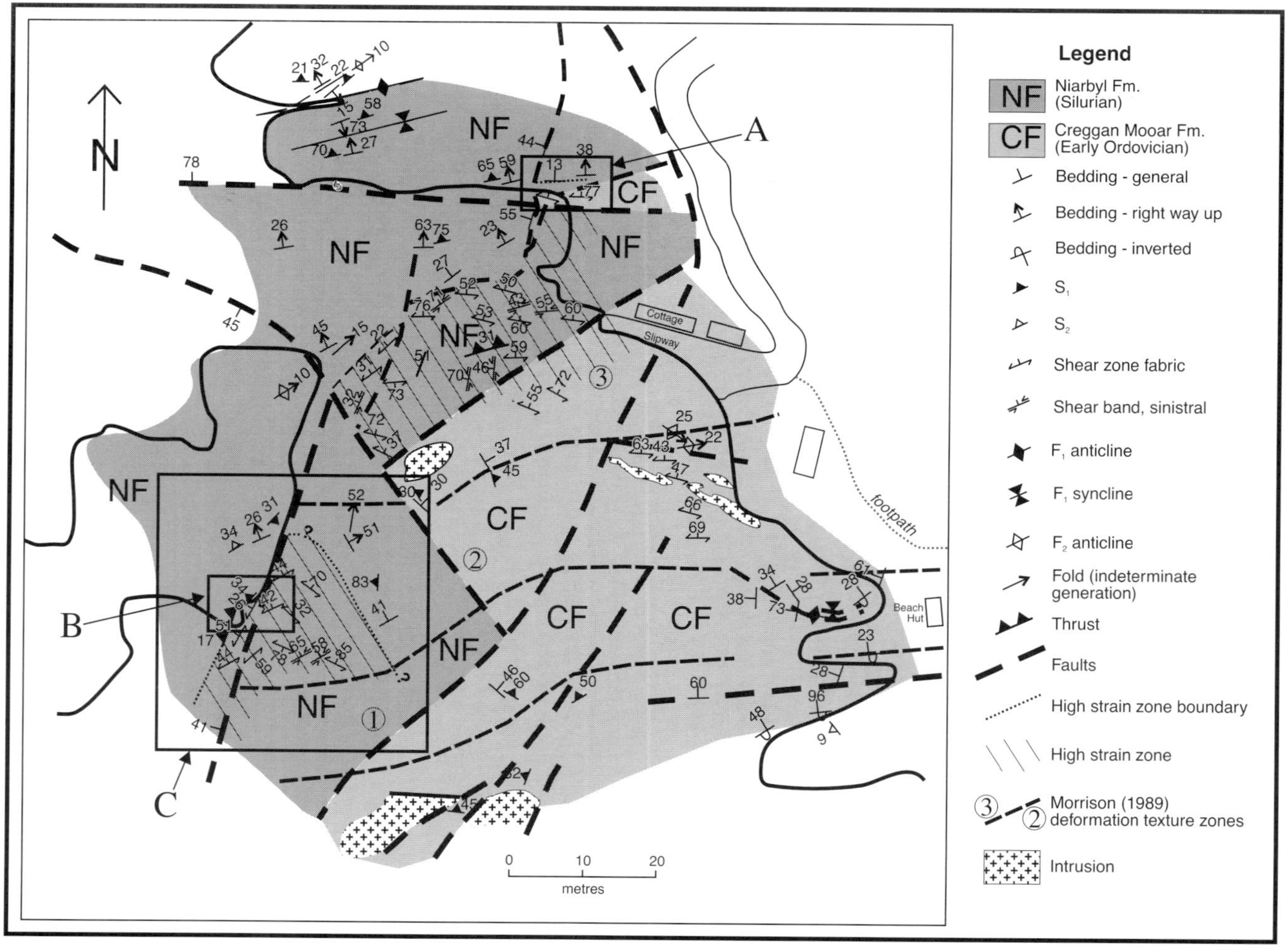

Fig. 14. Alternative geological map of the Niarbyl Fault Zone. A, Niarbyl Thrust and contact between the Niarbyl and Creggan Mooar Formations; B, site of southeast directed ramp-flat thrusts and thrust duplex (see also Fig. 15b); C, area of macroscopic fold in phyllonitic fault-zone fabric.

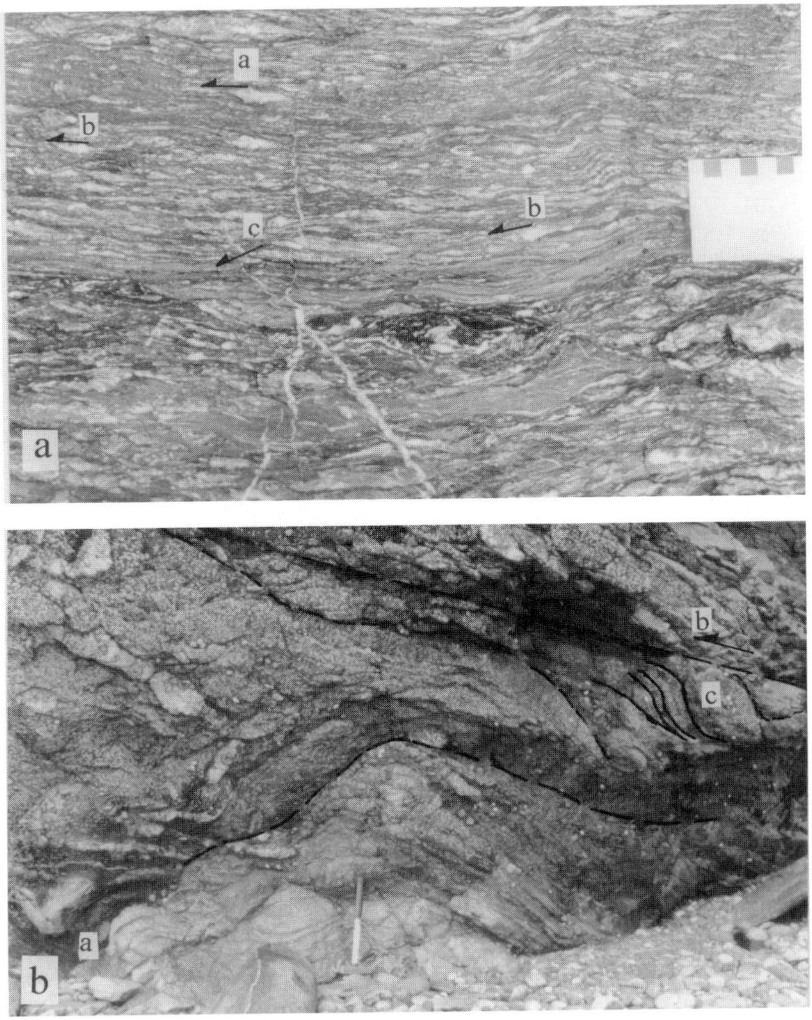

Fig. 15. Illustrations of structures and fabrics in the Niarbyl High-Strain Zone, just west of the slipway (location on Fig. 14). (**a**) Sinistral sense of shear defined by bedding (a) and quartz vein (b) boudins and shear bands (c). (**b**) Contact zone between high strain phyllonite (a) and top-to-the-southeast brittle thrust (b) and thrust duplex (c).

considered that many of the small shears, kinks and crenulations of the phyllonite fabric were produced by the D2 and D3 deformation, consistent with Morrison's views on timing.

The ductile shear zone is dissected by numerous brittle faults, some of which have a near north–south strike and steep dip, whilst others are moderately north to northwest dipping. The latter include two faults which truncate the phyllonite: the Niarbyl Thrust (Fig. 13) exposed at the east end of the cliff, 40 m north of the slipway, and another along the base of the south face of the sea-stack (Fig. 14, C). The fault at the sea-stack is considered here to be the likely continuation of the Niarbyl Thrust, although interrupted by numerous minor late faults.

Morrison (1989) considered that evidence for thrust movement rather than low-angle normal displacement on the Niarbyl Thrust is insubstantial. However, the fault zone at the sea-stack includes a 50 cm thick duplex, several thrusts with ramp-flat morphology, quartz vein arrays and a hanging wall anticline (Figs 14, B and 15b). Using ramp orientation as a guide, movement was towards the southeast, although the amount of displacement has not been determined. These thrusts are post-D1 on the basis that S1 has been rotated with bedding in thrust horses; the late or post-D1 phyllonite fabric is also deformed by them.

This complex occupies part of the northwest limb of a mesoscopic fold in the underlying Niarbyl High-Strain Zone phyllonite (Fig. 14, B and C). Just to the west of the brittle fault, the Niarbyl High-Strain Zone swings to a northeast strike, dips *c.* 25° northwest and, over a distance of *c.* 2 m, passes up from a high-strain zone, with stratiform quartz veins, through less intensely deformed, but folded, disrupted bedding and quartz veins, into relatively coherently bedded Niarbyl Formation turbidites (Fig. 15b). The top of this zone is marked by the top to the southeast brittle ramp-flat thrust complex described above.

Glen Maye. In the Glen Maye Gorge (Fig. 9), a north-northeast striking, gently dipping phyllitic fabric is developed in pelites attributed to the Creggan Mooar Formation in a zone exposed for 750 m across-strike, possibly representing a continuation of the Niarbyl High-Strain Zone. Its eastern boundary, under the road bridge in Glen Maye, is marked by an abrupt termination of outcrop. This is presumed to be an unexposed, north trending, brittle fault contact with Glen Rushen Formation slates, exposed *c.* 300 m upstream to the east. The western boundary, with the Niarbyl Formation, is defined by a felsite intrusion, presumed to occupy another late north–south brittle fault. To the west, in the lowest exposures of Niarbyl Formation, bedding is disturbed by prominent pinch-and-swell structures, morphologically comparable to those present in the formation in the brittle thrust system at Niarbyl (Fig. 14, B).

Discussion

The preceding paragraphs provide an objective record of the structures and fabrics in the Niarbyl Fault Zone with minimal interpretation. The origins of the phyllonite belt are discussed further in the next section. However, several aspects of the zone which have important bearing on the tectonic evolution of the Isle of Man remain the subjects of controversy and ongoing research. Three of the most significant problems are given below.

Kinematic and age relationships between the ductile shear zone and brittle thrusts. The ductile and brittle components of the Niarbyl Fault Zone may be: different expressions of a single, progressive, non-coaxial tectonic event; kinetically unrelated to one another and of different ages.

- The first interpretation suggests early sinistral ductile shear, followed immediately by brittle thrusting towards the southeast. This occurred during the later stages of the regional D1 events or shortly afterwards, on the basis that the phyllonite fabrics were imposed on D1 structures. The phyllonite fabric is deformed by S2 and co-planar S2 occurs in the formations above and below the Niarbyl Fault Zone, indicating a pre-D2 age, although the timing of brittle thrust–quartz vein systems in both formations relative to D2 is debatable. The Niarbyl Fault Zone in this interpretation might be viewed as a product of large-scale, sinistral transpression.
- The collective features of the late-D1–pre-D2 phyllonite belt, notably its width, west-northwest strike and mylonitic fabrics, are unusual in the Manx context, and its origin and regional tectonic significance remain enigmatic. However, the brittle deformation, which produced the Niarbyl Thrust cutting obliquely across the ductile shear, can be regarded as a much younger event, probably post-D2; the brittle thrust zone at the sea-stack in Niarbyl Bay has no clear genetic relationships with the essentially ductile D2 folds and S2 cleavage. The thrusting, in this interpretation, is attributed to the very late Caledonian (Acadian) compression which also locally deformed the late Silurian–early Devonian Peel Sandstone.

Protolith of the Niarbyl High-Strain Zone. The exposure of the Niarbyl High-Strain Zone is situated at the boundary between the Niarbyl and Creggan Mooar Formations; the protolith of much of it is the latter as indicated by the characteristic iron-rich laminae. The extent of involvement of the Niarbyl Formation distinguishes two possibilities:

- the two sequences were juxtaposed prior to or during formation of the Niarbyl High-Strain Zone such that rocks of both formations have been sheared and converted to phyllonite; the Niarbyl Thrust, and related minor top to the southeast structures, lie immediately at the top of the shear zone, within the Niarbyl Formation;
- the ductile shear zone is confined to the Creggan Mooar Formation; the younger Niarbyl Thrust carries the rocks of Niarbyl Formation, unaffected by the ductile shear zone, over the Niarbyl High-Strain Zone.

Confirming the presence of material from the Niarbyl Formation in the sheared rocks of the Niarbyl High-Shear Zone is particularly difficult. However, resolution of this problem may be achieved by geochemical analysis of relict sandstone within the shear zone, since the sandstone of the Niarbyl Formation has a distinct signature (Barnes *et al.* 1999).

The inland trace of the Niarbyl High-Shear Zone. Neither the phyllonite belt nor the Niarbyl Thrust can be traced immediately inland beneath the blanket of superficial deposits, so their positions inland are open to interpretation:

- One possibility is that the composite fault zone swings northeastwards and re-emerges in Glen Maye (Fig. 9). The fabric of the rocks of the Creggan Mooar Formation in that area would be interpreted as phyllonitic, rather than phyllitic. The inland extent of the zone here is controlled by north striking brittle faults. Similar structures truncate the inland extent of various units along the east coast, including anomalous southeast striking units such as the Lady Port Formation.
- The alternative is to consider that the east-southeast strike of the phyllonite in the Niarbyl High-Strain Zone is a reasonable guide to the orientation of the zone as a whole. In which case the Niarbyl High-Strain Zone extends east-southeast beneath the superficial deposits until it is truncated/offset by the fault between the Creggan Mooar and Glen Rushen Formations, and/or the tract-bounding structure. The phyllonite might be repeated to the north of Niarbyl, in the hanging wall of the Niarbyl Thrust, but any outcrops will be offshore. Considering the Niarbyl Thrust as a distinct structure, its position beneath the drift is unpredictable because it is dictated by small variations in its dip and later fault offsets, as demonstrated by its sinuous outcrop in Niarbyl Bay. However, if the thrust maintains its position as the lower boundary of the Niarbyl Formation, its trace is defined by the edge of the outcrop (Fig. 1).

Discussion of Manx structures and their regional context

This section considers each of the main stages of deformation in the Isle of Man described above in terms of timing and in the context of events recognized elsewhere in southern parts of the British Caledonides.

Tract structure

The structures inferred to separate the tectono-stratigraphic tracts from each other have yet to be characterized and they remain largely hypothetical. It is possible that better control may demonstrate that some of them are essentially stratigraphical boundaries. However, the tract 1–3 boundary, the Windy Corner Fault (Fig. 8), is arguably a north-west dipping, syn-D1 thrust, although its relationships with other structures is unclear. A similar inference may be drawn for the Niarbyl High-Strain Zone, by one interpretation. An alternative is that they are coeval, larger analogues of the small late thrusts, such as the Niarbyl Thrust, seen on the Isle of Man. Therefore, if faulted, the tract boundaries could be results of early and/or late collisional processes, i.e. latest Silurian–early Devonian to intra-Devonian.

D1 deformation

The age of D1 in the Wenlock strata of the Niarbyl Formation is constrained to the late Silurian–earliest Devonian, broadly late Caledonian (Acadian), because it does not affect the Peel Sandstone, inferred to be latest Silurian–early Devonian by Piper & Crowley (1999). It is likely that the D1 in the Ordovician Manx Group, similar in all respects to that in the Niarbyl Formation, is the same age, although an older, possibly Ordovician, age cannot be ruled out (cf. Gallagher *et al.* 1994; Tietzsch-Tyler 1996). The D1 structures are very similar to the main structures described from other parts of the southern Caledonides of Britain, folds having a 'Caledonoid' trend, generally upright axial planes and horizontal plunge. They are, however, subtly different from the more variably orientated folds controlled by reactivated Late Proterozoic basement in the Welsh Basin (e.g. Fitches in Treagus 1992). The uniformity on the Isle of Man may imply less basement influence, a different type of basement, basement with dominantly northeast–southwest structures, or a combination of these factors (cf. Kimbell & Quirk 1999).

As in other parts of the southern Caledonides, the F1 folds in the Isle of Man are commonly transected by S1 cleavage. In the Southern Uplands, the southern Lake District and parts of Wales, the transection is clockwise, a relationship interpreted to be a consequence of sinistral transpression at the western margin of the Midlands microcraton (Eastern Avalonia) during collision with Laurentia (Soper *et al.* 1987). However, as in parts of Wales, both clockwise and anticlockwise transection occurs on the Isle of Man, pointing to local block rotations during deformation (cf. Pratt & Fitches 1993).

High-strain zones

Two high-strain zones exposed on the west coast of the Isle of Man are described above, but their origin and regional significance are uncertain. The Lag ny Keeilley high-strain zone appears to be a Caledonoid D1 structure, but it may be a product of strain partitioning at the boundary between rigid and ductile rocks. Whether it is a zone of rotational (strike-slip shear) or coaxial (flattening) deformation is open to interpretation. Phyllonitic fabrics were formed in the Niarbyl High-Strain Zone during or shortly after the later stages of D1, possibly by ductile sinistral transpression. No other belt of such intense deformation with its unusual east-southeast trend, strongly discordant to the regional Caledonoid trend, has been documented in the Isle of Man, although this may be no more than a local effect. The origin of the early ductile fabric remains enigmatic, with a range of possible explanations:

- enhanced tectonism in a zone of pre-lithification disruption within the Creggan Mooar Formation, juxtaposed against the Wenlock Niarbyl Formation by later brittle faults; similar processes and structures have been identified in the Hawick Group in southwest Scotland (e.g. Kemp 1987; Lintern & Floyd 1999);
- a zone of post-lithification ductile deformation in the Creggan Mooar Formation, formed prior to either juxtaposition by faulting against, or being unconformably overlain by, the Niarbyl Formation;
- a zone of ductile deformation associated with the tectonic emplacement of the Niarbyl Formation into its present situation;
- strain partitioning, adjacent to the more rigid sandstone sequence of the Niarbyl Formation, into the more ductile rocks of the Creggan Mooar Formation, at either a sedimentary (unconformable) or faulted junction.

D2 deformation

D2 structures have not been detected in the late Silurian–early Devonian Peel Sandstone, but are present in the Silurian rocks of the Niarbyl Formation which crop out nearby at Peel Castle. By implication, the deformation took place after the Wenlock and before the early Devonian, and is assigned to late Caledonian events. The time gap between the D1 and D2 deformations has not been determined, so they may represent distinct events or stages in a deformation continuum.

In many parts of the Caledonides of southern Britain, flat-lying folds and cleavage have been imposed on the main, upright Caledonian structures. These late structures, commonly assigned to D2 events in local tectonic schemes, are small and are generally only sporadically developed: e.g. in Wales they are attributed to localized movement at various times in fault zones (e.g. Wilkinson & Smith 1988; Pratt 1992) and probably have little regional significance.

Flat-lying late structures also occur in the Skiddaw Group of the northern Lake District (Fraser *et al.* in Treagus 1992), approximately along-strike from the Isle of Man. Those F2 folds commonly are chevron in style, wavelengths are on a centimetre to metre scale and they have an axial-planar crenulation cleavage (Roberts 1971, 1976, 1977). In these respects, they resemble the D2 structures of the Isle of Man, although no large-scale structures have been identified. The D2 deformation in the northern Lake District took place during or after the emplacement of the *c.* 400 Ma Skiddaw Granite, on the basis that S2 cleavage has deformed minerals produced in the thermal aureole (Soper & Roberts 1971). Roberts (1977) attributed the D2 structures to gravitational collapse or to vertical compression caused by the rise of late components of the Lake District batholith. Hughes *et al.* (1993), on the other hand, associated them with south directed thrusts, although precise relationships and kinematics have not been defined.

Small-scale recumbent F2 folds and a gently dipping axial-planar crenulation cleavage are also locally well developed in the Southern Uplands in southwest Scotland (Barnes 1999; Barnes & Stone 1999). Those folds verge consistently down-dip and appear to be unrelated to larger scale structures, Here, they pre-date early Devonian granite intrusions and are probably late Silurian in age, as in the Isle of Man.

It is not yet determined whether the D2 structures in the Isle of Man have regional tectonic significance or represent minor localized structures of various ages, as in the Welsh Basin. Nevertheless, they closely resemble the widespread flat-lying structures of the northern Lake District, which were produced shortly after the main Caledonian deformation. The Manx D2 structures record a near-vertical flattening event, rather than regional shear associated with thrusting at the present level of exposure; there is no consistent D2 vergence direction and no clear spatial or kinematic association with thrust faults. The cause of this flattening remains enigmatic, although various hypotheses may be explored, such as: extensional relaxation and gravitational collapse following D1; vertical compression and lateral distension above the roof of a rising batholith; or flattening caused by thrust sheets emplaced above the present level of erosion.

D3 deformation

The ages of D3 structures are uncertain. However, for reasons to be discussed in detail by WRF in a later publication, the steep north striking folds are correlated with similarly oriented Variscan folds and cleavage in the Dinantian limestones of the Castletown area in the south of the Isle of Man.

Others were probably caused by the northwest–southeast to north–south compression which led to the formation of late Caledonian thrusts discussed below.

Thrusts

The Niarbyl Thrust, whether interpreted as part of a single, progressive ductile to brittle deformation episode or as the later of two discrete events, was produced late in the deformation sequence, certainly post-D1 and arguably D2 or post-D2. Whichever interpretation is accepted, it is regarded here as an end-Caledonian structure. Small, late thrusts occur sporadically elsewhere in the Manx Group, e.g. in the northwestern coastal section near Kirk Michael and at Langness. Similar structures (to be described and discussed elsewhere by WRF) are common in the rocks of the late Silurian–early Devonian Peel Group in White Strand Bay (Fig. 1). All show small southerly to southeasterly displacement and are almost certainly pre-Carboniferous structures because they are absent from the southern Manx Dinantian limestones.

These thrusts are probably members of a family of structures which also affected regions to the south shortly after the deposition of early Devonian red beds. Other, mostly much larger, members of this family include the south vergent folds and thrusts at Lligwy Bay in northeast Anglesey (Scott in Treagus 1992), the Carmel Head Thrust of northwest Anglesey (Bates in Treagus 1992) and the inferred South Irish Sea Lineament (Brewer *et al.* 1983).

In the Northern Fells of the Lake District, south directed thrusts cut early upright Caledonoid folds and cleavage, but it is not clear if these can be correlated with those on the Isle of Man. Hughes *et al.* (1993) considered them to be genetically related to small-scale recumbent folds and crenulation cleavages, whereas the recumbent D2 structures on the Isle of Man are inferred to be unrelated to thrusting. The Isle of Man is approximately along-strike and little more than 50 km from the Southern Fells of the Lake District. The younger Lower Palaeozoic rocks there were deformed by upright Caledonoid folds but were also caught up in the Westmoreland Monocline mountain front, a belt of steep northward backthrust deformation (Kneller 1990*a*, *b*; Kneller & Bell 1993). No northerly directed thrusts have been identified in the part of the Isle of Man discussed here but in other respects the deformation styles are similar.

A recurrent theme throughout the East Irish Sea region is that the main late Caledonian (Acadian) upright folds and cleavage are cut by thrusts. Some thrusts, as in the Lake District Northern Fells, may be older than others and related to recumbent folding. Those in the Southern Fells are northward directed, largely because of buttressing by the Lake District Batholith. Elsewhere in the region, including the Isle of Man, most thrusts appear to be southward directed and formed after deposition of Devonian red beds. These relatively young structures probably record the final stages and southernmost advance of the compressive deformation associated with the Iapetus suture zone, as envisaged by Jackson *et al.* (1995).

Summary remarks on the tract structure model

The large-scale tract structure of the Isle of Man proposed here was a necessary expedient to allow the development of a lithostratigraphical model of the Manx Group which recognizes the high level of uncertainty which still exists in relation to some boundaries between otherwise mappable rock units. This approach has two major advantages:

- it avoids the need experienced by previous workers (e.g. Lamplugh 1903; Simpson 1963) to construct large-scale tectonostratigraphical models based on a poor understanding of both the stratigraphy and the large-scale structure;
- it allows freedom in current interpretations to explore the full range of possible correlations between the fragments of lithostratigraphy which can reasonably be defined [e.g. cf. Woodcock *et al.* (1999), Woodcock & Barnes (1999) and discussion in Barnes *et al.* (1999)].

On a regional scale, it is interesting to note that, although the Manx Group has been broadly correlated with the Skiddaw Group in northwest England for a long time, detailed correlation has never been confirmed [e.g. Cooper *et al.* (1995) and references cited therein)]. Modern work seems to support lithological correlation between age equivalent (early Arenig) rocks of the sand-rich turbidite sequence of tract 1 (the Lonan and Santon Formations) with the Loweswater Formation in the Skiddaw Group [e.g. Stone & Evans (1997); Barnes *et al.* (1999)]. However, sequences in central and western tracts contain abundant quartz arenite for which there is no well-developed equivalent in the Skiddaw Group and may show closer lithological similarity with the Cambrian–early Ordovician sequence in southeast Ireland [e.g. Kennan & Morris (1999); McConnell *et al.* (1999)]. Quartz arenite interbedded in the tract 1 sequence provides a possible link with the sequence in other parts of the island, but in the absence of more precise biostratigraphical control there are many possible relationships (e.g. Barnes *et al.* 1999). It is possible on this basis that the Isle of Man preserves the tectonically 'concertina'd' lateral transition from the southeast Irish sequence to that exposed in northwest England. The relative structural simplicity of the tract 1 sequence, similar to that of much of the Skiddaw Group, allows the possibility that this may be essentially 'allochthonous' Skiddaw Group against which the more intensely deformed eastern margin of the southeast Irish sequence was juxtaposed during D1–D2. The occurrence beneath the Isle of Man of a major structural break in the crust (Kimbell & Quirk 1999) allows the possibility that this is a fundamental zone of convergence between two essentially 'Avalonian' crustal units rather than the 'Iapetus suture'.

We are indebted to colleagues involved in the recent investigations into the geology of the Isle of Man for discussion and shared experience in the field. Nigel Woodcock and David Quirk made their data from the southeast of the island available for presentation in Fig. 2. The paper was substantially improved by the helpful comments from the reviewers, including John Mendum, Barry Webb and Bob Holdsworth. Particular thanks to Greg Power who helped us to keep our act together. WRF publishes by permission of Robertson Research International; RPB publishes by permission of the Director, British Geological Survey (NERC); JHM publishes by permission of the Director, Geological Survey of Ireland.

References

BARNES, R. P. 1999. Geology of the Whithorn, Kirkcowan and Wigtown districts. *Memoir of the British Geological Survey*, sheets 2, 4W & 4E (Scotland), in press.

—— & STONE, P. 1999. Trans-Iapetus contrasts in the geological development of southern Scotland (Laurentia) and the Lakesman terrane (Avalonia). *This volume*

——, POWER, G. M. & COOPER, D. C. 1999. The definition of sandstone-bearing formations in the Isle of Man and correlation with adjacent areas – evidence from sandstone chemistry. *This volume.*

BREWER, J. A., MATTHEWS, D. H., WARNER, M. R., HALL, J., SMYTHE, D. K. & WHITTINGTON, R. J. 1983. BIRPS deep seismic reflection studies of the British Caledonides – the WINCH profile. *Nature*, **305**, 206–210.

GALLAGHER, V., O'CONNOR, P. J. & AFTALION, M. 1994. Intra-Ordovician deformation in southeast Ireland: Evidence from the geological setting, geochemical affinities and U–Pb zircon age of the Croghan Kinshelagh granite. *Geological Magazine*, **131**, 669–684.

GEOLOGICAL SURVEY. 1898. *Isle of Man. Solid and Drift geology. 1:63 360 scale*. Reprinted at 1:50,000 scale by Institute of Geological Sciences 1975 (Ordnance Survey, Southampton).

GILLOTT, J. E. 1955. Metamorphism of the Manx Slates. *Geological Magazine*, **92**, 141–154.

HOWE, M. P. A. 1999. The Silurian fauna (graptolite and nautiloid) of the Niarbyl Formation, Isle of Man. *This volume.*

HUGHES, R. A., COOPER, A. H. & STONE, P. 1993. Structural evolution of the Skiddaw Group (English Lake District) on the northern margin of eastern Avalonia. *Geological Magazine*, **130**, 621–629.

JACKSON, D. I., JACKSON, A. A., EVANS, D., WINGFIELD, R. T. R., BARNES, R. P. & ARTHUR, M. J. 1995. *United Kingdom Offshore Regional Report: The Geology of the Irish Sea.* HMSO for the British Geological Survey.

KEMP, A. E. S. 1987. Tectonic development of the Southern Belt of the Southern Uplands accretionary complex. *Journal of the Geological Society, London*, **144**, 827–838.

KENNAN, P. S. & MORRIS, J. H. 1999. Manganiferous ironstones in the early Ordovician Manx Group, Isle of Man: a protolith of coticule?. *This volume.*

KIMBELL, G. S. & QUIRK, D. G. 1999. Crustal magnetic structure of the Irish Sea region: evidence for a major basement boundary beneath the Isle of Man. *This volume.*

KNELLER, B. C. 1990*a*. *The Wenlock Rocks of Sheet 38.* British Geological Survey Technical Report WA/90/64.

—— 1990*b*. *The Ludlow Rocks of Sheet 38 (Ambleside).* British Geological Survey Technical Report WA/90/62.

—— & BELL, A. M. 1993. An Acadian mountain front in the English Lake District; the Westmorland Monocline. *Geological Magazine*, **130**, 203–213.

LAMPLUGH, G. W. 1903. *The Geology of the Isle of Man.* Memoir of the Geological Survey, UK. HMSO.

LINTERN, B. C. & FLOYD, J. D. 1999. *The Kirkcudbright–Dalbeattie district – a concise account of the geology*. Memoir of the British Geological Survey. Sheets 5W, 5E and part of 6W (Scotland).

MCCONNELL, B. J., MORRIS, J. H. & KENNAN, B. S. 1999. A comparisonof the Ribband Group (southeastern Ireland) to the Manx Group (Isle of Man) and Skiddaw Group (northwestern England). *This volume.*

MERRIMAN, R. J., ROBERTS, B., PEACOR, D. R. & HIRONS, S. R. 1995. Strain-related differences in the crystal growth of white mica and chlorite: a TEM and XRD study of the development of metapelitic microfabrics in the Southern Uplands thrust terrane, Scotland. *Journal of Metamorphic Geology*, **13**, 559–576.

MOLYNEUX, S. G. 1979. New evidence for the age of the Manx Group, Isle of Man. *In*: HARRIS, A. L., HOLLAND, C. H. & LEAKE, B. E. (eds) *Caledonides of the British Isles: Reviewed.* Geological Society, London, Special Publications, **8**, 415–421.

MOLYNEUX, S. 1999. A reassessment of Manx Group acritarchs. *This volume.*

MORRIS, J., WOODCOCK, N. H. & HOWE, M. 1999. The Niarbyl Formation: remnant of a Silurian anoxic turbidite basin on the Isle of Man. *This volume.*

MORRISON, C. W. K. 1989. *A study of the Anchizone–*

Epizone metamorphic transition. PhD Thesis, St Andrews University, Scotland.
ORR, P. & HOWE, M. P. A. 1999. Macrofauna and ichnofauna of the Manx Group (early Ordovician), Isle of Man. *This volume*.
PIPER, J. D. A. & CROWLEY, S. F. 1999. Palaeomagnetism of (Palaeozoic) Peel Sandstones and Langness Conglomerate Formation, Isle of Man: implications for the age and regional diagenesis of Manx red beds. *This volume*.
PLATT, J. P. & VISSERS, R. L. M. 1980. Extensional structures in anisotropic rocks. *Journal of Structural Geology*, **2**, 397–410.
POWELL, C. MC A. 1979. A morphological classification of rock cleavage. *Tectonophysics*, **58**, 21–34.
POWER, G. M. & BARNES, R. P. 1999. Relationships between metamorphism and structure on the northern edge of eastern Avalonia in the Manx Group, Isle of Man. *This volume*.
PRATT, W. T. 1992. The use of kink bands to constrain fault displacements: an example from the Bala Lineament, Wales. *Geological Magazine*, **129**, 625–632.
—— & FITCHES, W. R. 1993. Transected folds from the western part of the Bala Lineament, Wales. *Journal of Structural Geology*, **15**, 55–68.
QUIRK, D. G. & BURNETT, D. 1999. Lithofacies of Lower Palaeozoic deep marine sediments in the Isle of Man: a new map and stratigraphic model for the Manx Group. *This volume*.
——, ——, KIMBELL, G. S., MURPHY, C. A. & VARLEY, J. S. 1999. Shallow geophysical and geological evidence for a regional-scale fault duplex in the Lower Palaeozoic rocks of the Isle of Man. *This volume*.
ROBERTS, B., MORRISON, C. & HIRONS, S. R. 1990. Low grade metamorphism of the Manx Group, Isle of Man: a comparative study of white mica 'crystallinity' techniques. *Journal of the Geological Society, London*, **147**, 271–277.
ROBERTS, D. E. 1971. Structures of the Skiddaw Slates in the Caldew valley, Cumberland. *Geological Journal*, **7**, 225–238.
—— 1976. Cleavage formation in the Skiddaw Slates of the Northern Lake District, England. *Geological Magazine*, **113**, 377–382.
—— 1977. The structure of the Skiddaw Slates in the Blencathra–Mungrisdale area, Cumbria. *Geological Journal*, **12**, 33–58.
RUSHTON, A. W. A. 1993. Graptolites from the Manx Group. *Proceedings of the Yorkshire Geological Society*, **49**, 259–262.
SIMPSON, A. 1963. The stratigraphy and tectonics of the Manx Slates Series. *Quarterly Journal of the Geological Society, London*, **119**, 367–400.
—— 1964. Metamorphism of the Manx Slate Series, Isle of Man. *Geological Journal*, **4**, 415–434.
SOPER, N. J. & ROBERTS, D. E. 1971.Age of cleavage in the Skiddaw slates in relation to the Skiddaw aureole. *Geological Magazine*, **108**, 293–303.
——, WEBB, B. C. & WOODCOCK, N. H. 1987. Late Caledonian (Acadian) transpression in north-west England: timing, geometry and geotectonic significance. *Proceedings of the Yorkshire Geological Society*, **46**, 175–192.
STONE, P. & EVANS, J. A. 1997. A comparison of the Skiddaw and Manx groups (English Lake District and Isle of Man) using neodymium isotopes. *Proceedings of the Yorkshire Geological Society*, **51**, 343–347.
TIETZSCH-TYLER, D. 1996. Precambrian and early Caledonian orogeny in southeast Ireland. *Irish Journal of Earth Sciences*, **15**, 19–30.
TREAGUS, J. E. (ed.) 1992. Caledonian Structures in Britain South of the Midland Valley, *Geological Conservation Review Series*, **3**. Chapman & Hall.
WILKINSON, I. & SMITH, M. 1988. Basement fractures in North Wales: their recognition and control on Caledonian deformation. *Geological Magazine*, **125**, 301–306.
WOODCOCK, N. H. & BARNES, R. P. 1999. An early Ordovician turbidite system on the Gondwana margin: the southeastern Manx Group, Isle of Man. *This volume*.
—— & MORRIS, J. H. 1999. Debris flows on the Ordovician margin of Avalonia: Lady Port Formation, Manx Group, Isle of Man. *This volume*.
——, ——, QUIRK, D. G. ET AL. 1999. Revised lithostratigraphy of the Manx Group, Isle of Man. *This volume*.

Relationships between metamorphism and structure on the northern edge of eastern Avalonia in the Manx Group, Isle of Man

G. M. POWER[1] & R. P. BARNES[2]

[1]*School of Earth, Environmental and Physical Sciences, University of Portsmouth, Burnaby Road, Portsmouth PO1 3QL, UK*

[2]*British Geological Survey, Murchison House, West Mains Road, Edinburgh EH9 3LA, UK*

Abstract: The early Ordovician rocks of the Isle of Man, comprising fine-grained clastic sedimentary rocks at low regional metamorphic grade, contain a range of variably developed porphyroblasts. These are used to establish the relationship between the metamorphism and early Devonian deformation. Evidence for pre-deformation mineral growth is limited to diagenetic chlorite–mica stacks which may occur in most lithostratigraphical units. The first deformation (D_1) is marked by a pervasive cleavage, related to folding at various scales. Subsequent events formed crenulation cleavages in association with generally minor folding, including a gently dipping fabric developed relatively widely in response to the second phase of deformation (D_2). The main metamorphic peak, represented by growth of ilmenite, Mn-rich garnet, chloritoid and cordierite in muddy facies in the Manx Group, occurred after D_1 and before the onset of D_2. Local heating contributions are also recognized in restricted metamorphic aureoles around dioritic dykes and intermediate size granitic intrusions of syn-D_1 (Dhoon Granodiorite), syn-D_2 (Black Hut Dyke) and post-D_2 (Crosby and Foxdale Granite) ages.

Porphyroblast phases are best developed along the spine of the island and previous workers have suggested a metamorphic pattern related to underlying granitic intrusions. However, the timing of the growth of porphyroblasts within and outside the aureoles around exposed granites in this zone are very different, militating against any simple relationship between porphyroblast growth and igneous activity. Metamorphic mineral maps reveal a strong lithological control on the distribution of porphyroblasts. Chloritoid, in particular, is largely restricted to two prominent carbonaceous black mudstone units, the Barrule and Glen Rushen Formations, which strike northeast–southwest along the central part of the island. Cordierite is developed in the mudrocks of the Injebreck and Maughold Formations which crop out either side of the Barrule Formation. Further out across-strike, the absence of porphyroblasts (other than ilmenite) in the turbidite sandstone-rich parts of the sequence may again be, at least in part, related to composition. Here, however, low grades in the outermost tectonic tracts may be a real feature resulting from early D2 extensional reactivation and subsidence relative to the elevated central tracts.

The Lower Palaeozoic rocks of the Isle of Man are dominated by the Tremadoc–late Arenig Manx Group, comprising mudstone or silty mudstone with varying proportions of interbedded, generally fine-grained turbidite sandstone (Woodcock *et al.* 1999). These rocks, together with the broadly equivalent Skiddaw Group in north-west England (Cooper *et al.* 1995; Stone *et al.* 1999), were laid down on the northern edge of the Avalonian micro-continent before its separation from Gondwana (Woodcock & Barnes 1999). Tectonic deformation of the sequence (Fitches *et al.* 1999) appears to have occurred much later, probably in the early Devonian, as it affects Wenlock rocks on the Isle of Man and Pridoli rocks in northwest England (cf. Soper & Roberts 1971). This probably took place during collision of Avalonia with Laurentia, although Soper *et al.* (1992) have suggested that the impingement of Armorica on the southern edge of Avalonia may have been an important factor. Relationships between metamorphism and deformation in the Manx Group could be valuable in constraining regional models for the development of Avalonia during its collision with Laurentia.

In the Isle of Man, the first and principal deformation (D_1) of the Manx Group caused major folding and formed a pervasive cleavage (S_1), broadly parallel to the northeast striking, generally steeply dipping, bedding. Strike-parallel tracts (Fitches *et al.* 1999), each containing a

From: Woodcock, N. H., Quirk, D. G., Fitches, W. R. & Barnes, R. P. (eds) 1999.
In Sight of the Suture: the Palaeozoic geology of the Isle of Man in its Iapetus Ocean context.
Geological Society, London, Special Publications, **160**, 289–305. 1-86239-046-0/99/$15.00

recognizable lithostratigraphy (Woodcock *et al.* 1999) which cannot easily be correlated between tracts, may represent moderate to steep northwest dipping tectonostratigraphical units formed essentially during D_1. Later deformation is marked by variably developed crenulation cleavages in finer grained lithologies. Two of these are important, S_2 and S_3. The flat-lying second fabric (S_2) is relatively widely developed in relation to minor open recumbent folds and may be pervasive, locally combining with S_1 to produce a composite fabric. S_3 is a steep west-northwest dipping cleavage that is developed more locally.

Previous studies

The metamorphic history of the Lower Palaeozoic rocks of the Isle of Man has been the subject of several studies since the 'tentative discussion' by Lamplugh (1903, p. 106). Gillott (1955) and Simpson (1964*a*) both looked at the relationships between deformation and the growth of metamorphic minerals, and Roberts *et al.* (1990) constructed white mica crystallinity maps for the island. All these authors arrived at rather different conclusions regarding the sequence of metamorphic events. It is the aim of this paper to use new observations to resolve these differences and to produce a more complete metamorphic history in the context of the new geological model of the Manx Group (Fig. 1).

Gillott (1955) mainly used relationships observed in thin sections to deduce the relative timing of deformation as defined by cleavage development and growth of metamorphic minerals. He concluded that cordierite and spessartine–almandine garnet grew after the formation of the first cleavage but before the formation of a second cleavage. His observations did not include any mention of chloritoid and, on the basis of its apparently random orientation, he proposed that ilmenite was formed after both cleavages, but did not suggest how this might have been achieved. In summary (see Table 1), Gillott was proposing that the metamorphic thermal peak occurred early in the deformational history and certainly before the formation of a second cleavage.

Simpson (1964*a*) suggested that neither Gillott (1955) nor Lamplugh (1903) had adequately defined the second cleavage used as their time markers and that, as three cleavages are not uncommonly developed in these rocks, confusion may have occurred over which one of the two later cleavages was developed. His sequence of events (see Table 1) was to have growth of muscovite and quartz lenses to define the first cleavage and then all the other metamorphic minerals to have grown either during his second deformation event or after the formation of the second cleavage. Thus, Simpson (1964*a*) was stating that the metamorphic peak occurred during and after the second deformation. He recognized the presence of chloritoid but suggested that it was later than the second cleavage because of the apparently random orientation of the grains, an argument similar to that used by Gillott (1955) for ilmenite.

Development of cleavages

The structural history of the Lower Palaeozoic rocks of the Isle of Man is considered in detail by Fitches *et al.* (1999). The first cleavage is generally pervasive, striking northeast, and is clearly associated with D_1 folds widely developed on a variety of scales. It takes the form, in mudrocks, of a strong shape-preferred orientation of micas, commonly closely parallel to bedding. In psammites (Fig. 2a), S_1 is variably refracted and may be defined by a slight shape-preferred orientation of quartz grains and new growth of micas. The second cleavage may take a number of different forms depending on its orientation relative to S1. Where the two are similar in orientation a strong combined S_1–S_2 fabric is produced and may be difficult to distinguish from S_1 alone. When S_1 and S_2 are inclined to each other a crenulation cleavage of varying forms may be developed (e.g. Fig. 2). The third cleavage (S_3) is usually relatively steeply inclined but its strike is variable. In parts of the island, such as the area of Snaefell, it may strike around northwest. Typically, it may be observed as asymmetric microfolds in thin section or simply as a spaced fracture cleavage with the cleavage planes defined by concentrations of opaque minerals.

Other than their orientation, S_2 and S_3 do not have unique identifying features and it is critical that each is identified beyond doubt before making any interpretations. Ideally, they should be seen together in the same section and their relative age established (e.g. Fig. 2b). However, there are areas where only one of the two cleavages is developed and in non-oriented samples the unwary observer could misidentify the cleavage, as Simpson (1964) warned.

Growth of main metamorphic minerals

This section considers the relationships between the cleavages described above and the development of the main porphyroblastic metamorphic minerals, chlorite, ilmenite, chloritoid, cordierite, muscovite and Mn-rich garnet.

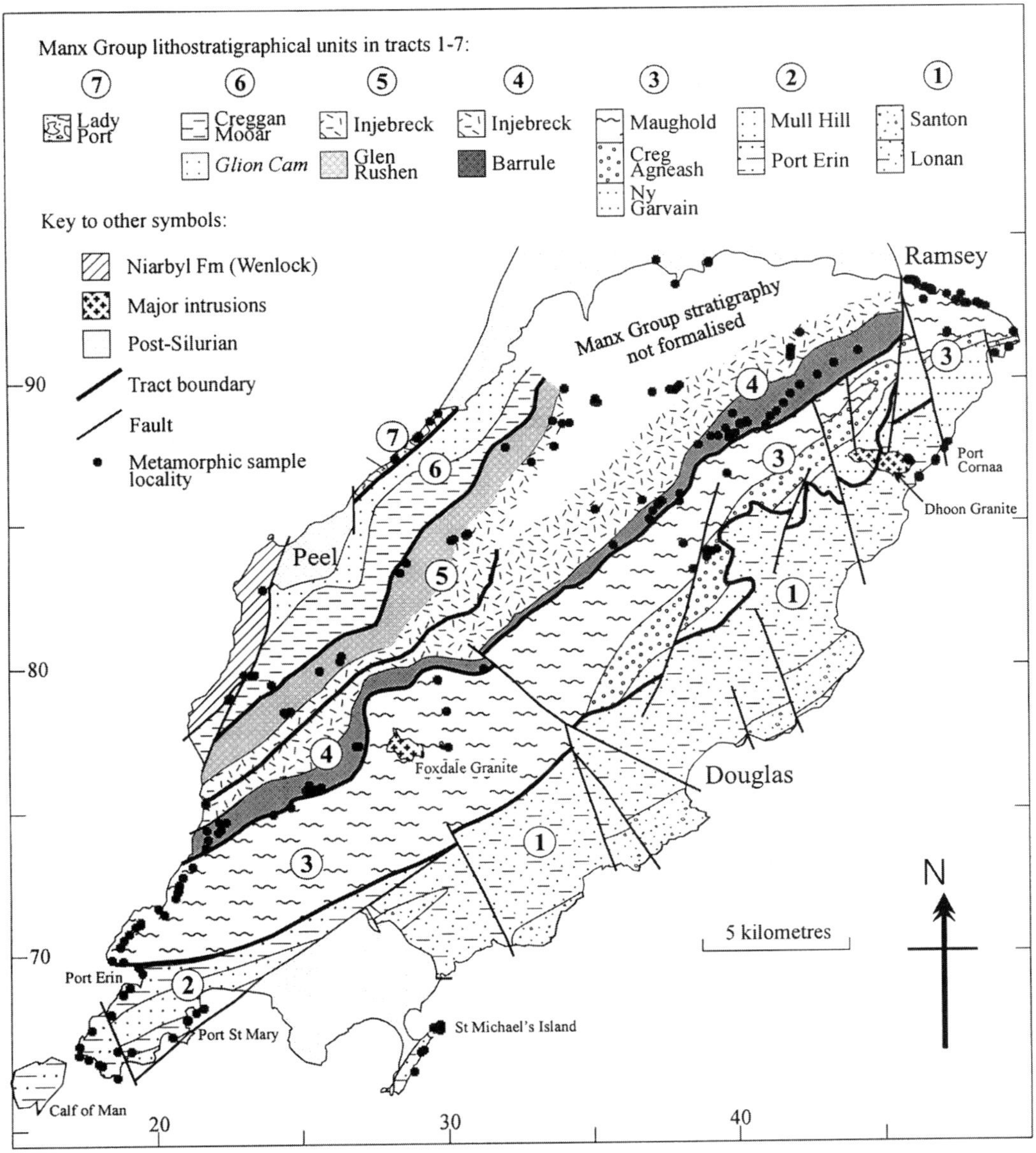

Fig. 1. Map showing the locations of samples on which this account of the metamorphic mineralogy of the Isle of Man is based, superimposed on the geological framework summarized by Woodcock *et al.* (1999).

Methods

The white mica crystallinity map of Roberts *et al.* (1990) shows a zone of higher crystallinity running northeast–southwest down the central axis of the island; rock samples for this study have been collected mainly from this zone. Mud rocks were taken in preference and, wherever possible, they were carefully orientated in the field. This was found to be essential if unambiguous information was to be obtained. Thin sections were cut in more than one orientation for some critical rocks. About 150 samples have been studied and the descriptions that follow are necessarily generalized. This is always a potential danger in studies of this type but it is essential groundwork before more specialized studies can be made.

Table 1. *Comparison of the tectonometamorphic sequences proposed by Gillot (1955) and by Simpson (1964a) with that established in the present work*

	Gillott (1955)	Simpson (1964)	Power and Barnes 1999 (excluding contact metamorphism)
D_1	**muscovite, biotite and chlorite** define S_1 cleavage	Muscovite felt and quartz lenses define S_1 (but NOT ilmenite)	S_1 cleavage defined by: white mica shape orientation; quartz slight shape fabric; wisps of carbon; chlorite. Ilmenite has shape preferred orientation in S_1 but encloses S_1 quartz and complex trails. **Ilmenite growth commenced late-D_1.** Chloritoid encloses straight S_1 wisps of carbon. Some chloritoid forms radiating clusters across S_1. **Chloritoid interpreted as post-D_1.**
Post- D_1	Cordierite has straight inclusion trails continuous with external S_1 S_2 deflected around cordierite or is continuous into edge of grains. **Cordierite interpreted post-D_1 pre-D_2.** S_1 not deflected by garnets, S_2 deflected around garnets. **Garnet interpreted post-S_1 and pre-S_2.**		Cordierite contains straight inclusion trails which are continuous with external S_1. Cordierite porphyroblasts usually terminate at S_2. Some are deformed in limbs of D_2 crenulations. **Cordierite post-D_1 and pre-D_2.** Mn-rich garnet has S_2 deflected around it and contains S_1 inclusion trails. **Mn-rich garnet post-D_1 and pre-D_2.**
D_2		Growth of muscovite plates, some parallel to S_2. **Ilmenite stated to be syn-D_2 because it only occurs along the axis of the Isle of Man syncline, a D_2 fold.**	**Muscovite flakes grow parallel to S_2 and enclose S_1 carbon trails.** Some rotation of ilmenite into S_2. Rotation of chloritoid into S_2. Chloritoid that cannot be rotated into S_2 is unstable and undergoes pressure solution. Muscovite 'wings' grow on long edges of chloritoid during D_2. **Chloritoid must be pre-D_2; Muscovite syn-D_2.**
Post-D_2	**Ilmenite interpreted as post-D_2** on the basis of its random orientation.	Biotite encloses D_2 folds. Cordierite helicitic to S_2. D_2 microfolding can easily be followed through the cordierite porphyroblasts. **Cordierite interpreted as growing during static conditions at the end of D_2.** Chloritoid has random orientation and intersects S_1 and S_2 at all angles. It cuts slip planes, microfolds and strain bands of D_2 but remains completely undeformed. **Bulk of chloritoid interpreted as post-D_2.** Garnet sits squarely astride S_2. **Garnet growth in static period between D_2 and D_3.** ?Muscovite random orientation. Cuts S_2 in all directions. Tourmaline grows with a random orientation. **Both post-D_2.**	
D_3			Pressure solution of D_2 muscovite in S_3 cleavage.

Chlorite–mica stacks

Chlorite and white mica are the main fabric-forming minerals defining the S_1 cleavage in pelitic beds. However, chlorite also occurs commonly as chlorite–mica aggregates or 'stacks' (e.g. Fig. 3) in all of the pelitic rocks in the Isle of Man, with the exception of the Barrule Formation.

Chlorite–mica stacks have been studied in slates from many parts of the world (e.g. Craig *et al.*

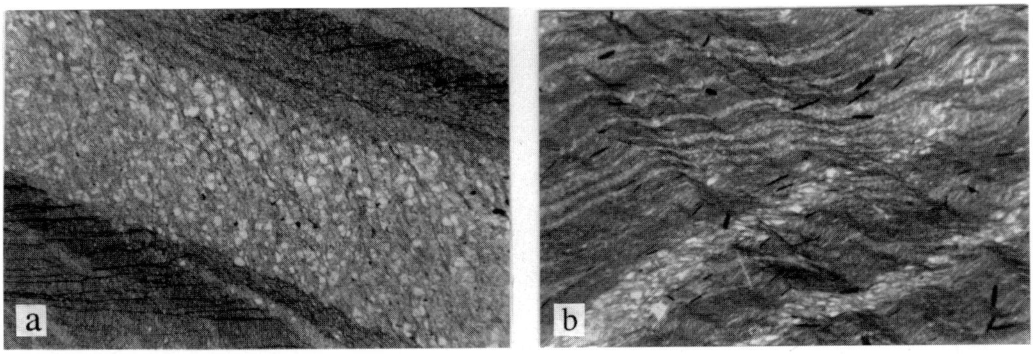

Fig. 2. (**a**) Psammite and pelite beds, sample 343, Creignish Wireless station [SC 1908 6673]. Psammite *c.* 2 mm thick. S_1 cleavage at a high angle to bedding is defined, in the psammite, by slight shape-preferred orientation of the detrital quartz grains and by new mica growth and, in the pelites, by a felt of orientated micas. S_2 is a near-horizontal crenulation cleavage with enhancement of the hinge zones and off-set of thin psammitic bed suggesting some mass transfer. (**b**) Pelite, sample 493, Bungalow Quarry [SC 400 866]. Width of field: 5.5 mm. Alternations of pelite and psammite define bedding inclined at *c.* 45° from middle right to bottom left. In the upper left of the picture a series of thin light stripes intersect the more steeply inclined psammite bed at a shallow angle. These are S_2 crenulations enhanced by mass transfer. S_2 may also be seen to cut almost horizontally through the pair of psammite beds. Several open, asymmetric microfolds have axial planes inclined from top left to bottom right. These define the S_3 cleavage and may be seen to fold the S_2 banding.

1982; Gregg 1986; Milodowski & Zalasiewicz 1990; Li *et al.* 1994). The more recent explanations for their origin invoke diagenetic formation of intergrowths of chlorite with some white mica that has overgrown and replaced volcanogenic biotite or other ferromagnesian minerals (e.g. Li *et al.* 1994). In the examples from the Isle of Man, the 0 0 1 planes of the mica intergrowths appear to be orientated close to bedding. However, there is no evidence of abrasion or other features that would support a purely detrital origin and the mica stacks are significantly larger than the grains in associated fine sandstone laminae. They are deformed by D_1 (Fig. 3), hence they are early and the most likely explanation is that they are products of diagenetic alteration of detrital minerals although relicts of these precursor minerals, have not been observed. It is suggested that, following Li *et al.* (1994), they may be used as an indication that the original sediments contained a volcanic component (cf. Barnes *et al.* 1999) which provided the constituents essential for chlorite growth. Local variations in the volcanogenic contribution might account for the apparently random occurrence of the chlorite–mica stacks throughout the sedimentary succession of the Isle of Man.

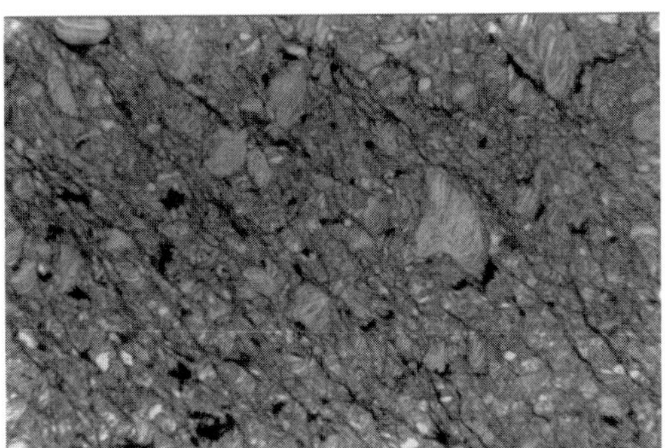

Fig. 3. Siltstone, sample GY881, St Michael's Island [SC 296 675]. Width of field, 2.2 mm. The S_1 cleavage is steeply inclined, top left to bottom right. Abundant chlorite–mica stacks pre-date D_1 and are modified by it.

Ilmenite

Ilmenite (e.g. Fig. 4a) is commonly developed in the pelitic rocks which crop out along the axis of the island and all the previous workers have commented on its presence. It may be detected in the field as tiny platelets that give a 'sparkle' to the rock. Examination of thin sections, particularly those where only S_1 is developed, gives the impression that the ilmenite has a preferred orientation close to S_1. However, in sections with a second and third cleavage present, the orientation of ilmenite is more random, possibly due to rotation by the later deformation, although some platelets are aligned in the crenulation cleavage planes (e.g. Fig. 4b). Figure 4a shows the typical relationship of the ilmenite grains to the S_2 crenulation cleavage, with chlorite pressure shadows and cleavage terminated at grains, or displaced around their tips,

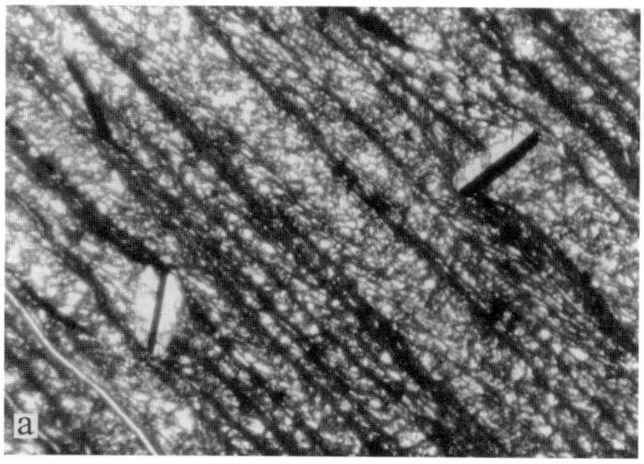

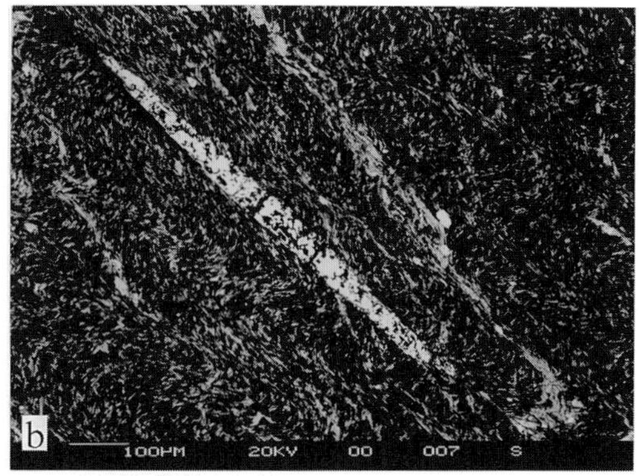

Fig. 4. (**a**) Semi-pelite, sample E66487, Port Mooar [SC 491 908]. Width of field, 2.2 mm. Ilmenite platelets lie at high angles to S_2, showing displacement of S_2 around tips of grains and chlorite pressure shadows. (**b**) Semi-pelite, sample E66487, Port Mooar [SC 491 908]. Width of field, 1 mm. Back-scattered electron image of ilmenite grain lying parallel to S_2. The complex patterns of the inclusions that appear as black grains within the ilmenite suggest more than one period of ilmenite growth.

supporting a pre-D_2 origin for the ilmenite. Quartz with a shape-preferred orientation may be detected within the ilmenite using an optical microscope. Back-scattered electron microscopy images (e.g. Fig. 4b) reveal that the inclusion trails within ilmenite may be complex and suggest more than one period of growth. For the present, until further investigations have been completed, it is concluded that the start of ilmenite growth must have been pre-D_2, probably late or post-D_1, although some syn-D_2 development cannot be ruled out.

Chloritoid

Chloritoid is a common constituent of many of the carbonaceous slates of the Isle of Man. It occurs as tiny colourless prisms, usually < 50 μm in length but up to 150 μm in some rocks. It cannot be detected in hand-specimen and is only revealed in thin sections. It overgrows S_1, defined by aligned wisps of carbonaceous matter within the grains (Fig. 5a and b). The angle between the wisps of carbonaceous matter and the edges of the prisms is not constant and this observation, coupled with the presence of radiating clusters of crystals across S_1 (Fig. 5b), are interpreted to indicate that chloritoid grew post-D_1 with no strong preferred orientation.

Chloritoid prisms may be partly rotated into S_2. This is suggested by evidence such as that shown in Fig. 5a where a spaced S_2 crenulation is partly defined by elongate prisms of chloritoid. In the S_2 fold-hinge domains between the cleavage planes, chloritoid may still appear at a high angle to S_2, depending on the original orientation of the chloritoid to the developing S_2. Where S_2 cleavage is strongly developed (Fig. 5c and d) chloritoid at a high angle to S_2 becomes unstable and acts as a nucleus for the growth of muscovite. The tips of chloritoid prisms abutting against S_2 undergo pressure solution and muscovite grows along the sides of the prisms (Fig. 5d). These 'mica-wings', developed during D_2, are clear evidence that chloritoid growth must have pre-dated D_2.

Cordierite

Cordierite is sporadically developed in pelitic rocks in the central zone of the Isle of Man without any obvious spatial relationship to igneous intrusions, although it is also directly associated with igneous bodies (see below). The cordierite occurs as spots, often < 1 mm in diameter. A common characteristic of all these spots is that they contain straight inclusion trails of the S_1 cleavage. Figure 6a shows a slate from near Porth e Vullen [SC 480 926], where it may be established from field observations that S_2 is prominently developed but S_3 could not be detected. Cordierite spots, 0.1 mm in diameter, have straight S_1 inclusion trails. The S_2 cleavage either truncates spots or deviates around them. The cordierite has often been replaced by mica aggregates. The most straightforward interpretation of these observations is that cordierite post-dates D_1 and pre-dates D_2.

Muscovite

White micas are major fabric-forming minerals in the S_1 cleavage. However, the extent of later recrystallization to muscovite is difficult to assess. The muscovite 'wings' on chloritoid (see above; Fig. 5c and d) indicate growth during D_2. In addition,

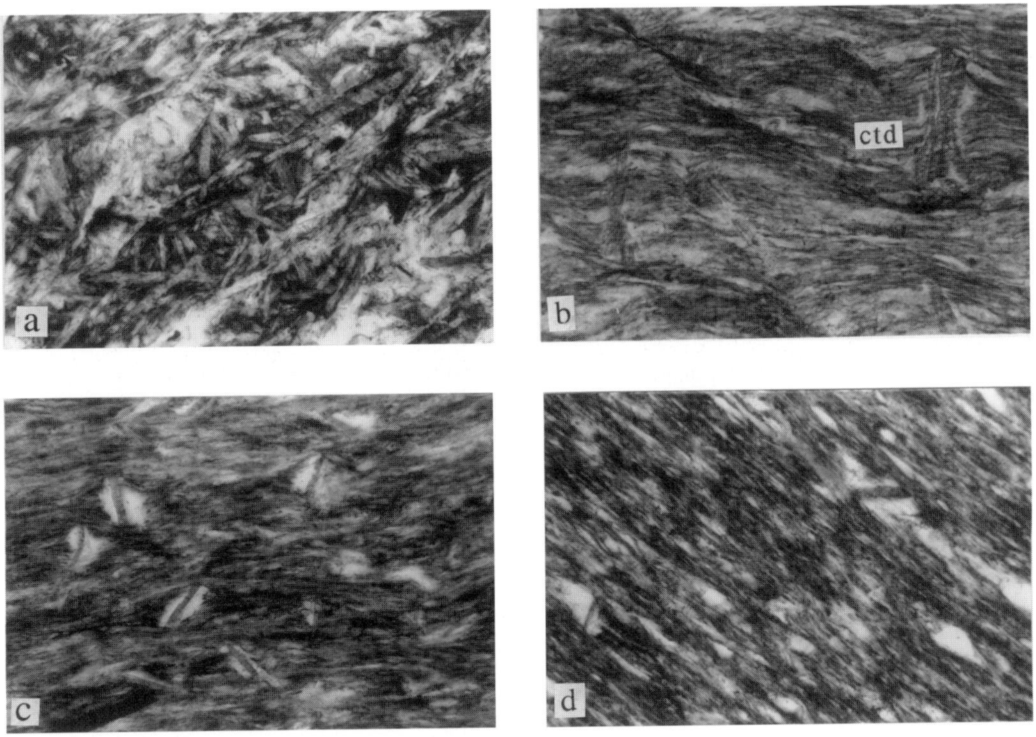

Fig. 5. (**a**) Carbonaceous pelite, sample 158, South Barrule [SC 2551 7581]. Width of field, 0.55 mm. A spaced S_2 cleavage runs across the field towards the upper right and is defined by aligned prisms of chloritoid. In intervening domains, chloritoid and mica flakes lie oblique to S_2 in the hinges of F_2 folds. Close observation reveals wisps of carbonaceous matter aligned within the chloritoid preserving remnants of the S_1 cleavage indicating that chloritoid overgrew S_1. Some grains have been rotated into the S_2 orientation whilst others remain at an angle to it, depending partly on their original orientation. The angle between the enclosed wisps of carbonaceous matter and the prism edges varies from grain to grain suggesting that the grains originally grew with a range of orientations. (**b**) Semi-pelite, sample 376, Sartfell [SC 3298 8717]. Width of field, 0.85 mm. Radiating clusters of chloritoid (ctd) up to 150 μm long overgrow S_1 cleavage (horizontal) within domains defined by a later crenulation cleavage. (**c**) Carbonaceous pelite, sample 370, Black Hut [SC 4016 8849]. Width of field, 0.55 mm. Composite S_1–S_2 cleavage with relict chloritoid lying at a high angle to the cleavage and showing development of 'mica wings'. Chloritoid growth must have been pre-D_2. (**d**) Carbonaceous pelite, sample 464, Glen Rushen Slate Quarry [SC 2450 7845]. Width of field, 0.85 mm. 'Mica wings' on chloritoid in a composite S_1–S_2 cleavage.

idiomorphic laths of muscovite up to 0.2 mm in length are present in some pelitic rocks. These flakes have a shape-preferred orientation parallel to S_2 and enclose S_1 wisps of carbonaceous matter. Figure 6b shows idiomorphic muscovite laths orientated at an angle to S_3, overgrowing cordierite and breaking down within the zones of intense S_3 cleavage development. This is interpreted as additional evidence for the syn-D_2 growth of muscovite.

Manganese-rich garnet

Lamplugh (1903) recorded the patchy occurrence of garnets commonly closely associated with quartz veins and confirmed their Mn-rich nature by chemical analysis. Gillott (1955) quoted further chemical analyses of these garnets, showing that they are quite variable in Mn content. Textural relationships for the garnets are shown in Fig. 7, together with examples of the sectorial growth patterns typically developed by Mn-rich garnets in low-grade metamorphism (Burton 1986). Mn-rich garnets contain S_1 inclusion trails and pre-date D_2 because S_2 cleavage is displaced around them (Fig. 7b) with the formation of pressure shadows.

Tourmaline

Previous workers mention tourmaline as one of the porphyroblastic phases developed during

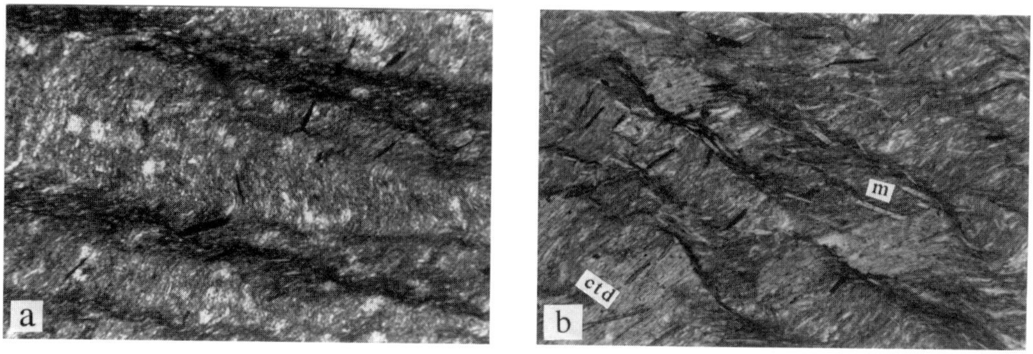

Fig. 6. (**a**) Pelite, sample 315, Cliff path, Port e Vullen [SC 4802 9258]. Width of field, 2.2 mm. Crossed polars and sensitive tint plate. S_2 crenulation cleavage near horizontal. Cordierite porphyroblasts (50 μm), seen as light areas, overgrow S_1 (near vertical). They are preserved in domains beween S_2 cleavages and deformed within the S_2 cleavages. (**b**) Pelite, sample 494, Bungalow Quarry [SC 400 866]. Width of field, 2.2 mm. S_3 crenulations are strongly developed, top left to bottom right, with S_1 orientated bottom left to top right. Cordierite spots 0.3 mm in length overgrow S_1 but are truncated by S_3. Prisms of chloritoid (ctd) may be detected apparently enclosed in the cordierite (e.g. bottom left). A muscovite lath (m) appears to overgrow the edge of one cordierite grain (lower right). The muscovite, which is broken down within the S_3 cleavage, is likely to be syn-S_2 and pre-S_3.

metamorphism. Most pelitic rocks were found to contain a few grains of tourmaline and textural relationships suggested a fairly early growth during the metamorphic history. Metasomatic and vein tourmaline will not be considered here. Kennan & Morris (1999) describe the presence of Mn-carbonate beds in the Creggan Mooar and Lady Port Formations, in one example associated with tourmalinite, which they interpret as syn-sedimentary volcanic exhalative deposits.

Metamorphic minerals in the thermal aureoles

Certain minerals and textures show a spatial association with igneous bodies suggesting that they formed directly as a result of thermal alteration related to intrusion. Thermal effects around four intrusions, the two larger granite bodies of Dhoon and Foxdale (Simpson 1964*b*, 1965) and two thick dykes, are considered in this section.

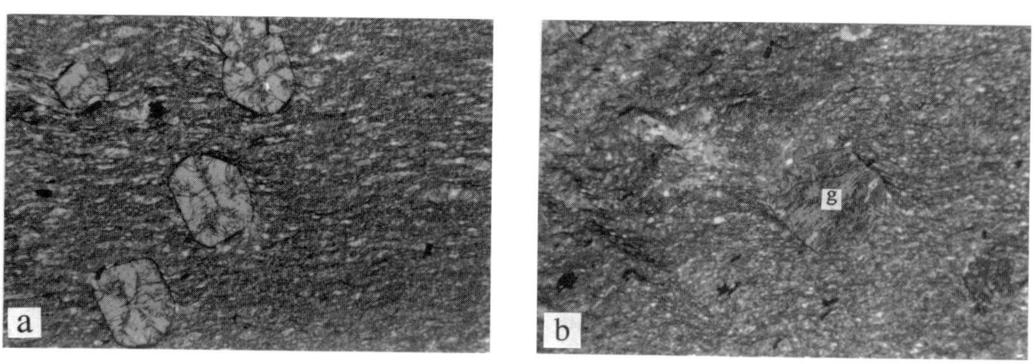

Fig. 7. (**a**) Semi-pelitic bed, sample 496, Bungalow Quarry [SC 400 866]. Width of field, 5.5 mm. Euhedral Mn-rich garnets *c.* 1 mm in diameter display prominent sectorial growth patterns that are characteristically associated with such garnets in low-grade metamorphism. Strong S_2 cleavage is slightly displaced around the garnets. (**b**) Alternating pelitic and semi-pelitic beds, sample 507, Bungalow Quarry [SC 400 866]. Width of field, 5.5 mm. Bedding is orientated from top right to bottom left, S_2 is horizontal and a very open S_3 crenulation runs from top left to bottom right. Weakly preserved S_1 within the domains bounded by S_2 in the more pelitic bed is near parallel to bedding. The Mn-rich garnet crystal (g) overgrows S_1 but S_2 is displaced around it. Growth of Mn-rich garnet is post-S_1 and pre-S_2.

The Dhoon Granodiorite

The form of the Dhoon granodiorite was mapped by Lamplugh (1898) although exposure around the periphery is poor and the contacts are nowhere seen. He showed it as an elongate body, *c.* 1.5 km long east–west, and not more than 0.5 km wide with a narrow (200 m) metamorphic aureole. This cannot be substantiated from the poor exposure and it must be largely schematic. Simpson (1964*b*) argued for a syn-tectonic emplacement of the Dhoon intrusion near the beginning of D_2.

The granodiorite is variably deformed, with cleavage apparent locally and a well-developed east–west trending zone of mylonite exposed in the main quarry at Dhoon [SC 459 872]. The granodiorite has a thoroughly metamorphosed character. Plagioclase grains are always cloudy and contain well-formed micas and epidote group minerals within their boundaries. Biotite flakes, in places, have a sufficiently strong shape-preferred orientation that it is possible to measure a shallow, spaced cleavage.

The best exposures in the thermal aureole occur in the coastal section between the Dhoon Granite and Port Cornaa (Fig. 1), where spotting is visible in more pelitic beds. Here the S_1 cleavage, associated with small-scale D_1 folds, dips *c.* 25° SE and is cut by a steeper cleavage, dipping *c.* 60° SSE. In thin section, the porphyroblasts, replaced by chlorite and biotite, are clearly flattened in the S_1 cleavage (Fig. 8). The aureole must therefore have been formed before D_1 was completed. Coupled with the deformation within the granodiorite, this indicates that the intrusion is pre- or syn-D_1. That it appears to pin the Windy Corner Fault as mapped (Fitches *et al.* 1999) would suggest that the fault is syn-D_1.

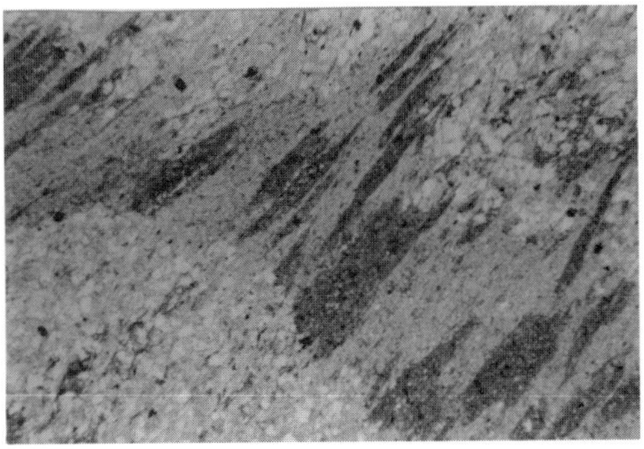

Fig. 8. Pelite and psammite beds, sample 427, South of Port Cornaa [SC 4717 8753]. Width of field 5.5 mm. Cross-section of a small-scale D_1 fold of a thin pelitic bed containing flattened and replaced dark spots in the aureole of the Dhoon Granodiorite. This relationship implies that the aureole of the Dhoon Granodiorite was formed pre- or possibly syn-D_1.

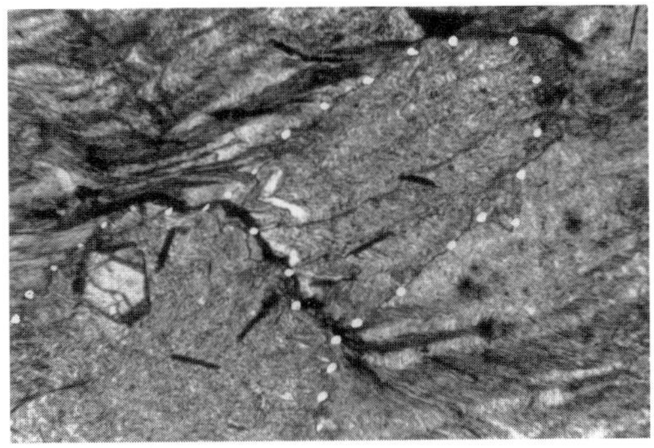

Fig. 9. Cordierite, chlorite rock, sample 373, Black Hut Quarry [SC 4042 8853]. Width of field, 5.5 mm. S_2 crenulations enclosed by contact metamorphic cordierite (centre and lower right, outlined by white spots). An idioblastic Mn-rich garnet appears to be enclosed by one cordierite crystal.

Black Hut Dyke

In the rocky outcrops above Black Hut [SC 402 885], on the northeast slopes of Snaefell, there is an old quarry in an east trending, intermediate dyke *c.* 15 m in thickness. The dyke cuts across the S_1 cleavage but has a strong, gently dipping fabric within it. This fabric is parallel to S_2 and developed axial planar to a small-scale S_2 fold in the Barrule Slate immediately to the north of the dyke contact. Intense spotting is developed close to the dyke, particularly on the southern side, and thermal metamorphic effects may be traced up to 10 m from its contacts. In thin section (Fig. 9) the rock in the aureole is seen to be composed principally of cordierite and chlorite. The cordierite has inclusion trails which preserve S_2 crenulations. Rare Mn-rich garnets also overgrow S_2 crenulations. The dyke is interpreted as a syn-D_2 intrusion, as this is the only way to reconcile the observations that the contact metamorphic minerals overgrow S_2 crenulations but the dyke itself has a strong S_2 cleavage developed within it.

Crosby granite Dyke

An old quarry [SC 325 791] near Crosby exploited a 10 m wide granitic dyke classified by Lamplugh (1903) as a 'Foxdale elvan', implying an association with the Foxdale Granite. Simpson (1966), noting that biotite in the contact aureole of this dyke overgrows S_2 but is deformed by D_3,

Fig. 10. Semi-pelite, sample CQ2, Crosby Quarry [SC 325 791]. Width of field, 5.5 mm. S_1 is near parallel to bedding and S_2 is at a high angle to it. Biotite, at the contact of the Crosby Granite, overgrows minute, open, D_2 crenulations.

suggested that the dyke had been emplaced in his 'F_2–F_3 static interval'. Our observations (Fig. 10) of biotite overgrowing S_2 crenulations support Simpson's (1966) interpretation.

The Foxdale Granite

Lamplugh's map (1898) shows an exposed granite *c.* 1 km × 0.5 km in size. He depicts the metamorphic aureole as 200 m wide to the south and several kilometres wide to the north. At best, this aureole can only be schematic. The rocks at the immediate contact of the Foxdale Granite are very poorly exposed now and there is no new evidence from this part of the contact zone. Two kilometres to the north, in the Archallagan Plantation [SC 30 79], loose blocks of a banded gneissic and phyllitic appearance may be found. Outcrops of metamorphic rocks occur around the old mine at Cornelly [SC 298 798], where granite was proved within 50 m of the surface (Lamplugh 1903), and in the Glion Darragh Valley [SC 297 804] further north. Lamplugh (1903) was puzzled by these rocks and commented that none of them had the appearance of contact hornfelses but rather of regional metamorphic rocks. Simpson (1965) proposed that the Foxdale–Archallagan Granite was emplaced as a syn-D_2 intrusion. He argued that coarse mica fabrics developed near parallel to bedding and to S_1 during the initiation of D_2 were formed under the influence of the intruding granite. He considered that this was followed by a static phase in which almandine garnet, staurolite, cordierite, brown biotite, green chlorite and muscovite porphyroblasts overgrew S_2. His photomicrographs (Simpson 1964*a*, plate 24D,–F) show convincing evidence of garnet, staurolite and cordierite grains growing helicitically over a crenulation cleavage such that crenulations may be traced into, through and out of grains without deviation. He proposed a renewal of D_2 stresses to account for a new fabric which is displaced around the porphyroblastic phases in the north of the aureole around Archallagan.

Thin sections of rocks from the Archallagan area have been examined and a number of general points may be made. The most prominent cleavage is commonly defined by shape-preferred orientation of chlorite. Garnet and biotite are common as porphyroblastic phases. Garnet overgrows S_2 crenulations (Fig. 11) and these crenulations are only preserved in the garnets and in some isolated domains in more pelitic beds. Elsewhere they have been overprinted by a strong new fabric which is displaced around the garnets giving them an augen-like appearance. Garnet has been extensively replaced by chlorite. Quartz-rich bands show polygonal textures that suggest annealing. Biotite porphyroblasts are nearly always enclosed by the new cleavage. A plausible explanation of these observations would be that garnet and biotite might be spatially related to the subsurface Archallagan granite (Lamplugh 1903) and that the peak of contact metamorphism is post-D_2. It is then not easily established whether the new cleavage is 'renewed movement on S_2', as proposed by Simpson (1966), or evidence of a later near-east trending shear zone. The annealing of quartz and abundant new chlorite growth close to this time might have been responsible for some of the coarsening interpreted by Simpson (1966) as syn-D_2. If this is true then there would be no reason to require that the Foxdale Granite is syn-D_2 in age. The fact that the granite itself lacks any pervasive

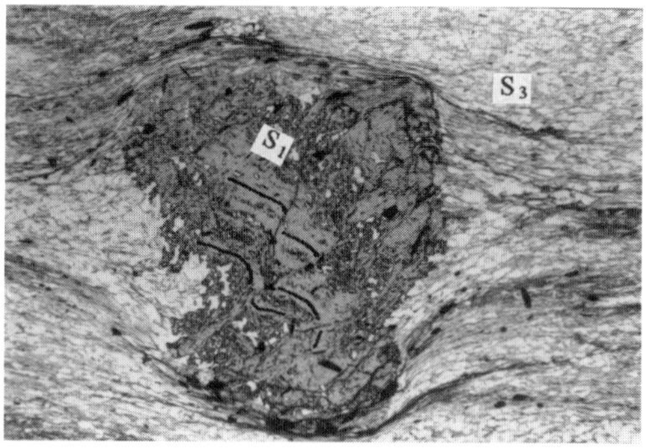

Fig. 11. Semi-pelite, sample 514, Glion Darragh [SC 295 798]. Width of field, 5.5 mm. Contact metamorphic garnet crystals with inclusion trails showing D_2 crenulations defined by S_1 (indicated by black lines) and both at a high angle to an external S_3 foliation which is displaced around the garnet. Newly grown mica and chlorite is present in S_3.

fabric cannot alone be taken as evidence against a syn-deformational age, but when considered together with the relationships of the metamorphic minerals it does tend to support a post-D_2 age.

Sequence of igneous activity

In summary, the evidence presented in this paper supports a sequence where the Dhoon Granodiorite is syn-D_1 and the Black Hut Dyke is syn-D_2, but the Crosby and Foxdale Granites are post-D_2. This is a greater time span than that proposed by Simpson (1965), who preferred to make the emplacement of both Dhoon and Foxdale syn-D_2 in age but with very different times taken for their consolidation.

Metamorphic map of the Isle of Man

The distribution of metamorphic minerals across the Isle of Man, based on the samples examined for this study, is shown in Fig. 12. This is generally similar to the distribution of metamorphic minerals illustrated by Simpson (1964*a*) but extends his findings in some important ways.

Chlorite–mica stacks (Fig. 12a) are, as expected, most obvious in the 'lower grade' rocks of the southeast and northwest of the island, and they also occur in the Silurian Niarbyl Formation. However, there is not a sharp cut-off and some rocks in which chloritoid, or even cordierite, is present may still contain some chlorite–mica stacks. Thus, they are seen as valuable indicators of diagenetic changes but must be considered in context with the complete mineral assemblage for determination of grade.

The cordierite occurrences shown (Fig. 12b) are only those that cannot be linked directly with thermal alteration close to exposed igneous intrusions (see above). The distribution of cordierite is the opposite to that of chloritoid (Fig. 12c), being largely restricted to the Maughold and Injebreck Formations and very rare in the carbonaceous slates of the Barrule and Glen Rushen Formations, suggesting that lithological composition is an important control. Three main clusters of cordierite, all well away from any known outcrop of igneous intrusions, stand out on Fig. 12, between Port Erin [SC 195 695] and Fleshwick Bay [SC 202 714] in the southwest, between Porth e Vullen [SC 476 927] and Maughold Head [SC 499 913] in the northeast, and more scattered occurrences around Snaefell [SC 394 881]. Additional cordierite localities recognized by Simpson (1964*a*) within the Maughold Formation may extend the Snaefell cluster towards the Foxdale Granite.

The possibility that this cordierite might represent the thermal aureoles of concealed igneous bodies should be considered. Cornwell (1972) presented a Bouguer gravity anomaly map for the Isle of Man, which has been combined with offshore data by Kimbell & Quirk (1999) but essentially maintaining Cornwell's interpretation. That is, in the southwest quadrant of the island the large, South Barrule, gravity low is centred close to the Foxdale Granite and it is considered to be almost certainly the indication of a large concealed granitic body. Similarly, in the northeast quadrant of the island, the Glen Mona gravity low intersects the east coast and extends from south of the Dhoon Granite, possibly as far as Ramsey, and is likely to correspond to a smaller concealed granite. In the central part of the island between these two gravity lows, Cornwell (1972) considers that 'if granite exists at all here it must lie at great depth'. Kimbell & Quirk (1999) allow that the data might be interpreted to allow a narrow granite (0.5 km wide) linking the two main anomalies at depth. In the northwest quadrant of the island there is a gravity high and a concealed granite may be completely ruled out.

The cordierite of the Port Erin to Fleshwick Bay section is unlikely to be related to the South Barrule gravity low to the north but it might be the surface expression of a thermal aureole associated with a concealed granite to the south. Cornwell (1972) suggested that the gravity low centred on the Calf of Man might indicate a concealed granite but he did not attempt to model it. However, no spotting has been recorded south of Port Erin closer to the centre of the gravity low. If it is contact metamorphic cordierite then the granite would have to be a post-D_1, pre-D_2 granite and earlier than the post-D_2 Foxdale Granite.

The cordierite on the coastal section east of Porth e Vullen might just possibly be spatially related to the granitic rocks represented by the Glen Mona gravity low. If this is the case, it would have to be a granite of a different age than the Dhoon Granodiorite. The spotting in the Dhoon aureole is not later than syn-D_1, whereas that at Porth e Vullen is post-D_1, pre-D_2. It is interesting to note here that Cornwell (1972) was unable to account for the Glen Mona anomaly using density values obtained from samples of the Dhoon Granodiorite and had to assume a lower density value.

The cordierite around Snaefell occurs either in the gap between the South Barrule and Glen Mona gravity lows or on the southeast edge of the gravity high in the northwest quadrant of the island. Again, it is post-D_1, pre-D_2 in age and is in an area for which the gravity data has not been interpreted as indicating a concealed igneous body.

A simple model of higher grade being directly associated with emplacement of igneous bodies (e.g. Roberts *et al.* 1990) is therefore difficult to

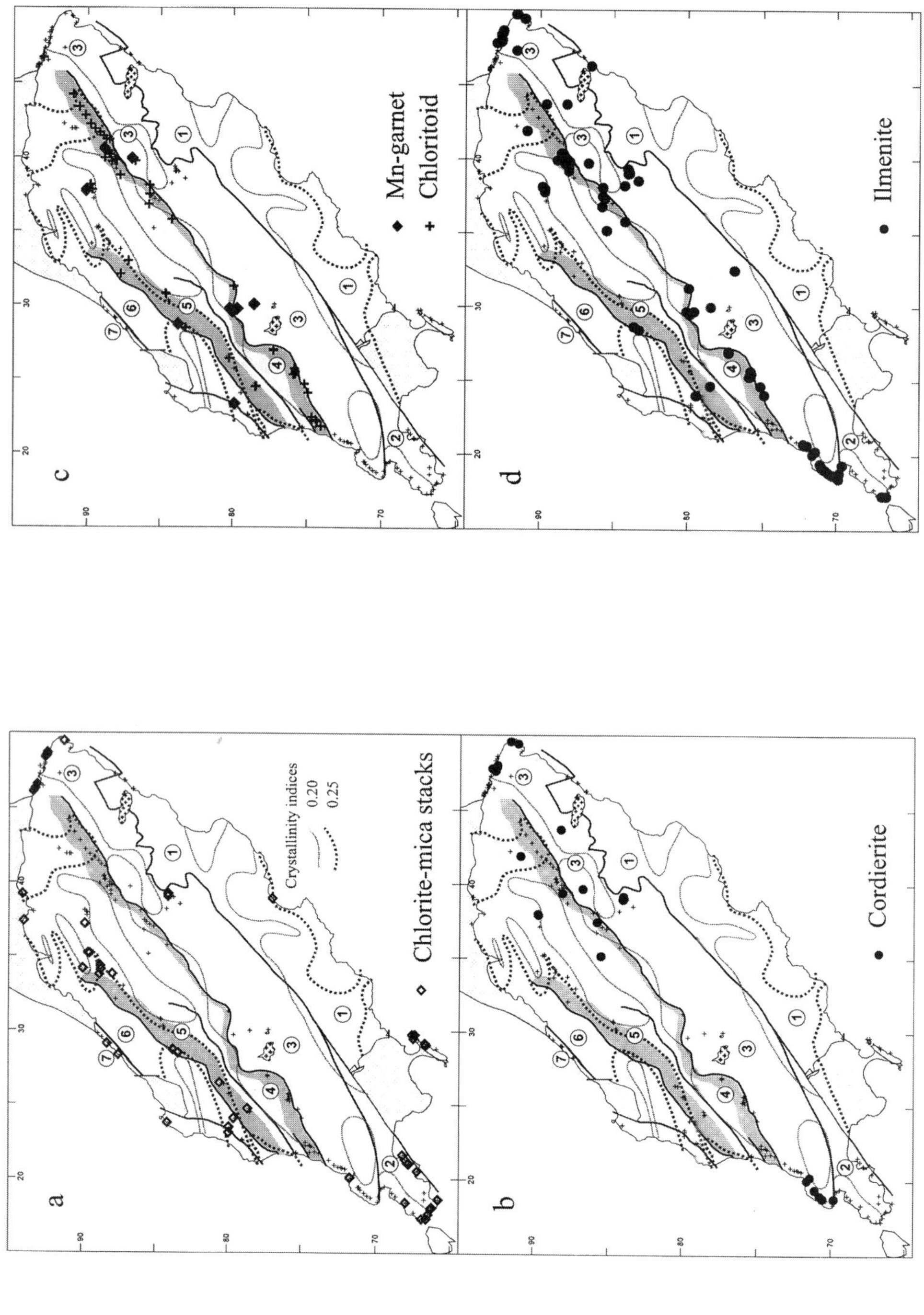

Fig. 12. Maps showing the distribution of metamorphic minerals (+ indicate sample sites) across the Isle of Man together with the outline of the geological framework shown in Fig. 1 and illite crystallinity contours from Roberts *et al.* (1990).

sustain. This is particularly true when the cordierite distribution is considered together with that of chloritoid where similar difficulties are encountered if distribution is considered to be controlled by the presence of thermal aureoles. A regional origin for the post-D_1 and pre-D_2 cordierite and chloritoid, with occurrences limited by rock composition, is therefore preferred.

A similar conclusion may be drawn for the Mn-rich garnet (Fig. 12c) which occurs sporadically in several units. It is very localized in distribution and almost certainly reflects the Mn content of the rock (Wang & Spear 1991).

Lithological/compositional control is most apparent in the distribution of chloritoid (Fig. 12c), which is common in the black mudstones of the Barrule and Glen Rushen Formations but relatively rare in the cordierite-bearing mudstone in the adjacent, but lithologically more varied, sequences of the Maughold and Injebreck Formations. It is known that chloritoid may only occur in rocks of quite particular compositions, rich in aluminium and iron [see Leslie (1988) for discussion]. It is therefore very likely that the carbonaceous slates of the Isle of Man satisfy these requirements and that this is as important as temperature variation for determining the presence or absence of chloritoid.

Important new evidence shows that the Glen Rushen Formation contains chloritoid. This western outcrop of black mudstone was previously considered to be a repetition by folding of the Barrule Formation (Lamplugh 1903; Simpson 1964*a*), but Woodcock *et al.* (1999) give it provisional status as a separate formation because of uncertainties in the structural interpretation. Work in progress (GMP) on the mineralogy and chemistry of the mudstones supports this division into two separate formations. New chloritoid localities have been recorded all along the length of the outcrop from Sartfell [SC 330 872] (Fig. 5b) in the north, to Glen Helen [SC 307 845], Slieu Whallian [SC 265 804] and, to the south, Glen Rushen slate quarries [SC 246 781] (Fig. 5d). In places it is well preserved, as at Sartfell, but commonly it is virtually overprinted by strong D_2 fabrics and the growth of new muscovite. These observations broaden the belt of porphyroblasts defined previously by Simpson (1964*a*, fig. 2) by several kilometres further to the northwest. This extended belt of chloritoid occurrences makes any direct relationship between concealed igneous bodies and metamorphic grade even more difficult to sustain than the case for cordierite, particularly as all the chloritoid is post-D_1, pre-D_2 in age.

It is also interesting that these new chloritoid localities occur to the west of the belt of higher grade rocks defined by Roberts *et al.* (1990) using a boundary at an illite crystallinity index of $\Delta°2\theta = 0.20$. However, the internationally recognized index value for the anchizone–epizone boundary is now $\Delta°2\theta = 0.25$ (Kisch 1990). Using this value, virtually all the new localities for chloritoid, often taken as an epizone index mineral (Bevins & Robinson 1988), falls within the epizone defined by the illite crystallinity isocrysts.

The spatial distribution of chloritoid and cordierite does not appear to follow any pattern that might correspond to a simple prograding sequence. A few rocks such as those at the Bungalow Quarry [SC 400 866] have cordierite + chloritoid + Mn-rich garnet, apparently in stable equilibrium together. Others, such as at Tholt y Will [SC 3812 8985], have Mn-rich garnet + chloritoid in one rock and cordierite in another in close proximity. The general impression is that there may not have been much variation in temperature required for the growth of any of these porphyroblast phases and that their distribution is primarily controlled by rock composition.

Relationship between metamorphism and larger scale structures

In addition to the lithological/compositional control discussed above, the distribution of metamorphic minerals across the Isle of Man, summarized on Fig. 12, may also suggest some relationship with the structural tracts proposed by Fitches *et al.* (1999). In the northwest, the Lady Port Formation in tract 7 is uniformly of low grade with chlorite–mica stacks preserved and only the development of quartz + chlorite + white mica defining the S_1 cleavage. The Silurian rocks of the Niarbyl Formation also appear to be of similarly uniform low grade, although few samples have been examined. Very limited sampling of tracts 1 and 6 suggests that these too are of lower grade, with chlorite–mica stacks well preserved. Tract 2, and the possibly equivalent sandstone-bearing formations in the lower part of the tract 3 sequence in the northeast of the island (Barnes *et al.* 1999), may occupy a transitional situation, with chlorite–mica stacks variably preserved and only ilmenite developed locally as a porphyroblast phase. The area of definite higher grade along the axis of the island, marked by abundant porphyroblasts, coincides with most of tract 3, and tracts 4 and 5.

Estimates of temperature and pressure of metamorphism

Determination of temperatures and pressures of metamorphism at these low grades is difficult and

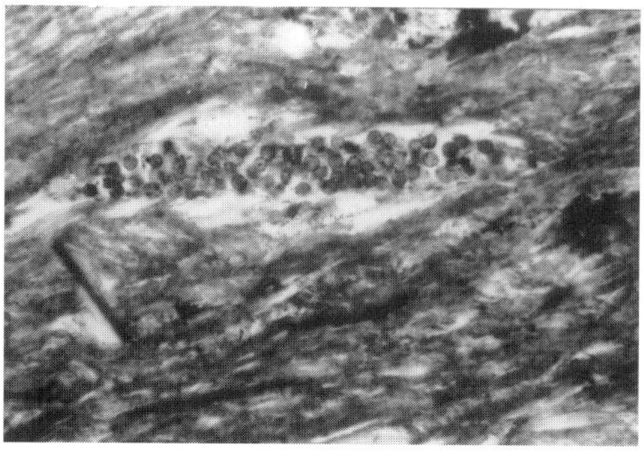

Fig. 13. Pelite, sample 337, Snaefell Tramway near Summit Station [SC 3943 8807]. Width of field, 0.85 mm. Acritarchs enclosed in filamentous tube. A 'bloom' of acritarchs may have been the main source for the carbonaceous matter now present in the carbonaceous slates of the Barrule Formation.

often imprecise. Information about rock and mineral composition is essential and knowledge of variables such as the composition of the fluid phase and the oxygen fugacity are also required; this is beyond the scope of the present paper. A few general comments may be made.

Many of the rocks with the porphyroblastic phases contain carbonaceous matter. This almost certainly originated from micro-organisms, mainly from the organic cell walls of acritarchs present in the original sediments. Thin-section study of the carbonaceous slates from the Barrule Formation reveals some well-preserved examples of acritarchs (Fig. 13), together with much finely divided carbonaceous material. It is apparent that many of these fragments of carbonaceous matter have shapes corresponding to parts of the more complete acritarchs and that they are the main source of carbonaceous matter in the rocks. The presence of carbonaceous matter in a rock will buffer the oxygen fugacity at lower levels and also influence the proportions of carbon species, such as carbon dioxide and methane, in the fluid phase.

The temperature of formation of chloritoid is not well defined. Bevins & Robinson (1988) proposed a temperature of close to 400°C at low pressures. This does not allow for the departure of the fluid phase from pure water or for the effect of low oxygen fugacities (Grieve & Fawcett 1974). Both of these factors will lower the temperature of the initial formation of chloritoid so that it may not have been greatly in excess of the 300°C usually taken for the beginning of the epizone at the isocryst for $\Delta°2\theta = 0.25$.

Seifert (1970) determined a temperature *c.* 500°C for the formation of cordierite at 2–3 Kb. At these temperatures, even highly aluminous pelitic rocks might be expected to contain biotite (Spear 1993) but this is rare in the Manx rocks. Pattison (1989) gives an example of the formation of cordierite in graphitic pelites taking place further out from the contact of the Ballachulish Granite than the first appearance of cordierite in non-graphitic pelites. Thus, it seems likely that the presence of carbonaceous material would lower the temperature for the first formation of cordierite.

Mn-rich garnets may form at temperatures $< 300°C$ depending on the Mn content of the rock (Wang & Spear 1991). Their formation and growth may well be influenced by the presence of carbonaceous matter in the rock (Burton 1986). Minimum temperature estimates of 350–400°C for the metamorphic peak may not be too unrealistic but it must be emphasized that there are many uncertainties involved. If a geothermal gradient of $35°C\ km^{-1}$, similar to recent estimates used for parts of the Welsh Basin (Roberts *et al.* 1996), is adopted this would imply burial of 10–13 km. This estimate is not greatly at variance with estimates of thicknesses for the Manx Group suggested by Woodcock *et al.* (1999), taking into account any additional cover that may have been present at the time of metamorphism.

Discussion

From the foregoing description, in the context of the illite crystallinity map produced by Roberts *et al.* (1990) and the recent tectonostratigraphical interpretation (Woodcock *et al.* 1999), the distribution of the metamorphic mineral phases developed on the Isle of Man may be the result of three possible processes:

- predominant compositional control within a fairly uniform, low grade of regional metamorphism established contemporaneously with the large-scale structure;
- post-metamorphic modification to the large-scale structure, with uplift of the tracts in the centre of the island relative to outer tracts;
- thermal effects of igneous bodies underlying the spine of the island superimposed on low grade regional metamorphism.

At least some compositional control is evident in the restriction of chloritoid to the black mudstone units of the Barrule and Glen Rushen Formations, as described above, and may also be apparent in the distribution of chlorite–mica stacks. Nevertheless, the heat input for this may have stemmed from underlying igneous intrusions, emplacement and/or underpinning of which could have facilitated relative uplift of the tracts in the centre of the island. However, the different timing of the development of cordierite in the thermal aureoles of the Dhoon (syn-D_1) and Foxdale (post-D_2) Granites, and the 'regional' post-D_1 and pre-D_2 cordierite and chloritoid, is convincing evidence that a complex sequence of igneous bodies would have to be invoked. Even then it would be difficult to explain the observed distribution of porphyroblasts. Furthermore, the Dhoon and Foxdale Intrusions are different in character and lie off the axial zone of porphyroblast-rich rocks, below the centre of which there is no geophysical evidence for a large concealed intrusion at shallow depth (Cornwell 1972).

The distribution of metamorphic minerals on the Isle of Man appears to some extent, to be independent of the grade of the rocks as represented by the crystallinity map of Roberts *et al.* (1990). The porphyroblast-bearing units are offset to the northwest within the main area of epizone grade ($\Delta°2\theta > 0.25$) and rocks in the northeastern part of that zone, in tracts 1 and 3, are devoid of porphyroblasts. Epizone grade rocks in the north of the Lower Palaeozoic outcrop, north of the confirmed extent of the Glen Rushen Formation in the area where the lithostratigraphy remains unresolved (Woodcock *et al.* 1999), contain only chlorite–mica stacks. Conversely, chloritoid is consistently found in the Barrule Formation in the ridge that leads to North Barrule [SC 910 443], cutting across an embayment of lower grade rocks as defined by the illite crystallinity.

In general, therefore, it seems likely that the present distribution of metamorphic minerals outside the limited thermal aureoles of the exposed granitic intrusions is largely a consequence of the primary composition of the rocks. In particular, it appears that the turbiditic sandstone-bearing units (in tracts 1 and 2; the lower part of tract 3; tract 7 and the Silurian Niarbyl Formation), despite containing interbedded mudstone throughout and being 'mud'-dominated in parts, were generally less susceptible to the development of porphyroblasts.

There may be abrupt changes in metamorphic grade across tract boundaries or cross-strike faults but these are difficult to resolve satisfactorily from the current sample distribution. The tract 6–7 boundary, for example, may relate to a marked change in porphyroblast character and lies close to the northwestern extent of the main area of epizonal rocks based on illite crystallinity. Therefore, post-metamorphic tectonic modification of the structure, reflected as variations in grade at the present outcrop level, cannot be ruled out. These may have occurred during D_2–D_3 deformation or in response to Upper Palaeozoic or younger tectonic events.

Conclusions and regional perspective

The main conclusions of this paper are:

- Manx Group rocks have undergone low grade regional metamorphism with the main metamorphic peak during a period of low tectonic stress between D_1 and D_2;
- igneous activity is established as syn-D_1 (Dhoon Granodiorite), syn-D_2 (Black Hut Dyke) and post-D_2 (Crosby and Foxdale Granites) on the basis of metamorphic mineral growth in restricted contact metamorphic aureoles;
- lithological/chemical control is very important in the determination of the distribution of the porphyroblast phases, chloritoid and cordierite;
- simple models of concealed igneous bodies acting as the heat source for metamorphism are unable to account for the observed distribution of the metamorphic rocks.

Regional perspective

The first regional review of metamorphism in Lower Palaeozoic rocks of the British Isles associated with the closure of the Iapetus Ocean was that of Oliver *et al.* (1984). Since that time a series of major papers have appeared on the metamorphism of the Lake District (e.g. Fortey 1989; Fortey *et al.* 1993) and Wales (Robinson & Bevins 1986; Bevins & Robinson 1988; Roberts *et al.* 1991, 1996), many using the illite crystallinity technique. A recurring problem is the distinction of early metamorphism associated with burial accompanied by basinal extension, and a consequently elevated geothermal gradient from later metamorphism associated with deformation. The Manx Group rocks, partly because of their unusual chemistry, have given rise to porphyroblastic phases and it has proved possible to distinguish the early growth of diagenetic chlorite–mica stacks from the subsequent development of metamorphic minerals that are clearly associated with the formation of cleavages. The differentiation of porphyroblast growth directly associated with contact metamorphism from the more regional development of porphyroblasts has helped to resolve the conflict between previous descriptions of these rocks (Gillott 1955; Simpson 1964*a*).

Fortey *et al.* (1993), in their account of the relationship between metamorphism and structure in the Skiddaw Group, interpret the patterns of their illite crystallinity isocrysts in terms of three factors. An early stage of metamorphism attributable to burial was followed by upper anchizone–epizone metamorphism during the early orogenic phase and formation of S_1 following the closure of Iapetus. Then, late tectonic uplift of metamorphosed rocks by southerly directed thrusts took place. There are striking similarities but also differences in detail between that sequence and the sequence which is inferred here for the Manx Group rocks.

Neil Fortey and John Jacques are thanked for constructive reviews. The skilled support of all the technical staff at Portsmouth is gratefully appreciated. Steve Crowley and Sean Mullins provided samples and advice. Fieldwork was funded by NERC grant No. GR9/01834. RPB publishes with the permission of the Director, British Geological Survey (NERC).

References

BARNES, R. P., POWER, G. M. & COOPER, D. C. 1999. The definition of sandstone-bearing formations in the Isle of Man and correlation with adjacent areas – evidence from sandstone chemistry. *This volume.*

BEVINS, R. E. & ROBINSON, D. 1988. Low grade metamorphism of the Welsh Basin Lower Palaeozoic succession: an example of diastathermal metamorphism. *Journal of the Geological Society, London,* **145,** 363–366.

BURTON, K. W. 1986. Garnet–quartz intergrowths in graphitic pelites: the role of the fluid phase. *Mineralogical Magazine,* **50,** 611–620.

COOPER, A. H., RUSHTON, A. W. A., MOLYNEUX, S. G., HUGHES, R. A., MOORE, R. M. & WEBB, B. C. 1995. The stratigraphy, correlation, provenance and palaeogeography of the Skiddaw Group (Ordovician) in the English Lake District. *Geological Magazine,* **132,** 185–211.

CORNWELL, J. D. 1972. A gravity survey of the Isle of Man. *Proceedings of the Yorkshire Geological Society,* **39,** 93–106.

CRAIG, J., FITCHES, W. R. & MALTMAN A. J. 1982. Chlorite–mica stacks in low-strain rocks from Central Wales. *Geological Magazine,* **119,** 243–256.

FITCHES, W. R., BARNES, R. P. & MORRIS, J. H. 1999. Geological structure and tectonic evolution of the Lower Palaeozoic rocks of the Isle of Man. *This volume.*

FORTEY, N. J. 1989. Low grade metamorphism in the Lower Ordovician Skiddaw Group of the Lake District, England. *Proceedings of the Yorkshire Geological Society,* **47,** 325–337.

——, ROBERTS, B. & HIRONS, S. R. 1993. Relationship between metamorphism and structure in the Skiddaw Group, English Lake District. *Geological Magazine,* **130,** 631–638.

GILLOTT, J. E. 1955. Metamorphism of the Manx Slates. *Geological Magazine,* **92,** 141–154.

GREGG, W. J. 1986. Deformation of chlorite–mica aggregates in cleaved psammitic and pelitic rocks from Isleboro, Maine, U. S. A. *Journal of Structural Geology,* **8,** 59–68.

GRIEVE, R. A. F. & FAWCETT, J. J. 1974. The stability of chloritoid below 10kb PH_2O. *Journal of Petrology,* **15,** 113–139.

HUGHES, R. A., COOPER, A. H. & STONE, P. 1993. Structural evolution of the Skiddaw Group (English Lake District) on the northern margin of eastern Avalonia. *Geological Magazine,* **130,** 621-629.

KENNAN, P. S. & MORRIS, J. H. 1999. Manganiferous ironstones in the early Ordovician Manx Group, Isle of Man: a protolith of coticule? *This volume.*

KIMBELL, G. S. & QUIRK, D. G. 1999. Crustal magnetic structure of the Irish Sea region: evidence from regional aeromagnetic data for a major basement boundary beneath the Isle of Man. *This volume.*

KISCH, H. J. 1990. Calibration of the anchizone: a critical comparison of illite 'crystallinity' scales used for definition. *Journal of Metamorphic Geology,* **8,** 31–36.

LAMPLUGH, G. W. 1898. 1-inch Geological maps: Sheet 100 (Solid and Drift) The Isle of Man. British Geological Survey.

—— 1903. *The Geology of the Isle of Man.* Memoir of the Geological Survey, UK. HMSO.

LESLIE, A. G. 1988. A chloritoid-bearing paragenesis in the Macduff Slates of Central Buchan. *Scottish Journal of Geology,* **24,** 223–232.

LI, G., PEACOR, D. R., MERRIMAN, R. J., ROBERTS, B. & VAN DER PLUIJM, B. A. 1994. TEM and AEM constraints on the origin and significance of chlorite–mica stacks in slates: an example from Central Wales, UK. *Journal of Structural Geology,* **16,** 1139–1157.

MILODOWSKI, A. E. & ZALASIEWICZ, J. A. 1990. The origin and sedimentary, diagenetic and metamorphic evolution of chlorite–mica stacks in Llandovery sediments of central Wales, U.K. *Geological Magazine,* **128,** 263–278.

MOLYNEUX, S. G. 1999. A reassessment of Manx Group acritarchs, Isle of Man. *This volume.*

OLIVER, G. J. H., SMELLIE, J. L., THOMAS, L. J. *ET AL.* 1984. Early Palaeozoic metamorphic history of the Midland Valley, Southern Uplands–Longford Down massif and the Lake District, British Isles. *Transactions of the Royal Society of Edinburgh: Earth Sciences,* **75,** 245–258.

PATTISON, D. R. M. 1989. P–T conditions and the influence of graphite on pelitic phase relations in the Ballachulish aureole, Scotland. *Journal of Petrology,* **30,** 1219–1244.

ROBERTS, B., MERRIMAN, R. J. & PRATT, W. 1991. The influence of strain, lithology and stratigraphical depth on white mica (illite) crystallinity in mudrocks from the vicinity of the Corris Slate Belt, Wales: implications for the timing of metamorphism in the Welsh Basin. *Geological Magazine,* **128,** 633–645.

——, MORRISON, C. & HIRONS, S. 1990. Low grade metamorphism of the Manx Group, Isle of Man: a comparative study of white mica 'crystallinity' techniques. *Journal of the Geological Society, London,* **147,** 271–277.

——, MERRIMAN, R. J., HIRONS, S. R., FLETCHER, C. J. N. & WILSON, D. 1996. Synchronous very low-grade metamorphism, contraction and inversion in the central part of the Welsh Lower Palaeozoic Basin.

Journal of the Geological Society, London, **153,** 277–285.

ROBINSON, D. & BEVINS, R. E. 1986. Incipient metamorphism in the Lower Palaeozoic marginal basin of Wales. *Journal of Metamorphic Geology,* **4,** 101–113.

SEIFERT, F. 1970. Low-temperature compatibility relations of cordierite in haplopelites of the system, K_2O – $MgO–Al_2O_3–SiO_2–H_2O$. *Journal of Petrology,* **11,** 73–99.

SIMPSON, A. 1964*a*. Metamorphism of the Manx Slate Series, Isle of Man. *Geological Journal,* **4,** 415–434.

—— 1964*b*. Deformed acid intrusions in the Manx Slate Series, Isle of Man. *Geological Magazine,* **101,** 20–36.

—— 1965. The syntectonic Foxdale–Archallagan granite and its metamorphic aureole, Isle of Man. *Geological Journal,* **4,** 415–434.

—— 1966. Summer field meeting in the Isle of Man. *Proceedings of the Geological Association,* **77,** 217–227.

SOPER, N. J. & ROBERTS, D. E. 1971. Age of cleavage in the Skiddaw Slates in relation to the Skiddaw aureole. *Geological Magazine,* **108**, 293–302.

——, STRACHAN, R. A., HOLDSWORTH, R. E., GAYER, R. A. & GREILING, R. O. 1992. Sinistral transpression and the Silurian closure of Iapetus. *Journal of the Geological Society, London,* **149,** 871–880.

SPEAR, F. S. 1993. Metamorphic phase equilibria and pressure–temperature–time paths. *Mineralogical Society of America Monograph.*

STONE, P., COOPER, A. H. & EVANS, J. A. 1999. The Skiddaw Group (English Lake District) reviewed: a model for early Palaeozoic sedimentation and tectonism at the northern margin of Avalonia. *This volume.*

WANG, P. & SPEAR, F. S. 1991. A field and theoretical analysis of garnet + chlorite + chloritoid + biotite assemblages from the tri-state (MA, CT, NY) area, USA. *Contributions to Mineralogy and Petrology,* **106,** 217–235.

WOODCOCK, N. H. & BARNES, R. P. 1999. An early Ordovician turbidite system on the Gondwana margin: the southeastern Manx Group, Isle of Man. *This volume.*

——, MORRIS, J. H., QUIRK, D. G. *ET AL.* 1999. Revised lithostratigraphy of the Manx Group, Isle of Man. *This volume.*

Trans-Iapetus contrasts in the geological development of southern Scotland (Laurentia) and the Lakesman Terrane (Avalonia)

R. P. BARNES & P. STONE

British Geological Survey, Murchison House, West Mains Road, Edinburgh EH9 3LA, UK

Abstract: The Iapetus Ocean was a major feature separating Laurentia and Avalonia in the early Ordovician. The early Palaeozoic, Laurentian margin of the ocean is preserved in the UK in inliers in the Midland Valley of Scotland and in the Southern Uplands Terrane. The oldest rocks, of similar age to the deep-marine facies of the Skiddaw and Manx Groups opposite on the Avalonian margin, form the fragmentary ophiolitic sequences preserved in the Ballantrae and Highland Border complexes. These, together with equivalents displayed more extensively in Newfoundland, show that a succession of volcanic arcs and back arc basins formed and were accreted onto the Laurentian margin as a result of subduction during the late Cambrian and early Ordovician. In contrast, a suprasubduction zone extensional regime may have dominated the Avalonian margin at that time.

Evidence for the progressive destruction of the ocean by northward directed subduction is preserved in the Caradoc–Wenlock rocks of the Southern Uplands. The tectonostratigraphic configuration, with fault-bound slices containing thick turbidite sandstone sequences, of individually restricted duration, resting southwards on progressively younger oceanic mudrocks, is suggestive of an accretionary complex, although the precise situation is debated. Interpretation of the structural complexity of the Southern Uplands is reliant upon extensive biostratigraphical data, prior to the availability of which the understanding was at a level comparable to that of the Manx Group at the present time.

The late Llandovery–Wenlock Hawick and Riccarton Groups in the Southern Uplands continue the general tectonostratigraphical pattern. These younger turbidites are, however, of distinctive lithological character and show similarities with Wenlock to Ludlow sandstone-dominated sequences in the Windermere Supergroup in the southern Lake District and the Niarbyl Formation on the Isle of Man. This correlation between the Laurentian and Avalonian margins confirms that the Iapetus Ocean was no longer a significant feature by the mid-Silurian. The Southern Uplands accretionary thrust front migrated southwards on to the Avalonian foreland during the late Silurian as the Avalonian plate was subducted beneath Laurentia. Deformation in the Southern Uplands was largely complete prior to emplacement of *c.* 400 Ma granite plutons, whereas the Acadian deformation of the Skiddaw and Manx Groups was concentrated in the early Devonian at *c.* 390 Ma. However, similarities in structural style in the two areas seem to arise from the operation of similar mechanisms.

In terms of present geography, the Lower Palaeozoic sequences of the Isle of Man and the Lake District (the Lakesman Terrane: Gibbons & Gayer 1985; Bluck *et al.* 1992) are juxtaposed against sequences of similar age which crop out in southern Scotland and northern Ireland (the Southern Uplands Terrane, Fig.1). However, comparison of the rock records in these areas suggests that their geological histories were very different during the Ordovician and the early part of the Silurian; these then converge with a number of common characteristics apparent from the late Llandovery into the Wenlock. Convergence was complete before the emplacement of early Devonian granite plutons in both areas.

The regional pattern is due to the well-documented separation of northwest England and the Isle of Man from Scotland and northern Ireland by the Iapetus Ocean until the late Ordovician or early Silurian (e.g. Cope *et al.* 1992). Evidence for this separation is seen in the faunal differences between Laurentia and Avalonia, north and south of the ocean, respectively (e.g. Cocks & Fortey 1982) and in palaeomagnetic data which provide some quantification of the separation (e.g. Torsvik *et al.* 1990; Torsvik & Trench 1991). Convergence of shallow-water faunas from the late Ordovician onwards (Cocks & Fortey 1990) give the first indication of a narrowing of the ocean with final closure taking place during the Silurian so that non-marine fish faunas were mixed by the Ludlow (Young 1990). The established pattern of

From: Woodcock, N. H., Quirk, D. G., Fitches, W. R. & Barnes, R. P. (eds) 1999.
In Sight of the Suture: the Palaeozoic geology of the Isle of Man in its Iapetus Ocean context.
Geological Society, London, Special Publications, **160**, 307–323. 1-86239-046-0/99/$15.00

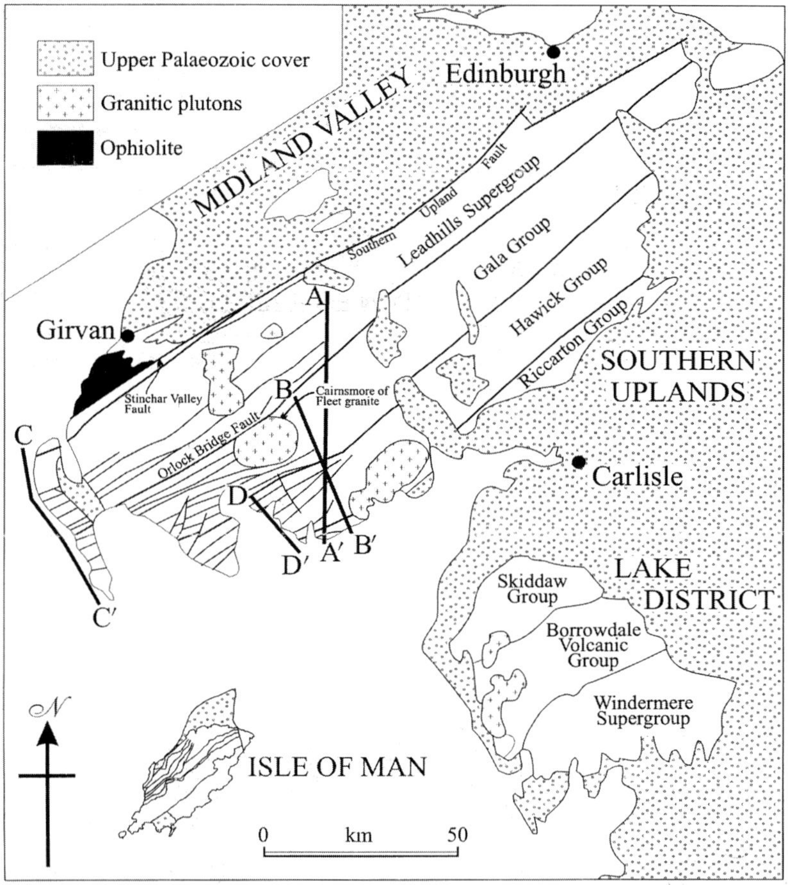

Fig. 1. General location map showing the Southern Uplands and Midland Valley Terranes in their modern situation with respect to the Isle of Man and the Lake District in northwest England. The detailed tract subdivision of the southwestern part of the Southern Uplands, interpreted during recent British Geological Survey mapping (e.g. Stone 1995; Barnes 1999) is shown for comparison with the scale of tectonostratigraphical tracts proposed on the Isle of Man (Fitches *et al.* 1999). Lines of cross-section refer to Fig. 2.

diachronous deformation throughout the late Ordovician and Silurian in the Southern Uplands of Scotland (e.g. Barnes *et al.* 1989) can be related to destruction of the Iapetus Ocean by northward subduction beneath Laurentia. Early Palaeozoic deformation at the Avalonian margin, on the other hand, as manifested in the Skiddaw Group of the English Lake District, followed a very different pattern (Hughes *et al.* 1993; Stone *et al.* 1999). There, extensive slump-and-fault disruption of the stratigraphy was followed by suprasubduction zone basin inversion; penetrative deformation with cleavage formation did not occur until the early Devonian (Soper & Kneller 1990), well after the juxtaposition of Laurentia and Avalonia into, more or less, their present relative positions.

The principal aim of this contribution is to place the early Palaeozoic geology of the Isle of Man in its regional context by considering the parallel evolution of the northern margin of the Iapetus Ocean, as shown in southern Scotland. No direct stratigraphical correlations are possible, although in the final phases of the depositional history the mid-Wenlock rocks in both areas have characteristics in common. Subsequently, after the juxtaposition of the opposing continental margins in the late Silurian, the same tectonic regime should have affected both areas.

As a prelude to this discussion, the historical perspective of geological research in the Southern Uplands, which has experienced many of the problems currently constraining interpretation of the geology of the Isle of Man (cf. Woodcock *et al.* 1999), is considered. The way in which the geological model has changed in response to the solution of those problems is an apposite illustration of the uncertainty in the present understanding of the geology of the Isle of Man.

Evolution of the geological model of the Southern Uplands

Interpretation of stratigraphy and large-scale structure are interrelated problems and usually information pertaining to one (generally the stratigraphy) is critical for solution of the other. The present difficulties faced in understanding the stratigraphy and the structure of the Lower Palaeozoic rocks of the Isle of Man are not dissimilar to those which faced the primary geological survey of the Southern Uplands in the 1870s:

- coastal exposure showing complex structure but relatively poor exposure inland, which generally allows only the principal lithological units to be mapped;
- little or no information about the nature of the junctions between rock units;
- little or no constraint on the relative age of different units;
- no knowledge of sedimentary way-up (generally not a problem now in the sandstone-bearing formations but still difficult in mudrock-dominated units), which precluded the facing direction of steeply inclined strata being used to control structural interpretation.

With subsequent work having overcome some of these problems, the evolution of the interpretation of the stratigraphy and structure of the Southern Uplands is instructive in that it indicates the degree of uncertainty in current interpretations of the Isle of Man. Three major stages can be identified in this evolution: early 'lithological' mapping, a leap in biostratigraphical control and advances which allowed greater structural control.

Early, lithological mapping-based interpretation

The primary geological survey of the Southern Uplands (e.g. Irvine 1878) recognized a number of units of broadly similar lithological character and derived a lithostratigraphy, but without biostratigraphical or structural control. The succession was assumed to young from south to north with the 'Ardwell Group' (Hawick Group), cropping out over a large area in the south of the Southern Uplands, considered the oldest part of the succession. This was thought to be succeeded by the 'Lower' or 'Moffat Black Shale Group' from its first occurrence northwards, and then the 'Queensberry Grit Group' (Gala Group). Inliers of the black shale within the latter were thought to be the Lower Black Shale Group in the cores of large folds. Divisions of the strata to the north mainly reflected variations in character of the turbidite sequence but, from further occurrences of black shale, a unit termed the 'Upper Black Shales' was included near the top of the succession. It was noted, however, that the latter 'bears so strong a resemblance to the lower unit that, but for the evidence of superposition, it might readily be identified with that band' (Irvine 1878). The whole of this sequence was assigned to the Lower Silurian (Ordovician in modern terminology) following the predilection of Murchison, then Director of the Survey.

Advances in biostratigraphical control

More or less in parallel with the work of the survey in the Southern Uplands, Lapworth (1876, 1878, 1889) developed a new biostratigraphical framework based on evolution of the graptolite faunas. Upper Silurian graptolite species had already been recorded from the southernmost parts of the Southern Uplands in the 'Riccarton Beds', which were thus considered to lie unconformably on, or be faulted against, the older rocks (Lapworth & Wilson 1871). However, Lapworth's biostratigraphy proved that the survey's stratigraphy was the wrong way up and that the black mudstones were in fact a single sequence. This was then adopted by the survey as described by Peach & Horne (1899) who, 'for the sake of convenience', divided the Southern Uplands into three strike-parallel belts reflecting the age of the sandstone succession:

- Northern Belt: volcanic rocks overlain by Moffat Shale passing up into sandstone of Ordovician age;
- Central Belt: Moffat Shale, as to the north but extending into the Llandovery and passing up into sandstone of Tarannon (Upper Llandovery) age divided into the 'massive grits and greywacke' of the Queensberry Grits (mainly the current Gala Group) in the north and the 'brown crusted flags' of the Hawick Rocks in the south;
- Southern Belt: sandstone of Wenlock age (Riccarton and Raeberry Castle beds).

The overall stratigraphy defined by Peach & Horne (1899), and their recognition that the 'oceanic deposits' of the Moffat Shale had been deposited continuously as the overlying coarse-grained terriginous materials were carried progressively further south (relative to the present geography) during the Ordovician and Silurian, have remained essentially unchanged to the present day.

Advances in structural understanding

Despite the radical revision of the stratigraphy, the perception of the structure of the Southern Uplands portrayed by Peach & Horne (1899) was developed from previous ideas (Fig. 2, A–A′). A large-scale anticlinorium–synclinorium model fitted the steep dips and the folding observed in coastal sections. The discontinuous Moffat Shale outcrops, interpreted as the cores of periclinal anticlines, were consequently mapped as lenticular inliers with subdivisions forming concentric zones, locally around a core of basic igneous rocks.

Major advances came in the 1950s when the application of newly recognized criteria for determining sedimentary way-up and petrographical characterization of the sandstone-dominated sequences provided new insight into the large-scale structure. Craig & Walton (1959) demonstrated from sedimentary way-up that the predominant northward younging of strata in the Kirkcudbright area is inconsistent with the old anticlinorium–synclinorium model. A lack of stratigraphic continuity over the perceived large-scale folds was also demonstrated by observations that the detrital components of the sandstone may change markedly from one side of a shale outcrop to the other (e.g. Walton 1955; Kelling 1961; Floyd 1976, 1982). However, further biostratigraphical work confirmed the overall southeast younging trend, with the oldest rocks cropping out in the northwest and progressively younger strata appearing southwards (e.g. Toghill 1970).

The apparent structural contradiction of a sequence which at outcrop is dominantly northward younging but which overall becomes younger southwards, was first explained by Craig & Walton (1959) as arising from major northeast–southwest strike faults cutting large-scale monoclinal folds (Fig. 2, B–B′). This paved the way for a radical new large-scale model, first presented by McKerrow *et al.* (1977) and developed by Leggett *et al.* (1979), which viewed the Southern Uplands Terrane as a supra-subduction zone accretionary prism. They suggested a series of strike-parallel, fault-bounded tracts, within each of which the volumetrically dominant, sand-rich turbidite sequence rests stratigraphically on a thin sequence of black shale situated at its southern margin. The southern margin of each black shale outcrop was interpreted to be a major strike-parallel fault (e.g. Fig. 2, C–D′). The age of the mudstone–turbidite transition within each tract becomes progressively younger to the southeast (Fig. 3). In early versions of this model the age range of the turbidite sequence in each tract was unconstrained, although it was suggested that it may span many graptolite zones. However, more recent work (e.g. Barnes *et al.* 1989; Stone 1995) has shown that the turbidite sequence is usually entirely within the youngest biozone seen in the mudstone or extends into the biozone above; an exception occurs in the northeast of the terrane where the younger tracts of the Gala Group commonly span several biozones in the late Llandovery (Rushton *et al.* 1996). Overall, notwithstanding this exception and despite their volumetric predominance, the turbidites in each tract occupy a relatively small time interval compared with the Moffat Shale Group which may represent up to 25 Ma in its southernmost outcrops (Fig. 3). More so than the variation in the age of the base of the turbidite sequence, this truncation of the top of the sequence from tract to tract requires syn-depositional deformation (Barnes *et al.* 1989; Barnes 1999) consistent with some form of sequential accretionary mechanism. Modern interpretations of the Southern Uplands Terrane have yet to achieve a complete consensus but require thrust-dominated structural configurations (e.g. Fig. 2, C–D′) to have been developed in an active margin, plate tectonic setting.

Implications for the Manx Group, Isle of Man

Following the early survey work of Lamplugh (1903), an interpretation of the geology of the Lower Palaeozoic rocks of the Isle of Man was developed by Simpson (1963) based on lithological mapping. As in the Southern Uplands, the resultant stratigraphy was shown to have severe problems when limited micropalaeontological control was produced by Molyneux (1979). However, even with a revised microfaunal interpretation (Molyneux 1999) and recently discovered graptolite and orthocone faunas (Rushton 1993; Howe 1999; Orr & Howe 1999), the stratigraphical and structural implications of full biostratigraphical control have yet to be realized in the Isle of Man. Recognition of compositionally distinct sandstone suites (Barnes *et al.* 1999) suggests the presence of fault-bounded tracts in the Isle of Man (Fitches *et al.* 1999) with similar implications for the overall 'tectonostratigraphy' as the petrographically discriminated sandstone suites in the Southern Uplands. Separate stratigraphies are identified within each tract in the Manx Group (Woodcock *et al.* 1999) but there are uncertainties relating to both the stratigraphical continuity and, in some tracts, the way-up of each 'sequence'. There is currently little or no clear means of relating the different sequences to one another or, with the exception of the Lonan sequence, to those in southeast Ireland

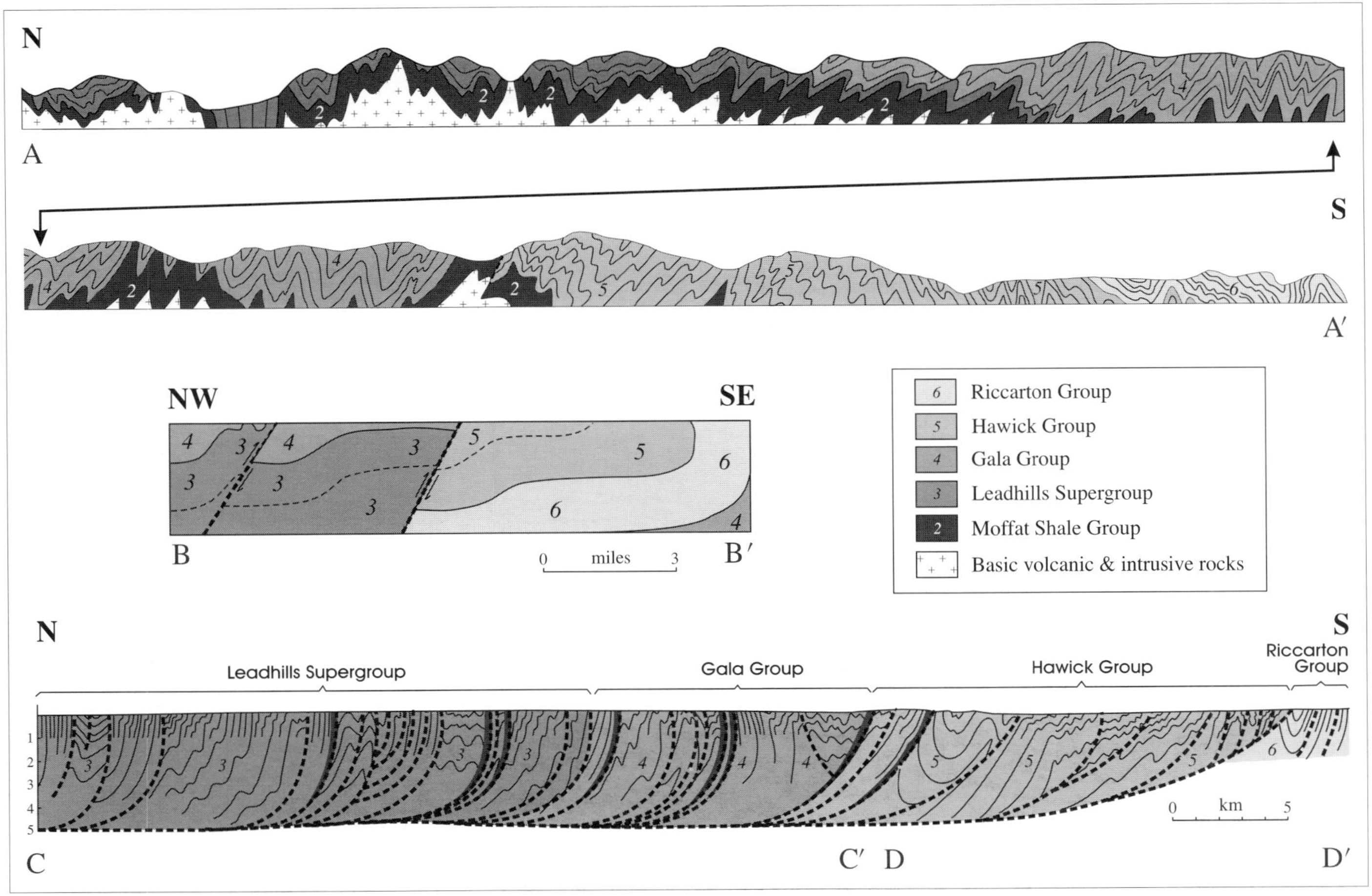

Fig. 2. Cross-sections across the strike of the Southern Uplands illustrating evolution of the detailed structural model following recognition of the biostratigraphical sequence: A–A′ [after Peach & Horne (1899)] is typical of early fold models; B–B′ early reconciliation of the contradiction of northerly sedimentary younging with progressive decrease in age of the sequence southwards [after Craig & Walton (1959); note that this section illustrates the sheet dip and that, at this stage, the rocks of the Hawick Group were considered to be younger than those of the Riccarton Group]; C–C′ plus D–D′ composite section illustrating recent imbricate thrust models, based on BGS work in southwest Scotland (British Geological Survey 1992, 1993; Stone 1995; Lintern & Floyd 1999). Lines of cross-section are shown on Fig. 1.

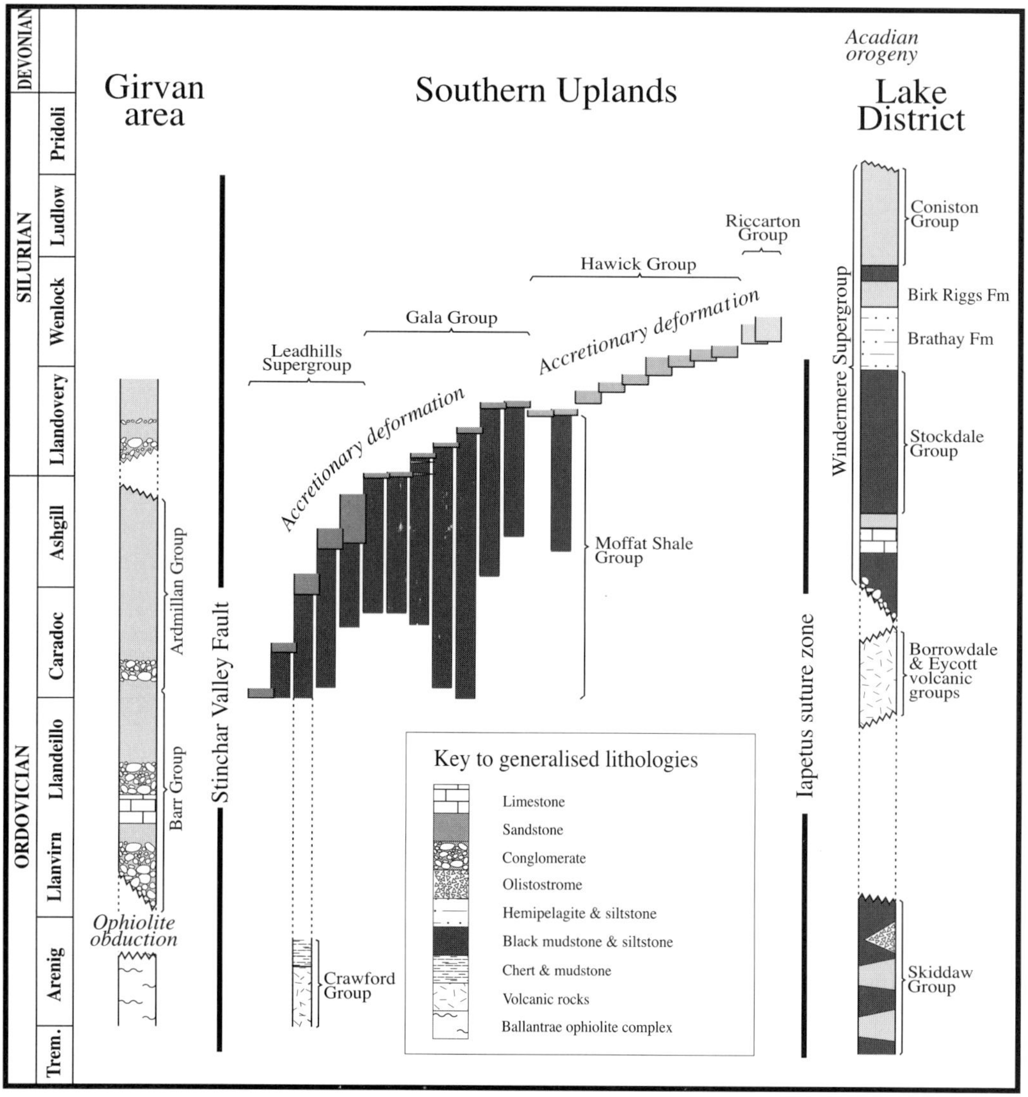

Fig. 3. Lithological successions in the Girvan area in southern Scotland and the Lake District in northwest England compared with the modern biostratigraphical/tectonostratigraphical interpretation of the Southern Uplands in southwestern Scotland (Barnes *et al.* 1989; Stone 1995).

and the Lake District. Until this can be achieved the large-scale structure of the Isle of Man will remain cryptic, as did that of the Southern Uplands prior to the availability of biostratigraphical detail.

The Laurentian margin: Tremadoc–Llanvirn ophiolite development

This phase of activity at the northern margin of the Iapetus Ocean is represented within the Midland Valley Terrane (Fig. 1), situated in southern Scotland between the Southern Upland Fault to the south and the Highland Boundary Fault to the north. Although extensively obscured by younger sedimentary cover, inliers of Lower Palaeozoic rocks show structurally disrupted early Ordovician ophiolite sequences adjacent to both faulted margins. These are temporally equivalent to the Manx Group but, from the biostratigraphical evidence of overlying sedimentary sequences discussed above, they record the separate development of the northern margin of the Iapetus Ocean at a time when there was wide separation of Avalonia and Laurentia.

In the north of the terrane the Highland Border Complex is structurally very fragmentary (Curry *et al.* 1984). Ophiolitic igneous rocks, largely serpentinites, are overlain by early Arenig limestone and conglomerate. A separate sequence of basaltic volcanic rocks, chert and black shale is though to be Llanvirn in age. From the character of the sedimentary lithologies and the geochemistry of the lavas, Robertson & Henderson (1984) suggested that these sequences formed in a small marginal basin.

A more complete sequence is preserved in the Ballantrae Complex, adjacent to the Southern Upland Fault in southwest Scotland (Church & Gayer 1973; Bluck *et al.* 1980; Stone & Smellie 1988). Serpentinized ultramafic rocks, a range of basaltic lava and lava breccia, together with minor chert and volcaniclastic lithologies, are structurally imbricated. Geochemical studies (Thirlwall & Bluck 1984; Stone & Smellie 1990 and refs cited therein) have shown that within-plate lava and two distinct types of arc tholeiite are present; boninitic rocks at one locality confirm the oceanic character of the arc lavas (Smellie & Stone 1992). Sedimentary interbeds in one of the within-plate sequences contain an early–middle Arenig graptolite fauna and a late Arenig fauna has also been reported but is of uncertain relationship to the rest of the complex (Stone & Rushton 1983). Radiometric dates from various components (Bluck *et al.* 1980; Thirlwall & Bluck 1984) suggest formation of the rock assemblage between *c.* 500 and 485 Ma, followed by obduction at *c.* 480 Ma. The detailed evolution of the sequence is obscured by the structural complexity but Smellie & Stone (1992) suggested that a sequence related to arc rifting and back-arc basin development was succeeded by ocean island tholeiites. The structural imbrication of the sequence then occurred as a result of obduction on to the Laurentian margin in late Arenig to early Llanvirn times. The suprasubduction components were believed to have been generated above a south (oceanward) dipping zone which facilitated subsequent arc–continent collision and ophiolite obduction. This was followed by a flip in subduction polarity and the establishment, from the late Llanvirn onwards, of northwards subduction of Iapetus Ocean crust.

A late Llanvirn to early Wenlock marine cover sequence resting unconformably on the ophiolitic rocks in the Midland Valley Terrane is best preserved in the Girvan area. Here, the lower part of the sequence, dominated by fan deltas with conglomerates containing much ophiolitic debris, transgresses northwards over the ophiolitic rocks from the Llanvirn to the late Caradoc (Williams 1962). From the Ashgill onwards the sequence is dominated by turbidites (Cocks & Toghill 1973) but passes up, near the top, into red subaerial deposits of early Wenlock age.

Laurentia–Avalonia comparisons

There are clear similarities between the fragmentary evidence of the early evolution of the Laurentian margin preserved in Scotland in the Midland Valley Terrane and the much more complete record in central Newfoundland (Colman-Sadd *et al.* 1992*a*). Some close correlations are possible between the two areas. The most striking is the early Arenig island-arc to back-arc transition recorded in the Betts Cove–Snooks Arm and associated ophiolites in Newfoundland (e.g. Swinden *et al.* 1989) and in the Ballantrae Complex in Scotland. These are so similar that Colman-Sadd *et al.* (1992*a*) considered that they may have formed in the same back-arc basin which closed, causing obduction of oceanic lithosphere in the late Arenig. Other ophiolites and volcanic sequences preserved in Newfoundland, ranging from Cambrian to mid-Ordovician in age, are not seen in the less deeply eroded Midland Valley Terrane. The polarity of subduction is, in most cases, obscure, although the Betts Cove–Snooks Arm and Ballantrae ophiolites have both been related to east or southeast directed subduction (e.g. Stockmal *et al.* 1990; Stone & Smellie 1990). Taken together, however, the full range of evidence suggests that the Laurentian margin of the Iapetus Ocean was characterized by several overlapping episodes of arc rifting and back-arc basin extension and closure.

This complex Laurentian margin, possibly resembling the present-day geology of the western Pacific, contrasts markedly with the passive/extensional nature of the early Ordovician Avalonian margin as preserved in the Skiddaw Group of the English Lake District (Fig. 3; Hughes *et al.* 1993; Stone *et al.* 1999). However, it should not be overlooked that elsewhere along the Avalonian margin there is evidence for a more active regime. Probable subduction-related magmatism commenced in Wales in the Late Tremadoc (Rhobell Volcanic Complex; Kokelaar 1986). Volcanic rocks of early Arenig age also occur in the Manx Group in the Isle of Man (e.g. Peel volcanics; Woodcock *et al.* 1999). Late Arenig ophiolite obduction along the southern side of the Iapetus Ocean has been reported from Newfoundland (Colman-Sadd *et al.* 1992*b*). A range of Caradoc–Ashgill arc systems that developed marginal to Avalonia but accreted to

Laurentia before final closure of Iapetus is reviewed by Cocks *et al.* (1997); they are preserved in Ireland (Grangegeeth terrane) and the northern Appalachians (Popelogan–Victoria arc assemblage). Against this background, the early Ordovician development of the British sector of the Avalonian margin could be viewed as anomalous.

Laurentian margin: Caradoc–Llandovery active margin accretion

South of the Midland Valley Terrane, the Southern Uplands Terrane (Fig. 1) is bounded to the north by the Southern Upland Fault and passes southwards beneath unconformably overlying Devonian and Carboniferous strata. It is dominated by a Caradoc–Wenlock sandstone rich turbidite sequence overlying a condensed sequence of grey to black mudstone and siltstone (the Moffat Shale Group) which ranges from Caradoc to mid-Llandovery in age, as described above. The boundary with the Midland Valley Terrane was probably established during the late Ordovician because debris in many of the older sandstone formations was apparently derived from the ophiolitic rocks to the north. Given the wide spread of ophiolitic rock along the Laurentian margin this provides little provenance control and some conglomerate units have apparently exotic clast assemblages from which McKerrow & Elders (1989) have argued for a provenance in or around Newfoundland and sizeable (*c.* 1500 km) post-Caradoc movement on the Southern Upland Fault. Conversely, derived shelly faunas in the deep-water strata of the northern part of the Southern Uplands bear a close resemblance to *in situ* Caradoc faunas in the Girvan succession and, from this association, Clarkson *et al.* (1992) deduced that movement could not have exceeded a few hundred kilometres.

Outcrops of the Moffat Shale Group in the northern part of the Southern Uplands are locally associated with basic volcanic and intrusive rocks. The relationship between them is difficult to determine due to tectonic disruption but the igneous rocks have generally been considered to be the 'basement' to the Southern Uplands sequence (e.g. Fig. 2, A–A′), representing the Iapetus oceanic crust in early accretionary prism models (e.g. Leggett *et al.* 1979). The most precise age control stems from one locality (Raven Gill; Hepworth 1981) where interbedded mudstone and chert have yielded conodonts of late Arenig age (Armstrong *et al.* 1990). This suggests a possible time equivalence with the Ballantrae ophiolite complex and, like the Ballantrae complex, the basic rocks exposed in the Southern Uplands do not represent a geochemically uniform association. From the Southern Uplands, Phillips *et al.* (1995) recognized alkaline within-plate basalts, possibly associated with extensional development of the depositional basin, and tholeiitic lavas of possible mid-ocean ridge affinity and island-arc or transitional character, the latter two types perhaps representing a back-arc basin. It seems likely that these fragments of older rocks within the Southern Uplands represent a complex basement, possibly composed, at least in part, of amalgamated fragments of pre-existing volcanic terranes much as may be inferred in the Midland Valley Terrane as discussed above.

Apart from the faunal comparisons, there is no tangible relationship between the late Ordovician and Llandovery turbidite cover sequence to the ophiolitic rocks in the Midland Valley Terrane and turbidites of the same age in the Southern Uplands. In the Leggett *et al.* (1979) model it is possible that the Midland Valley turbidites were deposited in a fore-arc basin behind an emergent trench slope break in the developing Southern Uplands accretionary prism. Alternative interpretations of the Southern Uplands terrane envisage a more direct link between the Midland Valley and Southern Uplands successions, either in a back-arc basin (Morris 1987; Stone *et al.* 1987) or in an extensional basin developed marginal to a narrow Laurentian continental shelf (Armstrong *et al.* 1996). Whichever initial model is preferred, the fundamental point is that the Caradoc to Llandovery development of the Southern Uplands Terrane was as an accretionary complex of some sort formed by sequential underthrusting at the northern, active margin of the Iapetus Ocean.

Laurentia–Avalonia comparisons

Temporally, the clastic sequence preserved in the Southern Uplands is younger than the Ordovician Manx Group as presently understood (Woodcock *et al.* 1999). The younger turbidites in the Southern Uplands, ranging up to the early *lundgreni* Biozone, may overlap in age with the Niarbyl Formation (Morris *et al.* 1999). However, all of these Wenlock rocks essentially post-date the closure of the Iapetus Ocean, and direct stratigraphical and structural comparisons with the Lake District and Isle of Man may be possible, as discussed more fully below. In the Lake District, the Caradoc saw the onset of a brief but intense episode of subduction-related volcanism, producing the Borrowdale Volcanic Group. This has been linked to the subduction of the Iapetus mid-ocean ridge (Pickering & Smith 1995) which would have effectively transferred the Avalonian margin on to the northwards subducting plate and initiated its inevitable collision with Laurentia. Sedimentation at the Avalonian margin following subduction shut-down there, as represented in the lower Caradoc to Llandovery part of the Windermere Supergroup, is suggestive of marine transgression across a subsiding shelf.

Avalonia–Laurentia collision: late Llandovery–Wenlock evolution of the Southern Uplands

Estimates of the timing of impingement of Avalonia with the Laurentian margin of the Iapetus Ocean vary according to the criteria used. The first signs of faunal convergence from either side of Iapetus begin in the late Ordovician and the faunal mixing even extends to non-marine fish by the Ludlow (Cocks *et al.* 1997 and refs cited therein). Volcanism, which may be broadly related to subduction processes, may have continued into the early Devonian in the Midland Valley of Scotland (Thirlwall 1981). The first clear signs of sedimentary links between the Southern Uplands and Avalonian Terranes are seen in the late Llandovery and early Wenlock (e.g. Barnes *et al.* 1989; Lintern *et al.* 1992), although to some extent reflecting wider, regional trends. Significant amounts of red mudstone first appear in the late Llandovery, in the Hawick Group in the Southern Uplands and in the Browgill Formation in the Lake District. The succeeding early Wenlock Brathay Formation sequence in the Lake District comprises finely interlaminated siltstone and carbonaceous mudstone which forms the background to interbedded turbidites (Birk Riggs Formation) higher in the Wenlock. This distinctive hemipelagite lithology also occurs interbedded with the Wenlock turbidite sequences of the Southern Uplands and has a very wide distribution throughout the residual Iapetus basins at that stratigraphical level (Kemp 1991).

There is no clear indication of a collisional event in the Southern Uplands sequences, although Rushton *et al.* (1996) present evidence for the interruption of the forward progress of the Southern Uplands accretionary thrust belt in the late Llandovery. They show that extended, apparently conformable, greywacke sedimentation occurred across up to five graptolite biozones coincident with back- and out-of-sequence thrusting in the hinterland. The latest Llandovery and earliest Wenlock was then a period of intense transpression before a more orthogonal, forward-breaking thrust pattern was re-established in the mid-Wenlock. This combination of stratigraphical and structural events may reflect the initial blocking of the Southern Uplands thrust belt, at the leading edge of Laurentia, by its first encounter with Avalonian continental crust, followed by the accommodation of the obstacle and the advance of the thrust belt on to Avalonia.

Another line of evidence may be drawn from the distribution of the Moffat Shale Group. This 'pelagic' black mudstone, the oceanic deposits in the McKerrow *et al.* (1977) and Leggett *et al.* (1979) model, appears beneath the turbidite sequence in almost every tract up to about the *crispus* Biozone (Fig. 3). The most southerly vestiges of Moffat Shale Group thus occur beneath the older parts of the Hawick Group but it is not seen in the more southerly tracts. This led Stone *et al.* (1987) to suggest that the 'oceanic' basin effectively ceased to exist in the mid-Llandovery, the Hawick Group representing deposition in a foreland basin setting as the Avalonian margin was thrust beneath the Laurentian margin. There is, however, little if any change in the overall structural style of the Southern Uplands, as far as can be determined, with the tract structure continuing southwards.

The loss of the Moffat Shale in southern parts of the Southern Uplands may have been due to the advance of the thrust front beyond the maximum southerly transgression of the Moffat Shale facies on to what was originally the Avalonian side of the Iapetus Ocean. Against this argument is the fact that shale sedimentation is apparent on the Avalonian margin for much the same time-span as that of the Moffat Shale Group. It began with the Caradoc Drygill Formation above the Borrowdale and Eycott Volcanic Groups (Ingham *et al.* 1978); Ashgill mudstone is preserved in the Cautley and Dent Inliers southeast of the outcrop of the volcanic rocks, whilst the Stockdale Group (Kneller *et al.* 1994) of the southern Lake District comprises a condensed Llandovery sequence dominated by black graptolitic shale. Hence, during the Llandovery, black shale was being deposited not only as the Moffat Shale Group in the relict Iapetus Ocean, but well southwards on the Avalonian continental margin.

An alternative explanation is that the loss of the Moffat Shale simply reflects the raising of the basal décollement into the turbidite sequence. Despite the uncertainties discussed above, it seems likely that this is significant in the regional context because it more-or-less coincides with a marked change in sandstone composition. Sandstone in the older part of the sequence is petrographically rather variable, with intermediate to basic igneous material forming a major component of some formations (e.g. Floyd 1982; Styles *et al.* 1989) which tectonically alternate, or are stratigraphically interbedded, with other formations derived from an ancient quartzo-feldspathic provenance (e.g. Stone & Evans 1995). Sandstone chemistry (Barnes 1998) suggests, however, a smoother trend, with mafic input increasing into the late Ordovician and then declining through the early part of the Llandovery in the dominantly quartzo-feldspathic Gala Group. A relatively subtle change in sandstone composition is apparent in the mid-Gala Group as the sandstone settles towards a uniform composition through the younger part of the sequence. From drainage chemistry, Stone *et al.* (1993) and Plant *et al.* (1999) note significant changes in the abundance of many element associations (particularly Pb–As, Rb–Sr–Ba, B–Li, Cr–Ti) across the Gala Group outcrop perpendicular to strike. However, whole rock REE abundances have been interpreted by Williams *et al.* (1996) in terms of a fundamental provenance switch in mid-Gala Group, suggesting a change from the intermittent volcanic provenance to a probably sedimentary source rich in heavy minerals. The Hawick Group sandstone is compositionally remarkably uniform throughout the sequence, but with a marked change from the Gala Group due to unusually high matrix carbonate content (up to 20%). It is, however, compositionally generally similar to the sandstone which became dominant during the late Wenlock and Ludlow in the Windermere Supergroup in the Lake District, consistent with the other indications of a closely linked system (e.g. Barnes *et al.* 1999).

The youngest strata preserved in the Southern Uplands, the mid-Wenlock Riccarton Group, are relatively weakly deformed compared with the Hawick Group but still form near-vertical, fault-bound packages that may be as little as 200 m in thickness (Kemp 1991; Lintern & Floyd 1999). The sedimentological links with coeval parts of the Windermere Supergroup have been discussed above and underlie a growing consensus that both sequences were deposited in a foreland basin which developed ahead of the Southern Uplands thrust front as it advanced on to the Avalonian margin, by that time being thrust beneath Laurentia. This tectonic situation was presaged in the Stone *et al.* (1987) sequential back-arc to foreland basin model for the Southern Uplands. More recently, detailed modelling of the Ludlow and Pridoli parts of the Windermere Supergroup by Kneller (1991) have established its foreland basin style of sedimentation, and the broader, trans-Iapetus implications have been discussed by Kneller *et al.* (1993) and Hughes *et al.* (1993). The regional significance of this model is that the Wenlock strata in the Niarbyl Formation (Morris *et al.* 1999), and the Wenlock and younger parts of the Kilcullen Group (southeast Ireland), could all have been deposited in a common, southwards migrating foreland basin developed above the sutured remains of the Iapetus Ocean. The propagation of the basal thrust system into the Cambro-Ordovician, Avalonian Skiddaw Group has been proposed by Stone *et al.* (1999) as the mechanism initiating the widespread resetting of Rb–Sr systems to the 430–420 Ma interval, i.e. broadly Wenlock (Tucker & McKerrow 1995).

Structural evolution of the Southern Uplands

The Ordovician and Silurian turbidite sequences of the Southern Uplands are typically steeply dipping to vertical, northeast to east-northeast striking and generally young northwards in a series of fault bounded tracts. In northern and central parts of the Southern Uplands the bounding fault-traces are marked by discontinuous slivers of the thin, but often fossiliferous, Moffat Shale Group preserved in stratigraphical continuity beneath the turbidite sequence (Fig. 3). Early movement occurred on thrusts propagated at a low angle to stratigraphy and was associated with the only phase of ductile deformation (D_1) to have affected many of the rocks in the Southern Uplands. Thrust propagation, and hence D_1, was diachronous, becoming younger southwards. Later phases of deformation were associated either with accommodation in the thrust hinterland, commensurate with D_1 deformation at the thrust front, or with intermittent sinistral shear imposed across the entire belt but focused into major strike-fault zones. These post-D_1 deformation phases have been referred to as D_2 (co-axial with gently plunging D_1) and D_3 (sinistral, steeply plunging), but their relationship is not the same everywhere (Barnes *et al.* 1989; Stone 1995; Barnes 1999). In parts of the Hawick Group, a significant component of sinistral shear during D_1, possibly equivalent to the first stages of D_3 in the thrust hinterland, produced steeply plunging or downward facing D_1 folds.

Thrust-related (D_1) deformation

D_1 folds, typically gently plunging and tight to isoclinal, were developed very variably throughout the Southern Uplands. Across-strike, highly folded zones occur interspersed with long homoclinal sections, usually of steeply inclined, north younging strata. This variation in structural style is, at least in part, related to the nature of the strata, with thickly bedded, massive greywacke less likely to be intensely folded than more thinly bedded strata. Slickensides or slickenfibres in thin veins along bedding surfaces, perpendicular to the fold axial orientation, demonstrate early fold growth by flexural slip, although these were themselves folded in the later stages of fold development. Individual tectonostratigraphical tracts are often marked by subtle variations in the style, orientation and intensity of D_1 folding (e.g. Barnes *et al.* 1986). D_1 deformation was particularly intense in the generally finer grained, more calcareous rocks of the Hawick Group in the southern part of the Southern Uplands, associated with the highest grades of regional metamorphism developed in the terrane.

The most widespread, regional manifestation of D_1 is a penetrative cleavage (S_1). In northern and central parts of the Southern Uplands, S_1 is best developed in the fine-grained, muddy lithologies, although even there it can be quite weak. It is rarely apparent macroscopically in sandstone, being represented only as a rough anastomosing fabric apparent in thin section. In parts of the Hawick Group, however, the foliation is more pervasive and is well developed in sandstone, where it is strongly refracted through graded beds, and is also commonly developed in felsic and lamprophyre dykes. The cleavage tends to be congruous with the D_1 folds, but may vary either in dip or strike from truly axial planar. Other than in the immediate vicinity of fold hinge zones, significant variation in the dip of the cleavage from that of the fold axial surface may cause bedding to be downward facing. This is particularly apparent in overturned, south dipping beds where cleavage commonly dips more steeply than bedding. As a consequence of this effect, the assessment of way-up or vergence from bedding cleavage relationships is generally unreliable in the Southern Uplands.

As mentioned above, some downward facing folds are locally present in the southern parts of the Hawick Group. S_1 cleavage typically contains the fold axial orientation in the northern part of the Southern Uplands but begins to transect the fold axis, by up to 20° clockwise locally, in central parts. This effect becomes commonplace throughout the Hawick Group due to systematic variation between the strike of the cleavage and that of the fold axial surface (cf. Anderson 1987). The cleavage transection has been explained in various ways as resulting from the evolution of the D_1 stress system (Stringer & Treagus 1980; Gray 1981; Sanderson *et al.* 1985).

Post-D_1 deformation (D_2 and D_3)

The extent and character of post-D_1 deformation varies widely across the Southern Uplands. In northern and central parts, post-D_1 structures coaxial with D_1 tend to occur only very locally and are difficult to correlate. To the south, in central and southern parts of the Hawick Group outcrop, coaxial post-D_1 deformation is widespread. Gently plunging, minor to mesoscale folds, coaxial with but refolding D_1 structures, occur in two styles, upright to inclined and recumbent. These have conjugate geometry and occur together locally as open box folds, suggesting that they formed together and consequently both are classified as D_2. Small recumbent folds, verging down the dip of bedding, are most common and are associated with a widely developed, gently dipping, S_2 crenulation cleavage. The orientation and geometry of the D_2 structures suggests that they formed either as a continuation of D_1 after locking of the D_1 folds, or by subsequent renewal of shortening on the tract-bounding faults. Alternatively, the recumbent folds may have been formed by subvertical shortening of bedding in more-or-less its present attitude, causing down-dip vergence, rather than by a consistent sense of shear on tract-bounding or other faults.

The steeply plunging sinistral folds (D_3) developed locally throughout the Southern Uplands are usually in narrow zones of shearing adjacent to tract-bounding faults and may therefore be associated with reactivation of these structures (e.g. Barnes *et al.* 1995). The relationships of the putative D_3 folds to D_2 are ambiguous (e.g. Barnes *et al.* 1989; Stone 1995), with indications of sinistral shear or refolding co-axial with development of D_1 folds at various times. It also seems likely that there were several episodes of sinistral shear superimposed on the diachronous D_1 and D_2 folding at different times.

The Moniaive Shear Zone

One particularly important example of strike-parallel sinistral shear is the Moniaive Shear Zone (Phillips 1994; Barnes *et al.* 1995; Phillips *et al.* 1995), named from the area around Moniaive, northeast of the Cairnsmore of Fleet Granite. It is a zone of high strain, kinematically similar to, but much wider than, the narrow shear zones associated with most tract-bounding faults. It has been recognized over a strike length of *c.* 100 km through the central part of the Southern Uplands where it is up to 5 km wide, generally truncated abruptly to the north by the Orlock Bridge Fault but dying out southwards within the northern tract of the Gala Group. It is characterized by the intermittent development of a pervasive foliation near-parallel to bedding, locally with a strong linear component, which commonly transposes all original structure. Strain within the shear zone is very variable (e.g. Phillips 1992, 1994) but a variety of kinematic indicators consistently show a sinistral sense of shear. Because the shear zone fabric is subparallel to the relatively weak S_1 cleavage outwith the shear zone, the two can only rarely be differentiated and unequivocal relative age relationships are difficult to establish. Cordierite porphyroblasts, widely distributed throughout the thermal metamorphic aureole of the early Devonian Cairnsmore of Fleet Pluton (*c.* 392 Ma; Halliday *et al.* 1980), are deformed by the shear zone foliation but the latter is generally overprinted by the biotite hornfelsing and later stages of the thermal metamorphism, closely constraining the timing of the final part of its development. Relatively high grades of regional metamorphism in the zone indicate that it formed at substantial depth. Barnes *et al.* (1995) suggested that the Moniaive Shear Zone is a composite feature, representing progressive but intermittent deformation over a long time period from its initiation during D_1, possibly in the early Silurian, until the early Devonian. Despite the possible long duration of intermittent deformation there are no grounds for assuming very large lateral displacement.

Overall, the style of deformation seen in the Moniaive Shear Zone is similar to the most intensely deformed part of a broad sinistral shear zone which locally marks the Orlock Bridge Fault in Ireland (the Slieve Glah Shear Zone; Anderson & Oliver 1986). However, suggestions made by Anderson & Oliver (1986) that this fault may be a terrane-bounding structure are refuted by Barnes *et al.* (1995) on the grounds that its effect diminishes progressively along-strike in the Southern Uplands until it becomes indistinguishable from the other tract-bounding faults. In addition, the stratigraphic break across the fault deduced by Anderson & Oliver (1986) has subsequently proved to be a considerable overestimate (Floyd *et al.* 1987; Floyd & Rushton 1993).

Structural relationships of intrusive rocks

Minor intrusions are abundant in parts of the Southern Uplands and many display key relationships which show that lamprophyric and felsic dykes were emplaced over a long period of time. In the Hawick Group tracts, dykes are particularly abundant and were emplaced during D_1, D_2 and D_3 deformation. They generally post-date D_1 deformation, although one or two dykes are demonstrably F_1 folded and many are S_1 cleaved. Throughout the Southern Uplands, later (but also mostly pre-dating the larger granite plutons) dykes were generally emplaced in north-northwest to north-northeast striking orientations associated with 'late' faulting, again with some associated deformation of the intrusive rocks showing syn-kinematic emplacement.

During thrust-related D_1 shearing and subsequent formation of conjugate D_2 folds, the direction of maximum shortening (σ_1) lay close to the dip direction of bedding. At an early stage, when the thrust slices were gently dipping, the overburden pressure would have been substantial and σ_2 therefore perpendicular to bedding. As the imbricate stack steepened, however, it appears that σ_2 and σ_3 switched, allowing bedding to be utilized for emplacement of the dykes. Subsequently, a change in the D_1 stress regime led to a larger component of sinistral shear in the more southerly Hawick Group tracts. This accompanied continued dyke emplacement to the north in areas actively undergoing D_3 deformation. Brittle deformation associated with the later stages of sinistral transpression caused along-strike (northeast–southwest) extension and formed conjugate fault sets into which dykes were emplaced (Barnes 1999).

Larger intrusions in the Southern Uplands include dioritic and granodioritic bodies, tens to hundreds of metres in size, and the much larger granitic plutons. The latter are generally post-tectonic, although the thermal aureole of the Cairnsmore of Fleet Pluton (cooling date *c.* 392 Ma) has an overlapping relationship with deformation related to the Moniaive Shear Zone since, as explained above, biotite in the thermal aureole overprints the shear zone foliation whereas earlier formed cordierite porphyroblasts are deformed.

Structural comparison with the Avalonian margin

The principal contrast which emerges in any trans-Iapetus structural comparison between the Southern Uplands and Lakesman terranes is the marked difference in timing of penetrative ductile deformation. In the Southern Uplands the thrust-related D_1 deformation was systematically diachronous from the late Ordovician in the north of the terrane to Wenlock or younger in the south. The overlapping deformation episodes, D_2 and D_3, were also diachronous, in as much as they occurred in different places at different times but have no clear spatial organization through time.

In sharp contrast to this pattern is the structural sequence seen in the Skiddaw Group where no tectonic cleavage was imposed until the early Devonian. At that time, only the D_3 sinistral shear was still operative in the Southern Uplands, although at the northern margin of that terrane there is evidence for folding and north directed thrusting on to the Midland Valley Terrane from the late Silurian. The clearest examples come from the Girvan area where the mid-Ordovician to mid-Silurian sedimentary cover to the early Ordovician ophiolite has been thrust northwards and structurally imbricated (Williams 1959). The youngest strata involved are late Wenlock so that the thrust movement was probably post-Wenlock in age, more or less contemporary with the initial collision of Laurentia and Avalonia further south.

The scale of this thrusting is uncertain but some authors (e.g. Bluck 1983) consider that the Southern Uplands Terrane is allochthonous and was emplaced northwards on to the Midland Valley basement at this time. Whatever its extent, this tectonic episode still appears to precede the first, early Devonian, penetrative tectonic deformation in the Lake District.

The contrast between the largely pre-Wenlock tectonism in the Southern Uplands and the apparently post-Pridoli first tectonic cleavage in the Lake District is the more remarkable in that continental collision certainly occurred at some time during the middle–late Silurian (e.g. Soper *et al.* 1992). This event appears to have left no orogenic imprint of any sort in these two terranes. However, crustal shortening across the suture zone did continue, as evinced by the Windermere Supergroup foreland basin (Kneller *et al.* 1993) and the resetting of Lake District Rb–Sr dates to the 420–430 Ma interval (Stone *et al.* 1999). A major structural décollement must have separated the converging terranes to act as a barrier to the northwards propagation of the subsequent early Devonian 'Acadian' deformation in the Lake District.

The 'tract' architecture resolvable through much of the southern part of the Southern Uplands (Fig. 1) is on a similar scale to that proposed on the Isle of Man by Fitches *et al.* (1999). The early tract-bounding faults, with at least tens of kilometres of reverse displacement and probable significant repeated lateral and normal reactivation, may be associated with intense deformation and imbrication of the attendant Moffat Shale, but little, if any, unusual deformation is usually apparent in the adjacent greywacke sequences. On the other hand, broad zones of localized high strain, such as developed locally on the Orlock Bridge Fault, the Moniaive Shear Zone and numerous zones associated with sedimentary disruption in parts of the Hawick Group (e.g. Kemp 1987; Lintern & Floyd 1999) may not represent zones on which a large amount of movement has occurred regionally. The significance of localized high strain zones with regard to major structural boundaries is debated in the Isle of Man (e.g. Fitches *et al.* 1999), but it is clear that in the Southern Uplands the strain preserved in the rocks is not necessarily a good indicator of the amount of movement on strike faults.

The style and geometry of the D_2 folds variably developed in the Southern Uplands is remarkably similar to those seen in the Ordovician rocks of the Manx Group and the Silurian Niarbyl Formation in the Isle of Man, particularly in their propensity to always verge down the dip of bedding (cf. Fitches *et al.* 1999). However, this presumably relates to similarity of processes rather than a correlatable 'event', particularly as they were demonstrably developed prior to the thermal aureole of the Cairnsmore of Fleet Granite in the Southern Uplands, dating them prior to D_1 as currently understood in the Lake District. Many elements of the relationship between dyke emplacement and deformation are also common in the Southern Uplands and the Isle of Man, with abundant S_1 cleaved dykes generally apparently post-D_1 folding in both areas. There is, however, a suite of very altered basic dykes with peperitic margins in the Isle of Man which may have equivalents in the Skiddaw Group (Hughes & Kokelaar 1993), which are presumably pre-tectonic but which have no obvious counterpart in the Southern Uplands.

Conclusions

The lack of any obvious lithostratigraphical correlation between southern Scotland and northwest England and the Isle of Man for most of their Ordovician and Silurian histories, as preserved in the now closely juxtaposed rocks, is as compelling evidence as that from the faunas for their separation by a major ocean for most of that period. During the late Cambrian and early Ordovician the Laurentian margin of the Iapetus Ocean was characterized by several overlapping episodes of arc rifting and back-arc basin extension and closure on a complex margin, possibly resembling the present-day western Pacific. This contrasted markedly with the passive, or perhaps suprasubduction, zone extensional nature of the early Ordovician Avalonian margin suggested by the Skiddaw and Manx Group sequences. The Caradoc–Wenlock sequence of the Southern Uplands is dominated by large volumes of sandstone-rich turbidite overlying a condensed mud-rich Moffat Shale sequence. Its tectonostratigraphical configuration suggests an accretionary complex and, although its precise setting is uncertain, points to sustained northward subduction throughout its depositional history. By contrast, in the Lake District, the Caradoc Borrowdale Volcanic Group records a volcanic episode which began with plateau andesite lavas and then developed thick but localized sequences of andesitic to rhyolitic pyroclastic rocks. The volcanic activity shut off abruptly in the late Caradoc and was succeeded by deposition of an unconformable and deepening shelf sequence in the early parts of the Windermere Supergroup. The early dyke suite and larger basic intrusions in the Isle of Man (e.g. Power & Crowley 1999) may be representative of the volcanic rocks but otherwise there is no evidence from the island of rocks younger than the Manx Group but older than the Wenlock Niarbyl Formation.

Major influxes of sandstone turbidites which first appear in the Windermere Supergroup in the mid-Wenlock and continue through the Ludlow were interpreted by Kneller *et al.* (1993) to mark rapid subsidence as a foreland basin developed in the early stage of overthrusting of Avalonia by Laurentia. The Niarbyl Formation (Morris *et al.* 1999) in the Isle of Man may be equivalent to one of the first of these pulses of sandstone, the Birk Riggs Formation (Barnes *et al.* 1999). In general, these Wenlock turbidites on the Avalonian margin continue a compositional trend established in the mid–late Llandovery in the Hawick Group in the Southern Uplands. This fits with other sedimentological similarities in the late Llandovery and Wenlock, suggesting effective closure of the intervening ocean by the mid-Silurian. The Moffat Shale sequence in the Southern Uplands, taken to be indicative of pelagic sedimentation in the open Iapetus basin, is very poorly preserved beneath northern parts of the Hawick Group and is nowhere seen to be younger than the mid-Llandovery (*cyphus* Biozone), again consistent with the passage to a different kind of basin architecture. There is otherwise, however, a seamless transition in the Southern Uplands in which the tectonostratigraphical configuration remains essentially unchanged into the youngest rocks preserved with no indication of a discrete collision event.

Deformation was essentially completed in the Southern Uplands by the time of emplacement of the late Silurian–early Devonian granite plutons. To the south, however, ductile deformation and cleavage formation was only just beginning in the Skiddaw and Manx Groups, although the 'tract' architecture of the Isle of Man is on a similar scale to that resolvable through much of the southern part of the Southern Uplands (Fig. 1). The zones of major early movement in the Southern Uplands, the tract-bounding faults, are not usually associated with broad high strain zones, although the attendant Moffat Shale may be intensely imbricated. Broad zones of localized high strain may not necessarily, however, represent zones on which a large amount of movement has occurred regionally. The style and geometry of the D_2 folds variably developed in the Southern Uplands, and many elements of the relationships between dyke emplacement and deformation seen there, are remarkably similar to those apparent in both the Ordovician and Silurian rocks of the Isle of Man. However, this must relate to common processes rather than correlatable 'events' with D_2 in the Southern Uplands being significantly older than D_2 in the Isle of Man, as deduced from the timing of deformation currently understood in the Lake District.

We are indebted to the many colleagues in BGS, universities and the Newfoundland Department of Mines with whom we have collaborated in work in southern Scotland. The paper was improved by helpful comments from the reviewers, J. M. Horák and T. Pharaoh, and the editor W. R. Fitches. The paper is published with the permission of the Director, British Geological Survey (NERC).

References

ANDERSON, T. B. 1987. The onset and timing of Caledonian sinistral shear in County Down. *Journal of the Geological Society, London*, **144**, 817–825.

—— & OLIVER, G. J. H. 1986. The Orlock Bridge Fault: A major late Caledonian sinistral fault in the Southern Uplands terrane, British Isles. *Transactions of the Royal Society of Edinburgh: Earth Sciences,* **77**, 203–222.

ARMSTRONG, H. A., CLARKSON, E. N. K. & OWEN, A. W. 1990. A new Lower Ordovician conodont faunule from the Northern Belt of the Southern Uplands. *Scottish Journal of Geology*, **26**, 47–52.

——, OWEN, A. W., SCRUTTON, C. T., CLARKSON, E. N. K. & TAYLOR, C. M. 1996. Evolution of the Northern Belt, Southern Uplands: implications for the Southern Uplands controversy. *Journal of the Geological Society, London*, **153**, 197–206.

BARNES, R. P. 1998. *Graphical display of sandstone geochemical data from the Southern Uplands, southern Scotland.* British Geological Survey Technical Report

—— 1999. Geology of the Whithorn, Kirkcowan and Wigtown districts. *Memoir of the British Geological Survey*, Sheets 2, 4W & 4E (Scotland), in press.

——, ANDERSON, T. B. & MCCURRY, J. A. 1987. Along-strike variation in the stratigraphical and structural profile of the Southern Uplands Central Belt in Galloway and Down. *Journal of the Geological Society, London*, **144**, 807–816.

——, COOPER, D. & POWER, G. M. 1999. The definition of sandstone-bearing formations in the Isle of Man and correlation with adjacent areas – evidence from sandstone chemistry. *This volume*.

——, LINTERN, B. C. & STONE, P. 1989. Timing and regional implications of deformation in the Southern Uplands of Scotland. *Journal of the Geological Society, London*, **146**, 905–908.

——, PHILLIPS, E. R. & BOLAND, M. P. 1995. The Orlock Bridge Fault in the Southern Uplands of SW Scotland, a terrane boundary? *Geological Magazine*, **132**, 523–529.

——, ROCK, N. M. S. & GASKARTH, J. W. 1986. Late Caledonian dyke-swarms in Southern Scotland: new field, petrological and geochemical data for the Wigtown Peninsula, Galloway. *Geological Journal,* **21,** 101–125.

BRITISH GEOLOGICAL SURVEY 1992. The Rhins of Galloway, Scotland, Sheets 1 & 3 with parts of 7 and 4W. Solid geology. 1:50 000.

—— 1993. Kirkcudbright, Scotland, Sheet 5W. Solid geology. 1:50 000.
BLUCK, B. J. 1983. Role of the Midland Valley of Scotland in the Caledonian orogeny. *Transactions of the Royal Society of Edinburgh: Earth Sciences*, **74**, 119–136.
——, GIBBONS, W. & INGHAM, J. K. 1992. Terranes. *In*: COPE, J. C. W., INGHAM, J. K. & RAWSON, P. F. (eds) *Atlas of Palaeogeography and Lithofacies*. Memoir of the Geological Society, London, **13**, 1–4.
——, HALLIDAY, A. N., AFTALION, M. & MACINTYRE, R. M. 1980. Age and origin of Ballantrae ophiolite and its significance to the Caledonian orogeny and Ordovician time scale. *Geology*, **8**, 492–495.
CHURCH, W. R. & GAYER, R. A. 1973. The Ballantrae ophiolite. *Geological Magazine*, **110**, 497–510.
CLARKSON, E. N. K., HARPER, D. A. T., OWEN, A. W. & TAYLOR, C. M. 1992. Ordovician faunas in mass-flow deposits, Southern Scotland. *Terra Nova*, **4**, 245–253.
COCKS, L. R. M. & FORTEY, R. A. 1982. Faunal evidence for oceanic separations in the Palaeozoic of Britain. *Journal of the Geological Society, London*, **139**, 465–478.
—— & —— 1990. Biogeography of Ordovician and Silurian faunas. *In*: MCKERROW, R. S. & SCOTESE, C. R. (eds) *Palaeozoic Palaeogeography and Biogeography*. Memoir of the Geological Society, London, **12**, 97–104.
—— & TOGHILL, P. 1973. The biostratigraphy of the Silurian rocks of the Girvan district, Scotland. *Journal of the Geological Society, London*, **129**, 209–243.
——, MCKERROW, W. S. & VAN STAAL, C. R. 1997. The margins of Avalonia. *Geological Magazine*, **134**, 627–636.
COLMAN-SADD, S. P., DUNNING, G. R. & DEC, T. 1992*b*. Dunnage–Gander relationships and Ordovician orogeny in central Newfoundland: a sediment provenance and U/Pb age study. *American Journal of Science*, **292**, 317–355.
——, STONE, P., SWINDEN, H. S. & BARNES, R. P. 1992*a*. Parallel geological development of Newfoundland and the Lower Palaeozoic terranes of southern Scotland: an assessment. *Transactions of the Royal Society of Edinburgh: Earth Sciences*, **83**, 571–594.
COPE, J. C. W., INGHAM, J. K. & RAWSON, P. F. (eds) 1992. *Atlas of Palaeogeography and Lithofacies*. Memoir of the Geological Society, London, **13**.
CRAIG, G. Y. & WALTON, E. K. 1959. Sequence and structure in the Silurian rocks of Kirkcudbrightshire. *Geological Magazine*, **96**, 209–220.
CURRY, G. B., BLUCK, B. J., BURTON C. J., INGHAM, J. K., SIVETER, D. J. & WILLIAMS, A. 1984. Age, evolution and tectonic history of the Highland Border Complex, Scotland. *Transactions of the Royal Society of Edinburgh: Earth Sciences*, **75**, 113–133.
FITCHES, W. R., BARNES, R. P. & MORRIS, J. H. 1999. Geological structure and tectonic evolution of the Lower Palaeozoic rocks of the Isle of Man. *This volume*.
FLOYD, J. D. 1976. *The Ordovician rocks of West Nithsdale*. PhD Thesis, University of St Andrews.
—— 1982. Stratigraphy of a flysch succession: the Ordovician of W Nithsdale, SW Scotland. *Transactions of the Royal Society of Edinburgh: Earth Sciences*, **73**, 1–9.
FLOYD, J. D. & RUSHTON, A. W. A. 1993. Ashgill greywackes in the Southern Uplands of Scotland: an extension of the Ordovician succession in the Northern Belt. *Transactions of the Royal Society of Edinburgh: Earth Sciences*, **84**, 79–85
——, STONE, P., BARNES, R. P. & LINTERN, B. C. 1987. Constraints on the significance of the Orlock Bridge Fault within the Scottish Southern Uplands. *Transactions of the Royal Society of Edinburgh: Earth Sciences*, **78**, 219–221.
GIBBONS, W. & GAYER, R. A. 1985. British Caledonian Terranes. *In*: GAYER, R A (ed). *The Tectonic Evolution of the Caledonian–Appalachian Orogen*. Friedr Vieweg and Sohn, 3–16.
GRAY, D. R. 1981. Cleavage-fold relationships and their implications for transected folds: an example from south-west Virginia, USA. *Journal of Structural Geology*, **3**, 265–277.
HALLIDAY, A. N., STEPHENS, W. E. & HARMON, R. S. 1980. Rb–Sr and O isotopic relationships in 3 zoned Caledonian granitic plutons, Southern Uplands, Scotland: evidence for varied sources and hybridization of magmas. *Journal of the Geological Society, London*, **137**, 329–348.
HEPWORTH, B. C. 1981. *Geology of the rocks between Leadhills and Abington, Lanarkshire*. PhD Thesis, University of St Andrews.
HOWE, M. P. A. 1999. The Silurian fauna (graptolite and nautiloid) of the Niarbyl Formation, Isle of Man. *This volume*.
HUGHES, R. A. & KOKELAAR, P. 1993. The timing of Ordovician magmatism in the English Lake District and Cross Fell inliers. *Geological Magazine*, **130**, 369–377.
——, COOPER, A. H. & STONE, P. 1993. Structural evolution of the Skiddaw Group (English Lake District) on the northern margin of eastern Avalonia. *Geological Magazine*, **130**, 621–629.
INGHAM, J. K., MCNAMARA, K. J. & RICKARDS, R. B. 1978. The Upper Ordovician and Silurian rocks. *In*: MOSELEY, F. (ed.) *The Geology of the Lake District*. Occasional Publication of the Yorkshire Geological Society, **3**, 121–145.
IRVINE, D. R. 1878. *Explanation of Sheet 2, Wigtownshire*. Memoir of the Geological Survey, Scotland.
KELLING, G. 1961. The stratigraphy and structure of the Ordovician rocks of the Rhinns of Galloway. *Quarterly Journal of the Geological Society of London*, **117**, 37–75.
KEMP, A. E. S. 1987. Tectonic development of the Southern Belt of the Southern Uplands accretionary complex. *Journal of the Geological Society, London*, **144**, 827–838.
—— 1991. Mid Silurian pelagic and hemipelagic sedimentation and palaeoceanography. *In*: BASSETT, M. G., LANE, P. D. & EDWARDS, D. (eds) *The Murchison symposium. Proceedings of an international conference on the Silurian system*. Special Papers in Palaeontology, **44**, 261–299.
KNELLER, B. C. 1991. A foreland basin on the southern

margin of Iapetus. *Journal of the Geological Society, London*, **148**, 207–210.
——, KING, L. M. & BELL, A. M. 1993. Foreland basin development and tectonics on the northwest margin of eastern Avalonia. *Geological Magazine*, **130**, 691–697.
——, SCOTT, R. W., SOPER, N. J., JOHNSON, E. W. & ALLEN, P. M. 1994. Lithostratigraphy of the Windermere Supergroup, Northern England. *Geological Journal*, **29**, 219–240.
KOKELAAR, B. P. 1986. Petrology and geochemistry of the Rhobell Volcanic Complex: amphibole-dominated fractionation at an early Ordovician arc volcano in North Wales. *Journal of Petrology*, **27**, 887–914.
LAMPLUGH, G. W. 1903. *The Geology of the Isle of Man.* Memoir of the Geological Survey, UK. HMSO.
LAPWORTH, C. 1876. On Scottish Monograptidæ. *Geological Magazine*, **3**, 308, 350, 499 and 544.
—— 1878. The Moffat Series. *Quarterly Journal of the Geological Society of London*, **34**, 240–346.
—— 1889. On the Ballantrae rocks of the south of Scotland and their place in the upland sequence. *Geological Magazine*, **26**, 20–24 and 59–69.
—— & WILSON, J. 1871. On the Silurian Rocks of the counties of Roxburgh and Selkirk. *Transactions of the Edinburgh Geological Society*, **2**, 46.
LEGGETT, J. K., MCKERROW, W. S. & EALES, M. H. 1979. The Southern Uplands of Scotland: a Lower Palaeozoic accretionary prism. *Journal of the Geological Society, London*, **136**, 755–770.
LINTERN, B. C. & FLOYD, J. D. 1999. The Kirkcudbright–Dalbeattie district – a concise account of the geology. *Memoir of the British Geological Survey*, Sheets 5W, 5E and part of 6W (Scotland), in press.
——, BARNES, R. P. & STONE, P. 1992. Discussion on Silurian and early Devonian sinistral deformation of the Ratagain Granite, Scotland: constraints on the age of Caledonian movements on the Great Glen system. *Journal of the Geological Society, London*, **149**, 858.
MCKERROW, W. S. & ELDERS, C. F. 1989. Movements on the Southern Uplands fault. *Journal of the Geological Society, London*, **146**, 393–395.
——, LEGGETT, J. K. & EALES, M. H. 1977. Imbricate thrust model of the Southern Uplands of Scotland. *Nature*, **267**, 237–239.
MOLYNEUX, S. G. 1979. New evidence for the age of the Manx Group, Isle of Man. *In:* HARRIS, A. L., HOLLAND, C. H. & LEAKE, B. E. (eds) *Caledonides of the British Isles: reviewed.* Geological Society, London, Special Publications, **8**, 415–421.
—— 1999. A reassessment of Manx Group acritarchs, Isle of Man. *This volume.*
MORRIS, J. H. 1987. The Northern Belt of the Longford-Down Inlier, Ireland and Southern Uplands, Scotland: an Ordovician back-arc basin. *Journal of the Geological Society, London*, **144**, 773–786.
——, WOODCOCK, N. H. & HOWE, M. P. A. 1999. The Silurian succession of the Isle of Man: the late Wenlock Niarbyl Formation, Dalby Group. *This volume.*
ORR, P. J., HOWE, M. P. A. 1999. Macrofauna and ichnofauna of the Manx Group (Early Ordovician) Isle of Man. *This volume.*
PEACH, B. N. & HORNE, J. 1899. *The Silurian Rocks of Britain, Volume 1. Scotland.* Memoir of the Geological Survey of the United Kingdom.
PHILLIPS, E. R. 1992. *Microfabric analysis of a series of sheared metasandstones exposed within the Moniaive Shear Zone, Southern Uplands, Scotland.* British Geological Survey Technical Report WH/92/45.
—— 1994. *Microstructural study of the Moniaive Shear Zone, Southern Uplands, Scotland.* British Geological Survey Technical Report WG/94/2.
PHILLIPS, E. R., BARNES, R. P., MERRIMAN, R. J. & FLOYD, J. D. 1995. The tectonic significance of Ordovician basic igneous rocks in the Southern Uplands, southwest Scotland. *Geological Magazine,* **132**, 549–556.
——, ——, BOLAND, M. P., FORTEY, N. J. & MCMILLAN, A. A. 1995. The Moniaive Shear Zone: a major zone of sinistral strike-slip deformation in the Southern Uplands of Scotland. *Scottish Journal of Geology*, **31**, 139–149.
PICKERING, K. T. & SMITH, A. G. 1995. Arcs and backarc basins in the Early Paleozoic Iapetus Ocean. *The Island Arc*, **4**, 1–67.
PLANT, J. A., STONE, P. & MENDUM, J. R. 1999. Regional geochemistry, terrane analysis and metallogeny in the British Caledonides. *In*: RYAN, P. & MACNIOCAILL, C. (eds) *Continental Tectonics.* Geological Society, London, Special Publications, in press.
POWER, G. M. & CROWLEY, S. F. 1999. Petrological and geochemical evidence for the tectonic affinity of the (?)Ordovician Poortown Basic Intrusive Complex, Isle of Man. *This volume.*
ROBERTSON, A. H. F. & HENDERSON, W. G. 1984. Geochemical evidence for the origins of igneous and sedimentary rocks of the Highland Border, Scotland. *Transactions of the Royal Society of Edinburgh: Earth Sciences*, **75**, 135–150.
RUSHTON, A. W. A. 1993. Graptolites from the Manx Group. *Proceedings of the Yorkshire Geological Society*, **49**, 259–262.
——, STONE, P. & HUGHES, R. A. 1996. Biostratigraphical controls of thrust models for the Southern Uplands of Scotland. *Transactions of the Royal Society of Edinburgh: Earth Sciences*, **86**, 137–152.
SANDERSON, D. J., ANDERSON, T. B. & CAMERON, T. D. J. 1985. Strain history and the development of transecting cleavage, with examples from the Caledonides of the British Isles (Abstract). *Journal of Structural Geology*, **7**, 498.
SIMPSON, A. 1963. The stratigraphy and tectonics of the Manx Slates Series. *Quarterly Journal of the Geological Society, London*, **119**, 367–400.
SMELLIE, J. L. & STONE, P. 1992. Geochemical control on the evolutionary history of the Ballantrae Complex, SW Scotland, from comparisons with recent analogues. *In*: PARSON, L. M., MURTON, B. J. & BROWNING, P. (eds) *Ophiolites and their Modern Oceanic Analogues.* Geological Society, London, Special Publications, **60**, 171–178.
SOPER, N. J. & KNELLER, B. C. 1990. Cleaved microgranite dykes of the Shap swarm in the

Silurian of NW England. *Geological Journal*, **25**, 161–170.
——, STRACHAN, R. A., HOLDSWORTH, R. E., GAYER, R. A. & GREILING, R. O. 1992. Sinistral transpression and the Silurian closure of Iapetus. *Journal of the Geological Society, London*, **149**, 871–880.
STOCKMAL, G. S., COLMAN-SADD, S. P., KEEN, C. E., MARILLIER, F., O'BRIEN, S. J. & MCQUINLAN, G. M. 1990. Deep seismic structure and plate tectonic evolution of the Canadian Appalachians. *Tectonics*, **9**, 45–62.
STONE, P. 1995. Geology of the Rhins of Galloway district. *Memoir of the British Geological Survey*, sheets 1 and 3 (Scotland).
—— & EVANS, J. A. 1995. Nd isotope study of provenance patterns across the British sector of the Iapetus suture. *Geological Magazine*, **132**, 571–580.
——, COOPER, A. H. & EVANS, J. A. 1999. The Skiddaw Group (English Lake District) reviewed: early Palaeozoic sedimentation and tectonism at the northern margin of Avalonia. *This volume*.
—— & RUSHTON, A. W. A. 1983. Graptolite faunas from the Ballantrae ophiolite complex and their structural implications. *Scottish Journal of Geology*, **19**, 297–310.
—— & SMELLIE, J. L. 1988. *Classical areas of British geology: the Ballantrae area: description of the solid geology of parts of 1:25 000 sheets NX08, 18 and 19*. HMSO (for British Geological Survey).
—— & —— 1990. The Ballantrae ophiolite, Scotland: an Ordovician island arc – marginal basin assemblage. *In*: MALPAS, J., MOORES, E. M., PANAYIOTOU, A. & XENOPHONTOS, C. (eds) *Ophiolites: Oceanic Crustal Analogues*. Procedings of the symposium 'TROODOS 1987', Geological Survey Department, Nicosia, Cyprus, 536–536.
——, FLOYD, J. D., BARNES, R. P. & LINTERN, B. C. 1987. A sequential back-arc and foreland basin thrust duplex model for the Southern Uplands of Scotland. *Journal of the Geological Society, London*, **144**, 753–764.
——, GREEN, P. M., LINTERN, B. C., SIMPSON, P. R. & PLANT, J. A. 1993. Regional geochemical variation across the Iapetus Suture zone: tectonic implications. *Scottish Journal of Geology*, **29**, 113–121.
STRINGER, P. & TREAGUS, J. E. 1980. Nonaxial planar S1 cleavage in the Hawick Rocks of the Galloway area, Southern Uplands, Scotland. *Journal of Structural Geology*, **2**, 317–331.
STYLES, M. T., STONE, P. & FLOYD, J. D. 1989. Arc detritus in the Southern Uplands of Scotland: mineralogical characterisation of a 'missing' terrane. *Journal of the Geological Society, London*, **146**, 397–400.
SWINDEN, H. S., JENNER, G. A., KEAN, B. F. & EVANS, D. T. W. 1989. Volcanic rock geochemistry as a guide for massive sulphide exploration in central Newfoundland. *Current Research*. Newfoundland Department of Mines, Geological Survey of Newfoundland, Report 89–1, 201–219.
THIRLWALL, M. F. 1981. Implications for Caledonian plate tectonic models of chemical data from volcanic rocks of the British Old Red Sandstone. *Journal of the Geological Society, London*, **138**, 123–138.
—— & BLUCK, B. J. 1984. Sr–Nd isotope and geochemical evidence that the Ballantrae 'ophiolite', SW Scotland, is polygenetic. *In*: GASS, I. G., LIPPARD, S. J. & SHELTON, A. W. (eds) *Ophiolites and Oceanic Lithosphere*. Geological Society, London, Special Publications, **13**, 215–230.
TOGHILL, P. 1970. The south-east limit of the Moffat Shales in the upper Ettrick Valley region, Selkirkshire. *Scottish Journal of Geology*, **6**, 233–242.
TORSVIK, T. H. & TRENCH, A. 1991. The Ordovician history of the Iapetus Ocean in Britain: new paleomagnetic constraints. *Journal of the Geological Society, London*, **148**, 423–425.
——, SMETHURST, M. A., BRIDEN, J. C. & STURT, B. A. 1990. A review of Palaeozoic paleomagnetic data from Europe and their palaeogeographic implications. *In*: MCKERROW, W. S. & SCOTESE, C. R. (eds) *Palaeozoic Palaeogeography and Biogeography*. Memoir of the Geological Society, London, **12**, 25–41.
TUCKER, R. D. & MCKERROW, W. S. 1995. Early Palaeozoic chronology: a review in light of new U–Pb zircon ages from Newfoundland and Britain. *Canadian Journal of Earth Sciences*, **32**, 368–379.
WALTON, E. K. 1955. Silurian greywackes in Peeblesshire. *Proceedings of the Royal Society of Edinburgh, Section B*, **65**, 327–357.
WILLIAMS, A. 1959. A structural history of the Girvan district, south-west Ayrshire. *Transactions of the Royal Society of Edinburgh*, **63**, 629–667.
—— 1962. The Barr and Lower Ardmillan Series (Caradoc) of the Girvan district, south-western Ayrshire, with descriptions of the brachiopoda. *Memoir of the Geological Society of London*, **3**.
WILLIAMS, T. M., HENNEY, P. J., STONE, P. & LINTERN, B. C. 1996. Rare earth element geochemistry of Lower Palaeozoic turbidites in the British trans-Iapetus zone: provenance patterns and basin evolution. *Scottish Journal of Geology*, **32**, 1–8.
WOODCOCK, N. H., MORRIS J. H., QUIRK, D. G. ET AL. 1999. Revised lithostratigraphy of the Manx Group, Isle of Man. *This volume*.
YOUNG, G. C. 1990. Devonian vertebrate distribution patterns and cladistic analysis of palaeo-geographic hypotheses. *In*: MCKERROW, W. S. & SCOTESE, C. R. (eds) *Palaeozoic Palaeogeography and Biogeography*. Memoir of the Geological Society, London, **12**, 243–255.

The Skiddaw Group (English Lake District) reviewed: early Palaeozoic sedimentation and tectonism at the northern margin of Avalonia

P. STONE[1], A. H. COOPER[2] & J. A. EVANS[3]

[1]*British Geological Survey, Murchison House, West Mains Road, Edinburgh EH9 3LA, UK*

[2]*British Geological Survey, Keyworth, Nottingham NG12 5GG, UK*

[3]*NERC Isotope Geoscience Laboratory, Keyworth, Nottingham NG12 5GG, UK*

Abstract: The Skiddaw Group is a 5 km thick sequence of Tremadoc–Llanvirn turbiditic mudstone and sandstone, including a major olistostrome, which occupies the northern part of the Lake District Lower Palaeozoic Inlier. Sporadic outcrop and borehole records indicate that similar strata extend beneath other parts of northern England. To the west, the Manx Group of the Isle of Man is a regional correlative. The Skiddaw Group was deposited on the Avalonian margin of the Iapetus Ocean, with constituent sediment derived largely from an earlier, possibly Precambrian, continental margin volcanic arc. Nd isotope data confirm the absence of juvenile detritus. Olistostrome emplacement in the late Arenig preceded subduction-related uplift of the deep-marine Skiddaw Group to form the subaerial basement to the mainly Caradoc, Borrowdale and Eycott volcanic groups. The scale of the unconformity beneath the volcanic rocks requires considerable pre-volcanic disruption and erosion of the Skiddaw Group prior to structural disturbance by volcanotectonic faulting. Volcanism ended in the late Caradoc when thermal re-equilibration, coupled with possible further extension, allowed marine transgression through the early Silurian. Ultimately, convergence of Avalonia with Laurentia initiated thrust imbrication of the Skiddaw Group as the Southern Uplands thrust belt extended across the sutured Iapetus Ocean. Thrust-related hydration caused widespread resetting of Rb–Sr isotope systems during the 430–420 Ma interval. A penetrative slaty cleavage with a broadly Caledonian trend was imposed during the Early Devonian, Acadian Orogeny and cuts an earlier, bedding-parallel (compaction) fabric. Later phases of Acadian compression probably involved reactivation of thrusts within the Skiddaw Group with associated strain partitioning resulting in domainal crenulation cleavage. Granite intrusion at *c.* 400 Ma coincided with the final cleavage episode.

The small, early Palaeozoic, palaeocontinent of Avalonia rifted from the northern margin of the larger Gondwana continent in the early Ordovician and drifted north as the early Palaeozoic Iapetus Ocean closed. Its extent has been reviewed recently by Cocks *et al.* (1997 and refs cited therein). The northeast corner of Avalonia now forms a triple junction with Laurentia and Baltica, westwards from which its northern margin defines the southern side of the Iapetus Suture Zone, the line of collision with Laurentia and the surface trace of the vestigal Iapetus Ocean. The suture zone can be traced from northern England and across Ireland (the Isle of Man lies immediately to the south of the suture; Fig. 1) to reappear in the Canadian provinces of Newfoundland, Nova Scotia and New Brunswick. Further southwest, in the USA, it clips the coast of Massachusetts and Connecticut.

In northwest England, the Lake District Lower Palaeozoic Inlier reveals part of the northern margin of the Avalonian continent including the Skiddaw Group, a sequence of possibly Cambrian–early Llanvirn (*sensu* Fortey *et al.* 1995) turbiditic mudstone and sandstone up to 5 km thick (Cooper *et al.* 1995). The largest Skiddaw Group outcrop (*c.* 480 km^2) is in the northern part of the Lake District with smaller inliers in the southern and eastern Lake District, and further east at Cross Fell and Teesdale. There are borehole records of similar rock beneath Upper Palaeozoic strata over a wide part of northern England. The distribution of the Skiddaw Group is illustrated in Fig. 1. Correlative strata may occur to the south, in the Ingleton Group of the Craven inliers, although significant differences in sandstone provenance between the Skiddaw and Ingleton Groups have been reported

From: Woodcock, N. H., Quirk, D. G., Fitches, W. R. & Barnes, R. P. (eds) 1999.
In Sight of the Suture: the Palaeozoic geology of the Isle of Man in its Iapetus Ocean context.
Geological Society, London, Special Publications, **160**, 325–336. 1-86239-046-0/99/$15.00

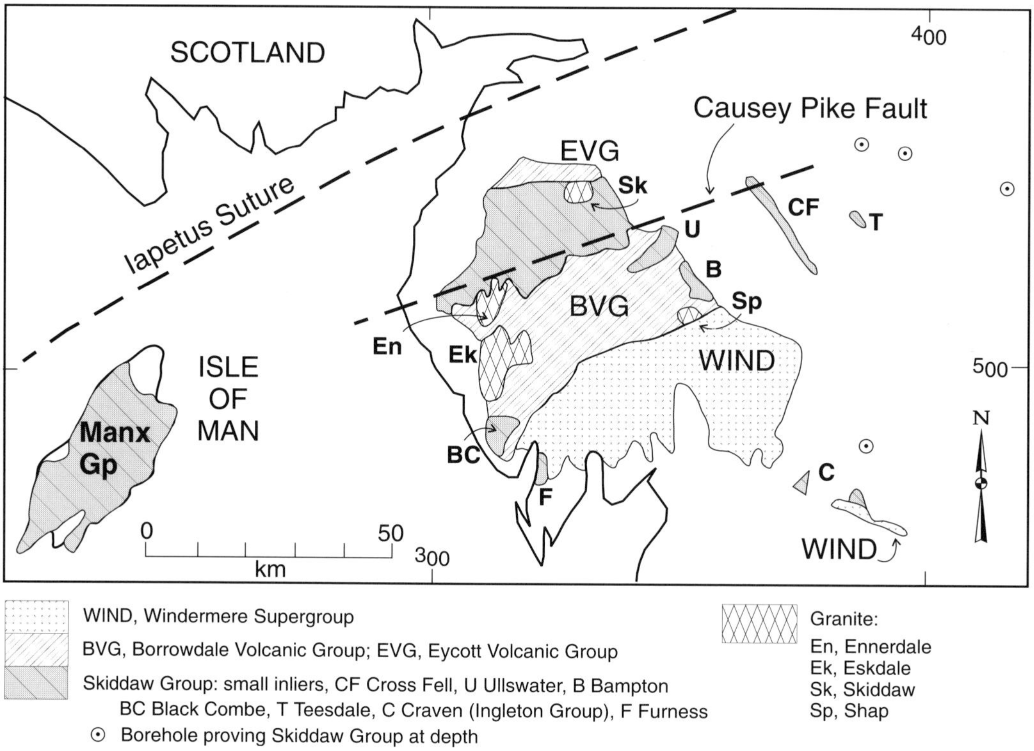

Fig. 1. Outline geology for the Lake District, northwest England, showing the distribution of the Skiddaw Group and related rocks.

(Stone & Evans 1997). The Manx Group in the Isle of Man shares many characteristics with the Skiddaw Group, as do various Lower Palaeozoic sequences in the Bellewstown and Leinster Terranes of southeast Ireland (Cooper *et al.* 1995 and refs cited therein). In the Canadian Atlantic provinces most of the Iapetean sedimentary sequences preserved at the Avalonian margin are older than the Skiddaw Group but one example, the St John Group of southern New Brunswick, ranges up to the early Ordovician with the black shales of the Reversing Falls Formation. These shales were deposited on a deep-marine shelf (Tanoli & Pickerill 1988) and so the age, lithology and likely depositional environment invite broad comparison with the Skiddaw and Manx Groups.

The Skiddaw Group has been the focus for much recent survey and research work and a coherent bio- and lithostratigraphy has now been established (Cooper *et al.* 1995). This in turn has allowed the structural understanding to advance (Hughes *et al.* 1993) so that the geological evolution of the group is now known in some considerable detail. This paper identifies and discusses those large-scale geotectonic events that can be confidently identified in the Skiddaw Group and which are likely to reflect the regional development of the Avalonian margin. Their effects should be recognizable in strata correlative with the Skiddaw Group and so they may have application as an interpretational template for nearby sequences such as the Manx Group of the Isle of Man.

The depositional basin

The key to development of a coherent lithostratigraphy for the Skiddaw Group (Fig. 2) was the establishment of reliable biostratigraphical control (Cooper *et al.* 1995 and refs cited therein). Two distinct stratigraphies are apparent in the main Lake District inlier, on either side of the Causey Pike Fault. To the north of this structure, in the Northern Fells Belt of Cooper *et al.* (1995), are preserved some 5 km of mainly mudstone turbidites that were deposited between the Tremadoc and the early Llanvirn, although there is some evidence that sedimentation may have commenced in the Cambrian

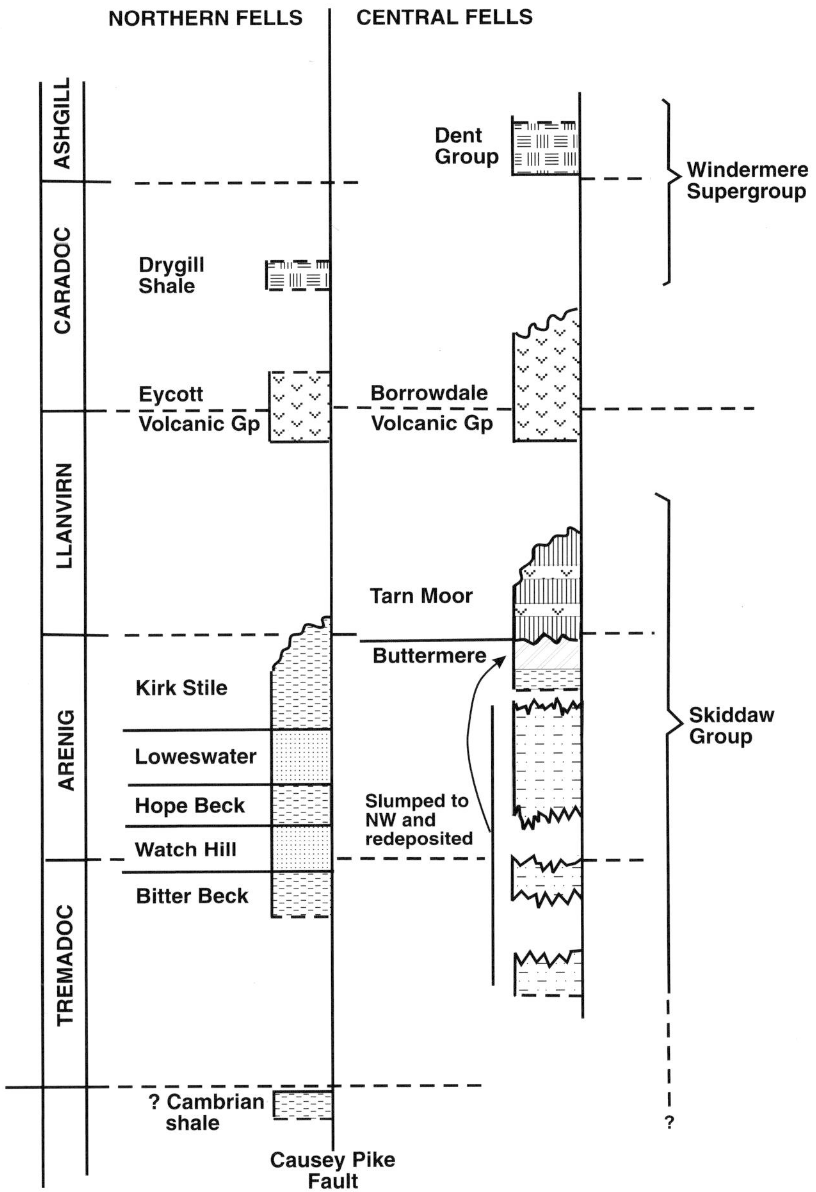

Fig. 2. A summary of Skiddaw Group stratigraphy. Unit names with no hierarchical status shown are formations. Modified from Cooper *et al.* (1995).

(Millward & Molyneux 1992). Wacke-type sandstone beds occur sporadically throughout the succession but become dominant at two levels, the Loweswater and Watch Hill Formations. South of the Causey Pike Fault, in the Central Fells Belt of Cooper *et al.* (1995), a major olistostrome, the Buttermere Formation, was emplaced from the south during the late Arenig and is overlain by the late Arenig–Llanvirn siltstones of the Tarn Moor Formation.

The thickness, duration and geographical extent of the Skiddaw Group successions suggest deposition in a large basin with a long history of subsidence. The provenance was deduced by Cooper *et al.* (1995) to be largely an old, inactive, continental volcanic arc lying to the southeast; Nd isotope results reported by Stone & Evans (1997) confirm the absence of a large juvenile component in the sandstones (εNd = –4.1 to –9.3). An extensional passive margin therefore seems a more likely site for deposition than an inter-arc or back-arc zone. In this apparent absence of coeval volcanism until late in its depositional history, the Skiddaw Group contrasts with the Manx Group which contains the interbedded Peel volcanic assemblage, described by Simpson (1963) as andesitic lava, tuff and agglomerate. The Manx Group siltstones adjacent to the volcanic rocks contain an Arenig microflora (Molyneux 1979; Cooper *et al.* 1995). One Nd isotope result of εNd = +2.1 from a sandstone (Stone & Evans 1997) has been interpreted as indicating a juvenile contribution to part of the Manx Group [referable to the Maughold Banded Group of Simpson (1963)]. However, this record proves to have been obtained from the vicinity of a large and highly altered felsic dyke and the result should be regarded as provisional until confirmed by further work.

Some idea of the possible basin configuration is given by Webb & Cooper (1988) from an examination of the widespread slump folds. Within the Northern Fells belt these are predominantly orientated towards the southeast and, although the origin of the larger folds has been debated (Hughes *et al.* 1993), there remains evidence of a likely southeast palaeoslope, at least during the Tremadoc and early Arenig. Conversely, south of the Causey Pike Fault in the Central Fells Belt, the Buttermere Formation is a large olistostrome emplaced towards the northwest; only relatively thin debris flow beds are seen in the Northern Fells Belt at an equivalent stratigraphical level. From the contrasting scale, style and orientation of the slump movements, Webb & Cooper (1988) deduced a steeper, faulted southerly margin to the depositional basin.

An extensive analysis of palaeocurrent evidence led Moore (1992) to propose two periods of submarine fan development (essentially the Watch Hill and Loweswater Formations) with different basin configurations for each. During deposition of the first (Watch Hill), axial flow was inferred along a trough orientated approximately east–west. Subsequent deposition (Loweswater) was spatially more complex, with greater influence by seafloor topography. The latter was thought to arise from syn-depositional extensional faults trending northwest–southeast within the depositional basin, with the intervening fault blocks tilted to the northeast.

The overall picture that emerges is of a north facing extensional half-graben system with the major boundary fault (or faults) on the southeastern side. The stratigraphical contrast across the Causey Pike Fault would have been aided if the basin was composite and so perhaps that structure was initiated as a northwest downthrowing normal fault partitioning the Skiddaw Group depositional basin.

Further evidence in support of an extensional basin model is provided by the illite crystallinity and clay mineral assemblages of the Skiddaw Group mudstones. Fortey *et al.* (1993) and Merriman & Frey (1999) suggest that early burial metamorphism, characterized by late diagenetic to low anchizonal grades, occurred under a higher geothermal gradient (> 35°C km^{-1}) than would normally be expected in an ensialic basin. They related this to high heat flow arising from extension and crustal thinning. Similar results have been reported from the Manx Group (Roberts *et al.* 1990). However, there must remain some uncertainty as to the relationship of burial metamorphism in the depositional basin during its putative extension (Tremadoc–Llanvirn), and that during subsequent suprasubduction zone basin uplift and volcanicity (Llanvirn–Caradoc; discussed below) when a high geothermal gradient would also be expected.

Finally, it is worth noting the similarities described by Plant *et al.* (1991) between some aspects of the regional geochemical characteristics of the Skiddaw Group and those of the Argyll and Southern Highland Groups in the late Proterozoic–early Palaeozoic Dalradian Supergroup of the Scottish Highlands. Both the Skiddaw Group and the upper Dalradian show enhanced levels of the gold pathfinder elements (As, Sb and Bi) and both sequences were considered to have acted as crustal reservoirs for later ore-forming processes. The geotectonic setting of the Dalradian is well established as an ensialic, tectonically controlled extensional basin (Anderton 1982) and a similar environment was deduced, by analogy, for the Skiddaw Group. Nevertheless, subsequent differences in terrane evolution were stressed: the Dalradian basin continued to extend until oceanic-style volcanism occurred, whereas the Skiddaw Group basin was closed and inverted above a developing subduction zone to become the foundations of a continental margin volcanic arc.

Basin uplift and volcanism

The first indications of volcanic activity during Skiddaw Group deposition are seen in the early Llanvirn (Cooper *et al.* 1995), with sporadic interbeds of volcaniclastic turbidite sandstone and bentonite ash in the Tarn Moor Formation of the Central Fells Belt (Fig. 2). The abundance and thickness of volcanic interbeds appear to increase upwards but Nd isotope data for the mudstones show that the juvenile component must be restricted to the discrete volcaniclastic beds since the background provenance remained relatively ancient. The values of εNd (calculated at a depositional age of 480 Ma) range from –8 to –10.6, equivalent to depleted mantle model ages of 1.52–2.01 Ga, which are compatible with derivation from a Proterozoic source. It must be stressed that these isotope data give an average value for all the components of the rock and do not relate to a single unique source. Nevertheless, they rule out any significant juvenile contribution to the Tarn Moor Formation mudstones. The Llanvirn initiation of volcanism therefore seems to have been either at some distance from the depositional basin or to have involved relatively small-scale and localized eruptions.

The onset of volcanicity in the early Llanvirn followed the late Arenig emplacement of the major Buttermere Formation olistostrome (Hughes & Kokelaar 1993; Cooper *et al.* 1995). It confirms that subduction of oceanic crust was initiated beneath the Avalonian continental margin by Llanvirn times and seismic activity, as a precursor to the volcanic episode, may have triggered the mass-flow movements. However, the olistostrome was clearly emplaced by downslope movement into a still-extant basin and so it may equally have been instigated by normal, extensional movement on the basin boundary fault(s). Whatever the trigger mechanism, the extensive slumping throughout much of the Skiddaw Group succession caused considerable stratigraphical disruption. Further disruption was inevitable during subduction-related uplift of the continental margin and inversion of the Skiddaw Group basin. Uplift was most likely caused by the generation and rise of andesitic melts above the subducted oceanic crust (Branney & Soper 1988; Hughes *et al.* 1993). It resulted in the deep-marine basinal strata being converted into the subaerial basement to the ensuing, mainly Caradoc, Borrowdale and Eycott volcanic groups.

The magnitude of the pre-volcanic stratigraphical disruption, caused by the combination of gravity driven, mass-slump movement and the subsequent basin inversion, may be gauged by the wide range of biostratigraphical zones determined immediately subjacent to the overstepping unconformity. At the southern margin of the Skiddaw Group outcrop, below the Borrowdale Volcanic Group, various Arenig–Llanvirn biostratigraphic levels occur close to the unconformity cut across the olistostrome and the overlying Tarn Moor Formation (Cooper & Hughes 1993). Further north, the Skiddaw Group strata immediately subjacent to the Eycott Volcanic Group range in age from possibly Cambrian to Llanvirn (Millward & Molyneux 1992), although there is stratigraphical coherence over wide areas of the outcrop. Much of the variation in the north therefore seems likely to have been caused by fault-block rotation prior to volcanicity. Rotation could have been either an extensional or compressional effect and was probably superimposed both on disrupted zones caused by the earlier slump movements and on areas which had escaped such disruption.

It should be stressed that there is no evidence for a compressive tectonic event producing penetrative deformation prior to the eruption of the volcanic rocks. This has been a long-running controversy in Lake District geology and is reviewed by Hughes *et al.* (1993). The earliest regional cleavage cuts both the Skiddaw Group and overlying volcanic and sedimentary formations which range up to Pridoli in age. It is a product of the early Devonian, Acadian Orogeny, which is discussed in more detail below. The only pre-volcanic, but post-depositional, fabric present in the Skiddaw Group is a widespread bedding-parallel compaction cleavage. This is developed particularly well in the most argillaceous lithologies and probably results from burial accentuation of the original bedding lamination either prior to, or during, basin uplift. The high heat flow thought to have been prevalent at the time (Merriman & Frey 1999; see also above) may have assisted the formation of this fabric by accelerating recrystallization of the clay minerals (Merriman & Peacor 1999). There is no indication that any penetrative tectonic fabric was imposed during pre-volcanic uplift and stratal disorganization. This situation has, in the past, given rise to dispute; Simpson (1967) and Helm (1970) proposed mid-Ordovician orogenesis, but Soper (1970) and Soper & Roberts (1971) established an early Devonian age for formation of the first tectonic cleavage.

It is now thought that the volcanic rocks that occur on either side of the main Skiddaw Group inlier, the Eycott Volcanic Group to the north and the Borrowdale Volcanic Group to the south (Figs 1 and 2) are penecontemporaneous and largely Caradoc in age (Millward & Molyneux 1992). Previously, the Eycott Volcanic Group was thought to be significantly older than the Borrowdale Volcanic Group on the basis of an acritarch flora described from sedimentary interbeds at the base of the former by Downie & Soper (1972). Both of the volcanic groups had a continental margin, supra-subduction zone origin which requires an arc-trench gap, probably exceeding 100 km of fore-arc, extending to the north; a situation supportive of an ensialic setting for the Skiddaw Group basin. A network of volcanotectonic, caldera-related faults disrupts the Borrowdale Volcanic Group (Branney & Soper 1988) with proved displacement commonly in excess of 400 m (Branney & Kokelaar 1994). It is highly probable that similar structures would have affected those parts of the Skiddaw

Group currently exposed with reactivation of the existing fault framework. Major structures, such as the Causey Pike Fault, would have been a likely focus for such movement.

The same pre-existing fault network may then have experienced further, post-volcanic reactivation as thermal re-equilibration, possibly coupled with renewed regional extension, allowed marine transgression across the eroded and largely extinct volcanic field. The oldest overlying marine strata are Longvillian (late Caradoc) in age (Ingham *et al.* 1978); the Drygill Formation shales, which form a faulted outlier against the Eycott Volcanic Group, and the equivalent Corona and Melmerby beds to the east in the Cross Fell Inlier. Further south, the basal strata of the Dent Group (Kneller *et al.* 1994), which unconformably overlie the southern margin of the Borrowdale Volcanic Group outcrop, are of Ashgill age (Fig. 2). These relationships imply an overall north–south transgression, although on the local scale the pattern is much more irregular (Ingham *et al.* 1978).

Post-volcanic convergence of Avalonia and Laurentia

The late Ordovician and early Silurian rocks of the Windermere Supergroup are of shallow-water to deep-shelf facies and indicate a return to passive margin conditions. Nevertheless, Avalonia continued to drift northwards until its Wenlock collision with Laurentia. This apparent paradox has been linked with the relative brevity, but great intensity, of the Borrowdale–Eycott volcanic episode by suggestions of Caradoc ridge subduction at the Avalonian margin of the Iapetus Ocean by Pickering & Smith (1995). These authors point out that such a situation could have two important outcomes. Firstly, it would allow creation of a window in the subducting slab which would explain the abrupt cessation of volcanicism at the Avalonian margin. Secondly, it would effectively transfer Avalonia on to a north moving plate which was being subducted beneath the Laurentian margin of the ocean.

Continued subduction at the northern, Laurentian margin of the Iapetus Ocean effected continental convergence by the late Llandovery or Wenlock (e.g. Soper *et al.* 1992). The accretionary complex that had developed at the leading edge of Laurentia overrode the Avalonian margin and continued southwards as a foreland fold and thrust belt preceded by a foreland basin. The initiation of the foreland basin in southern Scotland has been discussed by Stone *et al.* (1987); its progression on to Avalonia and across the Lake District is detailed by Kneller (1991) and Kneller *et al.* (1993). The Skiddaw Group was caught up in this process, in which south directed thrusts were the dominant structure, and the principal structures within the main outcrop may have been initiated at this stage (the Watch Hill, Loweswater and Gasgale thrusts; thrusts within the Causey Pike Fault System: Hughes *et al.* 1993, figs 1 and 3).

The timing of thrust propagation through the Skiddaw Group is indicated by the widespread resetting of the Rb–Sr isotopic systems in adjacent igneous rocks. This has been studied in an analogous situation in the Welsh Lower Palaeozoic basin by Evans *et al.* (1995), who concluded that the resetting was chemically controlled during hydration. In the northern Lake District, uplift associated with the thrust development would have opened fractures, increased permeability and facilitated the formation of secondary hydrated minerals, thus resetting the Rb–Sr system. The important age determinations in this debate are summarized in Fig. 3; some are also discussed by Hughes *et al.* (1996).

Biostratigraphical evidence for a Caradoc eruption age for the upper part of the Borrowdale Volcanic Group (Molyneux 1988) and a Sm–Nd garnet–whole rock age of 457 ± 4 Ma (Thirlwall & Fitton 1983) are in broad agreement. Palaeomagnetic results have been interpreted (Piper *et al.* 1997) as showing that the volcanic sequences were erupted during a single normal-polarity chron occupying the early Caradoc. These data contradicted precise Rb–Sr ages of 423 ± 3 and 432 ± 3 Ma which had previously been obtained by Rundle (1987). Contradictory results were also given by the Eskdale and Ennerdale granitic intrusions; Rundle (1979) obtained precise Rb–Sr results of 429 ± 4 and 420 ± 4 Ma, respectively, whereas U–Pb zircon dates obtained by Hughes *et al.* (1996) were 450 ± 3 and 452 ± 4 respectively. The U–Pb dates associated the plutons firmly with the later stages of volcanicism. Such contradictory age determinations have not been obtained from the younger, post-cleavage, Shap and Skiddaw granites. A Rb–Sr age from Shap of 394 ± 3 Ma (Wadge *et al.* 1978) compares with a U–Pb discordia zircon age of 390 ± 4 Ma (Pidgeon & Aftalion 1978). The Skiddaw Granite has given a K–Ar age of 399 ± 8 Ma [Shepherd *et al.* 1976; recalculated by Rundle (1982)] and a Rb–Sr age of 393 ± 5 Ma (Shepherd & Darbyshire 1981); both are within error of a poorly constrained preliminary U–Pb zircon age of 406 ± 12 Ma. Thus, there is no evidence for resetting of the younger granites by either thermal or hydration effects.

A Rb–Sr isochron of 427 ± 34 Ma (Fig. 4), reported by Evans (1996) from late Arenig Skiddaw Group mudstone (Kirk Stile Formation) at Dodd Wood [NY 2363 2785], is also relevant to the

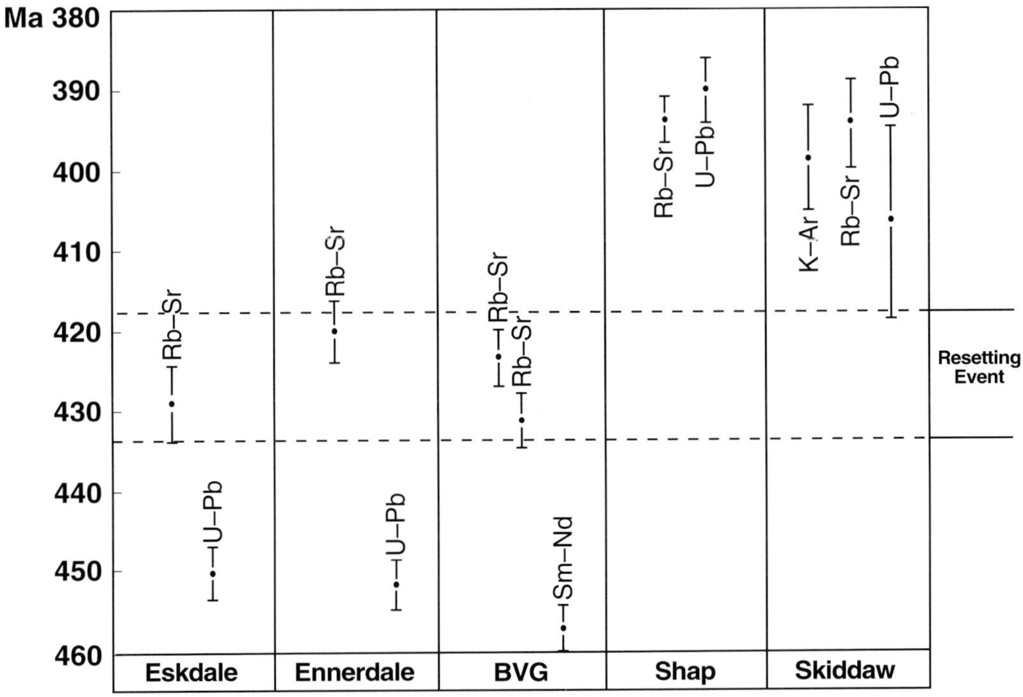

Fig. 3. A summary of radiometric age determinations from igneous rocks in the northern and central Lake District which define the 420–430 Ma resetting event. Sources of data are acknowledged in the text.

resetting debate. Despite the large error there is no overlap with the depositional age of *c.* 480 Ma. In mudstones such resetting is a dehydration reaction and has been linked to re-equilibration of the Rb–Sr system during the illite–smectite transition (Evans *et al.* 1995). Movement of fluids out of the mudstone during a dehydration reaction is thus coincident with fluid movement into the igneous rocks during hydration, both phenomena resetting the Rb–Sr systems to the 420–430 Ma interval.

In the southern Lake District, Rb–Sr resetting has also affected the Stockdale Rhyolite (Yarlside Volcanic Formation, Dent Group; Fig. 2), which crops out over several kilometres strike length near the northern margin of the Windermere Supergroup outcrop, a little to the southwest of the Shap Granite (Fig. 1). The Stockdale Rhyolite has given a Rb–Sr isochron age of 423 ± 4 Ma [Gale *et al.* 1979; reassessed by Rundle (1987)], although it is demonstrably an extrusive ignimbrite (Millward & Lawrence 1985) interbedded with Ashgill strata deposited in the 443–449 Ma interval (Tucker & McKerrow 1995). The Rb–Sr age was used, controversially, by Gale *et al.* (1979) to constrain the Palaeozoic timescale and led to much debate [e.g. Compston *et al.* (1982)]. In common with Hughes *et al.* (1996), the Stockdale Rhyolite is considered here to be another example of 420–430 Ma resetting (Fig. 4), which establishes that the process responsible extended into the southern Lake District.

If the resetting event in the 420–430 Ma interval was indeed caused by a thrust-induced uplift mechanism, that age is important since the process would have post-dated collision between Avalonia and Laurentia, which is thus fixed as Wenlock or earlier. It also means that thrust propagation through the Skiddaw Group immediately preceded the abrupt foreland basin deepening recorded by the Ludlow part of the Windermere Supergroup in the southern Lake District (Kneller 1991). The late Ludlow initiation of thrust detachment at the base of the Windermere Supergroup proposed by Kneller *et al.* (1993) was linked by Hughes *et al.* (1996) to the isotopic resetting but now seems more likely to be a slightly later manifestation of the southward propagating thrust system. Nevertheless, the late Ludlow timing does overlap with the younger end of the 420–430 Ma resetting range and the large-scale dewatering involved could have contributed to the fluid mobility necessary for resetting. A coherent tectonic model thus emerges,

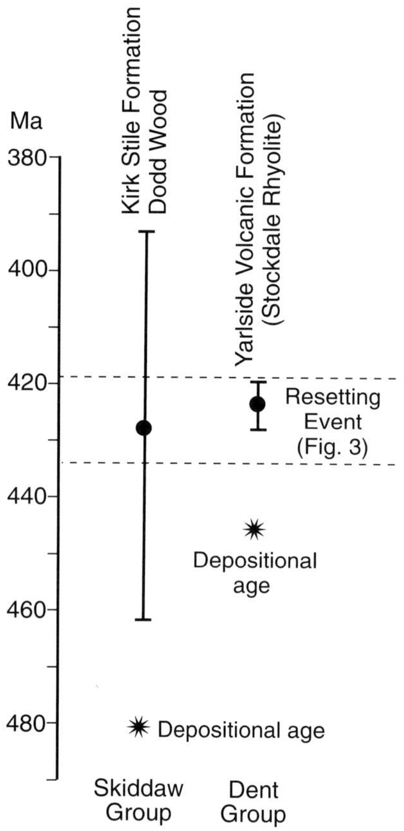

Fig. 4. A comparison of depositional and Rb–Sr ages from well-constrained stratigraphical units in both the northern and southern Lake District. Sources of data are acknowledged in the text.

linking several phases of geological development at the northern margin of Avalonia.

The Acadian Orogeny

The main regional cleavage affecting all rocks from the possibly Cambrian and Tremadoc parts of the Skiddaw Group up to the Pridoli beds at the top of the Windermere Supergroup was imposed during the Acadian Orogeny. The Pridoli ranges from 419 to 417 Ma according to Tucker & McKerrow (1995), so the Acadian event is likely to be 417 Ma or younger in age. The regional cleavage is cut by the Skiddaw Granite, which itself is dated at 399 ± 8 Ma (Rb–Sr; Rundle 1981) and 392 ± 4 Ma (K–Ar; Shepherd *et al.* 1976) (Fig. 3), although subsequent crenulation cleavage post-dates the granite (Soper & Roberts 1971). Formation of the regional slaty cleavage has also been demonstrated as broadly synchronous with intrusion of the Shap Granite (Soper & Kneller 1990) dated at 390 ± 6 Ma (Pidgeon & Aftalion 1978) (Fig. 3). This accumulation of evidence was used by Soper *et al.* (1987) to confirm that the main Acadian slaty cleavage was imposed during the Emsian Stage of the early Devonian. Since then, the main Acadian cleavage has been directly dated further south in the Craven (Ribblesdale) inliers where metamorphic white mica from a Ludlow bentonite gave ages of 397 ± 7 and 418 ± 3 Ma (K–Ar and Ar–Ar, respectively; Merriman *et al.* 1995), extending to somewhat older ages than the Emsian age range of 391–400 Ma proposed by Tucker & McKerrow (1995). Further evidence has been obtained from within the Causey Pike Fault zone where a concealed granite has produced the Crummock Water Aureole, dated at 401 ± 3 Ma (Rb–Sr; Cooper *et al.* 1988), which post-dates the main cleavage (Hughes *et al.* 1993). This also suggests that the age of the regional slaty cleavage formation may have been a little earlier than Emsian. The cause of the orogenic deformation has been suggested by Soper *et al.* (1992) to be the initial impingement of Armorica – Iberia on the southern margin of eastern Avalonia.

The Acadian slaty cleavage forms a regional arc in the Skiddaw Group with a west to east variation in trend between northeast–southwest and east–west (Soper *et al.* 1987). The cleavage is axial planar to gently plunging, steeply inclined, open to isoclinal folds, with amplitudes of hundreds of metres. Its fabric is of penetrative, pressure-solution type in some areas but of spaced, fracture type in others. Where the bedding-parallel compaction fabric is well developed, the regional Acadian cleavage may have the appearance of a crenulation fabric. Transecting fold–cleavage relationships are rare in the Skiddaw Group, in contrast to their widespread occurrance in the neighbouring Lower Palaeozoic slate belts of central Wales, the Southern Uplands of Scotland and the southern Lake District.

In the Skiddaw Group of the northern Lake District the regional Acadian cleavage is commonly crenulated by fabrics which are axial-planar to open, gently plunging minor folds with gently inclined axial planes. The attitude of these folds is variable and more than one generation is apparent with a widespread, gently dipping crenulation plane itself crenulated by later, more variable fabrics. Some of the crenulation fabrics are associated with minor folds related to a set of south directed thrusts (Roberts 1992). Further south, in the Black Combe Inlier (Rushton & Molyneux 1989), the Skiddaw Group has been severely deformed within a major, south directed thrust zone. On the north side of this zone the rocks are intensely cleaved, sheared and metasomatized with much quartz–tourmaline veining (Johnson 1992); south of the thrust zone the combination of slaty and crenulation cleavages is very similar to that seen in the main, northern inlier. However, only a short distance south from Black Combe, in the Furness Inlier, only the regional slaty cleavage is present (Soper 1970). The overall impression is of highly domainal crenulation fabrics imposed on an earlier, regional slaty cleavage during the later phases of the Acadian Orogeny. The pre-existing thrust system would have been reactivated at this time to form the principal domainal boundaries (Hughes *et al.* 1993), but evidence from the Crummock Water Aureole establishes that both sinistral strike-slip and south directed thrust movement continued on the Causey Pike Fault after *c.* 401 Ma, i.e. after the end of cleavage development.

In summary, the products of the earliest, broadly northwest–southeast orientated, Acadian crustal shortening in the Skiddaw Group are the regional cleavage and associated folds. At later stages of the same, approximately Emsian, event further strain increments reactivated a set of southward directed thrusts which formed the domainal boundaries for the development of further minor folds and associated crenulation cleavages. For the most part, the thrusts failed to propagate through the rigid mass of the Borrowdale Volcanic Group and its underpinning batholith, causing increased strain in the Skiddaw Group which was accommodated by more crenulation-related deformation.

Conclusions

This assessment of the regional geotectonic events responsible for shaping the Skiddaw Group has identified four principal phases of activity, discussed below.

The depositional basin (?Cambrian–Llanvirn)

An asymmetric extensional basin developed on the continental margin of Avalonia, probably from the late Cambrian onwards. An initial east–west basin trend may have become more complex later as northwest–southeast extensional faults varied the submarine topography. Slump folds converged from opposite sides of the basin and a large olistostrome was emplaced from the southern (steeper?) margin in the late Arenig. Provenance was from an ancient, deeply incised, continental volcanic arc; there is no evidence for a juvenile component, even in the youngest (early Llanvirn) mudstones which are interbedded with sporadic bentonite volcanic ash layers. At least 5 km of strata accumulated, locally disrupted by the extensive mass-flow movements. Support for sedimentation in an actively extensional environment comes from the illite crystallinity evidence for burial metamorphism in a high-heat flow environment (perhaps responsible for the widespread bedding-parallel compaction fabric) and from some aspects of the geochemical characteristics of the sedimentary succession.

Basin uplift and volcanism (Llanvirn and Caradoc)

During the late Llanvirn the deep-marine Skiddaw Group basinal strata were uplifted to provide the subaerial erosion surface on to which the Caradoc volcanic rocks were erupted. There is a substantial unconformity at the base of the volcanic groups but no evidence to suggest that it is of orogenic proportions. Instead, the stratal disruption seems likely to have been caused by a combination of the large-scale slump movements and fault-block rotation during either extension or subsequent basin inversion. Extensional and volcanotectonic faulting was a widespread feature of the volcanic episode and must have had a profound effect on the underlying Skiddaw Group, further complicating the structural architecture. The intensity and brevity of the volcanism has been linked to the Caradoc subduction of an Iapetus spreading ridge.

Convergence of Avalonia and Laurentia (Ashgill and Silurian)

The ridge subduction process effectively transferred the Avalonian margin to the Iapetus Ocean plate being subducted northwards beneath Laurentia. Following mid-Silurian collision of the two continents, the accretionary thrust complex at the Laurentian margin advanced into the Avalonian foreland. As the thrust front crossed the Skiddaw Group and the adjacent volcanic and intrusive rocks, uplift induced fluid movement which caused resetting of the Rb–Sr isotope systems to the 430–420 Ma interval. This timing is compatible with the development further south, ahead of the thrust front, of the Ludlow foreland basin in the southern Lake District, in which was deposited much of the Windermere Supergroup. The resetting mechanism extended southwards at least as far as the northern margin of the Windermere Supergroup outcrop.

Acadian Orogeny (early Devonian)

A regional slaty cleavage, axial-planar to upright folds, was imposed across the whole region, including the Skiddaw Group, during the early

Devonian. Thereafter, subsequent strain increments were taken up domainally within the Skiddaw Group, between the main thrust planes, by further folding and the development of crenulation cleavages. Renewed south directed thrust movement is apparent during this phase of the deformation, and localized increases in strain may have been caused by the failure of the thrusts to propagate readily through the volcanic massif and underlying batholith. Several generations of crenulation may be present at any one locality in both the main Skiddaw Group outcrop in the northern Lake District and in the Black Combe Inlier to the south. A little further south, in the Furness Inlier, only the regional slaty cleavage is seen, suggesting that the thrust systems did not extend that far.

Each of the events outlined above was a large-scale process which would have affected a considerable length of the Avalonian continental margin. Their effects should therefore be apparent in strata equivalent to the Skiddaw Group, but deposited and deformed at some distance from it. Such rocks comprise the Manx Group of the Isle of Man where sparse stratigraphical control has restricted any definitive geological interpretations so that it remains, arguably, the least understood of the Avalonian margin sequences. The Skiddaw Group provides an evolutionary template which may go some way towards resolving those outstanding difficulties of Manx geology.

We are indebted to many colleagues for their contributions to discussions of Lake District geology, in particular to Richard Hughes; and to the referees, David Millward and Brian McConnell, for their helpful reviews of the text. The paper is published by permission of the Director, British Geological Survey (NERC). NIGL publication number 307.

References

ANDERTON, R. 1982. Dalradian deposition and the late Precambrian–Cambrian history of the N Atlantic region: a review of the early evolution of the Iapetus Ocean. *Journal of the Geological Society, London*, **139**, 421–431.

BRANNEY, M. J. & KOKELAAR, B. P. 1994. Volcanotectonic faulting, soft-state deformation and rheomorphism of tuffs during development of a piecemeal caldera, English Lake District. *Geological Society of America Bulletin*, **106**, 507–530.

—— & SOPER, N. J. 1988. Ordovician volcano-tectonics in the English Lake District. *Journal of the Geological Society, London*, **145**, 367–376.

COCKS, L. R. M., MCKERROW, W. S. & VAN STAAL, C. R. 1997. The margins of Avalonia. *Geological Magazine*, **134**, 627–636.

COMPSTON, W., MCDOUGALL, I. & WYBORN, D. 1982. Possible two-stage ^{87}Sr evolution in the Stockdale Rhyolite. *Earth and Planetary Science Letters*, **61**, 297°302.

COOPER, A. H. & HUGHES, R. A. 1993. Discussion of Millward, D. and Molyneux, S. G. 1992 'Field and biostratigraphic evidence for an unconformity at the base of the Eycott Volcanic Group in the English Lake District'. *Geological Magazine*, **130**, 271–272.

——, RUSHTON, A. W. A., MOLYNEUX, S. G., HUGHES, R. A., MOORE, R. M. & WEBB, B. C. 1995. The stratigraphy, correlation, provenance and palaeogeography of the Skiddaw Group (Ordovician) in the English Lake District. *Geological Magazine*, **132**, 185–211.

COOPER, D. C., LEE, M. K., FORTEY, N. J., COOPER, A. H., RUNDLE, C. C., WEBB, B. C. & ALLEN, P. M. 1988. The Crummock Water aureole: a zone of metasomatism and source of ore metals in the English Lake District. *Journal of the Geological Society, London*, **145**, 523–540.

DOWNIE, C. & SOPER, N. J. 1972. Age of the Eycott Volcanic Group and its conformable relationship to the Skiddaw Slates in the English Lake District. *Geological Magazine*, **109**, 259–268.

EVANS, J. A. 1996. Basin development across Iapetus, Project No. 40079. Final report on Rb–Sr whole rock dating of mudrock suites from the Lake District and Southern Uplands. *NERC Isotope Geosciences Laboratory Report Series*, No. 87.

——, MILLAR, I. L. & NOBLE, S. R. 1995. Hydration during uplift is recorded by reset Rb–Sr whole-rock ages. *Journal of the Geological Society, London*, **152**, 209–212.

FORTEY, N. J., ROBERTS, B. & HIRONS, S. R. 1993. Relationship between metamorphism and structure in the Skiddaw Group, English Lake District. *Geological Magazine*, **130**, 631–638.

FORTEY, R. A., HARPER, D. A. T., INGHAM, J. K., OWEN, A. W. & RUSHTON, A. W. A. 1995. A revision of Ordovician series and stages from the historical type area. *Geological Magazine*, **132**, 15–30.

GALE, N. H., BECKINSALE, R. D. & WADGE, A. J. 1979. A Rb–Sr whole-rock isochron for the Stockdale Rhyolite of the English Lake District and a revised Mid-Palaeozoic time-scale. *Journal of the Geological Society, London*, **136**, 235–242.

HELM, D. G. 1970. Stratigraphy and structure in the Black Combe inlier, English Lake District. *Proceedings of the Yorkshire Geological Society*, **38**, 105–148.

HUGHES, R. A. & KOKELAAR, P. 1993. The timing of Ordovician magmatism in the English Lake District and Cross Fell inliers. *Geological Magazine*, **130**, 369–377.

——, COOPER, A. H. & STONE, P. 1993. Structural evolution of the Skiddaw Group (English Lake District) on the northern margin of eastern Avalonia. *Geological Magazine*, **130**, 621–629.

——, EVANS, J. A., NOBLE, S. R. & RUNDLE, C. C. 1996. U–Pb chronology of the Ennerdale and Eskdale intrusions supports sub-volcanic relationships with the Borrowdale Volcanic Group (Ordovician, English Lake District). *Journal of the Geological Society, London*, **153**, 33–38.

INGHAM, J. K., MCNAMARA, K. J. & RICKARDS, R. B. 1978. The Upper Ordovician and Silurian rocks. *In:* MOSELEY, F. (ed.) *The geology of the Lake District.* Occasional Publication of the Yorkshire Geological Society, **3**, 121–145.

JOHNSON, E. W. 1992. *Geology of the Stoupdale area of Black Combe, south-west Cumbria.* British Geological Survey Technical Report, No. WA/92/71.

KNELLER, B. C. 1991. A foreland basin on the southern margin of Iapetus. *Journal of the Geological Society, London*, **148**, 207–210.

——, KING, L. M. & BELL, A. M. 1993. Foreland basin development and tectonics on the northwest margin of eastern Avalonia. *Geological Magazine*, **130**, 691–697.

——, SCOTT, R. W., SOPER, N. J., JOHNSON, E. W. & ALLEN, P. M. 1994. Lithostratigraphy of the Windermere Supergroup, Northern England. *Geological Journal*, **29**, 219–240.

MERRIMAN, R. J. & FREY, M. 1999. Patterns of very low grade metamorphism in metapelitic rocks. *In:* FREY, M. & ROBINSON, D. (eds) *Low-grade Metamorphism.* Blackwell.

—— & PEACOR, D. R. 1999. Very low-grade metapelites: mineralogy, microfabrics and measuring reaction progress. *In:* FREY, M. & ROBINSON, D. (eds) *Low-grade Metamorphism.* Blackwell.

——, REX, D. C., SOPER, N. J. & PEACOR, D. R. 1995. The age of Acadian cleavage in northern England, UK: K–Ar and TEM analysis of a Silurian metabentonite. *Proceedings of the Yorkshire Geological Society*, **50**, 255–265.

MILLWARD, D. & LAWRENCE, D. J. D. 1985. The Stockdale (Yarlside) Rhyolite – a rheomorphic ignimbrite? *Proceedings of the Yorkshire Geological Society*, **45**, 299–306.

—— & Molyneux, S.G. 1992. Field and biostratigraphic evidence for an unconformity at the base of the Eycott Volcanic Group in the English Lake District. *Geological Magazine*, **129**, 77–92.

MOLYNEUX, S. G. 1979. New evidence for the age of the Manx Group, Isle of Man. *In*: HARRIS, A. L., HOLLAND, C. H. & LEAKE, B. E. (eds) *The Caledonides of the British Isles Reviewed.* Geological Society, London, Special Publications, **8**, 415–421.

—— 1988. Micropalaeontological evidence for the age of the Borrowdale Volcanic Group. *Geological Magazine*, **125**, 541–542.

MOORE, R. M. 1992. *The Skiddaw Group of Cumbria: Early Ordovician turbidite sedimentation and provenance on an evolving microcontinental margin.* PhD Thesis, University of Leeds.

PICKERING, K. T. & SMITH, A. G. 1995. Arcs and backarc basins in the Early Paleozoic Iapetus Ocean. *The Island Arc*, **4**, 1–67.

PIDGEON, R. T. & AFTALION, M. 1978. Co-genetic and inherited zircon U–Pb systems in granites: Palaeozoic granites of Scotland and England. *In:* BOWES, D. R. & LEAKE, B. E. (eds) *Crustal Evolution in Northwestern Britain and Adjacent Regions.* Geological Journal, Special Issue, **10**, 183–220.

PIPER, J .D. A., STEPHEN, J. C. & BRANNEY, M. J. 1997. Palaeomagnetism of the Borrowdale and Eycott volcanic groups, English Lake District: primary and secondary magnetization during a single late Ordovician polarity chron. *Geological Magazine*, **134**, 481–506.

PLANT, J. A., COOPER, D. C., GREEN, P. M., REEDMAN, A. J. & SIMPSON, P. R. 1991. Regional distribution of As, Sb and Bi in the Grampian Highlands of Scotland and English Lake District: implications for gold metallogeny. *Transactions of the Institution of Mining and Metallurgy (Section B: Applied earth science)*, **100**, B135–B147.

ROBERTS, B., MORRISON, C. & HIRONS, S. 1990. Low grade metamorphism of the Manx Group, Isle of Man: a comparative study of white mica 'crystallinity' techniques. *Journal of the Geological Society, London*, **147**, 271–277.

ROBERTS, D. E. 1992. Raven Crags, Mungrisdale. *In:* TREAGUS, J. E. (ed.) *Caledonian Structures in Britain south of the Midland Valley* (Joint Nature Conservation Committee). Chapman & Hall, 70–73.

RUNDLE, C. C. 1979. Ordovician intrusions in the English Lake District. *Journal of the Geological Society, London*, **136**, 29–38.

—— 1981. The significance of isotopic dates from the English Lake District for the Ordovician–Silurian time-scale. *Journal of the Geological Society, London*, **138**, 569–572.

—— 1982. *The chronology of igneous intrusion in the English Lake District.* PhD Thesis, University of London.

—— 1987. *Isotopic analyses of rocks from the English Lake District. Preliminary report and summary of results from the 1986 rock collection.* Report of the Isotope Geology Unit of the British Geological Survey, No. 87/6.

RUSHTON, A. W. A. & MOLYNEUX, S. G. 1989. The biostratigraphic age of the Ordovician Skiddaw Group in the Black Combe Inlier, English Lake District. *Proceedings of the Yorkshire Geological Society*, **47**, 267–276.

SHEPHERD, T. J. & DARBYSHIRE, D. P. F. 1981. Fluid inclusion Rb–Sr isochrons for dating mineral deposits. *Nature*, **290**, 5807.

——, BECKINSALE, R. D., RUNDLE, C. C. & DURHAM, J. 1976. Genesis of Carrock Fell tungsten deposits Cumbria: fluid inclusion and isotopic study. *Transactions of the Institution of Mining and Metallurgy*, **85**, B63–B73.

SIMPSON, A. 1963. The stratigraphy and tectonics of the Manx Slate Series, Isle of Man. *Quarterly Journal of the Geological Society of London*, **119**, 367–400.

—— 1967. The stratigraphy and tectonics of the Skiddaw Slates and the relationship of the overlying Borrowdale Volcanic Series in part of the Lake District. *Geological Journal*, **5**, 391–418.

SOPER, N. J. 1970. Three critical localities on the junction of the Borrowdale Volcanic rocks with the Skiddaw

Slates in the Lake District. *Proceedings of the Yorkshire Geological Society*, **37**, 461–493.

—— & KNELLER, B. C. 1990. Cleaved microgranite dykes of the Shap swarm in the Silurian of NW England. *Geological Journal*, **25**, 161–170.

—— & ROBERTS, D. E. 1971. Age of cleavage in the Skiddaw Slates in relation to the Skiddaw aureole. *Geological Magazine*, **108**, 293–302.

——, WEBB, B. C. & WOODCOCK, N. H. 1987. Late Caledonian (Acadian) transpression in north-west England: timing, geometry and geotectonic significance. *Proceedings of the Yorkshire Geological Society*, **46**, 175–192.

——, STRACHAN, R. A., HOLDSWORTH, R. E., GAYER, R. A. & GREILING, R. O. 1992. Sinistral transpression and the Silurian closure of Iapetus. *Journal of the Geological Society, London*, **149**, 871–880.

STONE, P. & EVANS, J. A. 1997. A comparison of the Skiddaw and Manx groups (English Lake District and Isle of Man) using neodymium isotopes. *Proceedings of the Yorkshire Geological Society*, **51**, 343–347.

——, FLOYD, J. D., BARNES, R. P. & LINTERN, B. C. 1987. A sequential back-arc and foreland basin thrust duplex model for the Southern Uplands of Scotland. *Journal of the Geological Society, London*, **144**, 753–764.

TANOLI, S. K. & PICKERILL, R. K. 1988. Lithostratigraphy of the Cambrian–Lower Ordovician Saint John Group, southern New Brunswick. *Canadian Journal of Earth Sciences*, **25**, 669–690.

THIRLWALL, M. F. & FITTON, J. G. 1983. Sm–Nd garnet age for the Ordovician Borrowdale Volcanic Group, English Lake District. *Journal of the Geological Society, London*, **140**, 511–518.

TUCKER, R. D. & MCKERROW, W. S. 1995. Early Palaeozoic chronology: a review in light of new U–Pb zircon ages from Newfoundland and Britain. *Canadian Journal of Earth Sciences*, **32**, 368–379.

WADGE, A. J., GALE, N. H., BECKINSALE, R. D. & RUNDLE, C. C. 1978. A Rb–Sr isochron age for the Shap Granite. *Proceedings of the Yorkshire Geological Society*, **42**, 297–305.

WEBB, B. C. & COOPER, A. H. 1988. Slump folds and gravity slide structures in a Lower Palaeozoic marginal basin sequence (the Skiddaw Group), North West England. *Journal of Structural Geology*, **10**, 463–472.

A comparison of the Ribband Group (southeastern Ireland) to the Manx Group (Isle of Man) and Skiddaw Group (northwestern England)

B. J. McCONNELL[1], J. H. MORRIS[1] & P. S. KENNAN[2]
[1]*Geological Survey of Ireland, Beggar's Bush, Dublin 4, Ireland*
[2]*Department of Geology, University College, Belfield, Dublin 4, Ireland*

Abstract: The lithostratigraphy of the Ribband Group of southeastern Ireland is revised through comparison to the equivalent Avalonian margin sequences of the Manx(Isle of Man) and Skiddaw (northern England) Groups. Four tracts are recognized in the Ribband Group, within which fossil age control and the 'coticule package' marker horizon are used to constrain lithofacies comparisons. Volcanic arc rocks in the Ribband and Manx Groups contrast with the passive margin provenance of the Skiddaw Group.

Ordovician palaeogeographic reconstructions place southeastern Ireland, the Isle of Man and the Lake District at the Iapetus Ocean margin of micro-continental Avalonia (e.g. McKerrow *et al.* 1991). The early Ordovician rock sequences in the three areas, the Ribband, Manx and Skiddaw Groups, consist predominantly of sedimentary rocks deposited in deep-marine environments (Cooper *et al.* 1995; Stone *et al.* 1999; Woodcock *et al.* 1999*b*). It is worthwhile, therefore, to attempt to correlate from the well-studied Skiddaw and Manx Groups to the less known Ribband Group, and so broaden the understanding of Avalonian margin evolution.

The starting point for correlations between the Ribband and Manx Groups is the new Manx Group lithostratigraphy, detailed by Woodcock *et al.* (1999*b*). It is attempted to 'fit' the Ribband Group to the Manx Group. The Skiddaw Group is used as an additional control, as it is the best known of the three sequences. Four tracts are identified in the Ribband Group, each with a separate, previously defined lithostratigraphy (Figs 1 and 2); (1) west of the Leinster Granite; (2) between the Leinster Granite and the Wicklow Fault Zone; (3) between the Wicklow Fault Zone and the late Ordovician Duncannon Group; (4) southeast of the Duncannon Group. Use of the term 'tract' does not necessarily imply recognized bounding tectonic structures. Tract 1 is bounded to the northwest by the Hollywood Shear Zone, and tracts 2 and 3 are separated by the Wicklow Fault Zone, but the Leinster Granite and the crop of the late Ordovician Duncannon Group are also limits of Ribband Group tracts in this sense. Comparison with the Manx and Skiddaw Group sequences suggests a tentative correlation of the lithostratigraphy of each of these tracts.

The Ribband Group and the Leinster Terrane

The Ribband Group (Crimes & Crossley 1968) is the early Ordovician, predominantly metasedimentary sequence, stratigraphically between the Cambrian Bray and Cahore Groups below and the late Ordovician Duncannon Group above (Brück *et al.* 1979). It lies within the main Leinster massif inlier of the Leinster Terrane (Murphy *et al.* 1991) (Fig. 1). Llanvirn mudstones and siltstones occur in the Kildare Inlier and presumed early Ordovician fine-grained metasedimentary rocks occur in the Balbriggan Inlier at the northern edge of the Leinster Terrane, but these are not included in the Ribband Group. The Leinster Terrane is bounded to the north by the Lowther Lodge Fault, which marks its junction with the intra-Iapetan Bellewstown Terrane. To the south, it is bounded by the Ballycogly mylonites at the junction with the Precambrian Rosslare Terrane. Within the Leinster Terrane, the Ribband Group is bounded to the west by the reverse dip-slip Hollywood Shear Zone, on which turbidites of the Silurian Kilcullen Group (Brück *et al.* 1979) have been thrust eastwards. This relationship is comparable to that between the Silurian Niarbyl Formation thrust over the early Ordovician Manx Group (Morris *et al.* 1999.

From: Woodcock, N. H., Quirk, D. G., Fitches, W. R. & Barnes, R. P. (eds) 1999.
In Sight of the Suture: the Palaeozoic geology of the Isle of Man in its Iapetus Ocean context.
Geological Society, London, Special Publications, **160**, 337–343. 1-86239-046-0/99/$15.00

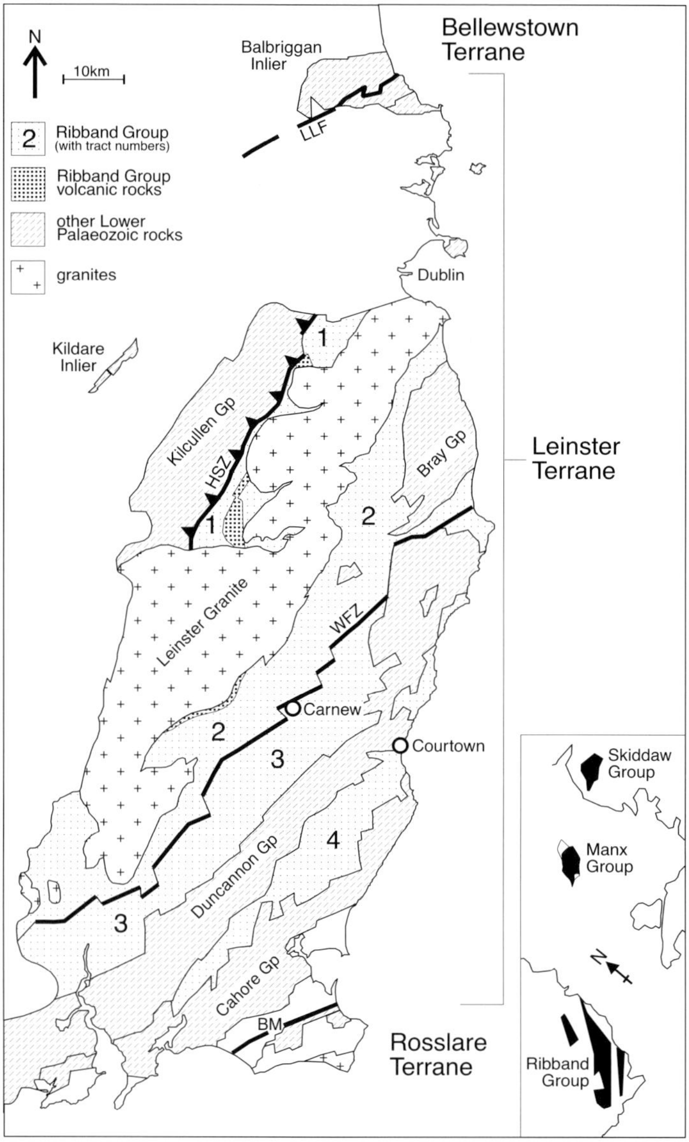

Fig. 1. Geological sketch map of the Leinster Terrane showing the tract subdivision of the Ribband Group. BM, Ballycogly mylonites; HSZ, Hollywood Shear Zone; LLF, Lowther Lodge Fault; WFZ, Wicklow Fault Zone.

The Ribband Group has been the subject of many local studies and many local stratigraphies have been erected [summarized by Brück *et al.* (1979)]. An attempt was made to unify these on the Geological Survey of Ireland 1:100 000 Bedrock Map Series (Tietzsch-Tyler & Sleeman 1994, 1995; McConnell *et al.* 1995), although this may have pushed units beyond their natural limits. The resultant stratigraphy (Fig. 3) considered the group in three belts; west of the Leinster Granite (our tract 1), east of the Leinster Granite (our tracts 2 and 3) and south of the Duncannon Group (our tract 4). The base of the group is the subject of some confusion. Traditionally, the Ribband Group has included the Upper Cambrian Booley Bay Formation (Moczydlowska & Crimes 1995). However, Tietzsch-Tyler & Sleeman (1994) considered that the Ballyhoge Formation of the Ribband Group conformably overlay the Booley Bay Formation of the Cahore Group, although that

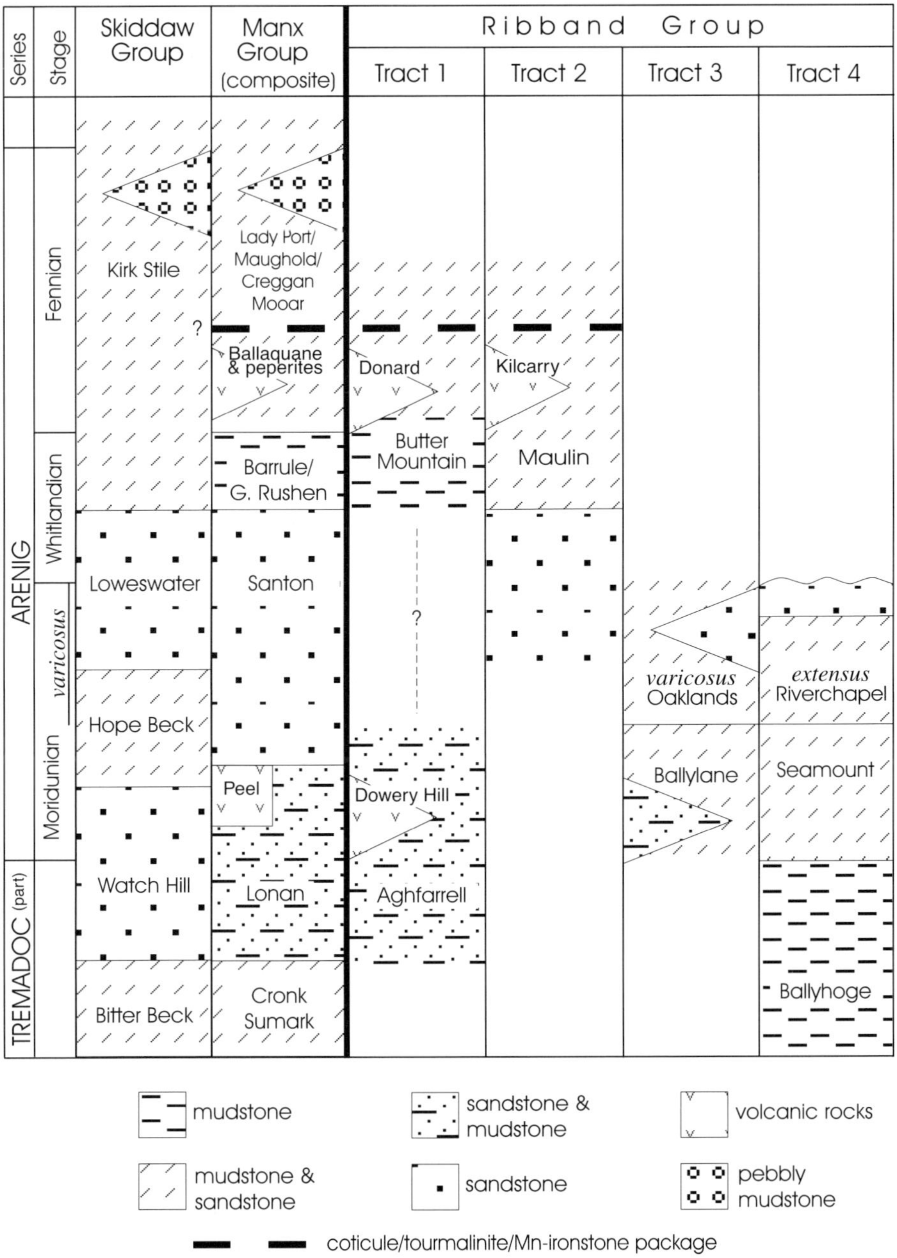

Fig. 2. Possible lithostratigraphical correlation of the four tracts of the Ribband Group and the Manx and Skiddaw Groups. The Manx Group column is composite from Woodcock *et al.* (1999*b*, fig. 9); note that the Peel volcanics are placed on the basis of their age but do not occur within the Lonan Formation.

relationship is not evident on their map, on which the apparent along-strike continuity of one grey to black mudstone and siltstone unit to the other is notable. The Llandeilian–Caradoc Courtown Limestone of the Duncannon Group unconformably overlies the Ribband Group at Courtown (Crimes & Crossley 1968; Brenchley & Treagus 1970).

The Riverchapel Formation of tract 4 has yielded graptolites of the *extensus* Biozone, probably the lower part [Skevington in Brenchley & Treagus

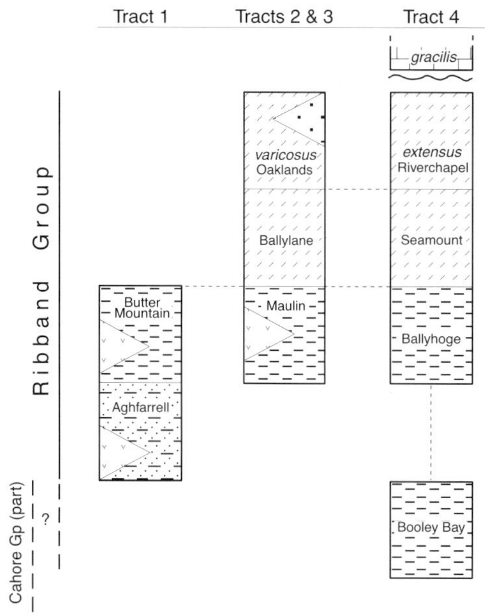

Fig, 3. Previous interpretation of Ribband Group stratigraphy (summarized from Tietzsch-Tyler & Sleeman 1994, 1995, McConnell *et al.* 1995). Ornament for dominant lithological character as on Fig. 2, except brick pattern indicates limestone.

(1970)]. Its correlative (Tietzsch-Tyler & Sleeman 1995) in the 'east-of-granite' belt (tract 3), the Oaklands Formation, contains graptolites corresponding to the *varicosus* Biozone or the poorly characterized strata below (Rushton 1996). The *varicosus* Biozone of the English Lake District and the lower part of the *extensus* Biozone are approximately equivalent, corresponding approximately to the Bendigonian (Cooper *et al.* 1995; Fortey *et al.* 1995, fig. 1). The Riverchapel and Oaklands Formations were believed to be the youngest part of the Ribband Group, suggesting that the group had a minimum preserved age of early Arenig. Attempts to use microfossils for dating have so far yielded poor results (e.g. Brück *et al.* 1974).

Comparison with the Manx Group

Biostratigraphical control

The *varicosus* Biozone age for the Oaklands Formation (Rushton 1996) and the lower *extensus* Biozone age for the Riverchapel Formation (Crimes & Crossley 1968) place them as approximate age equivalents of the Santon Formation of the Manx Group (Orr & Howe 1999) (Fig. 2). The varicoloured laminated siltstone and mudstone that comprise most of these Ribband Group units contrast with the medium- to thick-bedded greywacke and quartz wacke sandstones of the Santon Formation. However, the Oaklands Formation contains a greywacke sandstone member, the Palace Member, and the Riverchapel Formation is capped by feldspathic sandstone. The *varicosus* Biozone of the Skiddaw Group includes the upper part of the Hope Beck Formation of mudstone and siltstone, and the overlying lower part of the Loweswater Formation of quartz-rich feldspathic wackes. All three areas, therefore, contain a sandstone incursion in the *varicosus* Biozone, although this dominated the sedimentary record in the Manx Group and, as far as the preserved sequence indicates, was more limited in the Ribband Group.

The Riverchapel Formation passes down through the Seamount Formation of similar but less varicoloured sediments into the Ballyhoge Formation (Fig. 2), which is mentioned above as being similar to and possibly in vertical, or lateral, continuity with the Upper Cambrian Booley Bay Formation. It is possible, therefore, that the Ballyhoge Formation equates with the Tremadoc mudstone in the oldest known parts of the Manx and Skiddaw Groups (Cronk Sumark and Bitter Beck Formations, respectively).

Coticule package

The Oaklands Formation passes down to the green and grey siltstone and mudstone of the Ballylane Formation (Fig. 2), which Tietzsch-Tyler & Sleeman (1994, 1995) considered to stratigraphically overlie the dark slates, phyllites and schists of the Maulin Formation. Thus, there were apparently equivalent tripartite stratigraphies on either side of the Duncannon Group (Fig. 3). However, the gross younging direction of the Maulin Formation appears to be away from its contact with the Ballylane Formation. The Maulin Formation includes the distinctive 'coticule package' of coticule (spessartine-bearing quartzite) and tourmalinite, the product of an exhalative event and believed to be a stratigraphically confined package of large lateral extent within the Caledonian–Appalachian Orogen (Kennan & Kennedy 1983). In the Isle of Man, stratiform tourmalinite occurs in the Injebreck Formation and thinly bedded Mn–ironstone within the Maughold, Creggan Mooar and Lady Port Formations are considered to be the low-grade protolith of the coticule package (Kennan & Morris 1999). Accepting the 'coticule package' as a stratigraphic marker, late Arenig acritarchs from the Lady Port Formation (Molyneux 1999) appear to provide age control on all of the units containing the coticule package in the Ribband and Manx Groups. The coticule-bearing Maulin Formation would thus appear to be younger than the Oaklands Formation (*varicosus* Biozone) (Fig. 2), contrary to previous interpretations (Fig. 3).

The contact between the Maulin and Ballylane Formations equates with the Wicklow Fault Zone of Max *et al.* (1990), which those authors interpreted as a terrane boundary, without implication of an allochthonous relationship. Sheared serpentinites lie along it in the Carnew area (Gallagher 1989) and an aeromagnetic lineament continues from these to the south. The contact subdivides the Ribband Group between the Leinster Granite and Duncannon Group, giving the four tracts here (Figs 1 and 2) rather than the previous three belts (Fig. 3).

The coticule package can be traced from the Maulin Formation through a schist septum into the Butter Mountain Formation on the west side of the Leinster Granite, establishing the equivalence of the dark slates and schists of the two formations. The Butter Mountain Formation has a generally more pelitic lower part which appears to equate with the variation seen in the Maughold Formation of the Manx Group, the lower part of which may be equivalent to the Glen Rushen and Barrule Formations (Fig. 2; Woodcock *et al.* 1999*b*, fig. 9). This more pelitic lower part is not apparent in the Maulin Formation. Instead, the Maulin Formation phyllites and schists along the eastern margin of the Tullow Pluton [the Ballybeg Pelite of McArdle (1981)] pass down through an interbedded transition into the psammitic Ballybeg Greywackes. These coarser grained sediments possibly equate to the *varicosus* Biozone sandstone units of the southern Ribband Group tracts (Palace Member and top Riverchapel Formation) and the Manx Group (Santon Formation) (Fig. 2).

The Butter Mountain Formation appears to stratigraphically overlie the Aghfarrell Formation (McConnell *et al.* 1994), a sequence of thinly bedded greywacke siltstones and shales (Brück *et al.* 1979). Lithologically, the Aghfarrell Formation is most similar to the Lonan and Port Erin Formations in the Manx Group, but such a correlation would require a longer time range for the Aghfarrell Formation in order to bring its top up to the Glen Rushen-equivalent base of the Butter Mountain Formation, or a tectonic break between the two (Fig. 2).

The Avalonian margin–early Ordovician volcanic rocks

Apart from the many specific problems of goodness-of-fit of our correlations, there is a question over the equivalence of tectonic settings in which the three Avalonian margin sequences were deposited. The Skiddaw Group has been ascribed to an inactive continental or passive margin setting (Cooper *et al.* 1995). The oldest volcanogenic strata in the Skiddaw Group are distal volcaniclastic turbidites in the early Llanvirn Tarn Moor Formation (Hughes & Kokelaar 1993). In contrast, the Ribband Group contains several Arenig horizons of **subduction-related** volcanism (McConnell & Morris 1997; Brück 1976). The Dowery Hill basalts of the Aghfarrell Formation were interpreted by McConnell & Morris (1997) as recording volcanism early in the history of a volcanic arc. The plagioclase- and pyroxene-phyric Donard and Kilcarry andesites (Butter Mountain and Maulin Formations, respectively) probably record increasing maturity of this phase of arc magmatism.

There are two known occurrences of volcanic rocks in the Manx Group (Fig. 2), the Peel volcanics and an outcrop of tuff at Ballaquane Farm (Lamplugh 1903). The outcrops of Peel volcanics are isolated from their enclosing strata but an early Arenig age is indicated by acritarchs (Molyneux 1999). The Ballaquane tuff appears to occur in the Creggan Mooar Formation (Morris *et al.* 1999). In addition, peperitic shallow intrusions occur in the pelitic units considered by Woodcock *et al.* (1999*b*) to correlate to the Creggan Mooar Formation in the upper part of the preserved Manx Group. The correlations proposed above (Fig. 2) suggest a possible time equivalence of the Peel volcanics to the Dowery Hill basalts, and the Ballaquane tuff and peperites to the Donard and Kilcarry andesites. Three geochemical analyses of the Ballaquane tuff (Power, pers. comm.) are generally similar to data for the Donard andesites (Gallagher, pers. comm.) (Fig. 4). Both groups of data have a volcanic arc signature, but the small number of data and the altered state of the rocks makes it unsafe to interpret further.

Summary

The following tentative correlations can be made between the Ribband Group and the Manx and Skiddaw Groups (Fig. 2). The mudstones of the Ballyhoge Formation may equate in age and lithology with the Cronk Sumark and Bitter Beck Formations; the sandstones and mudstones of the Aghfarrell Formation with the Lonan and Watch Hill Formations; the mudstones and sandstones of lower Oaklands and Riverchapel Formations with the Hope Beck Formation (not apparent in the Manx Group where the time-equivalent Santon Formation may be an earlier development of the sandstones of the Ballybeg Greywackes and upper

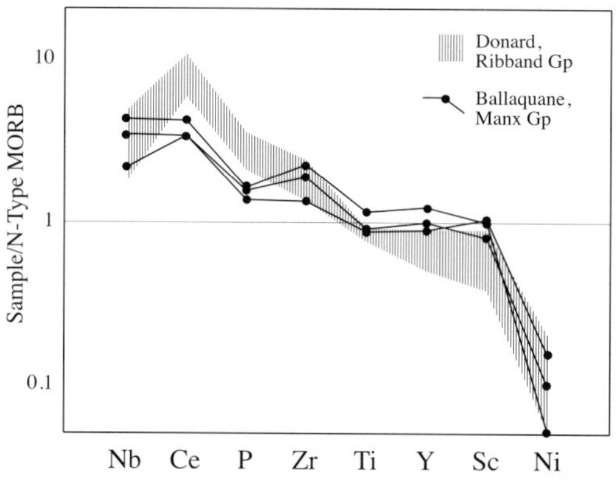

Fig. 4. Geochemical profiles of Donard (Ribband Group) and Ballaquane (Manx Group) volcanic rocks normalized to N-type MORB [normalizing values as in McConnell & Morris (1997)].

Oaklands and Riverchapel Formations); the pelitic base of the Butter Mountain Formation with the Barrule and Glen Rushen Formations; the mudstones, sandstones and coticule of the Maulin and Butter Mountain Formations with the Injebreck, Lady Port, Creggan Mooar and Maughold Formations.

The proposed correlations of the Ribband and Manx Group lithostratigraphies are tentative and model driven. Graptolite biozones are a secure criterion for comparison but the coticule 'marker' is of dubious accuracy. Lithofacies can vary laterally across the different environments within a basin and vertical changes can occur diachronously. However, there are gross similarities in the stratigraphy of the Manx, Skiddaw and Ribband Groups, and comparison to the Manx Group has produced a stratigraphic and structural model for the Ribband Group that can be tested.

Whether the volcanics of the Manx and Ribband Groups can be shown to be stratigraphically or geochemically comparable or not, the question remains how to relate the passive margin setting of the Skiddaw Group to the Ribband and Manx Groups' arc volcanism. It is hoped that the model of the Ribband Group proposed in this paper will stimulate further study to progress understanding of the early Ordovician Avalonian margin.

We thank Peadar McArdle and Matthew Parkes for discussion, Greg Power and Vincent Gallagher for permission to use their unpublished data, and Tony Cooper and anonymous referees for constructive reviews. BJM and JHM publish with permission of the Director, Geological Survey of Ireland.

References

BRENCHLEY, P. J. & TREAGUS, J. E. 1970. The stratigraphy and structure of the Ordovician rocks between Courtown and Kilmichael Point, Co. Wexford. *Proceedings of the Royal Irish Academy,* **69B**, 83–102.

BRÜCK, P. M. 1976. The andesitic and doleritic igneous rocks of west Wicklow and south Dublin. *Geological Survey of Ireland Bulletin,* **2**, 37–51.

——, POTTER, T. L. & DOWNIE, C. 1974. The Lower Palaeozoic stratigraphy of the northern part of the Leinster massif. *Proceedings of the Royal Irish Academy,* **74B**, 75–84.

——, COLTHURST, J. R. J., FEELY, M. *ET AL.* 1979. South-East Ireland: Lower Palaeozoic Stratigraphy and depositional history. *In*: HARRIS, A. L., HOLLAND, C. H. & LEAKE, B. E. (eds) *The Caledonides of the British Isles – reviewed.* Geological Society, London, Special Publications, **8**, 533–544.

COOPER, A. H., RUSHTON, A. W. A., MOLYNEUX, S. G., HUGHES, R. A., MOORE, R. M. & WEBB, B. C. 1995. The stratigraphy, correlation, provenance and palaeogeography of the Skiddaw Group (Ordovician) in the English Lake District. *Geological Magazine,* **132**, 185–211.

CRIMES, T. P. & CROSSLEY, J. D. 1968. The stratigraphy, sedimentology, ichnology and structure of the Lower Palaeozoic rocks of part of north-eastern Co. Wexford. *Proceedings of the Royal Irish Academy,* **67B**, 185–215.

FORTEY, R. A., HARPER, D. A. T., INGHAM, J. K., OWEN, A. W. & RUSHTON, A. W. A. 1995. A revision of Ordovician series and stages from the historical type area. *Geological Magazine*, **132**, 15–30.

GALLAGHER, V. 1989. The occurrence, textures, mineralogy and chemistry of a chromite-bearing serpentinite, Cummer, Co. Wexford. *Geological Survey of Ireland Bulletin,* **4**, 89–98.

HUGHES, R. A. & KOKELAAR, P. 1993. The timing of Ordovician magmatism in the English Lake District and Cross Fell inliers. *Geological Magazine,* **130**, 369–377.

KENNAN, P. S. & KENNEDY, M. J. 1983. Coticules - a key to correlation along the Appalachian-Caledonian orogen. *In*: SCHENK, P. E. (ed.) *Regional trends in the geology of the Appalachian-Caledonian-Hercynian-Mauritanide orogen.* Riedel, Dordrecht, 355-361.

KENNAN, P. S. & MORRIS, J. H. 1999. Manganiferous ironstones in the early Ordovician Manx Group, Isle of Man: a prolith of coticule. *This volume.*

LAMPLUGH, G. W. 1903. *The Geology of the Isle of Man.* Memoir of the Geological Survey, UK. HMSO.

MAX, M. D., BARBER, A. J. & MARTINEZ, J. 1990. Terrane assemblage of the Leinster Massif, SE Ireland, during the Lower Palaeozoic. *Journal of the Geological Society, London,* **147**, 1035–1050.

MCARDLE, P. 1981. The country rocks flanking the Leinster Granite between Aughrim and

Ballymurphy. *Geological Survey of Ireland Bulletin*, **3**, 85–95.

McConnell, B. & Morris, J. 1997. Initiation of Iapetus subduction under Irish Avalonia. *Geological Magazine*, **134**, 213–218.

——, Philcox, M. E., MacDermot, C. V. & Sleeman, A. G. 1995. *Bedrock Geology 1:100 000 scale Map Series, Sheet 16, Kildare – Wicklow.* Geological Survey of Ireland.

——, ——, Sleeman, A. G., Stanley, G., Flegg, A. M., Daly, E. P. & Warren, W. P. 1994. *Geology of Kildare – Wicklow; a geological description to accompany the Bedrock Geology 1:100 000 Map Series, Sheet 16, Kildare – Wicklow.* Geological Survey of Ireland.

McKerrow, W. S., Dewey, J. F. & Scotese, C. R. 1991. The Ordovician and Silurian development of the Iapetus Ocean. *Special Papers in Palaeontology*, **44**, 165–178.

Moczydlowska, M. & Crimes, T. P. 1995. Late Cambrian acritarchs and their age constraints on an Ediacaran-type fauna from the Booley Bay Formation, Co. Wexford, Eire. *Geological Journal*, **30**, 111–128.

Morris, J. H., Woodcock, N. H. & Howe, M. P. A. 1999. The Silurian succession of the Isle of Man: the late Wenlock Niarbyl Formation, Dalby Group. *This volume*.

Molyneux, S. G. 1999. A reassessment of Manx Group acritarchs, Isle of Man. *This volume*.

Murphy, F. C., Anderson, T. B., Daly, J. S. *et al.* 1991. An appraisal of Caledonian suspect terranes in Ireland. *Irish Journal of Earth Sciences*, **11**, 11–41.

Orr, P. J. & Howe, M. P. A. 1999. Macrofauna and ichnofauna of the Manx Group (early Ordovician), Isle of Man. *This volume*.

Rushton, A. W. A. 1996. *Trichograptus* from the Lower Arenig of Kiltrea, County Wexford. *Irish Journal of Earth Sciences*, **15**, 61–69.

Stone, P., Cooper, A. H. & Evans, J. A. 1999. The Skiddaw Group (English Lake District) reviewed: early Palaeozoic sedimentation and tectonism at the northern margin of Avalonia. *This volume*.

Tietzsch-Tyler, D. & Sleeman, A. G. 1994. *Bedrock Geology 1:100 000 scale Map Series, Sheet 23, South Wexford.* Geological Survey of Ireland.

—— & —— 1995. *Bedrock Geology 1:100 000 scale Map Series, Sheet 19, Carlow–Wexford.* Geological Survey of Ireland.

Woodcock, N. H., Quirk, D. G., Fitches, W. R. & Barnes, R. P. 1999*a*. In sight of the suture: the Lower Palaeozoic geological history of the Isle of Man. *This volume*.

——, Morris, J. H., Quirk, D. G. *et al.* 1999*b*. Revised lithostratigraphy of the Manx Group, Isle of Man. *This volume*.

A bibliography of the geology of the Isle of Man

EVA WILSON

The Lifeboat House Castletown, Isle of Man IM9 1LD, UK and
Centre for Manx Studies, 6 Kingswood Grove, Douglas, Isle of Man IM1 3LX, UK

Introduction

The bibliography presented here was initiated by the Centre for Manx Studies in 1995 as a part of their ongoing support for research on the Isle of Man. Lamplugh (1903) was a major source of reference for publications on the subject of Manx geology preceeding his work, amounting to one-third of all the entries listed here. Existing bibliographies, Thorpe (1972), Thomas (1977) and Burt *et al.* (1988) were included. Other important sources were the indexes and catalogues in the Manx National Heritage Library in the Manx Museum & National Trust and the British Geological Survey (BGS) Library. Some unpublished and unsorted British Geological Survey (BGS) material has not been listed in the bibliography, including data held or produced as part of the publication process for memoirs and maps, e.g. Lamplugh's field notebooks and correspondence.

The bibliography as a whole reflects the changing interests, economic as well as academic, on the geology of the island. Overall, the largest number of entries concerns the mines and the mining industry which survived until the 1920s. Numerous published articles from the nineteenth century were generated by the discussions and controversies on the nature of drift and on glaciation, a debate which has reopened in the latter half of this century. Research into the Lower Palaeozoic rocks is discussed in detail in Ford *et al.* (1999).

The author wishes to thank R. Sims and A. Franklin in the Manx National Heritage Library, and G. McKenna in the BGS Library. Help and advice was gratefully received from E. Brunton of the Mineralogy and Palaeontology Libraries of the Natural History Museum, and from G. Ryback who contributed references from his private database on mining and minerals. The author is greatly indebted to E. Bimpson in the Library of The Geological Society of London for assistance and for monitoring current literature and on-line sources. I am grateful to members of the Centre for Manx Studies and to P. Tomlinson in particular for her guidance in the control of the computer. Finally, my thanks to D. Quirk and D. Burnett for checking the bibliography.

AHLBERG, P. E. & COATES, M. I. 1997. There's a ratfish in our cellar! *Geology Today*, **13**, 22–23.

ALLEN, D. E. 1978. The present-day fauna and flora of Man as indicators of the date of the Flandrian severence. *In*: DAVEY, P. J. (ed.) *Man and Environment in the Isle of Man. British Series*, **54**. British Archaeological Reports, Oxford, 9–14.

ALLEN, J. R. L. & CROWLEY, S. F. 1983. Lower Old Red Sandstone fluvial dispersal systems in the British Isles. *Transactions of the Royal Society of Edinburgh*, **74**, 61–68.

ALLEN, T. 1904. Application for leave to build a shelter ... for dressing of flagstones. *In*: MOORE, A. W. (ed.) *Notes and Documents from the Records of the Isle of Man*. The Manx Sun, Douglas, 58.

ANON. 1880–1892. *Megaceros Hibernicus*. *Yn Lioar Manninagh*, **I**, **ii**, 23.

—— 1880–1892. Geological photography. *Yn Lioar Manninagh*, **1**, **ii**, 90.

—— 1821. Discovery of the Fossil Elk of Ireland in the Isle of Man. *Edinburgh Philosophical Journal*, **5**, 227.

—— 1823. Fossil Elk of the Isle of Man. *Edinburgh Philosophical Journal*, **8**, 198.

—— 1854. Antiquities of Mona – igneous rock in Douglas Bay. *Mona's Herald*, **7th January**.

—— 1874. Geology. *In*: *Jenkinson's Practical Guide to the Isle of Man*. Edward Stanford, 239–248.

—— 1887. Visit of the British Association to the Isle of Man. *Isle of Man Times*, **17th September**.

—— 1895. An Auriferous Quartz-vein near Douglas, Isle of Man; first record of Gold from the Island. *Nature*, **51**, 299.

—— 1895–1900. The British Association Excursion to the Isle of Man 24th–29th September 1896. *Yn Lioar Manninagh*, **III**, 211–223.

—— 1895–1900. Report of the Irish Elk Committee, 1898. *Yn Lioar Manninagh*, **III**, 26, 267, 327–330.

—— 1896. Arctic plants and Apus remains at Kirk

From: WOODCOCK, N. H., QUIRK, D. G., FITCHES, W. R. & BARNES, R. P. (eds) 1999. *In Sight of the Suture: the Palaeozoic geology of the Isle of Man in its Iapetus Ocean context*. Geological Society, London, Special Publications, **160**, 345–361. 1-86239-046-0/99/$15.00

Michael in the Isle of Man. *The Naturalist*, **August**, 244.

—— 1900. The mineral industry of the United Kingdon, iv. The Isle of Man. *The Quarry and Builders' Merchant*, **4**, 9–23.

—— 1903. The Barrule Granite Quarries of the Isle of Man. *The Quarry*, **1**, 205–209.

—— 1906–1912. Visit of the Yorkshire Geological Society. *Proceedings and Transactions of the Isle of Man Natural History and Antiquarian Society*, New Series, **I**, 101–114.

—— 1956. Island Hopes (Snaefell Mine). *Mine and Quarry Engineering*, **22**, 139.

—— 1957. China stone from the Isle of Man. *Mineralogical Magazine*, **97**, 78–82.

—— 1964–1972. Field Section Report, 1966–67. *Proceedings of the Isle of Man Natural History and Antiquarian Society*, New Series, **VII**, 270–276.

—— 1966. The Great Laxey Wheel in its prime. *Industrial Archaeology*, **3**, 305.

—— 1970. 'Lady Isabella' Wheel, Isle of Man. *Industrial Archaeology*, **7**, 337–338.

—— 1972–1980. Field Section Report 1968–1969. *Proceedings of the Isle of Man Natural History and Antiquarian Society, New Series*, **VIII**, 478–490.

ARONSON, R. 1987. A murder mystery from the Mesozoic. *New Scientist*, **116**, 56–59.

ARX VON, R. 1996. A glimpse of Snaefell Mine. *British Mining*, **57**, 34–46.

ASHTON, W. 1920. *The Evolution of a Coast Line, Barrow to Aberystwyth and the Isle of Man*. Ashton.

AUSTIN, R. L. & ALDRIDGE, R. J. 1973. Conodonts from horizons with *Goniatites crenistria* Phillips, in North Wales and the Isle of Man. *Geological Magazine*, **110**, 37.

BANNOCK, D. 1952. *A condensed report of mineral investigations in the Isle of Man*. Laxey. Unpublished in Manx National Heritage Library.

BARNE, J. H., ROBSON, C. F., KAZNOWSKA, S. S., DOODY, J. P. & DAVIDSON, N. C. (eds) 1996. *Coasts and seas of the United Kingdom Region 13. Northern Irish Sea Colwyn Bay to Stranraer, including the Isle of Man*. (Coastal Directories Series). Joint Nature Conservation Committee, Peterborough.

BARNES, J. & HOLROYD, W. F. 1898. Fossils from the Carboniferous Limestones of the Isle of Man, Derbyshire etc. exhibited at a meeting of the Manchester Geological Society. *Transactions of the Manchester Geological Society*, **25**, 394–396.

BASSLER, R. S. 1950. Faunal lists and Descriptions of Palaeozoic Corals. *Geological Society of America Memoir*, **44**.

BATHER, F. A. 1916–1917. *Hydreionocrinus Verrucosus* n.sp., Carboniferous, Isle of Man and some British specimen of Ulocrinus. *Transactions of the Geological Society of Glasgow*, **16**, 203–219.

BATTEN, R. L. 1966. *A Monograph of the Lower Carboniferous Gastropod fauna from the Hotwell Limestone of Conyston Martin, Somerset*. Palaeontographical Society, London.

BAWDEN, T. A. 1976. Mona–Erin. The Story of the Mines around Glen Maye. *Journal of the Manx Museum*, **VII**, 217–220.

——, GARRAD, L. S., QUALTROUGH, J. K. & SCATCHARD, W. J. 1967. *A Preliminary Study of the Industrial Archaeology of the Isle of Man*. Victoria Press, Douglas.

BEDFORD, J. E. 1889. Notes on the Isle of Man. *Transactions of the Leeds Geological Association*, **4**, 177–179.

BELL, A. 1897. *Tertiary deposits in North Manxland*. Report of the British Association for the Advancement of Science for 1896, 783.

—— 1912–1925. On an Irish Manx Pliocene Sea. *Proceedings of the Isle of Man Natural History and Antiquarian Society, New Series*, **II**, 372–383.

—— 1912–1925. The cliffs of North Ramsey and their fossil contents. *Proceedings of the Isle of Man Natural History and Antiquarian Society, New Series*, **II**, 383–392.

—— 1915. The fossiliferous molluscan deposits of Wexford and North Manxland. *Geological Magazine, Decade 6*, **2**, 164–169.

—— 1919. Fossil shells from Wexford and Manxland. *The Irish Naturalist*, **28**, 109–114.

BELT, T. 1874. Letter on the Glacial Period discussing the Isle of Man. *Nature*, **10**, 63.

BERGER, J. F. 1814. Mineralogical account of the Isle of Man. *Transactions of the Geological Society of London*, **2**, 29–65.

—— 1824. Reply to Mr Henslow's observations on Dr Berger's account of the Isle of Man. *Annals of Philosophy, Series 2*, **8**, 367.

BINNEY, E. W. 1867. Note on the plant remains in the Carboniferous of the Isle of Man. *Proceedings of the Manchester Literary and Philosophical Society*, **2**, 27.

—— 1876. A notice of some organic remains from the schists of the Isle of Man. *Proceedings of the Manchester Literary and Philosophical Society*, **16**, 1–8.

—— 1878. Notice of a fossil plant found at Laxey in the Isle of Man. *Proceedings of the Manchester Literary and Philosophical Society*, **17**, 85–88.

BIRCH, J. W. 1964. *The Isle of Man, A Study in Economic Geography*. University of Bristol at the University Press, Cambridge.

BIRD, R. H. 1974. *Britain's Old Metal Mines: A Pictorial Survey*. Bradford Barton.

BIRDS, J. A. 1875. Post-Pliocene formations in the Isle of Man. *Geological Magazine, Decade 2*, **2**, 80–85, 226–228, 428–430.

BISAT, W. S. 1934. The goniatites of the Beyrichoceras Zone in the north of England. *Proceedings of the Yorkshire Geological Society*, **22**, 280–309.

BLACKFORD, J. J. & INNES, J. B. 1996. *The Peel Bay area palynological assessment*. Report for Centre for Manx Studies.

BLAKE, J. F. 1905. On the order of succession of the Manx Slates. *Quarterly Journal of the Geological Society of London*, **61**, 358–373.

BOLTON, H. 1893. On a trilobite from the Skiddaw slate of the Isle of Man. *Geological Magazine, Decade 3*, **10**, 29–31.

—— 1894. On a trilobite from the Skiddaw slate of the Isle of Man. *Report of the British Association for Advancement of Science for 1893*, 770–771.

—— 1895. On the metamorphism of coal. *Transactions of the Manchester Geological Society*, **24**, 56–67.

—— 1899. The palaeontology of the Manx Slate of the Isle of Man. *Memoirs & Proceedings of the Manchester Literary & Philosophical Society*, **43**, 1–15.

BOND, G. 1941. Species and variation in British and Belgian Carboniferous schizophoridae. *Proceedings of the Geologists' Association,* **52**, 285–303.

BOTT, M. H. P. 1964. Gravity measurements in the North Eastern part of the Irish Sea. *Quarterly Journal of the Geological Society of London*, **120**, 369–396.

—— 1968. The geological structure of the Irish Sea Basin. *In*: DONOVAN, D. T. (ed.) *Geology of Shelf Seas. Proceedings of the Inter-University Geological Congress 1968*. Oliver & Boyd, 93–115.

BOWEN, D. Q. 1973. The Pleistocene succession of the Irish Sea. *Proceedings of the Geologists' Association*, **84**, 249–273.

BOWIE, S. H. U., PARKER, A. & RAYNOR, E. J. 1979. Uranium reconnaissance survey of British Lower Palaeozonic shales; trace elements related to clay mineralogy. *Transactions of the Institution of Mining and Metallurgy*, **88**, B61–B64.

BRACEGIRDLE, R. C. 1959–1964. Post-glacial invasions of Man by the animal kingdom. *Proceedings of the Isle of Man Natural History and Antiquarian Society, New Series*, **VI**, 388–401.

BRADY, G. S., JONES, T. R. & KIRBY, J. W. 1874–1884. *A Monograph of the British Fossil Bivalved Entomostraca*. Palaeontographical Society, London.

BRAND, P. J. 1972. Some British Carboniferous species of the Genus *Leptagonia* McCoy. *Bulletin of the Geological Survey of Great Britain*, **39**, 57.

BREWER, F. M., COX, J. D. & MORRIS, D. F. G. 1955. The occurrence of germanium in blende. *Geochimica et Cosmochimica Acta*, **8**, 131–136.

BRITISH GEOLOGICAL SURVEY. 1982. *54N–06W solid: Isle of Man*. Ordnance Survey for the Institute of Geological Sciences 1:250 000 series.

—— 1985. *54N–06W including part of 54N–08W sea bed: Isle of Man including part of Ulster, sea bed sediments and quaternary geology*. Ordnance Survey for the British Geological Survey 1:250 000 series.

BRITISH MUSEUM (NATURAL HISTORY). 1966. *British Palaeozoic Fossils*. 2nd editon. Trustees of the British Museum (Natural History), London.

BROWN, J. C. 1951. *Coast Erosion*. Report to the Isle of Man Harbour Commissioners.

BROWN, P. E., MILLER, J. A. & GRASTY, R. L. 1968. Isotopic ages of late Caledonian granitic intrusions in the British Isles. *Proceedings of the Yorkshire Geological Society*, **36**, 251–276.

BRUCE, J. R. 1925–1932. Animal and plant remains in the Manx post-Glacial deposits and Forest Bed. *Proceedings of the Isle of Man Natural History and Antiquarian Society, New Series*, **III**, 154–162.

—— 1961–1962. John Murrey's Smelt-House, Derbyhaven. *Journal of the Manx Museum*, **VI**, 152–154.

BURT, R. & WAITE, P. 1988. *Bibliography of the history of British metal mining: books, theses and articles published on the history of metal mining in England, Wales, Scotland and the Isle of Man since the Second World War*. University of Exeter, in association with the National Association of Mining History Organisations, Exeter.

——, BURNLEY, R., ALKINSON, M. & WAITE, P. 1983. *The Lancashire and Westmorland mineral statistics with the Isle of Man: metalliferous and associated minerals 1845–1913. 4. The Mineral statistics of the United Kingdom, 1845–1913*. Department of Economic History, University of Exeter, in association with the Northern Mines Research Society and the Peak District Mines Historical Society, Exeter.

CAIN, R. 1994. *The use of geotechnical tests to determine the glacial origin of Irish Sea Basin quaternary deposits*. BSc Thesis, University College of North Wales, Bangor.

CANNELL, W. 1843. *A New Guide and Visitor's Companion*. William Cannell.

CANTRILL, T. C., SHERLOCK, R. L. & DEWEY, H. 1919. Iron ores continued: sundry unbedded ores of Durham, East Cumberland, North Wales, Derbyshire, the Isle of Man, Bristol district and Somerset, Devon and Cornwall. *Memoirs of the the Geological Survey, Special Reports on Mineral Resources of Great Britain*, **9**, 87.

CARRUTHERS, R. G. & STRAHAN, A. 1923. Lead and zinc ores of Durham, Yorkshire, Derbyshire with notes on the Isle of Man. *Memoirs of the Geological Survey, Special Reports on Mineral Resources of Great Britain*, **26**.

—— & —— 1986. Lead and zinc ores of Durham, Yorkshire, Derbyshire with notes on the Isle of Man. *Memoirs of the Geological Survey, Special Reports on Mineral Resources of Great Britain*, **26**. (Facsimile reprint. Mining Facsimiles, Sheffield.)

CARTER, P. A. 1977. Lough Cranstal. *In*: TOOLEY, M. J. (ed.) *The Isle of Man, Lancashire Coast and Lake District. Guide for excursion A4. X INQUA Congress, Birmingham*. Geo Abstracts Ltd, University of East Anglia, 28–29.

CHALLIS, P. J. 1983. The Snaefell Mine Accident 1897. *British Mining*, **23**, 20–23.

CHALONER, J. 1656. *A Short Treatise of the Isle of Man*. John Streater.

—— 1864. A short treatise of the Isle of Man, digested into six chapters. Edited, with an introductory notice and copious notes, by the Rev. J. G. Cumming. *Manx Society*, **X**.

CHAPMAN, A. J., RICKARDS, R. B. & GRAYSON, R. F. 1993. The Carboniferous dendroid graptolites of Britain and Ireland. *Proceedings of the Yorkshire Geological Society*, **49**, 295–319.

CLAGUE, D. 1884. The geology of the Isle of Man. *Transactions of the Liverpool Geological Association*, **4**, 78–82.

—— 1887. The conglomerate beds of the Isle of Man. *Transactions of the Liverpool Geological Association*, **8**, 31, 64, 71.

—— 1895. Further notes on the geology of the Isle of Man. *Journal of the Liverpool Geological Association*, **15**, 19–29.

CLARK, S. D. 1982. Collecting on the Isle of Man. *Gems*, **1**, 17–18.

—— 1985. Collecting on the Isle of Man, part 2. *Gems & Mineral Realm*, **16**, 28–29.

CLARKE, C., HORTON, M. & STRATTON, M. 1985. *A survey of mining remains at Laxey, Isle of Man*. Institute of Industrial Archaeology, Research Paper 2.

CLAYTON, T. T. G. 1995. *A palaeo-environmental reconstruction of the Holocene climates of the Isle of Man through a palynological study of Close Sartfield, the Ballaugh Curragh*. BSc Thesis, Worcester College of Higher Education.

CLEMENS, F. M. A., DIXON, D. J. & WOODWARD, C. H. 1960. The Great Laxey Wheel. *Architectural Association Journal*, **75**, 210–225.

COLE, G. A. J. 1892. On occurence of Riebeckite in Britain. *Mineralogical Magazine and Journal of the Mineralogical Society*, **9**, 222–226.

COOPE, G. R. 1971. Coastal sections between Glen Mooar and Orrisdale (West Coast). 2. The fossil Coleoptera from Glen Ballyre and their bearing upon the interpretation of Late Glacial environments. *In*: THOMAS, G. S. P. (ed.) *Field Guide to the Quaternary of the Isle of Man*. Quaternary Research Association, Liverpool, 13–15.

—— 1981. Episodes of local extinction of insect species during the Quaternary as indicators of climatic change. *In*: FLENLEY, J. & NEALE, J. (eds) *The Quaternary in Britain. Essays, Reviews and Original Work on the Quaternary Published in Honour of Lewis Penny on his Retirement*. Pergamon Press, 216–221.

—— & GARRAD, L. S. 1988. The petrological identification of stone implements from the Isle of Man. *In*: CLOUGH, T. H. MCK. & CUMMINS, W. A. (eds) *The Petrology of Prehistoric Implements from the British Isles*. Stone Axe Studies 2. Council for British Archaeology, Research Report **67**, 67–70.

COOPER, A. H. & MOLYNEUX, S. G. 1990. The age and correlation of the Skiddaw Group (Early Ordovician) sediments in the Cross Fell inlier, northern England. *Geological Magazine*, **127**, 147–157.

——, RUSHTON, A. W. A., MOLYNEUX, S. G., HUGHES, R. A., MOORE, R. M. & WEBB, B. C. 1995. The stratigraphy, correlation, provenance and palaeogeography of the Skiddaw Group (Ordovician) in the English Lake District. *Geological Magazine*, **132**, 185–211.

CORKILL, W. H. 1921. Manx mines and megaliths. *Memoirs & Proceedings of the Manchester Literary & Philosophical Society*, **65**, 1–8.

CORNWELL, J. D. 1972. A gravity survey of the Isle of Man. *Proceedings of the Yorkshire Geological Society*, **39**, 93–106.

COWIN, F. 1973. *Laxey Mines Trail*. Manx Conservation Council.

COWLEY, C. H. 1932–1942. The caves of Peel Hill. *Proceedings of the Isle of Man Natural History and Antiquarian Society, New Series*, **IV**, 26.

CRAINE, D. 1932–1942. Some notes on the Parish of Ballaugh. *Proceedings of the Isle of Man Natural History and Antiquarian Society, New Series*, **IV**, 449–470.

CRICK, G. C. 1895. On a new species of Prolecanites from the Carboniferous limestone of Haw Bank Tunnel, Skipton. *Transactions of the Manchester Geological Society*, **23**, 80–88.

—— 1904. On a new form of Carboniferous nautiloid (*Amphoreopsis paucicamerata*) from the Isle of Man. *Proceedings of the Malacological Society*, **6**, 134–137.

CROMPTON, N. V. 1942–1956. Report of the Field Section for the year 1952–1953. *Proceedings of the Isle of Man Natural History and Antiquarian Society, New Series*, **V**, 470–471.

CROWLEY, S. F. 1985. Lithostratigraphy of the Peel Sandstones, Isle of Man. *Mercian Geologist*, **10**, 73–76.

—— & BOTTRELL, S. H. 1997. *An isotopic investigation of base metal mineralisation on the Isle of Man and its relationship to fluid flow events and geotechnic evolution of the Irish Sea: initial results*. Abstract from Geo-fluids Conference, Belfast.

CRUTTWELL, C. (ed.) 1781. *The Life of the Right Reverend Thomas Wilson D. D., Lord Bishop of Sodor and Man*. Compiled by the Rev. C. Cruttwell; to which is added his History of the Isle of Man. Bath.

CUBBON, A. M. 1942–1956. The Ice Age in the Isle of Man: A reconsideration of the evidence. *Proceedings of the Isle of Man Natural History and Antiquarian Society, New Series*, **V**, 499–512.

—— 1952. The Island in the Making. *In*: TAYLOR, R. (ed.) *Isle of Man Log*. Merseyside Youth Hostels Ltd, 13–16.

CUMMING, J. G. 1846. On Posidonian Schist amidst Trappean Beds, and on Traces of Drift-ice in the South of the Isle of Man. *Report of the British Association for the Advancement of Science for 1845*, 60.

—— 1846. On the geology of the Isle of Man. *Quarterly Journal of the Geological Society of London*, **2**, 317–348.

—— 1847. On the geology of the Calf of Man. *Quarterly Journal of the Geological Society of London*, **3**, 179–185.

—— 1848. *The Isle of Man; its History, Physical, Ecclesiastical, Civil, and Legendary*. Van Voorst.

—— 1854. On the superior limits of the glacial deposits in the Isle of Man. *Quarterly Journal of the Geological Society of London*, **10**, 211–232.

—— 1855. On some of the more recent changes in the Area of the Irish Sea. *Edinburgh New Philosophical Journal, Series 2*, **1**, 57–62.

—— 1861. *A Guide to the Isle of Man*. Edward Stanford, London.

—— *c*. 1861. List of the fossils collected by Cumming in the museum at King William's College. *In*: *Leech's New Illustrated Tourists' Guide*. F. Leech, Ramsey, 108–111.

DACKOMBE, R. V. 1977. *Lithology and genesis of Manx till*. PhD Thesis, University of Liverpool.

—— 1990. Solid geology. *In*: ROBINSON, V. & MCCARROLL, D. (eds) *The Isle of Man – Celebrating a Sense of Place*. Liverpool University Press, 18–39.

—— & MCCARROLL, D. 1990. The Manx landscape. *In*: ROBINSON, V. & MCCARROLL, D. (eds) *The Isle of Man – Celebrating a Sense of Place*. Liverpool University Press, 10–17.

—— & THOMAS, G. S. P. (eds) 1985. *Field Guide to the Quaternary of the Isle of Man*. Quaternary Research Association, Cambridge.

—— & —— 1991. The glacial deposits and Quaternary stratigraphy of the Isle of Man. *In*: EHLERS, J., GIBBARD, P. & ROSE, J. (eds) *Glacial deposits of Great Britain and Ireland*. Balkema, 333–344.

DARWIN, C. 1848. On the transportal of erratic boulders from a lower to a higher level. *Quarterly Journal of the Geological Society of London*, **4**, 315–323.

DAVEY, P. J. (ed.) 1978. *Man and Environment in the Isle of Man*. Two volumes, British Series, **54**. British Archaeological Reports, Oxford.

DAVIDSON, C. F. & BOWIE, S. H. U. 1951. On thucholite and related hydrocarbon–uraninite complexes, with a note on the origin of the Witwatersrand gold ores. *Bulletin of the Geological Survey of Great Britain*, **3**, 1–19.

DAVIDSON, J. 1906–1912. Langness: Its Geology, etc. *Proceedings and Transactions of the Isle of Man Natural History and Antiquarian Society, New Series*, **I**, 645–648.

DAVIDSON, T. 1858–1863. *A Monograph of the British Fossil Brachiopoda, V. The Carboniferous Brachiopoda, 3–4*. Palaeontographical Society Monographs, London.

DAVIES, B. E. & KEAR, B. S. 1974. Environmental factors and the soil pattern on the Isle of Man. *In*: ANON. (ed.) *Soils of the World*. Transactions of the 10th International Congress of Soil Science, Moscow, **VIII**, 93–99.

DAVIES, D. C. 1888. *A Treatise on Metalliferous Minerals and Mining*. 4th Edition. Lockwood & Son.

DAWKINS, W. BOYD. 1880–1892. On the geology of the Isle of Man. Part I. On the conglomerates of the south of the island. *Yn Lioar Manninagh*, **I**, **ii**, 16–18.

—— 1888. On the clay slates and phyllites of the south of the Isle of Man, and a section of the Foxdale mines. *Transactions of the Manchester Geological Society*, **20**, 53.

—— 1894. On the geology of the Isle of Man. Part 1. The Permian, Carboniferous and Triassic Rock and the New Saltfield of the North. *Transactions of the Manchester Geological Society*, **22**, 590–613.

—— 1895. The geology of the Isle of Man. Part 2. *Transactions of the Manchester Geological Society*, **23**, 147–159.

—— 1896. *On the geology of the Isle of Man*. Report of the British Association for the Advancement of Science, **66**, 155–169.

—— 1901–1905. Excursion – St John's to Foxdale. *Yn Lioar Manninagh*, **IV**, 16–18.

—— 1901–1905. Excusion to Langness. *Yn Lioar Manninagh*, **IV**, 77–78.

—— 1901–1905. Excursion to the Salt Works, Bride. *Yn Lioar Manninagh*, **IV**, 119–120.

—— 1902. The red sandstones of Peel, Isle of Man. *Quarterly Journal of the Geological Society of London*, **58**, 633–646.

—— 1902. The Carboniferous, Permian and Triassic Rocks under the Glacial Drift in the North of the Isle of Man. *Quarterly Journal of the Geological Society of London*, **58**, 647–660.

—— 1913. Coal and salt in the Isle of Man. *Mannin*, **1**, 28–32.

DAWSON, J. 1966. Beryllium in the Foxdale Granite. *Bulletin of the Geological Survey of Great Britain*, **25**, 55–58.

DEVOY, R. J. 1983. Possible landbridges between Ireland and Britain: A geological appraisal. *In*: DEVOY, R. J., SLEEMAN, D. P. & WOODMAN, P. C. (eds) *Proceedings of the Postglacial Colonization Conference, University College, Cork 1983*. Irish Biogeographical Society, Occasional Paper No. 1, 15–26.

—— 1985. The problem of a Late Quaternary landbridge between Britain and Ireland. *Quaternary Science Reviews*, **4**, 43–58.

DEWEY, H. & EASTWOOD, T. 1925. Copper ores of the Midlands, Wales, the Lake District and the Isle of Man. *Memoirs of the Geological Survey. Special Reports on Mineral Resources of Great Britain*, **30**, 78–87.

—— & —— 1986. Copper ores of the Midlands, Wales, the Lake District and the Isle of Man. *Memoirs of the Geological Survey. Special Reports on Mineral Resources of Great Britain*, **30**, 78–87. (Facsimile reprint. Mineral Facsimiles, Sheffield.)

DICKSON, C. A., DICKSON, J. H. & MITCHELL, G. F. 1970. The Late Weichselian Flora of the Isle of Man. *Philosophical Transactions of the Royal Society of London*, **B258**, 31–79.

DICKSON, E. & HOLLAND, P. 1888–1889. An examination of some volcanic rocks of the Isle of Man. *Proceedings of the Liverpool Geological Society*, **6**, 123–131.

—— & —— 1894. Notes on shell breccia dredged off the coast of the Isle of Man, and on the source of the carbonate of lime of the shells of marine organisms. *Proceedings of the Liverpool Geological Society*, **7**, 164.

DICKSON, J. A. D. 1967. *The structure and sedimentation of the Carboniferous rocks of the Castletown area, Isle of Man*. PhD Thesis, University of London.

—— & BARBER, C. 1976. Petrography, chemistry and origin of early diagenetic concretions in the Lower Carboniferous of the Isle of Man. *Sedimentology*, **23**, 189–211.

—— & —— 1977. Chemical variation in a partially dolomitised Visean limestone bed, Isle of Man. *Mineralogical Magazine*, **41**, 145–153.

——, FORD, T. D. & SWIFT, A. 1987. The stratigraphy of the Carboniferous rocks around Castletown, Isle of Man. *Proceedings of the Yorkshire Geological Society*, **46**, 203–229.

DOBSON, M. 1986. The Beckwiths Push. *Manx Mines Research Group. Mann-Underground*, **5**, 6–7.

—— 1988. On an investigation of a walled up adit at Dhyrnane Maughold. *Manx Mines Research Group. Mann-Underground*, **8**, 5–7.

—— 1989. Glencherry. Garth Mine. Mount Dalby Siver Lead Mine. Niarbyl. Kerroo Mooar Silver–Lead Mine. *Manx Mines Research Group. Mann-Underground*, **9**, 1–11, 16–26.

—— 1991. Douglas Head Adit. *Manx Mines Research Group. Mann-Underground*, **10**, 1–2.

—— & HOLLIS, D. B. 1993. Lead mines and trials in Glen Auldyn, Isle of Man. *Bulletin of the Peak District Mines Historical Society*, **12**, 31–37.

DOUGHTY, P. S. no date. The state and status of geology in United Kingdom museums. *Geological Society Miscellaneous Papers (London)*, **13**, 118.

DOUGLAS, I. 1987. Plate tectonics, palaeoenvironment and limestone geomorphology in west-central Britain. *Earth Surface Processes and Landforms*, **12**, 481–495.

DOWNIE, C. & FORD, T. D. 1966. Microfossils from the Manx Slate Series. *Proceedings of the Yorkshire Geological Society*, **35**, 307–322.

DUNLOP, R. 1912. Notes on the geology of the Isle of Man. *Transactions of the Edinburgh Geological Society*, **10**, 242.

—— 1923. Notes on four Nautili. *Transactions of the Geological Society of Glasgow*, **17**, 78–79.

DURANT, G. P. & GRANT, C. 1985. The Scarlett Volcanic Complex, Isle of Man: a shallow water submarine volcano. *Geological Society Newsletter (London)*, **14**, 46.

DWERRYHOUSE, A. R. 1895. On terminal curvature in the Isle of Man. *Glacialists Magazine*, **2**, 164.

EAKIN, P. A. & GIZE, A. P. 1992. Reflected-light microscopy of uraniferous bitumens. *Mineralogical Magazine*, **56**, 85–99.

EASTWOOD, T. 1963. *British Regional Geology. Northern England.* 3rd Edition. HMSO.

ECCLES, P. G. & KAYE, K. A. 1984. Sulby reservoir, Isle of Man: design and construction. *Journal of the Institution of Water Engineers and Scientists*, **38**, 197–216.

EDEN, R. A., DEEGAN, C. E., RHYS, G. H., WRIGHT, J. E. & DOBSON, M. R. 1973. *Geological investigations with a manned submersible in the Irish Sea and off western Scotland.* Institute of Geological Sciences Report, No. **73/2**, 28.

EDWARDS, M. H. & HAIME, J. 1850–1855. *A Monograph of the British Fossil Corals.* Palaeontographical Society Monographs, London.

EHLERS, J., GIBBARD, P. & ROSE, J. (eds) 1991. *Glacial Deposits of Great Britain and Ireland.* A.A. Balkema.

ERDTMAN, G. 1924–1925. Pollen statistics from the Curragh and Ballaugh, Isle of Man. *Proceedings of the Liverpool Geological Society*, **14**, 158–163.

—— 1928. Studies in the postarctic history of the forests of north-western Europe. I. Investigations in the British Isles. *Geologiska Föreningens i Stockholm Förhandlingar*, **50**, 123–192.

EVANS, J. W. & STUBBLEFIELD, C. J. 1929. *Handbook of the Geology of Great Britain.* Murby.

EVENS, E. D., GRINSELL, L. V., PIGGOTT, S. & WALLIS, F. S. 1962. Complete list of specimens examined by the south-western implement petrology committee. *Proceedings of the Prehistoric Society*, **28**, 242–251.

EYLES, C. H. 1986. *Sedimentation on glacially-influenced continental shelves.* PhD Thesis, University of Toronto.

—— & EYLES, N. 1984. Glaciomarine sediments of the Isle of Man as a key to the late Pleistocene stratigraphic investigations in the Irish Sea basin. *Geology*, **12**, 359–364.

——, —— & MCCABE, M. 1985. Reply to comment on 'Glaciomarine sediments of the Isle of Man as a key to Late Pleistocene stratigraphic investigations in the Irish Sea Basin'. *Geology*, **13**, 446–447.

EYLES, N. & MCCABE, A. M. 1989. The late Devensian (<22,000 BP) Irish Sea Basin: the sedimentary record of a collapsed ice sheet margin. *Quaternary Science Reviews*, **8**, 307–351.

—— & —— 1991. Glaciomarine deposits of the Irish Sea Basin: The role of Glacio-Isostatic disequilibrium. *In*: EHLERS, J., GIBBARD, P. & ROSE, J. (eds) *Glacial Deposits of the British Isles.* A.A. Balkema, 311–331.

——, —— & WINGFIELD, R. T. R. 1992. The late Devensian (<22,000 BP) Irish Sea Basin: the sedimentary record of a collapsed ice sheet margin. Discussion and reply. *Quaternary Science Reviews*, **11**, 377–379.

FERGUSON, J. 1966. Variation in two species of the Carboniferous brachiopod *Pleuropugnoides*. *Proceedings of the Yorkshire Geological Society*, **35**, 353–374.

FLENLEY, J. & NEALE, J. (eds) 1981. *The Quaternary in Britain. Essays, Reviews and Original Work on the Quaternary Published in Honour of Lewis Penny on his Retirement.* Pergamon Press.

FOORD, A. H. & CRICK, G. C. 1888–1897. *Catalogue of the Fossil Cephalopoda in the British Museum.* London. Trustees of the British Museum (Natural History).

—— & —— 1894. On the identity of *Ellipsolites compressus* with *Ammonites Henslowi*. *Geological Magazine, Decade 4*, **1**, 11–17.

FORBES, D. 1867. Researches in British Mineralogy. *Philosophical Magazine, Series 4*, **34**, 329–354.

FORBES, E. 1831–1834? *Geology and flora of the Isle of Man, with recent additions by James F. Robinson.* Manuscript in the Royal Botanic Garden Library, Edinburgh.

—— 1839. Natural history. *In*: *Illustrated Guide and Visitor's Companion through the Isle of Man.* 2nd Edition. J.Quiggin.

—— 1841. *On a Pleistocene tract in the Isle of Man, and the relations of its fauna to that of the neighbouring sea.* Report of the British Association for the Advancement of Science for 1840, 104.

—— 1842. Geology. *In*: *Quiggin's Illustrated Guide and Visitor's Companion through the Isle of Man.* Quiggin (now Glover), 55–58.

—— 1846. List of fossils from the Pleistocene Marine Formation of the Isle of Man, collected by the Rev. J. G. Cumming and named by Prof. E. Forbes. *Quarterly Journal of the Geological Society of London*, **2**, 346–347.

—— 1846. On the connexion between the distribution of the existing fauna and flora of the British Isles, and the geological changes which have affected their area, especially during the epoch of the Northern Drift. *Geological Survey Memoir*, **1**, 336–432.

FORD, T. D. 1964–1972. Slump structures in the Peel Sandstone Series, Isle of Man. *Proceedings of the*

Isle of Man Natural History and Antiquarian Society, New Series, **VII**, 440–448.
—— 1974. *Geological field course guide*. Student field course handout, University of Leicester.
—— 1981. The stones of St Paul's. *Geologists' Association Circular*, **825**, 8.
—— 1984. Field meeting report: Isle of Man. *Proceedings of the Yorkshire Geological Society*, **45**, 133–136.
—— 1985. Field excursion to the Isle of Man: Report. *Mercian Geologist*, **9**, 243–244.
—— 1987. Geological Excursion Guide No. 4: Isle of Man. *Geology Today*, **3**, 64–69.
—— 1993. *The Isle of Man*. Geologists' Association Guide, **46**. Geologists' Association.
—— 1994. Mining on the Isle of Man. *Dips, Angles and Spurs*, **13**, 6–9.
—— 1995. A tale of Manx Geology. *University of Leicester Students Geological Society Journal*, **33**, 22–24.
—— 1999. The geology of Laxey mine. *Mercian Geologist*. in press.
FOSTER, C. LEN. 1883. On an emanation of carbonic acid gas at Foxdale Mine in the Isle of Man. *Transactions of the Royal Geological Society of Cornwall*, **10**, 175–176.
—— 1894. *A Text-Book of Ore and Stone Mining*. Charles Griffin.
—— & MILLER, A. E. 1898. *Report to HM Secretary of State for the Home Department on the circumstances attending an underground Fire which occurred at the Snaefell Lead Mine, Isle of Man, in the month of May 1897*. HMSO.
FULLEN, M. A. 1994. *Soils of the Isle of Man*. Student handout.
——, HARRIS, J. & KEAR, B. S. 1996. *Soils of the Isle of Man. Research Report, 5*. Centre for Manx Studies, Douglas.
GACIRI, S. J. & INESON, P. R. 1981. Hydrogeochemical lead–zinc–copper anomalies in the Isle of Man. *Transactions of the Institution of Mining and Metallurgy*, Section B, **90**, 120–125.
GARNER, R. 1878. Professor Edward Forbes and his Country. *Midland Naturalist* **1**, 67–70, 90–94.
GARRAD, L. S. 1978. The Carboniferous fossil collection in the Manx Museum. *Newsletter of the Geological Curators' Group*, **2**, 3.
—— 1978. John Kelly's Ballawillan Carding Mill and the Ohio Mine, East Baldwin, Isle of Man. *Industrial Archaeology*, **13**, 331–334.
—— 1978. Stone implements and their users in the Isle of Man. *In*: DAVEY, P. J. (ed.) *Man and Environment in the Isle of Man, Part 1. British Series, 54*. British Archeological Reports, Oxford, 165–170.
—— 1979. The Manx Museum shells from the Manxland drift. *Newsletter of the Geological Curators' Group*, **2**, 5.
—— 1980–1989. Items in the collection of the Manx Museum likely to be from horizontal mills. *Proceedings of the Isle of Man Natural History and Antiquarian Society, New Series*, **IX**, 251–256.
—— 1981. Snaefell Mine, Lonan, Isle of Man, with particular reference to the use of water power: Part I. *Industrial Archaeology*, **16**, 116–125.
——, BAWDEN, T. A., QUALTROUGH, J. K. & SCATCHARD, W. J. 1972. *The Industrial Archaeology of the Isle of Man*. David and Charles.
GARWOOD, E. J. 1900. *Life-zones in the British Carboniferous rocks*. Report of the British Association for the Advancement of Science for 1899, 371-375.
GASKING, S. 1880–1892. On the geology of the south of the Island. *Yn Lioar Manninagh*, **I**, **i**, 119–122.
—— 1880–1892. The so-called 'Old Red Sandstone' formation of the Isle of Man. *Yn Lioar Manninagh*, **I**, **i**, 135–138.
—— 1887–1888. With the British Association in the Isle of Man. *Transactions of the Liverpool Geological Association*, **8**, 21–28.
GEDDES, P. 1985. Laxey Trail. *Manx Mines Research Group. Mann-Underground*, **1**, 1–6.
—— 1986. A first for the Manx research group. Glen Darragh level. *Manx Mines Research Group. Mann-Underground*, **5**, 3–5.
—— 1989. Rebuilding the Welsh Shaft. *Manx Mines Research Group. Mann-Underground*, **9**, 3–10.
—— 1994. Laxey Mine: Excavation at mouth of adit. The Capping of Glen Cherry. The Reclaiming of Dixons. *Bulletin of Laxey Mines Research Group. Mann-Underground*, **12**, 1–15.
GEE, S. 1969. Derbyshire Caving Club. Report on the Easter 1969 Expedition to the Mines of the Isle of Man. *Isle of Man Natural History and Antiquarian Society: Field Section. Notes and Queries*, **1**, 21–25.
GEFFEN, A. J. 1991. *Irish Sea Study Group: Report of the Irish Sea Conference, Isle of Man, October, 1990*. Liverpool University Press.
GEIKIE, A. 1897. *The Ancient Volcanoes of Great Britain*. Macmillan.
—— 1912–1925. Address to the London Manx Society. *Proceedings of the Isle of Man Natural History and Antiquarian Society, New Series*, **II**, 71–86.
GEIKIE, J. 1893. The intercrossings of erratics in glacial deposits. *In*: *Fragments of Earth Lore etc*. John Bartholomew.
GEOLOGICAL SURVEY OF GREAT BRITAIN & GEIKIE, A. 1893–1897. *Annual Report of the Geological Survey and Museum of Practical Geology to the Department of Science and Art*. HMSO.
GEORGE, I. E. 1888–1889. Notes on some Manx Lavas. *Journal of the Liverpool Geological Association*, **9**, 41–47.
GEORGE, T. N. 1932. The British Carboniferous Reticulate Spiriferidae. *Quarterly Journal of the Geological Society of London*, **88**, 516–575.
—— 1958. Lower Carboniferous paleogeography in the British Isles. *Proceedings of the Yorkshire Geological Society*, **31**, 227–318.
GIBB, J. O. T., ALLEN, J. R. & HAWKINS, S. J. 1996. The application of bio-monitors for the assessment of mine-derived pollution on the west coat of the Isle of Man. *Marine Pollution Bulletin*, **32**, 513–519.
GILBERTSON, W. 1835. A notice of the result of a visit to the Isle of Man in search of fossils. *Magazine of Natural History*, **8**, 119–120.
GILL, E. L. 1903. Keisley limestone–pebbles from the Isle of Man. *Quarterly Journal of the Geological Society of London*, **59**, 307–310.

GILLINGS, A. M. 1981. Recollections of a Laxey miner: The story of Bobby Kelly as told to A. M. Gillings. *Bulletin of the Peak District Mines Historical Society*, **8**, 43–48.
GILLOTT, J. E. 1954. *Sedimentary petrology, metamorphism and structure of the Manx slates*. PhD Thesis, University of Liverpool.
—— 1955. Metamorphism of the Manx Slates. *Geological Magazine*, **92**, 141–154.
—— 1956. Breccias in the Manx Slate; their origins and stratigraphic relations. *Liverpool and Manchester Geological Journal*, **1**, 370–380.
—— 1956. Structural geology of the Manx Slate. *Geological Magazine*, **93**, 301–313.
GORDON, M. O. 1900. Similar geological structures in South Tyrol and the Isle of Man. *Nature*, **6l**, 490.
—— 1900. Rock structures in South Tyrol and the Isle of Man. *Nature*, **62**, 7.
GOULD, S. J. 1974. The origin and function of `bizarre' structures: antler size and skull size in the `Irish Elk' *Megaloceras giganteus*. *Evolution*, **28**, 191–220.
GRANT, R. F. M. 1955. The Snaefell Mine. *Mining Magazine*, **92**, 79–86.
GRAY, F. J. 1897–1898. The Rocks of the Isle of Man. *Hardwicke's Science Gossip, New Series*, **4**, 199-203, 219–222, 259–261.
GREG, R. P. & LETTSOM, W. G. 1858. *Manual of the Mineralogy of Great Britain and Ireland*. Van Voorst.
—— & —— 1977. *Manual of the mineralogy of Great Britain and Ireland*. (Facsimile reprint with supplementary list of British minerals by L. J. Spencer and a fourth supplementary list together with a foreword by P. G. Embrey.) Lapidary Publications, Broadstairs.
GREGORY, J. M. 1920. The red rocks of a deep bore at the end of the Isle of Man. *Transactions of the Institute of Mining Engineers of Scotland*, **59**, 156–168.
GRESSWELL, R. K. 1956. Note on pro-glacial lake deposits on the western side of the Isle of Man. *Liverpool and Manchester Geological Journal*, **1**, 381–384.
—— 1966. *Report of coastal conditions around Ayre, Isle of Man, with special reference to advisability of winning gravel at and near the Point of Ayre*. Report to Messrs B.S.G. (Aggregates), Ramsey.
—— 1967. *Physical Geography*. Longmans.
GRINDLEY, T. 1862. Geology of the Isle of Man. *The Geologist*, **5**, 171–183.
—— 1863. Footprints in the Cambrian? slates. *The Geologist*, **6**, 315.
—— 1865. On certain tracks in the Manx Slates. *Geological Magazine*, **2**, 542.
—— 1888–1890. On some recent errors in Manx Geology. *Transactions of the Manx Geological Society*, **1**, 1–12.
GROSSMAN, K. 1892. *Photographs of Geological Features*. Manx National Heritage Library.
GROVES, A. W. (ed.) 1953. *Wartime investigations into the haematite and manganese ore resources of Great Britain and Northern Ireland*. Permanent Records of Research and Development No. 20.703. Ministry of Supply.
HARDMAN, W. A. & LOMAS, J. 1898. On the floor deposits of the Irish Sea. *Proceedings of the Liverpool Geological Society*, **8**, 205–232.
HARKER, A. 1888. The Igneous Dykes of the North of England. *The Naturalist*, **161**, 349–353.
—— 1894. The Foxdale Granite. *Midland Naturalist*, **March**, 68.
—— 1895. Carrock Fell, part 3. The Grainsgill Greisen. *Quarterly Journal of the Geological Society of London*, **51**, 143–144.
HARKNESS, R. & NICHOLSON, H. 1866. On the Lower Silurian Rocks of the Isle of Man. *Quarterly Journal of the Geological Society of London*, **22**, 488–491.
HARPER, C. T. 1966. Potassium–argon ages of slates from the southern Caledonides of the British Isles. *Nature*, **212**, 1339–1341.
HARRIS, A. L., HOLLAND, C. H. & LEAKE, B. E. (eds) 1980. *The Caledonides of the British Isles – Reviewed*. Geological Society, London, Special Publications, **8**.
HARRISON, A. 1968. Economic opportunities, 1700. Some observations from the Derby Papers. *Journal of the Manx Museum*, **VII**, 81–83.
HARRISON, S. N. 1880–1892. Boulders around the Maughold Coast. *Yn Lioar Manninagh*, **I**, **ii**, 208–212.
—— 1892–1895. Striated rock surfaces in Maughold. *Yn Lioar Manninagh*, **II**, 52.
—— 1895–1900. Report of the Geological Section. *Yn Lioar Manninagh*, **III**, 21–22, 120–121.
—— 1895–1900. Some remarkable boulders noticed in 1897. *Yn Lioar Manninagh*, **III**, 323–324.
—— 1901–1905. Report of the Geological Section. *Yn Lioar Manninagh*, **IV**, 41–42.
—— 1912–1925. Isle of Man Fossils: List of shells from the Shellag Brooghs. *Proceedings of the Isle of Man Natural History and Antiquarian Society*, New Series, **II**, 167–169.
HARRISON, W. (ed.) 1871. The old historians of the Isle of Man: Camden, Speed, Dugdale, Cox, Wilson, Willis and Grose. *Manx Society*, **18**.
HARRISON, W. J. 1883. Geology of the Isle of Man. *Knowledge*, **3**, 277–278.
HAVILAND, A. 1880–1892. The necessity for collecting and arranging the ascertained facts relating to the glaciation of Man. *Yn Lioar Manninagh*, **I**, **i**, 57–64.
—— 1880–1892. The aspects of the 'Crags' of the Manx Mountains in relation to the glaciation of the island. *Yn Lioar Manninagh*, **I**, **i**, 77–82.
—— 1886. On the study of the glacial period in the Isle of Man. *Isle of Man Times*, **16th January**.
—— 1888–1890. Inaugural address by the President with a Manx geological bibliography. *Transactions of the Manx Geological Society*, **1**, 7th January, 1–16.
—— 1888–1890. The President's retiring Address. *Transactions of the Manx Geological Society*, **1**, 28th November, 1–8.
—— 1889. The physical geography and scenery of the Isle of Man. *Isle of Man Times*, **20 February**.
HELM, D. G., ROBERTS, B. & SIMPSON, A. 1963. Polyphase folding in the Caledonides south of the Scottish Highlands. *Nature*, **200**, 1060–1062.
HENSLOW, J. S. 1821. Supplementary observations to Dr Berger's account of the Isle of Man. *Transactions of*

the Geological Society of London, Series 1, **5**, 482–505.

—— 1824. Remarks upon Dr Berger's Reply to Mr Henslow's Observations etc. *Annals of Philosophy, Series 2*, **8**, 407.

Herdman, W. A. 1894. Notes on the submarine deposits of the Irish Sea. *Proceedings of the Liverpool Geological Society*, **7**, 171–182.

—— 1900. *The Geological Succession and Origin of Man*. C. Tinling.

Hewitt, W. 1885. Notes on glacial deposits and markings in the south of the Isle of Man. *Proceedings of the Liverpool Geological Society*, **5**, 352–358.

Heywood, S. A. 1987. An exotic local. *Manx Life*, **XVI**, 36–38.

Hibbert, S. 1825. Account of the circumstances connected with the discovery of the fossil Elk in the Isle of Man, etc. *Edinburgh Journal of Science*, **3**, 15–28.

—— 1825. Notice of the remains of an animal resembling the Scandinavian Elk, recently discovered in the Isle of Man, etc. *Edinburgh Journal of Science*, **3**, 129.

Hill, D. 1940. *A Monograph on the Carboniferous Rugose Corals of Scotland*. Palaeontographical Society, London.

Hind, W. 1896–1905. *A Monograph of the British Carboniferous Lamellibranchiata*. Palaeontographical Society, London.

—— 1907. The palaeontological succession of the Carboniferous rocks in the south of the Isle of Man. *Proceedings of the Yorkshire Geological Society*, **16**, 137–154.

Hobson, B. 1880–1892. On the igneous rocks of the South of the Isle of Man. *Yn Lioar Manninagh*, **I**, **ii**, 337–348.

—— 1891. *Olivine–dolerite (or basalt) dykes of the Isle of Man*. Manx National Heritage Library.

—— 1891. On the igneous rocks of the south of the Isle of Man. *Quarterly Journal of the Geological Society of London*, **47**, 432–450.

—— 1892. An Irish Augite. *Geological Magazine, Decade 3*, **9**, 348–350.

Hollis, D. B. 1986–1987. The Great Laxey winding engine. *Bulletin of the Peak District Mines Historical Society*, **9**, 306–312.

—— 1986–1987. Great Laxey Mine, Isle of Man. *Bulletin of the Peak District Mines Historical Society*, **10**, 17–23.

—— 1987. Uranium in the Isle of Man: a review of the history and geology of uranium prospects at Snaefell and Laxey mines. *British Mining*, **34**, 44–45.

—— 1987. The forgotten iron mines of Kirk Maughold, Isle of Man. *British Mining*, **34**, 46–54.

—— 1988. The Kirk Maughold iron mines, Isle of Man: Recent explorations. *British Mining*, **37**, 4–15.

—— 1988. Hydrocarbons in sulphide ore deposits. *British Mining*, **37**, 16.

—— 1988. The 'Ohio' ('Great East Baldwyn', or 'Man') mine, the West Baldwyn mine, and the Abbeylands mine, Isle of Man: a case study in small scale nineteenth century speculative mining. *British Mining*, **37**, 63–78.

—— 1989. The search for iron, lead and gold in Douglas Bay, Isle of Man; old mining trials and new insight into the geology. *Bulletin of the Peak District Mines Historical Society*, **10**, 291–304.

—— 1990. Bell Hole, Foxdale – the history and geology of a typical small mine of the Isle of Man. *Bulletin of the Peak District Mines Historical Society*, **11**, 35–42.

—— 1993. The geology of the Isle of Man – Report on the expedition, August 1991, of the Glasgow Geological Society. *Proceedings of the Glasgow Geological Society, Session 129–133*, 54–63.

Hopson, T. & Lamb, J. (eds) 1995. *Manx hill – land report*. Manx Nature Conservation Trust, Tynwald Mills, St Johns.

Horn, D. P. 1992. A numerical model for shore-normal sediment size variation on a macrotidal beach. *Earth Surface Processes and Landforms*, **17**, 755–773.

—— 1992. A review and experimental assessment of equilibrium grain size and the ideal wave-graded profile. *Marine Geology*, **108**, 161–174.

—— 1993. Sediment dynamics on a macrotidal beach; Isle of Man, UK. *Journal of Coastal Research*, **9**, 189–208.

—— & Hardistry, J. 1990. The application of Stokes' wave theory under changing sea levels in the Irish Sea. *Marine Geology*, **94**, 341–345.

—— & Mason, T. 1994. Swash zone sediment transport modes. *Marine Geology*, **120**, 309–325.

Horne, J. 1874. Geology of the Isle of Man. *Transactions of the Edinburgh Geological Society*, **2**, 323–347.

—— 1875. The Post-Pliocene Formations of the Isle of Man. *Geological Magazine, Decade 2*, **2**, 329–331.

Howorth, H. H. 1877. Geology of the Isle of Man. *Geological Magazine, Decade 2*, **4**, 410–413, 456–459.

—— 1894. The most recent changes of level and their teaching. *Geological Magazine, Decade 4*, **1**, 405–413.

Hoyle, D. M. 1983. Beckwiths mine, Isle of Man. *Mineral Realm*, **3**, 26–28.

Hughes, R. A., Cooper, A. H. & Stone, P. 1993. Structural evolution of the Skiddaw Group (English Lake District) on the northern margin of eastern Avalonia. *Geological Magazine*, **130**, 621–629.

Humphreys, P. E. 1942–1956. Prehistoric Man. *Proceedings of the Isle of Man Natural History and Antiquarian Society, New Series*, **V**, 479–483.

Hunt, R. 1884. *British Mining*. Croxby Lockwood.

Hutchings, W. M. 1896. Clays, Shales and Slates. *Geological Magazine, Decade 4*, **3**, 309.

Huxley, T. H. 1869. The anniversary address of the President. Memoir of the Rev. J. G. Cumming 1812–1868. *Quarterly Journal of the Geological Society of London*, **25**, xxxvii–xxxviii.

Ineson, P. R. & Mitchell, J. G. 1979. K–Ar ages from the ore deposits and related rocks of the Isle of Man. *Geological Magazine*, **116**, 117–128.

Innes, J. B. 1995. *The Dhoo Valley, Isle of Man: a palaeoenvironmental assessment*. Research Report, **2**. Centre for Manx Studies, Douglas.

HM Inspectors of Mines 1887. *Mining and mineral statistics of the United Kingdom of Great Britain and Ireland, including lists of mines and mineral works, for the year 1886 and a list of plans of abandoned mines*. HMSO.

—— 1895–1896. *List of quarries in the United Kingdom of Great Britain and Ireland and the Isle of Man*. Home Office.

—— 1897–1921. *List of Mines in the United Kingdom of Great Britain and Ireland and the Isle of Man*. HMSO.

—— 1897–1921. *List of Quarries in the United Kingdom of Great Britian and Ireland and the Isle of Man*. HMSO.

—— 1898–1920. *Mines and Quarries: General Report with Statistics by the Chief Inspector of Mines*. HMSO.

—— 1922–1948. *List of Quarries in Great Britain and the Isle of Man*. HMSO.

—— & ATKINSON, J. B. 1895–1897. *List of Mines in the United Kingdom of Great Britain and Ireland and the Isle of Man*. Home Office.

—— & FOSTER, C. LEN. 1882. *Mining reports. Report upon the inspection of metalliferous mines in the counties of Anglesey, Brecon, Cardigan, Carnavon, Denbigh, Flint, Merioneth, Montgomery, Radnor, and Shropshire, and in the Isle of Man, for ... 1881*. Home Office.

—— & —— 1889–1899. *Report of the North Wales and Isle of Man District (No. 9) to Her Majesty's Secretary of State*. HMSO.

INSTITUTE OF GEOLOGICAL SCIENCES 1977. *54N–06W: Isle of Man*. Institute of Geological Sciences, London 1:250 000 series, Bouguer gravity anomaly map.

—— 1978. *54N–06W: Isle of Man*. Institute of Geological Sciences, London 1:250 000 series, aeromagnetic anomaly map..

IRISH SEA STUDY GROUP 1990. *The Irish Sea – An Environmental Review*. Liverpool University Press.

ISLE OF MAN FIELD GUIDES 1958. *The Geology and Archaeology of Scarlett*. Manx Museum and National Trust.

JACKSON, D. I. 1992. Hydrocarbon exploration in the Manx Basin. *Proceedings of Cumberland Geological Society*, **5**, 382–383.

—— & MULHOLLAND, P. 1993. Tectonic and stratigraphic aspects of the East Irish Sea Basin and adjacent areas: contrasts in their post-carboniferous structural styles. *In*: PARKER, J. R. (ed.) *Petroleum Geology of Northwest Europe: Proceedings of the 4th Conference*. The Geological Society, London, 791–808.

——, JACKSON, A. A., EVANS, D., WINGFIELD, R. T. R., BARNES, R. P. & ARTHUR, M. J. 1995. *The Geology of the Irish Sea*. United Kingdom Offshore Regional Report British Geological Survey. HMSO.

JENKINSON, H. I. 1874. *Jenkinson's Practical Guide to the Isle of Man*. Edward Stanford.

JESPERSEN, A. 1970. *The Lady Isabella Waterwheel of the Great Laxey Mining Company, Isle of Man, 1854–1954*. 3rd Revised edition. Private Publication, Virum.

JOACHIM, K. 1977. Glen Ballyre. *In*: TOOLEY, M. J. (ed.) *The Isle of Man, Lancashire Coast and Lake District. Guide for Excursion A4. X INQUA Congress, Birmingham*. Geo Abstracts Ltd, University of East Anglia, 33–36.

—— 1978. *Late Glacial coleopteron assemblages from the west coast of the Isle of Man*. PhD Thesis, University of Birmingham.

JOHNSON, R. J. (ed.) 1985. *The Geomorphology of North-West England*. Manchester University Press.

JOLLIFFE, I. P. 1979. *Interim report on coastal instability at Ballure Park, Isle of Man*. Report to the Isle of Man Harbour Board and the Isle of Man Railways Board.

—— 1981. *An investigation into coastal erosion problems in the Isle of Man: causes, effects, and remedial strategies*. Report to the Isle of Man Harbour Board and the Isle of Man Government.

JONES, N. S., KAIN, J. K. & STRIDE, A. H. 1965. The movement of sand waves on Warts Bank Isle of Man. *Marine Geology*, **3**, 329–336.

JONES, R. L. & KEEN, D. H. 1993. *Pleistocene Environments in the British Isles*. Chapman & Hall.

JONES, T. A. 1927. The geological history of the Isle of Man. *Proceedings of the Liverpool Geological Society*, **14**, 285–305.

JONES, T. R. & KIRBY, J. W. 1886. Notes on the distribution of the ostracoda of the carboniferous formations of the British Isles. *Quarterly Journal of the Geological Society of London*, **42**, 496–514.

JONES, W. 1980. *The nature and movement of beach shingle around the Point of Ayre, Isle of Man*. BSc Dissertation, Bedford College, London.

JUDD, J. W. 1897. William Smith's Manuscript maps. *Geological Magazine, Decade 4*, **4**, 439–447.

KEAR, B. S. 1971. *Agriculture soil survey. Report on the examination of soils in the first phase areas of development around the main urban centres in the Isle of Man*. Department of Town and Country Planning, Isle of Man Local Government Board.

—— 1972–1980. Soils of the Isle of Man. *Proceedings of the Isle of Man Natural History and Antiquarian Society, New Series*, **VIII**, 38–50.

—— 1979. *Soils in the Isle of Man*. Additional Special Meeting, North of England Soils Discussion Group, Field Excursion Guide, 2–5.

—— 1980. Soils in the Isle of Man: pattern and process. *Manchester Geographer, New Series*, **1**, 21–34.

—— 1982. *The Isle of Man. Its soils and their significance*. North of England Soils Discussion Group, Special Publications, **18**, 53.

KEEGAN, P. Q. 1888. In the Isle of Man (section on geology and botany). *Hardwicke's Science Gossip*, **280**, 73–75.

KENDALL, J. D. 1883. The structure of the Cumberland Coal Field. *Transactions of the North of England Institute of Mining Engineers*, **32**, 319–356.

KENDALL, P. F. 1880–1892. Manx minerals. *Yn Lioar Manninagh*, **I, ii**, 167.

—— 1880–1892. The glacial phenomena of the Isle of Man. *Yn Lioar Manninagh*, **I, ii**, 227–228.

—— 1880–1892. On the glacial geology of the Isle of Man. *Yn Lioar Manninagh*, **I, ii**, 397–438.

—— 1889. On the occurrence of *Nassa serrata* Brocchi in the glacial drift of the North Shore, Ramsey, Isle of Man. *Transactions of the Manx Geological Society*, **1, 4th February**, 5–8.

—— 1891. On the source of some remarkable boulders in the Isle of Man. *Memoirs & Proceedings of the*

Manchester Literary & Philosophical Society, Series 4, **4**, 217–220.

Kenyon, P. 1965. The Great Laxey Mining Company, Isle of Man: surviving evidence. *Industrial Archaeology*, **2**, 154–167.

Kermode, P. M. C. 1880–1892. Fossil shells from the boulder clay and sand, North Ramsey. *Yn Lioar Manninagh*, **I**, **ii**, 96–98.

—— 1895–1900. Excursion to Poortown, St German's. *Yn Lioar Manninagh*, **III**, 318–319.

—— 1897. Occurrences of Irish Elk remains in the Isle of Man. *Glacialists Magazine*, **5**, 101–112.

—— 1898. The 'Irish Elk' *Cervus giganteus* in the Isle of Man. *Geological Magazine, Decade 6*, **5**, 116–119.

—— 1898. *Reports of the committee of the British Association on the finding of the remains of the Irish Elk in the Isle of Man*. Report of the British Association for the Advancement of Science for 1897, 548–551.

—— 1899. *Reports of the committee of the British Association on the finding of the remains of the Irish Elk in the Isle of Man*. Report of the British Association for the Advancement of Science for 1898, 376.

—— 1900. *Reports of the committee of the British Association on the finding of the remains of the Irish Elk in the Isle of Man*. Report for the British Association for the Advancement of Science for 1899, 349–350.

—— 1925–1932. Address of retiring president – Apologia. *Proceedings of the Isle of Man Natural History and Antiquarian Society, New Series*, **III**, 406–417.

Kidson, C. & Tooley, M. J. (eds) 1977. *The Quaternary History of the Irish Sea*. Geological Journal Special Issue, **7**.

Kinahan, G. H. 1896. On possible land connections, in recent geological times, between Ireland and Great Britain. *Transactions of the Manchester Geological Society*, **24**, 113–130.

King, C. A. M. 1976. *Northern England. In:* Brown, E. H. & Clayton, K. (eds) *The Geomorphology of the British Isles*. 1st Edition. Methuen.

—— & Andrews, J. T. 1967. Radiocarbon date and significance from the Bride moraine, Isle of Man. *Geological Journal*, **5**, 305–308.

King, R. J. 1995. Minerals explained 19: The dolomite group. *Geology Today*, **11**, 105–107.

Kinvig, R. H. 1944. *A History of the Isle of Man. A Social, Cultural, and Political History*. 1st Edition. Liverpool University Press.

Kitto, W. H. 1892–1895. (1) Note on the History of Foxdale Mines. (2) Feather Ore (Plumosite). *Yn Lioar Manninagh*, **II**, 32–33.

Ladds, E. F. (ed.) 1976. Industrial Archaeology Group study of limekilns. *Isle of Man Natural History and Antiquarian Society Notes and Queries*, **1**, 7–8.

Laffoley, N. 1990. Iron, lead and gold in Douglas Bay. *Bulletin of the Peak District Mines Historical Society*, **11**, 20.

Lambreck, K. 1996. Glaciation and sea-level change for Ireland and the Irish Sea since Late Devensian/Midlandian time. *Journal of the Geological Society, London*, **153**, 853–872.

Lamplugh, G. W. 1895. The Crush-conglomerates of the Isle of Man. *Quarterly Journal of the Geological Society of London*, **51**, 564–597.

—— 1896. An outline of the Geology of Man. *In*: *An Appendix to the Handbook for the Liverpool Meeting of the British Association in 1896*. Philip, Son & Nephew, 165–182.

—— 1898. The Glacial Period and the Irish fauna. *Nature*, **57**, 245.

—— 1898. *Isle of Man, Solid and Drift*. Geological Survey geological maps, Map Sheet 100.

—— 1900. Rock structures in the Isle of Man and in South Tyrol. *Nature*, **61**, 612.

—— 1900. On some effects of Earth-movement on the Carboniferous volcanic rocks of the Isle of Man. *Quarterly Journal of the Geological Society of London*, **56**, 11–25.

—— 1903. *The Geology of the Isle of Man*. Memoir of the Geological Survey, of the United Kingdom. HMSO.

—— 1903. *Economic Geology of the Isle of Man, With Special Reference to the Metalliferous Mines*. Geological Survey of Great Britain, HMSO.

—— 1906. Address to the Geological Section on British Drifts and the Interglacial Problem. *British Association for the Advancement of Science York 1906*, 1–27.

Landless, J. 1976. An introduction to the mines of the Isle of Man. *British Mining*, **3**, 24–41.

Laxey Heritage Trust 1989. *Laxey Mine Trail*. Leaflet.

Lee, A. 1888. Marble and Marble Workers, A Handbook for Architects, Artists, Masons and Students. Croxby Lockwood & Son.

Leech *c*. 1861. *New Illustrated Tourists' Guide*. F. Leech.

Lewis, F. J. 1907. *The peat moss deposits in the Cross Fell, Caithness and Isle of Man districts*. Report of the British Association for the Advanvement of Science for 1907, 410–417.

Lewis, H. C. 1894. *Papers and Notes on the Glacial Geology of Great Britain and Ireland*. Longmans, Green and Co.

Lewis, H. P. 1927. '*Auloclisia*' a new coral genus from the Carboniferous Limestone. *Proceedings of the Yorkshire Geological Society*, **21**, 29–46.

—— 1928. *The zoning of the Avonian rocks in the south of the Isle of Man*. Report of the British Association for the Advancement of Science for 1927, 322.

—— 1929. The Avonian sucession in the south of the Isle of Man. *Abstracts of the Proceedings of the Geological Society of London*, **1206**, 26–30.

—— 1929. On the Avonian coral *Caninophyllum, gen. nov.* and *C. archiaci* (Edwards and Haime). *Annals and Magazine of Natural History*, **10**, 456–468.

—— 1930. The Avonian succession in the south of the Isle of Man. *Quarterly Journal of the Geological Society of London*, **86**, 234–290.

—— 1931. A sandstone with fluorspar cement and other sandstones from West Cumberland [Isle of Man]. *Geological Magazine*, **68**, 543–557.

—— 1933. The occurrence of fossiliferous pebbles of Salopian age in the Peel Sandstone [Isle of Man]. *Summary of Progress of the Geological Survey*, **2**, 91–108.

Lomas, J. 1892–1895. Notes on a geological excursion to Man. *Yn Lioar Manninagh*, **II**, 22–29.

—— 1903. Quartz dykes near Foxdale, Isle of Man. *Geological Magazine, Decade 4*, **10**, 34–36.

LONGSTAFF, J. D. 1933. A revision of the British Carboniferous Loxonematidae with descriptions of new forms. *Quarterly Journal of the Geological Society of London*, **89**, 87–124.

LOUIS, H. 1923. The production of lead in Britain. *Chemistry & Industry, New Series*, **42**, 252–255, 286–290.

MACARTHUR, C. W. P. 1990. Dr Jean-François Berger of Geneva (1779–1833): From the Travelling Fund the Wollaston Donation. *Archives of Natural History*, **17**, 97–119.

MACKAY, R. A. & SCHNELLMAN, G. A. 1963. *The mines and minerals of the Isle of Man ... submitted to the Industrial Officer for the Government of the Isle of Man*. Isle of Man Government.

MCCARROLL, D. 1990. The Quaternary Ice Age in the Isle of Man: an historical perspective. *In*: ROBINSON, V. & MCCARROLL, D. (eds) *The Isle of Man: Celebrating a Sence of Place*. Liverpool University Press, 40–54.

——, GARRAD, L. S. & DACKOMBE, R. V. 1990. Late glacial and postglacial environmental history. *In*: ROBINSON, V. & MCCARROLL, D. (eds) *The Isle of Man: Celebrating a Sense of Place*. Liverpool University Press, 55–76.

MCCORRY, M. 1995. Petrological examination of five samples from the Isle of Man. *In*: BARTON, R. S. *Manx granite tempered ware*. MA Thesis, University College, Dublin, 142–146.

MCCOY, F. 1851. Description of some new Mountain Limestone Fossils. *Annals and Magazine of Natural History, Series 2*, **7**, 167.

MCCRONE, J. 1941–1946. Mines and quarries in 1826. *Journal of the Manx Museum*, **V**, 26.

MCCULLOCH, J. 1819. *A Description of the Western Islands of Scotland, Including the Isle of Man: Comprising an Account of Their Geological Structure, With Remarks on Their Agriculture, Scenery and Antiquities*. Three volumes. Constable & Co.

MCKERROW, W. S. & SOPER, N. J. 1989. The Iapetus suture in the British Isles. *Geological Magazine*, **126**, 1–8.

MANX CONSERVATION COUNCIL. 1972. *Scarlett Geological Nature Trail*. Leaflet.

MANX MUSEUM AND NATIONAL TRUST. 1989. *Lady Isabella and the Great Laxey Mine*. Official guide.

MANX NATURE CONSERVATION TRUST *c*. 1992. *Scarlett Nature Trail*. Leaflet.

MARR, J. E. 1898. *The Principles of Stratigraphical Geology*. Cambridge University Press.

MARROW, P. H. 1886. A geological ramble to the Isle of Man. *Transactions of the Liverpool Geological Association*, **7**, 56–62.

MATHIESON, N. 1942–1956. Manx mines during the Athol period. *Proceedings of the Isle of Man Natural History and Antiquarian Society*, **New Series**, **V**, 559–560.

MEADOWS, N. S., TRUEBLOOD, S. P., HARDMAN, M. & COWAN, G. (eds) 1997. *Petroleum Geology of the Irish Sea and Adjacent Areas*. Geological Society, London, Special Publications, **124**.

MEGAW, B. 1957–1965. 'Bakenaldwath' and the medieval lead mines. *Journal of the Manx Museum*, **VI**, 105–107.

METEOROLOGICAL OFFICE 1963. Rainfall over the Isles of Man, Jersey and Guernsey 1916–1950. *Hydrological Memoranda*, **23**.

MITCHELL, G. F. 1958. A late-glacial deposit near Ballaugh, Isle of Man. *New Phytologist*, **57**, 256–263.

—— 1960. The Pleistocene history of the Irish Sea. *Advancement of Science*, **17**, 313–325.

—— 1965. The Quaternary deposits of the Ballaugh and Kirk Michael districts, Isle of Man. *Quarterly Journal of the Geological Society of London*, **121**, 359–381.

—— 1970–1973. The Pleistocene history of the Irish Sea: a second approximation. *Scientific Proceedings of the Royal Dublin Society, Section A*, **4**, 181–199.

—— 1971. Coastal sections between Glen Mooar and Orrisdale (West Coast). 1. Glen Mooar to Orrisdale. *In*: THOMAS, G. S. P. (ed.) *Isle of Man – Easter 1971, Field Guide*. Quaternary Research Association, Liverpool, 3–12.

—— 1971. Coastal sections between Ramsey and Point of Ayre. 5. Lough Cranstal. *In*: THOMAS, G. S. P. (ed.) *Isle of Man – Easter 1971, Field Guide*. Quaternary Research Association, Liverpool, 43.

—— & PARKES, H. M. 1948–1950. The Giant Deer in Ireland. Studies in Irish Quaternary Deposits No. 6. *Proceedings Royal Irish Academy, Section B*, **52**, 291–314.

MITCHELL, M. & WHITE, D. E. 1966. Catalogue of figured, described and cited Carboniferous corals in the collection of the Geological Survey and Museum, London. *Bulletin of the Geological Survey of Great Britain*, **24**, 19–56.

MOLYNEUX, S. 1980. New evidence for the age of the Manx Group, Isle of Man. *In*: HARRIS, A. L., HOLLAND, C. H. & LEAKE, B. E. (eds) *The Caledonides of the British Isles – Reviewed*. Geological Society, London, Special Publications, **8**, 415–421.

MOORE, A. W. 1900. *A History of the Isle of Man*. two volumes. Fisher Unwin.

—— 1904. Manx mines in the old times. *In*: MOORE, A. W. (ed.) *Notes and Documents from the Records of the Isle of Man*. The Manx Sun, 2–5.

—— 1904. Land taken away by the Sea 1658. *In*: MOORE, A. W. (ed.) *Notes and Documents from the Records of the Isle of Man*. The Manx Sun, 36–37.

MOORE, E. J. 1931. The ecology of the Ayreland of Bride, Isle of Man. *Journal of Ecology*, **19**, 115–136.

MOORE, R. M. 1992. *The Skiddaw Group of Cumbria: early Ordovician turbidite sedimentation and provenance of an evolving micro-continental margin*. PhD Thesis, University of Leeds.

MORTON, G. H. 1879. Geology of the Isle of Man. *Geological Magazine, Decade 2*, **6**, 211–213.

MUIR-WOOD, H. M. & COOPER, G. A. 1960. Morphology, classification and life habits of the Productoidea (Brachiopoda). *Geological Society of America Memoir*, **81**.

MURRAY, J. 1818. On phosphate of iron. *Annals of Philosophy*, **2**, 29–65.

MUSSON, R. M. W. 1993. *Earthquakes in the Isle of Man*. British Geological Survey, Global Seismology Series, Technical Report No. WL/93/10.

——, RITCHIE, M. E. A. & INNES, I. 1988. *The Isle of Man earthquake of 9 November 1987*. British Geological Survey, Global Seismology Series, Technical Report No. WL/88/9.

NEELY, G. J. H. 1933–1942. Excursion on 13th June 1935. To the Braddan gravel pits, Poortown Quarry and Peel. *Proceedings of the Isle of Man Natural History and Antiquarian Society, New Series*, **IV**, 130–131.

NIVEN, J. 1973. The Manx beaches. *Gems*, **5**,(5), 25–28.

NOCKOLDS, S. R. 1931. The Dhoon (Isle of Man) granite: a study in contamination. *Mineralogical Magazine*, **22**, 494–509.

OSWALD, H. R. 1823. *On the stratification of alluvial deposits, and the crystalization of Calcareous stalactites, in a letter to Dr John Maccullough*. Pamphlet, Douglas.

—— 1825. Observations relative to the Fossil Elk of the Isle of Man. *Edinburgh Journal of Science*, **3**, 28–31.

OWENS, R. M. 1984. *A Monograph of the Carboniferous Trilobites of Britain, Part 1*. Palaeontographical Society, London.

PANTIN, H. M. 1978. Quaternary sediments from the north-east Irish Sea: Isle of Man to Cumbria. *Bulletin of the Geological Survey of Great Britain*, **64**.

PARKER, J. G. & TAYLOR, P. M. (eds) 1993. *The Coast of North Wales and North West England. An Environmental Appraisal*. Hamilton Oil Company Ltd.

PARKES, T. 1988. Douglas Head. *Manx Mines Research Group. Mann-Underground*, **8**, 1–4.

PARNELL, J. 1988. Mineralogy of uraniferous hydrocarbons in carboniferous-hosted mineral deposits, Great Britain. *Uranium*, **4**, 197–218.

PEARCE, A. & ROSE, G. 1979. Mining remains in the Isle of Man. *Bulletin of the Peak District Mines Historical Society*, **7**, 216–230.

PHILLIPS, B. A. M. 1967. The post-glacial raised shoreline around the North Plain, Isle of Man. *Northern Universities Geographical Journal*, **8**, 43–48.

—— 1969. *Cliff and shore platform development in the Isle of Man*. PhD Thesis, University of Aberystwyth.

—— 1971. *Geomorphological observations at the sites of Ballure and Phurt, Isle of Man*. Field guide.

PHILLIPS, J. A. 1884. *A Treatise on Ore Deposits*. 1st Edition. Macmillan.

—— & LOUIS, H. 1896. *A Treatise on Ore Deposits*. 2nd Edition. Macmillan.

PICKIN, J. 1990. Stone tools and early mining in England and Wales. *In*: CREW, P. & CREW, S. (eds) *Early Mining in the British Isles*. Plas Tan y Bwlch, 39–42.

—— & WORTHINGTON, T. 1989. Prehistoric mining hammers from Bradda Head, Isle of Man. *Bulletin of the Peak District Mines Historical Society*, **10**, 274–275.

POCOCK, Y. P. 1968. Carboniferous schizophoriid brachiopods from Western Europe. *Palaeontology*, **11**, 64.

PRIDHAM, C. 1997. *The Story of the Ayres*. Manx Nature Conservation Trust, Isle of Man.

QUALE, B. 1794. *General View of the Agriculture of the Isle of Man*. Board of Agriculture, London.

—— & QUALE, T. 1992. *General View of the Agriculture of the Isle of Man by Basil Quale 1794 and Thomas Quale 1812*. Facsimile edition, Douglas.

QUALE, T. 1812. *General View of the Agriculture of the Isle of Man*. Board of Agriculture, London.

QUINE, J. 1911. *The Isle of Man*. Cambridge County Geographies.

QUIRK, D. G. 1982. Mine Survey – Glenn Foss level, Isle of Man. *Newsletter of the Leicester University Potholing Club*, **Summer**, 24–25.

—— 1994. *Petroleum Geology of the Isle of Man. Petroleum Exploration Society of Great Britain. Field Trip Guide*. Oxford Brookes University.

—— 1995. *Oil and gas exploration around the Isle of man*. Centre for Manx Studies, Day School, June 30th 1995.

—— 1996. *Geology of the Isle of Man*. Field Guide. Oxford Brookes University.

—— & FORD, T. D. 1994. The early Ordovician Manx Group and its relationship to the Iapetus Suture. *In*: *Caledonian Terrane relationships in Britain, 24th September 1994. (Programme with abstracts.)* British Geological Survey, Keyworth, 20.

—— & KIMBELL, G. S. 1997. Structural evolution of the Isle of Man and central part of the Irish Sea. *In*: MEADOWS, N. S., TRUEBLOOD, S. P., HARDMAN, N. & COWAN, G. (eds) *Petroleum Geology of the Irish Sea and Adjacent Areas*. Geological Society, London, Special Publications, **124**, 135–159.

——, KING, J., ROBERTS, I., POSTANCE, R. A. & ODELL, I. 1987. Basin ideas on boulder conjecture; microtectonics and collapsed limestones in the Carboniferous of the Isle of Man. *AAPG Bulletin*, **71**, 604.

——, FORD, T. D., KING, J. A., ROBERTS, I. L., POSTANCE, R. B. & ODELL, I. 1990. Enigmatic boulders and syn-sedimentary faulting in the Carboniferous Limestone of the Isle of Man. *Proceedings of the Yorkshire Geological Society*, **48**, 99–113.

RADCLIFFE, F. J. 1990. *Langness. Field Guide, 1*. Isle of Man Natural History and Antiquarian Society Field Section, Douglas.

—— 1994. *Manx Mines Rocks and Minerals*. Manx Heritage Foundation, Douglas. Resource book.

RADCLIFFE, J. W. 1964–1972. Mines in Maughold. *Proceedings of the Isle of Man Natural History and Antiquarian Society, New Series*, **VII**, 343–355.

RALFE, P. G. 1912–1925. A Manx lead mine in the 13th century. *Proceedings of the Isle of Man Natural History and Antiquarian Society, New Series*, **II**, 414–417.

READE, T. M. & WRIGHT, J. 1906. The Pleistocene clays and sands of the Isle of Man. *Proceedings of the Liverpool Geological Society*, **17**, 103–117.

REED, F. R. C. 1900. Woodwardian Museum Notes: A new Carboniferous Cephalopod, *Pleuronautilus? scarlettensis*, sp. nov. *Geological Magazine, Decade 4*, **7**, 105–106.

—— 1942. Some new Carboniferous Trilobites. *Annals and Magazine of Natural History, Series 11*, **9**, 649–671.

REEVES, G. M. 1972–1980. On the remains of a skeleton of *Cervus giganteus* (Irish Elk) in a late-glacial kettle hole deposit near Kirk Michael, Isle of Man. *Proceedings of the Isle of Man Natural History and Antiquarian Society, New Series*, **VIII**, **4**, 416–422.

RENSHAW, J. C. 1976. *Coast erosion in the north of the Isle of Man*. BSc Dissertation, Bedford College, London.

REYNOLDS, S. H. 1928. Breccias [Isle of Man]. *Geological Magazine*, **65**, 97–107.

—— 1929. *A Monograph on the British Pleistocene Mammalia: The Giant Deer*. The Palaeontographical Society, London.

ROBERTS, B., MORRISON, C. & HIRONS, S. 1990. Low grade metamorphism of the Manx Group, Isle of Man: a comparative study of white mica 'crystallinity' techniques. *Journal of the Geological Society, London*, **147**, 271–277.

ROBERTSON, D. 1985. North Bradda Mines. *Manx Mines Research Group. Mann-Underground*, **1**, 1–2.

—— 1989. The Great Laxey Mine. *Manx Mines Research Group. Mann-Underground*, **9**, 13–15.

ROBINSON, E. 1984. *London, Illustrated Geological Walks. Book One*. Scottish Academic Press.

ROBINSON, V. 1990. Extractive and manufacturing industries. *In*: ROBINSON, V. & MCCARROLL, D. (eds) *The Isle of Man – Celebrating a Sense of Place*. Liverpool University Press, 219–237.

—— & MCCARROLL, D. (eds) 1990. *The Isle of Man – Celebrating a Sense of Place*. Liverpool University Press.

ROUSE, C. 1990. The Isle of Man's unstable coast. *In*: ROBINSON, V. & MCCARROLL, D. (eds) *The Isle of Man – Celebrating a Sense of Place*. Liverpool University Press, 77–83.

RUSHTON, A. W. A. 1993. Graptolites from the Manx group. *Proceedings of the Yorkshire Geological Society*, **49**, 259–262.

RUSSELL, G. 1978. The structure and vegetation history of the Manx hill peats. *In*: DAVEY, P. (ed.) *Man and Environment in the Isle of Man. British Series, 54*. British Archaeological Reports, 39–49.

SCARFFE, A. 1987. The Building of the Laxey Wheel. *Bulletin of the Peak District Mines Historical Society*, **10**, 129–140.

—— 1987. The Snaefell Mine disaster. *Manx Mines Research Group. Mann-Underground*, **7**, 1–3.

SCATCHARD, W. J. 1971. Excavations at Beckwith Mine, Patrick 1969–1970 – a preliminary report. *Isle of Man Natural History and Antiquarian Society: Field Section. Notes and Queries*, **1**, 145–149.

SCHWEITZER, J. & MUELLER, G. 1985. Anomale Diferenzlaufzeiten und Amplitudenverhältnisse von SKS und SKKS bei Tonga–Fidschi Tieferdbeben [Different traveltime anomalies and amplitude ratios of SKS and SKKS-waves near Tonga–Fiji deep-focus earthquakes]. *Jahrestagung der Deutschen Geophysikalischen Gesellschaft e.V.*, **45**, 123.

SENIOR, J. R. 1996. Appendix 2: Pottery thin-section analysis. *In*: DAVEY, P. J., FREKE, D. J. & HIGGINS, D. A. (eds) *Excavations in Castletown, Isle of Man 1989–1992*. Liverpool University Press, 160–162.

SHEPPARD, A. J. 1996. *An environmental impact assessment of the Snaefell Mines, Isle of Man*. Undergraduate dissertation, St Andrews University.

SHIMMIN, J. P. 1972–1980. The Port of Laxey. *Proceedings of the Isle of Man Natural History and Antiquarian Society, New Series*, **VIII**, 378–384.

SHIPMAN, J. 1887. *Holiday Letters of a Geologist*. Carrick & Young.

SIMPSON, A. 1961. *The stratigraphy and tectonics of the Manx Slate Series, Isle of Man*. PhD Thesis, University of London.

—— 1963. F2 quartz veining in the Manx Slate Series. *Nature*, **199**, 900–901.

—— 1963. The stratigraphy and tectonics of the Manx Slate Series, Isle of Man. *Quarterly Journal of the Geological Society of London*, **119**, 367–400.

—— 1964. Deformed acid intrusions in the Manx Slate Series, Isle of Man. *Geological Journal*, **4**, 189–206.

—— 1964. The metamorphism of the Manx Slate Series. *Geological Magazine*, **101**, 20–36.

—— 1965. Discussion of an exhibit: microfossils of the Manx Slate Series. *Proceedings of the Geological Society of London*, **1625**, 125–126.

—— 1965. F1 cross-folding in the Manx Slate Series, Isle of Man. *Geological Magazine*, **102**, 440–444.

—— 1965. The syn-tectonic Foxdale–Archallagan granite and its metamorphic aureole. *Geological Journal*, **4**, 415–434.

—— 1966. Summer field meeting in the Isle of Man. *Proceedings of the Geologists' Association*, **77**, 217–227.

—— 1968. The Caledonian history of the north-eastern Irish Sea region and its relation to surrounding areas. *Scottish Journal of Geology*, **4**, 135–163.

SKELTON, R. H. 1955. The Manx mines. *Mining Magazine*, **92**, 9–18.

SKIPSEY, E. 1989–1990. Aspects of the geology of the Isle of Man. *Proceedings of the Cumberland Geological Society*, **5**, 237–238.

—— 1992. Geology of the Isle of Man. *Proceedings of the Westmorland Geological Society*, **5**, 12–15.

SLATER, G. 1931. The structure of the Bride moraine, the Isle of Man. The section of the north east coast between Port Cranstal and Point Cranstal. *Proceedings of the Liverpool Geological Society*, **15**, 284–296.

SLEIGHT, J. 1995. *The geology of the Castletown area, Isle of Man*. Undergraduate dissertation, University of Durham.

SMITH, B. 1927. On the carboniferous limestone series of the northern part of the Isle of Man. *Summary of Progress of the Geological Survey 1926*, 108–119.

—— 1931. Borings through the glacial drifts of the northern plain of the Isle of Man. *Summary of Progress of the Geological Survey*, **3**, 14–23.

SMITH, J. 1909–1912. Carboniferous limestone rocks of the Isle of Man. *Transactions of the Geological Society of Glasgow*, **14**, 119–164.

SMYTH, W. W. 1876. Metallic mining. *In*: BEVAN, G. P.

(ed.) *British Manufacturing Industries*. Edward Stanford.

—— 1879–1884. List of minerals found in the Isle of Man. *Transactions of the Isle of Man Natural History and Antiquarian Society*, **I**, 143–147.

—— 1881. On the occurence of feather-ore (Plumosite) in Foxdale mines, Isle of Man. *Transactions of the Royal Geological Society of Cornwall*, **10**, 82–89.

SOLLAS, W. J. 1895. An experiment to illustrate the mode of flow of a viscous fluid. *Quarterly Journal of the Geological Society of London*, **51**, 361–368.

SOPER, N. J., ENGLAND, R. W., SNYDER, D. B. & RYAN, P. D. 1992. The Iapetus suture zone in England, Scotland and eastern England: a reconciliation of geological and deep seismic data. *Journal of the Geological Society, London*, **149**, 697–700.

——, STRACHAN, R. A., HOLDSWORTH, R. E., GAYER, R. A. & GREILING, R. O. 1992. Sinistral transgression and the Silurian closure of the Iapetus. *Journal of the Geological Society, London*, **149**, 871–880.

SOUTHGATE, T., SLINN, D. J. & EASTHAM, J. F. 1983. Mine-derived metal pollution in the Isle of Man. *Marine Pollution Bulletin*, **14**, 137–140.

STENNING, E. H. 1912–1925. The life and work of the Reverend J. G. Cumming. *Proceedings of the Isle of Man Natural History and Antiquarian Society, New Series*, **II**, 402–414.

—— 1930–1934. Some geological influences on Manx scenery. *Journal of the Manx Museum*, **II**, 179.

—— 1932–1942. Excursion in Castletown district 20th June 1932. *Proceedings of the Isle of Man Natural History and Antiquarian Society, New Series*, **IV**, 1–6.

—— 1932–1942. Ancient Structures uncovered at Derbyhaven May 1935. *Proceedings of the Isle of Man Natural History and Antiquarian Society, New Series*, **IV**, 145–151.

—— 1932–1942. Excursion to Santan, 19th June 1941. *Proceedings of the Isle of Man Natural History and Antiquarian Society, New Series*, **IV**, 579–581.

—— 1950. *The Isle of Man. The County Books*. Robert Hale.

—— 1958. *Portrait of the Isle of Man. Portrait Books.* Robert Hale.

STONE, P. & EVANS, J. A. 1997. A comparison of the Skiddaw and Manx Groups (English Lake District and Isle of Man) using neodymium isotopes. *Proceedings of the Yorkshire Geological Society*, **51**, 343–347.

STRAHAN, A. 1879. Geology of the Isle of Man. *Geological Magazine, Decade 2*, **6**, 286.

STRICKLAND, H. E. 1842–1843. Some remarkable concretions in the Tertiary beds of the Isle of Man. *Quarterly Journal of the Geological Society of London*, **4**, 8–10.

SUGGIT, S. C. 1981. *Genesis of some Manx outwash deposits*. MSc Thesis, University of Liverpool.

SWIFT, A. 1993. Mantle-derived heat recorded by conodont colour alteration in the Carboniferous of the Isle of Man. *Geological Journal*, **28**, 171–177.

SWYNNERTON, F. 1912–1925. Notes on some ancient trees on the seashore below Strand-Hall, Bay-ny-Carricky. *Proceedings of the Isle of Man Natural History and Antiquarian Society, New Series*, **II**, 159.

SYLVESTER-BRADLEY, P. C. & KING, R. J. 1963. Evidence for abiogenic hydrocarbon. *Nature*, **198**, 728–730.

TAYLOR, B. J., BURGESS, I. C., LAND, D. H., MILLS, D. A. C., SMITH, D. B. & WARREN, P. T. 1971. *British Regional Geology. Northern England*. 4th Edition. HMSO.

TAYLOR, J. E. 1862. Supposed imprints in the Lower Cambrian beds of the Isle of Man. *The Geologist*, **5**, 321.

—— 1864. The Cambrian strata of the of the Isle of Man. *Transactions of the Manchester Geological Society*, **4**, 70–107.

TAYLOR, J. H. & GAMBA, E. A. 1933. The Oatland igneous complex. *Proceedings of the Geologists' Association*, **44**, 355–377.

TEACHERS LOCAL STUDIES PROJECT (ed.) 1976. *Manx Geography Series, No. 1 Geology of the Coastline*. Isle of Man Board of Education.

TELLET, F. S. 1888–1890. Inaugural Address by the President on Manx Glacial Geology. *Transactions of the Manx Geological Society*, **1**, 1–16.

TEMPLE, P. H. 1956. *Aspects of the geomorphology of the Isle of Man*. MA Thesis, Liverpool University.

THOMAS, G. S. P. 1971. Coastal sections between Ramsey and Point of Ayre (East Coast). 1. Ballure section: relationship between foreign and local drift. *In*: THOMAS, G. S. P. (ed.) *Isle of Man – Easter 1971, Field Guide*. Quaternary Research Association, Liverpool, 22–24.

—— 1971. Coastal sections between Ramsey and Point of Ayre (East Coast). 2. The Ayre drift plain. *In*: THOMAS, G. S. P. (ed.) *Isle of Man – Easter 1971, Field Guide*. Quaternary Research Association, Liverpool, 27–29.

—— 1971. Coastal sections between Ramsey and Point of Ayre (East Coast). 4. The Phurt section.*In*: THOMAS, G. S. P. (ed.) *Isle of Man – Easter 1971, Field Guide*. Quaternary Research Association, Liverpool, 41–42.

—— 1971. Coastal sections between Ramsey and Point of Ayre (East Coast). 6. The Ayre raised beach.*In*: THOMAS, G. S. P. (ed.) *Isle of Man – Easter 1971, Field Guide*. Quaternary Research Association, Liverpool, 44–45.

—— 1971. Coastal sections between Ramsey and Point of Ayre (East Coast). 7. The Kimmeragh pits. *In*: THOMAS, G. S. P. (ed.) *Isle of Man – Easter 1971, Field Guide*. Quaternary Research Association, Liverpool, 46–47.

—— 1971. Coastal sections between Glen Mooar and Orrisdale (West Coast). 3. Cliff sections south of Glen Mooar. The local deposits.*In*: THOMAS, G. S. P. (ed.) *Isle of Man – Easter 1971, Field Guide*. Quaternary Research Association, Liverpool, 16–21.

—— (ed.) 1971. *Field Guide to the Quaternary of the Isle of Man*. Quaternary Research Association, Liverpool.

—— 1976. The Quaternary stratigraphy of the Isle of Man. *Proceedings of the Geologists' Association*, **87**, 307–323.

—— 1976. *The Quaternary stratigraphy of the Isle of*

Man. PhD Thesis, University College of Wales, Aberystwyth.

—— 1977. Guide for the Isle of Man excursion. *In*: TOOLEY, M. J. (ed.) *The Isle of Man, Lancashire Coasts and the Lake District. Guide for Excursion A4. X INQUA Congress, Birmingham*. Geo Abstracts Ltd, University of East Anglia, 19–37.

—— 1977. The Isle of Man. *In*: KIDSON, C. (ed.) *Bibliography of the Quaternary of the Irish Sea Coasts. Proceedings X INQUA Congress, Birmingham*. Geo Abstracts Ltd, University of East Anglia, 52–57.

—— 1977. The Quaternary of the Isle of Man. *In*: KIDSON, C. & TOOLEY, M. J. (eds) *The Quaternary History of the Irish Sea*. Geological Journal Special Issue, 7, 155–179.

—— 1984. The origin of the glacio-dynamic structure of the Bride Moraine, Isle of Man. *Boreas*, **13**, 355–364.

—— 1985. *Guide Book to the Isle of Man Excursion*. Quaternary Research Association, Cambridge.

—— 1985. The Quaternary of the northern Irish Sea. *In*: JOHNSON, R. J. (ed.) *The Geomorphology of North-West England*. Manchester University Press, 143–158.

—— & DACKOMBE, R. V. 1985. Comments and reply on 'Glaciomarine sediments of the Isle of Man as a key to late Pleistocene stratigraphic investigations in the Irish Sea Basin' by C. H. Eyles and N. Eyles. *Geology*, **13**, 445–447.

—— & GELL, W. A. 1971. Coastal sections between Ramsey and Point of Ayre (East Coast). 3. Cliff sections between Dog Mills and Phurt. *In*: THOMAS, G. S. P. (ed.) *Field Guide to the Quaternary of the Isle of Man*. Quaternary Research Association, Liverpool, 30–40.

—— & TAPP, R. E. 1971. Coastal sections between Ramsey and Point of Ayre (East Coast). 8. The Bride drainage channels. *In*: THOMAS, G. S. P. (ed.) *Isle of Man – Easter 1971, Field Guide*. Quaternary Research Association, Liverpool, 48–49.

——, CONNAUGHTON, M. & DACKOMBE, R. V. 1985. Facies variation in a Late-Pleistocene supraglacial outwash sandur from the Isle of Man. *Geological Journal*, **20**, 193–213.

THORPE, J. A. (ed.) 1972. *North and west Lancashire and the Isle of Man; a bibliography of the geology and physical geography*. University of Lancaster Occasional Papers, **5**.

TILSLEY, J. W. 1977. Trilobites (Proetacea) from the Visean reef limestone at Treak Cliff, Castleton, Derbyshire. *Mercian Geologist*, **6**, 155–170.

TODD, J. 1895–1900. On the borings for coal and the experimental salt works in the north. *Yn Lioar Manninagh*, **III**, 65–72.

TODD, S. P., MURPHY, F. C. & KENNAN, P. S. 1991. On the trace of the Iapetus suture in Ireland and Britain. *Journal of the Geological Society, London*, **148**, 869–880.

TOOLEY, M. J. 1969. *Sea-level changes and the development of coastal plant communities during the Flandrian in Lancaster and adjacent areas*. PhD Thesis, University of Lancaster.

—— 1974. Sea level changes during the last 9000 years in N.W. England. *Geographical Journal*, **140**, 18–42.

—— 1976. Flandrian sea level changes in West Lancashire and their implications for the Hillhouse coastline. *Geological Journal*, **11**, 137–152.

—— (ed.) 1977. *The Isle of Man, Lancashire Coasts and the Lake District. Guide for Excursion A4. X INQUA Congress, Birmingham*. Geo Abstracts Ltd, University of East Anglia.

—— 1978. Flandrian sea-level changes and vegetational history on the Isle of Man. *In*: DAVEY, P. (ed.) *Man and Environment on the Isle of Man, Part 1*. British Archaeological Reports, **54**, 15–24.

—— 1982. Sea-level changes in northern England. *Proceedings of the Geologists' Association*, **93**, 43–51.

—— 1985. Sea-level changes and coastal morphology in north-west England. *In*: JOHNSON, R. H. (ed.) *The Geomorphology of North-west England*. Manchester University Press, 94–121.

TOWNLEY, R. 1791. *A Journal Kept in the Isle of Man: With Observations on the Soil, Clime, and Natural Production of that Island, also Antiquities of Various Kinds Now Extant There*. Two volumes. J. Ware & Son, Whitehaven.

TRENHAILE, A. S. & LAYZELL, M. G. J. 1981. Shore platform morphology and the tidal duration factor. *Transactions of the Institute of British Geographers*, **6**, 82–102.

TURNER, J. S. 1953. On the Carboniferous nautiloids: Some Middle Visean species from the Isle of Man. *Liverpool and Manchester Geological Journal*, **1**, 298–325.

—— 1960–1963. Some reflections on the medium scale cartography of the British Lower Carboniferous strata. *Transactions of the Leeds Geological Association*, **7**, 151–173.

WALDEN, J. & SLATTERY, M. C. 1993. Verification of a simple gravity technique for preparation of particle size fractions suitable for mineral magnetic analysis. *Earth Surface Processes and Landforms*, **18**, 829–833.

——, SMITH, J. P. & DACKOMBE, R. V. 1987. The use of mineral magnetic analyses in the study of glacial diamicts; a pilot study. *Journal of Quaternary Science*, **2**, 73–80.

——, —— & —— 1992. Mineral magnetic analyses as a means of lithostratigraphic correlation and provenance indication of glacial diamicts; intra- and interunit variation. *Journal of Quaternary Science*, **7**, 257–270.

——, —— & —— 1992. The use of simultaneous E- and Q-Mode Factor Analysis as a tool for assisting interpretation of mineral magnetic data. *Mathematical Geology*, **24**, 227–247.

WARD, C. 1970. The Ayres raised beach. *Geological Journal*, **7**, 217–220.

WARD, J. C. 1880. Notes on the geology of the Isle of Man. *Geological Magazine*, *Decade 2*, **7**, 1–9.

WARRINER, D. & GILLINGS, A. M. 1983. Exploration and survey of the Great Laxey Mine, Isle of Man. *Bulletin of the Peak District Mines Historical Society*, **8**, 373–382.

WATSON, E. 1971. Remains of pingos in Wales and the Isle of Man. *Geological Journal*, **7**, 381–387.

WATSON, J. 1916. *British and Foreign Marbles and other Ornamental Stones, A Descriptive Catalogue of the Specimen in the Sedgewick Museum, Cambridge*. Cambridge University Press.

WEBB, J. S. 1959. Notes on geochemical prospecting for lead–zinc deposits in the British Isles. *In*: *The Future of Non-ferrous Mining in Great Britain and Ireland: A Symposium*. Institution of Mining and Metallurgy, London.

WEEKS, A. & SIMPSON, J. H. 1991. The measurement of suspended particulate concentrations from remotely-sensed data. *International Journal of Remote Sensing*, **12**, 725–737.

WEIR, J. 1925. On new carboniferous gastropoda from the Isle of Man. *Geological Magazine*, **62**, 433–438.

—— 1932. The British and Belgian Carboniferous Bellerophontidae. *Transactions of the Royal Society of Edinburgh*, **56**, 767–865.

WHITE, F. W. 1909. The complex of igneous rocks at Oatland, Santon, Isle of Man. *Proceedings of the Yorkshire Geological Society*, **17**, 110–131.

WHITE, J. 1883. Remarks upon the Geology of the Isle of Man. *Transactions of the Geological Society of Glasgow*, **7**, 81–86.

WILKS, J. 1785. Remarks on Bog Timber in the Isle of Man. *Gentleman's Magazine* **1785**, 503.

WILLIAMS, K. 1932–1942. Entomological Report For 1939. *Proceedings of the Isle of Man Natural History and Antiquarian Society*, **IV**, 490–501.

WILLIAMSON, A. 1987. The Laxey Mines Tramways. Douglas Bay Mine. Maughold. *Manx Mines Research Group. Mann-Underground*, **6**, 1–9.

WILSON, E. 1978. The J. G. Cumming Collection of Fossils from the Limestone in the Isle of Man. *Newsletter of the Geological Curators' Group*, **2**, 99–101.

—— 1980. The 'Manx' Steps of St Paul's – Fact or Fiction? *Manx Life* **X**, number 3, 21–23.

—— 1981. Manx limestone and the steps of St Paul's. *Geologists' Association Circular*, **823**, 10–11.

—— 1981. The Stones of St Paul's. *Geologists' Association Circular*, **825**, 8–9.

—— 1982. Manx limestone and the steps of St Paul's. *Proceedings of the Geologists' Association*, **93**, 309–310.

—— 1999. Report on the stones in the standing masonry of the Lord's house. *In:* FREKE, D. *Excavations on St Patrick's Isle, Peel, Isle of Man; Prehistoric, Medieval and Later, 1982–1988.* Liverpool University Press. In press.

WILSON, G. & GEIKIE, A. (ed.) 1861. *Memoir of Edward Forbes*. Edmonston Douglas, Edinburgh.

WILSON, BISHOP THOMAS. 1722. A new Survey and Description. Account of the Isle of Man. *In*: Camden, W. *Britannia: or, a Chorographical Description of Great Britain and Ireland, together with the adjacent Islands.* Gibson, E. (ed.). 2nd Edition.

WOODS, G. 1811. *An Account of the Past and Present State of the Isle of Man ... including a sketch of its mineralogy*. Baldwin.

WRIGHT, B. M. 1874. Mineralogy. *In*: *Jenkinson's Practical Guide to the Isle of Man.* Edward Stanford, 217–38, 265.

WRIGHT, J. 1895–1900. The Foraminfera of the Pleistocene Clay, Shellag. *Yn Lioar Manninagh*, **III**, 627–629.

—— 1953. *Monograph of the British Carboniferous Crinoidea.* Palaeontographical Society, London.

WRIGHT, J. E., HULL, J. M., MCQUILLIN, R. & ARNOLD, S. E. 1971. *Irish Sea Investigations.* Institute of Geological Sciences Report, No. 71/19.

ZIEGLER, P. A. 1990. *Geological Atlas of Western and Central Europe.* 2nd Edition. Shell Internationale Petroleum Maatschappij B.V., The Hague.

Index